Productive Multivocality in the Analysis of Group Interactions

COMPUTER-SUPPORTED COLLABORATIVE LEARNING

The *Computer-Supported Collaborative Learning Book Series* is for people working in the CSCL field. The scope of the series extends to 'collaborative learning' in its broadest sense; the term is used for situations ranging from two individuals performing a task together, during a short period of time, to groups of 200 students following the same course and interacting via electronic mail. This variety also concerns the computational tools used in learning: elaborated graphical whiteboards support peer interaction, while more rudimentary textbased discussion forums are used for large group interaction. The series will integrate issues related to CSCL such as collaborative problem solving, collaborative learning without computers, negotiation patterns outside collaborative tasks, and many other relevant topics. It will also cover computational issues such as models, algorithms or architectures which support innovative functions relevant to CSCL systems.

The edited volumes and monographs to be published in this series offer authors who have carried out interesting research work the opportunity to integrate various pieces of their recent work into a larger framework.

For other titles published in this series, go to
http://www.springer.com/series/5814

Daniel D. Suthers • Kristine Lund
Carolyn Penstein Rosé • Chris Teplovs
Nancy Law
Editors

Productive Multivocality in the Analysis of Group Interactions

Springer

Editors
Daniel D. Suthers
Department of Information
& Computer Sciences
University of Hawai'i at Manoa
Honolulu, HI, USA

Kristine Lund
ICAR Research Lab
CNRS — University of Lyon
Lyon, France

Carolyn Penstein Rosé
Language Technologies Inst. & HCI Inst.
Carnegie Mellon University
Pittsburgh, Pennsylvania, USA

Chris Teplovs
Problemshift, Inc.
Windsor, ON, Canada

Nancy Law
University of Hong Kong Centre for Info.
Technology in Education
Hong Kong
Hong Kong SAR

ISBN 978-1-4614-8959-7 ISBN 978-1-4614-8960-3 (eBook)
DOI 10.1007/978-1-4614-8960-3
Springer New York Heidelberg Dordrecht London

Library of Congress Control Number: 2013954412

Printed on acid-free paper

Springer is part of Springer Science+Business Media (www.springer.com)

Contents

Part VII Reflections

Contributors

David Adamson Carnegie Mellon University, Pittsburgh, PA, USA

Michael Baker CNRS—Telecom ParisTech, Paris, France

Karine Bécu-Robinault ENS of Lyon—IFE—University of Lyon, Lyon, France

Patrick Brown DuBray Middle School, Fort Zumwalt School District, St Peters, MO, USA

Wenli Chen National Institute of Education, Singapore, Singapore

Ming Ming Chiu University at Buffalo—State University of New York, Buffalo, NY, USA

Ulrike Cress Knowledge Media Research Center, Tuebingen, Germany

Gregory Dyke University of Lyon, Lyon, France

Regina Frey Washington University in St. Louis, St. Louis, MO, USA

Nobuko Fujita University of Windsor, Windsor, ON, Canada

Sean P. Goggins The University of Missouri, Columbia, MO, USA

Cindy E. Hmelo-Silver Indiana University, Bloomington, IN, USA

Iris Howley Carnegie Mellon University, Pittsburgh, PA, USA

Heisawn Jeong Hallym University, Chuncheon, South Korea

Joachim Kimmerle Department of Applied Cognitive Psychology and Media Psychology, University of Tuebingen, Tuebingen, Germany

Timothy Koschmann Department of Medical Education, Southern Illinois University, Springfield, IL, USA

Rohit Kumar Carnegie Mellon University, Pittsburgh, PA, USA

Therese Laferriere Laval University, Quebec City, QC, Canada

Nancy Law University of Hong Kong, Hong Kong, Hong Kong

Chee Kit Looi National Institute of Education, Singapore, Singapore

Kristine Lund CNRS—University of Lyon, Lyon, France

Yoshiaki Matsuzawa Shizuoka University, Shizuoka, Japan

Elijah Mayfield Carnegie Mellon University, Pittsburgh, PA, USA

Richard Medina University of Hawaii, Honolulu, HI, USA

Yusuke Niihara InfoCom, Cambridge, MA, USA

Claire O'Malley School of Psychology, University of Nottingham, Nottingham, UK

Ritsuko Oshima Shizuoka University, Shizuoka, Japan

Jun Oshima Shizuoka University, Shizuoka, Japan

Carolyn Penstein Rosé Carnegie Mellon University, Pittsburgh, PA, USA

Keith Sawyer University of North Carolina, Chapel Hill, Chapel Hill, NC, USA

Hajime Shirouzu National Institute for Educational Policy Research, Tokyo, Japan

Yanjie Song Hong Kong Institute of Education, New Territories, Hong Kong

Gerry Stahl Drexel University, Philadelphia, PA, USA

Jan-Willem Strijbos Ludwig-Maximilians-Univerity Munich, Munich, Germany

Daniel D. Suthers University of Hawaii, Honolulu, HI, USA

Chris Teplovs Problemshift Inc., Windsor, ON, Canada

Stefan Trausan-Matu University "Politehnica" of Bucharest, Bucharest, Romania

Yun Wen National Institute of Education, Singapore, Singapore

On-Wing Wong University of Hong Kong, Hong Kong, Hong Kong

Part I
Introduction

Section Editor: Dan Suthers,
University of Hawai'i

This book is about a long-term collaborative effort known as the "productive multivocality project" that sought to engage researchers from different analytic traditions (i.e., multiple "voices") in productive dialogue with each other while analyzing shared data from group interactions in collaborative learning settings. It will be of interest to persons who want to understand and collaborate with colleagues from other traditions, to students who want to broaden their understanding of theoretical and methodological traditions available to them and how they might be brought into coordination, and to researchers who are interested in the particular learning settings and analytic results found in the five data corpora and associated analyses that make up the body of the book. Most significantly, the book offers a vision of how fields of study (such as the learning sciences) that are comprised of diverse traditions can counter tendencies towards fragmentation and achieve some level of coherence.

This first section of the book introduces the reader to the project on which this book is based and provides a guide to the book. Chapter 1, "The Productive Multivocality Project: Origins and Objectives," introduces the concept of multivocal analysis and why it is needed in the learning sciences, provides a brief historical account of the collaborations (series of workshops) out of which this work arose, and previews some of the major lessons learned in the form of problems encountered and strategies that we found useful for avoiding these problems while engaging analysts from multiple traditions with each other. Chapter 2, "Methodological Dimensions" details dimensions we used in our project and use throughout the book for describing different approaches to the analysis of interaction. Finally, Chap. 3, "A Readers' Guide to the Productive Multivocality Project," provides a preview of the book and a guide to using it as a resource for different purposes (e.g., for researchers who want to undertake multitradition collaborations themselves or students who want to learn about different analytic traditions). Perusal of Chaps. 1 and 3 should enable the reader to make effective use of the rest of the volume.

Chapter 1
The Productive Multivocality Project: Origins and Objectives

Daniel D. Suthers

The key idea of this volume is that scientific and practical advances in an area of study can be obtained if researchers working in multiple traditions—including traditions that have been assumed to be mutually incompatible—make a concerted effort to engage in dialogue with each other, comparing and contrasting their understandings of a given phenomenon and how these different understandings can either complement or mutually elaborate each other. Incompatibilities may remain but at least are reduced to essential and possibly testable differences once the noise of nonessential differences has been reduced. This key idea potentially applies to many fields, particularly in the social and behavioral sciences in which no single tradition has established primacy. The present volume offers case studies and insights of interest to anyone concerned with understanding the coordinated use of multiple *methods* but goes beyond mixed methods to address the coordinated joint work of diverse *methodologists* or the discourse within a diverse or "multivocal" discipline.

The researchers involved as editors and authors in the present volume work in the areas of collaborative learning, technology-enhanced learning, and cooperative work. We share an interest in understanding group interactions, including interactions mediated by various technologies ranging from paper and pencil to online environments. We approach this topic from a variety of traditional disciplinary homes and theoretical and methodological traditions that converge in a "field" known as computer-supported collaborative learning (CSCL) (Koschmann, Hall, & Miyake, 2001), the study of how interaction leads to learning with the support of designed artifacts. CSCL is situated more generally in the learning sciences (Sawyer, 2006), the interdisciplinary study of human learning and of the design and implementation of innovations and methods in support of learning and instruction.

D.D. Suthers (✉)
Department of Information and Computer Sciences, University of Hawai'i at Manoa,
1680 East West Road, POST 309, Honolulu, HI 96822, USA
e-mail: suthers@hawaii.edu

D.D. Suthers et al. (eds.), *Productive Multivocality in the Analysis of Group Interactions*, Computer-Supported Collaborative Learning Series 16,
DOI 10.1007/978-1-4614-8960-3_1, © Springer Science+Business Media New York 2013

In addition to the methodological project behind the key idea, this volume also offers research contributions within CSCL and the learning sciences.

The diversity of CSCL is salient to anyone involved in the conference series or journal that bears this name. The CSCL community is an international community (Kienle & Wessner, 2006) consisting of researchers, designers, and practitioners from computer science, education, educational psychology, human–computer interaction, and psychology as well as linguistics and other educational, information, learning, and social sciences (Wessner & Kienle, 2007). Hence numerous theoretical frameworks and methodological traditions drive work in this community to the extent that one can question whether it can be called a single field of study.

We take the term *multivocal* from Bahktin (Bakhtin, 1981; Koschmann, 1999), who used it to describe the presence of multiple "voices" that can be discerned in texts. Here the "text" is the collective discourse of those who identify with the CSCL community and its core values. This multivocality is a strength only to the extent that there is sufficient commonality to support dialogue between the voices and reach some degree of coherence in the discourse of CSCL (Suthers, 2006). The learning sciences and CSCL are too diverse (theoretically and methodologically) for unification to be possible. Moreover, unification is not at present even desirable—diversity is our strength in exploring alternate approaches to understanding learning in interaction. However, we would benefit from *boundary objects* (Star & Griesemer, 1989) that form the basis for dialogue between theoretical and methodological traditions applied to the analysis of learning in and through interaction. The question at hand is what constitutes effective boundary objects and how they may be leveraged.

Motivated by these considerations, the authors of this volume and other colleagues collaborated over a period of 5 years through a series of workshops and online interaction, seeking appropriate boundary objects and strategies for supporting productive multivocality between multiple analytic traditions in CSCL. This collaboration has become known as the "productive multivocality project." With this book we offer to colleagues in our own and other fields the insights of our activities. This chapter provides an overview of the project and summarizes its lessons. After a brief history of the project, the chapter summarizes dimensions for describing analytic approaches (discussed further in Chap. 2, Lund & Suthers, 2014), the composition of our data corpus, and strategies for productive multivocality (see also Chaps. 32–34: Dyke, Lund, Suthers, & Teplovs, 2014; Lund, Rosé, Suthers, & Baker, 2014; Rosé & Lund, 2014). Readers interested primarily in an executive summary of our insights are encouraged to read the present chapter with Chap. 31 (Suthers, Lund, Rosé, & Teplovs, 2014), which provides a more comprehensive post hoc summary of what we have learned. But the accounts in these summary chapters are given in the abstract: the case studies through which our work was conducted provide concrete examples. The body of this volume consists of five sections, each using a case study to investigate specific barriers to multivocal analyses, strategies to overcome these barriers, and benefits that may accrue from leveraging theoretical and methodological diversity. These case studies also offer other potential value to readers beyond the productive multivocality objectives. They serve as examples to students learning about new methods (see also Chap. 32), provide examples of how multiple methods may be combined in approaching one's own data (complementing volumes such as Tashakkori & Teddle, 2003), and yield research results that may be of interest to researchers studying the specific settings and phenomena we analyzed. The reader is referred to Chap. 3 (Suthers, Rosé, Lund, & Teplovs, 2014) for a guide to selective reading of the rest of the volume under these various reading objectives. The final section of the book discusses various issues encountered and lessons learned, offering implications for research programs and fields of study. Let us now begin our story.

Origins and Development of the Productive Multivocality Project

This project received inspiration from and emerged out of various earlier efforts, including a video analysis workshop at CSCL 2009 (Suthers, Christie, Goldman, & Hmelo-Silver, 1999), Tim Koschmann's "data fest" workshops at several CSCL and Winter Text Conferences, and various workshops and collaborations organized by Gerry Stahl around the Virtual Math Teams data (culminating in Stahl, 2009). The present Productive Multivocality project developed through a series of workshops at the International Conference on the Learning Sciences (ICLS) in 2008 and 2010, the CSCL conference in 2009, and the STELLAR Alpine Rendez-Vous (ARV) in 2009 and 2011. An interim report was also presented at a CSCL 2011 symposium (Suthers et al., 2011). Below we describe the motivations for each workshop and how major lessons learned led to changes in our strategy in each subsequent workshop. This historical account is relevant because it explains how our findings are based on what went wrong or was found to be insufficient as well as what worked.

A Common Framework for CSCL Interaction Analysis (ICLS 2008)

A premise of our first workshop was that common conceptions, representations, and tools are needed to support and bridge between multiple theoretical perspectives as well as to facilitate the application of different analytical methods and tools to complex data sets. Progress in any scientific discipline requires that practitioners share common objects such as instrumentation, data sources, and analytic methods that enable researchers to replicate or challenge results. Shared instruments and representations mediate the daily work of scientific discourse (e.g., Latour, 1990; Roth, 2003), and advances in other scientific disciplines have been accompanied with representational advances. Similarly, we reasoned that researchers studying learning in distributed and networked environments need shared ways of conceptualizing and representing what takes place in these environments to serve as the common foundation for our scientific and design discourse.

The goal of our first workshop (organized by volume editors Suthers, Law, and Rosé, with Nathan Dwyer) was to establish requirements for a common conceptual and representational framework to support collaborative learning process analysis, by (a) demonstrating our analytic tools to one another in the context of analyses we had conducted, (b) identifying commonalities among these tools and analyses along four dimensions, and (c) generating requirements for a common conceptual model and abstract transcript that might also form the bases for shared analytic software. The dimensions are as follows:

- *Purpose of analysis*. What is the analyst trying to find out about interaction? (In our context, some aspect of learning or meaning-making in interaction is usually a focus.)
- *Units of action, interaction, and analysis.* In terms of what fundamental relationships between actions do we conceive of interaction? What is the relationship of these units to the unit of analysis? The unit of interaction should not be confused with the unit of action or unit of analysis: units of action (e.g., chat messages or a discussion postings) are put into relation to each other by units of interaction (e.g., uptake of others' contributions) in a manner that constructs a model of interaction informative for the desired unit of analysis.
- *Representations of data and analytic interpretations*. What representations of data and representations of analytic constructs and interpretations capture these units in a manner consistent with the purposes and theoretical assumptions? Specifically, what requirements does the analytic method place on the representation of the original trace of activity? How are units of action interaction represented in terms of this trace representation (if they are)? What subsequent interpretations are layered on top of these representations, and how are they in turn expressed?
- *Analytic manipulations taken on those representations*. What are the analytic moves that transform a data representation into successive representations of interaction and interpretations of this interaction? How do these transformations lead to insights concerning the purpose of analysis?

These dimensions are described further in Chap. 2. At the workshop, we found that the dimensions were helpful for characterizing diversity (i.e., they described ways in which our approaches *differed* from each other), but we realized that our multivocality presented challenges in identifying a single common conceptual and representational framework for analysis. Yet, we felt that we were gaining some understanding from looking at each other's analyses. A software "tool fair" also generated considerable interest, and we noted the need to make our theoretical assumptions explicit.

Common Objects for Productive Multivocality in Analysis (CSCL 2009)

In our second workshop (organized by editors Suthers, Law, Lund, Rosé, and Teplovs), we decided to tackle multivocality head on by having analysts from different traditions assigned to analyze the same data set, a strategy that many others have tried (e.g., Koschmann, 2011). Two corpora were used, from the Virtual Math Teams (Stahl, 2009) and Knowledge Forum (Scardamalia, 2004) projects. We continued to use the four dimensions to characterize different analyses and added the following dimension.

- *Theoretical assumptions underlying the analysis*. What ontological and epistemological assumptions are made about phenomena worth studying, and how can we come to know about them? (Here we assume that such phenomena broadly include interaction.)

This dimension was needed to warrant the decisions expressed in the first four dimensions. Theoretical assumptions permeate the other methodological dimensions. For example, representations of data embody implicit theoretical commitments (Ochs, 1979).

As the analyses were presented, we tried to use our dimensions to discover commonalities ("common objects") that can support productive multivocality. We also sought to determine whether the analytic differences are complementary (potential sources of richer understanding) or incompatible (potential barriers to a common discipline). Again, we found that the dimensions highlighted how the analyses differed rather than their commonalities. Asking ourselves what we did have in common, we agreed that we shared (a) learning through collaborative interaction as our topic of study and (b) the desire and willingness to engage in this activity together. These are key prerequisites for productive multivocality. Although we had hoped that multiple analyses of shared data corpora would provide a basis for dialogue, the analyses presented were disconnected in part because the analysts were approaching these corpora with entirely different questions: they were "talking past" each other. This observation led to the objective of identifying "pivotal moments" in the next workshop.

Pinpointing Pivotal Moments in Collaboration (ARV 2009)

Our third workshop (organized by Lund, Law, Rosé, Suthers, and Teplovs) continued the prior strategy of having researchers from different theoretical and methodological traditions analyze shared data corpora. We used a different Knowledge Forum corpus (the basis of the case study in Chaps. 20–24 of this volume) and a Japanese primary school mathematics class (Chaps. 4–8 of this volume). As before, we assigned analysts to data, deliberately pairing up analysts from different methodological traditions. We also assigned an analyst to data from a setting he did not normally study (the textual analysis of Bakhtin being applied to multimodal data) and grappled with the question of how data-hungry quantitative methods can inform microanalysis. We addressed the prior mismatch in analytic objectives by asking analysts to identify the *pivotal moments* in the interactions recorded in the data. The definition of pivotal moments was purposefully left unspecified, providing a projective stimulus that drew out different researchers' assumptions and insights and led to exciting comparative and integrative discussion.

As expected, analysts differed in their conception and identification of pivotal moments, but these differences (as well as some congruencies) generated productive discussion of how learning arises from interaction. In this workshop we first articulated our core strategy for multivocality: assign diverse analysts to shared corpora and charge them with analytic objectives that are deliberately open to interpretation (e.g., "pivotal moments"). During this and the prior workshop, our own objectives shifted: we talked less about sharing the *same* concepts or representations and more about *boundary objects* (such as the corpora and pivotal moments) supporting dialogue between different traditions. Boundary objects "have different meanings in different worlds but their structure is common enough to more than one world to make them recognizable, a means of translation" (Star & Griesemer, 1989, p. 393). We found that it is useful to align analytic results (e.g., to find overlaps and differences in pivotal moments identified) and so wanted to explore further how shared analytic frameworks (e.g., Howley, Mayfield, & Rosé, 2013; Suthers, Dwyer, Medina, & Vatrapu, 2010) and shared analytic software tools (e.g., Tatiana; Dyke, Lund, & Girardot, 2009) could serve as or produce appropriate boundary objects.

Productive Multivocality in the Analysis of Collaborative Learning (ICLS 2010)

In our fourth workshop (organized by Lund, Suthers, Law, Rosé, and Teplovs), we sought to build on the success of the third workshop, replicating the strategy of having deliberately diverse analysts identify pivotal moments in shared corpora. There were two novelties. First, we brought in new data corpora and new analysts. Corpora included a Group Scribbles mathematics classroom in Singapore (subsequently replaced) and university-level chemistry study groups in the USA (Chaps. 9–13 of this volume). Second, we wanted to revisit the possibility that a shared software tool and its data and analytic representations would help support more detailed comparisons between analyses, by providing all the data and analyses within the common tool. This latter effort enabled analyses to be shared ahead of the workshop and is reported in Dyke et al. (2011).

The primary strategy again proved to be productive, surfacing issues and exemplifying insights by the case studies. In the chemistry case, analysts discovered that they had different conceptions of "leadership," leading to refinement of this concept and its analytic manifestations. With the exception of one analyst who emphasized implicit interaction via nonverbal means, most analysts concluded that there was not much collaborative learning taking place in the Group Scribbles mathematics corpus. Although we recognized that educators must deal with failed collaboration all the time and therefore research could examine these missed opportunities, we decided that analysts and (subsequently) readers of this volume would not be very motivated to put time into an "uninteresting" case (in fact, one analyst on this corpus dropped out of the project). However, many other interesting examples were available from the Singapore Group Scribbles setting.

Leveraging Researcher Multivocality for Insights on Collaborative Learning (ARV 2011)

The final formal workshop of this collaboration (organized by editors Rosé, Lund, Suthers, Law, and Teplovs, with Gregory Dyke) brought in two more data corpora that are represented in the present volume. At our request, our Singapore colleagues replaced the mathematics corpus with another Group Scribbles corpus, this one on learning about electric circuits. This corpus has features not found in the prior corpora, including use of technology to support face-to-face interaction, use of physical manipulatives (batteries, wires, and light bulbs), and the multimodality that results from this combination. It forms the basis of Chaps. 14–19. A final corpus along with three new analysts was introduced, involving the use of a software agent in discovery learning of 9th-grade biology (Dyke, Adamson, Howley, & Rosé, 2013). This corpus is unique in two ways: the use of agents in support of collaborative learning and the role that the analyses are playing in iterative design and improvement of this software environment. It forms the case study of Chaps. 25–30 of the present volume. The end of the 2-day workshop was structured to identify themes common across the case studies and thus surface practical, methodological, and theoretical issues and strategies for productive multivocality that are highlighted in the present volume (especially in Chaps. 31–34).

Subsequent collaborations continued beyond ARV 2011 with numerous individual and small group meetings at conferences and each others' institutions and resulting in a number of papers (e.g., Chiu & Fujita, 2014; Dyke, Howley, Adamson, & Rosé, 2012; Dyke, Kumar, Ai, & Rosé, 2012; Dyke et al., 2011; Dyke et al., 2013; Howley et al., 2013; Jeong, Chen, & Looi, 2011; Medina & Suthers, 2013; Oshima, Matsuzawa, Oshima, Chan, & van Aalst, 2012; Oshima, Oshima, & Matsuzawa, 2012; Oshima, Oshima, Matsuzawa, van Aalst, & Chan, 2011; Reynolds & Chiu, 2012; Schwarz et al., 2010; Suthers et al., 2011; Wise & Chiu, 2011a, 2011b). The remainder of the chapter discusses the diversity of our data and methods and summarizes issues and strategies that will be revisited throughout the book and discussed further in Chap. 31 onwards.

The Corpora and Analytic Traditions

In selecting the data corpora (case studies) and analysts for this project, we were cognizant of the need to bring multiple theoretical and methodological traditions to bear on a diversity of interactional settings. Diversity of data and traditions helps ensure that we encounter the range of issues present in a multivocal research community and helps make a more convincing case for the generality of our conclusions. Of course, we also worked within the constraints of the available data and analysts and had to consider the motivations of our project participants.

Data Corpora for Case Studies

Data corpora for case studies were subject to two individual criteria (i.e., criteria that are applied independently of what other data corpora were under consideration): the data must have the potential to show learning through interaction, and must be compelling as evidenced by the desire and willingness of multiple analysts to spend time analyzing that data. The corpus was also subject to collective criteria of achieving diversity, deliberately sampling various interactional and learning settings of interest. We wanted to achieve diversity of age levels, diversity of settings (formal and informal learning in schools, workplaces, and elsewhere), diversity of interactional media (face-to-face, synchronous, and asynchronous computer-mediated communication), and diversity of domains or topics of study.

In the end, we were able to obtain and perform multiple analyses of the corpora shown in Table 1.1, listed by domain, population and setting, and interactional media. As one can see from Table 1.1, we were successful in obtaining various topics, age groups, and interactional media within formal educational settings.

Table 1.1 Summary of data corpora

Chapters	Topic	Age and institutional setting	Interactional setting and media
4–8	Mathematics	6th-grade Japanese classroom	Face to face with origami paper and blackboard
9–13	Chemistry	Undergraduate peer-led team learning	Face to face with paper and whiteboard
14–19	Electricity	Primary school in Singapore	Primarily face to face with circuit components and Group Scribbles software
20–24	Education	Graduate level in Toronto	Asynchronous discussions in Knowledge Forum
25–30	Biology	Secondary school in Pittsburgh	Mixed face to face and online with concert chat and conversational agents in support of collaborative learning

The emphasis is on science and mathematics, and we are missing case studies in informal settings or workplaces.

Analytic Traditions

A project on productive multivocality requires sufficient diversity of theoretical and methodological traditions. There is a "sampling bias" in this project in that the traditions represented are those brought by persons who were willing to commit the effort to either share their data or analyze others' data and participate in the workshops. The persons we were able to recruit use methods as diverse as various forms of content analysis, conversation analysis, polyphonic analysis, semiotic and multimodal analysis, social network analysis, statistical discourse analysis, computational linguistics, and uptake analysis. Theoretical traditions include cognitivism, constructivism, dialogism, ethnomethodology, group cognition or intersubjective meaning-making, knowledge building, progressive inquiry, semiotics, and systemic functional linguistics.

Reflecting on the corpora and traditions represented, there are clearly gaps. We particularly would have liked to include data from outside of formal schooling, such as a workplace setting, and in conjunction with this to have included sociocultural traditions of analysis (attempts were made to recruit relevant data and participants but were unsuccessful). Also, our case studies are biased towards small group interaction and hence microanalysis rather than large-scale networks of learners. Yet, we believe that we have sufficient diversity to have encountered and grappled with major issues in achieving productive multivocality in the analysis of interaction. Our attempts to bring the analytic traditions listed above into conversation with respect to the various corpora encountered difficulties that we overcame with the strategies discussed in the next section.

Issues and Strategies for Productive Multivocality

As suggested in the preceding account of the historical development of the project, our series of workshops was an iterative process in which we refined our shared objectives, encountered issues and problems, and developed strategies for meeting these objectives. Our objectives shifted from one of identifying common representations and practices that would enable the specification of requirements for shared data and tools to one of enabling productive dialogue between multiple traditions through whatever boundary objects served this purpose. Following is a preview of some of the strategies we developed for making our dialogue productive. These strategies, along with the issues they are intended to address, are discussed in greater detail in Part VII of this volume, with a summary in Chap. 31 and more detailed discussion of methods for achieving productive multivocality in Chaps. 32–34.

Use Standards, Metadata, and Repositories to Share Data and Tools

There is a great redundancy in the software efforts behind analysis. Many research groups develop their own tools, and there are technical barriers to applying these tools to data gathered in multiple settings. The first workshop began with the objective of developing standards that would enable a suite of software tools developed at different labs to interoperate on common data and analytic representations. These solutions have been the focus of a number of other efforts. For example, Harrer, Monés, and Dimitracopoulou (2009) have developed standards for representing data and analyses, and Reffay, Betbeder, and Chanier (2012) have proposed standards for a data repository. Ontologies have long been a focus in the Artificial Intelligence and Education community (e.g., Mizoguchi, Ikeda, & Sinista, 1997).

Our project did not culminate in the development or the adoption of standards across the project, but methods of sharing data and tools were critical to each case study. An exception was that the Tatiana analytic tool (Dyke et al., 2009) served a useful role as a common tool in several of the case studies. Tatiana provided a medium within which to share synchronized replayable data traces (e.g., video, transcripts, and log files) and to construct analytic representations (e.g., coded segments) on top of these traces that are also synchronized with them. The case studies in Part II (Case Study 1, Fractions), Part IV (Case Study 3, Electric Circuits), and Part VI (Case Study 5) in particular made use of Tatiana for sharing data and/or comparing analytic results.

Technical solutions that enable researchers in different settings to reuse the software developed and data gathered elsewhere are useful but not sufficient: to bring multiple traditions into productive dialogue they must share an object of study.

Analyze the Same Data

An obvious and well-known strategy for engaging researchers in dialogue is to have them analyze the same data and discuss their results so that different perspectives on and results obtained concerning the same object of study may be compared. This strategy has been found to be useful within single traditions. For example, in quantitative content coding multiple coders are used to achieve reliability, and similarly collaborative interaction analysis reaches a richer understanding of interaction through group review of video data (Jordan & Henderson, 1995). Work within education and CSCL has taken this strategy: recent examples are Koschmann (2011) and Stahl (2009).

This strategy was introduced in our second workshop and continued throughout the project. Some of the multivocal dialogue that takes place actually precedes the analysis of the data, as participants need to agree on what data is worth considering and how it should be selected and represented. Data selection and preparation will expose assumptions. We found that this strategy can productively be augmented with an auxiliary strategy of a shared analytic objective, considered shortly.

Pair Up Diverse Traditions

If the analysts assigned to a data corpus work in similar theoretical or methodological traditions, it will be easier for them to talk to each other. They will share basic assumptions and will be able to focus on the nuances of their results and fine-tune their analytic practice. Such work is valuable but does not address the objective of fostering dialogue between representatives of theoretically and methodologically diverse communities who are working within a given area of study (such as learning through social interaction).

We have found that it is useful to pair up analysts from quite distinct traditions. This approach surfaces otherwise implicit assumptions concerning what data is suitable for study and what questions are worth asking, and once these questions are resolved (and with the application of a strategy described below), comparisons of results can lead to productive dialogue concerning analytic concepts and results. For example, in Part II (Case Study 1), analysts from three traditions compared the points of interaction that they considered to be the most significant, finding agreement on some but non-overlap on others. This discrepancy led one analyst to reconsider how he was defining these "pivotal moments." In Part III (Case Study 2), the concept of "leadership" was refined through juxtaposition of linguistic and conceptual coding methods. In Part VI (Case Study 5), analysts from several traditions problematized a core design assumption behind the data provider's software.

Push Methods Outside of Their Comfort Zone

The next strategy is related to (and perhaps inevitable given) the strategy of pairing up diverse analytic traditions, as in any deliberately diverse pair one analyst may feel closer to the data than the other. We found it useful to give an analyst data that is not of the type they normally analyze. This of course must be done with care, as too great of a mismatch would not be productive. The objective from a research community perspective is not merely to challenge individual researchers but rather to explore how analytic traditions might be applicable beyond the scope of data to which they have been usually applied. The benefits for the community are that analytic traditions are brought out of their isolation, coming into contact with each other, and also we discover unanticipated ways in which they might contribute to understanding new phenomena.

In our project, a clear example of the success of this strategy was when we asked an analyst who had been doing conversation analysis of texts (written conversation) informed by Bahktin to analyze video data that included gestures and manipulation of paper and blackboard diagrams (see Part II, Case Study 1). A potential issue is whether the analytic method is also pushed outside of its zone of validity. For example, in the same case study a statistical breakpoint analysis was applied to a sample that might be considered too small for this method. Yet the exercise has utility as long as it is understood that a different scientific game is being played: rather than generalizing to a population from a sample, statistical analysis was used to expose features of the data that other analysts might consider from their standpoints.

Address a Shared Analytic Objective

As we found in our second workshop, it is not sufficient to have diverse analysts take on the same data. There is no guarantee that their analyses (or even how they construe the object of study) will be comparable, and given that they come from different traditions they are likely to "talk past" each other. Identification of this problem led to our most crucial strategy: to request analysts to approach the data with a shared analytic objective so that the different analyses can be compared and hence the traditions brought into dialogue with respect to these traditions. In our case, we asked analysts to identify the "pivotal moments" of the interaction found in the data: What events were most crucial for the collaboration?

The concept of a "pivotal moment" is deliberately vague. Vagueness can be understood as advantageous if we consider the concept of a "shared analytic objective" with respect to the objectives of our project. We cannot ask analysts to address the same research question at the usual level of specificity found within a given analytic tradition, because a research question that is well specified within one tradition may not be interesting to or make sense within another tradition or may even violate its assumptions. We need to offer analytic objectives that are interpretable by each tradition involved so that they can be brought into dialogue with each other around this object. An analytic objective that only makes sense within one tradition is not "shared." An analytic objective that is sharable across traditions acts as a boundary object (Star & Griesemer, 1989)—one that is interpretable by all traditions involved, perhaps differently, but this is what makes the exercise interesting! In a sense, vagueness is a great advantage. To draw an analogy in which analytic traditions are psychodynamic persons, the objective of finding pivotal moments serves as a "projective stimulus" in which each tradition sees, or upon which it projects, what is important in the given data. This strategy is exemplified well in Part II (Case Study 1).

Eliminate Gratuitous Differences in Data Considered

In some cases, we found that analysts came to different conclusions merely because they looked at different aspects of the data. This was the case in our first Group Scribbles study, discussed in Chap. 19, in which it was found that analysts differed on whether they analyzed private (as well as public) activity and whether verbal acts, nonverbal acts, and the states of artifacts that resulted were considered. Once gratuitous differences are eliminated, the differences in results and interpretations that remain are more likely to be essential to the dialogue needed between traditions.

An issue discussed previously arises again: analytic traditions may differ in the data considered because they differ in what is considered relevant or in the "amount" of data needed to meet validity requirements for the tradition (e.g., inferential statistics vs. conversation analysis). This problem has been dealt within the author's laboratory through an overlapping technique: a focal session is chosen for microanalysis, but analysts who have larger data requirements (e.g., to study role development or relationship formation over time) analyze the data they require and report the implications of the results for understanding the results of session microanalysis.

Align Analytic Representations

Having eliminated (to the extent possible) gratuitous differences in the scope of data considered, we have found that it is extremely helpful to represent analytic results in some form that can be brought into alignment with each other for comparison. The most obvious basis for such an alignment is time: different interpretations of the same sequence of events are given a visual representation along a common timeline. Such representations highlight congruences and discrepancies and serve as excellent prompts and resources for conversation between analysts. Chapter 33 discusses the role of representations and tools for achieving productive multivocality in greater detail.

Iterate

The above strategies imply that iteration is required. For example, even if analysts have agreed on a data corpus and a shared analytic objective, in the first meeting they may discover that they have examined different aspects or scopes of the data. Inconsequential differences should be eliminated and the analyses repeated to focus on essential (e.g., conceptual and epistemological) differences and convergences.

Step Back from Methods

None of the above strategies will help if participants remain within their methodological boxes. Ultimately we want to bring theoretical ideas into dialogue, but this can be prevented if the methods in which one is trained are taken as fundamental to how the phenomenon is viewed. The researchers who will be most successful in achieving productive multivocality in a community are those who can take off their methodological eyeglasses and *dialogue about methods as object-constituting, evidence-producing, and argument-sustaining tools*. This dialogue requires careful consideration of what methods as inscriptions and means of operating on inscriptions bring with them intrinsically as well as what teleological and theoretical commitments are made in the practices of applying these tools to a domain.

Conclusions

Sharing analyses has benefits both for the individual analysts and the community. Analysts are confronted with aspects of the data highlighted by others that they might not have themselves considered; epistemological assumptions are challenged; analytic concepts are fine-tuned; and a multidimensional understanding of the phenomenon being investigated and analytic constructs used to approach it is gained. The process leads to greater dialogue and mutual understanding in our community. Yet, these benefits do not accrue merely by putting analysts together in the same room or even by having them analyze the same data. Productive multivocality is facilitated by strategies such as eliminating gratuitous differences in the scope and representation of data considered and deliberately pairing diverse analysts charged with a common yet flexible analytic objective.

The collaboration constituting this project is, we believe, unprecedented and significant in our field. Many volumes result from one-shot workshops, but sustained collaboration over a period of years is rare, particularly in the face of academic incentive structures that provide greater rewards to solo efforts and self-promotion. The researchers we worked with on this project are large in number and represent diverse disciplines and analytic traditions, yet all shared a commitment to the project and were congenial colleagues to work with. This volume is a testament to their dedication to finding ways to bring the individual and collective needs of research in CSCL and the learning sciences into congruence with each other.

Acknowledgements The author gratefully acknowledges the collaboration and hard work of the editors and many other researchers who made this project possible and rewarding. The author was partially supported by NSF Award 0943147 during this project. The views expressed herein do not necessarily represent the views of NSF.

References

Bakhtin, M. (1981). Discourse in the novel. In M. Holquist (Ed.), *The dialogic imagination* (pp. 259–422). Austin: University of Texas.

Chiu, M. M., & Fujita, N. (2014). Statistical discourse analysis: Knowledge creation during online discussions. In S.-C. Tan, H.-J. So & J. Yeo (Eds.), *Knowledge creation in education*. New York: Springer

Dyke, G., Adamson, A., Howley, I. K., & Rosé, C. P. (2013). Enhancing scientific reasoning and discussion with conversational agents. *IEEE Transactions on Learning Technologies, 6*(3), 240–247.

Dyke, G., Howley, I. K., Adamson, D., & Rosé, C. P. (2012). Towards Academically Productive Talk Supported by Conversational Agents. In S. A. Cerri, W. J. Clancey, G. Papadourakis & K. Panourgia (Eds.), *Proceedings of the 11th International Conference on Intelligent Tutoring Systems (ITS 2012)*, Chania, Crete, Greece (pp. 531–540). Berlin: Springer

Dyke, G., Kumar, R., Ai, H., & Rosé, C. P. (2012). Challenging Assumptions: Using sliding window visualizations to reveal time-based irregularities in CSCL processes. In J. van Aalst, K. Thompson, M. J. Jacobson & P. Reimann (Eds.), *Proceedings of the 10th International Conference of the Learning Sciences* (Vol. 1, pp. 363–370). Sydney, Australia: International Society of the Learning Sciences

Dyke, G., Lund, K., & Girardot, J.-J. (2009). Tatiana: an environment to support the CSCL analysis process. In C. O'Malley, P. Reimann, D. Suthers & A. Dimitracopoulou (Eds.), *Computer Supported Collaborative Learning Practices: CSCL 2009 Conference Proceedings* (pp. 58–67). Rhodes, Greece: International Society of the Learning Sciences

Dyke, G., Lund, K., Jeong, H., Medina, R., Suthers, D. D., van Aalst, J., et al. (2011). Technological affordances for productive multivocality in analysis. In: H. Spada, G. Stahl, N. Miyake, & N. Law (Eds.), *Connecting Computer-Supported Collaborative Learning to Policy and Practice: Proceedings of the 9th International Conference on Computer-Supported Collaborative Learning (CSCL 2011)* (Vol. I, pp. 454–461). Hong Kong: International Society of the Learning Sciences

Dyke, G., Lund, K., Suthers, D. D., & Teplovs, C. (2014). Analytic representations and tools for productive multivocality. In D. D. Suthers, K. Lund, C. P. Rosé, C. Teplovs & N. Law (Eds.), *Productive multivocality in the analysis of group interactions*, Chapter 33. New York: Springer

Harrer, A., Monés, M., & Dimitracopoulou, A. (2009). Users' data: Collaborative and social analysis. In N. Balacheff, S. Ludvigsen, T. de Jong, A. Lazonder, & S. Barnes (Eds.), *Technology enhanced learning. Principles and products* (pp. 175–193). New York: Springer.

Howley, I. K., Mayfield, E., & Rosé, C. P. (2013). Linguistic analysis methods for studying small groups. In C. E. Hmelo-Silver & A. M. O'Donnell (Eds.), *International handbook of collaborative learning*. New York: Routledge, Taylor and Francis Group.

Jeong, H., Chen, W., & Looi, C.-K. (2011). Analysis of group understanding in artifact-mediated discourse. In H. Spada, G. Stahl, N. Miyake, & N. Law (Eds.), *Connecting Computer-Supported Collaborative Learning to Policy and Practice: Proceedings of the 9th International Conference on Computer-Supported Collaborative Learning (CSCL 2011)* (Vol. II, pp. 786–790). Hong Kong: International Society of the Learning Sciences

Jordan, B., & Henderson, A. (1995). Interaction analysis: Foundations and practice. *The Journal of the Learning Sciences, 4*(1), 39–103.

Kienle, A., & Wessner, M. (2006). The CSCL community in its first decade: Development, continuity, connectivity. *International Journal of Computer Supported Collaborative Learning, 1*(1), 9–33.

Koschmann, T. (1999). Toward a dialogic theory of learning: Bakhtin's contribution to understanding learning in settings of collaboration. In C. Hoadley & J. Roschelle (Eds.), *Proceedings of the 1999 Conference on Computer Support for Collaborative Learning (CSCL 1999)*. Palo Alto, CA: International Society of the Learning Sciences

Koschmann, T. (2011). *Theories of learning and studies of instructional practice*. New York: Springer.

Koschmann, T., Hall, R., & Miyake, N. (Eds.). (2001). *CSCL II. Carrying forward the conversation*. Mahwah, NJ: Lawrence Erlbaum Associates.

Latour, B. (1990). Drawing things together. In M. Lynch & S. Woolgar (Eds.), *Representation in scientific practice* (pp. 19–68). Cambridge, MA: MIT Press.

Lund, K., Rosé, C. P., Suthers, D. D., & Baker, M. (2014). Epistemological encounters in multivocal settings. In D. D. Suthers, K. Lund, C. P. Rosé, C. Teplovs & N. Law (Eds.), *Productive multivocality in the analysis of group interactions*, Chapter 34. New York: Springer

Lund, K., & Suthers, D. D. (2014). Methodological dimensions. In D. D. Suthers, K. Lund, C. P. Rosé, C. Teplovs & N. Law (Eds.), *Productive multivocality in the analysis of group interactions*, Chapter 2. New York: Springer

Medina, R., & Suthers, D. D. (2013). Juxtaposing practice: Uptake as modal transposition. In *Proceedings of the 10th International Conference on Computer Supported Collaborative Learning (CSCL '13)*. Madison, WI

Mizoguchi, R., Ikeda, M., & Sinista, K. (1997). Roles of shared ontology in AI-ED research—intelligence, conceptualization, standardization, and reusability. In B. D. Boulay & R. Mizoguchi (Eds.), *Artificial intelligence in education: Knowledge and media in learning systems* (pp. 537–544). Amsterdam: Ios Press.

Ochs, E. (1979). Transcription as theory. In E. Ochs & B. B. Schieffelin (Eds.), *Developmental pragmatics* (pp. 43–72). New York: Academic.

Oshima, J., Matsuzawa, Y., Oshima, R., Chan, C. K. K., & van Aalst, J. (2012). Social Network Analysis for Knowledge Building: Establishment of Indicators for Collective Knowledge Advancement. In J. van Aalst, K. Thompson, M. J. Jacobson & P. Reimann (Eds.), *The Future of Learning: Proceedings of the 10th International Conference of the Learning Sciences (ICLS2012), Volume 2, Short Papers, Symposia, and Abstracts* (pp. 465–466). Sydney, NSW, Australia: International Society of the Learning Sciences

Oshima, J., Oshima, R., & Matsuzawa, Y. (2012). Knowledge building discourse explorer: A social network analysis application for knowledge building discourse. *Educational Technology Research & Development, 60*:1–19.

Oshima, J., Oshima, R., Matsuzawa, Y., van Aalst, J., & Chan, C. K. K. (2011). Network structure analysis for knowledge building: A macroscopic view of collaborative learning discourse, *Annual Meeting of Educational Research Association*. New Orleans, LA

Reffay, C., Betbeder, M.-L., & Chanier, T. (2012). Multimodal learning and teaching corpora exchange: Lessons learned in five years by the Mulce project. *International Journal of Technology Enhanced Learning, 4*(1/2), 11–30.

Reynolds, R., & Chiu, M. M. (2012). Contribution of motivational orientations to student outcomes in a discovery-based program of game design learning. In J. van Aalst, K. Thompson, M. J. Jacobson & P. Reimann (Eds.), *Proceedings of the 10th International Conference of the Learning Sciences* (Vol 1, pp. 356–360). Sydney, Australia: International Society of the Learning Sciences

Rosé, C. P., & Lund, K. (2014). Methods for multivocality. In D. D. Suthers, K. Lund, C. P. Rosé, C. Teplovs & N. Law (Eds.), *Productive multivocality in the analysis of group interactions*, Chapter 32. New York: Springer

Roth, W.-M. (2003). *Toward an anthropology of graphing: Semiotic and activity-theoretic perspectives*. The Netherlands: Springer (formerly Kluwer Academic Publishers).

Sawyer, R. K. (Ed.). (2006). *The Cambridge handbook of the learning sciences*. New York: Cambridge University Press.

Scardamalia, M. (2004). CSILE/Knowledge Forum®. In *Education and technology: An encyclopedia* (pp. 183–193). Santa Barbara: ABC-CLIO.

Schwarz, B., Wang, C., Chiu, M. M., Ching, C. C., Walker, E., Koedinger, K. R., et al. (2010). Adaptive human guidance of computer-mediated group work. In K. Gomez, L. Lyons & J. Radinsky (Eds.), *Proceedings of the 9th International Conference of the Learning Sciences* (Vol. 2, pp. 149–156). Chicago, IL: International Society of the Learning Sciences

Stahl, G. (Ed.). (2009). *Studying virtual math teams*. New York: Springer.

Star, S. L., & Griesemer, J. R. (1989). Institutional ecology, "translations" and boundary objects: Amateurs and professionals in Berkeley's museum of vertebrate zoology. *Social Studies of Science, 19*(3), 387–420.

Suthers, D. D. (2006). Technology affordances for intersubjective meaning-making: A research agenda for CSCL. *International Journal of Computer Supported Collaborative Learning, 1*(3), 315–337.

Suthers, D. D., Rosé, C. P., Lund, K., & Teplovs, C. (2014). A readers' guide to the productive multivocality project. In D. D. Suthers, K. Lund, C. P. Rosé, C. Teplovs & N. Law (Eds.), *Productive multivocality in the analysis of group interactions*, Chapter 3. New York: Springer

Suthers, D. D., Christie, M. A., Goldman, S., & Hmelo-Silver, C. E. (1999). Collaborative use of representations: Analyzing learning interactions: A workshop held on Dec 11, 1999 in conjunction with the 3rd computer supported collaborative learning conference. Stanford University

Suthers, D. D., Dwyer, N., Medina, R., & Vatrapu, R. (2010). A framework for conceptualizing, representing, and analyzing distributed interaction. *International Journal of Computer Supported Collaborative Learning, 5*(1), 5–42.

Suthers, D. D., Lund, K., Rosé, C. P., Dyke, G., Law, N., Teplovs, C., et al. (2011). Towards productive multivocality in the analysis of collaborative learning. In H. Spada, G. Stahl, N. Miyake, & N. Law (Eds.), *Connecting Computer-Supported Collaborative Learning to Policy and Practice: Proceedings of the 9th International Conference on Computer-Supported Collaborative Learning (CSCL 2011)* (Vol. III, pp. 1015–1022). Hong Kong: International Society of the Learning Sciences

Suthers, D. D., Lund, K., Rosé, C. P., & Teplovs, C. (2014). Achieving productive multivocality in the analysis of group interactions. In D. D. Suthers, K. Lund, C. P. Rosé, C. Teplovs & N. Law (Eds.), *Productive multivocality in the analysis of group interactions*, Chapter 31. New York: Springer

Tashakkori, A., & Teddle, C. (Eds.). (2003). *Handbook of mixed methods in social and behavioral research.* Thousand Oaks, CA: Sage Publications.

Wessner, M., & Kienle, A. (2007). Interdisciplinarity in the CSCL community: an empirical study. In C. A. Chinn, G. Erkens & S. Puntambekar (Eds.), *Proceedings of the 8th International Conference on Computer Supported Collaborative Learning* (pp. 776–784). New Brunswick, NJ, USA: International Society of the Learning Sciences

Wise, A. F., & Chiu, M. M. (2011). Analyzing temporal patterns of knowledge construction in a role-based online discussion. *International Journal of Computer Supported Collaborative Learning, 6*, 445–470.

Wise, A. F., & Chiu, M. M. (2011b). Knowledge construction patterns in online conversation. In H. Spada, G. Stahl, N. Miyake, & N. Law (Eds.), *Connecting Computer-Supported Collaborative Learning to Policy and Practice: Proceedings of the 9th International Conference on Computer-Supported Collaborative Learning (CSCL 2011).* Hong Kong: University of Hong Kong

Chapter 2
Methodological Dimensions

Kristine Lund and Daniel D. Suthers

The Productive Multivocality Project brought together analysts from different theoretical and methodological traditions to learn whether and how our approaches can complement each other and where essential differences lie. As a conceptual aid, we developed a set of five dimensions along which to describe analytic methods. This chapter discusses these dimensions, which are then used throughout this volume to briefly characterize the various analytic methods when introducing them in the case studies and also as a conceptual tool in our summary discussions of the project. The dimensions essentially take a distributed cognition view on analysis, by describing how analyses are achieved through transformations of representations in a system of analysts and analytic representations (Hutchins, 1995). Briefly stated, the dimensions as they were introduced in Chap. 1 of this volume are as follows:

1. *Theoretical assumptions*: What ontological and epistemological assumptions are made about phenomena worth studying, and how can we come to know about them?
2. *Purpose of analysis*: What is the analyst trying to find out about interaction?
3. *Units of action, interaction, and analysis*: In terms of what fundamental relationships between actions do we conceive of interaction? What is the relationship of these units to the unit of analysis?
4. *Representations*: What representations of data and representations of analytic constructs and interpretations capture these units in a manner consistent with the purposes and theoretical assumptions?

K. Lund (✉)
ICAR Research Lab, CNRS—University of Lyon, 15 parvis René Descartes, 69007 Lyon, France
e-mail: kristine.lund@ens-lyon.fr

D.D. Suthers
Department of Information and Computer Sciences, University of Hawai'i at Manoa, 1680 East West Road, POST 309, Honolulu, HI 96822, USA
e-mail: suthers@hawaii.edu

D.D. Suthers et al. (eds.), *Productive Multivocality in the Analysis of Group Interactions*, Computer-Supported Collaborative Learning Series 16,
DOI 10.1007/978-1-4614-8960-3_2, © Springer Science+Business Media New York 2013

5. *Analytic manipulations*: What are the analytic moves that transform a data representation into successive representations of interaction and interpretations of this interaction? How do these transformations lead to insights concerning the purpose of analysis?

The dimensions taken as a whole are methodological in the sense that they aid us in our study (ology) of methods, and as such they invite consideration of how theory and method are linked and influence each other. Exploring the relations between theory and method in studies of group interaction is a central theme of this volume. Below we consider methodological issues associated with each of the above dimensions in turn.

Theoretical Assumptions

What ontological and epistemological assumptions are made about phenomena worth studying, and how can we come to know about them?

Researchers carry out their work within a particular paradigm, although they might not explicitly articulate this. Some researchers may not critically examine their ontological stance (what is the nature of reality?) or their epistemological stance (how can we come to know about the nature of reality?) in relation to their methods, but whether implicit or explicit, these stances make a difference in how one carries out research (Guba & Lincoln, 1982; Tuli, 2011). For example, whether one believes that reality exists independent of ourselves and that existing laws can be discovered (i.e., positivism) or whether one believes that reality is socially constructed and therefore subjective (e.g., social constructionism) has implications for acceptable methods of evaluating claims. Yet, often young researchers are taught methods without ever being asked to consider the underlying ontological or epistemological issues, and experienced researchers may not consider these issues. Bryman (2007) notes that some researchers—especially those employing mixed methods (e.g., both quantitative and qualitative)—avoid the ontological divide by labeling themselves as "pragmatists" (e.g., Johnson & Onwuegbuzie, 2004) and thinking of their research in terms of what can be done with outcomes instead of attempting to resolve a millennia old philosophical dilemma (see also Onwuegbuzie & Leech, 2005). Finally, some authors argue (e.g., Guba & Lincoln, 1982; Johnson & Onwuegbuzie, 2004) that both qualitative and quantitative methods may be used appropriately with any research paradigm. According to Guba and Lincoln, the debate should take place in relation to the implications of assumptions inherent in the overarching paradigms and not on the relative utility of qualitative versus quantitative methods.

Since we agree with Guba and Lincoln, let's take a closer look at the implications of epistemological assumptions, which concern the relationship between the knower or would-be knower and what can be known. The answer to the ontological question constrains the answer to the epistemological one (Guba & Lincoln, 1994). For example, if there is a reality "out there," independent of our observing it, then our posture is one of objective observation. And conversely, if we claim objectivity, then we are implying that a "real" world exists about which we can be objective. Indeed, questions of method are secondary to and dependent upon questions of paradigm, the latter being the belief system or world view (based on ontological and epistemological positions) that guides the investigator in choices of method (Guba & Lincoln, 1994).

In the setting of this project, we expected that everyone would include "interaction" among the phenomena worth studying and possibly some version of "learning." Rather than simply naming phenomena, it is more illuminating to identify what the method assumes about the forms interaction and learning take and the aspects of phenomena worth attending to. In what follows, we use learning as an example. How is it defined? What exactly about learning is being focused on? Researchers conceptualize group interaction and learning in different ways, depending on the researcher's framework (Greeno, Collins, & Resnick, 1996; Suthers, 2006).

One definition of learning might be the permanent modification, due to interactions with the environment, of the disposition of an individual to carry out a behavior or perform a mental activity (Le Ny & Sabah, 2002). Within the behaviorist view of this definition, an example is operant conditioning, in which a learner changes behavior that operates upon the environment in order to maximize rewards and minimize punishment. Psychology as the behaviorist views it is a purely objective experimental branch of natural science (Watson, 1913) and is therefore aligned with the positivist ontological stance. The cognitivist view of this definition of learning—like behaviorism—understands learning as resulting from experience within a stable, objective world, but instead of focusing on direct contingencies between stimuli and responses, it uses models of mental processes to mediate the stimulus–response relationship (Kirschner & Whitson, 1997). In either case, these theoretical orientations lead naturally to methods that quantify relationships between environmental stimuli or conditions and measurable aspects of behaviors on relatively moderate time scales.

Alternative views of learning still consider the individual as the agent of learning but attempt to apprehend learning in the context of social interaction, with other individuals, groups, or communities. The Vygotskian approach radically reoriented learning theory from an individualistic to a sociocultural perspective, but social can refer to both an interaction between two people (e.g., adult–child) or to wider interactions within culturally defined structures (Kozulin, 2003). Each psychological function that is to be learned is seen as appearing twice during development, once in the form of interaction with others and a second time as an inner internalized form of this function (Vygotsky, 1978). In a similar socially oriented view, Tomasello (1999) argues that human cultural learning is possible because as individuals, we have the ability to understand others as beings like us, who have intentional and mental lives like our own. In order to socially learn the conventional use of a tool or a symbol, children must understand why (to what end?) someone else uses that tool: What is its intentional significance? These sociocultural views on learning do not fit into the positivistic stance, long the dominant view in science. Tongue in cheek, Kozulin (op. cit.: 435) notes the difficulties for Vygotsky: his "samples are small, data are unclear and/or ambiguous, advanced statistics are absent, and it is not clear how he controlled the independent variables." But since we can safely infer that these are not measures for success in Vygotsky's ontological and epistemological view, it doesn't matter. From Tomasello's (1999) evolutionary perspective, much can be accomplished culturally in a quarter of a million years, and young children have countless learning experiences by actively engaging with their cultural environments over the course of several years, days, or even hours. As Tomasello's goal is to explain the universal features of what is unique to human cognition (e.g., the creation and use of material, symbolic, and institutional artifacts with accumulated histories) but also the particularities of specific cultures, he focuses in "Vygotskian fashion" (p. 10) on the kinds of evolutionary, historical, and ontogenetic processes that might have transformed the fundamental skills shared with primates (e.g., perception, memory, attention, categorization) into what is specific about human cognition. Thus, these theoretical perspectives lead to methods that examine a much broader range of time scales and relevant objects (e.g., the role of cultural histories and artifacts).

But what if we want to talk about the group of the agent of learning instead of individual learning as influenced by external social or cultural influences? Stahl (2010) argues that there are distinct phenomena and processes at the individual, small-group, and community levels, and analyses at each level reveal different insights. He gives an alternative to (1) theories with a psychological view of mental processes at the individual level but that still acknowledge social and cultural influences and (2) theories at the community level (e.g., Engeström, 1999; Lave & Wenger, 1991; Suchman, 1987). Stahl (2006) introduced the term *group cognition* to refer to processes at the small-group level that are neither reducible to processes of individual minds nor imply the existence of a group mind. They are processes like "interpersonal trains of thought, shared understandings of diagrams, joint problem conceptualizations, common references, coordination of problem-solving efforts, planning, deducing, designing, describing, problem solving, explaining, defining, generalizing, representing, remembering and reflecting as a group" (Stahl 2010, p. 35). Suthers (2006) prefers to dispense with the cognitive metaphor, calling processes at this level of agency *intersubjective meaning-making* and points out that these processes involve compositions of interpretations of aspects of prior contributions that are taken up by participants. Intersubjective meaning-making is similar to distributed cognition (Hutchins, 1995), but the focus is on interpretations of meaning that have generative power rather than transformations of representations that implement a computation in a socio-technical system. Methodological consequences of this theoretical conception of learning include the need foreground the interactional processes by which groups accomplish learning and to derive explanatory accounts from these actual processes (Koschmann, Stahl, & Zemel, 2004; Koschmann et al., 2005).

Although we did not originally mean for the theoretical assumption dimension to also include methodological assumptions, such assumptions could well fall under this dimension if stated in epistemological terms (how we come to know about the phenomenon of interest). For example, ethnomethodology (Garfinkel, 1967) and arguably to a lesser extent conversation analysis (Goodwin & Heritage, 1990) are based on the theoretical assumption (if we may attempt a brief gloss of Garfinkel's complex prose) that no sociological entities (norms, rules, etc.) external to actual instances of behavior are needed to explain the organized nature of that behavior, as this ordered nature is accomplished by the very methods that participants use to make their behavior organized for and to themselves. Therefore, the constructs used to describe participants, action, and context must be used by or at least recognizable in the orientations of the participants themselves. This stance has radically emic implications for researchers' methods. For example, it excludes hypothesis testing, application of coding schemes, or generalization beyond the situated accomplishment of the participants. Even interviewing informants, normally considered appropriate for emic anthropological research, is excluded, as the methods by which participants organize their interview behavior are not the same as their methods of participation in their culture (Goodwin & Heritage, 1990). Essentially, ethnomethodological inquiry is a process of uncovering participants' analysis of their own behavior.

Another example of a methodological assumption, but this time stemming from a positivistic paradigm, is the idea that only experimental inquiries allow you to determine whether a treatment causes an outcome to change (Light, Singer, & Willet, 1990; cited by Maxwell, 2004). Maxwell explains that this view of causality stems from Hume, who argues that we cannot directly perceive causal relationships, and thus, we can have no knowledge of causality beyond the observed regularities in associations of events (Maxwell, op. cit.: 244). Holding this assumption about causality implies that causal inferences require a systematic comparison of situations in which the presumed causal factor is present or absent (or perhaps varies in strength) as well as being able to control for other possible explanatory factors. On the other hand, realism (as opposed to positivism and some aspects of constructivism) gives an alternative view of causal explanation that sees "causation as fundamentally a matter of processes and mechanisms rather than observed regularities" (Maxwell, op. cit.: 246). Maxwell goes on to explain that realism asserts that some causal processes can indeed be directly observed (contrary to what Hume argued), that context is intrinsically involved in causal processes (and is not just reduced to a set of extraneous variables), that mental events and processes are real phenomena that can be causes of behavior, and that causal explanation does not inherently depend on preestablished comparisons.

These examples all illustrate how methodological assumptions depend upon overarching ontological and epistemological viewpoints. Assumptions about the nature of reality, about context, language, or knowledge, collectively constitute a mechanism for investigation that produces or reflects interpretations framed in its own terms and not neutral descriptions and explanations (Yanchar & Williams, 2006). In the following sections, we show how the other methodological dimensions also depend on ontological, epistemological, and their associated methodological assumptions. They are purpose of analysis, units of interaction, representations, and analytic manipulations.

Purpose of Analysis

What is the analyst trying to find out about interaction?

Some example purposes of analysis were already stated in the theoretical assumptions section. The reader will recall that Tomasello's far-reaching goal is to explain the universal features of what is unique to human cognition, but he attempts to accomplish this through the study of how intentional tool use is socially learned. A major goal for Vygotsky (Kozulin, 2003: 436) was to draw a developmental path of a given phenomenon (e.g., mediated memory, scientific concepts). To achieve this goal, he carefully investigated the developmental phases of the phenomenon in question in every study. Vygotsky's objective of studying "not only the final effect of the operation, but its specific psychological structure" led to the method of double stimulation, in which secondary stimuli are offered that the learner can incorporate as auxiliary means to problem solving (Vygotsky, 1978). Some other examples of purposes of analysis within computer-supported collaborative learning, expressed at different levels of granularity, are (1) descriptively characterizing the phenomenon by making interaction apparent; (2) finding causal relationships between variables, e.g., how to link process quality and knowledge construction; (3) design-oriented purposes, such as how to mediate and transform learning and teaching with technology; (4) practice-oriented purposes, such as how to support instructors; (5) seeking metrics to use in other research or applications, such as how to measure the quality of collaboration; and (6) methodological purposes, such as how to define the process of interaction analysis (derived from Lund, 2011).

For understanding specific analytic methods, it is more informative to consider "near" purposes (e.g., "the recognition of inter-animation patterns among voices," to take an example from Trausan-Matu, this volume) rather than ultimate "far" purposes (e.g., to understand how learning takes place in small groups). Thus we will generally characterize analyses in terms of near purposes. Of course, the connection to the larger purpose can be made as well (e.g., stating how understanding interanimation of voices might bear upon understanding learning in small groups). This dimension serves as a nice bridge between what has been foregrounded under theoretical assumptions to what relationships the analysis will actually attend to.

Units of Action, Interaction, and Analysis

In terms of what fundamental relationships between actions do we conceive of interaction? What is the relationship of these units to the unit of analysis?

Originally, this dimension was called simply "Unit of Interaction," as the relational structure that makes an analysis an analysis of interaction (rather than some other kind of analysis) is important for understanding our methods. However, over the course of the project, we found that (1) unit of interaction is easily confused with unit of observation, action, or analysis and that (2) it is informative to identify these other units as well as the unit of interaction. Therefore we discuss all of these units explicitly.

In some paradigms, the unit of observation is the smallest entity for which data is gathered. For example, the unit of observation may be student's response to a single question in a student-test administration (and there are many students and several tests). In conversation analysis, the units over which we work can be below the utterance level. The unit of observation is the smallest data available to be coded, quantified, or interpreted.

But often the unit of observation is at a finer grain than the unit you are interested in making a claim about. For example, you might be making observations at the individual student level, but you are interested in comparing performance of students who work with an intelligent tutoring system versus performance of those using a textbook. Your analysis would aggregate students across these two groups, and the groups become your units of analysis. Unit of analysis is relative to the analysis: different analyses can take the same data and operate with different units of analysis. Hierarchical analysis explicitly works with multiple nested units.

Interaction is *inter*-action: something between actions. There are more than just two actions; there is also some kind of relationship between them. We therefore assumed that any analysis of interaction would work with a relationship between actions as one of its fundamental units. The way one characterizes interaction is a crucial difference between methods.

We asked the analysts in this book to include the unit of observation and other units of analysis in their description, but we requested that their description of unit of interaction clearly state what relationships between actions are taken as fundamental to the analysis. If interaction is related sets of actions, then the analyst should specify what that relation is and whether units of action are logically prior to the interaction or can only arise after identifying the unit of interaction. For some methods the unit of interaction may be obvious as it is very explicit in the method, such as in polyphony (Trausan-Matu, Chap. 6, this volume), uptake analysis (Looi, Song, Wen, & Chen, Chap. 15, this volume; Medina, Chap. 16, this volume), or relevancies between adjacency pairs (Stahl, this volume). For others it may require more thought, for example, while a statistical breakpoint analysis in statistical discourse analysis (Chiu, 2008; Chap. 7, Chiu, this volume-a) does not explicitly ask about relationships between individual acts, it seeks to group acts by discontinuities in variables between sets of acts within two contiguous time spans. As it turned out, some analyses, such as Jeong (Chap. 18, Jeong, this volume), did not work with an explicit relationship of interaction.

Inclusion of this dimension was partly influenced by conversation analysis. CA was developed in order to analyze "practices of reasoning and inference that inform the production and recognition of intelligible courses of action. Central to the achievement of this objective has been the development of a theory of context that links processes of interpretation to action within a reflexive, time-bound process" (Goodwin & Heritage, 1990). Contrary to former linguistic approaches that worked on isolated or invented sentences, CA sought to treat the stream of speech actually uttered by a speaker in conversation as forms of action that were situated within specific contexts. The analysis of any utterance should therefore begin from the action (talk or other forms of action) and other aspects of the setting that it emerges from. In CA, the emblematic notion of the unit of interaction is the "adjacency pair" (e.g., such as question–response or greeting–greeting), developed by Sacks and Schegloff (Sacks & Schegloff, 1979; Schegloff & Sacks, 1973) where a current action requires the production of a reciprocal action at the first possible opportunity (Goodwin & Heritage, 1990). When a reciprocal action does not occur, participants (and hence CA analysts) attempt to understand why this was the case. This particular definition of the unit of interaction is supported by an ontological assumption, namely, that such adjacency pairs are not a description of statistical regularities in patterns of action nor are they a specification of some internalized rule that drives behavior. Rather, they illustrate how participants constrain one another and analyze each other's actions in order to produce the appropriate reciprocal action and develop coherent interactional sequences (Goodwin & Heritage, op. cit). However, in this volume we intend "unit of interaction" to allow for other ontological assumptions and also to extend to nonconversational media. Although CA originally focused on audiotaped and transcribed talk, it later extended the notion of action and reciprocal action to include multimodality (e.g., gestures, gaze, posture, and coordination of technological artifacts), as is particularly evident in the work of Goodwin (2000, 2003). In the CSCL context, Suthers and colleagues have been inspired by ethnomethodology and conversation analysis in order to also argue that not only the meanings of utterances are contextual and negotiated in order to support action, but also the same is true for nonlinguistic representations that support action (Suthers, Dwyer, Medina, & Vatrapu, 2010; Suthers, Dwyer, Vatrapu, & Medina, 2006). They use the term "uptake" instead of "adjacency pair" as a generalized building block of interaction that can be constructed of relations between nonadjacent events and found in diverse media.

Representations

What representations of data and representations of analytic constructs and interpretations are used to capture these units in a manner consistent with the purposes and theoretical assumptions?

Analyses of interaction (as undertaken by researchers rather than ethnomethodological participants) almost always include the construction of representations of the interaction—the "data" record such as a video or audio recording and practices of constructing and interpreting successive analytic representations, sometimes beginning with a "transcript" and possibly including representations of segments (units of analysis), annotations, codes, links, aggregations of units or of metrics, summaries, etc. Thus, analysis can be characterized in part by what representations are constructed.

The ability to create and manipulate visual representations is a cognitive skill that scientists acquire as they become accomplished participants in the methods that define a particular domain. Gooding (2010) argues that the important feature of a representation is its plasticity and integrative power, enabling its adaptation to the changing social and cognitive demands of the creative process (see also "cognitive dimensions of notations," Blackwell & Green, 2003). He also argues that this adaptability of representations is managed in the context of three constraints: (1) theories about the domain and problem-solving methods regarding it (in our case, group interaction); (2) "imaging conventions" or notations (two examples for group interaction are social network analysis and transcriptions); and (3) "material resources" of imaging technologies (an example for group interaction is synchronizing multiple streams of data: videos, transcriptions, and traces of computer-mediated human interaction) (Dyke, Lund, & Girardot, 2009). Using the terminology of Suthers (2001), "representational tools" are a form of material resource that make the imaging conventions of "notations" available in software settings; and these notations may offer variable affordances for individual and group interaction.

We can understand the process of analysis, particularly multivocal analysis, as a form of *distributed cognition* (Hutchins, 1995). Distributed cognition is neither solely internal nor solely external but takes place through transformations of a system of representations that are distributed between the two. The social and cognitive acts of analysis, like other such acts, involve translations between representations. To take an example from Suthers and colleagues (Suthers et al., 2010; Suthers & Medina, 2011), a time-ordered representation of individual contributions and their characteristics such as actor, linguistic content, and medium can be translated into a relational graph based on how words, phrases, and ideas are echoed across contributions and how actors address each other (polyphonic analysis does something similar); and this graph of observable contingencies can be converted into a summary representation of uptake evidenced by such contingencies, which in turn is folded into a sociogram of who uptakes from whom with what frequency (Suthers & Rosen, 2011; Suthers & Desiato, 2012).

The representations we use say a lot about our methods. They may also suggest implicit theoretical assumptions (although not in a deterministic manner: the researcher also has agency). Consider, for example, transcripts. Some analyses may require different information than others, and part of the value in transcripts is that they are selective, making some aspects of the data salient at the cost of others. Gail Jefferson (2004) compares unelaborated transcripts by Harvey Sacks with her own notational conventions that capture the nuances of prosody and timing. She illustrates how some questions of interpretation do not even arise, let alone can be resolved, without the information her notation includes. Yet, in making prosody and timing salient, the salience of the interaction as a verbal conversation is somewhat obscured. Also, her notation focuses primarily on verbal acts and relegates nonverbal acts to annotations or parenthetical comments, implying that nonverbal acts are merely contextual or play a subordinate role. One might use separate columns for verbal and nonverbal acts, but this implies that there is non-overlap and does not highlight the coordination across multiple verbal and nonverbal semiotic fields (Goodwin, 2003). Ochs (1979) provides a detailed discussion of how the notational format of transcripts has biases that can be derived from or have theoretical implications, with examples in the transcription of interaction between an adult and a very young child. When transcripts are written in sequential order, as is common for conversation analysis, there is bias towards reading contributions as contingent upon immediately prior contributions and setting up expectations (preferences) for immediately following contributions. However, very young children do not necessarily attempt to make their contributions relevant to the immediately prior contribution. They may engage in running narratives where their contributions are more relevant to their own prior contributions. As a fix for this, Ochs suggests placing participants in their own separate columns, aligned horizontally for time, but enabling one to read each participants' narrative independently. This may then lead to a new bias: in languages in which we read in the left-to-right direction, the interlocutor placed on the left may be seen as dominant or as the initiator of all interactions. To counter this bias, Ochs suggests placing the adult on the right-hand side.

Once the transcript is constructed, we then construct other analytic representations from it that offer restricted and selected narratives about what the world was like at a particular moment through a combination of symbolic, iconic, and indexical signs (Duranti, 2006). As Duranti points out, both a transcript's evolution and the evolution of the transcript's interpretations can provide us with a record of our epistemological and theoretical changes. We will see examples throughout this volume, including how graphs of relationships between events make interactional structure explicit under concepts of adjacency, polyphony, transformations, and uptake (Looi et al., Chap. 15, this volume; Lund & Bécu-Robinault, Chap. 17, this volume; Medina, Chap. 16, this volume; Stahl, Chap. 28, this volume; Trausan-Matu, Chap. 6, this volume); how interaction can be differentially understood through representations of changes in values of collections of variables (Chiu, Chap. 7, this volume; Chap. 23, this volume) or is understood primarily through the physical artifacts that it produces (Jeong, this volume); and how it can be abstracted to networks of relations between concepts and/or persons (Goggins & Dyke, Chap. 29, this volume; Teplovs, Chap. 21, this volume). Here we have only touched on a few ways in which representations of different facets of human interaction show a variety of ways of portraying and understanding interactional phenomena. Many more examples are possible when considering other analytic representations: see Chap. 33 (Dyke, Lund, Suthers, & Teplovs, this volume) for further discussion.

Analytic Manipulations

What are the analytic moves that transform a data representation into successive representations of interaction and interpretations of this interaction? How do these transformations lead to insights concerning the purpose of analysis?

The foregoing account has already noted that the act of analysis can be viewed as consisting of certain manipulations and transformations of representations, presumably beginning with data representations and then deriving analytic representations and interpretations. The manipulations operate on the representations described by the previous dimension, translating one to another. In the process, the unit of interaction is involved, being identified and either interpreted directly or transformed in other ways into what is worth interpreting. The final representation(s) should make salient something relevant to the identified purpose of analysis. Just as we understand interaction as not consisting of isolated acts but rather as acts being understood in relation to each other, analysis is not understood as isolated representations, but rather the representations are understood in relation to each other and the practices through which they are transformed and interpreted. These practices will reflect theoretical assumptions, particularly the epistemology of the tradition within which notations become representations. As a simple example, a tradition in which learning is a matter of uncovering participant practices for doing learning will "transform" (they would not put it this way) records of participant interaction into rich accounts of how particulars of coordinated vocalization, gaze, gesture, etc. offer and affirm interpretations of meaning among the group, while another tradition that seeks accounts of regularities between theoretical constructs across the "noise" of multiple settings may take the same transcript and generate counts of codes related to these constructs and aggregate them numerically for statistical characterization. We will see many examples of different kinds of manipulations throughout this volume.

Conclusions

We end this chapter with an anecdote by Richards (1995) illustrating how the methodological dimensions of two researchers from different disciplines guide what aspect of a phenomenon of interest they focus on. Richards was at a faculty party where researchers discovered that one batch of homemade beer was less bubbly than another one. A biologist suggested that it was because there was less air in the bottle, and decreased oxygen meant that the yeast would die sooner, thereby converting less sugar to alcohol and producing fewer bubbles. A physicist countered that it was instead crucial to calculate how much pressure was building up in the bottle and that the increased pressure was what was probably killing the yeast and that what should be examined was what the effect of more fluid and less air would be on the amount of pressure in the bottle.

As Richards tells it, the party quickly formed into two groups: one of biologists and one of physicists, each discussing the theory that made sense within their respective scientific traditions. Neither group talked to each other, and it was clear that they were not going to compare results. Neither group was posing more interesting or more relevant questions, but perhaps if they had conversed and worked together, they would have discovered ways of converging. It may be safe to say that both groups were operating in positivistic paradigms, with their associated theoretical assumptions of discovery of objective universal laws and indeed both were trying to understand the bubbliness of the batch of beer (purpose of analysis). However, each had a different unit of analysis (e.g., relation of oxygen quantity to yeast life vs. relation of pressure to yeast life) and therefore different representations and analytic manipulations. Richards doesn't give the solution to the enigma, but both hypotheses can be tested by first keeping constant pressure and decreasing oxygen level and then keeping constant oxygen level and increasing pressure and, in both cases, checking to see if the beer is equally less bubbly in both cases than a "control" batch of beer, from which the experimental values of oxygen and pressure varied.

Although in this particular case one or both theories may be true (they are not necessarily incompatible) and this result is verifiable by experiment, such an example helps us to see how some disciplinary views on what constitutes explanation of phenomena may be more difficultly reconciled. If we consider an experimental cognitive psychologist and a conversation analyst, it is already difficult to converge on a similar purpose of analysis. The former is most likely in a positivistic paradigm, using quantitative analyses in an attempt to discover causal connections between isolated variables, whereas the latter will be in a constructivist paradigm, using qualitative analyses in order to describe the details of participants' negotiations of events in a particular context. Both may be interested in human interaction but will focus on different aspects of it and employ different units of interaction and therefore different representations and analytical transformations. As Richards (op. cit.) asks (p. 59): "As we give up truth or nature as the ultimate determinant, and assume some degree of incommensurability between traditions, how do I, as a scientist, make a rational decision to accept or join a new tradition?" We hope this book gives researchers, both new to and experienced in their fields, a means to answering this question while they examine more critically the tradition(s) they have been educated in.

References

Blackwell, A., & Green, T. (2003). Notational systems—the cognitive dimensions of notations framework. In J. M. Carroll (Ed.), *HCI models, theories, and frameworks: Toward a multidisciplinary science* (pp. 103–134). San Francisco, CA: Morgan Kaufmann Publishers.

Bryman, A. (2007). Barriers to integrating quantitative and qualitative research. *Journal of Mixed Methods Research, 1*(1), 8–22.

Chiu, M. M. (2008). Flowing toward correct contributions during group problem solving: A statistical discourse analysis. *Journal of the Learning Sciences, 17*(3), 415–463.

Chiu, M. M. (this volume-a). Social metacognition, micro-creativity and justifications: Statistical discourse analysis of a mathematics classroom conversation. In D. D. Suthers, K. Lund, C. P. Rosé, C. Teplovs & N. Law (Eds.), *Productive multivocality in the analysis of group interactions*, Chapter 7. New York: Springer

Chiu, M. M. (this volume-b). Statistical discourse analysis of an online discussion: Cognition and social metacognition. In D. D. Suthers, K. Lund, C. P. Rosé, C. Teplovs & N. Law (Eds.), *Productive multivocality in the analysis of group interactions*, Chapter 23. New York: Springer

Duranti, A. (2006). Transcripts, like shadows on a wall. *Mind, Culture & Activity, 13*(4), 301–310.

Dyke, G., Lund, K., & Girardot, J.-J. (2009). Tatiana: An environment to support the CSCL analysis process. In C. O'Malley, P. Reimann, D. Suthers & A. Dimitracopoulou (Eds.), *Computer Supported Collaborative Learning Practices: CSCL 2009 Conference Proceedings* (pp. 58–67). Rhodes, Greece: International Society of the Learning Sciences

Dyke, G., Lund, K., Suthers, D. D., & Teplovs, C. (this volume). Analytic representations and tools for productive multivocality. In D. D. Suthers, K. Lund, C. P. Rosé, C. Teplovs & N. Law (Eds.), *Productive multivocality in the analysis of group interactions*, Chapter 33. New York: Springer

Engeström, Y. (1999). Activity theory and individual and social transformation. In Y. Engeström, R. Miettinen, & R.-L. Punamäki (Eds.), *Perspectives on activity theory* (pp. 19–38). New York, NY: Cambridge University Press.
Garfinkel, H. (1967). *Studies in ethnomethodology*. Englewood Cliffs, NJ: Prentice-Hall.
Goggins, S. P., & Dyke, G. (this volume). Network analytic techniques for online chat. In D. D. Suthers, K. Lund, C. P. Rosé, C. Teplovs & N. Law (Eds.), *Productive multivocality in the analysis of group interactions*, Chapter 29. New York: Springer
Gooding, D. C. (2010). Visualizing scientific inference. *Topics in Cognitive Science, 2*, 15–35.
Goodwin, C. (2000). Action and embodiment within situated human interaction. *Journal of Pragmatics, 32*, 1489–1522.
Goodwin, C. (2003). The semiotic body in its environment. In J. Coupland & R. Gwyn (Eds.), *Discourses of the body* (pp. 19–42). New York: Palgrave/Macmillan.
Goodwin, C., & Heritage, J. (1990). Conversation analysis. *Annual Review of Anthropology, 19*, 283–307.
Greeno, J. G., Collins, A. M., & Resnick, L. B. (1996). Cognition and learning. In D. C. Berliner & R. C. Calfee (Eds.), *Handbook of educational psychology* (pp. 15–46). New York: Simon & Schuster Macmillan.
Guba, E. G., & Lincoln, Y. S. (1982). Epistemological and methodological bases of naturalistic inquiry. *Educational Communication and Technology Journal, 30*(4), 233–252.
Guba, E. G., & Lincoln, Y. S. (1994). Competing Paradigms in Qualitative Research. In N. Denzin & Y. Lincooln (Eds.), *Handbook of qualitative research* (1st Ed. 1994) (pp. 105–117). Thousand Oaks, CA: Sage.
Hutchins, E. (1995). *Cognition in the wild*. Cambridge, MA: The MIT Press.
Jefferson, G. (2004). Glossary of transcript symbols with an introduction. In G. H. Lerner (Ed.), *Conversation analysis: Studies from the first generation* (pp. 13–31). Amsterdam/Philadelphia: John Benjamins.
Jeong, H. (this volume). Development of group understanding via the construction of physical and technological artifacts. In D. D. Suthers, K. Lund, C. P. Rose, C. Teplovs & N. Law (Eds.), *Productive multivocality in the analysis of group interactions*, Chapter 18. New York: Springer
Johnson, R. B., & Onwuegbuzie, A. J. (2004). Mixed methods research: A research paradigm whose time has come. *Educational Researcher, 33*(7), 14–26.
Kirschner, D., & Whitson, J. A. (1997). Editor's introduction to situated cognition: Social, semiotic, and psychological perspectives. In D. Kirschner & J. A. Whitson (Eds.), *Situated cognition: Social, semiotic, and psychological perspectives* (pp. 1–16). Mahwah, NJ: Lawrence Erlbaum Associates.
Koschmann, T., Stahl, G., & Zemel, A. (2004). The video analyst's manifesto (or the implications of Garfinkel's policies for studying practice within design-based research). In Y. Kafai, N. Sandoval, N. Enyedy, A. Nixon & F. Herrera (Eds.), *Proceedings of the Sixth International Conference of the Learning Sciences* (pp. 278–385). Mahwah, NJ: Lawrence Erlbaum Associates
Koschmann, T., Zemel, A., Conlee-Stevens, M., Young, N., Robbs, J., & Barnhart, A. (2005). How do people learn: Member's methods and communicative mediation. In R. Bromme, F. W. Hesse, & H. Spada (Eds.), *Barriers and biases in computer-mediated knowledge communication (and how they may be overcome)* (pp. 265–294). Amsterdam: Kluwer Academic Press.
Kozulin, A. (2003). Psychological tools and mediated learning. In A. Kozulin, B. Gindis, V. S. Ageyev, & S. M. Miller (Eds.), *Vygotsky's educational theory in cultural context* (pp. 15–38). New York: Cambridge University Press.
Lave, J., & Wenger, E. (1991). *Situated learning: Legitimate peripheral participation*. Cambridge: Cambridge University Press.
Le Ny, J.-F., & Sabah, G. (2002). Apprentissage (learning). In G. Tiberghien, H. Abdi, J.-P. Desclés, N. Georgieff, M. Jeannerod, J.-F. Le Ny, P. Livet, J. Pynte, G. Sabah, (Eds.), *Dictionnaire des sciences cognitives* (*Dictionary of Cognitive Sciences*). (pp. 30–34) Paris: Armand Colin
Light, R. J., Singer, J., & Willet, J. (1990). *By design: Conducting research on higher education*. Cambridge, MA: Harvard University Press.

Looi, C.-K., Song, Y., Wen, Y., & Chen, W. (this volume). Identifying pivotal contributions for group progressive inquiry in a multi-modal interaction environment. In D. D. Suthers, K. Lund, C. P. Rose, C. Teplovs & N. Law (Eds.), *Productive multivocality in the analysis of group interactions*, Chapter 15. New York: Springer

Lund, K. (2011). Analytical frameworks for group interactions in CSCL systems. In S. Puntambekar, G. Erkens, & C. E. Hmelo-Silver (Eds.), *Analyzing collaborative interactions in CSCL: Methods, approaches and issues* (pp. 391–411). New York: Springer.

Lund, K., & Bécu-Robinault, K. (this volume). Conceptual change and sustainable coherency of concepts across modes of interaction. In D. D. Suthers, K. Lund, C. P. Rose, C. Teplovs & N. Law (Eds.), *Productive multivocality in the analysis of group interactions*, Chapter 17. New York: Springer

Maxwell, J. A. (2004). Causal explanation, qualitative research, and scientific inquiry in education. *Educational Researcher, 33*(2), 3–11.

Medina, R. (this volume). Cascading inscriptions and practices: Diagramming and experimentation in the Group Scribbles classroom. In D. D. Suthers, K. Lund, C. P. Rose, C. Teplovs & N. Law (Eds.), *Productive multivocality in the analysis of group interactions*, Chapter 16. New York: Springer

Ochs, E. (1979). Transcription as theory. In E. Ochs & B. B. Schieffelin (Eds.), *Developmental pragmatics* (pp. 43–72). New York: Academic.

Onwuegbuzie, A. J., & Leech, N. L. (2005). On becoming a pragmatic researcher: The importance of combining quantitative and qualitative research methodologies. *International Journal of Social Research Methodology, 8*(5), 375–387.

Richards, J. (1995). Construct[ion/iv]ism: Pick One of the Above. In L. P. Steffe & J. Gale (Eds.), *Constructivism in Education*, (pp. 57–63). New Jersey: Laurance Erlbaum Associates.

Sacks, H., & Schegloff, E. A. (1979). Two preferences in the organization of reference to persons in conversation and their interaction. In G. Psathas (Ed.), *Everyday language: Studies in ethnomethodology* (pp. 15–21). New York, NY: Irvington Press.

Schegloff, E. A., & Sacks, H. (1973). Opening up closings. *Semiotica, 8*, 289–327.

Stahl, G. (2006). *Group cognition: Computer support for collaborative knowledge building*. Cambridge, MA: MIT Press.

Stahl, G. (2010). Group cognition as a foundation for the new science of learning. In M. S. Khine & I. M. Saleh (Eds.), *New science of learning: Cognition, computers and collaboration in education* (pp. 23–44). New York, NY: Springer.

Stahl, G. (this volume). Interaction analysis of a biology chat. In D. D. Suthers, K. Lund, C. P. Rosé, C. Teplovs & N. Law (Eds.), *Productive multivocality in the analysis of group interactions*, Chapter 28. New York: Springer

Suchman, L. A. (1987). *Plans and situated actions: The problem of human-machine communication*. Cambridge: Cambridge University Press.

Suthers, D. D. (2001). Towards a systematic study of representational guidance for collaborative learning discourse. *Journal of Universal Computer Science, 7*(3), http://www.jucs.org/jucs_7_3/towards_a_systematic_study

Suthers, D. D. (2006). Technology affordances for intersubjective meaning-making: A research agenda for CSCL. *International Journal of Computer Supported Collaborative Learning, 1*(3), 315–337.

Suthers, D. D., & Desiato, C. (2012). Exposing chat features through analysis of uptake between contributions. Proceedings of the Hawaii International Conference on the System Sciences (HICSS-45), January 4–7, 2012, Grand Wailea, Maui, Hawai'i (CD-ROM). New Brunswick: Institute of Electrical and Electronics Engineers, Inc. (IEEE).

Suthers, D. D., Dwyer, N., Medina, R., & Vatrapu, R. (2010). A framework for conceptualizing, representing, and analyzing distributed interaction. *International Journal of Computer Supported Collaborative Learning, 5*(1), 5–42.

Suthers, D. D., Dwyer, N., Vatrapu, R., & Medina, R. (2006). Analyzing interactional construction of meaning in online learning. In *Proceedings of the 39th Hawai'i International Conference on*

the System Sciences (HICSS-39), January 4–7, 2006, Poipu, Kauai (CD-ROM). Poipu, Hawai'i: IEEE Computer Society Press
Suthers, D. D., & Rosen, D. (2011). A unified framework for multi-level analysis of distributed learning. Proceedings of the First International Conference on Learning Analytics & Knowledge, Banff, Alberta, February 27-March 1, 2011.
Suthers, D. D., & Medina, R. (2011). Tracing interaction in distributed collaborative learning. In S. Puntambekar, G. Erkens, & C. E. Hmelo-Silver (Eds.), *Analyzing interactions in CSCL: Methods, approaches and issues* (pp. 341–366). New York: Springer Science+Business Media, LLC.
Teplovs, C., & Fujita, N. (this volume). Socio-dynamic latent semantic learner models. In D. D. Suthers, K. Lund, C. P. Rosé, C. Teplovs & N. Law (Eds.), *Productive multivocality in the analysis of group interactions*, Chapter 21. New York: Springer
Tomasello, M. (1999). *The cultural origins of human cognition.* Cambridge, MA: Harvard University Press.
Trausan-Matu, S. (this volume). Collaborative and differential utterances, pivotal moments, and polyphony. In D. D. Suthers, K. Lund, C. P. Rosé, C. Teplovs & N. Law (Eds.), *Productive multivocality in the analysis of group interactions*, Chapter 6. New York: Springer
Tuli, F. (2011). The basis of distinction between qualitative and quantitative research in social science. *Ethiopian Journal of Education and Sciences, 6*(1), 97–108.
Vygotsky, L. S. (1978). *Mind in society: The development of higher psychological processes.* Cambridge, MA: Harvard University Press.
Watson, J. B. (1913). Psychology as the behaviorist views it. *Psychological Review, 20*, 158–177.
Yanchar, S. C., & Williams, D. D. (2006). Reconsidering the compatabilty thesis and eclecticism: Five proposed guidelines for method use. *Educational Researcher, 35*(9), 3–12.

Chapter 3
A Reader's Guide to the Productive Multivocality Project

Daniel D. Suthers, Carolyn Penstein Rosé, Kristine Lund, and Chris Teplovs

This chapter serves as a guide to a book reporting on a 5-year collaboration among researchers exploring the basis for productive dialogue between multiple theoretical and methodological traditions in the analysis of group interaction. Following a description of the overall format of the book, several reading strategies are described, and the chapters are outlined with annotations to help the reader implement these strategies.

Organization

The seven sections of this book include the present introductory section, five sections focused on case studies in which multiple analysts analyze the same data, and a final section summarizing lessons learned and implications. The case study

D.D. Suthers (✉)
Department of Information and Computer Sciences, University of Hawai'i at Manoa, 1680 East West Road, POST 309, Honolulu, HI 96822, USA
e-mail: suthers@hawaii.edu

C.P. Rosé
Carnegie Mellon University, Language Technologies Inst. & HCI Inst., Gates Hillman Center, 5415, 5000 Forbes Avenue, Pittsburgh, PA 15213, USA
e-mail: cprose@cs.cmu.edu

K. Lund
ICAR Research Lab, CNRS—University of Lyon, 15 parvis René Descartes, 69007 Lyon, France
e-mail: kristine.lund@ens-lyon.fr

C. Teplovs
Problemshift, Inc., Windsor, ON, Canada, N9E 4M5
e-mail: chris.teplovs@gmail.com

D.D. Suthers et al. (eds.), *Productive Multivocality in the Analysis of Group Interactions*, Computer-Supported Collaborative Learning Series 16,
DOI 10.1007/978-1-4614-8960-3_3, © Springer Science+Business Media New York 2013

sections each has the same internal structure. They each begins with a chapter written by the persons providing the data that describes the setting in which the data was gathered and the nature of the data. Several chapters providing alternative analyses of the data follow this data description chapter. Some of these analyses were iterated after being influenced by other analyses in the sections, resulting in some cross talk between analysis chapters, but the bulk of the discussion of productive multivocality issues is in the final chapter of each section, the discussant chapter. Depending on what was most salient for the given case study, the discussant chapters identify challenges that came up in achieving productive multivocality and how they were addressed, compare the results obtained, and summarize theoretical and methodological issues that were exposed.

Reading Strategies

Any reading of this book should begin with Chap. 1, to understand the key motivations and insights of the project. Then various reading strategies are envisioned, according to the reader's goals. Of course, one strategy is to simply read the entire book in sequence, and with this in mind the book is organized to provide a comprehensive survey of diverse analytic approaches as applied to equally diverse interactional settings, age levels, and topics in five case studies, culminating in what we learned from the entire enterprise. Each of the five case study sections ends with a discussion of issues in and strategies for achieving productive multivocality that are illustrated by the section. The final section of the book aggregates and abstracts these issues and lessons from the case studies. If for whatever reason the reader cannot or need not read the entire book, then one of the following strategies may apply according to the reader's goal.

Reading Goal: Understanding Productive Multivocality

Chapter 31 (Suthers, Lund, Rosé, & Teplovs, this volume) was written to stand alone as a summary of the productive multivocality project and its lessons and implications. It can be used, for example, as an executive summary for those who can read only one chapter or as an introductory reading in a graduate seminar. Readers who wish to go into more depth concerning what we have learned about productive multivocality and implications for a research field might skim the final discussion chapter of each of the five case study sections (Chaps. 8, 13, 19, 24, and 30) to first encounter the issues and lessons in context and then concentrate on the final section of the book where we discuss various aspects of productive multivocality that could apply to other fields of inquiry. References to the case studies in the case study discussions and final reflection chapters might inform such readers where to dive into the case studies for informative examples.

Reading Goal: Figuring Out How to Approach One's Own Data

Some readers may be faced with a dataset similar to one of ours and want to learn about different analytic approaches and what they have to offer, possibly with the intention of using multiple methods. Such a reader may begin with the guide to chapters that follows below to identify the case study that is closest to their interests according to their interactional setting, grade/age level, and topic, and then read the corresponding section in detail. Chapter 31 and the Methods for Multivocality chapter (Rosé & Lund, Chap. 32, this volume) also provide some practical pointers for thinking about how to approach the task.

Reading Goal: Learning About the Range of Analytic Approaches Available

Other readers may also be interested in learning about analytic approaches but not with any particular data in mind. Students and researchers who have been trained in one tradition may want to broaden their perspective, or early career readers may want to explore alternative traditions to pursue. The book can also support a graduate methods course through this strategy. These readers should begin with the description of dimensions along which analytic methods may be described in Chap. 2 (Lund & Suthers, this volume) and then use the guide to chapters that follows to either identify case studies that bring together the methods they want to explore or to construct their own reading trajectory that follows particular methods of interest (e.g., ethnomethodological, network analysis, statistical discourse analysis (SDA)) threaded through the case studies. Chapter 32 compares the experience of the expert analysts whose work is represented in this book with the experiences of graduate students just learning about multivocal analysis and therefore may provide some useful guidance for newcomers.

Reading Goal: Identifying Results for Research and Practice in Application Areas

Since this book includes various specific studies by reputable researchers, each with their own results and insights, this book can also serve as a resource for researchers or practitioners who are not interested in methodological issues but rather are most interested in research questions or issues of practice in one or more of the particular settings we studied. For example, a researcher may be interested in software agents in an intelligent tutoring system context, or a mathematics educator may want to examine a case study of how conceptual issues in mathematics may be addressed with group exercises using simple tools such as paper and a blackboard. Again, such a reader can use the remainder of this chapter to identify the case study that is closest to their interests according to their interactional setting, grade/age level, and topic.

Data Section 1: Pivotal Moments in Origami Fractions

Section Editor: Kristine Lund CNRS

In this section, learning fractions in a 6th-grade Japanese classroom provide the focus for three analytical approaches, each identifying "pivotal moments" within the interaction. The data consists of an English-subtitled video in Japanese of six students folding origami paper and of one teacher monitoring their progress on the blackboard and an accompanying transcription of their talk and gestures. One analyst (Shirouzu, Chap. 5, this volume) sought to identify where the personal foci of learners originate; what happens in the interaction once a learner focuses on, for example, shapes or production methods; and how learner outcomes are related to such foci. Another analyst (Trausan-Matu, Chap. 6, this volume) identified the semantic content of "voices" and their interanimation patterns in a polyphony framework. A third analyst (Chiu, Chap. 7, this volume-a) applied SDA to the dataset in order to see whether recent sequences of utterances affected the likelihood of creating utterances categorized as new ideas, correct ideas, micro-creativity, or justifications. As a consequence of our multivocal approach, all three analysts revisited their methods and modified them in light of discussion with the others. An analysis of the methodological dimensions (cf. Lund & Suthers, Chap. 2) across the three researchers is presented, and lessons learned are summarized in the discussant chapter (Lund, Chap. 8, this volume).

Chapter 4 (Data): Learning Fractions Through Folding in an Elementary Face-to-Face Classroom

Hajime Shirouzu

Shirouzu introduces the fractions dataset in Chap. 4, entitled "Learning Fractions Through Folding in an Elementary Face-to-Face Classroom," a dataset he collected while visiting and teaching students twice in a remote area in Japan. In his chapter, he clarifies the rationale behind his data selection, the design principles of the class he taught, and the learning task he presented to the students as well as its objectives.

Chapter 5 (Analysis): Focus-Based Constructive Interaction

Hajime Shirouzu

If we can analyze the diversity of both the paths learners take and the goals that they reach in a collaborative situation, we will be able to utilize such diversity for further enriching learning. This chapter proposes the model of "focus-based constructive interaction," which hypothesizes that the intramental interaction of each individual creates a personal focus affecting how he verbalizes and acts in collaborative moments and that the verbalization leads to his learning outcome. By applying this model to the origami fraction data, the chapter demonstrates that, even in a shared situation involving the six children, each child deepened his or her own understanding by asking his or her own questions and searching the external world for answers along his or her own focus, which remained relevant for several months. It also shows that the difference in foci produced different interpretations and promoted social interactions among them. The analytic devices of focus and role were discussed and contrasted with individual attributes for explaining individuals' diverse progressions through social interaction.

Chapter 6 (Analysis): Collaborative and Differential Utterances, Pivotal Moments, and Polyphony

Stefan Trausan-Matu

This chapter presents a multivocal analysis method of collaborative learning and its application on the origami fractions dataset, considering several dimensions: spoken dialogue, body language, the visual dimension, internal dialogue (at an intramental level), and echoes. The analysis is performed starting from the polyphonic model, which was previously used for instant messenger conversations and discussion forums and was extended for the face-to-face (F2F) classroom interactions in this dataset. The analysis includes the identification of the voices, in an extended sense, interanimation patterns among them, collaborative and differential utterances, changes in the rhythm (chronotopes), and pivotal moments of the interactions.

Chapter 7 (Analysis): Social Metacognition, Micro-Creativity, and Justifications: Statistical Discourse Analysis of a Mathematics Classroom Conversation

Ming Ming Chiu

This analysis shows how SDA can identify the locations and consequences of pivotal moments and how characteristics of recent turns of talk such as questions and evaluations (social metacognition) are linked to characteristics of subsequent turns of talk, such as correct ideas, new ideas, or justifications. Along with the other studies in this unit, this analysis shows how multivocality can suggest cycles of analyses and help develop further statistical methods.

Chapter 8 (Discussion): A Multivocal Analysis of Pivotal Moments for Learning Fractions in a 6th-Grade Classroom in Japan

Kristine Lund

This chapter compares the pivotal moments each analyst described using the five methodological dimensions discussed in Chap. 2 (Lund & Suthers, this volume): theoretical assumptions, purpose of analysis, unit of analysis/unit of interaction, data representations, and manipulation of data representations. Conclusions are drawn on how redefining the unit of analysis and the unit of interaction in light of other researchers' analyses, interpreting other researchers' pivotal moments in one's own framework, and comparing the semantics of and the relations between analytical concepts all contribute to helping an analyst surpass the limits of a particular method.

Data Section 2: Peer-Led Team Learning for Chemistry

Section Editor: Carolyn P. Rosé

In this section, we use a multivocal leadership construct as a lens for viewing and comparing the dynamics of two different peer-led teams as they solve a chemistry problem related to de Broglie's equation. Four different analysts offer their interpretation of the data, resulting in three analysis chapters: one providing an ethnographic analysis (Sawyer, Frey, & Brown, Chap. 10, this volume); another comparing and contrasting two multidimensional coding and counting approaches, each including a cognitive, relational, and motivational dimension (Howley, Mayfield, Rosé, & Strijbos, Chap. 11, this volume); and finally, a content-focused network analysis chapter (Oshima, Matsuzawa, Oshima, & Niihara, Chap. 12, this volume). Chapter 13 (Carolyn P. Rosé, this volume) discusses how the juxtaposition of the three distinct lenses reveals new insights into the intricate nature of complex constructs like leadership that argue strongly that a multivocal analysis is more than the sum of its parts.

Chapter 9 (Data): Peer-Led Team Learning in General Chemistry

Keith Sawyer, Regina Frey, and Patrick Brown

Peer-led team learning (PLTL) is a collaborative learning technique that has been used on many college campuses, particularly in large lecture classes in departments of chemistry. Several studies have shown that PLTL results in improved learning.

However, researchers have not investigated the discourse practices used by peer leaders and students, and among students themselves, that give rise to this enhanced understanding of chemistry content. To better understand the interactional mechanisms that make PLTL effective, three PLTL sessions for each of 15 veteran peer leaders were videotaped over the course of one semester. The dataset presented here contains transcripts of two PLTL groups as they solved the same problem.

Chapter 10 (Analysis): Knowledge Building Discourse in Peer-Led Team Learning Groups in First-Year General Chemistry

Keith Sawyer, Regina Frey, and Patrick Brown

To better understand the interactional mechanisms that make PLTL effective, we closely examined videotapes of two PLTL groups as they both solved the same chemistry problem. In one group, students engaged in group knowledge building: intellectual conversations where they asked each other questions, provided procedural and conceptual explanations, and closely monitored each others' understanding of the problem. This led to an increasingly accurate understanding of the problem. In the contrasting group, their conversations focused on rote application of formulas as they worked to calculate a "correct" solution. Our analyses help us to understand what effective collaborative discourse looks like and have practical implications for how peer leaders are trained and how peer groups are organized.

Chapter 11 (Analysis): A Multivocal Process Analysis of Social Positioning in Study Groups

Iris K. Howley, Elijah Mayfield, Carolyn P. Rosé, and Jan-Willem Strijbos

This chapter compares two multidimensional analyses of the PLTL chemistry dataset, which each includes a cognitive, relational, and motivational dimension. These multidimensional analyses serve to highlight the ways in which the complementary perspectives on collaborative processes offered by each dimension can be integrated in a way that offers deep insights into social positioning within collaborative groups. Differences revealed particularly along the relational and motivational dimensions raise important questions regarding the operationalization of interaction style as displayed through language and highlight the value of multivocality for the purpose of refining important constructs in ways that work towards theory building through integration of findings across research groups that employ different analytic frameworks coming from a common theoretical foundation.

Chapter 12 (Analysis): Application of Network Analysis to Collaborative Problem Solving Discourse: An Attempt to Capture Dynamics of Collective Knowledge Advancement

Jun Oshima, Yoshiaki Matsuzawa, Ritsuko Oshima, and Yusuke Niihara

This chapter presents an analysis of collaborative knowledge building in the PLTL corpus using a social network analysis approach. The goal is to present an analysis of collective knowledge advancement that goes beyond what has been accomplished using existing methods and offers a unique bird's eye view of how knowledge advancement proceeds over time.

Chapter 13 (Discussion): A Multivocal Analysis of the Emergence of Leadership in Chemistry Study Groups

Carolyn P. Rosé

This chapter reflects on the three analysis chapters describing the PLTL chemistry dataset in two different stages. The first stage focuses on the concept of leadership and contrasts the three quantitative analyses presented by Oshima, Rosé, and Strijbos at the workshop on Multivocality at ICLS 2010. Based on these reflections, a multi-faceted image of ideal leadership emerges that would not be visible in any single one of the frameworks investigated. This chapter integrates the perspectives discussed within these three chapters, illustrating how this multivocal separation between different leadership constructs allows us to view how it is possible to present one's views as standing on their own without denying others the right to have their own voice. Following up on this integration, a second wave of reflection focuses on the subsequent, more detailed written analyses, including a new qualitative analysis by Sawyer and colleagues, that enables a more in-depth comparison across analytic approaches at both the individual level and the group level. Questions are raised related to assessment of collaborative problem solving that must be addressed in future work.

Data Section 3: Multimodality in Learning About Electricity with Diagrammatic and Manipulative Resources

Section Editor: Daniel D. Suthers

The data for this section is from an innovative primary school science classroom in Singapore (Chen & Looi, Chap. 14, this volume). Group Scribbles collaborative sketching software (Brecht et al., 2006) is used in conjunction with physical manipulatives (batteries, light bulbs, and wires) in an exercise to understand how basic electric circuits work. The corpus was analyzed by Looi, Song, Wen, and Chen (Chap. 15, this volume) using uptake and content analysis guided by a theory of progressive inquiry; Medina (Chap. 16) using uptake analysis with an ethnomethodological orientation towards unpacking group accomplishments; Lund and Bécu-Robinault (Chap. 17) focusing on coherence and conceptual change in translations between media and modes motivated by a theory of semiotic bundles; and Jeong (Chap. 18) using content analysis under her conception of "group understanding." The discussion chapter by Suthers (Chap. 19, this volume) identifies two major themes across the analyses: what evidences understanding and practices of multimodal interaction across various media. Suthers describes a related Group Scribbles case study that preceded the present one, discusses pragmatic issues concerning transcript sharing, and then compares the analyses in various ways summarized in his abstract below.

Chapter 14 (Data): Group Scribbles-Supported Collaborative Learning in Primary Grade 5 Science Class

Wenli Chen and Chee Kit Looi

This chapter describes the setting and context of a group of primary grade 5 (about age 11) students doing a collaborative learning activity in a science class. Data from this setting are analyzed in subsequent chapters in this book section. Students, in groups of four, used a networked technology called Group Scribbles (GS) to jointly complete a learning task—how to connect a circuit with batteries, wire, and a light bulb so that the bulb would light up. They shared information, negotiated meaning, and constructed knowledge through both GS interaction and F2F discussion. The lesson designers attempted to optimize the use of GS and F2F interaction in real classrooms to support students' collaborative learning, with the aim of harnessing the specific features of each medium.

Chapter 15 (Analysis): Identifying Pivotal Contributions for Group Progressive Inquiry in a Multimodal Interaction Environment

Chee Kit Looi, Yanjie Song, Yun Wen, and Wenli Chen

This chapter adopts an interaction analysis method using the notion of uptake to investigate the development of progressive inquiry learning in a classroom setting using GS. In progressive inquiry learning, students work together on elaborating a shared object such as a research problem, products in a shared digital space like GS, or experimental practices to be reflected on and transformed. An uptake analytical framework is applied to code different facets of interactions in a small group, comprising verbal interactions (utterances and gestures), artifacts created in GS, and hands-on experimental practices into events as coordination acts and to identify uptakes and pivotal contributions (a contribution that plays the role of shifting the direction of the subsequent events seamlessly or abruptly) from such interactions. The analysis illuminates how the pivotal contributions influenced the direction of the group progressive inquiry and led the group to developing progressive understanding of the science concepts.

Chapter 16 (Analysis): Cascading Inscriptions and Practices: Diagramming and Experimentation in the Group Scribbles Classroom

Richard Medina

The analysis discussed in this chapter draws attention to the interactional and inscriptional practices observed in Group Scribbles science classroom. The critical finding is the identification of a pivotal sequence of interaction occurring in the later half of the activity in which one member of the group proposes an innovation for illuminating two light bulbs in a single circuit. The proposal and its subsequent endorsement by the other members are contingent on an immediately prior interaction in which the group appropriates another group's circuit diagram. Together, this pair of adjacent sequential structures exposes multiple instances of uptake between participants. These uptake relations are realized through an ensemble of contingencies consisting of persistent diagrams, tabletop materials, and a locally situated interactional practice.

Chapter 17 (Analysis): Sustainable Coherency of Concepts Across Modes of Interaction

Kristine Lund and Karine Bécu-Robinault

Our analyses illustrate nine instances of what we call multimodal and multimedial reformulations of content beginning either with drawings of physics experiments and going to the manipulation of the physics experiments themselves or beginning with the experiments and going to the drawings. We postulated that each time one of these reformulations occurred, it was a potential (yet rare) pivotal moment for conceptual change because content was being transformed across modes and media. Within the nine instances of reformulation, we found two types of pivotal moments (three instances in all). The first type was changing one's conception from an intuitive everyday view on physics to a canonical view of physics. The second was maintaining a canonical view of physics while also integrating more complexity in terms of experiments constructed, drawings made, or concepts talked about. In addition, the notion of the semantic bundle enabled us to show how the ongoing interaction supplied building blocks that illustrated either sustained conceptual change coherent with canonical physics or difficulties that students faced.

Chapter 18 (Analysis): Development of Group Understanding via the Construction of Physical and Technological Artifacts

Heisawn Jeong

The analyses reported in this chapter analyzed the development of group understanding along the dimensions of domain understanding and intersubjectivity based on the artifacts that a student group constructed during learning. In terms of the domain knowledge development, the analyses identified a progression of four circuit understandings, showing that the group's understanding of electrical circuits became more sophisticated over time as the group considered additional ways to light the bulb(s). The four group understandings also differed in terms of the extent to which they were interactively constructed so that some were constructed mainly by pooling individual ideas while others were more or less co-constructed in the process of collaborative artifact construction.

Chapter 19 (Discussion): Agency and Modalities in Multimediated Interaction

Daniel D. Suthers

As reported in previous chapters, four teams each analyzed traces of a group of students in a Singapore primary school science classroom, interacting F2F and with the aid of a shared whiteboard (Group Scribbles) while manipulating electrical circuits. The four analyses, undertaken from various theoretical and methodological traditions, identified pivotal events that changed the direction of the group's activity, gave accounts of how activity in multiple modes was coordinated simultaneously to enact innovations, examined how translations between different media evidenced changes in conceptual understanding, and characterized the group's understanding through the artifacts they produced. The present chapter summarizes the origins of this work in a prior analysis of Group Scribbles, discusses challenges encountered in producing shared transcripts or otherwise bringing our analytic artifacts into alignment for purposes of comparison, discusses analytic conflicts that led to synthetic agreement in one case and agreement to disagree in another, and characterizes how the analytic approaches expose different theoretical conceptions of the distribution of agency across individuals and of activity across modalities. The chapter concludes with advice for future efforts at productive multivocality.

Data Section 4: Knowledge Building Through Asynchronous Online Discourse

Section Editor: Chris Teplovs

In this section we investigate data from an online graduate level course in education that used Knowledge Forum (Scardamalia, 2004) as its principal communication medium. Three analyses seek to identify and explore "pivotal moments" in the context of a broader analysis of the dynamics of group processes that support knowledge building and investigate the potential of automated analyses for use by learners, teachers, and researchers. The chapters highlight different approaches to analysis of asynchronous discourse data. Teplovs and Fujita (Chap. 21, this volume) analyze social and semantic networks derived from the discussions; Chiu (Chap. 23, this volume-b) applies SDA to analyze how prior messages influence a given message; and Law and Wong (Chap. 22, this volume) explore simple visualizations of student activity that may be usable by teachers managing knowledge-building classrooms. The section closes with a critical reflection on some of the advantages and problems of multivocal analyses and presents a model of iterative design-based research (DBR) that capitalizes on some of the unique affordances of multivocality (Fujita, Chap. 24, this volume).

Chapter 20 (Data): Online Graduate Education Course Using Knowledge Forum

Nobuko Fujita

Progressive discourse is a kind of collaborative discourse for inquiry in which participants share, question, and revise their ideas to deepen understanding and build knowledge. Although progressive discourse is central to knowledge building pedagogy, it is not known whether it is possible to detect its emergence in the participation patterns in asynchronous conferencing environments or what kinds of instructional interventions are most effective to support its development. To characterize episodes of discourse in which participants honor the commitments for progressive discourse and to refine designs of peer and software-based scaffolding, the data used in this section was collected in the context of a study that examined student interactions on the asynchronous online discussion platform, Knowledge Forum®, in an online graduate educational technology course.

Chapter 21 (Analysis): Sociodynamic Latent Semantic Learner Models

Chris Teplovs and Nobuko Fujita

In this chapter we present a framework for learner modelling that combines latent semantic analysis and social network analysis of online discourse. The framework is supported by newly developed software, known as the Knowledge, Interaction, and Social Student Modelling Explorer (KISSME), that employs highly interactive visualizations of interactions and semantic similarity among learners. Our goal is to develop, use, and refine KISSME to generate and test predictive models of learner interactions to optimize learning.

Chapter 22 (Analysis): Exploring Pivotal Moments in Students' Knowledge Building Progress Using Participation and Discourse Marker Indicators as Heuristic Guides

Nancy Law and On-Wing Wong

This chapter sets out to identify pivotal moments in students' knowledge building progress for an online asynchronous corpus generated by a class of master's-level students in the context of a totally online course. The main motivation for this study is to develop a methodology that can be effectively automated to aid teachers and/or researchers to quickly gain a good overview of students' progress in understanding at an overall class level from a very large, semantically rich, and complex discourse corpus. The methodology incorporates the use of participation and discourse marker indicators to provide an overview of the nature and depth of students' engagement in relation to key concepts targeted for student learning and to support the heuristic selection of a small sample of notes for use by the teacher and/or researcher for further in-depth qualitative analysis. This methodology has the potential of being developed into a teacher's pedagogical aid to more effectively facilitate students' collaborative inquiry and knowledge building. As a researcher's productivity tool in understanding students' developmental trajectory in learning through discourse, it offers a distinct possibility for developing and validating knowledge building theory on the basis of empirical discourse analysis of large sets of corpus.

Chapter 23 (Analysis): Statistical Discourse Analysis of an Online Discussion: Cognition and Social Metacognition

Ming Ming Chiu

This study revised a statistical method (SDA) designed for linear sequences of turns of talk to apply to branches of messages in asynchronous online discussions. The revised

SDA was used to test for cognitive and social metacognitive relationships among 17 students' 1,330 asynchronous messages during a 13-week online graduate educational technology course. Multivocality benefits included enhancing a statistical method to expand its scope, exposure to other analytic methods' simpler user-interfaces, and potential integration of multiple methods into a computer program capable of semiautomatic analyses.

Chapter 24 (Discussion): Critical Reflections on Multivocal Analysis and Implications for Design-Based Research

Nobuko Fujita

This chapter presents critical reflections on the multivocal analyses presented in the preceding chapters in this volume by Teplovs and Fujita, Law and Wong, and Chiu on the asynchronous discussion data collected in an online graduate education course using Knowledge Forum. The multivocal analyses are discussed along five dimensions: theoretical assumptions, purpose of analysis, unit of analysis/unit of interaction, data representations, and manipulations on data representations. The diverse interpretations and findings of pivotal moments are explicated in light of broader dynamic group processes that support knowledge building in online graduate course contexts. The implications of multivocal analysis for DBR are discussed.

Data Section 5: A Data-Driven Design Cycle for 9th-Grade Biology

Section Editor: Carolyn P. Rosé

The unique focus of this section is on using multivocality to enhance a data-driven design process by offering a multifaceted understanding of how interventions under development interact with group functioning. Four analysts offer their interpretation of what went right and what went wrong in a pilot evaluation of a new form of software agent-based support for scientific discovery learning in 9th-grade biology (Dyke, Howley, Kumar, & Rosé, Chap. 25, this volume). The four distinct analytic approaches include ethnographic analysis (Cress & Kimmerle, Chap. 27), ethnomethodological interaction analysis (Stahl, Chap. 28), network analysis (Goggins & Dyke, Chap. 29), and linguistic analysis from a systemic functional linguistic perspective (Howley, Kumar, Mayfield, Dyke, & Rosé, Chap. 26). Each methodological lens identifies unique opportunities to refine and improve the intervention, which illustrates how a multivocal iterative development process enables each design iteration to suggest a wider breadth of opportunities for improvement in DBR (Hmelo-Silver, Chap. 30).

Chapter 25 (Data): Towards Academically Productive Talk Supported by Conversational Agents

Gregory Dyke, Iris K. Howley, David Adamson, Rohit Kumar, and Carolyn P. Rosé

In the past 6 years, technology for dynamic support for collaborative learning has matured in terms of its ability both to monitor online interaction through automatic collaborative learning process analysis as well as to offer context-appropriate support for effective participation in groups, such as using conversational agent technology. In recent years, we have been exploring an approach called academically productive talk (APT) as scaffolding for online collaborative learning discussions. In this form of agent-based support, the computer agent poses as an APT facilitator who asks questions that call for a relatively elaborated response (e.g., both a solution and a reason for the solution) and then presses the group to build on or challenge these ideas, with the purpose of keeping student reasoning at center stage and increasing student ownership of ideas. This study reports on an iterative design process for developing the concept of APT agents for supporting online collaborative learning. This effort extended over 2 years during which we have conducted two complete cycles of design development, deployment, and analysis, with the second-year design drawing on lessons learnt from the multivocal analyses presented in the chapters within this section, which were conducted after the first-year study.

Chapter 26 (Analysis): Gaining Insights from Sociolinguistic Style Analysis for Redesign of Conversational Agent-Based Support for Collaborative Learning

Iris K. Howley, Rohit Kumar, Elijah Mayfield, Gregory Dyke, and Carolyn P. Rosé

Data from an early stage of development of conversational agent-based support for collaborative learning provides an ideal resource for demonstrating the value of sociolinguistic style analysis paired with time series visualizations as part of an iterative design process. The method illustrated in this chapter was introduced in earlier publications focusing separately on the sociolinguistic style analysis and the time series visualization using the Tatiana tool. However this chapter is unique in its application to data that is at such an early stage in a development process. The data is admittedly raw and contains many examples of interaction gone awry. Nevertheless, the value in this analysis is in a demonstration of what insights can be gained through detailed stylistic analysis of conversational behavior that informs the next steps of intervention development.

Chapter 27 (Analysis): Successful Knowledge Building Needs Group Awareness: Interaction Analysis of a 9th-Grade CSCL Biology Lesson

Ulrike Cress and Joachim Kimmerle

This chapter presents an analysis of chat protocols from four 9th-grade biology classrooms with 50 students at a public school in Pittsburgh, PA. Particular aspects of knowledge building processes in small computer-supported groups are described and explained. We provide examples from the chat protocols that hint at successful knowledge building and from which we can learn something about how the development of knowledge takes place. Moreover, we provide examples that illustrate why four types of group awareness (social, action, activity, and knowledge awareness) are crucial for collaboration, why a lack of group awareness may be detrimental to CSCL, and which strategies students will apply in order to establish group awareness and common ground. Concluding, we point to implications for future design processes of CSCL scenarios.

Chapter 28 (Analysis): Interaction Analysis of a Biology Chat

Gerry Stahl

This is an analysis of data from initial attempts to combine (a) technology from the Virtual Math Teams (VMT) Project, (b) helping agents, (c) collaborative small groups, and (d) accountable-talk prompting in order to scaffold biology student online chats about videotaped results of a biology experiment. Analysis of the response structure of the chat log of a student group reveals characteristics of their interactions in terms of building collaborative knowledge. In particular, the mediation by the VMT technology, helping agents, and accountable-talk training is analyzed to determine their influences in promoting productive learning-oriented interaction. A DBR analytic perspective provides suggestions for redesign of the socio-technical approach based on the findings from the interaction analysis. Redesign in response to the analysis results in clear improvement, as seen in analysis of the response structure of a chat log from a second test cycle.

Chapter 29 (Analysis): Network Analytic Techniques for Online Chat

Sean P. Goggins and Gregory Dyke

Multivocal analysis applies two or more research methods to the same dataset and then applies reflexivity in a joint analysis to achieve greater insights than would be possible with a single method. In this pilot study, we demonstrate how the application of specific methods is influenced by the ordering of the methods and present a guideline for future multivocal analysis of online chat data using network analytic techniques. We do this in two phases. First, we use Stahl's ethnomethodological analysis of one session of biology chat discourse to inform decisions about how to identify and weight implicit connections between participants. Implicit connections are useful because they can be easily automated and presented in real time. We then contrast Stahl's analysis with the networks we derive from those implicit connections, showing some similarities. Second, we use Tatiana to construct ethnomethodologically informed networks for the full corpora and perform network analysis on the resulting explicit connections. The results are not aligned with our first-phase analysis of network position and roles for members. Further inquiry illustrates that the session chosen for ethnomethodological analysis by Stahl has different characteristics than the other six sessions, drawing our use of that analysis for building implicit connections in the corpora into question. We conclude with a clear vision for applying the group informatics methodological approach to corpora prior to the performance of time-consuming qualitative methods like ethnomethodologically informed analysis. Weaving methods together in the right order, we argue, will lead to more rapid and deeper insight.

Chapter 30 (Discussion): Multivocality as a Tool for Design-Based Research

Cindy E. Hmelo-Silver

This chapter provides an integrated perspective of the discussions and analyses related to the DBR process enacted in a multivocal way in this 9th-Grade Biology section of the book. The focus of the work is iterative development of what are referred to as accountable-talk agents to support collaborative learning in an urban high school science laboratory. This discussant chapter provides an interpretation of the multivocal process, how it sometimes worked and sometimes didn't, and what lessons were learned along the way. This early stage in a DBR program is timely for understanding how a complex socio-technical intervention affected collaboration.

Reflections

The final section steps back from the specific data corpora and their attendant issues to return to the question of when and how productive multivocality can be achieved. Chapter 31 (Suthers et al., this volume) summarizes the entire project and the major lessons learned, and can be used as a self-contained reading. Chapter 32 (Carolyn

Penstein Rosé & Lund, this volume) uses a conceptual model of how multivocality relates to methodological traditions to consider pathways for approaching multivocality and possible pitfalls. They compare our experience in this project with the experience of a class of graduate students in their attempt at productive multivocality. Chapter 33 (Dyke, Lund, Suthers, & Teplovs, this volume) examines how data and analytic representations are used and given meaning in analysis, with examples derived from the case studies of this volume, and discusses the implications of representational affordances for multivocality. It concludes with strategies for effective use of representations in support of productive multivocality. Chapter 34 (Lund, Rosé, Suthers, & Baker, this volume) examines what happened when different epistemologies encountered each other in the case studies and discusses what could or should have happened (e.g., when the epistemologies did not engage with each other, or the engagement was not productive). The chapter shows how epistemological encounters can help to bridge between isolated traditions that work on similar objects of study. Chapter 35 (Law & Laferriere, this volume) takes a critical look at which aspects of this project may have meaningful implications for educational practitioners such as teachers. While some of our work may only be of interest to researchers, the authors find types of relevance to practice: informing immediate pedagogical decision-making and providing more general insight and understanding to the processes and outcomes of learning and knowledge building in collaborative contexts. Finally, in Chap. 36 (Koschmann & O'Malley, this volume), two prominent researchers from different methodological traditions who were not involved in the project discuss the implications of this research collaboration and the relation of multivocality to other literatures. Their chapter takes the form of a dialogue, constituting their own productive multivocality.

Chapter 31: Achieving Productive Multivocality in the Analysis of Group Interactions

Daniel D. Suthers, Kristine Lund, Carolyn P. Rosé, and Chris Teplovs

This chapter reports on the productive multivocality project, a 5-year collaboration among researchers exploring the basis for productive dialogue between multiple analytic traditions in the analysis of group interaction, focusing on educational settings. The project was motivated by the need to bring cohesion to multidisciplinary fields such as the learning sciences in a manner that respects and leverages their diversity. Five data corpora were each analyzed by several analyst teams representing various theoretical and methodological traditions, and we explored strategies for engaging these teams in productive dialogue. This chapter offers a self-contained summary of the project and its major insights and lessons and can serve as a starting point for further reading. After briefly reviewing the motivations and history of the project, we then summarize the five data corpora, the analyses done on them, and the challenges for productive multivocality that we encountered and what we learned from these case studies. The chapter concludes with a discussion of strategies for productive multivocality.

Chapter 32: Methodological Pathways for Avoiding Pitfalls in Multivocality

Carolyn Penstein Rosé and Kristine Lund

This chapter explores multivocality from a methodological perspective. A conceptual model is presented for thinking about multivocality and how it relates to methodological traditions. We reflect back on what we have learned through experimentation with multivocality through the five data sections of the book and draw principles for best practices that we offer to the broader research community. As a running theme throughout the chapter and as an invitation to disseminate multivocality to the next generation of researchers in our field, we contrast the experience of expert analysts whose work is presented in the preceding data sections with the experience of students working in groups on their first discourse analysis project in the context of a computational models of discourse analysis (CMDA) class.

Chapter 33: Analytic Representations and Affordances for Productive Multivocality

Gregory Dyke, Kristine Lund, Daniel D. Suthers, and Chris Teplovs

This chapter describes and reflects upon the analytic representations used in the analyses presented in this book and the roles they played in multivocal analysis. As shown in other chapters, multivocality across analyses based on shared datasets can be productive in a variety of ways and for a variety of reasons. From a pragmatic perspective this productivity is also dependent on the ability of analysts to share datasets, perform analyses, inscribe new analytic knowledge into representations, and use these representations as a basis for discussion. In this chapter, we examine how representations are used and given meaning in analysis. We catalogue the types of entities and attributes inscribed in representations, the notational systems by which they are encoded, and the kinds of moves that result in the creation of new representations. We then discuss the opportunities for multivocality afforded by the representations present in the different data sections and discuss the properties desirable in a framework for coordinating analytic representations. We describe instances of representation-based productive multivocality found in this volume, presenting nine strategies for researchers seeking to engage in productive multivocality. This chapter will be of interest to tool designers but also provides guidance to researchers in reflectively choosing representations (and their affordances for interpretation and manipulation) so as to maximize their ability to engage in productive multivocality.

Chapter 34: Epistemological Encounters in Multivocal Settings

Kristine Lund, Carolyn Rosé, Daniel D. Suthers, and Michael Baker

Researchers usually work and evolve in the scientific frameworks in which they were trained, without questioning their epistemological foundations. However, this may be required when researchers coming from different disciplines and paradigms try to work together on the same object of study. This chapter reflects on epistemological encounters in a 5-year project of multidisciplinary collaborations in the analysis of interaction. We argue for maintaining diversity of epistemological traditions while either achieving complementarity within explanatory frameworks on different levels or maintaining productive tension. We then present the extent to which researchers in our project and a similar project encountered each other's epistemologies when they compared their analyses of shared corpora. The majority of comparisons in various contexts led to engagement between epistemologies, and some of these epistemological encounters were productive and glitch free, others had difficulties, but still led to productivity, while still others led to missed opportunities and in one case to radicalizing incommensurable stances. A minority of comparisons in other contexts did not lead to engagement but could either still be fruitful or not productive at all. In conclusion, we summarize the consequences of engaging with epistemologies through the comparisons researchers make of their analyses in multivocal contexts, showing how epistemological encounters can help to bridge between isolated traditions that work on similar objects of study.

Chapter 35: Implications for Practice

Nancy Law and Therese Laferriere

While the focus of this book is generally to explore whether multivocal analysis of the same dataset can lead to productive interactions among researchers and possible theoretical and/or methodological developments that this may bring about, this chapter explores whether such multivocality would have meaningful implications for practice. Our analysis demonstrates that irrespective of the analysts' theoretical or methodological constructs, whether the work has pedagogical relevance depends largely on the purpose and focus of the analysis. A meaningful analysis from the practice perspective can be made by researchers who do not themselves generate the data and using analytical methods that are grounded on theoretical frameworks different from the ones underpinning the pedagogical practice contexts from which the data were collected. Pivotal moments that are directly linked to the subject matter domain being studied are likely to be easily appreciated by teachers as relevant to their practice. However, not all pivotal moments have direct relevance to pedagogical practice. Further, this preliminary study provides substantial evidence that the multivocality in interaction analysis can be productive in providing valuable insight and pedagogical support to teachers interested in implementing collaborative learning in their everyday practice. Overall, we find that multivocal interaction analysis can contribute to two types of relevance to practice: those that can inform more immediate pedagogical decision-making and those that provide more general insight and understanding to the processes and outcomes of learning and knowledge building in collaborative contexts.

Chapter 36: A Dialog on "Productive Multivocality"

Timothy Koschmann and Claire O'Malley

This chapter presents a reflection on the whole productive multivocality project in the form of a dialogue between two researchers in the CSCL field who come from different analytic perspectives. The reflections include comparisons of the project with other attempts to bring to bear different analytic methods on common data as well as other attempts to aggregate findings over multiple datasets. The chapter also reflects upon the successes and challenges of the productive multivocality project as measured against the five overarching questions that they set themselves at the outset of the project.

References

Brecht, J., DiGiano, C., Patton, C., Davis, K., Chaudhury, S. R., Roschelle, J., et al. (2006). Coordinating networked learning activities with a general purpose interface. *The Proceedings of the 5th World Conference on Mobile Learning, Banff, Canada*

Chen, W., & Looi, C.-K. (this volume). Group Scribbles-supported collaborative learning in a primary grade 5 science class. In D. D. Suthers, K. Lund, C. P. Rose, C. Teplovs & N. Law (Eds.), *Productive multivocality in the analysis of group interactions*, Chapter 14. New York: Springer

Chiu, M. M. (this volume-a). Social metacognition, micro-creativity and justifications: Statistical discourse analysis of a mathematics classroom conversation. In D. D. Suthers, K. Lund, C. P. Rosé, C. Teplovs & N. Law (Eds.), *Productive multivocality in the analysis of group interactions*, Chapter 7. New York: Springer

Chiu, M. M. (this volume-b). Statistical discourse analysis of an online discussion: Cognition and social metacognition. In D. D. Suthers, K. Lund, C. P. Rosé, C. Teplovs & N. Law (Eds.), *Productive multivocality in the analysis of group interactions*, Chapter 23. New York: Springer

Cress, U., & Kimmerle, J. (this volume). Successful knowledge building needs group awareness: Interaction analysis of a 9th grade CSCL biology lesson. In D. D. Suthers, K. Lund, C. P. Rosé, C. Teplovs & N. Law (Eds.), *Productive multivocality in the analysis of group interactions*, Chapter 27. New York: Springer

Dyke, G., Howley, I. K., Kumar, R., & Rosé, C. P. (this volume). Towards academically productive talk supported by conversational agents. In D. D. Suthers, K. Lund, C. P. Rosé, C. Teplovs & N. Law (Eds.), *Productive multivocality in the analysis of group interactions*, Chapter 25. New York: Springer

Dyke, G., Lund, K., Suthers, D. D., & Teplovs, C. (this volume). Analytic representations and affordances for productive multivocality. In D. D. Suthers, K. Lund, C. P. Rosé, C. Teplovs & N. Law (Eds.), *Productive multivocality in the analysis of group interactions*, Chapter 33. New York: Springer

Fujita, N. (this volume). Critical reflections on multivocal analysis and implications for design-based research. In D. D. Suthers, K. Lund, C. P. Rosé, C. Teplovs & N. Law (Eds.), *Productive multivocality in the analysis of group interactions*, Chapter 24. New York: Springer

Goggins, S. P., & Dyke, G. (this volume). Network analytic techniques for online chat. In D. D. Suthers, K. Lund, C. P. Rosé, C. Teplovs & N. Law (Eds.), *Productive multivocality in the analysis of group interactions*, Chapter 29. New York: Springer

Hmelo-Silver, C. E. (this volume). Multivocality as a tool for design-based research. In D. D. Suthers, K. Lund, C. P. Rosé, C. Teplovs & N. Law (Eds.), *Productive multivocality in the analysis of group interactions*, Chapter 30. New York: Springer

Howley, I. K., Kumar, R., Mayfield, E., Dyke, G., & Rosé, C. P. (this volume). Gaining insights from sociolinguistic style analysis for redesign of conversational agent based support for collaborative learning. In D. D. Suthers, K. Lund, C. P. Rosé, C. Teplovs & N. Law (Eds.), *Productive multivocality in the analysis of group interactions*, Chapter 26. New York: Springer

Howley, I. K., Mayfield, E., Rosé, C. P., & Strijbos, J.-W. (this volume). A multivocal process analysis of social positioning in study groups. In D. D. Suthers, K. Lund, C. P. Rosé, C. Teplovs & N. Law (Eds.), *Productive multivocality in the analysis of group interactions*, Chapter 11. New York: Springer

Jeong, H. (this volume). Development of group understanding via the construction of physical and technological artifacts. In D. D. Suthers, K. Lund, C. P. Rose, C. Teplovs & N. Law (Eds.), *Productive multivocality in the analysis of group interactions*, Chapter 18. New York: Springer

Koschmann, T., & O'Malley, C. (this volume). A Dialog on "Productive Multivocality." In D. D. Suthers, K. Lund, C. P. Rosé, C. Teplovs & N. Law (Eds.), *Productive multivocality in the analysis of group interactions*, Chapter 26. New York: Springer

Law, N., & Laferriere, T. (this volume). Multivocality in interaction analysis: Implications for practice. In D. D. Suthers, K. Lund, C. P. Rosé, C. Teplovs & N. Law (Eds.), *Productive multivocality in the analysis of group interactions*, Chapter 35. New York: Springer

Law, N., & Wong, O.-W. (this volume). Exploring pivotal moments in students' knowledge building progress using participation and discourse marker indicators as heuristic guides. In D. D. Suthers, K. Lund, C. P. Rosé, C. Teplovs & N. Law (Eds.), *Productive multivocality in the analysis of group interactions*, Chapter 22. New York: Springer

Looi, C.-K., Song, Y., Wen, Y., & Chen, W. (this volume). Identifying pivotal contributions for group progressive inquiry in a multi-modal interaction environment. In D. D. Suthers, K. Lund, C. P. Rose, C. Teplovs & N. Law (Eds.), *Productive multivocality in the analysis of group interactions*, Chapter 15. New York: Springer

Lund, K. (this volume). A Multivocal analysis of pivotal moments for learning fractions in a 6th grade classroom in Japan. In D. D. Suthers, K. Lund, C. P. Rosé, C. Teplovs & N. Law (Eds.), *Productive multivocality in the analysis of group interactions*, Chapter 8. New York: Springer

Lund, K., & Bécu-Robinault, K. (this volume). Conceptual change and sustainable coherency of concepts across modes of interaction. In D. D. Suthers, K. Lund, C. P. Rose, C. Teplovs & N. Law (Eds.), *Productive multivocality in the analysis of group interactions*, Chapter 17. New York: Springer

Lund, K., Rosé, C. P., Suthers, D. D., & Baker, M. (this volume). Epistemological encounters in multivocal settings. In D. D. Suthers, K. Lund, C. P. Rosé, C. Teplovs & N. Law (Eds.), *Productive multivocality in the analysis of group interactions*, Chapter 34. New York: Springer

Lund, K., & Suthers, D. D. (this volume). Methodological dimensions. In D. D. Suthers, K. Lund, C. P. Rosé, C. Teplovs & N. Law (Eds.), *Productive multivocality in the analysis of group interactions*, Chapter 2. New York: Springer

Medina, R. (this volume). Cascading inscriptions and practices: Diagramming and experimentation in the Group Scribbles classroom. In D. D. Suthers, K. Lund, C. P. Rose, C. Teplovs & N. Law (Eds.), *Productive multivocality in the analysis of group interactions*, Chapter 16. New York: Springer

Oshima, J., Matsuzawa, Y., Oshima, R., & Niihara, Y. (this volume). Application of network analysis to collaborative problem solving discourse: An attempt to capture dynamics of collective knowledge advancement. In D. D. Suthers, K. Lund, C. P. Rosé, C. Teplovs & N. Law (Eds.), *Productive multivocality in the analysis of group interactions*, Chapter 12. New York: Springer

Rosé, C. P. (this volume). A multivocal analysis of the emergence of leadership in chemistry study groups. In D. D. Suthers, K. Lund, C. P. Rosé, C. Teplovs & N. Law (Eds.), *Productive multivocality in the analysis of group interactions*, Chapter 13. New York: Springer

Rosé, C. P., & Lund, K. (this volume). Methodological pathways for avoiding pitfalls in multivocality. In D. D. Suthers, K. Lund, C. P. Rosé, C. Teplovs & N. Law (Eds.), *Productive multivocality in the analysis of group interactions*, Chapter 32. New York: Springer

Sawyer, R. K., Frey, R., & Brown, P. (this volume). Knowledge building discourse in peer-led team learning (PLTL) groups in first-year general chemistry. In D. D. Suthers, K. Lund, C. P. Rosé, C. Teplovs & N. Law (Eds.), *Productive multivocality in the analysis of group interactions*, Chapter 10. New York: Springer

Scardamalia, M. (2004). CSILE/Knowledge Forum®. In *Education and technology: An encyclopedia* (pp. 183–193). Santa Barbara: ABC-CLIO.

Shirouzu, H. (this volume). Focus-based constructive interaction. In D. D. Suthers, K. Lund, C. P. Rosé, C. Teplovs & N. Law (Eds.), *Productive multivocality in the analysis of group interactions*, Chapter 5. New York: Springer

Stahl, G. (this volume). Interaction analysis of a biology chat. In D. D. Suthers, K. Lund, C. P. Rosé, C. Teplovs & N. Law (Eds.), *Productive multivocality in the analysis of group interactions*, Chapter 38. New York: Springer

Suthers, D. D. (this volume). Agency and modalities in multimediated interaction. In D. D. Suthers, K. Lund, C. P. Rosé, C. Teplovs & N. Law (Eds.), *Productive multivocality in the analysis of group interactions*, Chapter 19. New York: Springer

Suthers, D. D., Lund, K., Rosé, C. P., & Teplovs, C. (this volume). Achieving productive multivocality in the analysis of group interactions. In D. D. Suthers, K. Lund, C. P. Rosé, C. Teplovs & N. Law (Eds.), *Productive multivocality in the analysis of group interactions*, Chapter 31. New York: Springer

Teplovs, C., & Fujita, N. (this volume). Socio-dynamic latent semantic learner models. In D. D. Suthers, K. Lund, C. P. Rosé, C. Teplovs & N. Law (Eds.), *Productive multivocality in the analysis of group interactions*, Chapter 21. New York: Springer

Trausan-Matu, S. (this volume). Collaborative and differential utterances, pivotal moments, and polyphony. In D. D. Suthers, K. Lund, C. P. Rosé, C. Teplovs & N. Law (Eds.), *Productive multivocality in the analysis of group interactions*, Chapter 6. New York: Springer

Part II
Case Study 1: Pivotal Moments in Origami Fractions

Section Editor: Kristine Lund, CNRS, University of Lyon

In this section, learning fractions in a 6th grade Japanese classroom provides the focus for three analytical approaches, each identifying moments within the interaction that were "pivotal," in a specific way, depending on the researcher's approach. The data consists of an English-subtitled video in Japanese of six students folding origami paper and of one teacher monitoring their progress on the blackboard, an accompanying transcription of their talk and gestures as well as detailed explanations of how papers were folded by each child. As a consequence of our multivocal approach, all three analysts revisited their methods and modified them in light of discussion with the others.

Shirouzu introduces the fractions dataset in Chap. 4, entitled "Learning Fractions through Folding in an Elementary Face-to-Face Classroom," a dataset he collected while visiting and teaching students twice in a remote area in Japan. In his chapter, he clarifies the rationale behind his data selection, the design principles of the class he taught, and the learning task he presented to the students as well as its objectives.

In Chap. 5, entitled "Focus-Based Constructive Interaction," Shirouzu presents an analysis of his own dataset. His goal is to understand where the personal foci of learners originate, what happens in the interaction once a learner focuses on, for example, shapes or production methods, and how learner outcomes are related to such foci. He also shows how foci and roles students take on (i.e., active task-doer or reflective task monitor) provoke different interpretations of the objects and events discussed during the interaction between the children. Pivotal moments center on how foci emerge or are mobilized in the interaction.

Next, Trausan-Matu presents his analysis in Chap. 6, entitled "Collaborative and Differential Utterances, Pivotal Moments, and Polyphony." He identifies the semantic content of "voices" and their inter-animation patterns beginning from a polyphony framework that he extends to include gestures in the analysis. He considers several dimensions: spoken dialogue, body language, the visual dimension, internal dialogue (at an intra-mental level), and echoes. Here, pivotal moments center on collaborative and/or differential utterances.

The third analyst, Chiu gives us Chap. 7, entitled "Social Metacognition, Micro-creativity and Justifications: Statistical Discourse Analysis of a Mathematics Classroom Conversation." He applies statistical discourse analysis to the dataset in order to see whether recent sequences of utterances affected the likelihood of creating utterances categorized as new ideas, correct ideas, micro-creativity or justifications. Pivotal moments pinpoint where one description of activity changes to another.

Lund wraps up Sect. 2 with her chapter called "A Multivocal Analysis of Pivotal Moments for Learning Fractions in a 6th Grade Classroom in Japan" by comparing the pivotal moments each analyst described using the five methodological dimensions discussed in Chap. 2: theoretical assumptions, purpose of analysis, unit of analysis/unit of interaction, data representations, and manipulation of data representations. She shows how redefining the unit of analysis and the unit of interaction in light of other researchers' analyses, interpreting other researchers' pivotal moments in one's own framework, and comparing the semantics of and the relations between analytical concepts all contribute to helping an analyst surpass the limits of a particular method.

Chapter 4
Learning Fractions Through Folding in an Elementary Face-to-Face Classroom

Hajime Shirouzu

Introduction

This chapter describes in detail a data set used for analysis by Shirouzu, Tausan-Matu, and Chiu and discussion by Lund in the following chapters. The data set consists of a lesson that involves six children studying the multiplication of fractions in a sixth-grade classroom in Japan and their recall of its content after 5 months. The task for the children was to cut out three-fourths of two-thirds of a piece of origami paper and then discuss whether or not their solutions were the same. In summary, they created eight solutions of five types and reached the conclusion that these solutions were the same, because their area equaled one-half of the whole by the multiplication $2/3 \times 3/4$. Not all of the children, however, remembered this conclusion in their long-term recall. The three researchers analyzed the interaction that took place during the lesson as well as its relationship to recall in the next three chapters, which are integrated by Lund in her discussion chapter.

This section clarifies the rationale for data selection, the design principles of the class, and its learning task and objectives. Since this is a "data" chapter, I avoid writing my own research question or theoretical position here. This is a difficult task, because any selection of a learning event is theory-laden. At the end of the chapter, I attach the protocol transcribed from the lesson video, so readers may check my potential bias in explaining the data set.

Rationale for Data Selection

We need to find "boundary objects" (Star & Griesemer, 1989; Suthers, this volume) for collaborative analyses in order to clarify our multivocality in theoretical and

H. Shirouzu (✉)
National Institute for Educational Policy Research, Tokyo, Japan
e-mail: shirouzu@nier.go.jp

D.D. Suthers et al. (eds.), *Productive Multivocality in the Analysis of Group Interactions*, Computer-Supported Collaborative Learning Series 16,
DOI 10.1007/978-1-4614-8960-3_4,

methodological positions on learning in social settings. The boundary objects should be related to theoretically important issues and should have thick data that affords multiple revisits from various viewpoints.

A good candidate is the data for the whole class, which enables us to trace *all verbal and behavioral data* of *all children in a class* through *the entire lesson time*. Theoretically, such data can be analyzed from multiple viewpoints (e.g., the interplay between group cognition and individual cognition, cognitive trajectories of each learner in the collaborative situation, or collective emergence of new ideas). Methodologically, such data enables detailed analysis of every child in an exhaustive way.

I found no paper meeting the criteria italicized above in *The Journal of the Learning Sciences*, *Cognition & Instruction*, *Cognitive Science* or some journals on mathematics. For example, Barrett and Clements (2003), Engle (2006), Engle and Conant (2002), Izsak (2005, 2008), and Lobato, Burns, and Munoz (2003) included detailed analyses of conversations or drawings but traced only some of the class students or focused on interactions between teacher and students. In contrast, Sherin (2002); Strom, Kemeny, Lehrer, and Forman (2001); and Wortham (2001) analyzed the entire verbal and behavioral data of one lesson but treated students anonymously (i.e., did not trace each student consistently through the whole lesson). Since class size is the key for meeting the criteria, I chose a small but authentic class and transcribed almost all the verbal and behavioral data of all the children.

Design Principles

We need simple but general principles to design a class so that our findings can be related to fundamental issues. Conceptual understanding is one such issue, as its variant, conceptual change, remains a hot theme in the learning sciences and in CSCL (Roschelle, 1992; Vosniadou, 2008).

Japan was once famous for its lessons that promote children's conceptual understanding, called elaboration lessons (*neriage-jugyou*) by Stigler and Hiebert (1999). In a typical lesson, children work on a task for the day, present multiple solution methods, and engage in discussion to build a consensus on which method is good and why. The tasks presented at the beginning can often be accomplished by all learners by utilizing their prior knowledge, familiar procedures, or external aids at hand. When those variations have been collected, teachers often ask children conceptual questions to seek commonalities among them. Teachers believe that this kind of reflection and integration helps children move from procedural knowledge of how to do something, to declarative knowledge, and then to conceptual knowledge.

Three design principles are at work here from our learning scientists' perspectives (Miyake, 2008; Roschelle, 1992): (1) externalization of initial thoughts and solutions, (2) sharing for reflection on those variations, and (3) integration of those variations. Linn, Lewis, Tsuchida, and Songer (2000) also pointed out that *eliciting*, *exchanging*, and *reflecting* on students' ideas is an instructional feature of Japanese science education. Although whole-class consensus building discussion is believed to lead to conceptual understanding, accumulated observations tell us that only some of the children participate in the discussions (Sato, 2006; Shirouzu, 2008). The issue is how individual children deepen their understanding in the discussion. Therefore, I designed a lesson along these principles for the purpose of detailed analysis of the issue.

Learning Task and Objectives

It is necessary to prepare a learning task that can be solved in *multiple* ways. "Multiple" means not only procedurally different but also differing in the degree of abstraction (e.g., being solvable diagrammatically as well as algorithmically). Such a feature enables us to observe two things: how children produce different solution methods and reflect on those variations, and how they change their methods as they develop an understanding of the task through reflection.

The task I used here is a simple fraction calculation, cutting out three-fourths of two-thirds of a sheet of origami paper, following the classic example of de la Rocha's cottage cheese problem (Lave, Murtaugh, & de la Rocha, 1984). A dieter in de la Rocha's case responded to a recipe calling for "three-fourths of two-thirds cup of cottage cheese" by taking two-thirds cup of the cheese, flattening it into a uniformly thick circular disk on a cutting board, and drawing a cross on it with his finger so that he could save the desired amount by discarding the quarter. He never verified his procedure with a written algorithm, which would have produced $3/4 \times 2/3$ cup $= 1/2$ cup. Thus, this task can be solved with external materials and many different strategies, one of which relies solely on internal cognitive resources (algorithmic knowledge).

This dieter's action provoked two contrasting reactions: one viewed it as human active use of external resources (Brown, Collins, & Duguid, 1989; Pea, 1993) and the other as human passiveness to available methods, suppressing retrieval of even simple mathematical knowledge (Palinscar, 1989; Salomon, 1990; Wineburg, 1989). However, a closer inspection of the dieter's solution process suggests that he was not passive but an active user of external resources. The dieter first measured two-thirds cup of cheese and then laboriously removed it from the cup and made a circular disk so that he could divide it into four equal parts. The representation of the cheese as two-thirds of the cup was transformed to a "one-as-a-whole" disk. Thus, he utilized the intermediate result and re-represented it in the external world, which made completion of his first solution step clearer and made it easier for him to initiate the second step and verify its progress. In this sense, the dieter used external resources actively, with the *proto plan* of dividing the overall task into simpler subtasks to obtain the secondary amount of "three-fourths of two-thirds cup" and by interactively actualizing the plan in the external world.

In this study, I did not use cheese but rather a square sheet of origami paper for three purposes. First, I gave a sheet of paper to each child to observe how he or she solved this problem. Second, if the children used external resources to solve it, such actions would leave traces on the external world including origami paper so that each child could compare the intermediate and resultant solution states with the original state on his or her origami paper. This was impossible for the dieter in de la Rocha's study, since the original state of the whole cup ceased to exist when he moved the cheese onto the cutting board. Third, since each child was given a piece of paper, the children could compare their solutions.

The learning objective of using this task was to connect the children's hands-on experience to algorithmic knowledge to deepen their understanding of fraction multiplication. This objective could be divided into three sub-objectives:

1. Notice that the commonality among various solutions is the area.
2. Notice that the area is one-half of the whole, no matter how different their solutions look.
3. Explain it diagrammatically (e.g., by comparing the answer with the cut-out portion), or explain it algorithmically by fraction multiplication.

Meeting these objectives enables children to understand various things (see Izsák, 2008 for potential of this task for fraction learning). For example, the phrase "three-fourths of two-thirds" means fraction multiplication of "$2/3 \times 3/4$." This operation involves taking three-fourths out of two-thirds of the whole. The resultant area can also be acquired by taking two-thirds of three-fourths (the commutative law of multiplication). In this sense, the global objective can be the single one stated above; however, the children were allowed to notice various things, reach their own conclusions, and even express freely what they learned from their experiences. In this study, I led the children's discussion toward the objectives above, yet allowed them to solve the task and express their thoughts freely, and examined what emerged from such solving activities and discussion. I also collected long-term recall data of the lesson to see how each child memorized it, since such recall often leads to understanding the individual learner in a collaborative situation (Hatano & Inagaki, 1991; Miyake, 1986).

Method

The data comes from a sixth-grade classroom in a remote school in Japan, which had a total of seven children. The seven children had been brought up together from the first grade (age 6) in the same class; they knew each other very well and were able to express their opinions freely. They had already mastered fraction multiplication.

I visited there twice with a 5-month interval as a teacher to give two lessons, both of which were recorded by video and audio recorders for analyses. One boy was absent from the first lesson, so there were six children: two females (G and K) and four males (F, N, O, and Y; all pseudo-initials), seated around a teacher's desk (Fig. 4.1). I conducted the lesson on the first visit and collected recall data on the second visit.

Fig. 4.1 Image of the class: Six children, teacher, and blackboard at the end of the lesson

I prepared three activities, *solving*, *sharing*, and *comparing*, for the first lesson, according to the design principles above. I planned to ask all children to solve the task of cutting out 3/4 of 2/3 of origami paper at least once in the solution activity, to share their solutions by oral explanations with some demonstrations in the sharing activity, and to discuss whether their solutions were the same in the comparison activity. I sought to collect variations from all the children and to spend much time on sharing and comparing them with multiple chances to reflect on the solutions.

After 5 months, I asked the same children at the beginning of the second class to write down anything that they remembered about what happened in the first class. The question was printed on an A4 sheet of paper, with another question asking what mathematics unit they liked. Five minutes were devoted to this inquiry. After this, I debriefed the aim of the first lesson, explaining the origin of the task, situated theory of cognition, transfer of knowledge, and role of abstraction including mathematics, because the higher objective of the two lessons was to introduce cognitive science to children by taking advantage of their hands-on experience in the first class.

Results

In this section, I summarize the overall process of the first lesson, report results of recall in the second lesson, and then describe the first lesson in detail.

Table 4.1 Correspondence among phase, leaning activity, and design principle

Phase	Learning activity	Design principle
Phases 1 and 2	Solving and sharing	Externalization and sharing for reflection
Phases 3 to 7	Comparing	Integration

Brief Synopsis of the Lesson

The author, acting as a teacher, began the lesson by distributing sheets of origami and scissors and inviting the children to cut 3/4 of 2/3 of the origami paper. All children gathered around the teacher's desk closely enough to see each other's use of the origami paper. Two children, G and N, initially took the lead in solving the task. After the two presented their equivalent solutions, I wrote them on the blackboard, asking for Child N to explain his solution process and share it with the others. I also presented that process on the board with extra origami paper (the lesson up to this stage is referred to as "Phase 1"). In response to the teacher's comment that there might be other solutions, all six children successively solved the task and explained their solutions. Everyone used external resources, i.e., origami paper, to solve the task, but their ways had rich variety as shown later in Fig. 4.3 (Phase 2). I exhibited a total of eight solutions on the blackboard and asked whether they were all the same. No child responded to that question clearly (Phase 3). I made a paired comparison of two selected solutions a total of five times. The children verbalized responses to the comparisons, such as "exactly the same," "although the production methods differ, the shape is the same," and "though areas are the same, the shape and production method differ." I noted these commonalities on the blackboard, asking what the same was consistently, and obtained the response "area" (Phase 4). When asked, Child Y indicated that the areas of the solutions were one-half, explaining his reasoning using congruity. However, he withdrew this explanation in spite of others' consent (Phase 5). Then, Child Y tried again to explain why all answers were one-half by fraction multiplication. All the other children concurred with this explanation (Phase 6). I judged that the explanation by congruence was insufficient and asked if the solution obtained from the origami and the 1/2 derived from a calculation could be connected by an equal sign. This question was aimed at challenging their understanding of the commutative law. However, it was left as homework for the next class since we ran out of time (Phase 7). The class took 50 min in total.

The seven phases roughly include the three activities presented in Table 4.1. About 30 min was spent in Phases 1–2 and about 20 min in Phases 3–7. As you see in Table 4.1, the latter activity of less time was divided into more phases than the former ones, since it had qualitatively different components.

A transcript was prepared from the videotape (see the Appendix). Each line represents an utterance spoken in one breath or an important action taken. Lines are numbered from the beginning (1) to the end (584). For analysis of the following chapters, video data was supplied and the Japanese was translated into English and synchronized with the video as subtitles. Drawings of each student's folded origami solution were also provided.

Table 4.2 Contents of reports

Y	We made 3/4 of 2/3 using origami paper. Then the 2/3 × 3/4 made 1/2, and we thought why it resulted in 1/2.
K	Various types of 1/2 of origami paper were made. We thought why 2/3 × 3/4 equals 1/2.
G	3/4 of 2/3 of origami paper was expressed by shapes. We thought what is "=."
N	What is the shape of 3/4 of 2/3 of colored paper? Are various shapes produced the same or not?
O	To solve the problem of dividing origami paper into 3/4 of 2/3, we folded it into 2/3 and then 3/4 of 2/3. Various shapes were obtained.
F	Origami paper was used and folded to find 2/3 and 3/4.

Report After 5 Months

While the children seemed to be of one voice at the end of the Phase 6, the report 5 months later revealed individual differences among the children in their memory of the gist of the lesson. Table 4.2 presents the literal contents of every child's reports.

Detailed Description of the Lesson

Here, I describe children's speech and actions phase by phase by referring to the line numbers of the transcript.

Phase 1

Lines 1 to 42. Children were asked to solve the task of obtaining 3/4 of 2/3 of origami paper using provided origami paper and scissors by oral instruction that was also written on the blackboard. Children N and G reacted to it and completed their solution (lines 7–40). Their solutions happened to be the same (Fig. 4.2). They followed the instructions literally, cutting out the two-thirds part and then cutting three-quarters of it.

Lines 43 to 127. The teacher asked N, who turned in the answer first, to explain his solution to others, displaying it by new sheets of paper and some notations on the blackboard as shown in Fig. 4.2. The rest of the children listened to his explanation, including Child G, who indicated that she shared the same procedure with N by nodding in response to the teacher's question (line 93).

Phase 2

Lines 128 to 201. The teacher then encouraged all children to tackle the task, saying "there is more than one correct solution … Although N and G solved the problem

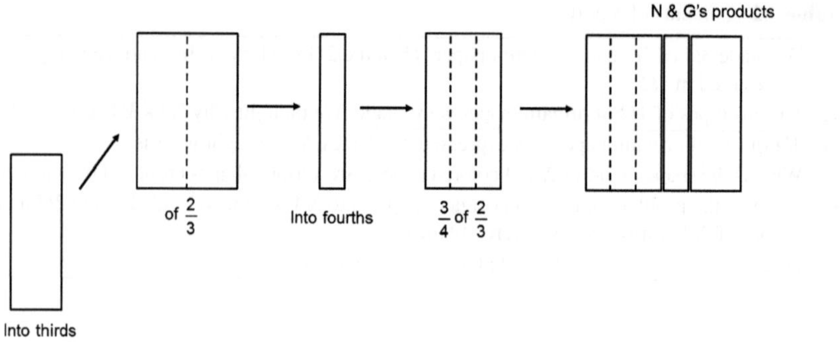

Fig. 4.2 Blackboard at line 127: Display of N's and G's solutions

Fig. 4.3 Solution processes

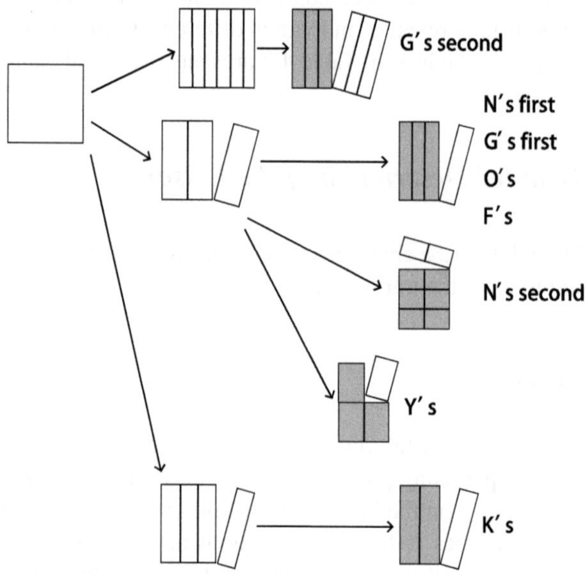

in this way, there may be other solutions" (lines 130–137). All six children successively solved the task in various ways.

Figure 4.3 represents all the solutions obtained during the lesson in a simplified manner, indicating folds by bold lines and cuts by detached rectangles. Let me explain the diagram with N's first solution as an example. It is represented as "N's first" path from an original square in the leftmost column, to the midpoint where some parts are cut out, and lastly to the resultant state represented by shading. The length of the arrow roughly corresponds to the efficiency of the solution steps. Next, I will describe each child's solution process in detail.

Child N: He changed his approach from the first trial by making a right-angled turn to fold the second step (see N's first and N's second in Fig. 4.3).

Child G: She also changed her approach by folding the paper into sixths, opening it, and then cutting out three-sixths. This solution was more efficient than her first one, as represented by the length of its arrow in Fig. 4.3. However, she hesitated to turn in her solution, taking time to look at the solutions on the blackboard and K's solution approach next to her, as if to confirm the correctness of her solution (lines 151–182).

Child O: He solved the task in the same way as N's and G's first ones without noticing it at this point (see line 236).

Child F: He first failed to find a solution but then completed it when supported by other children (lines 273–308).

Child K: She inverted the order of fractions of the task into "2/3 of 3/4," the solution of which essentially did not require the extra folding after cutting the three-quarter part because the two-thirds was already there in the creases (Fig. 4.3). Child K, however, did not appear to realize it, folding the part again helped by a teacher in charge (thus, the length of her solution process was the same with others' two-step solutions in Fig. 4.3).

Child Y: He had planned to solve the task in the same way as N's second solution (lines 140–167) but happened to notice at the midpoint that he did not have to fold the "2/3 rectangle" into fourths and instead only had to fold it into halves to take the 3/4 of 2/3 area (lines 168–178). At line 172, he was helped by Child N to complete the final step, taking one-fourth from the 2/3 rectangle. Figure 4.4 illustrates his actions, gestures, and origami states through this process. Numbers mean line numbers, shaded cells highlight his shift between two solutions, and thick bars mean pauses longer than 5 s.

Lines 202 to 405. The teacher asked all children to explain their own solution. A total of eight answers of five types (the shaded areas in Fig. 4.3) were posted on the blackboard with their solution processes as shown in Figs. 4.1 and 4.5. The answers differed from each other in shape or production method, providing the class with rich variations.

In explaining the solution, each child clearly expressed the features of his or her own method. Child Y articulated that he had obtained the three-fourths from the "2/3 rectangle" by folding it only once (lines 205–211). After some pause, Child O admitted that his solution was the same as the N's and G's first ones (line 236). Child N said that his was the same as the first solution until the mid-step but diverged from there (lines 243–256). Child K (lines 328–341) mentioned that she made "3/4" first and then "2/3 of it." Child G said that she reflected on her first solution and folded the paper into "han-bun (everyday expression of one-half)" of sixths from the start (lines 368–394).

Phase 3

Lines 406 to 437. The teacher asked the children to discuss whether the answers were all the same, but their responses were slow and no clear answer was obtained. The teacher changed how to express question three times but failed to foster discussion.

Origami State	Y's Actions and Gestures
140-	[Hesitates, glances at the blackboard,] [and starts to fold the origami paper into three equal portions.] [Slowly divides the paper into three equal parts.]
147	[Opens the origami paper (so as to fold it neatly)]
147	[and divides it into three equal parts again.]
149	[Divides the origami paper into three equal parts and leaves them as they are.] [Cuts 1/3 and places it on the thighs while leaving 2/3 as it is.]
160	[Halves 2/3 into 1/3,]
161	[Touches one-fourth from the bottom.] [Attempts to fold by turning outside in by his both thumbs.]
162	[Slides his both index fingers to one-half position.] [Attempts to fold, but...]
163	[Opens.] [Looks at the blackboard steadily.]
164	[Folds 2/3 vertically in half.] [Folds it in half while thinking and then opens it]
165	[Opens.] [and looks the four pieces doubtfully.]
167	[Attempts to fold 2/3 in half]
168-	[and then stops.] [Looks at the blackboard] [Looks at the blackboard and inclines his head.]
170	[Folds it in half at a dash while inclining head; N is watching.]
172-	[Opens.] [Attempts to cut off the lower right 1/4 part and peels it off] (at the same time, N's hand extends and appears to point the upper left 1/4 portion) [Sees N's face and nods each other]
176	[Cut the lower right portion by scissors.]

Fig. 4.4 Child Y's solution process

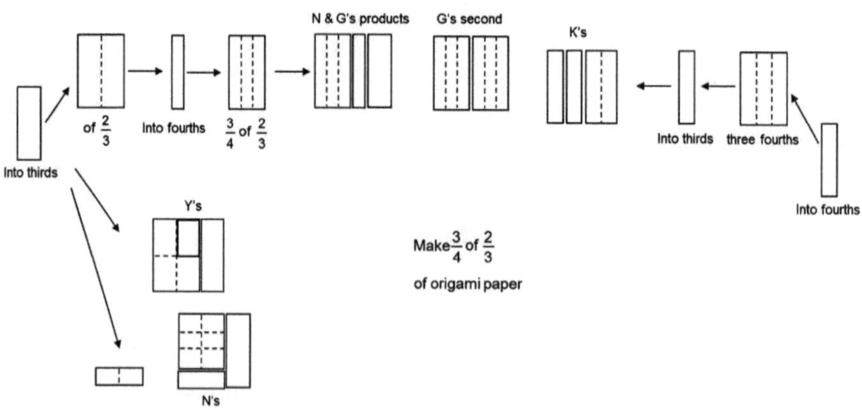

Fig. 4.5 Blackboard at line 405: Display of all children's solutions

Table 4.3 Paired comparisons and commonalities found

Solution 1	Solution 2	Commonality and difference
N's first	G's first	"The same." (Lines 448)
N's first	G's second	"Although the production methods differ, the shape is the same." (Lines 457–459)
N's first	N's second	"Though *areas* are equal, the shape and production method differ." (Lines 473–476)
N's first	K's	"Although the shapes are the same, the production methods differ." (Lines 481–482)
N's first	Y's	"Although the areas are the same, the shapes and production methods differ." (Lines 488)

Solution 1 and solution 2 mean solutions paired; see Fig. 4.3 for their details

Table 4.4 Conversation at third paired comparison: first reference to "areas"

Time	Line	Speaker	Speech and action
35:58	469	T	What do you think of N's two solutions? [Places N's two solutions on the teacher's desk.]
	470	Y	[Moves toward the teacher's desk by further raising his hip.]
	471	Anonymous	[Whispers] The shapes differ.
36:14	472	Y	Differ [with clear voice]
	473	Y	though areas are the same [with low voice]
	474	G	Though the areas are the same
	475	T	Yes
36:20	476	G	┌ the shapes and production methods differ.
		K	| The shape and production methods differ.
		N	└ The shape and production methods differ.

Phase 4

Lines 438 to 489. The teacher finally resorted to "paired comparison," that is, letting the children compare two pieces of the paper like N's first solution and G's second one. Repeating this five times in total provided a scaffold that enabled the children to compare the variations and see commonalities among them as shown in Table 4.3.

As you see in Table 4.3, the abstract commonality "area" first appeared in the third comparison. Table 4.4 transcribes the scene. Child Y was silent up to this point during the comparison activity but approached the desk at line 470. After the other children had responded (line 471), he said "*areas* are the same" in a low voice (line 473). Child G immediately followed this verbalization (line 474).

Lines 490 to 499. The teacher visualized the results of the comparisons and wrote commonalities on the blackboard (Fig. 4.6). When he asked, "What among these is constant?" (line 495), children said first quietly but then loudly, "the area" (lines 496–497).

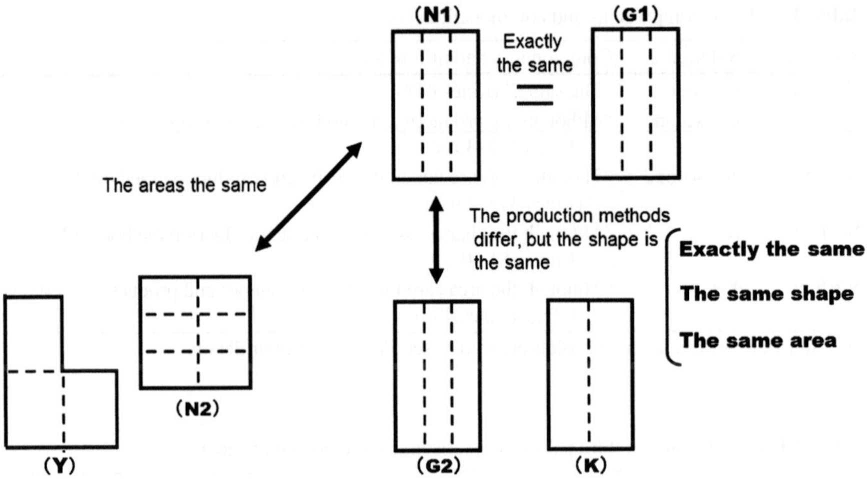

Fig. 4.6 Blackboard at line 495

Phase 5

Lines 500 to 534. When asked "How large is the area?" by the teacher, Child Y clearly answered "2-bun-no-1 (algorithmic expression of one-half)." He attempted to explain it by mating the portion of origami paper representing the answer to the rest of G's second solution, saying "This (the answer) and this (remaining part) return to the original form when they are put together in this way, so I think it is probably 1/2." However, he withdrew that idea despite consent from Children G and K (lines 500–519).

Phase 6

Lines 535 to 548. Finally, Y explained that all answers are 1/2 by the following calculation (lines 536–546): "Another (explanation) is, when these two (fractions) are multiplied, I think that the ratio (of the answer) to the whole can be obtained. When 2/3 is multiplied by 3/4, the product is 6/12 and it is equal to 1/2 after being reduced, all (answers) are 1/2 of the whole."

He took the floor from the teacher and asked the other children, "What do you all think?"(line 547) The others answered, "OK" (line 548).

Phase 7

Lines 549 to 584. The teacher resisted their jump from externally driven reasoning to an algorithmic one. For example, the teacher sought to let them back up their own explanation by using "a core square" of 1/12 area to demonstrate that every shape has six of them. The lesson time, however, ran out.

Conclusion

I hope this description has given you some sketch of the data set and can serve as a boundary object. In sum, the children solved the task in various ways and appeared to change their understanding of the task through discussion, but whether, how, and why the change took place should be explained. The following chapters will tackle this.

Appendix

This section shows the transcript in the form that it was originally shared, with analysts Trausan-Matu and Chiu and discussant Lund (cf. Table 4.5). Section authors use line numbers when referring to their analyses.

Table 4.5 Transcript provided by Shirouzu and shared with analysts and discussant

Time	Line	Person	Talk/action	Blackboard
START 0:00	1	T	Here we have a piece of origami paper, a pencil, and a pair of scissors. [Showing them.]	
	2	T	What I want you to do is …	
	3	T	to use these to make three-fourths of two-thirds of this origami paper. [Writing on the blackboard.]	"To make 3/4 of 2/3 of this origami paper."
0:27	4	T	Can anybody do that? [Putting tools on the teacher's desk.]	
0:30	5	N	Can I?	
	6	T	Oh, you need this? [Handing a piece of origami paper to N.]	
	7	N	[Starts to fold the paper into a rectangle of one-third of the total area.]	
	8	F	Of two-thirds …	
	9	G	Of two-thirds …	
	10	K	Three-fourths.	
1:00	11	T	Ummm? Do you want to try? [Slides a piece of origami paper in front of G.]	
	12	F	How about you? [Slides his hands toward G and K.]	
	13	G	[Acts like folding the paper into a rectangle.]	
		F	Two-fourths of two-thirds.	
	14	F	Huh? [Reaches toward the paper but stops.]	
		N	[Still folding the paper.]	
	15	F	Mm, of two-thirds.	

(continued)

Table 4.5 (continued)

Time	Line	Person	Talk/action	Blackboard
	16	K	Three-fourths of two-thirds. [Speaking to G.]	
	17	G	[Touching the paper.]	
		F	Ah, three. [After hearing K.]	
	18	G	Can I use a pair of scissors?	
		N	[Opening the folded paper of one-third area.]	
	19	T	Of course.	
1:30	20	N	[Folds paper into one-third area again.]	
	21	N	[Takes the pair of scissors in front of G.]	
2:00	22	G	[Neatly folds the paper into a rectangle of one-third area.]	
		N	[Cuts out a two-thirds area, leaving the other one-third area on the table.]	
	23	G	Three-fourths of two-thirds, right? [Speaking to K as if to confirm.]	
	24	G	I do not need this part. (One-third area.)	
	25	G	N, lend me those. [Reaches toward the scissors.]	
	26	F	[Gets the scissors in front of N and slides it toward G.]	
	27	G	[Cuts out a two-thirds area, leaving the other part on the table.]	
		N	[Folds the cut-out part of 2/3 area into one-fourth, opens it, and waits for the scissors.]	
	28	N	[Cuts out 3/4 of the precut part, leaving the remaining part on the table]	
		G	[Folds the cut-out part of 2/3 area into one-fourth, opens it, and waits for the scissors.]	
		F	[Looking around N's hands.]	
	29	N	[Puts his answer on the table.]	
	30	N	[Touches three cut-out parts on the table one by one.]	
	31	G	N, lend me those. [Reaches toward the scissors.]	
		N	[Handing the scissors.]	
	32	N	[Slides the answer slowly to T.]	
2:47	33	T	Okay? [To N]	
	34	T	Have you finished?	
	35	T	Thank you. [Takes N's solution and holds it in the air.]	
	36	T	Let me take this too. [Pulls N's remaining part to him.]	
		G	[Cuts 3/4 and places 1/4 on the desk.]	

(continued)

Table 4.5 (continued)

Time	Line	Person	Talk/action	Blackboard
	37	G	[Joins 1/3 and 1/4 (of 2/3) and then overturns them so that the colored side is up.]	
	38	G	[Simultaneously slides the answer as well as 1/3 and 1/4 (of 2/3) forward.]	
	39	T	Which one? [Confirms the answer that G wants to submit.]	
	40	G	[Pushes the answer slightly forward.]	
3:00	41	T	Is it this one? [Pulls the answer.]	
	42	O	[Has not touched the origami paper yet.]	
		Y	[Has not touched the origami paper yet.]	
3:10	43	T	[While picking out N's answer] Because this is the first answer.	Posts N's and G's answers to the left of the center.
	44	T	Then, this is the completed one.	
	45	T	Because it is special, let me know your name.	
	46	T	What is your name?	
	47	N	Yes, my name is N.	
	48	T	[Looks back and takes a glance while recording N's answer on the blackboard.]	
	49	T	Then, let's record this as N's answer.	
		Anonymous	[Laugh.]	
	50	T	[Picks up the remaining parts from the desk and posts them on the blackboard.]	Puts on N's and G's remaining parts to the right of their respective answers.
	51	T	What is your name?	
	52	G	G.	
	53	T	Then, let's record this as G's answer.	
4:06	54	T	Well ...	
	55	T	[While indicating the answers with both hands at a position slightly apart from the blackboard.]	
	56	T	This has been completed.	
	57	T	You gave me this one. If the answer alone is displayed,	
	58	T	other children may not understand how this was produced.	
	59	T	N, you gave the answer first,	
	60	T	so please explain how you made it. [Hands a new piece of colored origami paper to N.]	

(continued)

Table 4.5 (continued)

Time	Line	Person	Talk/action	Blackboard
		N	May I fold this? [Confirms the requested task.]	
	61	T	Yes.	
	62	N	First, using this origami paper,	
	63	T	Yes.	
	64	N	I fold this into three parts	
	65	N	so that the three parts overlap. [Starts to fold it.]	
4:48	66	T	Then? [Folds another piece of origami paper into 2/3 on the teacher's desk.]	
	67	N	Then I [open],	
	68	N	well …	
	69	N	from whichever end	
	70	N	I cut the portion up to the second line as viewed from this side. [Folds to 2/3.]	
	71	T	Yes.	
	72	N	Then,	
	73	N	I fold this cut-out portion.	
	74	T	Yes.	
		N	[Glances toward the blackboard.]	
	75	N	I next fold it again like this [folds 2/3 in half]	
	76	N	and	
	77	N	fold it into half again. [Folds 1/3 into half.]	
	78	N	[Opens 2/3.]	
	79	N	Then	
	80	N	this piece can be divided into four,	
	81	T	Yes.	
	82	N	and from these four pieces,	
	83	T	Yes.	
	84	N	the fold comes to the third position	
	85	N	from the end,	
	86	T	Indeed! [Snaps fingers.]	
	87	N	then [glances at the blackboard]	
	88	N	3/4 of 2/3 has been obtained.	
5:43	89	T	Yes, indeed.	
		G	[Whispers to K with smile.]	
	90	G	[Shows that her answer is the same by attitude before being asked by T.]	
	91	T	[Looks back while putting up N's answer.]	
	92	T	Is it the same?	
	93	G	[Nods explicitly.]	
	94	T	hh	
	95	T	Now look at how they got the answer.	
	96	T	Well, did you fold the origami paper	

(continued)

Table 4.5 (continued)

Time	Line	Person	Talk/action	Blackboard
6:05	97	T	into three equal parts?	
	98	T	After folding it into three equal parts,	Posts the green origami paper folded to 1/3 to the left of the N's answer. Writes "1/3" below it.
	99			
	100	T	did you fold 1/3 in this way?	
	101	T	In this way?	
	102	T	Or, in this way? [Confirms the direction of folding.]	
	103	T	Don't you care about it?	
	104	T	In this way?	
		N	[Nods.]	
	105	T	Then, this is the 2/3 part.	Places 2/3 of the pink origami paper to the right of the above answer. Writes "1/3" below it.
	106	T	After folding to 2/3,	
	107	T	[Draws 3/4 of 2/3 as made by N.]	
	108	T	did you cut 2/3	
	109	T	and fold it into four equal parts?	
	110	T	Then,	
	111	T	you cut it into four narrow strips.	Posts 1/4 of 2/3 of blue origami paper. Writes "dividing into four equal parts."
	112	T	And then,	
	113	T	there is no place for G. [Moves G's answer.]	Moves G's answer.
	114	T	Let's record as collaboration.	Writes "N's and G's answers" above it.
	115	T	Then [folds 2/3 of a new origami paper into four equal parts and opens it]	
	116	T	[Another teacher (Teacher H) brings a magnet.]	
	117	T	Open it in this way	
	118	T	and cut it [while bringing it to the blackboard]	
	119	T	this is a 3/4 of 2/3.	
	120	T	Thank you.	

(continued)

Table 4.5 (continued)

Time	Line	Person	Talk/action	Blackboard
	121	T	This piece was obtained by cutting these three parts.	Puts on 2/3 of 3/4 of yellow origami paper. Marks "3/4 of 2/3."
	122	T	This is 3/4 of 2/3.	
	123	T	This one can be obtained. [Moving the 1/3 part.]	Puts on 1/3 next to the answer in yellow.
	124	T	Another one can be obtained. [Moves 1/4.]	Posts 1/4 next to the answer in yellow.
	125	T	[Drops 1/4.]	
	126	T	These are the discarded pieces, aren't these?	Connects the pieces including the rest part to N's and G's first answers.
	127	T	Then, G's solution is the same.	
9:22	128	T	Then,	
	129	T	I said today that	
	130	T	there is no correct solution.	
	131	T	This means that	
	132	T	this (task) has not	
	133	T	yet been completed	
	134	T	[while touching the blackboard].	
	135	T	3/4 of 2/3.	
	136	T	Although N and G solved the problem in this way,	Underlines the history of paper folding.
	137	T	there may be other solutions.	
	138	T	Now then,	
9:42	139	T	I would like to ask you who did not do it before to try now. [Distributes colored origami papers to all children.]	
9:50	140	G	[Folds the origami paper into three equal parts and folds to the half to obtain a 1/6 part.]	
		K	[Folds the origami paper to the half immediately and then folds it into four equal parts.]	
		N	[Folds paper into 2/3 and folds the opposite side to 2/3 to divide the origami paper into three equal parts and then opens it.]	
		O	[Folds the origami paper into three equal portions.]	
		Y	[Hesitates, glances at the blackboard, and starts to fold the origami paper into three equal portions.]	

(continued)

Table 4.5 (continued)

Time	Line	Person	Talk/action	Blackboard
		Anonymous	[Looks like starting at a dash.]	
	141	G	[And opens it fully.]	
		K	[Appears to look at the instruction on the board.]	
10:05	142	G	[Folds the origami paper neatly to the half and folds its 1/3 (presumably aligned with the fold already produced).]	
		K	Please lend them to me. [Waits the scissors by patting N's shoulder across F.]	
		F	[Folds into 2/3 after attempting to fold three times.]	
		N	[Cuts the 2/3 portion and removes 1/3 by hand.]	
		O	[Opens to 2/3.]	
10:10	143	G	[Opens.]	
		F	[Opens paper and turns it over.]	
		N	[Places the scissors down; K picks it up.]	
		O	[Folds to 2/3 carefully.]	
	144	F	[Folds 2/3 in half to align with the folds for 2/3; produces folds for dividing it into three equal parts.]	
10:15	145	G	[Counts six folds by her finger.]	
		K	[Receives the scissors, cuts 3/4, and places 1/4 on the desk.]	
		F	[Opens.]	
		N	[Divides 2/3 vertically into four equal parts.]	
		O	[Divides into three equal parts and then opens it.]	
		Y	[Slowly divides the paper into three equal parts.]	
	146	F	[Appears to look at the instruction.]	
10:25	147	G	[Divides into six equal parts again.]	
		K	[Places the scissors. G takes the placed scissors.]	
		F	[Turns the paper over and divides 2/3 of it into four equal parts.]	
		N	[Opens the paper and looks at it.]	
		O	[Laughs at failing to take scissors; keeps the paper as 2/3, as if intending not to forget the state of 2/3.]	
		Y	[Opens the origami paper (so as to fold it neatly) and divides it into three equal parts again.]	
10:35	148	G	[By taking the scissors earlier than N]	
		K	[Holds the 3/4 portion lightly]	
		F	[While pointing to a fold on the origami paper.]	

(continued)

Table 4.5 (continued)

Time	Line	Person	Talk/action	Blackboard
		N	[Attempts to take the scissors but taken by G first.]	
	149	G	[cuts 3/6 and places the rest part on the desk (finished the task most early).]	
		Y	[Divides the origami paper into three equal parts and leaves them as they are.]	
10:40	150	Teacher H	Let's bring your own scissors.	
		G	[Finished by drawing the chair (gives a glance to N's hand).]	
		K	[and folds 3/4 into three equal parts again.]	
		F	[Turns the origami paper round and round, while muttering] of thirds ….	
		N	[Stands up and attempts to take his scissors.]	
		O	[Attempts to take scissors]	
		Y	[Goes to get scissors]	
10:50	151	G	[Picks up two of her own portions and compares them.]	
		K	[Folds the origami paper slowly into three equal parts.]	
		F	No. Wrong.	
11:00	152	K	[Opens the origami paper, holds it in the air, and looks at it with the hands placed on the face.]	
		F	No. Wrong.	
		N	[After returning, quickly cuts 3/4 of 2/3.]	
11:05	153	G	[Folds one side only neatly.]	
		K	[Glanced by T.]	
		N	[First, holding the solution by hand]	
		O	[but returns without scissors.]	
		Y	[but returns without them,]	
11:10	154	T	[To K] What happened?	
		K	[Stopping.]	
		N	Completed. [Submits it together with the remaining part.]	
		O	[Scissors on the teacher's desk is taken by Y.]	
		Y	[and takes scissors from O.]	
	155	T	[Receives N's solution, though not noticed yet.] Oh!	
		Y	[While holding scissors between thighs,]	
11:20	156	T	This is N's second solution.	
		G	[Picks up two pieces, places one of them, and retains it with a finger.]	
		K	[Stops her hand, and scratches her hair.]	
		F	You have finished quickly! [To N]	

(continued)

Table 4.5 (continued)

Time	Line	Person	Talk/action	Blackboard
		N	[Falls on the teacher's desk in a waiting posture.]	
		O	[Stands up again and goes to get scissors.]	
		Y	[folds 2/3 again.]	
	157	G	You made so quickly. [To N]	
		F	[Slightly distracted.]	
	158	T	[Asks K] What happened? [While pointing out her hand.]	
11:30	159	T	[Puts N's solution on the blackboard.]	Posts N's second answer on the lower left midmost.
		K	Let's take scissors. [Stands up, goes to take scissors,]	
		F	[Overturns the origami paper and divides it into four equal parts.]	
		O	[With no scissors at hand, using the scissors on the teacher's desk,]	
		Y	[Cuts 1/3 and places it on the thighs while leaving 2/3 as it is.]	
11:40	160	K	–	
		O	[cuts 2/3 and drops 1/3 on the teachers desk.]	
		Y	[Halves 2/3 into 1/3,]	
11:50	161	G	[Compares own solution with that on the blackboard.]	
		K	–	
		F	[Thinks deeply.]	
		Y	[touches 1/4 from the bottom of 1/3,]	
11:55	162	K	[and returns.]	
		F	[Thinks deeply.]	
		O	[Folds the 2/3 part into four equal parts]	
		Y	[and opens the origami paper in the middle of raising the finger to the half]	
12:05	163	G	[Compares own solution with that on the blackboard (seemingly not confident in her own solution this time)]	
		F	[Thinks deeply.]	
		N	[Looks at O's solution.]	
		O	[opens it, glances at the blackboard,]	
		Y	[Looks at the blackboard steadily.] [Folds 2/3 vertically in half.]	
12:15	164	G	[Looks at K's solution.]	
		K	[Stopping and G looks at it.]	
		F	What is this?	
		N	[Looks at O's solution.]	
		O	[cuts off 3/4,]	

(continued)

Table 4.5 (continued)

Time	Line	Person	Talk/action	Blackboard
		Y	[Folds it in half while thinking, then opens it,]	
	165	Y	[and looks the four pieces doubtfully.]	
	166	Y	[T steals a glance.]	
12:20	167	T	[Asks K] What happened?	
		G	[Returns to look at own hands.]	
		K	[Stopping.]	
		F	[Thinks with hands on chin and looking at the desk.]	
		N	[Looks at O's solution.]	
		O	[and after cutting,]	
		Y	[Attempts to fold 2/3 in half and then stops.]	
12:30	168	G	[Touches the solution.]	
		K	[While stopping, teacher H looks at it across her shoulder.]	
		N	[Follows O's solution with eyes while raising his right hand slightly and seemingly wanting to intervene O.]	
		O	[submits both 1/3 and 1/4 of 2/3 (seeming slightly unconfident).]	
		Y	[Looks at the blackboard.]	
12:35	169	G	[Touches the solution.]	
		K	[Teacher H advances by one step.]	
		N	[Looks at Y's solution.]	
		Y	[Looks at the blackboard and inclines his head.]	
12:40	170	G	[Looks at K's solution.]	
		K	[Teacher H looks at it by hanging out his body.]	
		N	[Looks at O's solution.]	
		Y	[Folds it in half at a dash while inclining head; N is watching.]	
	171	Y	[Opens.]	
12:45	172	K	[Teacher H looks at it by hanging out his body.]	
		F	Then, let's assume	
		N	[Appears to point the upper left 1/4 portion of Y's solution.]	
		Y	[Attempts to cut off the lower right 1/4 part and peels it off (at the same time, N's hand extends and appears to point the upper left 1/4 portion).]	
12:50	173	K	[Teacher H looks at it by hanging out his body.]	
		F	this portion does not exist.	

(continued)

Table 4.5 (continued)

Time	Line	Person	Talk/action	Blackboard
		N	[Nods to Y while looking at each other. Takes the 1/3 portion cut by Y to the teacher's desk.]	
		Y	[Looks at each other and nods.]	
	174	F	Well.	
		Y	[N brings 1/3 (already cut-out part) to the teacher's desk.]	
	175	F	Oh, this is wrong.	
12:55	176	G	[Picks up two pieces and, after looking at the blackboard, places one of the portions on the teacher's desk.]	
		K	Oh? [while rocking her legs]	
		F	It seems understandable but is really not understandable [while repeating folding].	
		N	[Leaving from Y a little]	
		Y	[Cut the lower right portion by scissors.]	
13:05	177	K	[holding scissors by hand]	
		F	This, this point ...	
13:10	178	G	[Looks at the scene where teacher H extends his hand.]	
		K	[Teacher H extends his hand and points out one portion of the origami paper.]	
		F	This is the point.	
		N	[Places the solution cut by Y on the teacher's desk.]	
		Y	[The 1/6 part sticking to the scissors was brought to the teacher's desk as it is. The solution was submitted by N.]	
13:15	179	K	[Cuts 2/3 and drops 1/3 on the desk.]	
		F	First what of thirds?	
	180	F	First what of fourths?	
	181	F	No? [Scratches head.]	
13:18	182	T	Is that OK? [To G]	
		G	[Submits the solution—T receives the answer and rest part.]	
		Y	[Laughs together with O while pointing to the blackboard (Laughed about the way of folding?).]	
13:25	183	F	Oh, I can't understand.	
		K	[Submits the solution—T receives the answer and rest part.]	
	184	T	Yes [to K].	
		F	Well ...	
13:45	185	F	4/3 or 3/4? [N attempts to intervene by chance.]	
13:55	186	T	You may do (your work) late [to F].	

(continued)

Table 4.5 (continued)

Time	Line	Person	Talk/action	Blackboard
14:00	187	F	Oh no! [being puzzled]	
	188	F	Well …	
14:10	189	T	[Attempts to touch F's origami paper.]	
	190	T	How far did you get? [To F]	
		F	[Takes scissors after being touched by T.]	
14:25	191	F	Wait a moment.	
	192	F	This maybe …	
14:35	193	F	What is	
	194	F	this?	
	195	F	I understood. [Cutting only the 2/3 part of 1/3 by scissors.]	
14:45	196	F	Well …	
	197	F	Oh? [Submits it while inclining his head—T receives the solution and other pieces.]	
	198	T	Yes. [Receives F's solution.]	
	199	T	[Arranges solutions on the desk.]	
	200	T	How far did you get? [All other children had finished by that time.]	
14:50	201			
	202	T	Now,	
	203	T	all of you submitted solutions, then let me ask a question to each of you.	
	204	T	First [while touching Y's answer on the desk], how do you make this one?	
	205	Y	[Standing in front of the blackboard] First, let's fold the origami paper into three equal parts in this way. [Folds the origami paper into three equal portions.]	
	206	T	Yes.	
	207	T	These are the same up to this point [while pointing out the first step of N's and G's first solutions].	
		Y	[Continues to fold after glancing at T's explanation.]	
	208	Y	Then,	
	209	T	Yes.	
		Y	let's cut this 1/3 part like this. [Cuts it by scissors.]	
	210	Y	This paper is now divided into four portions,	
	211	Y	and this shape was obtained by cutting necessary ones from them.	
15:57	212	T	Indeed.	
	213	T	This is the final shape, isn't it?	
	214	T	2/3 has been obtained	

(continued)

Table 4.5 (continued)

Time	Line	Person	Talk/action	Blackboard
	215	T	by dividing it into three equal portions.	
	216	T	So far, the process is the same as G's and N's first solution [using the history of the first solution].	Encloses 2/3 below pink origami paper in a circle.
	217	T	Then, from this point,	
	218	T	by the way, what is your name?	
	219	Y	Y	
	220	T	then this is Y's original idea.	Posts 1/4 of 2/3 in the lower right position and draws a branching arrow.
	221	T	Cut 2/3 of	
	222	T	the origami paper folded into three equal portions,	
	223	T	and then cut the 1/3 part in half.	
	224	T	And this is what you get	
	225	T	[By lifting the 1/4 (of 2/3) piece that was cut] by cutting into four equal parts and	
	226	T	by opening it.	
	227	T	This (answer) is splendidly made,	Posts Y's answer.
	228	T	and the remaining parts are attached	Posts Y's remaining part.
	229	T	like this.	Connects them by drawing an arrow.
	230	T	Can you understand it? [To the children]	
	231	K	[Nods several times.]	
		N	[Nods.]	
	232	T	Is this one OK?	
	233	T	There are many other solutions.	
	234	T	How do you make this one? [To O]	
17:23	235	T	Is this included in these solutions?	
		N	[Looks at O's face.]	
17:27	236	O	[Nods after pausing for four seconds.]	
17:34	237	O	I made the same solution as that.	
	238	T	Oh, this is the same as these solution.	
	239	T	Let's post this one also.	Posts O's answer and remaining parts to the right of N's and G's solutions.
	240	T	Many children solve in this way.	
17:52	241	T	This is N's second solution [while pointing out a solution on the blackboard].	

(continued)

Table 4.5 (continued)

Time	Line	Person	Talk/action	Blackboard
17:57	242	T	How did you obtain this solution? [To N]	
	243	N	First,	
	244	T	Yes,	
	245	N	Well, like in the	
	246	T	first time, fold this into three equal	
18:14	247	N	portions [while folding it],	
	248	N	[opening it and cutting 1/3]	
	249	N	fold it in half like this,	
18:37	250	N	[by folding 2/3 in half]	
	251	N	fold it again in this way [while folding in half again]	
	252	N	it is divided into four equal parts [while opening it]	
	253	N	cut out two portions [while cutting],	
19:01	254	N	and the center line is useless, but	
19:19	255	N	3/4 is obtained by these three lines, and thus	
	256	N	the solution was obtained [while pointing to his own solution on the blackboard].	
	257	T	Well, yes [by extending his hand to the blackboard],	Encloses the pink answer within a square.
	258	T	this portion is	
	259	T	2/3 in the same manner as before.	
	260	T	Although you folded this portion into four equal portions	
	261	T	[while folding 2/3] from this side last time,	
	262	T	you cut it vertically this time.	
	263	T	It is right, isn't it?	
		N	[Nods.]	
	264	T	Then, this separates	Draws an additional branched line from the pink answer
	265	T	from the 2/3 portion.	to below Y's answer.
	266	T	2/3 is divided into three (correctly four) equal portions,	Posts the four equal parts of 2/3 in yellow divided by transverse folds.
	267	T	and when it is opened completely,	
	268	T	3/4 of 2/3 is this portion,	Posts the part of the answer.
	269	T	this is the portion taken out,	

(continued)

Table 4.5 (continued)

Time	Line	Person	Talk/action	Blackboard
20:30	270	T	this is the portion cut off,	While posting 1/4 and the first cut-out 1/3, points them out.
	271	T	and this is the solution.	
	272	T	Then, yes, hh	
	273	T	Please.	
		F	Well, not yet [being about to give up redoing].	
	274	T	OK. [Returns the origami paper to F.]	
	275	T	Redo it.	
		F	It is this position. [Cuts 1/3 of 1/3 of the former origami paper and submits it.]	
	276	T	Yes, do you have something to say? [To girls G and K]	
	277	T	OK.	
		F	[Picks up new piece of origami paper.]	
	278	F	First, dividing it into three equal parts [while folding it into three equal parts]	
	279	T	Yes.	
	280	F	then, folding it in this way,	
	281	T	Yes.	
	282	F	[while opening the 2/3 portion only]	
	283	T	Yes.	
		F	uuh, then	
	284	F	Well, yes? [Fully opens it and then flaps it (to F, which is 2/3 seems unclear; cannot advance to dividing the paper into four equal parts)]	
	285	T	Yes? hh	
		Anonymous	[Laugh]	
	286	F	Then,	
	287	F	by folding in this way,	
	288	F	folding into three equal parts,	
	289	F	cut out this part [while cutting 2/3 with scissors.]	
	290	F	Then,	
22:09	291	F	[by cutting the 1/3 portion into half]	
	292	F	Then, it is completed.	
	293	G	[Thrusts her body forward.]	
	294	T	Yes, hh from girls …	
	295	T	Speak up.	
	296	G	You divided it into three equal portions,	
	297	G	with this and this.	
	298	G	Why here <inaudible>	
22:20	299	K	This is the portion to take [while repeatedly pointing out 2/3].	

(continued)

Table 4.5 (continued)

Time	Line	Person	Talk/action	Blackboard
	300	F	[Inclines his head.]	
	301	G	Let's take 3/4 of 2/3.	
	302	F	[Surrounded by G, K, and N.]	
		N	3/4 of this [while touching the origami paper]	
22:40	303	F	[Begins to divide 2/3 into four equal parts.]	
	304	F	[Folds, opens, and holds scissors.]	
	305	T	There is no secret.	
		K	[Points out the portion, 3/4 of 2/3, to take. Whispers something.]	
	306	T	You may speak up. hh	
		F	[Cuts the fold for 3/4 of 2/3 with scissors.]	
		Anonymous	[Laugh]	
	307	Teacher H	You may speak loudly.	
	308	T	Is your solution changed in some points after your friend helped you?	
		F	[Hurriedly folds the first solution.]	
23:40	309	G	[Points out.]	
		K	[Points out.]	
	310	T	This portion was further divided, wasn't it?	
	311			Posts solutions and the remaining parts of two sessions of F's trial.
	312	T	You made this break by yourself	
	313	T	and other breaks in cooperation with your friend.	
	314	T	And then [to K],	
	315	T	because I suppose you are eager to speak,	
24:20	316	T	[by taking up K's solution]	
	317	T	please look at this. [To all children]	
	318	T	There are three breaks,	
	319	T	no I am sorry, folds,	
	320	T	if we verify it again like Colombo,	
	321	T	[Places K's solution and other pieces on the desk] and look at them again.	
	322	T	[while tracing the lines on the origami paper with finger.]	
	323	T	And those are united like this	
	324	T	and divided into three and then four equal parts.	The way to fold the paper to take 3/4 of 2/3 is explained with chalk.

(continued)

Table 4.5 (continued)

Time	Line	Person	Talk/action	Blackboard
	325	T	There should be four folds.	
	326	T	Why are there only three folds?	
	327	T	Can you explain the reason? [To K]	
	328	K	First, I intended to prepare 3/4,	
	329	T	Yes.	
	330	K	so I folded the origami paper into four [dividing into four equal portions],	
	331	T	Yes.	
	332	K	and then I prepared 3/4 [by opening it and cutting 1/4 of it].	
	333	T	Yes.	
	334	K	Then I took 2/3 of it [divided 3/4 into three equal portions again].	
	335	T	Yes.	
	336	K	These are three equal parts	
	337	T	Yes.	
	338	K	and when this portion is cut,	
		N	[Nods slightly, spontaneously.]	
	339	T	Oh,	
	340	T	your friends nodded.	
	341	K	2/3 can be obtained.	
	342	T	You cut the paper into four equal portions first, didn't you? [Draws the new origami paper K cut.]	
	343	T	All others cut the paper into three equal portions first,	
	344	T	but this time, you first cut it into four equal portions.	
	345	T	I will write from the right this time.	Writes "Fold into four equal parts" on the right end, and posts 1/4 of the origami paper.
	346	T	These are the results of division into four equal parts.	
	347	T	Then 4/3 was produced.	Writes "3/4" and posts it on the blackboard.
	348	T	This was folded again	
	349	T	into three equal parts.	
	350	T	3/4 was further divided into three equal parts.	Writes "to divide into three equal parts" and posts it on the blackboard.
	351	T	Then,	
	352	T	There is no 2/3.	

(continued)

Table 4.5 (continued)

Time	Line	Person	Talk/action	Blackboard
	353	T	Let me take it.	Removes 3/4 and substitutes it by writing with chalk.
	354	T	Then,	
	355	T	what was produced?	
	356	K	Two-thirds.	
	357	T	Yes.	
	358	T	Well, is this 3/4?	Writes "3/4 of 2/3"
	359	K	Oh, 2/3 of 3/4.	
	360	T	2/3 of	then erases it and writes "2/3 of 3/4."
	361	T	3/4 was produced.	
	362	T	This is the answer.	Posts the solution.
28:20	363	T	These are the remaining parts.	Posts the remaining part.
	364	T	[Confirms the solution name.]	
	365	F	F	
	366	K	K	
28:45	367	T	How did you make this?	
	368	G	Well, I looked at the first solution,	
	369	T	Yes.	
	370	G	then, I thought that task was to make 4/3 of 3/2,	
	371	T	Yes.	
	372	G	and there was the remaining part.	
	373	T	Yes.	
	374	G	Well …	
	375	T	Yes.	
	376	G	yes, well …	
	377	T	Yes, you may come.	
		G	[goes to the blackboard]	
	378	G	if we viewed like this, I mean,	
	379	T	Yes.	
	380	G	this (1/4 of 2/3) and this (1/3),	
	381	T	Yes.	
		G	if I combine these,	
	382	G	there are three (equal portions) [pointing out the remaining part].	
	383	T	Yes.	
	384	G	And, because I have also three (portions) here [points out the answer],	
	385	T	Yes.	
	386	G	if I fold the origami paper into halves	
	387	T	Yes.	

(continued)

Table 4.5 (continued)

Time	Line	Person	Talk/action	Blackboard
	388	G	and	
	389	T	Yes.	
	390	G	draw a line here, to divide it into six equal parts,	
	391	T	Yes.	
	392	G	I thought that I could produce 3/4 of 2/3,	
	393	T	Yes.	
29:28	394	G	I mean, if I produce half of six equal parts, or 3/6 (I can produce 3/4 of 2/3).	
	395	T	Yes.	
	396	T	There are many creases, but you prepared	
	397	T	six pieces, didn't you?	
		G	[Nods.]	
	398	T	After looking at her solution [looks at the blackboard while holding her solution in his hand],	
	399	T	your solutions have one [counting], two breaks [pointing out G' and N's first solution],	
	400	T	this also has two breaks [O's solution],	
	401	T	K's solution also has two breaks [K's solution].	
	402	T	This type of solution also has one [counting], two breaks [N's second solution];	
	403	T	however, her solution has only one break,	
	404	T	like this.	Compares N's and G's solutions.
	405	T	This is G's second solution. [Writing her name he has forgot to do.]	Writes G's name.
		G	hhh	
30:12	406	T	Well,	
	407	T	many solutions have gathered, so where	
	408	T	should I begin?	
	409	T	Do you think that	
	410	T	N's and G's solution,	
	411	T	F's solution,	
	412	T	N's solution, and	
	413	T	K's solution are	
	414	T	all the same?	
	415	T	This is the next question.	Writes "Are all the same?" in the upper center.
30:53	416	T	Are all of these the same?	
	417	T	What do you think?	
	418	Anonymous	[Most children incline their heads.]	
31:01	419	T	Many incline your heads like this. hhh	

(continued)

Table 4.5 (continued)

Time	Line	Person	Talk/action	Blackboard
31:13	420	T	It is difficult to solve this kind of problem	
	421	T	with your head alone.	
	422			
	423	T	Let's use an object.	
	424	T	What do you want?	
31:32	425	T	The solution name is given there.	
		G	[Leans over the desk.]	
		F	[Shakes his head.]	
		Y	[Looks at the blackboard with mouth open.]	
	426	T	What do you want? [Leans over G on a seat near him.]	
	427	F	Uhh, well.	
		Anonymous	[Pause]	
	428	T	OK. Say anything you like.	
32:09	429	T	Yes. What do you want to say? [Turns to N.]	
	430	T	[Circles his hand holding the chalk.]	
		N	Well. [Straightens body a little]	
	431	T	I now ask the question, are these all the same?	
32:28	432	T	Who thinks they are all the same? [Raises his own hand.]	
	433	T	And who disagrees?	
		G	Well. [Rubs his eyes.]	
	434	T	Or,	
32:49	435	T	do you have any question, such as	
	436	T	they are the same when said like this or they differ when said like that?	
	437	G	[Whispers something to K.]	
		K	[Whispers something to G.]	
33:00	438	T	Do you want to compare them?	
	439	T	What do you think?	
		F	[Inclines head as if embarrassed.]	
33:05	440	T	Which should we begin with?	
33:09	441	T	N's and G's solution [removes the solutions from the blackboard and puts them on the desk]	
	442	T	are the ones completed first.	
	443	T	This is a complete one.	
	444	T	This is one example.	
	445	T	This is another one [places another solution on the desk].	
33:41	446	T	How do you compare?	
	447	Y	[Approaches the teacher's desk.]	
		Anonymous	[All children thrust their bodies forward.]	
33:46	448	G	The same.	

(continued)

Table 4.5 (continued)

Time	Line	Person	Talk/action	Blackboard
		K	The same.	
		N	The same.	
	449	T	The same.	
	450	T	Here, everyone agreed.	
	451	T	This one and this one are the same.	
	452	T	So,	While returns N1 and G1 to the blackboard, writes "=" between them.
34:04	453	T	how do you compare these [N1 and G2]?	
34:10	454	T	This one and this one.	
	455	T	Please.	
	456	F	Oh.	
		Anonymous	[Pause.]	
34:40	457	N	Although the production methods differ [starts quietly],	
	458	T	Yes.	
	459	G	The shape is the same.	
		K	The shape is the same.	
		N	The shape is the same.	
	460	T	Yes.	
	461	T	These are exactly the same.	Writes "Exactly the same" above the former "=."
	462	T	So,	
	463	T	where should I put these? [Searches for a space on the blackboard.]	
	464	T	Let's put them there.	Moves the two solutions [N1 and G1] to the lower right.
	465	T	These are exactly the same.	Posts them on the right and left and writes "Exactly the same."
	466	T	Where should I put these?	
35:36	467	T	The production methods differ, but the shapes are the same.	Posts G2 on the top, links it with the bottom, and writes "The methods differ, but the shapes are the same."
	468	T	Yes. Thank you.	
35:58	469	T	What do you think of N's two solutions? [Places N's two solutions on the teacher's desk.]	

(continued)

Table 4.5 (continued)

Time	Line	Person	Talk/action	Blackboard
	470	Y	[Moves toward the teacher's desk by further raising his hip.]	
	471	Anonymous	[Whispers] The shapes differ.	
36:14	472	Y	Differ [with clear voice]	
	473	Y	though areas are equal [with low voice].	
	474	G	The areas are the same,	
	475	T	Yes.	
36:20	476	G	but the shapes and production methods differ.	
		K	The shape and production method differ.	
		N	The shape and production method differ.	
		Anonymous	The shape and production method differ.	
	477	T	The areas are the same.	
	478	T	Because the areas are the same,	Connects diagonally and writes "Areas are the same."
	479	T	this is the last comparison [N's first solution and K's one].	
		Y	[Leans over the desk.]	
36:52	480	T	What do you think of these?	
37:04	481	G	Although shapes are the same,	
		K	Although shapes are the same,	
		N	Although the shapes are the same,	
		O	Although the shapes are the same,	
		Y	Although the shapes are the same,	
	482	G	the production methods differ.	Puts K's solution on the blackboard.
		K	the production methods differ.	
		N	the production methods differ.	
		O	the production methods differ.	
		Y	[Quickly goes back to his seat.]	
	483	T	Ok, let's compare this group.	
	484	T	Do you think this one and this one are the same or do they differ?	
	485	T	This is like this	
	486	T	and this is like this. [Talking about the results of comparisons.]	
37:24	487	T	Sorry to skip you, how do you compare [N's first one with Y]?	
37:27	488	G	Although the areas are the same, the shapes and production methods differ.	
		K	Although the areas are the same, the shapes and production methods differ.	
		N	Although the areas are the same, the shapes and the production methods differ.	

(continued)

Table 4.5 (continued)

Time	Line	Person	Talk/action	Blackboard
		O	Although the areas are the same, the shapes and production methods differ.	
		Y	Although the areas are the same, the shapes and production methods differ.	
	489	T	Let's put it here.	Puts Y's answer on the blackboard.
	490	T	This means that there are three types of relations:	
	491	T	exactly the same,	Draws relations on the blackboard.
	492	T	not exactly the same but the same in shape,	
	493	T	and the same in area.	
	494			
38:13	495	T	What among these is constant?	
38:14	496	Anonymous	[Whispers] Area.	
	497	Anonymous	[All together] Area.	
	498	T	Area.	
	499	T	The areas are the same.	
38:18	500	T	How large is the area?	
38:20	501	Anonymous	[Whispers] of 2 (halves), 2 …	
38:24	502	Y	1/2 [in low voice]	
	503	Y	[Slowly] of the whole.	
		T	[Following Y] 1/2 of the whole	
	504	G	Ah.	
		K	Ah [moves right hand].	
		N	Ah. [Nods.]	
38:34	505	T	Why? [To Y]	
38:38	506	Y	Well.	
	507	T	Come!	
		Y	Well. [Stands up and goes to the blackboard, looks around solutions, and]	
	508	T	Yes. Which do you want?	
	509	T	Do you want this? [points to blue, N2.]	
		Y	[Points out G2.]	
38:59	510	T	Or this? This is the remaining part.	
		Y	[Looks back on the teacher's desk.]	
	511	Y	Well. [Turns around to T together with other children while holding G's answer and rest part.]	
	512	Y	This (the answer) and this (rest part) return to	
	513	T	Yes.	
		Y	[Pause]	
	514	Y	to original form when they are put together in this way, so I think it is probably be 1/2.	

(continued)

Table 4.5 (continued)

Time	Line	Person	Talk/action	Blackboard
	515	T	Yes.	
	516	T	When this part and this part are put together,	
		Y	[Returns to his own seat mystified in the middle of T's explanation.]	
	517	T	the original form is obtained.	
	518	T	Oh, stay here.	
		Y	Ah? Oh?	
	519	Y	Is this wrong?	
	520	T	hhh	
		G	That's OK [in loud voice].	
		K	OK [loudly].	
		Anonymous	[Laughter.]	
	521	K	OK.	
		Y	[Laughing while inclining head.]	
	522	G	Yes.	
	523	T	Now.	
	524	T	What shall we do?	
	525	T	When this part and this part are put together, the original form is obtained [reproduces it on the teacher's desk].	
	526	T	The girls said something about (this Y's comment).	
	527	T	Then he thought, "This is wrong."	
	528	T	Please do it first. [To G and others]	
39:37	529	T	Do you have something to say?	
	530	Y	[Looking steadily at the blackboard.]	
40:10	531	T	Don't you have any thoughts about this?	
	532	T	Can't you explain it well?	
	533	T	Please. [To Y]	
	534	T	What do you mean, "it's wrong"?	
	535	Y	[Raises hand like a joke after a pause.]	
40:20	536	Y	[Goes to the blackboard again.] Another one is	
40:24	537	Y	Yes.	
		Y	[by pointing out the instruction] well, because we have to make 2/3 and its 3/4,	
	538	T	Yes.	
		Y	when these two are multiplied,	
	539	Y	well,	
	540	T	Yes.	
		Y	I think that the ratio to the whole can be obtained. [The tone quickens.]	
40:38	541	T	Yes.	
		Y	Well, then,	

(continued)

Table 4.5 (continued)

Time	Line	Person	Talk/action	Blackboard
	542	Y	when 2/3 is multiplied by 3/4,	
	543	T	Yes.	
		Y	the product is 6/12,	
	544	T	Yes.	
		Y	and because it is equal to 1/2 after being reduced,	
	545	Y	well,	
	546	T	Yes.	
		Y	I thought all (answers) are 1/2 of the whole.	
40:53	547	Y	What do you all think?	
time change	548	G	OK.	
		K	OK.	
		F	OK.	
		N	OK.	
		O	OK.	
45:06	549	T	Is it really OK?	
45:08	550	T	Are you sure?	
45:10	551	T	You tried various solution methods, didn't you? Are all of them really the same?	
		N	Ah?	
45:22	552	T	Here, these are exactly the same and have the same areas.	
45:26	553	T	In this case, the shapes are the same, but the methods differ.	
45:42	554	T	Or, do you mean that although the area are the same, shapes and methods differ?	
45:46	555	T	You said about such differences a little while ago, didn't you?	
45:52	556	T	But now, well, the product of 2/3 × 3/4 is 6/12, which is 1/2 after being reduced.	
46:15	557	T	These are the same.	
46:19	558	T	Is it true?	
46:20	559	T	What do you think?	
46:23	560	T	Is it true?	
46:29	561	T	I would now like to ask each of you what you did,	
46:35	562	T	because all of you have solved it.	
46:36	563	T	Well, G.	
	564	G	Well …	
46:55	565	G	Well …	
47:00	566	T	What is your product?	
47:15	567	T	[Draws a circle on the blackboard.]	
47:34	568	T	What is K's product?	
47:39	569	T	What do you think?	

(continued)

Table 4.5 (continued)

Time	Line	Person	Talk/action	Blackboard
48:00	570	T	Then, when saying again plainly,	
48:05	571	T	lastly, because here is 2/3 × 3/4,	
48:07	572	T	"OF" means multiplication,	
48:09	573	T	so the total is 1/2.	
48:11	574	T	However,	
48:13	575	T	among your solutions,	
48:20	576	T	some are exactly the same, some are same in shape but differ in production method, and some are same in area but differ in both shape and production method.	
48:33	577	T	Here, we consider these to be different,	
	578	T	and these are same.	
48:42	579	T	Is it okay?	
	580	T	What about this (gap)? [Looks at the blackboard and the hand.]	
49:21	581	T	How do you think? [To F and N]	
		N	Uumm	
49:29	582	T	This is the last question that I want you to think about in my lesson of today.	
	583	T	[Teacher H] Remember what you have done so far. Anything is OK.	
49:57	584	T	[Inclining toward G and K to prompt their response.]	
		T	TIME IS OVER. [Leaves this last question as a homework.]	

References

Barrett, J. E., & Clements, D. H. (2003). Quantifying length: Children's developing abstractions for measures of linear quantity in one-dimensional and two-dimensional contexts. *Cognition and Instruction, 21*, 475–520.

Brown, J. S., Collins, A., & Duguid, P. (1989). Situated cognition and the culture of learning. *Educational Researcher, 18*, 32–42.

Engle, R. A. (2006). Framing interactions to foster generative learning: A situative explanation of transfer in a community of learners classroom. *The Journal of the Learning Sciences, 15*, 451–498.

Engle, R., & Conant, F. R. (2002). Guiding principles for fostering productive disciplinary engagement: Explaining an emergent argument in community of learners classroom. *Cognition and Instruction, 20*, 399–483.

Hatano, G., & Inagaki, K. (1991). Sharing cognition through collective comprehension activity. In L. B. Resnick, J. M. Levine, & S. D. Teasley (Eds.), *Perspectives on socially shared cognition* (pp. 331–348). Washington, DC: American Psychological Association.

Izsák, A. (2005). "You have to count the squares": Applying knowledge in pieces to learning rectangular area. *The Journal of the Learning Sciences, 14*(3), 361–403.

Izsák, A. (2008). Mathematical knowledge for teaching fraction multiplication. *Cognition and Instruction, 26*, 95–143.
Lave, J., Murtaugh, M., & de la Rocha, O. (1984). The dialectic of arithmetic in grocery shopping. In B. Rogoff & J. Lave (Eds.), *Everyday cognition: Its development in social context* (pp. 67–94). Cambridge, MA: The Harvard University Press.
Linn, M., Lewis, C., Tsuchida, I., & Songer, N. B. (2000). Beyond fourth-grade science: Why do U.S. and Japanese students diverge? *Educational Researcher, 29*(3), 4–14.
Lobato, J., Burns, A., & Munoz, R. (2003). How "Focusing Phenomena" in the instructional environment support individual students' generalizations. *Mathematical Thinking & Learning, 5*, 1–36.
Miyake, N. (1986). Constructive interaction and the iterative process of understanding. *Cognitive Science, 10*, 151–177.
Miyake, N. (2008). Conceptual change through collaboration. In S. Vosniadou (Ed.), *International handbook of research on conceptual change* (pp. 453–478). New York, NY: Routledge.
Palinscar, A. S. (1989). Less charted waters. *Educational Researcher, 18*, 5–7.
Pea, R. D. (1993). Practices of distributed intelligence and designs for education. In G. Salomon (Ed.), *Distributed cognitions: Psychological and educational considerations* (pp. 47–87). Cambridge, MA: Cambridge University Press.
Roschelle, J. (1992). Learning by collaborating: Convergent conceptual change. *The Journal of the Learning Sciences, 2*, 235–276.
Salomon, G. (1990). *If intelligence is distributed: What about the cultivation of individuals' abilities?* Paper presented at a symposium in American Educational Research Association, Boston.
Sato, M. (2006). *Challenging for higher learning: Building communities of learners*. Tokyo, Japan: Shogakukan [in Japanese].
Sherin, M. G. (2002). A balancing act: Developing a discourse community in a mathematics classroom. *Journal of Mathematics Teacher Education, 5*, 205–233.
Shirouzu, H. (2008). Observing the lesson as "outsiders": Lesson study from learning scientists' perspective. *Proceeding of Annual Meeting of Japanese Association of Educational Psychology*, s58–s59. [in Japanese]
Star, S. L., & Griesemer, J. R. (1989). Institutional ecology, 'translations' and boundary objects: Amateurs and professionals in Berkeley's Museum of Vertebrate Zoology. *Social Studies of Science, 19*(3), 387–420.
Stigler, J. W., & Hiebert, J. (1999). *The teaching gap: Best ideas from the world's teachers for improving education in the classroom*. New York, NY: The Free Press.
Strom, D., Kemeny, V., Lehrer, R., & Forman, E. (2001). Visualizing the emergent structure of children's mathematical argument. *Cognitive Science, 25*, 733–773.
Suthers, D. D. (this volume, Ch. 1). The productive multivocality project: Origins and objectives. In D. D. Suthers, K. Lund, C. P. Rosé, C. Teplovs, & N. Law (Eds.), *Productive multivocality in the analysis of group interactions*, Chapter 1. New York, NY: Springer.
Vosniadou, S. (2008). *International handbook of research on conceptual change*. New York, NY: Routledge.
Wineburg, S. S. (1989). Remembrance of theories past. *Educational Researcher, 18*, 7–10.
Wortham, S. (2001). Interactionally situated cognition: A classroom example. *Cognitive Science, 25*, 37–66.

Chapter 5
Focus-Based Constructive Interaction

Hajime Shirouzu

Introduction

We need to analyze learning processes more deeply and explain mechanisms of how and why collaborative learning works more precisely in order to optimally support it. For example, we do not yet know how diverse the paths learners take are and how diverse the goals are that they reach in a collaborative situation. If we can actually utilize such diversity, we could enrich their learning. In this chapter, I propose the theoretical and analytic model, "focus-based constructive interaction." It clarifies the existence and mechanism of diversity by documenting divergent paths of six learners and their interactions in the origami fraction data (Chap. 4).

In order to tackle this issue, I integrate two lines of research that use different units of analysis to reveal mechanisms of collaboration. One line takes a group as its unit of analysis to reveal the convergent nature of the collaborative process and its summative benefit (Roschelle, 1992). The other line takes each individual as its unit to clarify the divergent, individualistic process of conceptual change through social interaction (Miyake, 1986; Shirouzu, Miyake, & Masukawa, 2002). While the former claims that the difference in members' thoughts forces collaborative conceptual change through consensus-building toward "convergence," the latter maintains that the difference promotes "constructive interaction," driving each member to construct more persuasive explanations for themselves as well as for the others.

Although the two lines have different theoretical claims, they have complementary methodological strengths that we can exploit for renovating our analytic methods. The former makes full use of conversation analysis (Sacks, Schegloff, & Jefferson, 1974) and pinpoints common grounds constructed by members' conversation, action, or use of external resources. I use the analytic device "collaborative utterances" (Sacks, 1962/1995) in this chapter, the conversational type in which two

H. Shirouzu (✉)
National Institute for Educational Policy Research, Tokyo, Japan
e-mail: shirouzu@nier.go.jp

D.D. Suthers et al. (eds.), *Productive Multivocality in the Analysis of Group Interactions*, Computer-Supported Collaborative Learning Series 16,
DOI 10.1007/978-1-4614-8960-3_5, © Springer Science+Business Media New York 2013

Table 5.1 Example conversation of a collective pivotal moment

Time	Line	Speaker	Speech and action
35:58	469	T	What do you think of N's two solutions? [Places N's two solutions on the teacher's desk.]
	470	Y	[Moves toward the teacher's desk by further raising his hip.]
	471	Anonymous	[Whispers] The shapes differ.
36:14	472	Y	Differ. [In a clear voice.]
	473	Y	Though the areas are the same. [In a low voice.]
	474	G	Though the areas are the same.
	475	T	Yes.
36:20	476	G	But the shapes and production methods differ.
		K	The shape and production methods differ.
		N	The shape and production methods differ.

or more persons build a sentence together. The latter line, being founded on the tradition of cognitive science or protocol analysis (Newell & Simon, 1972) in particular, separates a joint situation into threads of each member's expressions in order to trace the personal knowledge structure through verbal and behavioral expressions. Combining these strengths, we can pinpoint different *pivotal moments* (Suthers, Chap. 1, this volume) depending on our choice of units of analysis and examine the interplay of common ground and individual learning. One goal of this chapter is to use this combined method to reveal hidden diversity in individual learning, given the apparent convergence in the class.

I explain the method by an example presented in Table 5.1, taken from the origami fraction data. There, six children compared two of their solutions of cutting 3/4 of 2/3 of a square sheet of origami paper and found various commonalities and differences. A normative answer to the targeted solutions in Table 5.1 was "although the shapes and production methods differ, the areas are the same." It was the first time that the comparison required children to refer to the abstract commonality "area." As shown in the table, this was first difficult for the children, but they collaboratively formed the answer by saying "the shapes differ" (lines 471–472), "though the areas are the same" (lines 473–474), and "the shapes and production methods differ" (line 476). They struggled for the commonality (lines 471–473), and once found, the word "area" was revoiced (O'Connor & Michaels, 1996) for sharing in the class (lines 473–474). Combined with a task analysis, I defined this collaborative utterance of lines 471–476 as a collective pivotal moment, in the same way as Trausan-Matu (Chap. 6). However, a closer look at each speaker of each utterance reveals subtle individual differences. Child Y referred to the "areas" sooner than the others at line 473, Child G immediately reacted to it, but Children K and N did not refer to the area. Instead, Children K and N referred to the shapes and production methods at line 476.

If we design a learning environment that utilizes this diversity, we should be able to explain such diversity in more specific ways, for example, on what each child focused, why he or she referred to particular attributes, what he or she learned from the others' expressions, and to what outcome such unique expression led. In Chap.6, Trausan-Matu explains the diversity by individual attributes like Child Y as the "divergent thinker" and Child G as the "mirror." Although I admit that such individual attributes as thinking modes, characteristics, and abilities contribute to the results, it is more interesting to see to what extent we can explain things happening in a class by observing what happens there. If we want to see how each learner differs from the others and how he or she changes in the interaction, we should take a more individualized unit of analysis. Taking an individualized unit also requires taking a cognitive approach: how each learner interprets the problem, what solutions he or she explores in interaction with the outside world, to what extent he or she is satisfied with his or her solution, and how he or she summarizes his or her experience through verbalization. Another goal of this chapter is to explain individual differences by the cognitive approach, including changes in knowledge structure.

In this chapter, I first assumed that the children made explicit their personal knowledge structure or *personal foci* through reference or non-reference to particular attributes. I then analyzed preceding parts of the lesson to seek *individual pivotal moments*, because the preceding interactions of each child with others and with external resources could prepare them to verbalize particular things. Finally, I explain where the personal foci had come from, what interaction they caused to take place, and what outcomes they led to. By this analysis, I will show how Children Y and G's interaction with external resources and with others prepared them to focus on the word "area" but in different ways.

The analysis will reveal two implications. First, the combined method helps us to pinpoint achievements of the class as collective/collaborative pivotal moments as well as to clarify individual differences in contributing to the achievements. Second, by introducing the construct of personal focus, we can better explain how individuals are deepening their understanding in diverse ways. Specifically, we hypothesize that the intramental interaction of each individual creates a personal focus affecting how he or she verbalizes and acts in collaborative moments and that such verbalization leads to his or her learning outcome. It also explains how pivotal moments of individuals lead to collaborative moments of the class.

Interplay of Group and Individual Cognition

A hot theme in the learning sciences and CSCL is how to clarify the interplay of group cognition and individual cognition, that is, how each individual mind changes through interaction with others' minds while also contributing to changes in others (Baker, Hansen, Joiner, & Traum, 1999; Ford & Forman, 2006; Hatano & Inagaki, 1991, 1994; Miyake, 2008; Stahl, 2006; Suthers, Medina, Vatrapu, & Dwyer, 2007). The rapid growth of information technology also enables us to keep finer records of individual trajectories in collaborative situations (Miyake, Oshima, & Shirouzu, 2011; Roy, 2011), requiring newer and suitable analytic methods.

Convergence-Oriented Approach

One analytic approach assumes that "collective action must be seen as more than just a sum of isolated individual acts" (Baker et al., 1999, p. 31) and analyzes a collaborative situation as a whole, including participants' conversations, actions, and usage of tools as well as its context. It focuses on *grounding*, the interactive process by which common ground (mutual understanding) is constructed and maintained. Roschelle (1992) analyzed how a pair of high-school students acquired the concept of acceleration by using computer simulation software in order to claim that the crux of collaboration is *convergence* to mutual understanding. When he looked at the pair as a whole and analyzed their language based on his framework for the scientific understanding of acceleration (e.g., various metaphors to represent the velocity and acceleration), a pattern emerged in which "convergent conceptual change is achieved incrementally, interactively, and socially through collaborative participation in joint activity" (Roschelle, 1992, p. 238). This convergence-oriented approach has advantages for detecting collective phenomena as well as for pointing out advancement in collective understanding.

Divergence-Oriented Approach

The convergence-oriented approach has two disadvantages. First, it has difficulty in explaining individual differences in learning outcomes because it assumes that the overall outcome of collaboration is "convergence." In reality, we often observe that learning outcomes differ from learner to learner even though learners appear to converge on a shared understanding (Forman & McPhail, 1993; Hatano & Inagaki, 1991; Saito & Miyake, 2011). The approach has another difficulty in explaining what *role* each individual takes in the grounding process because it assumes that every member is uniformly motivated to achieve convergence. Yet, we often observe not uniform but heterogeneous relationships among members in collaboration. Without complementary approaches to explain those differences, we would not know the actuality of diversity and its positive or negative effects on learning.

The theory of constructive interaction (Miyake, 1986; Shirouzu et al., 2002) is one such approach that explains collective phenomena with the individual as its unit of analysis. This approach regards differences among members as a precious source for collaboration, in the sense that differences trigger members' reflection upon what they once considered "understood" and make them know what was not known. Such differences come not only from initial diversity (prior experience, knowledge, or expertise) but also from role division and exchange. When two members are put in a joint work situation, role division occurs naturally: one member takes the initiative in performing a task and the other monitors it. The solution process of the former (task-doer) is externalized, with traces of that process being observed by the latter (monitor). The monitor often has a slightly broader view of such traces because he or she does not fully share the doer's plan or problem consciousness and is thus free to offer a different interpretation of the externalization. When the monitor begins to provide comments, he or she serves as the doer in turn, being bound to his or her own interpretation, while the first doer becomes the monitor, gaining a broader perspective of the comments he or she received. This role division and exchange lead a pair to an iterative chain of reinterpretations.

Clarifying the Relation Between the Two Approaches

Norman's constructs (1991), the *system view* and the *personal view*, make it easier to understand the relationship between the convergence-oriented and divergence-oriented approaches. He proposed that when a person uses an artifact to accomplish some task, this situation can be viewed from two different viewpoints. The outside observer takes the system view, the total structure of person plus artifact that enhances the overall performance. The person, however, takes the personal view of an artifact that changes his or her task. For example, a calculator enables the user to solve a complicated problem more quickly and more correctly than without it, but it forces him or her to learn new things. The collaborative situation can also be seen from these two different points of view. From the system point of view, learners in collaboration achieve together what they cannot achieve individually. Their concepts may converge from idiosyncratic, naïve ones into an integrated, scientific one. Thus, if researchers impose what they see from the outside on each learner, the learner strives to converge with each other and construct common ground. From the personal point of view, however, each learner faces others and tools from his or her own point of view. What one learner sees inevitably differs from what another sees because they cannot take the same place. Even when they look at the same thing, they can focus on different parts. Even when one learner is satisfied with his or her explanation, another learner finds it incomplete and wants to go beyond, or vice versa. As a result, there may be more complex and diverse ways of understanding than can be seen from the outside. Of course, we cannot truly access personal views from the outside, but we can infer them from data and the appropriate model. By taking advantage of both cognitive and situated theories of cognition, many learning scientists now try to propose integrated models that clarify individual cognition in a situation (Engle, 2006; Lobato, 2006; Stahl, 2006). In this chapter, I try to infer what children think, especially when they do not talk (e.g., listening to others or manipulating things), based on the following model.

Focus-Based Constructive Interaction

I propose here a model called "focus-based constructive interaction," which is an expanded version of the model of constructive interaction (Miyake, 1986). The latter model explains how learners deepen their understanding through discussions, whereas the former model includes the process of how they enter and leave the discussions, explaining their individual understanding progressions with their personal focus as a key construct. In short, the model of focus-based constructive interaction assumes that when one finds an unexpected result or an incomplete solution in the outer world, one questions or becomes conscious of a problem. This question may not yet be in words but influences how he or she focuses on the relevant information when entering the discussion. Differences among the foci of members trigger constructive interactions, through which built-in mechanisms of role division and exchange make their foci clearer. Finally, each member deepens his or her own understanding according to his or her focus, which remains long after the discussion ends. Hereafter, I briefly explain why I think this is the case (theoretical background) and how I apply these ideas to the data (analytical framework).

Personal Focus

Stahl (2006) proposed an integrated model with two cycles: personal understanding and social knowledge building. Learning starts with each individual's tacit pre-understanding, but when some breakdown takes place through his or her interaction with the external world, this tacit understanding is *problematized* (Engle & Conant, 2002), creating a personal focus. When the learner's focus is verbalized and taken up in a social setting, a social knowledge building process then starts. Multiple alternatives are proposed, triggering conflicting interpretations. Through the discussion of these alternatives and interpretations, rationales for different points of view are created, and the interchange gradually converges on a shared understanding.

This notion of "focus" is resonant with the definition by cognitive psychologists: an activated part within one's knowledge structure in interaction with external information (Kuhn & Ho, 1980; Neisser, 1976). When one has a plan to carry out but finds an unexpected result, one naturally focuses on the gap and the reason for it. Humans take in an enormous amount of sensory information, but they use only part of this information in a top-down way. One's focus guides one's verbalization and reaction to others' verbalizations in a social setting. Thus, although Stahl's model predicts a convergence process in social knowledge building, learners with different foci can hardly converge on a shared understanding.

Constructive Interaction

We can combine the notion of focus with the model of constructive interaction in order to predict how the foci interact with roles taken in social settings. When a role division between task-doer and monitor emerges there, we can assume that the focus is closely related to how and when each learner takes the role of the doer. More specifically, we hypothesize that, when things (e.g., tasks, problems, or topics) come into his or her personal focus, the learner tends to initiate task-doing and tries to articulate his or her thoughts in words. However, he or she generally cannot fully express his or her thoughts by words alone and thus reflects on them, only to find flaws of reasoning or reconstruct the thoughts. This is the *effect of task-doing*. The others (listeners) not only observe the doer's verbalization objectively but also try to relate it to their own focus, if they have one. We can hypothesize that, when the doer's verbalizations are relevant to his or her focus, the listener becomes an *active monitor* who goes on to verbalize his or her newer thoughts. The monitor, having somewhat free cognitive resources when listening, can integrate the doer's verbalized results into his or her slightly broader schema by seeking what he or she can share, take away, add, or object to. This is the *effect of monitoring*. The effects of both task-doing and monitoring are accumulated in each individual's mind, through which the personal focus becomes clearer and tied with more critical information. Based on this, each person articulates his or her final thoughts, the articulation of which remains lengthy because his or her learning experience has been abstracted in the form of words. This series of hypotheses predicts that even a learner who monitors a discussion silently remembers its content well when he or she finally takes the role of the doer and articulates his or her thoughts. It also predicts that what each learner remembers differs from that of other learners, depending on what he or she articulates in a class.

Diversity

The model of focus-based constructive interaction predicts that there are as many diverse learning paths as there are learners. However, these are too complicated to report fully in one chapter. I instead introduce the types of foci and their effects on collaboration to capture and explain diversity concisely. Hatano and Inagaki (1986) proposed that there are two courses of expertise: adaptive and routine. The former is characterized by meta-cognitive reflection upon problem-solving procedures, producing a conceptual understanding that enables new ways of solving that are adaptable to new situations. The latter is characterized as efficient and accurate in repeatedly solving similar problems. This model can be applied to defining the focus. When learners solve the same task, one learner can reflect on its conceptual dimension, but another can think of its procedural dimension, both of which influence their foci. Solving problems via external resources, for example, often leaves external traces, which some people reflect upon and reinterpret to find another way of solving the problem, while others confirm their solution and repeat the same solving method. The former would focus on conceptual commonalities, but the latter would focus on procedural or visible similarities.

Analytical Framework

I apply the model to the particular task and lesson used in the Shirouzu data (Chap. 4). When Shirouzu et al. (2002) asked a 100 college students to get 2/3 of 3/4 of a sheet or 3/4 of 2/3 of a sheet of origami paper as their first trial and then the same problem

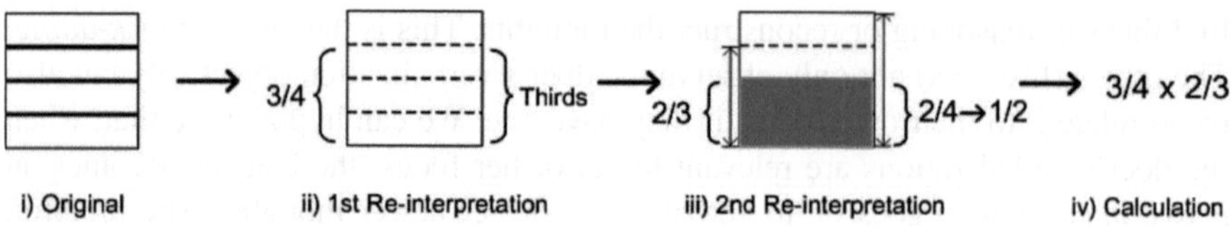

Fig. 5.1 Gradual reinterpretation of "getting 2/3 of 3/4"

with the order of fractions reversed, either in a solo condition or a paired one, they found three things:

1. More than 90 % of the time the participants, either as solos or in pairs, folded the paper to solve the task in the first trial.
2. The solos kept the same solution strategy in the second trial, but more than half of the pairs shifted to the arithmetic calculation.
3. The pairs' shift was a gradual one, in which the two members took turns in monitoring the partner's externalized results and reinterpreting them from a more global perspective.

The former two findings indicate that solutions to this task change from procedural and external resource-dependent to conceptual and internal resource-oriented one through collaboration. The last finding indicates that the change took place gradually when pairs solved the task in their first trial. Figure 5.1 illustrates three reinterpretations of the task "2/3 of 3/4." The most externally oriented, two-step strategy requires first folding the paper into four (level 1). Upon doing this, the pairs could reinterpret the just-completed 3/4 as already having three equal-size rectangles, which eliminates the physical necessity of a second folding (level 2). They could further reinterpret the 2/3 of the designated 3/4 as 2/4 of the original square or a half of the whole (level 3). This reinterpretation often led the pairs to realize that the problem was solvable by calculation (level 4). When analyzing this process with the unit of an individual, the monitor (who observed or listened to the doer at each moment) reinterpreted 100 % of the time from level 1 to 2 and 60 % of the time from level 2 to 3.

This process implies that the view is narrow when the doer reflects upon his or her own externalized traces, whereas the monitor grasps the situation from a slightly broader perspective and can change the view of the traces (from levels 1 to 2). Here, the first doer is pushed out of task-doing, only to listen to the partner's claims as a monitor, which contributes to widening the view (from levels 2 to 3) or abstracting the results (to level 4). This final abstraction, or *active verbalization*, leads this member to take the initiative in shifting to calculation in the next trial. In contrast, solos kept the strategy of level 1 during both trials, implying that their view remained narrow and their focus was on the procedural and not conceptual dimension.

This indicates that it is difficult even for college students to grasp the answer as a half of the whole, and that the key to the shift is reflection upon the external traces at level 2. The pairs often spent the longest time shifting from level 1 to 2 amongst all the shifts between levels and showed emotional expressions like surprise. In addition, the task is changed from drawing oblique lines on origami paper (Shirouzu et al., 2002) to cutting out the designated area in the origami fraction data, which increases the difficulty of grasping the answer in the frame of the paper. We thus expect that the shift would take place gradually in the lesson according to the framework of the four levels above. The remaining issue is how the shift occurs through collaboration and especially role-taking. The above result implies that any person who takes the role of monitor can cause an upward shift and that the person who takes the doer's role cannot do it by himself or herself. This is the strongest position insisting that the collaborative situation matters (i.e., the model of constructive interaction). The opposite position would claim that the individual matters: a person who has a particular ability or skill causes a shift. In this view, the shift occurs regardless of the roles, or the roles are fixed by person throughout the collaborative process (Chap. 6). There could be a third position: both the situation and the individual matter. The collaborative situation naturally allows roles to be divided among the participants, but anyone who monitors the situation does not always cause the point-of-view shift. Instead, an individual who experiences particular breakdowns causes the shift (i.e., the model of focus-based constructive interaction). This issue is an empirical question worth examining, as it is closely related to the design of the lesson.

The Five Dimensions Characterizing the Approach

I summarize the discussion above along five dimensions (see Lund & Suthers, Chap. 2, this volume) and posit it as a set of hypotheses and analytic methods for this study.

Theoretical Assumptions

I take the divergence-oriented (constructive interaction) approach theoretically and the individual as its unit of analysis methodologically. However, I also use the convergence-oriented approach and group-unit analysis. It is similar to Hatano and Inagaki's (1994) "two-level analysis approach," which proposed to investigate the target phenomenon of collective comprehension activity with individual outcomes as a collective/inter-mental process as well as an individual/intramental process (reflecting the inter-mental process). In addition, I take the strong cognitive approach to reveal the understanding processes of each individual using the model of focus-based constructive interaction. This explains not only what each individual takes away from collective activity, as Hatano and Inagaki suggest, but also what he or she contributes to the activity, by clarifying his or her personal focus created through intramental/embodied interaction prior to inter-mental/conversational interaction.

Purpose of Analysis

The main purpose of the analysis is to reveal and explain the diverse learning paths and outcomes of the six children. Methodologically, I use the combined method to reveal diversity in individual learning under convergence in the class and show that there could be different pivotal moments for the class and individuals. Cognitively, I explain the diversity by the growth of personal foci among the children through their social and physical interactions.

Units of Interaction

I analyze social interaction using as units of analysis both groups (more than two children) and individuals. Although the former concentrates on joint achievement and the latter on role-taking, both analyze the inter-person/mental interaction as a unit of interaction. The unique contribution of this chapter resides in the analysis of the intra-person/mental interaction between the inner and outer resources of each child. Here, inner resource refers to the prior knowledge, interest, and problem consciousness that the children have, and outer resource refers to the physical and social environments in which they act. The analysis therefore considers not only talking but also actions or gestures.

Representations and Analytic Interpretations

The first part of the Results section uses summary tables to show both collective achievement and individual differences. The second part sketches the intramental interactions and their juxtaposition to show how social interaction is caused by diversity.

Analytic Manipulations

The analyses go as follows.

1. Analyzing class achievement (collaborative utterances) based on a framework with four understanding levels for solution strategies by detecting keywords or key phrases:

   First level: Following the instruction procedurally using external resources.
   Second level: Eliminating physical steps by reinterpreting the external traces or instruction.
   Third level: Realizing that the answer's area is one-half diagrammatically.
   Fourth level: Realizing that it can be calculated ($2/3 \times 3/4 = 1/2$).

2. Clarifying individual differences in learning outcomes by content analysis of a report 5 months after the lesson.
3. Returning to analyze collaborative moments (found in step 1) and reveal who takes what roles to detect the personal focus of each child.
4. Analyzing the intramental interactions of each individual to determine where each learner's focus has come from and how it has grown, and reconstructing the collaborative moments as constructive interaction among learners having those foci.

Results

I briefly explain the Shirouzu dataset structure to show how the analyses follow from it. The lesson has two main segments: to cut out 3/4 of 2/3 of a piece of origami paper (Phases 1 and 2) and to discuss whether or not their solutions were the same (Phases 3–7). First, I analyze the levels of understanding through which the class proceeded through these seven phases and identify collective pivotal moments. Second, I analyze the content of reports 5 months later, when students were asked to write down what they remembered about the last class. Third, I analyze collective moments in Phases 3 and 4 to reveal types of personal foci because these phases were the first stages of sharing solutions and discussing commonalities. Fourth, I analyze how individual children solved the task in Phases 1 and 2 to postulate what focus and pivotal moment each child had and how he or she interacted with each other in Phases 3–7.

Collective Pivotal Moments

If we take the class as our unit of analysis, the class as a whole changed from a folding-and-cutting approach to an algorithmic one. It also accumulated a variety of solutions, found the commonality "area," and achieved an integrated explanation using math (i.e., $2/3 \times 3/4 = 1/2$). Specifically, I coded children's explanations and conversations to capture how their understanding shifted among the four levels above. An explanation could be a single child's articulation, more than two children's joint utterances, or one child's incomplete explanation with others' support. I define upward shifts among the levels as collective pivotal moments.

As a result, the children's verbalization of their solutions in Phases 1 and 2 clearly included explanations of the two-step solution strategies of level 1. There could be explanations at level 2, but children did not articulate them (e.g., "I saw the usable traces here and skipped the second folding"), which prevented us from coding it clearly.

In contrast, their shift from level 1 (or 2) to 3 was clear. With peculiarities in the children's discussion about multiple answers, level 3 took longer to happen: "*realize that the answers have the same area,*" "*it is one half,*" and "*it is so diagrammatically.*"

They did not answer the teacher's question in lines 406–437, but they gradually and collaboratively explained that the area was common between the two solutions, from lines 471 to 476 (Table 5.1). Child Y made the core statement at line 473, but this formed a complete sentence with other utterances in the rest of the lines. In addition, they revoiced to each other (from Y to the others in lines 471 and 472; from G to Y in lines 473 and 474). We thus identified lines 471–476 as the first collective pivotal moment. When several children searched for an appropriate word to represent the area in line 501 ("A" means anonymous), Child Y again completed the sentence by saying that the answer is 1/2 of the whole as below. This was followed by emotional expressions from the other children, indicating that this was what they had searched. We defined the lines 501–504 as the second collective pivotal moment.

500	T	How large is(.) the area?
501	A	[Whispers] 2-bun-no- (halves), 2:::
502–503	Y	2-bun-no-1 (algorithmic expression of one-half) [in low voice] of the whole.
503	T	[Following Y] 1/2 of the whole
504	G	Ah.
	K	Ah [moves right hand].
	N	Ah::: [Nods.]

Finally, when asked why it was 1/2, Child Y explained it diagrammatically with Child G's second answer as below. The core statements are the explanation from lines 511 to 513, but its correctness puzzled him. Instead, Children G and K gave strong support, which was followed by warm laughter. We defined lines 512–520 as the third collective moment.

512–514	Y	This (the answer) and this (rest part) return to [Pause] to original form when they are put together in this way, so I think it is probably 1/2.
515–517	T	Yes. When this part and this part are put together, the original form is obtained.
	Y	[Returns to his own seat mystified in the middle of T's explanation.]
518	T	Oh, (.)stay here.
518–519	Y	Ah?(.) Oh? Is this wrong?
520	G	That's OK [in loud voice]
	K	OK [loudly]
	A	[Laughter.]

In addition to these gradual shifts to level 3 understanding, the children went on to level 4 as below. After Child Y articulated an integrative explanation, he sought approval of the class and actually obtained it. We defined lines 536–548 as the fourth collective pivotal moment.

536–547	Y	Another is, when these two (fractions) are multiplied, I think that the ratio to the whole can be obtained. When 2/3 is multiplied by 3/4, the product is 6/12 and it is equal to 1/2 after being reduced, all (answers) are 1/2 of the whole. What do you all think?
548	All	OK

Table 5.2 Elements found in the reports

Child	"2/3×3/4" "1/2"	"Shape"	"3/4 of 2/3" "origami"
Y	Found		Found
K	Found		Found
G		Found	Found
N		Found	Found
O		Found	Found
F			Incompletely found

In sum, the children searched for appropriate words and, when found, gave support to each other with a warm atmosphere. Their search proceeded in the direction that the framework predicted, which might give them a sense of unity in gradually finding hidden answers.

Diversity in Learning Outcomes

When we look at the elements identified in students' reports, we can find individual differences as summarized in Table 5.2. I analyzed whether each report included the expressions in double quotation marks or not (see Table 4.2). Although everyone referred to the task and material ("3/4 of 2/3" and "origami"), there is a mutually exclusive pattern between "math" and "shape." The students who referred to the "shape" did not refer to the math ("2/3×3/4" and "1/2"). Although all of the children agreed with Child Y's algorithmic explanation at the end of the class, they differed in summarizing and memorizing their learning experience. We need analyses other than the convergence-oriented one in order to explain the diversity in learning processes and outcomes.

Role-Taking and Collective Moments

I returned to the lesson to analyze the roles taken and the expressions used by each child based on the constructive interaction framework. In Phase 1, Children N and G solved the task as the doers, and Child N explained his solution of level 1 strategy. The other five children, including Child G, monitored his explanation. In Phase 2, all six children assumed the role of the doers by solving the same task. As a result, four of six children presented different solutions, indicating that exchanging roles from monitor to doer served to enrich the solutions. In Phase 3, subtle exchanges of roles were observed. Table 5.3 indicates who referred to particular commonalities or differences at what times of comparison. The leftmost number represents the number of the comparison, and the "solutions 1 and 2" columns represent targeted

Table 5.3 Role-taking in each comparison

	Solution 1	Solution 2	Utterance	Task-doer	Active monitor
1	N's first	G's first	"The same"	N, G, K	–
2	N's first	G's second	"Although the production methods differ,"	N	G, K
			"the shape is the same."	N, G, K	–
3	N's first	N's second	"The shapes differ …"	A (incl. N, G, K)	Y
			"Though areas are the same,"	Y←G	N, K
			"the shape and production method differ."	N, G, K	–

solutions for paired comparisons (see Fig. 4.3). The "task-doer" column lists who verbalized each "utterance," and the "active-monitor" column indicates who listened to that but verbalized the next "utterance" directly after. By this analysis, we can break collaborative utterances into particular utterances of a particular child.

As clearly shown, in the first and second comparisons, Children N, G, and K formed a "group of task-doers" (Saito & Miyake, 2011), but in the third comparison (the first collective moment), Child Y changed his role from monitor to doer and verbalized the abstract commonality, "area." After that, he led the class by explaining the area diagrammatically (the second and third moments) and algorithmically (the fourth moment), the task-doing of which enabled him to understand and memorize the gist of the lesson as revealed in his report (Table 5.2). Thus, the model of constructive interaction approach could explain individual differences in learning outcomes. Child Y, who took the lead, remembered the lesson from the algorithmic point of view, but other children who followed his lead did not.

However, this does not tell us why Child Y was able to take the lead. In addition, Child N always served as the doer in Phases 1 and 2 and took the lead in the third comparison in Phase 3 (Table 5.3) but did not refer to the area. The theory of constructive interaction might explain that Child N took the doer's role too much and had no time for reflection, yet this explanation is too coarse. This is why and where I introduce the concept of focus.

Table 5.4 rearranges the utterances of the three children in the first to third comparisons of Table 5.3. Child N referred to the production method sooner than any of the others but did not refer to the area at the third comparison, even though both compared solutions were his own. This implies that his focus was on the "production methods," not the "area." In contrast, Child Y had referred to nothing through the second comparison but referred to the area at the third comparison, indicating that his focus was on the "area." Child G did not refer to the production methods at the second comparison, although one of the compared solutions was hers and the other was the same as her first solution (thus, she could make this comparison intramentally). Instead, Child G echoed Y's reference to the area at the third comparison. Her focus seemed to be on the "area" rather than on the "production methods."

Table 5.4 Utterances of three children

Solution 1	Solution 2	N	Y	G
N's first	G's first	"The same"	–	"The same"
N's first	G's second	"The production methods differ,"	–	–
		"the shape is the same."	–	"The shape is the same."
N's first	N's second	"The shapes differ …"	(Differ.)	"The shapes differ …"
		–	"Areas are the same."	("Areas are the same.")

Individual Pivotal Moments

We need a further analysis that reveals the children's individual learning processes, how they had their own questions, expressed ideas, observed others' explanations, and deepened their own understanding. In other words, we should explain when their foci were created (individual pivotal moments) and how these foci grew to interact with each other to form collective moments. I analyzed the individual trajectories of Children Y and G, since Child G assumed the role of active monitor, often supporting Child Y's utterances, but she only referred to the "shape" in her report in contrast to Y's reference to the "math" (Table 5.2).

Child Y's Trajectory

Video analyses of the actions and gestures of Children Y and G in Phases 1 and 2 clarified the difference. After Child Y monitored N's solution and explanation in Phase 1, he had planned to solve the task by changing the direction of folding 3/4 out of 2/3 (i.e., the same as Child N's second solution). However, he happened to notice at the midpoint that he did not have to fold the "2/3 rectangle" into fourths and instead only had to fold it into halves to take the 3/4 of 2/3 area (see Figs. 4.3 and 4.4). We assume that this led him to the question, "Why was the answer obtained in a different shape?" or more intuitively, "Are the shapes  and  the same? What is common?" On this assumption, it is reasonable to say that Child Y tried to resolve his question and took advantage of the opportunity to verbalize the "area" and explain it as a half by its shape. Since the shape he used was G's second solution and its answer  was the same as the rest  in its form, his explanation was visually understandable. Yet, he gave up this explanation and resorted to the mathematical explanation, for which he asked others' consent, indicating his satisfaction. Why did he give up the diagrammatic explanation? What was the difference between it and the algorithmic one? One possibility is that if he had sought a commonality

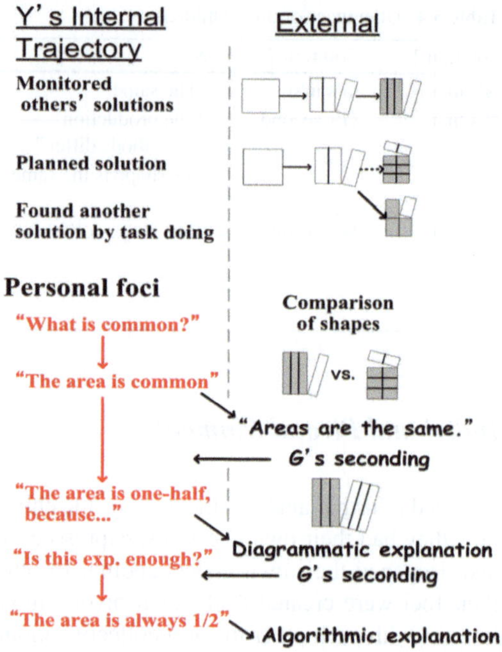

Fig. 5.2 Child Y's intramental interaction

between different shapes as we assumed, an explanation based on the shape of a single answer would account for only a part of the whole, and he felt the incompleteness. If so, we can conclude that Child Y searched for the answer to his question, tried to propose various explanations in the collaborative situation, and found the algorithmic explanation to be the most plausible one, an impression of which lasted for 5 months and appeared in his report.

If we summarize and sketch his understanding as the intramental interaction between internal trajectory and external resources (cf. Fig. 5.2), we notice two things. First, he experienced a shift from a level 1 solution to that of level 2, to which social interaction (monitoring others' solutions) as well as intramental interaction (his own solving actions) might have contributed. Second, if we consider that the shift caused him to have a personal focus, we can better understand his following speech and actions. For example, this explains why Child Y did not stop with his diagrammatical explanation, regardless of Child G's strong support, because he needed to integrate various shapes other than G's answer. The convergence-oriented theory would hardly explain this, since individuals do not have to search for new ideas after convergence. Thus, we defined line 172 as an individual pivotal moment for him.

Child G's Trajectory and Interaction with Child Y

Child G also monitored N's solution and explanation and reflected upon her solution in Phase 1. She then recreated the same shape by a new method in Phase 2 (see Fig. 4.3).

However, she hesitated to turn in her answer, implying her lack of confidence in its accuracy. She might have the question, "Did I arrive at the correct answer by this method?" or "Is the shape  really the same as ? Is my answer OK?" On this assumption, it is plausible that Child G sought a guarantee of the correctness of her answer beyond its visual properties and jumped at Child Y's verbalization of the commonality "area." When Child Y picked up her answer and explained that it was 1/2 based on its shape, that explanation could be clear enough to resolve her question and support her answer.

Figure 5.3 adds Child G's hypothetical internal trajectory to Y's. We can see that she often agreed with Y, but a closer look reveals that she might have had a different personal focus than Child Y (represented by different colors in Fig. 5.3). In particular, when Child Y explained G's answer as one-half, he was not satisfied with it, but Child G seemed to be satisfied with it as confirming the correctness of her own solution. This indicates their different conceptual levels of foci. Her personal focus influenced when she terminated her own understanding process and how she remembered her learning experience. In this sense, we defined lines 151–182 as her pivotal moment. One more important thing is that the two children could communicate with each other, even though they might have meant or searched for different things when saying the same thing (e.g., the word "area" might have meant "something common" to G but "something to be clarified" to Y). This contributes to the first and third collective pivotal moments.

Discussion

Two main results emerged from focus-based constructive interaction analysis. First, even in a shared situation, each learner deepens his or her own understanding by asking his or her own questions, pursuing his or her own goals, and searching the external world for answers or explanations. Thus, even if people share the same external information or data, they integrate it differently and to different degrees according to their questions or goals (the teacher's failure in Phase 7 shows that he also had a different goal in this lesson). Second, this difference itself produces different interpretations and promotes social interactions among learners. For example, Child G did not share Y's internal question but picked up his word "area" based on her own need, which socially displayed a new commonality to the class and supported Y's advancement. Children deepened their *individual understanding* through *social interactions*.

The implications for the design of the lesson are twofold. First, the results tell us that a shared context does not guarantee shared understanding among learners. Educators should be more cautious about what each individual learner takes away from a shared situation. Second, we can take advantage of such diversity. If we had let students exchange their delayed reports of the lesson, they would have noticed different understandings of even the same lesson, which could have further

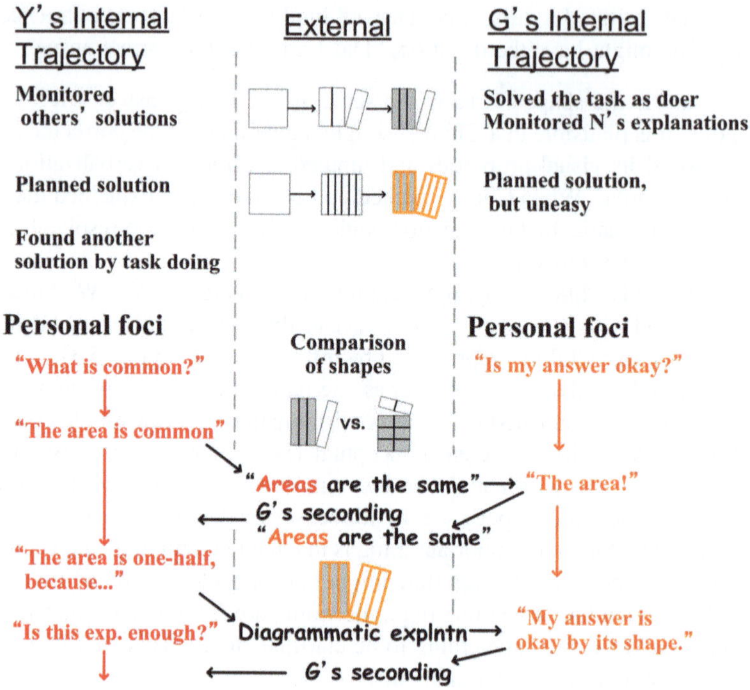

Fig. 5.3 Interaction of Children Y and G's intramental interactions

diversified their learning. The accumulation of such opportunities also engenders meta-cognitive reflection on one's own solutions or mental models, which could raise their adaptability to newer problem situations. Of course, the lack of challenges and justifications in the lesson (pointed out by Chap. 7) might have increased the diversity in their understanding. Thus, it remains an empirical question whether such encapsulation and its exchange promote divergence or convergence.

Space prevents me from documenting the other four children's trajectories and their interactions, but we are always able to treat a collaborative situation as an intersection of multiple trajectories of every learner. Figure 5.3 shows an image of such trials, that is, multiple ebbs and flows of talks and actions at the foreshore of each learner. Advanced technologies enable us to keep a record of interactions among any persons at any time and place (Dyke, Lund, & Girardot, 2009). The units of analysis can function as a zoom from the group as a whole through the emergent small group to the individual. With technological advancement, we should be more flexible in adopting complementary units of analysis or interaction.

This chapter proposes "role" and "focus" as analytic devices. The role enables us to untangle collaboration into interactions of individual processes. It also gives us a hint on how to redesign the lesson. For example, we can make the elaboration lesson more constructive by introducing small-group discussion in the discussion phases, which increases the opportunities for active task-doing by silent observers and reflective monitoring by impulsive leaders. The small group serves both as ground for role-exchange among individuals and as a task-doing/monitoring group within the class. Yet, there is another possibility that the monitor can learn even when he or she remains silent like Child K in this lesson. This is an empirical question.

The focus helps us hypothesize individual understanding process. Types of foci make our hypotheses plausible to some degree. For example, when asked to solve the same task repeatedly but in various ways in Phase 2, one can solve it differently using the solutions of level 1 (e.g., changing the direction of folding like Child N). This secures the correctness of his or her solution because an external resource-dependent strategy provides ample opportunities to confirm the solution through external traces. This solver does not have to get his or her solution problematized. He or she would have no personal focus or at most that of solution procedures. However, if one changes the solution from that of level 1 to that of level 2, the correctness of one's solution has to be problematized because the reinterpretation of traces is an internal process and its correctness cannot be confirmed externally (e.g., Child Y). He or she would seek some basis for solutions, which activates his or her personal focus. Even though he or she might not focus on the area directly, his or her focus would be on something other than concrete procedures.

We should examine the generalizability and usability of foci as an analytic device in other learning tasks and lessons. More importantly, we do not know how one gains a particular focus. For example, why did Child Y have the focus postulated here? As Trausan-Matu claimed, Child Y might be the "divergent thinker," who tends to focus on newer and higher dimensions. This is a question that cannot be answered by the result of just one lesson. We should have traced Children G or Y in various lessons longitudinally to examine whether their ways of focusing were constant enough to call them *cognitive styles* or were flexibly changing depending on the topic, the task, and their own thoughts. By doing that, we could ask how the accumulated experience of having foci in various situations engenders practices, styles, and attributes. Multivocal analysis of a huge database of learning processes enables us to build bridges across our strength to clarify the actuality and diversity of learning in the real world.

References

Baker, M., Hansen, T., Joiner, R., & Traum, D. (1999). The role of grounding in collaborative learning tasks. In P. Dillenbourg (Ed.), *Collaborative learning: Cognitive and computational approaches* (pp. 31–63). Oxford, UK: Pergamon.

Dyke, G., Lund, K., Girardot, J.-J. (2009). Tatiana: An environment to support the CSCL analysis process. *CSCL2009* (pp. 58–67), Rhodes, Greece.

Engle, R. A. (2006). Framing interactions to foster generative learning: A situative explanation of transfer in a community of learners classroom. *The Journal of the Learning Sciences, 15*, 451–498.

Engle, R. A., & Conant, F. R. (2002). Guiding principles for fostering productive disciplinary engagement: Explaining an emergent argument in a community of learners classroom. *Cognition and Instruction, 20*(4), 399–483.

Ford, M. J., & Forman, E. A. (2006). Redefining disciplinary learning in classroom contexts. In J. Green & A. Luke (Eds.), *Review of research in education* (Vol. 30, pp. 1–32). Washington, DC: American Educational Research Association.

Forman, E. A., & McPhail, J. (1993). Vygotskian perspective on children's collaborative problem solving activities. In E. A. Forman, N. Minick, & C. A. Stone (Eds.), *Contexts for learning: Sociocultural dynamics in children's development* (pp. 213–229). New York, NY: Oxford University Press.

Hatano, G., & Inagaki, K. (1986). Two courses of expertise. In H. Stevenson, H. Azuma, & K. Hakuta (Eds.), *Child development and education in Japan* (pp. 262–272). New York, NY: Freeman.

Hatano, G., & Inagaki, K. (1991). Sharing cognition through collective comprehension activity. In L. B. Resnick, J. M. Levine, & S. D. Teasley (Eds.), *Perspectives on socially shared cognition* (pp. 331–348). Washington, DC: American Psychological Association.

Hatano, G., & Inagaki, K. (1994, April). *A two-level analysis of collective comprehension activity*. Paper presented at the symposium, "Integrating the cognitive and social in the construction of mathematical and scientific knowledge", American Educational Research Association Meeting, New Orleans, LA.

Kuhn, D., & Ho, V. (1980). Self-directed activity and cognitive development. *Journal of Applied Developmental Psychology, 1*, 119–133.

Lobato, J. (2006). Alternative perspectives on the transfer of learning: History, issues and challenges for future research. *The Journal of the Learning Sciences, 15*(4), 431–449.

Miyake, N. (1986). Constructive interaction and the iterative process of understanding. *Cognitive Science, 10*, 151–177.

Miyake, N. (2008). Conceptual change through collaboration. In S. Vosniadou (Ed.), *International handbook of research on conceptual change* (pp. 453–478). New York, NY: Routledge.

Miyake, N., Oshima, J., & Shirouzu, H. (2011). *Robots as research partners for promoting young children's collaborative learning*. The pre-conference workshop at CSCL2011, Hong Kong

Neisser, U. (1976). *Cognition and reality: Principles and implications of cognitive psychology*. New York, NY: Freeman.

Newell, A., & Simon, H. A. (1972). *Human problem solving*. Englewood Cliffs, NJ: Prentice-Hall.

Norman, D. A. (1991). Cognitive artifacts. In J. M. Carroll (Ed.), *Designing interaction: Psychology at the human-computer interface* (pp. 17–38). Cambridge, UK: Cambridge University Press.

O'Connor, M. C., & Michaels, S. (1996). Shifting participant frameworks. In D. Hicks (Ed.), *Discourse, learning, and schooling* (pp. 63–103). Cambridge, UK: Cambridge University Press.

Roschelle, J. (1992). Learning by collaborating: Convergent conceptual change. *The Journal of the Learning Sciences, 2*, 235–276.

Roy, D. (2011). The birth of a word. Talk presented at TED (http://blog.ted.com/2011/03/10/deb-roy/).

Sacks, H. (1962/1995). *Lectures on conversation*. Oxford, UK: Blackwell.

Sacks, H., Schegloff, E. A., & Jefferson, G. (1974). A simplest systematics for the organization of turn-taking for conversation. *Language, 50*, 696–735.

Saito, M., & Miyake, N. (2011). Socially constructive interaction for conceptual change. In N. Law et al. (Eds.), *CSCL2011*, Hong Kong.

Shirouzu, H., Miyake, N., & Masukawa, H. (2002). Cognitively active externalization for situated reflection. *Cognitive Science, 26*, 469–501.

Stahl, G. (2006). *Group cognition: Computer support for building collaborative knowledge*. Cambridge, MA: MIT Press.

Suthers, D. D., Medina, R., Vatrapu, R., & Dwyer, N. (2007). Information sharing is incongruous with collaborative convergence: The case for interaction. In C. Chinn, G. Erkens, & S. Puntambekar (Eds.), *Computer supported collaborative learning 2007*. Rutgers, NJ: ISLS.

Chapter 6
Collaborative and Differential Utterances, Pivotal Moments, and Polyphony

Stefan Trausan-Matu

Introduction

A deep analysis of collaborative learning sessions should consider several facets. A first aspect is in what degree and how group interactions involved in joint learning provide a scaffold for the individual development. Students have a personal, individual learning trajectory, which interferes with that of the other students when they are learning in a group, like in polyphonic music, where voices have both longitudinal and transversal dimensions.

The dual individual–group perspectives are extremely important for students entering into a process with two cycles, in which they should interact with the others, debate, negotiate meaning in order to construct knowledge, and, meanwhile, internalize it (Stahl, 2006; Vygotsky, 1934/1962). Starting from Bakhtin's (1981) dialogism ideas we consider dialogue as being essential in both the group and individual cycles: Students enter in dialogues with other students in the first case and with themselves in the second case (for example, the "make problematic" link in Stahl's cycle of knowledge building (Stahl, 2006)). Moreover, we consider that there is an interaction between external and internal dialogues, for example, external dialogue utterances of one student may have as reaction an internal dialogue utterance at other (or even the same) student, which may be externalized later as an utterance with a loud voice.

Another issue to be considered in collaborative learning is the identification of all types of implied utterances, the role played by words and spoken or written communication but also by other types of communication acts, which may be similar in effect with textual utterances. Natural language is both a means of joint knowledge building and a way for professors towards monitoring the learning process. However, natural language is not the sole way of communication in collaborative learning.

S. Trausan-Matu (✉)
University "Politehnica" of Bucharest, Bucharest, Romania
e-mail: stefan.trausan@cs.pub.ro

D.D. Suthers et al. (eds.), *Productive Multivocality in the Analysis of Group Interactions*, Computer-Supported Collaborative Learning Series 16,
DOI 10.1007/978-1-4614-8960-3_6, © Springer Science+Business Media New York 2013

In addition to spoken or written language, other means for collaborative knowledge construction may be identified: visual communication, either using diagrams, drawings, images, and objects or body language. All of these may be considered as utterances, in a generalized way, and all may give indicators for learning. However, a big problem is that the set of utterances that may be taken into account is very large, even if we consider only the textual ones. Therefore we should have a means to identify the relevant ones, which are recurrent, have an influence on the learning process, and have an "echo" in the future. As we will later discuss in detail, we call such utterances "voices."

There were several approaches directed towards the analysis of collaborative learning sessions. Their vast majority considered textual utterances: transcriptions of spoken conversation, logs of instant messenger (chat), forum interventions, and even wikis, for example: CORDTRA (Hmelo-Silver, Chernobilsky, & Masto, 2006), COALA (Dowell & Gladisch, 2007), DIGALO and other tools used in the Argunaut system (Harrer, Hever, & Ziebarth, 2007), and ColAT (Avouris, Fiotakis, Kahrimanis, & Margaritis, 2007). Multimedia utterances were also considered, for example in TATIANA (Dyke, Lund, & Girardot, 2009). Some of these systems use several kinds of argumentation graphs, some of them in the idea of Toulmin (1958), or more elaborated structures like the contingency graphs (Suthers, Dwyer, Medina, & Vatrapu, 2007).

The analysis in existing approaches is usually focused on pairs of utterances: adjacency pairs (Schegloff & Sacks, 1973; Jurafsky & Martin, 2009), transacts (Joshi & Rosé, 2007), or, considering also longer distance connections, uptakes (Suthers et al., 2007). We consider that also another, more global unit of interaction than that of pairs of utterances should be considered: threads of utterances interanimating in a polyphonic framework (Trausan-Matu, Stahl, & Sarmiento, 2007). For example, even the significance of a pair of utterances may be totally different if they are singular or if the second utterance in the pair appears after a thread of repetitions of the first utterance. Moreover, repetitions of utterances, either singular or in pairs, may generate a rhythm.

Utterances which are influential become "voices" that means threads having a duration and/or echoes. In our vision, an utterance, in a generalized sense (and consequently a potential voice), may be a word, a sentence, a paragraph, a paper, a book, a turn in a conversation, a figure, a gesture, etc. Utterances may be not only individual, but they may also be generated by a group (for example, all students move their chairs as a chorus at the beginning of the origami fractions session).

Learning, either individual or collaborative, has duration (a longitudinal dimension in time) and can occur at different rhythms (even in the same session) of dialogue and in different settings. Consequently, a derived problem is what types of space–time situations or chronotopes (Bakhtin, 1981; Ligorio & Ritella, 2010) may be identified in the analyzed data, and in what degree they are well suited for achieving a good collaboration.[1] For example, in the beginning of collaborative problem solving a chronotope with few verbal utterances may be detected, in which

[1] In our approach, we consider a chronotope as "a genre of movement or pacing in the space that participants adopt over the temporal duration of an activity" (Ligorio & Ritella, 2010).

students explore the problem. When they reach to build collaboratively a solution, another chronotope may be detected, which may be called also a region of good collaboration (Banica, Trausan-Matu, & Rebedea, 2011), in which threads of verbal utterances occur in a rapid rhythm.

Changes in learning rhythm are the starting point for passing from one chronotope to another (Ligorio & Ritella, 2010) and may be considered pivotal moments in the learning session. Changes of rhythm are often associated with the presence of special utterances, for example, collaborative or differential (Trausan-Matu et al., 2007), which may be therefore considered as cues for detecting pivotal moments. Collaborative utterances, even if they sometimes don't mark a change of rhythm (a passage from a chronotope to another) are also candidates for pivotal moments because they are not frequent situations and they display moments in which the group behaves like a whole; it really collaborates, which is a desideratum in computer-supported collaborative learning.

A good professor is able to orchestrate utterances as voices: he/she gives texts to students to read, speaks, uses images and gestures, and even analyzes and directs the class's acts (or utterances as a group) in order to build a coherent thread of ideas. This process is similar to music not only by the existence of a polyphony of voices but also through the created rhythms.

The identification of the types of chronotopes, of collaborative moments, and of pivotal moments in a learning session are very important for a teacher in order to manage students' activity. A model which can provide a unifying view on the above facets is multivocality and polyphony (Trausan-Matu et al., 2007), which will be used in this chapter for analyzing the origami fractions data set. This model may also be used to implement semiautomatic analysis tools, which provide facilities for the visualization of voices and their interanimation and potential pivotal moments (Trausan-Matu & Rebedea, 2009; Chiru & Trausan-Matu, 2012).

The Five Dimensions Characterizing the Approach

The method of analysis of collaborative learning used in this chapter is based on considering small-group interactions from the perspectives of dialogism and polyphony (Bakhtin, 1981, 1984; Trausan-Matu et al., 2007), repetition and rhythm as an involvement provider (Tannen, 1989), interanimation (Wegerif, 2005; Trausan-Matu et al., 2007), conversation analysis (Sacks, 1962/1995)—collaborative utterances and adjacency pairs), and collaborative moments (Stahl, 2006). The five dimensions on which our approach may be understood are the following:

Assumptions Underlying the Analysis

Theoretical assumptions. Knowledge may be constructed in small groups (Stahl, 2006, 2009). In this process, interplays take place between the group discourse and the understanding of the participants as individuals (Stahl, 2006).

Small-group conversations for problem solving and collaborative learning often take the form of multi-threaded discourse that follows polyphonic patterns (Trausan-Matu et al., 2007). Both group discourse and individual thinking are characterized by dialogism and multivocality (Bakhtin, 1984; Trausan-Matu et al., 2007).

Methodological assumptions. Interanimation patterns (Trausan-Matu et al., 2007) may be detected in interactions, and they offer a glimpse on the collaborative learning processes of the group. Conversation analysis and ethnomethodology (Garfinkel, 1967) may be used for providing cues for detecting interanimation and collaboration (by identification of associated member methods). Integrating natural language processing (NLP) techniques (for the automatic identification of adjacency pairs, repetition, and discourse threads) with polyphony identification, social network analysis, and graphical visualizations may provide a way for analyzing the contributions of each participant and their interanimation.

Purpose of Analysis

A main purpose of analysis from the point of view of this chapter's approach is the recognition of interanimation patterns among voices (in particular considering participants and discussion threads) and, as a result, the inference of pivotal moments mentioned earlier and regions of good collaboration. Related purposes are the identification of collaborative and differential utterances, of adjacency pairs, of voices (discourse threads) and their interactions, and of the semantic and pragmatic content of the utterances. Eventually, starting from the above data, an evaluation of the contribution of each participant to the learning process may be also derived.

Units of Interaction

The most important units of interaction in our approach are voices, in a generalized sense, which means, from another perspective, discourse threads viewed in a polyphonic weaving. However, as units of interaction are also considered pairs of utterances. We remind that utterances, in a generalized sense may be: words, sentences, gestures, and images. All these may be seen also as units of action.

Representations of Data and Analytic Interpretations

Transcriptions of textual utterances are codified using a complex XML schema in order to be available for an automatic analysis. Graphical representations of some types of voices and their interanimation are generated automatically. Graphical representation of the evolution of the contribution of each participant may also be represented.

Analytic Manipulations

There are two main analysis directions. The first of them is the analysis of discourse for identifying voices, repetitions of generalized utterances (as defined above) in order to construct threads, and their interactions. This objective includes the analysis of speech acts, adjacency pairs, collaborative and differential utterances, co-references, argumentation chains, contrapuntal/polyphonic structure, etc. and (if available) nonverbal communication and individual/group body language. NLP tools are used as a support of the analysis. The second direction is the analysis of the social network of user links between their utterances.

The Polyphonic Model and the Interanimation Patterns

Polyphony is an example of a joint achievement of several independent participants acting sequentially (singing in music or emitting utterances in dialogues) starting from a common theme and meanwhile keeping coherence among them. It originated as a concept and practice in music, and it can be extrapolated to texts, as Bakhtin (1984) emphasized and even, in our opinion, to spoken and nonverbal artifacts. Polyphony may occur in musical pieces with more than one melodic line (or voice) at a time, in contrast with monophony, where a single voice (part) is present. Polyphony differs also from homophony because even if in both cases multiple voices are present, in the former they have a high degree of independence. However, even if they are independent, in order to achieve polyphony, the voices obey some implicit constraints, some so-called counterpoint rules, for example, in order to achieve a joint harmonic, pleasant musical piece. Polyphony may be seen as a model of group interaction and creativity, in which independent individuals (voices, in a metaphorical sense) achieve a joint activity during a period of time.

We consider a voice, in a generalized way, beyond the acoustic sense, as a distinctive presence in a group, influencing the other voices. An utterance or a sequence of utterances become a voice if they have a longitudinal dimension, they last, they have an echo in time, and they may be perceived as a coherent thread. Meanwhile, to have a distinctive presence, a voice should have a transversal dimension, opposing but also keeping coherence with the other voices.

One important feature of the polyphonic model is that if we consider the generalized perspective of a voice, it may be applied for an integrated analysis of different types of media for communication. Even if it was conceived by integrating ideas from music and text, it may be applied to analyzing video images, as will be the case in the analysis of the origami fractions data set presented in this chapter. A voice in our polyphonic model may be a spoken utterance, a written utterance on the blackboard, but also nonverbal utterances like a gaze, a movement, or the acts of the teacher, a student, or a group of students.

The polyphonic model of group interaction in collaborative learning considers that in a conversation different longitudinal threads (or “voices”) appear, composed of utterances and their echoes, each of them having independence but achieving a joint (a consonant) discourse (Trausan-Matu et al., 2007; Trausan-Matu & Rebedea, 2009). However, the interaction in a group inherently involves the solving of the dissonances appearing between voices. Therefore, as also Bakhtin noticed for texts, in general (Bakhtin, 1981), participants face both centrifugal (divergent, towards difference) and centripetal (convergent, towards unity) forces, along two directions: longitudinal and transversal, following constraints that are similar to the music counterpoint rules (Trausan-Matu et al., 2007). These forces have an important effect: they obligate the participants to perceive dissonances that put their utterances under question (they make them “problematic” in the personal cycle of the knowledge building (Stahl, 2006)), and they generate an interanimation phenomenon. The polyphonic analysis tries to identify interanimation patterns along the two dimensions while corresponding to the two types of forces.

The polyphonic analysis of a joint activity like those specific to collaborative learning combines the individual and group perspectives. Similarly to the case of participants in an improvising jazz quartet, each learner is listening to (and sometimes also looking at) the others and is also playing in the same time, achieving a joint musical piece. It is very important to consider the group as a whole—not just individual developments or dyadic interactions within the group. The joint achievement of the group, be it music or spoken or written dialogue, is constrained by the centripetal and centrifugal forces towards convergence/divergence, and it may be seen as a creative or a “thinking device” (Wegerif, 2005). The presence of centrifugal and centripetal forces may be discovered by the identification of interanimation patterns among participants’ utterances.

Interanimation patterns may be classified in unity-pursuing patterns, characterized by a trend towards continuity and achieving coherence in the interaction and differential interanimation patterns (Trausan-Matu et al., 2007). They may be identified, for example, in transcriptions or chat logs using conversation analysis (CA—Sacks, 1962/1995) or NLP, and they may be the starting point for analyzing the degree of collaboration and personal contributions (Trausan-Matu & Rebedea, 2010). Interanimation patterns occur also in face-to-face interaction, including nonverbal behavior, as will be discussed later in this chapter.

A very important case of unity patterns is the cumulative talk (Mercer, 2000) or, in Sacks’ words, collaborative utterances (Sacks, 1962/1995). This type of convergent interaction is characterized by the fact that two or more participants spontaneously build together a sentence, as if they were a single person. Two examples are found in Sacks (1962/1995):

Joe	(Coughs) We were in an automobile discussion,
Henry	discussing the psychological motives for
Mel drag	racing on the streets

and Trausan-Matu et al. (2007):

ModeratorSf	Could you guys tell templar what's going on?
Mathpudding	We're experimenting with circles
Mathman	and finding as many possible relations as we can

This kind of pattern occurs also in the data set, at several points, for example, at utterances 457–459, one of the pivotal moments is as follows:

34:40	457	N	Although the production methods differ [starts quietly],
	458	T	Yes.
	459	G	The shape is the same.
		K	The shape is the same.
		N	The shape is the same.

Collaborative utterances appear in several places in the origami fractions data set. They are rare, and they are generally related to pivotal moments (which might be related to what Stahl (2006) calls "collaboration moments") in which the group displays cohesion and sometimes understanding. Collaborative utterances may also be nonverbal, in body language, like the fact that everybody (excepting Y, see next section for details) moves the chairs as if it were a choreography, at moment 0:25[2] in the video.

If collaborative utterances might be considered examples of consonances in the polyphonic metaphor, differential patterns may be viewed as examples of dissonances, of something felt as unfinalized or wrong. They have a very important role in triggering further utterances of other participants as a result of incompleteness perception.

A differential pattern example is (taken from Stahl, 2006 and commented in Trausan-Matu & Rebedea, 2009) the following:

1:21:53	Teacher: And you don't have anything like that there?
1:21:56	Steven: I don't think so
1:21:57	Jamie: Not with the same engine
1:21:58	Steven: ┌ No
	Jamie: └ Not with the same
1:21:59	Teacher: With the same engine … but with a different (0.1) … nose cone?=
1:22:01	Chuck: ┌ =The same=
	Jamie: └ =Yeah,
1:22:02	Chuck: These are both (0.8) the same thing
1:22:04	Teacher: Aw ┌ right
1:22:05	Brent: └ This one's different

Remark that this differential pattern occurs after a series of repetitions of "the same" which becomes a thread or, in other perspective, a voice inducing a dissonance needing a resolution.

[2] The indicated times are those from the video file, not those in the transcription or subtitling, which are slightly different.

Differential patterns are also essential for the identification of pivotal moments in the origami fractions data set. An important fact to remark is that the below examples of differential patterns are connected to the collaborative utterance (473–474) marking the first pivotal moment. Note that almost the same words occur in the above and below excerpts from two different corpora:

35:58	469	T	What do you think of N's two solutions? [Places N's two solutions on the teacher's desk.]
	470	Y	[Moves towards the teacher's desk by further raising his hip.]
	471	Anonymous	[Whispers] The shapes differ.
36:14	472	Y	Differ [with clear voice]
	473	Y	though areas are equal [with low voice].
	474	G	The areas are the same,
	475	T	Yes.
36:20	476	G	but the shapes and production methods differ.

Differential patterns may occur also (as in the case of collaborative utterances) in body language, as it will be discussed in a section below.

The Analysis of the Origami Fractions Data Set

Many unity and differential interanimation patterns of different kinds, on different dimensions (verbal and nonverbal), may be identified in the origami fractions data set, occurring among different types of voices: participants' spoken utterances, body language utterances, solutions, opinions, threads of repeated words, etc. Some of the interanimation patterns are unprompted (for example, the collaborative utterances) and some are induced by the teacher (for example, threads of repeated differential patterns aimed at inducing the answer to the problem). In general, teachers should know how to handle voices and interanimation patterns. They have to be able to detect collaborative utterances that may be a sign of moments of collaboration. The repeating of difference patterns may induce understanding. Different kinds of additional voices like images or drawings may be used for inducing interanimation.

Pivotal moments in our perspective are generally associated to the presence of collaborative or differential utterances, which occur many times as a result of threads' (voices') interaction. As we mentioned previously, pivotal moments (and collaborative and differential utterances) also coincide sometimes with changes in the learning rhythm, marking the passing from one chronotope to another (Ligorio & Ritella, 2010).

I discovered some of the instances of interanimation patterns on a later, more thorough analysis, after seeing that Chiu's analysis (Chap. 7, this volume) contained more pivotal moments than mine. Moreover, his discussion on micro-creativity enforced me the unifying view of CSCL and computer-supported group creativity under the polyphonic model.

We analyze the origami fractions data set, according to several dimensions which can be considered intertwined, following the polyphonic model. The dimensions we consider in the following sections are spoken dialogue, body language, visual dimension, internal dialogue (at an intramental level), and echoes. In each of these dimensions, several voices, in a metaphorical way, and their polyphonic interactions may be detected.

Spoken Dialogue

The first and probably most important dimension consists of individual and collaborative utterances in the spoken dialogue. This dimension may be investigated by CA (Sacks, 1962/1995), discourse analysis (Tannen, 1989), interanimation (Trausan-Matu et al., 2007; Trausan-Matu & Rebedea, 2009), and NLP methods (Trausan-Matu & Rebedea, 2010).

As mentioned in the previous section, collaborative and differential patterns may be detected in the transcribed data of the origami fractions session. A very "dense" segment, with several collaborative and differential utterances, is between utterances 469 and 482. The segment starts with the first pivotal moment labeled by Shirouzu's analysis (Shirouzu, Chap. 5, this volume) which is also the fifth of Chiu (472–474) (Chiu, Chap. 7, this volume). We also identified this segment as a pivotal moment within our polyphonic perspective due to both differential interanimation patterns (at 471 and 476) and collaborative utterances (473–474, 476, and 481–482).

35:58	469	T	What do you think of N's two solutions? [Places N's two solutions on the teacher's desk.]
	470	Y	[Moves towards the teacher's desk by further raising his hip.]
	471	Anonymous	[Whispers] The shapes differ.
36:14	472	Y	Differ [with clear voice]
	473	Y	though areas are equal [with low voice].
	474	G	The areas are the same,
	475	T	Yes.
36:20	476	G	but the shapes and production methods differ.
		K	The shape and production method differ.
		N	The shape and production method differ.
		Anonymous	The shape and production method differ.
	477	T	The areas are the same.
	478	T	Because the areas are the same,
	479	T	this is the last comparison [N's first solution and K's one].
		Y	[Leans over the desk.]
36:52	480	T	What do you think of these?
37:04	481	G	Although shapes are the same,
		K	Although shapes are the same,
		N	Although the shapes are the same,
		O	Although the shapes are the same,
		Y	Although the shapes are the same,
	482	G	the production methods differ.
		K	the production methods differ.
		N	the production methods differ.
		O	the production methods differ.
		Y	[Quickly goes back to his seat.]

The second pivotal moment identified by Shirouzu (Shirouzu, Chap. 5, this volume) corresponds also to collaborative utterances 502–503:

38:13	495	T	What among these is constant?
38:14	496	Anonymous	[Whispers] Area.
	497	Anonymous	[All together] Area.
	498	T	Area.
	499	T	The areas are the same.
38:18	500	T	How large is the area?
38:20	501	Anonymous	[Whispers] of 2 (halves), 2 …
38:24	502	Y	1/2 [in low voice]
	503	Y	[Slowly] of the whole.
		T	[Following Y] 1/2 of the whole
	504	G	Ah.
		K	Ah. [Moves right hand.]
		N	Ah. [Nods.]

Other collaborative utterances occur in several places of the data set. For example, at utterances 8–10, a first verbal joint, collaborative utterance marks the beginning of the problem solving:

0:00	1	T	Here we have a piece of origami paper, a pencil, and a pair of scissors.
	2	T	What I want you to do is …
	3	T	to use these to make three-fourths of two-thirds of this origami paper.
0:27	4	T	Can anybody do that?
0:30	5	N	Can I?
	6	T	Oh, you need this? [Handing a piece of origami paper to N.]
	7	N	[Starts to fold the paper into a rectangle of one-third of the total area.]
	8	F	Of two-thirds …
	9	G	Of two-thirds …
	10	K	Three-fourths.

Differential patterns may also be considered for detecting pivotal moments, as mentioned above. They occur sometimes after a series of repetitions (as in the collaborative moment in the solving of the rocket nose problem in Stahl (2006)) and/or together a collaborative utterance, like at utterances 471–476 in the Shirouzu (Shirouzu, Chap. 4, this volume) data set.

Body Language

The second dimension of analysis that we consider is body language, which may contain individual or collective utterances. An example is the moment (at 0:25) when, after the teacher appears in front of the students, all of them move their chairs forward, excepting Y. Such a movement seemed like a collective spontaneous sign of their entering into the lesson space. Ethnomethodology may be used for analyzing such member methods (Garfinkel, 1967). The Y student's body language is in many moments independent, behaving like a distinct voice, on a differential pattern. He stays almost immobile for the majority of the first 30 min. An important moment is at about 36:37, when for several tens of seconds Y stands up, moves towards the table, and looks transversally. This is important because this moment coincides with another crucial moment (pivotal moment 1), when student Y has a very important contribution.

Collective body language is also displayed by some students (N, F, and K) when they avoid answering the teacher's question:

561	T	I would now like to ask each of you what you did,
562	T	because all of you have solved it.

They do this by putting their hands over their eyes, putting their heads on the table, or looking elsewhere (all different "methods" of avoiding an answer that are very well known by professors).

The reaction of the students may be viewed also as a voice saying "we don't want to answer any more," and it could even be considered as a pivotal moment in the lesson, possibly indicating students' fatigue and thus the beginning of another chronotope. As a consequence, the teacher does not insist and answers himself.

The Visual Dimension

What participants see is a third dimension of analysis. Visual data on the blackboard, what other participants do, and even others' body language are "voices" that may generate reactions that may be sparks triggering interanimation patterns. The actions of the teacher that writes on the table and displays the solutions may be seen as voices that are supposed to trigger students' internal reasoning and responses.

Shirouzu (Chap. 4, this volume) mentions that the origami fractions experiment had two phases. In the first 30 min, children were instructed to solve the problem of "obtaining 3/4 of 2/3 of colored paper (origami paper)" using provided colored paper and scissors. This process occurred mainly individually, although a joint component is present because they could look at each other and compare their utterances (including origami folding and cutting acts), seeing who solved the problem and how they went about it. The visual dimension was enforced by the teacher when displaying the solutions on the table.

Intramental Dimension

In the beginning part of the data set some students are folding and cutting origami (G and N) and some are watching (Y, K, and O). After others started to individually solve the problem by cutting and folding origami, Y proposed a solution (at 13.07) totally different from the others. He also has the major contributions at pivotal moments 2 and 3, and after 5 months he was the student having the best description of the origami session findings. One explanation of his achievements in spite of his predominant less active participation might be that he is probably rather an intramental than an intermental reasoner, a lurker positioning himself on differential positions (as discussed in the body language section) even regarding his own verbal utterances (doubting at utterance 519 that what he said before was right: "Is this wrong?"). At least in this data set, his actions show that he prefers to look and afterward to act. In polyphony terms, he prefers to develop a counterpoint while internalizing others' voices and to have inner dialogues rather than entering into polyphony with the others. Even the fact that he does not stay at the table for the majority of time is perhaps an argument for our idea. Based on what we can observe in the origami fractions data set we could say that Y is a divergent thinker.

A similar assertion may be said partially about K, who inversed the order of fractions (2/3 of 3/4 instead of 3/4 of 2/3). We may remark also that K, from the beginning, has a different position:

8	F	Of two-thirds …
9	G	Of two-thirds …
10	K	Three-fourths.

It is interesting to note that even if K seems to inverse the order, she also had one of the best rememberings of the idea of the session after 5 months. In another interpretation of utterance 10, K might be completing the previous utterances according to teacher's specification.

The ideas of inner speech and dialogue have an important role in the writings of Vygotsky (1934/1962) and Bakhtin (Voloshinov, 1929/1973). For example, Bakhtin says: "There are no ontological differences between inner and outer speech" (Clark & Holquist, 1984).

Stahl's personal understanding cycle contains also inner acts: "We may be able to repair our understanding by explicating the implications of that understanding and resolving conflicts or filling in gaps—by reinterpreting our meaning structures—to arrive at a new comprehension" (Stahl, 2006). He considers that what happens at the individual mind level is socially determined: "The process of interpretation that seems to be carried out at the level of the individual mind is already an essentially social process" (Stahl, 2006).

Some neurology researchers are also supporting the idea of inner speech, following the ideas of the Russian school initiated by Vygotsky and continued by Luria (DeBleser & Marshall, 2005). Neural correlates of inner speech are also mentioned (Jones & Fernyhough, 2007). I searched such evidences after the "polyphonic interanimation" of my, Shirouzu's, and Lund's ideas and opinions related to the intramental dimension, exchanged during our interactions around the data set.

Thinking—the intramental activity—is, in our vision dialogical, implying inner speech which, similarly to the outer speech, is composed of inner utterances. If we consider that there is such a dimension, at least two types of students' thinking may be supposed to be present in the origami data set. The first one is that occurring when they are individually trying to obtain the solution by folding origami following the verbalized goal specified by the professor: "to make 3/4 of 2/3 of colored (origami) paper." In support of this idea we remark that they have to achieve at least two sequential steps (obtain 2/3 and 3/4) and therefore to propose actions, to remember them, and to validate their correctness, all of these made without loud voice. We may say that they have to emit inner utterances like "I fold ..." or "I cut ...." Such utterances might not be linguistic; they might be generalized utterances and mental imagery of the folding, cutting, and comparing acts.

In order to solve the problem students should also emit inner utterances in a kind of inner dialogue with themselves, containing sentences as "the (partial) result is good/wrong" or adjacency pairs like question–answer.

A second type of utterance at the intramental dimension is generated by looking at others' solutions, at the teacher's writing, and at the display of solutions on the blackboard. Hearing teacher's and others' utterances probably also generates inner utterances (for example, "my solution is the same as ..." or "my solution is different from ..."), and interanimation patterns may occur (for example, we can consider adjacency pairs (Schegloff & Sacks, 1973) between external and internal utterances, which might also be uptakes (Suthers et al., 2007)). Other types of thinking may be identified, for example, to prepare an answer to teacher's questions and even the attempts to avoid an answer (N, F, and K after the teacher's utterances 561 and 562).

Echoes

The fifth analysis dimension in our approach is the long-term effect, the long-term echo of the voices, spoken, inner, or of another kind, which were present in the lesson. This dimension is very important because it is, in fact, the main goal of the teaching session. The analysis made after 5 months shows that either students forgot or did not initially understand the conclusion of the lesson. After 5 months, only Y, who proposed the solution, and K remembered the final conclusion (Y: "The $2/3 \times 3/4$ made 1/2 and we were taught why it resulted in 1/2"; K: "We thought why $2/3 \times 3/4$ equals 1/2"). An answer to the question "why was there a difference between Child Y and G, in spite of G's convergent moves to Y at pivotal moments 1 and 2" may be given starting from the idea of collaborative utterances. G acted as a member of a group, participating in collaborative utterances, but she didn't internalize the utterance; she only acted as a "mirror" (see Tannen, 1989).

Tools for Helping the Polyphonic Analysis

In the analysis presented in this chapter, the detection of pivotal moments was based primarily on a manual analysis towards the identification of interanimation patterns and the identification of changes in rhythm (passing from a chronotope to another), which sometimes co-occur. The automatic detection of voices, of the instances of interanimation patterns, and of polyphony would be extremely useful, but it is extremely difficult, even if only for textual utterances. An easier task is to assist a human analyst by trying to identify specific behaviors that may indicate the possible presence of voices, interanimation patterns, and changes of rhythm. For example, it is easy to detect repeating words or phrases which may signal a thread, a voice. Moreover, discourse markers, cue phrases, and particular speech acts may be used for detecting differential patterns.

The semiautomatic content-based analysis system PolyCAFe (Polyphonic Conversation Analysis and Feedback generation) proved helpful for the analysis of collaborative learning sessions using instant messenger chats (Trausan-Matu & Rebedea, 2010). This system is based on the polyphonic model (Bakhtin, 1981, 1984; Trausan-Matu et al., 2007) and assists human analysts in the detection and the visualization of the presence of voices, interanimation patterns, participation, contribution, semantic content, and collaboration in conversations (Trausan-Matu & Rebedea, 2010; Rebedea, Dascalu, Trausan-Matu, Armitt, & Chiru, 2011; see also http://www.ltfll-project.org/index.php/polycafe.html). The system uses techniques from NLP and social network analysis (Dascalu, Rebedea, & Trausan-Matu, 2010; Rebedea et al., 2011).

PolyCAFe is a module developed in the EU FP7-IST project "Language Technologies for Lifelong Learning" (LTfLL, see http://www.ltfll-project.org), and it provides textual feedback and interactive graphic visualization of instant messenger chats, transcribed conversations, forums, or other collaborative activities. The system offers (among other services) facilities for the identification of adjacency pairs, for identification of the most frequent used concepts, and for the visualization of threads (voices) and their interactions.

PolyCAFe was used for the analysis of the transcription of the discussions in the origami fractions data set. For this purpose, the transcription[3] was encoded into a specific XML schema, processed, and analyzed with the graphical facilities. Figure 6.1 shows the graphical visualization of the threading generated by the words "different," "same," "solution," and "number" in the origami fractions data set, which may be used for the identification of some interanimation patterns (each participant's utterances are small rectangles on a horizontal line, time flowing from left to right; the threads of appearance of concepts (words) are shown with distinct colors; the ruler shows the number of utterances). For example, after several rhythmical repetitions of the word "same," a joint appearance of "same" and "different" occurs after utterance 480.

[3]Of course such a tool loses many useful details of the face-to-face origami data set because it is not able to identify nonverbal utterances. However, as it will be seen below, it still provides some useful representation for analyzing rhythm and pivotal moments.

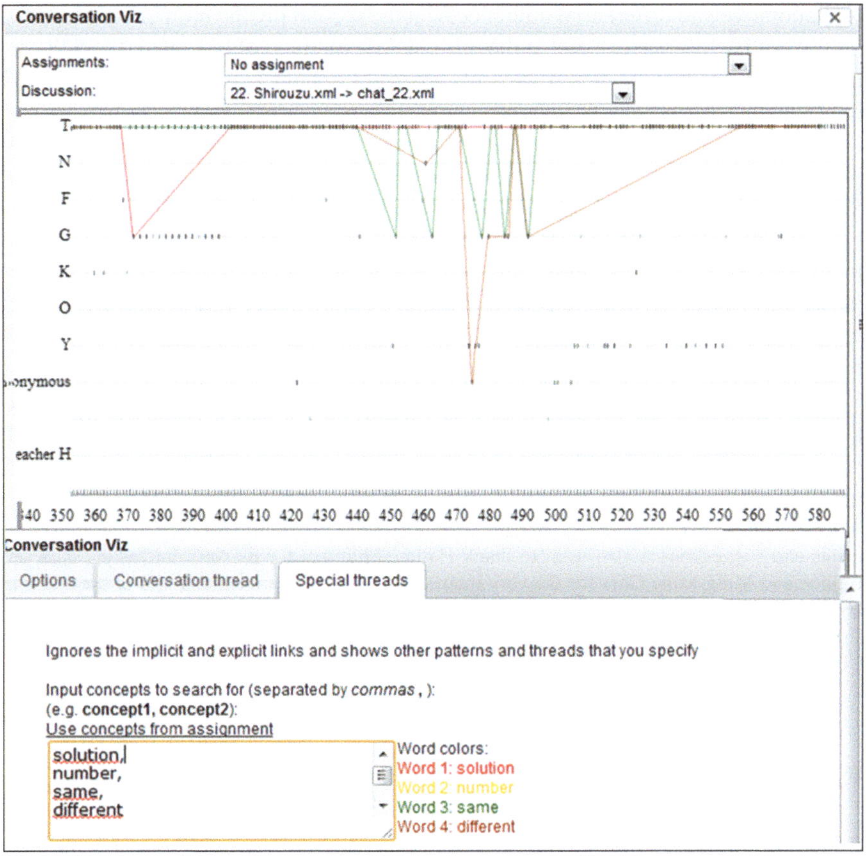

Fig. 6.1 The PolyCAFe visualization of a fragment of the conversation in the origami fractions data set

Conclusions

Pivotal moments in the approach presented in this chapter are related to collaborative moments (collaborative utterances), to other interanimation patterns (for example, differential utterances), and sometimes to changes in the rhythm (chronotope) of interacting voices. The analysis presented showed that the detection of pivotal moments in conversations may start from the identification of two types of interanimation patterns: collaborative and differential utterances and their succession.

In the origami fractions data set the pivotal moments that can be detected by the polyphonic approach are in the first minute (collaborative utterances both spoken 8–10 and collective body language), at the first pivotal moment of Shirouzu (collaborative and differential utterances), at the second pivotal moment of Shirouzu (collaborative utterances), and at the third pivotal moment of Shirouzu (the 548 collaborative utterance). Another possible pivotal moment is at utterances 561–562 (body language).

An important achievement of the analysis of the origami data set with the polyphonic model was the natural extension of its usage beyond textual utterances. Voices and interanimation patterns were identified also between verbal and nonverbal utterances. The concept of generalized utterances was introduced in order to include visual utterances, body language, and group utterances. Moreover, the existence of the intramental dimension that includes inner dialogue and inner utterances was asserted because it may explain some observed facts in the data set. The assertion of this dimension is also based on the ideas of inner speech (Vygotsky, 1934/1962; DeBleser & Marshall, 2005; Jones & Fernyhough, 2007), inner dialogue (Voloshinov, 1929/1973), and personal understanding cycle (Stahl, 2006). However, further investigations and evidence are needed for this latter dimension.

Acknowledgements I would like to thank Kris Lund and Dan Suthers for their very useful comments and suggestions. I also want to thank Hajime Shirouzu for the very interesting data set; Hajime and Ming Ming Chiu for the very interesting discussions on the analysis of the learning sessions; and Costin Chiru, Traian Rebedea, and Mihai Dascalu for their help on the automated support development. The work in this chapter was supported partially by the EU FP7 projects LTfLL and ERRIC.

References

Avouris, N., Fiotakis, G., Kahrimanis, G., & Margaritis, M. (2007). Beyond Logging of Fingertip Actions: Analysis of Collaborative Learning Using Multiple Sources of Data. *Journal of Interactive Learning Research, 18*, 231–50.

Bakhtin, M. (1981). *The dialogic imagination: Four essays*. London: University of Texas Press.

Bakhtin, M. (1984). *Problems of Dostoevsky's poetics (C. Emerson, Trans.)*. Minneapolis, MN: University of Minnesota Press.

Banica, D., Trausan-Matu, S., & Rebedea, T. (2011). Detecting collaboration regions in a chat session. In H. Spada, G. Stahl, N. Miyake, N. Law (Eds.), *CSCL 2011 Conference Proceedings* (Vol. 2, pp. 771–775). Hong Kong.

Chiru, C., & Trausan-Matu, S. (2012) Identification and Classification of the Most Important Moments from Students' Collaborative Discourses. In *Proceedings of ITS 2012*. Springer

Chiu, M. M. (this volume). Social metacognition, micro-creativity and justifications: Statistical discourse analysis of a mathematics classroom conversation. In D. D. Suthers, K. Lund, C. P. Rosé, C. Teplovs & N. Law (Eds.), Productive Multivocality in the Analysis of Group Interactions, Chapter 7. New York: Springer.

Clark, K., & Holquist, J. M. (1984). *Mikhail Bakhtin*. Cambridge, MA: Harvard University Press.

Dascalu, M., Rebedea, T., & Trausan-Matu, S. (2010). A Deep Insight in Chat Analysis: Collaboration, Evolution and Evaluation, Summarization and Search. *AIMSA 2010, Lecture Notes on Artificial Intelligence* (Vol. 6304, pp. 191–200). Springer.

DeBleser, R., & Marshall, J. C. (2005). Egon Weigl and the Concept of Inner Speech. *Cortex, 41*, 249–57.

Dowell, J., & Gladisch, T. (2007). Design of argument diagramming for case-based group learning. *ACM International Conference Proceeding Series* (Vol. 250, 99–105).

Dyke, G., Lund, K., & Girardot, J.-J. (2009). Tatiana: an environment to support the CSCL analysis process in O'Malley, C., Suthers D., Reimann P., Dimitracopoulou A. (Eds.) Computer Supported Collaborative Learning Practices - CSCL2009 Conference Proceedings, Rhodes, Greece, 58–67.

Garfinkel, H. (1967). *Studies in ethnomethodology*. Englewood Cliffs, NJ: Prentice-Hall.

Harrer, A., Hever, R., & Ziebarth, S. (2007). Empowering researchers to detect interaction patterns in e-collaboration. *Frontiers in Artificial Intelligence and Applications, 158*, 503–10.

Hmelo-Silver, C.E., Chernobilsky, E., & Masto, O. (2006). *Representations for Analyzing Tool-mediatí Collaborative Learning. Proceedings of ICLS 2006*. New Brunswick.

Jones, S. R., & Fernyhough, C. (2007). Neural correlates of inner speech and auditory verbal hallucinations: A critical review and theoretical integration. *Clinical Psychology Review, 27*, 140–54.

Joshi, M., & Rosé, C.P. (2007). Using transactivity in conversation summarization in educational dialog. In *Proceedings of the SLaTE Workshop on Speech and Language Technology in Education*

Jurafsky, D., & Martin, J.H. (2009) *Speech and Language Processing: An Introduction to Natural Language Processing, Computational Linguistics, and Speech Recognition* (2nd edn.), London: Pearson Prentice Hall

Ligorio, M. B., & Ritella, G. (2010). The collaborative construction of chronotopes during computer supported collaborative professional tasks. *International Journal on Computer-Supported Collaborative Learning, 5*(4), 433–45. Springer.

Mercer, N. (2000). *Words and minds. How we use language to think together*. London: Routledge.

Rebedea, T., Dascalu, M., Trausan-Matu, S., Armitt, G., & Chiru, C. (2011). Automatic Assessment of Collaborative Chat Conversations with PolyCAFe. In C. Delgado Kloos et al. (Eds.), *EC-TEL 2011* (LNCS, Vol. 6964, pp. 299–312). Berlin: Springer.

Sacks, H. (1962/1995). *Lectures on conversation*. Oxford: Blackwell.

Schegloff, E. A., & Sacks, H. (1973). Opening up closings. *Semiotica, 8*, 289–327.

Shirouzu, H. (this volume). Learning fractions through folding in an elementary face-to-face classroom. In D. D. Suthers, K. Lund, C. P. Rosé, C. Teplovs & N. Law (Eds.), Productive Multivocality in the Analysis of Group Interactions, Chapter 5. New York: Springer.

Stahl, G. (2006). *Group cognition: Computer support for building collaborative knowledge*. Cambridge, MA: MIT Press.

Stahl, G. (Ed.). (2009). *Studying virtual math teams*. Boston, MA: Springer.

Suthers, D. D., Dwyer, N., Medina, R., & Vatrapu, R. (2007). A framework for eclectic analysis of collaborative interaction. In C. Chinn, G. Erkens & S. Puntambekar (Eds.), *The Computer Supported Collaborative Learning (CSCL) Conference 2007* (pp. 694–703). New Brunswick: International Society of the Learning Sciences.

Tannen, D. (1989). *Talking voices: Repetition, dialogue, and imagery in conversational discourse*. Cambridge: Cambridge University Press.

Toulmin, S. (1958). *The uses of argument*. Cambridge: University Press.

Trausan-Matu, S., & Rebedea, T. (2009). Polyphonic Inter-Animation of Voices in VMT. In G. Stahl (Ed.), *Studying Virtual Math Teams* (pp. 451–473). Boston, MA: Springer.

Trausan-Matu, S., & Rebedea, T. (2010). A Polyphonic Model and System for Inter-animation Analysis in Chat Conversations with Multiple Participants. In A. Gelbukh (Ed.), *CICLing 2010* (LNCS, Vol. 6008, pp. 354–363). Heidelberg: Springer.

Trausan-Matu, S., Stahl, G., & Sarmiento, J. (2007). Supporting polyphonic collaborative learning. *E-service Journal* (Indiana University Press), *6*(1), 58–74.

Voloshinov, V. N. (1929/1973). *Marxism and the philosophy of language*. New York, NY: Seminar Press.

Vygotsky, L. S. (1934/1962). *Thought and language*. Cambridge, MA: MIT Press.

Wegerif, R. (2005). Towards a dialogic understanding of the relationship between CSCL and teaching thinking skills. In *Proceedings of th 2005 conference on Computer support for collaborative learning: learning 2005: the next 10 years!* (CSCL '05) (pp. 707–16). International Society of the Learning Sciences.

Chapter 7
Social Metacognition, Micro-Creativity, and Justifications: Statistical Discourse Analysis of a Mathematics Classroom Conversation

Ming Ming Chiu

In this chapter, I apply statistical discourse analysis (SDA, Chiu, 2008a) to Shirouzu's classroom data both to identify the locations and consequences of pivotal moments and to accompany other methodologies to yield multivocality insights. When asked to solve a novel problem, students try to create new ideas (*micro-creativity*) and assess their utility via explanations or justifications (Chiu, 2008a). [Micro-creativity occurs at specific moments, unlike the daily-life "small c" creativity of ordinary people and the "big C" creativity that affects societies (Sternberg & Lubart, 1999).] While micro-creativity provides grist for solving a problem, justifications support or refute an idea's usefulness by linking it to data, using a warrant, or supporting a warrant with backing (Toulmin, 2003). Hence, justifications are also a crucial component of the micro-creative process. A natural follow-up question is how classroom processes affect new ideas and justifications and whether their effects differ across time.

In this study, I address these issues by statistically modeling individual and conversation turn characteristics that affected micro-creativity or justifications as students solved a fraction problem in a Japanese classroom. This study contributes to the classroom process literature in four ways. First, I document when new ideas, correct ideas, and justifications occur, whether they occur uniformly during a lesson or more frequently in some time periods (*meso-time context*) than in others (Wise & Chiu, 2011). I statistically identify pivotal moments that divide the lesson into distinct time periods. Second, I test how the recent sequences of actions (*micro-time context*) affect the likelihoods of micro-creativity or justifications (Wise & Chiu, 2011). Third, I test whether the above effects differ across participants, classrooms, or time periods. Lastly, I discuss how other analysts' results and ideas have improved both SDA and its results. By understanding how the multivocality of several analyses informs one another, we can develop stronger methods and reap greater insights from a data set.

M.M. Chiu (✉)
University at Buffalo—State University of New York, Buffalo, NY, USA
e-mail: mingchiu@buffalo.edu

D.D. Suthers et al. (eds.), *Productive Multivocality in the Analysis of Group Interactions*, Computer-Supported Collaborative Learning Series 16,
DOI 10.1007/978-1-4614-8960-3_7, © Springer Science+Business Media New York 2013

Micro-creativity and Justification

Classroom participants' cognitive or social metacognitive processes might influence one another's thinking. This section focuses on how they might affect one another's micro-creativity and justifications.

Cognition

Classroom participants can build on one another's ideas to create new ideas through processes such as sparked ideas, error recognition, and jigsaw pieces (Paulus & Brown, 2003). Comments by one person (e.g., a key word) might spark another person to activate related concepts in his or her semantic network and propose a new idea (Nijstad, Diehl, & Stroebe, 2003). Specifically, a student might build on a correct idea to create another correct idea, or replace a flawed idea with a correct, new idea (Orlitzky & Hirokawa, 2001). Like fitting jigsaw pieces together, classroom participants can also put together different pieces to create a new idea (Milliken, Bartel, & Kurtzberg, 2003).

New ideas are often accompanied by justifications. Chiu and Khoo (2003) showed that classroom participants often supported their new ideas with justifications, especially before a disagreement. After a new idea, other classroom participants often evaluate its validity and give justifications to support their evaluation, especially if they disagree with a wrong idea (Orlitzky & Hirokawa, 2001).

Social Metacognition

Whereas individual metacognition is monitoring and controlling one's own knowledge, emotions, and actions, *social metacognition* is people's monitoring and control of one another's knowledge, emotions, and actions (Chiu & Kuo, 2009). Social metacognition can aid classroom problem solving through repetition, evaluation of one another's ideas, identification of problems (via disagreements or questions), or justification of different positions (Chiu & Kuo, 2009).

By repeating old information, students show shared understanding, common ground, and solidarity (Chiu, 2000a). Repetitions that organize and synthesize previous ideas can help classmates understand relationships among ideas, recognize gaps, and create a productive foundation for new ideas and correct ideas (Wise & Chiu, 2011). As repetitions review previously discussed ideas, they typically do not provoke new justifications.

Classroom participants often evaluate the previous speaker's action and problem-solving approach (Orlitzky & Hirokawa, 2001). For example, after one student says "three-sixths (3/6) is two," another student can respond by agreeing ("right"), using a neutral action ("what did you say?"), disagreeing ("no, that's wrong"), or changing the topic ("are you going to the party tonight?"). While agreements continue the current problem-solving path, disagreements and changes of topic (ignoring the previous action) try to change it (Chiu, 2001).

Evaluations can also be right or wrong in some contexts (such as simple mathematics problems). Correct evaluations support correct ideas ("three-sixths is one-half, uh-huh") or identify flawed ideas ("uh-uh, three-sixths is not two,"), thereby contributing to a foundation of partially shared understanding of correct ideas that group members can use to build new ideas, correct ideas, micro-creativity, or justifications. In contrast, incorrect evaluations reject correct ideas ("nope, three-sixths isn't a half,") or accept flawed ideas ("three-sixths is two, yeah"), embedding flaws in their partially shared understanding. Group members using this partially shared understanding can import these flaws into their new ideas, resulting in wrong ideas (Cobb, 1995).

Through their monitoring, classroom participants can recognize problems or difficulties (perturbations), express them through disagreements or questions, and address them with new ideas and justifications (Buchs, Butera, Mugny, & Darnon, 2004). Piaget (1985) defines two types of perturbations: (a) lacunae, gaps in understanding, often expressed through questions, and (b) obstacles, often expressed through negative feedback (disagreement).

A person asking a question (elicitation) typically shows a gap in his or her understanding (except for artificial teacher questions, Tsui, 1992). For example, a student asks, "how did you get half?" This gap can motivate the need for a new idea and suggest a direction for creating one and its accompanying justifications. Thus, questions might aid creation of new ideas, correct ideas, or justifications.

Meanwhile, disagreements can aid micro-creativity and justifications both directly and indirectly. Disagreements can correctly identify obstacles to be overcome (e.g., "no, three-sixths can't be two because it has to be smaller than one.") and directly stimulate justifications that support creation of new ideas (Chiu, 2000b; Coleman, 1998). Furthermore, a disagreement (even if wrong) can stimulate the attention of classroom participants, helping them consider more aspects of the situation from other perspectives to create further justifications and possibly new ideas—especially from social loafers who might stop relying on others (Nemeth, Personnaz, Personnaz, & Goncalo, 2004).

Disagreements can also indirectly encourage reluctant classmates to express their ideas, especially after agreements and repetitions of an existing idea suggest a majority view (Nemeth et al., 2004). Thus, a disagreement by another member, regardless of its validity, legitimizes the existence of different opinions, freeing all classroom participants to express new ideas, including those unrelated to the specific disagreement (Nemeth et al., 2004). Hence, disagreements can aid new ideas, correct ideas, and justifications.

After perturbations provoke new ideas, justifications often follow. Chiu and Khoo (2003) showed that members of successful groups often anticipated criticisms and justified their new ideas. Likewise, after a person disagrees with a proposal, the original proposer might try to justify it by linking it to data, using a warrant, or

Table 7.1 Hypothesized model of the effects of classroom problem-solving process on the outcome variable correct contributions (symbols in parentheses indicate expected direction of relationship with the outcome variables: positive [+], negative [–], or unknown [?])

	→ Dependent variables		
Classroom processes	New idea	Correct idea	Justify
Cognition			
New idea	+	+	+
Correct idea	?	+	?
Social metacognition			
Repeat	+	+	–
Evaluate correctly	+	+	+
Question	+	+	+
Disagree	+	+	+
Justify	+	+	+

supporting a warrant with backing (Toulmin, 2003). In response, other members can present new ideas and justifications (Chiu, 2008b). Similarly, when a student shows a gap in understanding by asking a question, other members can respond with explanations and justifications (Lu, Chiu, & Law, 2011). As justifications support an idea's validity, justifications might help create *correct* new ideas rather than *wrong* new ideas (e.g., Chiu, 2001).

Present Study

In sum, this study statistically models how cognitive and social metacognitive processes influence the likelihoods of new ideas, correct ideas, and justifications (see Table 7.1). To reduce omitted variable bias, I control for time (Chiu, 2008a) and demographic variables (gender, teacher vs. student). Learning of other analysts' results inspired me to improve my analysis, and I have noted them as changes to the original analysis below.

Method

In this study, I examine a lesson in which a teacher helps students learn multiplication of fractions by folding paper (see Shirouzu chapter). Their classroom processes were videotaped and transcribed. My content analyses (Krippendorff, 2004) yielded multidimensional coding of each conversation turn. While the conversation turn is the unit of analysis, the unit of interaction is a sequence of one type of action following another. The interaction as a whole is characterized by the probabilities of these sequences, which is modeled with SDA (Chiu & Khoo, 2005). See Shirouzu chapter for participants, data, and procedure.

Table 7.2 Coding of an artificial classroom discourse segment along five dimensions

Person	Action	EPA	KC	Validity	Justify	IF
Bob	Do three-sixths	*	N	√	[]	!
Lyn	Three-sixths is, um, is–	+	R	√	[]	–
Don	Three sixths is two	+	N	X	[]	–
Bob	Wrong, three sixes is eighteen	–	N	X	[]	–
Lyn	What?	n	0	0	0	?
Jan	It's three sixths, not three sixes. Three is half of six, so three sixths is one-half	–	N	√	J	–

Variables

In addition to individual (gender, teacher vs. student) and time period variables (discussed below), each conversation turn was coded along five dimensions. The dimensions were *evaluation of the previous action* (EPA: agree [+], neutral [n], disagree [–], ignore/new topic[*]), *knowledge content* (KC: new idea [N], repetition [R], null content [0]), *validity* (right [√], wrong [X], null content [0]), *justify* ([J], no justification [], null content [0]), and *form of invitation to participate* (IF: command [!], question [?], statement [_.]). See Table 7.2.

Some variables are created from combinations of the above variables. For example, a correct evaluation is either agreeing with a correct, previous idea or disagreeing with an incorrect, previous idea.

Analysis

This section specifies the assumptions underlying the analysis, its purpose, units of interaction, representations of the data, and analytic manipulations.

Assumptions Underlying the Analysis

Theoretical assumptions. SDA (Chiu & Khoo, 2005) has several theoretical assumptions. First, as with any statistics (e.g., count, mean, standard deviation), SDA assumes that instances of a category (e.g., justification) with the same value (e.g., is vs. is not [coded as 1 vs. 0]) are sufficiently similar to be treated as equivalent for the purpose of this analysis. This specific study has three additional theoretical assumptions. Characteristics of recent conversation turns, participating individuals, and time constitute a micro-context in which future talk emerges. Third, characteristics of recent conversation turns, their authors, and the time period can influence characteristics of later conversation turns. Fourth, residuals reflect attributes related to the dependent variables that are not specified in the theoretical model and not correlated with the explanatory variables.

Methodological assumptions. Like traditional regressions, SDA assumes a linear combination of explanatory variables. (Nonlinear aspects can be modeled as nonlinear functions of variables [e.g., age squared] or interactions among variables [question x correct].) SDA also requires independent and identically distributed residuals and a modest, minimum sample size.

Purpose of Analysis

SDA (1) identifies pivotal moments along specific dimensions that divide the data into distinct time periods, (2) tests whether variables are linked to greater or reduced likelihoods of dependent variables of interest, and (3) tests whether these links differ across time periods.

Units of Interaction That Are Taken as Basic in the Analysis

While the unit of analysis is a conversation turn, the unit of interaction is a sequence of one type of action following another. The interaction as a whole is characterized by the probabilities of these sequences, which is modeled with SDA.

Representations of Data and Analytic Interpretations

I used the standard representations of a database table, a summary statistics table, a table of breakpoints, a time series graph, and a path diagram. I converted the initial data representation of a database table with one utterance per row to one conversation turn per row, keeping the given attributes such as time, actor, and content. Next, I added columns (variables) for coding the argumentative attributes of each conversation turn as occurring or not. Then, I performed statistical analyses to test relationships across this table of vectors, resulting in a summary statistics table, a table of breakpoints, and a table of results of regression models (via SDA). To aid reader comprehension, I capitalize on readers' understanding of spatial relationships to convert the tables into graphs and path diagrams.

Analytic Manipulations

Addressing the above hypotheses with this data set requires modeling (1) differences across time (time periods, serial correlation); (2) three binary, infrequent, dependent variables; and (3) sequences of conversation turns that can differ across people, show indirect mediation effects, or yield false positives. See Table 7.3.

To address these difficulties, a simplified version of SDA is used (Chiu, 2008b; Chiu & Khoo, 2005). First, a breakpoint analysis statistically identifies pivotal moments that separate distinct time periods. Differences due to time periods or

Table 7.3 Statistical discourse analysis strategies to address each analytic difficulty

Analytic difficulty	Statistical discourse analysis strategy
Differences across time	
Different time periods	Breakpoint analysis (Chiu and Khoo 2005)
Differences across time/serial correlation	I^2 index of Q-statistics (Huedo-Medina, Sanchez-Meca, Marin-Martinez, and Botella 2006)
Dependent variables	
Binary	Logit (Kennedy 2004)
Infrequent	Bias estimator (King and Zeng 2001)
Multiple	Multivariate outcome analyses (Goldstein 1995)
Explanatory variables	
Sequences of conversation turns	Vector auto-regression (VAR, Kennedy 2004)
Differences across people	Interaction terms (Kennedy 2004)
Indirect, mediation effects	Mediation tests (Sobel 1982)
False positives	Two-stage linear step-up procedure (Benjamini, Krieger, and Yekutieli 2006)

people are tested with interaction terms (Kennedy, 2004). If not modeled properly, resemblances among adjacent conversation turns can result in serial correlation of errors (Kennedy, 2004). An I^2 index of Q-statistics can test conversation turns in many time periods for serial correlation, which can be modeled if needed (Huedo-Medina et al., 2006).

Furthermore, the three dependent variables were binary and infrequent (new idea, correct idea, and justification). To model a binary-dependent variable, Logit or Probit is used. When dependent variables occur far less than 50 % of the time, standard regressions will yield biased results. To remove this bias, I used King and Zeng's (2001) bias estimator. Multiple outcomes can have correlated residuals that underestimate standard errors. To model several dependent variables properly, a multivariate outcome analysis is needed (Goldstein, 1995).

The explanatory variables can include sequences, differ across people, yield indirect effects, or show false positives. Sequences of explanatory variables are modeled with vector auto-regression (VAR, Kennedy, 2004). Different effects across people are tested with interaction terms (Kennedy, 2004). To test for indirect effects, Sobel's (1982) mediation test was used. Testing many hypotheses via explanatory variables raises the likelihood of a false positive (Type I error). To control for this false discovery rate (FDR), the two-stage linear step-up procedure was used, which outperformed 13 other methods in computer simulations (Benjamini et al., 2006).

Identify Pivotal Moments and Time Periods

Some actions (e.g., correct ideas) might occur more often at the end of a session (e.g., close to a solution) than at the beginning (e.g., discussion of a problem).

I operationalize *pivotal moment* as *a conversation turn that separates a portion of the conversation into two distinct time periods* (before and after) *with substantially different likelihoods of the focal variable* (e.g., correct ideas). The different likelihoods of the focal variable in the *before* and *after* time periods suggest that the interactions in the two time periods differ substantially.

SDA can statistically identify pivotal moments that divide a session into time periods with more vs. fewer correct ideas. These pivotal moments can then be used to test whether the relationships between explanatory variables and correct ideas differ across time periods (Chiu, 2008b). Initially, a univariate time-series model (auto-regressive order 1 model) has no pivotal moments. In (7.1), Correct_t indicates whether a correct idea occurs at conversation turn t. The regression coefficient β indicates whether Correct_t is related in some way to whether a correct idea occurred in the previous utterance, Correct_{t-1}, with constant C_0 and residual ε_t:

$$Correct_t = \beta Correct_{t-1} + C_0 + \varepsilon_t \tag{7.1}$$

Next, we added pivotal moments. The number of potential pivotal moments (i) can range from 1 to p, with corresponding pivotal moment location dummy variables ($Break_i$) and regression coefficients (C_i):

$$Correct_t = \beta Correct_{t-1} + C_0 + \varepsilon_t + C_1 Break_1 + C_2 Break_2 + \ldots + C_p Break_p \tag{7.2}$$

For each number of pivotal moments (first 1 break, then 2 breaks, … 6 breaks), all possible locations of pivotal moments were modeled. (Only six pivotal moments were tested because current microcomputers require over a year to test seven pivotal moments.) For each model, the Bayesian information criterion (BIC) was computed from the log-likelihood function L, n observations, and k estimated parameters: BIC $= [-2L + \ln(n)\,k]/n$. Information criteria indicate whether a model suitably balances parsimony and goodness of fit. Unlike other information criteria, the BIC yields a consistent estimator for the number of lagged variables in the true model (Kennedy, 2004). The best model has the lowest BIC.

Explanatory Model

Next, the explanatory model was estimated with multivariate logit (Goldstein, 1995):

$$\mathbf{Action}_{jy} = \beta_{0y} + \mathbf{e}_{jy} \tag{7.3}$$

$\mathbf{Action}_{jy}$ is a vector of y dependent variables (new idea, correct idea, and justification) for turn j. β_{0y} are its grand mean intercepts, and its residuals are $\mathbf{e}_{jy}$. First, the statistically identified time period dummy variables (**Time**) were entered into the regression model:

$$\begin{aligned}&\mathbf{Action}_{jy} = \beta_{0y} + \mathbf{e}_{jy} + b_{ty}\mathbf{Time}_{jy} + \beta_{iy}\mathbf{Individual}_{jy} + \\ &\beta_{\mathbf{cy}}\mathbf{Current\_Conversation\_turn}_{jy} + \\ &\beta_{py}\mathbf{Previous\_Conversation\_turn}_{(j-1)y} + \\ &f_{py}\mathbf{Two\_Conversation\_turns\_ago}_{(j-2)y} + \ldots + \\ &\beta_{xy}\mathbf{Interactions}_{jy}\end{aligned} \tag{7.4}$$

Each set of predictors was tested for significance with a nested hypothesis test (χ^2 log likelihood, Kennedy, 2004), and nonsignificant variables were removed.

Then, individual characteristics were entered: teacher (vs. student) and girl (**Individual**). Next, characteristics of the current conversation turn were entered: repeat, correctly evaluate, agree, disagree, ignore, question, and command (**Current_Conversation_turn**). Then, characteristics of the previous turn were entered: justification (-1), correct (-1), new idea (-1), repeat (-1), correctly evaluate (-1), agree (-1), disagree (-1), ignore (-1), question (-1), and command (-1) (**Previous_Conversation_turn**). Next, the characteristics from two turns ago (**Two_Conversation_turns_ago**) were tested and so on until no variables were significant. To test for moderation, I added interactions of all significant variables (**Interactions**).

An alpha level of 0.05 was used. Testing many hypotheses raises the likelihood of a false positive (Type I error). To control for the FDR, the two-stage linear step-up procedure was used (Benjamini et al., 2006).

Path analysis estimated direct and indirect effects (Kennedy, 2004). As time constrains the direction of causality, the explanatory variables were ordered temporally in the path analysis. The odds ratio of each variable's total effect (E, direct plus indirect) was reported as the increase or the decrease (+E% or –E%) in the dependent variable (Kennedy, 2004).

This model was initially tested on the full data set. Upon learning that another analyst (Shirouzu) viewed the class discussion activity as the most important part of the lesson, I did a separate analysis on the class discussion activity subset of the data, delineated as occurring after the statistically identified primary pivotal moment (see details below). With 582 turns in the full data set, statistical power exceeded 0.99 for an effect size of 0.2 (Cohen, West, Aiken, & Cohen, 2003). With 134 turns in the subset, statistical power is 0.95 for an effect size of 0.3 (Cohen et al., 2003).

Sample Size

Green (1991) proposed the following heuristic sample size, N, for a multiple regression with M explanatory variables and an expected explained variance R^2 of the outcome variable:

$$N > \left(\left\{8 \times \left[\left(1 - R^2\right) / R^2\right]\right\} + M\right) \quad 1 \tag{7.5}$$

For a large model of 25 explanatory variables with a small expected R^2 of 0.10, the required sample size is 96 conversation turns: $= 8 \times (1 - 0.10)/0.10 + 25 - 1$. Less data are needed for a larger expected R^2 or for smaller models. In practice, two groups of students talking for half an hour will often yield more than 100 speaker turns, sufficient for SDA. In this data set, we converted the 582 utterances to 443 conversation turns, which exceeds the required sample size of 96. To aid comparisons across chapters, we use utterance identification numbers (rather than conversation turn identification numbers).

Results

Summary Statistics and Pivotal Moments

In the data subset in which the class compares student answers, the key variables occurred more often (especially new ideas, micro-creativity, correct evaluations, questions, and disagreements) than in the overall data set of the entire class (see Table 7.4). The percentages in the data subset are similar to those in other studies of face-to-face mathematics problem solving by groups of students (e.g., Chiu, 2008b).

SDA yielded five significant pivotal moments for micro-creativity and three pivotal moments for justifications (see Table 7.2 and Fig. 7.1). The micro-creativity pivotal moment at utterance 448 is strongly supported through its consistent identification in the optimal models of one, two, three, four, and five pivotal moments; hence, it is the *primary pivotal moment* (an idea raised through the three analysts' [Shirouzu, Trausan-Matu, and me] discussion of whether some pivotal moments are more important than others). Consider the pivotal moment at utterance 448, when several students recognize the equivalence of two different solutions. After students have solved the problem of finding 3/4 of 2/3 of a square sheet of paper, the teacher asks them to compare two students' solutions.

Table 7.4 Summary statistics of significant variables

	% in each data set	
Variable	Overall	Subset
Cognition		
Repeat	37	40
Correct idea	42	50
New idea	12	17
Micro-creativity	12	17
Social metacognition		
Evaluate correctly	24	43
Question	13	17
Disagree	1	3
Justify	7	9
Individual		
Teacher	27	31
Girl	28	21

Table 7.5 Micro-creativity and justification pivotal moments

# of pivotal moments	Micro-creativity						Justifications					
	BIC	At utterance # …					BIC	At utterance # …				
0	0.767						0.533					
1	0.766				**448**		0.464	**206**				
2	0.755			164	**448**		0.456	**206**			394	
3	0.750			164	**448**	476	0.433	**206**			394	537
4	0.750	20	39		**448**	476	0.445	**206**	310		394	537
5	0.747	20	39	164	**448**	476	0.447	**206**	310	359	394	537

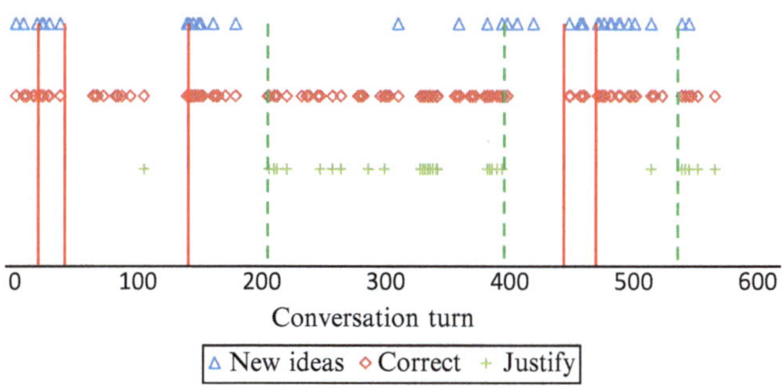

Fig. 7.1 Occurrences of three actions over time: new ideas, correct ideas, and justifications. *Red solid vertical lines* indicate pivotal moments for new ideas, and correct ideas. *Green dashed vertical lines* indicate pivotal moments for justifications

Utterance #	Person	Talk and actions
440–447	T:	Which should we begin with? N's and G's solutions [removes the solutions from the blackboard and puts them on the desk] are the ones completed first. This is a complete one. This is one example. This is another one [places another solution on the desk]. How do you compare?
448	**G, K, N:**	**The same**
449	T:	The same. Here, everyone agreed. This one and this one are the same. So, how do you compare these [N1 and G2]? This one and this one. Please

Students G, K, and N all say that the solutions are "the same," the first of many correct, new ideas (see Fig. 7.1). After listening to another analyst's (Trausan-Matu) discussion of collaborative utterances, I considered whether this pivotal moment consisted of more than this shared utterance and might include preceding or following utterances. As the teacher asked the question that elicited the student answers, the previous turn clearly contributes to this pivotal moment. Arguably, the teacher's confirmation of the students' shared answer in the following turn is also part of the pivotal moment. Hence, SDA only identifies the conversation turn at the heart of the pivotal moment, not its outer boundaries. Hence, the statistically identified conversation turn does not necessarily encapsulate all key aspects of the pivotal moment.

Qualitative methods are needed to examine both the boundaries of the pivotal moment and the mechanism by which it operates. For example, usage of a polyphonic framework (Trausan-Matu, Stahl, & Sarmiento, 2007) to identify the before and after threads of utterances separated by the pivotal moment can indicate where the conversation shifted from one thread to the next. Furthermore, ethnomethodologists (e.g., Sacks, 1995) might examine the detailed relationships among the words and actions near the pivotal moment to understand the mechanism(s) by which one thread becomes another.

The pivotal moments are not necessarily the same across variables. For example, SDA identifies three different pivotal moments for justifications (at utterances 206, 394, and 537). The *primary pivotal moment* at utterance 206 occurs in all the optimal models with one to five pivotal moments. Consider the pivotal moment at utterance 206. After several students have presented their initial solutions, the teacher demonstrates the common first step of several solutions.

Utterance #	Person	Talk and actions
205	Y:	[Standing in front of the blackboard.] First, let's fold the origami paper into three equal parts in this way. [Folds the origami paper into three equal portions.]
206	**T:**	**Yes. These are the same up to this point [while pointing out the first step of N's and G's first solutions].**
207–210	Y:	[Continues to fold after glancing at T's explanation.] Then, Yes. let's cut this 1/3 part like this. [Cuts it with scissors.] This paper is now divided into four portions, and this shape was obtained by cutting necessary ones from them.

The teacher folds the paper into three equal parts to justify his claim that the first step of the solutions of N and G are the same. This justification pivotal moment ignites a new time period with many justifications by students (see Fig. 7.1). As with the micro-creativity pivotal moment discussed above, the identified conversation turn does not encapsulate the entire pivotal moment, which is a continuation of an idea that started two turns earlier.

The analysts' discussion mentioned earlier also inspired a way to compare the relative importance of pivotal moments across dimensions; the reduction in BIC after adding a primary pivotal moment shows how much it alters the likelihood of its target phenomenon (e.g., justification) in the following time period. Comparing the primary pivotal moments of justifications and micro-creativity, the justification pivotal moment at utterance 206 has a larger impact on the likelihood of justifications in the subsequent time period compared to the impact of the pivotal moment at utterance 448 on subsequent micro-creativity (13 % > 0.1 %; 13 % = [0.533 − 0.464]/0.533; 0.1 % = [0.767 − 0.766]/0.767). In the next step of the analysis, all of these pivotal moments are entered into the explanatory model.

Explanatory Model

As shown in the explanatory models below, the significant relationships in the full data set and those of the subsample differed substantially. All reported results are from the final models with only significant variables.

Correct Idea, New Idea, and Justification in the Full Lesson

Time period, correct ideas, questions, correct evaluations, and agreement were linked to subsequent correct ideas (Fig. 7.2). A correct idea was 8 % more likely in the time period after the primary pivotal moment for micro-creativity. After a correct idea in the previous turn, a correct idea in the current turn was 6 % more likely. If a correct idea and question both occurred in the previous turn, a correct idea was 9 % less likely to follow. Examination of the classroom videotape showed that correct ideas accompanied by questions were often followed by acknowledgements.

Utterance #	Person	Talk and actions
547	Y:	I thought all (answers) are 1/2 of the whole. What do you all think?
548	N, F, K, O:	Ok.

After Y presents a correct idea (1/2) and asks for other's opinions ("What do you all think?"), four students (N, F, K, and O) simply agree ("Ok").

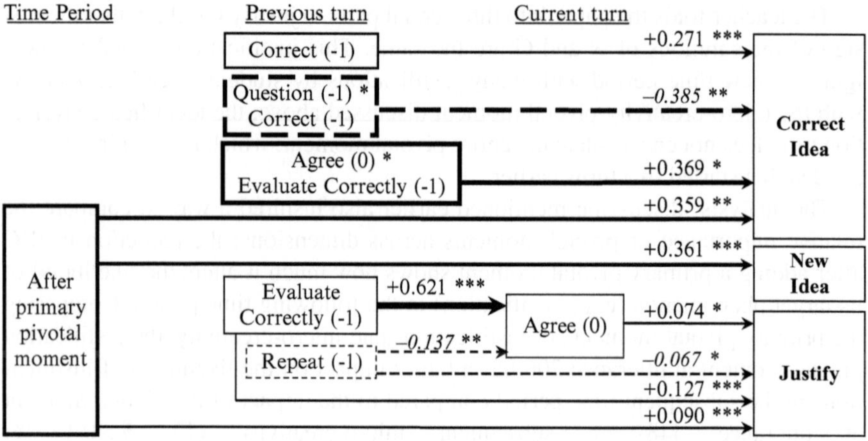

Fig. 7.2 Path diagram modeling correct idea, new idea, and justification in the full data set. *Solid lines* indicate positive links. *Dashed lines* indicate negative links. *Thicker lines* indicate stronger links. $*p<0.05$, $**p<0.01$, $***p<0.001$

In contrast, a correct idea was 10 % more likely after a correct evaluation in the previous turn and an agreement in the current turn. These variables accounted for 20 % of the variance of correct ideas.

New ideas were 9 % more likely after the primary pivotal moment for micro-creativity. No other variables were linked to new ideas for this data set. This time period variable accounted for 7 % of the variance of new ideas.

Time period, correct evaluations, repetitions, and agreements were linked to justifications. A justification was 2 % more likely in the time period after the primary pivotal moment for micro-creativity. After a correct evaluation, a justification was 5 % more likely. After a repetition however, a justification was 2 % less likely. Meanwhile, justifications were 7 % more likely to occur with an agreement in the same turn. While correct evaluations yielded 16 % more agreements in the next turn, repetitions yielded 3 % fewer agreements in the next turn. These variables accounted for 15 % of the variance of justifications.

Correct Idea, New Idea, and Justification in the Subsample

Time period and correct evaluations were linked to correct ideas. A correct idea was 7 % more likely after the secondary pivotal moment of micro-creativity in utterance 467. Moreover, a correct evaluation in the previous turn or the current turn raised the likelihood of a correct idea by 12 or 11 %, respectively (see Fig. 7.3). These variables accounted for 42 % of the variance of correct ideas in this data subset.

Time period, correct evaluations, repetitions, and justifications were linked to new ideas. A new idea was 9 % more likely after the secondary pivotal moment. After a correct evaluation three turns ago, a new idea was 9 % more likely. After a

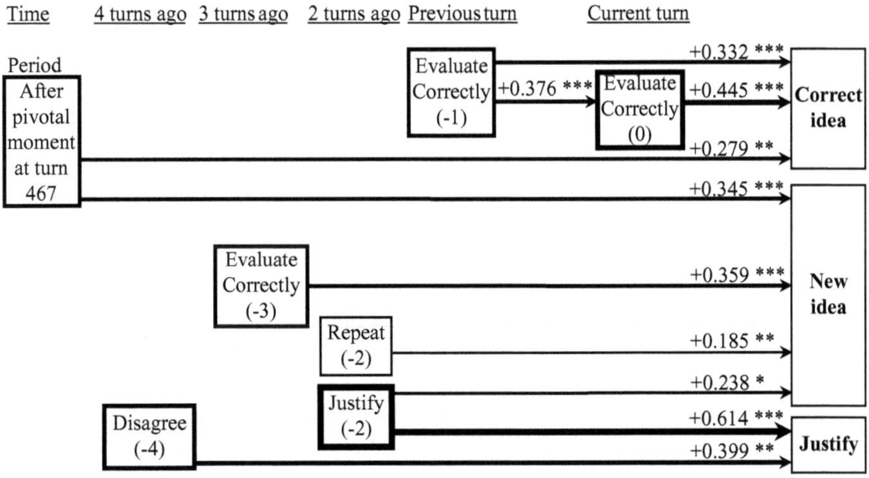

Fig. 7.3 Path diagram of correct idea, new idea, and justification in the data subset after the primary pivotal moment for micro-creativity. *Solid lines* indicate positive links. *Dashed lines* indicate negative links. *Thicker lines* indicate stronger links. $*p<0.05$, $**p<0.01$, $***p<0.001$

repetition or a justification two turns ago, a new idea was 5 or 6 % more likely, respectively. These variables accounted for 38 % of the variance of new ideas in this data subset.

Justification was 10 or 15 % more likely after a disagreement four turns ago or a justification two turns ago. These variables accounted for 54 % of the variance of justifications.

All other variables were not significant. Notably, gender, teacher (vs. student), and their interactions were not significant, showing that these variables and their relationship did not differ significantly with respect to gender or position in this data set.

Discussion

This study examines antecedents of students' micro-creativity, new ideas, correct ideas, and justifications as they solve a fraction problem under the guidance of their teacher. SDA statistically identified five pivotal moments and six distinct time periods of high vs. low micro-creativity but a different set of three pivotal moments and four time periods of frequent vs. infrequent justifications. The explanatory models provide support for some of the hypotheses but differ substantially across time, as shown in the different results of the full data set vs. its subset. Furthermore, other analysts' methods and results provided multivocality grist for further insights and methodological developments.

Pivotal Moments and Time Periods

The statistically identified pivotal moments showed distinct time periods and different degrees of importance. New ideas, correct ideas, and justifications occurred more frequently in some time periods than in others. Inspired by Trausan-Matu's discussion of collaborative utterances, I found that a pivotal moment can have boundaries beyond a single turn, incorporating aspects of both earlier and later turns. Also, the pivotal moments and time periods identified for micro-creativity and justifications differed, showing that a pivotal moment along one dimension is not necessarily a pivotal moment along another dimension.

Statistical identification of a pivotal moment is also an invitation to understand its mechanism(s) through a multivocality cycle of further qualitative and statistical analyses. A detailed, qualitative analysis of the actions, changes, and their relationships around the pivotal moment (e.g., via ethnomethodology, Sacks, 1995) can suggest a mechanism(s) that alters the interaction. Such examinations of multiple pivotal moments can provide comparative case studies to test whether these hypothesized mechanisms are idiosyncratic or not. After specifying these mechanisms through operationalized variables, SDA can test these mechanism hypotheses across the entire data set (e.g., Wise & Chiu, 2011).

A comparison of our three analysts' (Shirouzu, Trausan-Matu, and me) pivotal moment results also inspired a method to assess their relative importance. All three analysts identified one common pivotal moment, which suggested that it was more important than the others. Returning to my analyses, I saw that the pivotal moment with the largest corresponding reduction in the BIC would indicate the greatest impact on the target phenomena (e.g., justifications). Furthermore, this reduction in BIC measure applies across different target phenomena, so it serves as a general method for comparing the relative impact of different breakpoints.

Explanatory Models

The results of the explanatory models showed some support for many cognition hypotheses (correct ideas) and social metacognition hypotheses (repetitions, correct evaluations, disagreements, and justifications), but they differed across time periods. The results partially supported the correct idea hypotheses but did not support the new ideas or micro-creativity hypotheses, in part due to reflective practices. Only a correct idea unaccompanied by a question was often followed by another correct idea; a correct idea in the form of a question was often followed by a simple agreement rather than a correct idea. The significant interaction between correct idea and question also highlights the importance of the immediate temporal context in moderating the effect of a specific action. As shown above, students often reflected on ideas, especially new ones, which hindered chains of new ideas or micro-creativity. Whether these reflections follow ideas regularly in Japanese classrooms or other classrooms remain open questions.

The analyses of the full data set and the data subset show another benefit of multivocality. After Shirouzu indicated that the data subset (class discussion) was the most substantive part of the lesson, SDA was applied to only the data subset to test if the relationships among independent and dependent variables differed in both the data subset and in the full data set. The results showed that the explanatory model for the entire data set differed from that of the data subset for *all* significant explanatory variables, showing that the relationships among variables differed entirely across time periods. These different explanatory models across time periods highlight the time-dependent nature of the statistical relationships and suggest that statistical models without proper modeling of time periods can be incomplete. Statistical methods such as SDA are needed to test whether relationships among variables and their accompanying hypotheses are supported, rejected, or not significant in both the entire data set and in each time period.

Identification of these differences in relationships among variables across time periods raises the question of why these differences occur. The above multivocality cycle of qualitative and statistical analyses of pivotal moment mechanisms might help account for these differences. If the pivotal moment mechanisms do not account for them, then the SDA results suggest where to look; researchers can conduct qualitative analyses (e.g., ethnomethodology, Sacks, 1995) of instances in which an independent variable-dependent variable relationship occurs in one time period and instances in which it does not occur in another time period. Comparative case studies can then yield hypotheses regarding moderation variables, which in turn can be tested by SDA.

Conclusion

This analysis shows how SDA can both identify the locations and consequences of pivotal moments and accompany other methodologies to yield multivocality insights. SDA of a classroom lesson showed the impacts of the meso-time context of time periods within a lesson and the micro-time context of recent conversation turns. The statistically identified pivotal moments distinguished time periods for each dependent variable and yielded different relationships among variables across time periods. While the statistical analysis identified a conversation turn at the heart of each pivotal moment, detailed discourse analysis showed that the boundary of the pivotal moment often extended to earlier and later turns. The statistical analysis identified a set of pivotal moments and its time periods of higher vs. lower micro-creativity but a different set of pivotal moments and time periods for more vs. fewer justifications. Lastly, the explanatory models differed across time periods, highlighting the importance of the meso-time context.

Methodologically, this analysis shows how multivocality can suggest cycles of analyses and help develop further statistical methods. SDA identified pivotal moments, time periods, and differences in relationships across time periods that can ignite cycles of further qualitative and statistical analyses. Detailed qualitative analysis (e.g., ethnomethodology) of pivotal moments and contrast cases of relationships between variables (that occur in one time period but not in another) can specify candidate mechanism hypotheses that can be operationalized and tested with SDA. Furthermore, the three analysts' identification of one common pivotal moment inspired a statistical method (change in BIC) to test the relative importance of different pivotal moments. In short, this study showed not only how SDA can be applied but also how its use with other methods yields further benefits.

References

Benjamini, Y., Krieger, A. M., & Yekutieli, D. (2006). Adaptive linear step-up procedures that control the false discovery rate. *Biometrika, 93*, 491–507.

Buchs, C., Butera, F., Mugny, G., & Darnon, C. (2004). Conflict elaboration and cognitive outcomes. *Theory into Practice, 43*, 23–30.

Chiu, M. M. (2000a). Group problem solving processes: Social interactions and individual actions. *Journal for the Theory of Social Behavior, 30*(1), 27–50.

Chiu, M. M. (2000b). Status effects on solutions, leadership, and evaluations during group problem solving. *Sociology of Education, 73*(3), 175–195.

Chiu, M. M. (2001). Analyzing group work processes. In F. Columbus (Ed.), *Advances in psychology research* (Vol. 4, pp. 193–222). Huntington, NY: Nova Science.

Chiu, M. M. (2008a). Effects of argumentation on group micro-creativity. *Contemporary Educational Psychology, 33*, 382–402.

Chiu, M. M. (2008b). Flowing toward correct contributions during group problem solving: A statistical discourse analysis. *Journal of the Learning Sciences, 17*(3), 415–463.

Chiu, M. M., & Khoo, L. (2003). Rudeness and status effects during group problem solving. *Journal of Educational Psychology, 95*, 506–523.

Chiu, M. M., & Khoo, L. (2005). A new method for analyzing sequential processes: Dynamic multi-level analysis. *Small Group Research, 36*, 600–631.

Chiu, M. M., & Kuo, S. W. (2009). From metacognition to social metacognition: Similarities, differences, and learning. *Journal of Education Research, 3*(4), 1–19.

Cobb, P. (1995). Mathematical learning and small-group interaction: Four case studies. In P. Cobb & H. Bauersfeld (Eds.), *The emergence of mathematical meaning: Interaction in classroom cultures* (pp. 25–130). Hillsdale, NJ: Lawrence Erlbaum Assoc.

Cohen, J., West, S. G., Aiken, L., & Cohen, P. (2003). *Applied multiple regression/correlation analysis for the behavioral sciences*. Mahwah, NJ: Lawrence Erlbaum.

Coleman, E. (1998). Using explanatory knowledge during collaborative problem solving in science. *The Journal of Learning Sciences, 7*, 387–427.

Goldstein, H. (1995). *Multilevel statistical models*. Sydney: Edward Arnold.

Green, S. B. (1991). How many subjects does it take to do a regression analysis? *Multivariate Behavioral Research, 26*, 499–510.

Huedo-Medina, T. B., Sanchez-Meca, J., Marin-Martinez, F., & Botella, J. (2006). Assessing heterogeneity in meta-analysis. *Psychological Methods, 11*, 193–206.

Kennedy, P. (2004). *A guide to econometrics*. Cambridge: Blackwell.

King, G., & Zeng, L. (2001). Logistic regression in rare events data. *Political Analysis, 9*, 137–163.

Krippendorff, K. (2004). *Content analysis*. Thousand Oaks, CA: Sage.

Lu, J., Chiu, M. M., & Law, N. (2011). Collaborative argumentation and justifications. *Computers in Human Behavior, 27*, 946–955.

Milliken, F. J., Bartel, C. A., & Kurtzberg, T. R. (2003). Diversity and creativity in work groups. In P. B. Paulus & B. A. Nijstad's (Eds.), *Group creativity* (pp. 32–62). New York, NY: Oxford University Press.

Nemeth, C. J., Personnaz, B., Personnaz, M., & Goncalo, J. A. (2004). The liberating role of conflict in group creativity. *European Journal of Social Psychology, 34*, 365–374.

Nijstad, B. A., Diehl, M., & Stroebe, W. (2003). Cognitive stimulation and interference in idea generating groups. In P. B. Paulus & B. A. Nijstad (Eds.), *Group creativity: Innovation through collaboration* (pp. 137–159). New York, NY: Oxford University Press.

Orlitzky, M., & Hirokawa, R. Y. (2001). To err is human, to correct for it divine. *Small Group Research, 32*, 313–341.

Paulus, P. B., & Brown, V. (2003). Ideational creativity in groups. In P. B. Paulus & B. A. Nijstad (Eds.), *Group creativity* (pp. 110–136). New York, NY: Oxford University Press.

Piaget, J. (1985). *The equilibrium of cognitive structures*. Chicago: University of Chicago Press.

Sacks, H. (1995). *Lectures on conversation*. Oxford: Blackwell.

Sobel, M. E. (1982). Asymptotic intervals for indirect effects in structural equations models. In S. Leinhart (Ed.), *Sociological methodology* (pp. 290–312). San Francisco, CA: Jossey-Bass.

Sternberg, R. J., & Lubart, T. I. (1999). The concept of creativity. In R. J. Sternberg (Ed.), *Handbook of creativity* (pp. 3–15). New York, NY: Cambridge University Press.

Toulmin, S. E. (2003). *The uses of argument*. Cambridge: Cambridge University Press.

Trausan-Matu, S., Stahl, G., & Sarmiento, J. (2007). Supporting polyphonic collaborative learning. *E-service Journal, 6*(1), 58–74.

Tsui, A. (1992). A functional description of questions. In M. Coulthard (Ed.), *Advances in spoken discourse analysis* (pp. 89–110). London: Routledge.

Wise, A., & Chiu, M. M. (2011). Analyzing temporal patterns of knowledge construction in a role-based online discussion. *International Journal of Computer-Supported Collaborative Learning, 6*, 445–470.

Chapter 8
A Multivocal Analysis of Pivotal Moments for Learning Fractions in a 6th-Grade Classroom in Japan

Kristine Lund

Introduction

The Japanese 6th-grade fractions dataset was one of the first datasets we studied in the workshop series that led to this book. Data was initially provided by Shirouzu for the 2009 Alpine Rendez-Vous.[1] Amongst the researchers who had applied to be analysts at that workshop, we chose Trausan-Matu and Chiu to be analysts for the fractions dataset. As organizers, one of our first goals was to explore to what extent researchers could carry out their own analyses on a dataset that was not gathered for their own purposes (this would be the case for Trausan-Matu and Chiu but for different reasons). Another goal was to encourage researchers to move outside of their comfort zones when analyzing new datasets (also aimed at Trausan-Matu and Chiu). We thought if researchers were challenged in how they applied their analytical methods, this would help them to reflect on the limits of their methods. Reflecting on limits can lead first to being capable of explaining why those limits exist and thus be helpful to other researchers new to the method. Such reflection can also perhaps help to surpass those limits. In this chapter, we will see different ways to surpass limits of an initial analytical method as well as other findings made possible by comparing and contrasting three analytical approaches on the same dataset.

The data concerns six students studying the multiplication of fractions in a 6th-grade classroom in Japan (see Shirouzu, Chap. 4, this volume for details on the dataset). Their task was to cut out 3/4 of 2/3 of a piece of origami paper and then to discuss whether or not their solutions were the same. A teacher led and monitored

[1] The Alpine Rendez-Vous (2007, 2009, 2011) was initially supported by the Kaleidescope European network of excellence and then by the STELLAR network of excellence: http://www.stellarnet.eu/programme/wp3/rendez-vous.

K. Lund (✉)
ICAR Research Lab, CNRS—University of Lyon, 15 parvis René Descartes,
69007 Lyon, France
e-mail: kristine.lund@ens-lyon.fr

D.D. Suthers et al. (eds.), *Productive Multivocality in the Analysis of Group Interactions*, Computer-Supported Collaborative Learning Series 16,
DOI 10.1007/978-1-4614-8960-3_8, © Springer Science+Business Media New York 2013

activity, and work was carried out both on the blackboard and by folding pieces of the paper. Video data was supplied, and the Japanese was transcribed, translated into English, and synchronized with the video as subtitles. Drawings of each student's folded origami solution were also provided.

In the remaining part of this chapter, I first use five dimensions for reflecting on productive multivocality to discuss each of the approaches (Shirouzu's, Trausan-Matu's, and Chiu's). Second, I compare analysts' pivotal moments in relation to their respective analytical constructs: conceptual change, interanimation patterns, and frequency of new ideas. Third, I show how sharing a dataset and comparing analyses influenced each individual researcher's analytical trajectory. Finally, I summarize the lessons learned for productive multivocality.

Five Dimensions for Reflecting on Productive Multivocality

In this section we use the five dimensions refined during our series of workshops in order to compare and contrast the approaches of Shirouzu, Trausan-Matu, and Chiu. These dimensions are theoretical assumptions, purpose of analysis, unit of analysis/unit of interaction, data representations, and manipulations on data representations.

Theoretical Assumptions

The Consequences of Students' Focusing

Shirouzu (Chap. 5, this volume) argues that both intra- and inter-mental activities are either displayed by or can be inferred from the interaction being studied. "Inter-mental" interaction refers to social interaction among humans, while "intramental" interaction refers to the interaction between inner and outer resources. Inner resources refer to the learner's prior knowledge, interest, and problem consciousness, and outer resources refer to the learner's physical and social environment. Shirouzu used the theory of focus-based constructive interaction—based on the theory of constructive interaction (see his chapter, this volume)—to analyze the role of individuals within the collective. Since the theory of constructive interaction does not predict (1) which student will be the first doer, (2) which student will respond as active monitor to the first doer's externalization, (3) how the contents of task-doing and monitoring evolve, and (4) what each learner retains from the interaction, Shirouzu developed an analytical method to answer these questions.

If, as Shirouzu argues, role division and exchange between participants are responsible for constructive interaction where "constructive" means to understand what has not been previously understood and if a learner enters a social process with his or her own personal focus, based on priori knowledge, then when tasks, problems, or topics come into a learner's personal focus, he or she will tend to initiate task-doing. In addition, when a listener hears verbalizations and sees actions relevant to his or her focus, he or she will become an active monitor. Shirouzu argues that there are effects for both task-doing and task-monitoring involving reflecting on and assessing thoughts, leading to learning outcomes, but also that such thoughts lead to choice of personal focus, as the interaction progresses. Shirouzu specifically chose a fractions paper-folding task so that problem-solving steps could be visible to the observer as an interaction with external resources.

Discussion

I argue that the part of his analysis where activities are displayed carries more weight than the part of his analysis where they are inferred. Although observed activities can be interpreted differently, depending on the observer's viewpoint and framework, each person doing the interpreting can see the activities, so, depending on their perceptive capacities, all observers begin from the same data. On the other hand, since inferences about activities are not observable but made according to an observer's viewpoint and framework, it seems that they are more susceptible to controversy. Having said this, the specific nature of the task Shirouzu chose may sufficiently reduce the inferential space so that the inferences become convincing. In addition, Shirouzu mostly relies on dialogue interpreted in relation to external resources and knowledge about the task as given by the teacher (assumed to be acknowledged by the learners) to illustrate the inferences he makes about learners' thoughts.

Interanimation Patterns Illustrate Collaborative Learning

Trausan-Matu's (Chap. 6, this volume) main theoretical assumption is that if we accept a generalized perspective of "voice," the polyphonic model can be used effectively to perform an integrated analysis of different media types used in communication. He argues that group interaction must deal with dissonances appearing between voices and thus faces two forces: a centripetal force (going towards the center) that is convergent, towards unity, and a centrifugal force (going away from the center) that is divergent, towards difference. The former are unity-pursuing interanimation patterns, whereas the latter are differential interanimation patterns. When these dissonances put into question learners' own utterances, they can generate an interanimation pattern where learners are reacting to these forces. So, interanimation or transactivity may be detected in interactions, and they offer a glimpse of the collaborative learning processes of the group. In addition, he argues that natural language processing (NLP) techniques may be used for the automatic identification of discourse threads, and integrating NLP techniques with polyphony identification and social network analysis may provide a way for analyzing the contributions of each participant and their interanimation.

Discussion

Polyphony originated as a concept and practice in music and was extrapolated to texts (Bakhtin, 1984). We chose Trausan-Matu as an analyst for Shirouzu's dataset to see whether polyphony could be applied to other types of data (namely, nonverbal) in ways that further the understanding of the learning either occurring or that could occur in the interaction. Trausan-Matu considers the dataset through several intertwined dimensions in which polyphonic interactions are detected: (1) spoken dialogue, (2) body language, (3) visual aspects such as writing on the blackboard and folding paper, (4) internal dialogue at the intramental level, and (5) echoes—or in other words—the long-term effect of the voices, which can be related to actual learning outcomes. Like Shirouzu, Trausan-Matu also analyzes intramental activity and also defines it as the interaction between inner and outer resources. According to Trausan-Matu, intramental activity is dialogical, and two types of thinking can be detected in the dataset: (1) students try to obtain the solution by folding and by thinking (i.e., dialoguing with themselves) about solutions and (2) students look at others' solutions and at what the teacher draws and writes on the board and then this generates inner utterances (thinking). In both cases, interanimation patterns can be detected between external and internal "utterances." Trausan-Matu thus suffers from the same criticism as Shirouzu concerning inferences about learners' thoughts, although again, inferences seem logical in the specific interactive context. The main theoretical assumption difference between Shirouzu and Trausan-Matu is that Trausan-Matu chooses to attribute characteristics to individuals based on what he analyzes in the interaction and to use these characteristics as explanatory for what occurs, whereas Shirouzu prefers to avoid characterizing individuals, keeping to what is observed in or inferred from the interaction.

Theoretical and Methodological Assumptions

Chiu (Chap. 7, this volume) argues that when students are faced with a problem to which they do not know the solution, they try to create new ideas and assess these new ideas via explanations or justifications. If these new ideas are also correct, Chiu calls them micro-creativity. While new ideas are necessary for solving the problem—Chiu further argues—justifications support or refute an idea's usefulness by linking it to data, using a warrant, or supporting a warrant with backing. Based on the literature on cognitive and social meta-cognitive processes, he builds eight separate hypotheses (to be tested with statistical discourse analysis—SDA) about whether or not certain types of coded utterances (e.g., questions, disagreements, correct evaluations, repetitions) raise or reduce the likelihood of subsequent new ideas, correct ideas, micro-creativity, and/or justifications.

We chose Chiu as an analyst in order to see to what extent SDA could be applied beyond peer conversations—in this case, teacher–student interactions. Given his statistical method, Chiu had to make assumptions that may better be qualified as methodological rather than theoretical. He points out that SDA, like any statistical analysis, assumes that different instances of a category are sufficiently similar to be treated as equivalent during analysis.

Discussion

Chiu avoids attributing personality characteristics to individuals, nor does he postulate about intramental activities. However, he does look at the individual's role in the group interaction, as do Shirouzu and Trausan-Matu, but from a different angle. Chiu categorizes each individual's utterance over five dimensions (not to be confused with our five methodological dimensions for reflecting upon productive multivocality) and then checks to see if statistically a particular individual's utterance X generates more or less of the next utterances categorized as X, Y, Z, etc. that are coming from other individuals (he did not find any statistically significant difference between individuals for this dataset). Each of the five dimensions can take a limited number of values, and only one value is coded for each dimension.

Comparison of Theoretical Assumptions

Our three analyst's approaches can be compared on a theoretical level concerning how individual participation is characterized as influencing group interaction. Trausan-Matu attributes personal characteristics to learners based on the way they participate in the interanimation patterns he detects (where patterns are either sequences of collaborative or differential utterances). Shirouzu analyzes roles as indicative of personal focus and postulates where focus comes from, where it leads the interaction to in terms of new understandings for participants, and what it can predict for individual learner outcomes. Finally, in regard to the role of the individual in the group interaction, Chiu's analysis showed that particular individual characteristics (e.g., gender) did not significantly predict different interactional sequences. Granted, Shirouzu and Trausan-Matu see individual differences during qualitative analyses and Chiu does not during a statistical analysis, but the analysts are not looking at the same dimensions. Shirouzu searches for collaborative moments in which personal focus is made visible (often through comparisons of similar and different solution characteristics—e.g., production method or shape) and analyzes where that focus originates and where it takes the interaction. Trausan-Matu searches for collaborative and differential moments that can be considered, respectively, as either consonances or dissonances (in the polyphonic metaphor), and he argues that the latter play a role in triggering further utterances as a result of a need to complete the unfinished feature of the dissonances. On a more abstract level, both are looking at how moments of collective understanding and moments of socio-cognitive conflict come about and influence how the interaction plays out. Rather than focusing on these moments and what comes before and after (although he moved towards this later, as a result of our collaboration), Chiu focuses on breaking the interaction down into time periods, defined by breakpoints. One set of breakpoints corresponded to divisions of the interaction into time periods distinguishing between higher vs. lower micro-creativity, and another set of breakpoints corresponded to divisions of the interaction into time periods distinguishing between more vs. less justifications. As discussant I have the goal of attempting to relate collaborative and differential moments, personal focus, micro-creativity, and justifications. I will address this in the section § *Comparing analysts' pivotal moments.*

Purpose of Analysis

Shirouzu's purpose of analysis was to analyze (1) where personal foci of learners originate, (2) what happens next in the interaction once a learner focuses on something, and (3) what learner outcomes do such foci and interactions lead to.

Trausan-Matu's purpose of analysis was to identify discourse threads, the interactions in which they take place, and the semantic content of the utterances in chats (in the original application of his method) and of the interanimation among participants. Finally, starting from the above data, an evaluation of the contribution of each participant to the learning process was computed. Interaction takes place (1) between learners who interact through talk or actions, (2) between a single learner's utterances who dialogues with himself or herself, or even (3) between the talk/action of one student and the inner dialogue of another student. A first derived goal was to understand how group interactions scaffold individual learning. A second derived goal was to evaluate different collaborative situations in terms of what the interanimation patterns reveal about the quality of collaboration, thus giving us an indicator for choosing them. A third derived goal was to leverage these interanimation patterns as a way for teachers to manage students' activity.

The purpose of Chiu's analysis was to use SDA to statistically model individual and conversation turn characteristics that affected micro-creativity or justifications. He first documented when new ideas, correct ideas, micro-creativity, and justifications occurred. Second, he analyzed whether they occurred uniformly throughout the whole interaction or whether they occurred more frequently in particular time periods where pivotal moments divided the whole interaction into distinct time periods. Third, he tested how recent sequences of actions affected the likelihood of creating new ideas, correct ideas, micro-creativity, or justifications. In abstract terms, he examined variables that significantly increased or decreased the likelihoods of dependent variables of interest. Fourth, he tested whether the above effects differed across participants or across time periods. Chiu postulates that if educators can understand how classroom processes influence micro-creativity or justifications, they can help classroom participants act in ways that facilitate greater micro-creativity and reduce behaviors that hinder micro-creativity (e.g., evaluate ideas slowly and carefully to raise the likelihood of a correct evaluation which aids others' micro-creativity).

Comparison of Purposes of Analysis

Although the main purpose of analysis of all three researchers was to understand the role of the individual in the group and for Shirouzu and Trausan-Matu to understand the role of the group in individual learning, the way each researcher chose to attempt to qualify those roles was different. Shirouzu assessed the role of the individual in terms of task-doer or task monitor during group work. Trausan-Matu assessed the individual role as part of an adjacency pair of utterances—occurring in interaction with an interlocutor—and classified as either converging or diverging. Chiu assessed the role of the individual through how a particular utterance type can lead to other or the same utterance types coming from others in the group (or from the same individual). Taking these different ways of qualifying the nature of the individual contribution to the group together gives a more complete picture and incites an integrative approach that remains open to still other definitions of the role of the individual within the group. A similar effect exists in relation to how Shirouzu and Trausan-Matu attempted to qualify the role of the group in individual learning. Shirouzu described how students' inferred internal trajectories interact both with external traces of the student's own and group member's activity and with other students' inferred internal trajectories. He then used this to explain why learning was or was not persistent when students were questioned six months later. Trausan-Matu considered that the "voices" in interaction can be spoken dialogue, body language, and visual data such as objects and actions but also inner speech. If learning is favored by convergence and divergence, then identifying interanimation patterns among participants' "voices" in group interaction contributes to showing how individual learning occurs. In subsequent sections, I show how the different approaches to understanding both the role of the individual in the group and the role of the group in individual learning fostered analytical refinements or the elaboration of new methods. All analysts also had the purpose of finding pivotal moments, and I will define their approaches in the section § *Comparing analysts' pivotal moments*.

Unit of Analysis/Unit of Interaction

Multi-faceted Interventions and Sequences of Related Turns

Shirouzu's unit of analysis consisted of a participant's multi-faceted intervention. He considered the participant's verbalizations but also how they were spoken (prosody), to whom these verbalizations seemed to be addressed, body posture of participants, and their gaze and gestures but also their actions in relation to the problem (folding and cutting the origami paper). His unit of interaction was of two types. The first was not sequential in that it was not an excerpt of the interaction that was analyzed. Instead, Shirouzu compared a single student's internal processes (as illustrated by dialogue and inferences Shirouzu made) with (1) the student's manipulation of external resources and (2) what other students said about the manipulations or dialogue. The second type of unit of interaction used by Shirouzu was indeed a sequence of interaction where he noted which students were task-doers and which were monitors as well as what was being compared and contrasted in the dialogue. In other words, a unit of interaction of the second type was a sequence with action/talk labeled "task-doer" and other action/talk in relation to it, labeled "task monitor."

Sequences of Related Turns

Trausan-Matu's unit of interaction was dependent on his unit of analysis. Indeed, detecting interanimation patterns (are they unity-pursuing or differential?) requires understanding how individual utterances and interventions react to one another, so Trausan-Matu's pivotal moments were also sequences of turns or groupings of sequences of turns, not far from one another. In confrontation with the dataset, Trausan-Matu modified the analytical constructs of both "utterance" and "adjacency pair." Utterances were no longer just verbal as they were initially defined in the polyphony framework (although they had originally been transferred from music); they could now also be actions, text, figures, gesture, etc. In addition, utterances could be internal (e.g., inner speech). Finally, utterances were no longer solely individual but could be generated by a group. Trausan-Matu gave the example of all students moving their chairs as a chorus. During his analysis, the construct "adjacency pair" also took on a different meaning from its original definition (Schegloff & Sacks, 1973), both by the fact that the nature of "utterance" had changed and also because Trausan-Matu argued for considering external and internal utterances as adjacent. If this is to be the case, one can then question what defining characteristic makes a pair of utterances "adjacent." Should adjacent pairs be expressed in the same mode and medium in order to illustrate transparent shared ordering? Or can adjacency be determined on a semantic level across different modes and media and even across talk and actions on the one hand and thought on the other?

Sequences of Types of Turns

Chiu used the conversation turn as the unit of analysis but then used them to define a unit of interaction, which was a sequence of one type of action (conversation turn), followed by another. In other words, one utterance was a "breakpoint" in between a collection of utterances of type 1, followed by a collection of utterances of type 2. Chiu characterized the interaction as a whole by the probabilities of such sequences, modeled by SDA. After hearing about Trausaun-Matu's collaborative utterances, Chiu added a new unit of analysis by studying the utterances before and after what he had defined as a pivotal moment.

Data Representations

Form of Provided Data

Data was originally provided by Shirouzu in the form of (1) a video of the 6th-grade classroom in the original Japanese with English subtitles of talk and (2) an Excel file containing two versions of the transcript—a parallel version with the rows

Table 8.1 A sequence of the transcript transformed by Chiu

Time	Line	Person	Talk/action	Blackboard
36:14	*472*	*Y*	*Differ [with clear voice]*	*36:14*
	473	*Y*	*though areas are equal [with low voice]*	
	474	*G*	*The areas are the same*	
	475	T	Yes	
36:20	**476**	**G**	**but the shapes and production methods differ**	
		K	The shape and production method differ	
		N	The shape and production method differ	
		Anonymous	The shape and production method differ	
	477	T	The areas are the same	
	478	T	Because the areas are the same	Connects diagonally and writes, "Areas are the same."
	479	T	this is the last comparison [N's first solution and K's one]	
		Y	[Leans over the desk]	
36:52	480	T	What do you think of these?	
37:04	481	G	Although shapes are the same	

indicating interaction interventions and the columns indicating the time stamp, line number, state of the blackboard and participant, and a serial version with the rows indicating the participant and the columns indicating time, line number, participant, talk and action, and finally the state of the blackboard (cf. Shirouzu's data presentation chapter, this volume, and Table 8.1 for an extract).

In the parallel version of the transcript, when there is overlapping speech or overlapping actions, the overlapping elements appear on the same row, but we do not know exactly how they overlap. Comments on the transcription (cf. extract of serial version in Table 8.1) appear in square brackets and deal with prosody (lines 472 and 473), body postures (line 479), relations between solutions (also line 479), specifying deictic references (not shown), specifying to whom speech seems addressed (not shown), gaze (not shown), and pointing or other communicative gestures (line 504). When the transcriber was unable to hear who spoke, the speech was labeled as anonymous (line 476). We can assume that all of these elements were important to Shirouzu for understanding the dataset. Whatever its form (participants appearing in rows or columns), a shared transcript containing time stamps and line numbers that could be used to refer to interesting phenomena was of course crucial to facilitate the comparison of analyses.

Representations of Students' Talk, Gestures, and Actions

Chiu used the serial version of the transcript Shirouzu provided to record his choice of pivotal moments. Overlapping talk and actions were kept noted by multiple rows corresponding to one line number (line 476). At one point in our collaboration, Chiu harnessed the representational powers of colors by highlighting Shirouzu's three pivotal moments in red (Shirouzu's first pivotal moment is shown in Table 8.1 in italics: lines 472–474) and his own five pivotal moments in blue (Chiu's last breakpoint is shown here in bold: line 476, around which he defines a pivotal moment that includes the utterances in Shirouzu's first pivotal moment) in order to compare them. How each researcher defined pivotal moments will be explained in a subsequent section.

Shirouzu and Trausan-Matu used the parallel transcript form for their respective analyses and Chiu used the serial form, but they all watched the video in order to do their analyses. In addition, all of them used other representations of data. Shirouzu manually made drawings of students' successive attempts at solutions and associated students' actions, gaze, and gestures with each step of folding and cutting, using these elements to specify relations between internal and external trajectories (students Y and G are shown in his chapter, although he did this for his PhD for all students, but in Japanese). He also used a table to illustrate comparisons between pairs of students' solutions and what they say about their solutions because comparing is a principle design activity for his pedagogical sequence.

Trausan-Matu mainly did analysis manually and concentrated on locating interanimation patterns and marking them as he read through the transcription and watched the video, but he pointed out that the PolyCAFe system[2] could be used to combine locating such pattern couples with an automatic content-based analysis. The PolyCAFe system integrates content-based NLP with social network analysis and polyphonic analysis.

In addition to marking the transcription, Chiu performed analyses automatically by writing a computer program in PerlTM that sent commands to the Eviews statistics programTM and used a database table, a summary statistics table, a table of breakpoints, a time series graph, and a path diagram.

Analytic Manipulations on Data Representations

Shirouzu carried out three steps in his method to study focus-based constructive interaction: (1) pinpointing collaborative moments and detecting personal focus, (2) backtracking to determine where each learner's focus originated and how it has grown and reconstructing collaborative moments as constructive interaction, and

[2]The PolyCAFe system was developed in the European FP7 project LTfLL (http://www.ltfll-project.org/index.php/polycafe.html) for the visualization of the threading of utterances, for analyzing and assessing computer-supported collaborative learning chats. It is available online at http://ltfll-lin.code.ro/ltfll/wp5/login.php

(3) explaining what happens after the collaborative moments and predicting final learning outcomes for each learner. In order to carry out these steps, Shirouzu manipulates the data representations in a variety of ways. He manipulates drawings of students' solutions by showing the steps involved in each solution and then how they compare to each other. For example, some students share the same first step but then differ in the second step, and in order to show this, drawings must be juxtapositioned in a specific way. Shirouzu also re-created the state of the blackboard at various times in his paper. In one instance, he showed drawings of each student's solution(s) and put arrows between some of the solution with labels saying how the solutions were either same or different. He showed which student took which role (task-doer or active monitor) each time two solutions were compared in the interaction. Finally, he compared the interplay of internal (talk and inferred thought) and external trajectories (folding and cutting of paper, others' reactions) of particular students.

Trausan-Matu did not manipulate the data, per se—at least outside of his PolyCAF program, referred to only very briefly at the end of his chapter. His analytical method was accomplished by reading over the transcription, watching the video, and noting when interanimation patterns occurred.

Chiu created a statistical model for each possible combination of locations of breakpoints (up to a maximum of six [maximum allowable by current microcomputers]. Then, he computed the Bayesian information criterion (BIC) for each statistical model to determine which model accounts for the most variance with the fewest breakpoints. The model with the smallest BIC is the best model, so he used those breakpoint locations to divide the data into specific time periods. After the three analysts noticed that they had defined one common pivotal moment, Chiu realized that this pivotal moment had the greatest impact on producing justifications and that this corresponded to the largest BIC reduction. So, Chiu discovered that his method had new analytical power, after following up on the fact that analysts converged on defining a particular sequence as pivotal.

Comparing Analysts' Pivotal Moments

As organizers of the series of workshops culminating in this book, we chose the concept of pivotal moments as a boundary object that would be sufficiently shared by analysts to be comparable, yet remain sufficiently ambiguous to allow for each analyst to define it, according to his or her own framework. Below, I will first compare Shirouzu's pivotal moments to those of Trausan-Matu and then Shirouzu's pivotal moments to those of Chiu. Second, I will comment about possible relations between Trausan-Matu and Chiu's pivotal moments. The reader is referred to the dataset presentation chapter in this section in order to use the line numbers to follow identification of pivotal moments. This comparison, the discussion on the five methodological dimensions, and our general discussions throughout our collaboration have all enabled the analysts to reflect further on their own methods and motivated them to revisit their analyses.

Conceptual Change Compared to Interanimation Patterns

Trausan-Matu defined pivotal moments in collaboration by detecting the changes in the degree of interanimation of voices as illustrated by collaborative and differential utterances. Collaborative utterances illustrate a convergence pattern and correspond for example to the collective display of understanding. An example of a differential utterance is when an explanation given by one learner is perceived as incomplete, thus inciting a second learner to add to it. In the polyphonic view, this exemplifies a type of "dissonance" between the two learners that is remedied by the second learner's addition. Trausan-Matu used a polyphonic model of group interaction where a conversation contains different longitudinal threads (or "voices") composed of utterances, each of them having independence but achieving a joint discourse. Trausan-Matu found combinations of interanimation patterns (collaborative, differential, or both) at Shirouzu's pivotal moments (lines 472–474, lines 502–504, line 519, and lines 538–547). This latter pivotal moment respects Shirouzu's first and second definitions, but there are no strong interanimation patterns.

In addition to locating interanimation patterns as pivotal moments, Trausan-Matu initially suggested attributing characteristics to certain learners, based on how he interprets their role in the interaction:

Line 173 (Y's unique solution portrays him as a divergent thinker)
Line 179 (K may be a divergent thinker, she is the only one to change the order of fractions)
Line 470 (gestures that divide Y from group portray him as divergent thinker)

In the final version of his chapter, the fact that K may be a divergent thinker is not emphasized, perhaps because there are more reasons to infer from the dataset to hypothesize such an attribute for Y.

Finally, Trausan-Matu found two different instances of collective body language, one of which could be pivotal (learners hedging) and neither of which were located by Shirouzu or Chiu. Recall that the polyphonic model as it is typically applied to learning datasets focuses on talk and that we asked Trausan-Matu to analyze this dataset knowing that he would be confronted with analyzing gestures. Indeed, it was possible to extend the application of the polyphonic model from music to text to gestures and find both collaborative (learners hedging) and differential (Y staying physically back) interanimation patterns, the former of which seems more pivotal for teachers wanting to manage class participation.

Conceptual Change Compared to Frequency of New Ideas

Whereas Shirouzu qualitatively noted four pivotal moments of high-level thinking showing levels of conceptual change, Chiu identified six time periods of distinctly different frequencies of new ideas. Chiu makes five observations about Shirouzu's pivotal moments in comparison to his own. Firstly Shirouzu identified pivotal moments, whereas Chiu identified time periods divided by pivotal moments. Second, Shirouzu's pivotal moments are sequences of conversation turns, whereas Chiu's pivotal moments are statistically restricted to one conversation turn (requiring further qualitative analysis to identify a pivotal moment's boundaries). Third, locations of the pivotal moments reflect different scopes of interest. Shirouzu's pivotal moments focus on localized conceptual thinking, whereas Chiu's pivotal moments cover the entire classroom interaction. Chiu's first three breakpoints indicate shifts in the activity from teacher instructions to student folding to looking at one another's solutions. Much of the conceptual thinking displayed by the learners occurs later, so Shirouzu does not focus on this first part of the interaction. Fourth, Shirouzu and Chiu both identified a corresponding pivotal moment—his first (472–474) and breakpoint—his fifth (476). Chiu indicates a sharp increase in new ideas with his breakpoint at line 476. Fifth, Chiu's breakpoints do not recognize Shirouzu's other pivotal moments of conceptual thinking because they occur during a time period with a similar number of other new ideas. This shows that some pivotal moments (472–6) are pivotal in two ways: (1) the moment itself is important and (2) they elevate the conversation to elicit many more new ideas. Others are important conceptually but do not change the subsequent interaction (Shirouzu). Still other pivotal moments can hinder the subsequent interaction in negative ways (some of Chiu's).

Interanimation Patterns Compared to Frequency of New Ideas

As with Shirouzu, Chiu also shared one pivotal moment with Trausan-Matu (lines 472–474 where "the areas are the same," "but the shapes and production methods differ"). This is the one area in the dataset where all analysts coincided in their foci. It might be termed a super pivotal moment (to be distinguished from what Chiu called a primary pivotal moment) because it is pivotal in the three different frameworks. In this case, the group noticed conceptually that the solutions have the same areas, but their shapes and production methods are different. The differential interanimation pattern focused on "same" versus "differ," and this is where there was a drop in new ideas. Based on this one event, it is clear that we are not able to conclude on any relationship between interanimation patterns compared to frequency of new ideas, but we do know that this super pivotal moment is an important moment because the group has reached a new shared conceptual level.

Chiu defined a primary pivotal moment as a moment that all his statistical models identified as a pivotal moment (he statistically identified utterance 448 as the primary pivotal moment for micro-creativity and utterance 206 as the primary pivotal moment for justifications—each dimension can only have one primary pivotal moment). After being prompted by the teacher, students G, K, and N all say (in unison) in utterance 448 that N's and G's solutions are the same. In utterance 260, in response to student Y who is demonstrating a solution, it's the teacher who says,

"Yes. These are the same up to this point [while pointing out the first step of N's and G's first solutions]."

This discussion, emanating initially from a comparison of interanimation patterns and frequency of new ideas, has brought us a new insight. In addition to researchers defining a moment in time or a sequence of interaction as being pivotal based on their theoretical assumptions, purpose of analysis, unit of analysis, and unit of interaction, these pivotal moments can also be seen as important for two reasons. The first is when all three researchers agree that a moment is pivotal but for different reasons—it's a super pivotal moment. And the second is when a particular method shows a moment as being more important than other methods in relation to a particular criterion that evaluates the moment. For example, Chiu's primary pivotal moment at utterance 448 has greater influence than other identified pivotal moments on the subsequent conversation turns' micro-creativity (likewise the primary pivotal moment at 206 has the greatest influence on subsequent conversation turns' justifications).

In the final sections, I first comment on the effect sharing a dataset and analyses had on analysts' individual analyses and what that meant for multivocality. Then I summarize the lessons on how confronting analyses contributed to making multivocality productive.

Effects of Sharing a Dataset and Analyses on the Individual Analytical Process

As could possibly be expected from a data provider, Shirouzu showed great interest in deciphering other analysts' points of view on his own dataset, and as our collaboration progressed, he took on a kind of integrator stance, striving to see how analysts' viewpoints could fit together. Having witnessed how doing this changed the way he looked at his own pivotal moments and while writing this chapter, I asked him to comment on how his interaction with the other analysts led to these modifications. In what follows, I paraphrase what he told me and I summarize how a similar process occurred for Trausan-Matu and Chiu.

After the workshop at the Alpine Rendez-Vous in 2009, Shirouzu was impressed by how differently Trausan-Matu and Chiu analyzed his dataset, and he thus became more aware of the variety of possible interpretations. At this point, he took Chiu's pivotal moment breakpoints and reconsidered them as evidence of instructor intentions either being carried out or not, a characteristic of his pedagogical task that he wanted to evaluate. Shirouzu said he was cognizant of the fact that he was still taking in others' analyses within the context of his own framework and preoccupations, although he felt that his view on his data was already widening. In parallel, Shirouzu was still pondering his own analyses: in a laboratory experiment concerning the origami task, the monitor shifted levels of conceptualization of the task by reflecting upon the task-doer's traces, whereas in the dataset analyzed for this section, child Y shifted levels of conceptualization by reflecting upon his own traces. Shirouzu did not yet have an explanation for this phenomenon of being able to shift levels, both while reflecting on a task-doer's and on one's own work.

In preparation for the symposium at CSCL 2011, Shirouzu had understood in more detail how Chiu analyzed "change" as a form of micro-creativity and that Trausan-Matu's analyses employed both a conversation analysis approach (e.g., a collective view) but also affected fixed attributes (e.g., divergent thinker, mirror) to individuals in order to explain the interaction. At this point, Shirouzu decided to counter the fixed attributes aspect of Trausan-Matu's approach by emphasizing for his own analyses what was observable during the interaction. In addition, because Chiu dealt with conceptual "changes," Shirouzu was incited to look at "changes" that were not conceptual (e.g., student G used the word "area" to mean something that her solution had in common with other students' solutions and not as a concept to be clarified). In an effort to explain both how student Y was able to shift levels of conceptualization while reflecting upon his own traces and how student G came to be satisfied with her solution, Shirouzu introduced "internal cognitive processes" into his analyses and he integrated them into what he calls "focus-based constructive interaction." This was also when Shirouzu changed his definition of pivotal moments from "when the monitor reflects upon externalized traces" to "when the monitor or the doer reflects upon externalized traces." At this point in time, Shirouzu says he had a tendency to explain things by dichotomy (individual vs. collective, internal vs. external, procedural vs. conceptual), and he muses that this may have been because he was contrasting himself with Trausan-Matu and Chiu as well as with other researchers.

Shirouzu said that the summary I wrote about the dataset and analyses for the CSCL 2011 symposium helped him to see the relations between the analyses and the clarification questions I asked about the chapters in this section and the visualizations of (un)shared pivotal moments helped him to reflect upon his interpretations.

During the Alpine Rendez-Vous in 2011, Shirouzu tried to explain individual differences with mechanisms that were not based on either ability or character, leading to the concept of "personal focus." He also defined pivotal moments based on different units of analyses in order to delve deeper into the notion of personal focus, yet he still felt that the "individual unit" had stronger explanatory power than the "collective unit."

Shirouzu describes feeling his integrative stance more and more when our group discussed and presented at the CSCL symposium in 2011. Although he felt that the audience wanted a single "correct" interpretation of the data, we instead deepened our own different representations of it by interacting with each other. At this point, Shirouzu thought that there could be a relation between his "personal focus" cognitive construct and Trausan-Matu's "thinking style" in the sense that perhaps accumulated experiences of having foci in various situations could generate styles, practices, and attributes. If students G or Y could have been followed in a longitudinal study through a number of years in different classes, perhaps it would be possible to relate personal foci during problem solving to individual characteristics. Shirouzu was thus able to identify new research directions because he moved from considering his own interpretation as the definitive answer to taking into consideration alternative views. In sum, faced with multivocality, Shirouzu moved from a contrasting stance to an integrative stance where he recognized that different units of analysis and frameworks could be complementary.

Perhaps in response to Shirouzu's interrogations about individual attributes and our discussion about their justification, Trausan-Matu's final chapter shows less emphasis regarding them; his analyses instead concentrate on interanimation patterns, including verbal, nonverbal, inner-dialogue, and talk-in interaction. However, it seems that Trausan-Matu's main individual analytical process change was more influenced by the nature of the dataset with which he was confronted than by the other researchers' analyses. He modified the meanings of the analytical constructs of "utterance" and "adjacency pair" and extended the field of applicability of the polyphony method by analyzing nonverbal gestural aspects of interaction for the first time and, in doing so, illustrated insights about the pedagogical task.

Chiu performed new analyses focused on the class discussion activity phase of the pedagogical task after understanding that Shirouzu had a special interest in it. In light of Trausan-Matu's discussion of collaborative utterances, Chiu also expanded his window of analysis around what he had considered as pivotal moments. For example, he noted that SDA identifies the pivotal moment's heart (e.g., a shared utterance) and not a pivotal moment's boundaries (what provoked the utterance or happened because of it). Although he used quantitative methods, while discussing with Trausan-Matu, Chiu recognized the necessity of qualitative methods and iterative loops of both in order to examine the boundaries of pivotal moments and understand the mechanisms by which they operate. Thus, in a similar way to Shirouzu, Chiu was inspired by other analytical methods, giving him new insights and ways of refining his methods.

Lessons Learned for Productive Multivocality

All analysts measured the quality of the collaboration in some way (Lund, 2011) but with different indicators and units of analysis, using both qualitative and quantitative methods that were adapted to the small size of the dataset. In this section, I discuss nine lessons for multivocality that all deal with ways to surpass the limits of an initial analytical method.

Insights Gained as a Result of Changing Unit of Analysis and Unit of Interaction

Shirouzu and Trausan-Matu described sequences of turns as pivotal moments because they focused on moments of convergence or divergence whereas Chiu restricted his breakpoints to a single conversation turn, as his goal was to divide the interaction into distinct periods of frequency of new ideas. These differences in focus of attention reflect both how questions spurred by underlying theoretical frameworks (intramental interaction, micro-creativity) guide the eye and how criteria for applying particular analytical techniques influence choosing the unit of analysis, although such differences did not require changing the method for segmenting the interaction, initially defined by Shirouzu. However, two analysts changed their original definitions of unit of analysis and unit of interaction as a result of our collaboration and, in doing so, gained insights. Chiu decided to look at the context around which his pivotal moments occurred (thus enlarging his unit of interaction) after understanding more about collaborative utterances from Trausan-Matu, and this had two consequences. The first is that he broadened his understanding of why his breakpoints were pivotal using qualitative analyses to supplement his quantitative SDA approach, and the second is that he thus became more convinced of how quantitative and qualitative methods can be used in concert to obtain more complete understanding of group interactions. In confrontation with the data, Trausan-Matu changed the nature of both his unit of analysis and his unit of interaction by adding nonverbal aspects of the interaction to his conception of "utterance" and "adjacency pair," and this also had two consequences, one of them potential. The first is that he was able to extend his polyphony framework to the analysis of nonverbal data and through watching the video he defined a collective pivotal moment where students took on various body postures, moved, or gazed elsewhere in order to avoid answering the teacher. This was not originally transcribed and was thus not analyzed by the others. Perhaps it was Trausan-Matu's newfound sensitivity to gesture that enabled him to pinpoint this non-transcribed moment, useful for monitoring the class. The second consequence—potential, as it has not actually come about—is that modifying the meaning of analytical constructs by adding new phenomena to them could make one realize that the framework in which one is situated is in fact quite similar to another framework. Perhaps utterances could be now termed "interventions" as they are no longer only verbal and perhaps there is some discussion to be had on what exactly makes an "adjacency pair" adjacent and how this could be related to the notions of transactivity and uptake.

In summary, the lessons in productive multivocality that came out of redefining one's unit of analysis or unit of interaction are multiple. First, other researchers' methods, in completely different frameworks, can give us ideas on where to look to enrich our own analyses. Second, qualitative methods can be used in concert with quantitative methods both for explaining results and for generating hypotheses to be tested. Although this second lesson may be well known for some, it's always exciting to discover it anew in a specific context. Third, using an analysis technique on new types of data can first extend the field of application of the analysis technique and, second, do so in a way that is insightful for learning. Fourth, extending the field of application of a method leads to redefining analytical constructs, and this can lead to recognizing that similarities with other frameworks may merit further study.

Insights Gained by Reacting to Other Visions of Collaborative Interaction

A fifth lesson can be found in how analysts reacted to each other's respective analyses but outside of reframing unit of analysis and unit of interaction. Shirouzu demonstrated that he was able to match new meanings to Chiu's interpretations of pivotal moments (occurring in Chiu's framework) that were relevant to him in his own framework. For example, Shirouzu saw his first pivotal moment as a collective display of new understanding, whereas Chiu viewed it as indicating the end of a period of frequent ideas, occurring just after teacher acknowledgment. Indeed it is compatible that the moment when collective understanding is reached could correspond to the beginning of a drop in new ideas because learners are consolidating their knowledge in terms of concepts already expressed. Reexamining this moment in terms of Chiu's definition of ideas as "new" or "old" led Shirouzu to suggest that in his own framework, new ideas could correspond to conceptual or procedural changes of how to view the solutions, progressing potentially towards a collaborative pivotal moment. Shirouzu also noticed that Chiu's five breakpoints corresponding to frequency of new ideas corresponded to when and how the pedagogical designer's intentions were actualized by students' behavior. So, our fifth lesson in productive multivocality for this dataset shows that in two separate instances, Shirouzu was able to reinterpret Chiu's breakpoints in his own framework and further his own understanding.

Extending One's Own Method by Reacting to Other Analysis

There is a sixth, seventh, and eighth lesson in this same category—how reacting to other's analyses can extend the limits of one's method—and it is demonstrated by Shirouzu as he was influenced by Trausan-Matu. Although all analysts consider the individual role within the collective, only Shirouzu and Trausan-Matu draw inferences of individual thought processes from the interaction. For example Trausan-Matu explains that one learner's reaction to another learner's statement might be generated by internal dialogue. As already discussed, one could describe an "adjacency pair" within a sequence where the first element comes from a first learner, the second from the internal dialogue of a second learner, and the third from how the second learner reacts to the first. For example, if a "dissonance" results from the difference between an utterance and a learner's internal belief, a corrective reaction utterance may occur. Shirouzu would explain Tasaun-Matu's convergence and dissonance by (1) tracking the conceptual and procedural pivotal moments of each learner, (2) seeing how these moments form collaborative moments, and (3) seeing if the individual pivotal moments can explain the collaborative ones. After seeing Trausan-Matu's analysis, Shirouzu was encouraged to explore how conceptual change could be influenced not only by convergence but also by divergence. On the other hand, he disagreed with Trausan-Matu's method of explaining interanimation patterns by attributing personality characteristics to individuals (e.g., he is a divergent thinker). However, after some thought, he used Trausan-Matu's viewpoint to develop a direction for future research where he would study how personal foci, analyzed in different situations over time, could help in constituting individual attributes such as divergent thinker. Our sixth and seventh lessons in productive multivocality are that confronting one analysis with another can lead to fine-tuning of analytical concepts (such as conceptual change) but that it also can lead to rendering explicit beliefs about evidence-based ways of measuring individual participation in the collective. Our eighth lesson is that comparing analytical viewpoints with others can lead to the definition of new research questions. A ninth lesson, stemming from the entire collaboration, is that it influenced researchers to adopt an integrative stance, to actively search for ways in which the analyses of others could be complementary to one's own.

What About Learning Fractions?

Finally, how has this endeavor helped to understand how students learn fractions? How has it helped to understand the role of the individual in group interaction and how the group influences individual learning? In answer to the first question, Shirouzu shows that there exist "levels of conceptual understanding" for the fractions problem in the same way that there existed levels of conceptual understanding for how a sewing machine works (Miyake, 1986). His analysis shows how students progress in their levels of understanding and has implications for pedagogical design of teaching–learning sequences on fractions (e.g., how to get students on different conceptual levels to interact with each other and deepen their levels on their own pace). Perhaps "levels of conceptual understanding" should also be sought out for other tasks so that these pedagogical strategies can become useful in other contexts. In addition, the levels found in understanding fractions provide an argument in favor of the generality of the theory of constructive interaction based on the iterative process of understanding, even though it has now taken the name of focus-based constructive interaction. Chiu's analysis provided an indicator (frequency of ideas) for whether a teacher's pedagogical goals (e.g., changes in how to view solutions to the fractions problem) were being met by the students. This could serve as a monitoring mechanism for learning fractions.

In sum, our collaboration around the fraction dataset proved to be very productive on three different levels: better understanding and then surpassing the limits of particular analytical methods, better understanding of how students learn fractions, and finally hopefully becoming better researchers, with more broadened perspectives.

Acknowledgements I would like to warmly acknowledge Hajime Shirouzu, Stefan Trausan-Matu, and Ming Ming Chiu; Hajime's dataset and their analyses and reflections made this chapter possible. I would also like to acknowledge my co-organizers throughout the Productive Multivocality workshops: Dan Suthers, Carolyn Rosé, Nancy Law, Gregory Dyke, and Chris Teplovs as well as all the participants in the workshop series.

References

Bakhtin, M. (1984). *Problems of Dostoevsky's poetics* (C. Emerson, Trans.). Minneapolis, MN: University of Minnesota Press.

Chiu, M. M. (this volume). Social metacognition, micro-creativity and justifications: Statistical discourse analysis of a mathematics classroom conversation. In D. D. Suthers, K. Lund, C. P. Rosé, C. Teplovs & N. Law (Eds.), *Productive multivocality in the analysis of group interactions, Chapter 7*. New York, NY: Springer.

Lund, K. (2011). Analytical frameworks for group interactions in CSCL systems. In S. Puntambekar, G. Erkens, & C. Hmelo-Silver (Eds.), *Analyzing Collaborative Interactions in CSCL: Methods, Approaches and Issues* (pp. 391–411). New York, NY: Springer.

Miyake, N. (1986). Constructive interaction and the iterative process of understanding. *Cognitive Science*, 10, 151–177.

Schegloff, E. A., & Sacks, H. (1973). Opening up closings. *Semiotica, 8*, 289–327.

Shirouzu, H. (this volume). Learning fractions through folding in an elementary face-to-face classroom. In D. D. Suthers, K. Lund, C. P. Rosé, C. Teplovs & N. Law (Eds.), *Productive multivocality in the analysis of group interactions, Chapter 4*. New York, NY: Springer.

Shirouzu, H. (this volume). Focus-based constructive interaction. In D. D. Suthers, K. Lund, C. P. Rosé, C. Teplovs & N. Law (Eds.), *Productive multivocality in the analysis of group interactions, Chapter 5*. New York, NY: Springer.

Trausan-Matu, S. (this volume). Collaborative and differential utterances, pivotal moments, and polyphony. In D. D. Suthers, K. Lund, C. P. Rosé, C. Teplovs & N. Law (Eds.), *productive multivocality in the analysis of group interactions, Chapter 6*. New York, NY: Springer.

Part III
Case Study 2: Peer-Led Team Learning for Chemistry

Section Editor: Carolyn P. Rosé,
Carnegie Mellon University

In this section we use a multivocal leadership construct as a lens for viewing and comparing the dynamics of two different peer-led teams as they solve a chemistry problem related to de Broglie's equation. Four different analysts offer their interpretation of the data: one providing an ethnographic analysis and the other three providing different forms of quantitative analysis. The juxtaposition of the three distinct lenses reveal new insights into the intricate nature of complex constructs like leadership and role taking more generally.

Sawyer, Frey, and Brown kick off the PLTL Chemistry section with a chapter that describes the theoretical framework and research questions that prompted the collection of the dataset used in this section of the book. As described in that chapter, Peer-Led Team Learning (PLTL) is an approach to using collaborative learning as a mode for supporting learning in college-level chemistry, particularly where lectures are large and therefore limited with respect to opportunities for intensive interaction. While this approach has been demonstrated to be effective in improving learning, questions remain about the inner workings of PLTL groups. In order to address these questions, a larger dataset, from which the two interactions presented in this chapter as well as the other chapters within this section were selected.

In a second chapter, Sawyer, Frey, and Brown present a qualitative analysis of their own dataset that offers a perspective on how the two groups selected for sharing contrast with one another. A blow-by-blow analysis is presented that offers an up close view of the two groups and how their respective interactions played out. This chapter sets up a contrast between the first group, referred to as the Gillian group, as being more interactive and conceptually focused, where the second group, the Matt group, was portrayed as less interactive and more procedurally focused. Questions are raised about the two different peer leaders and how their interactions with the groups may have led to this contrast.

Next, Howley, Mayfield, Rosé, and Strijbos present two contrasting three-dimensional coding and counting analyses of the PLTL dataset, each consisting of a Cognitive, Relational, and Motivational dimension. These two multidimensional analyses highlight ways in which complementary perspectives on collaborative processes offered by each dimension can be integrated in a way that offers deep insights into social positioning within collaborative groups. Nevertheless, distinctions in operationalization of the three dimensions between coding schemes raise questions about what these dimensions can tell us about collaborative problem solving and illustrate the care with which interpretation of distributions of codes must be carried out. These quantitative analyses raise questions about some contrasts drawn within the qualitative analysis presented by Sawyer and colleagues.

As a final analytic perspective, Oshima, Matsuzawa, Oshima, and Niihara present a Social Network Analysis of the dataset. This analysis is unique in its ability to use a similar representation to view the interactions from an individual or group perspective, and how the collaborative processes unfolded over time.

In a final chapter, Rosé compares and contrasts these analyses, bringing into sharper focus the questions that remain for future work on assessment of collaborative problem solving.

Chapter 9
Peer-Led Team Learning in General Chemistry

Keith Sawyer, Regina Frey, and Patrick Brown

Introduction

There is now a consensus in science education research that the most effective learning environments are those in which students engage in productive, collaborative discourse to build knowledge (e.g., American Association for the Advancement of Science [AAAS], 1989; National Research Council [NRC], 1996). Knowledge building occurs when students engage in collaborative conversations intended to advance both individual understanding and the collective knowledge of the group in pursuit of a common goal (Bereiter, 2002; Engle & Conant, 2002; Rogoff, Matusov, & White, 1996). Many learning scientists have shown that engaging in collaborative discourse contributes to deeper conceptual understanding, greater transferability of knowledge, and better retention (Engle & Conant, 2002; Greeno, 2006; Sawyer, 2006; Scardamalia & Bereiter, 2006). However, promoting collaboration among undergraduate students is a challenge because many science courses are large lectures that are primarily focused on individual learning (Seymour & Hewitt, 1994). As a result, there has been very little research on students' collaborative discourse practices in college science settings.

This chapter describes the setting and context of the discourse that occurs during Peer-Led Team Learning (PLTL) in first-year General Chemistry at Washington University in St. Louis. PLTL is designed to facilitate chemistry literacy and success for all students, not only for chemistry majors, by supplementing the lecture with formalized study groups that provide opportunities for active and collaborative

K. Sawyer (✉)
University of North Carolina, Chapel Hill, NC, USA
e-mail: rksawyer@email.unc.edu

R. Frey
Washington University in St. Louis, St. Louis, MO, USA

P. Brown
DuBray Middle School, Fort Zumwalt School District, St Peters, MO, USA

D.D. Suthers et al. (eds.), *Productive Multivocality in the Analysis of Group Interactions*, Computer-Supported Collaborative Learning Series 16,
DOI 10.1007/978-1-4614-8960-3_9,

learning (Gosser et al., 2001; Gosser & Roth, 1998; Sarquis et al., 2001; Siebert & McIntosh, 2001). Peer leaders are selected from undergraduate students who have received an A in the class previously. A peer leader is assigned to a group of six to eight students who meet for 2 h each weekend to solve chemistry problems designed by the course instructors. Neither the peer leader nor the students are given the solutions to the problems, because the goal of the session is not to get the correct answer; rather, it is to provide opportunities for engaging in problem solving while discussing the concepts used in the problem. The PLTL structure implemented at Washington University in St. Louis has been previously described in Brown, Sawyer, and Frey (2010) and Hockings, DeAngelis, and Frey (2008).

Literature Review

PLTL was inspired by educational research demonstrating that cooperative learning environments help students learn (Bossert, 1988–1989; Johnson & Johnson, 1992; Slavin, 1995; Webb & Palincsar, 1996). For example, cooperative classroom groups result in greater learning than competitive or individualistically structured learning environments (Johnson & Johnson, 1989; Johnson, Johnson, & Smith, 1998). Additionally, peer teaching has been shown to provide enhanced learning to both the peer teacher and to the student (Bargh & Schul, 1980; Fuchs et al., 1997). Johnson et al. (1998) reported that in college settings cooperative learning is almost 150 % as effective as individual or competitive learning environments; several reviews of the literature describe the significant, positive effects of cooperative learning on student learning in science (Blosser, 1992; Johnson, Johnson, & Maruyama, 1983; Mullen, Johnson, & Salas, 1991; Slavin, 1995).

A number of studies show that cooperative learning in undergraduate chemistry courses has a positive effect on student learning (Basili & Sanford, 1991; Bowen, 2000; Cooper, 1995; Hockings et al., 2008; Lewis & Lewis, 2005, 2008; Lundeberg, 1990; Paulson, 1999; Tien, Roth, & Kampmeier, 2002). Lewis and Lewis (2008) compared the outcomes of students experiencing traditional instruction with students experiencing cooperative learning and inquiry. The researchers found that regardless of students' background knowledge (SAT sub-scores and SAT average), students experiencing cooperative learning and inquiry had higher performance. Basili and Sanford (1991) compared students who worked together in small groups with students who learned through direct instruction. They found that the students who worked together in small groups scored higher on content tests. Lundeberg (1990) evaluated the effectiveness of a peer-led, cooperative learning program that encouraged students to use think-aloud strategies to develop conceptual understanding of chemistry concepts. She found that the program increased students' final grades in the course.

PLTL in undergraduate chemistry courses has been designed based on the above findings (Gafney & Varma-Nelson, 2008; Gosser et al., 2001; Gosser & Roth, 1998; Hockings et al., 2008; Sarquis et al., 2001; Schray, Russo, Eglof, Lademan, & Gelormo, 2009), and PLTL has been the object of several empirical studies.

Hockings et al. (2008) investigated the influence of PLTL on student performance in General Chemistry. They found that PLTL participation statistically improved student performance by an average of one-third of a grade, after controlling for students' backgrounds. Schray et al. (2009) compared groups in Organic Chemistry in a community college setting led by fellow students (referred to as in-class peer leaders) with groups led by students who were 1 year older and had previously taken Organic Chemistry (referred to as standard peer leaders). They found no significant difference in the achievement between the groups led by in-class peer leaders with those led by standard peer leaders. However, students typically preferred having standard peer leaders facilitate their group sessions.

Methodology

As part of a larger study of discourse between peer leaders and students and among the students themselves, three PLTL sessions for each of 15 veteran peer leaders were videotaped over the course of one semester, resulting in approximately 60 h of video data. From this larger study, a primarily quantitative analysis (Brown et al., 2010) allowed us to identify significant differences across groups, in both peer-leader style and in student discourse. For this volume, we compare two PLTL groups as they both solved the same problem. We selected these two groups because they demonstrated dramatic differences in student discourse practices. The first group talked more and engaged with deeper conceptual issues; in the second group, students primarily worked alone quietly, focused narrowly on solving the problem, and did not engage with the underlying concepts.

Both groups were videotaped and transcribed verbatim. The written transcripts were annotated with relevant gestures to include what students were doing when they were not talking.

The Chemistry Problem and the Learning Goals

There are three "key ideas" that underlie the chemistry problem that these two groups are working on, and four equations that correspond to these main concepts. The students are learning about the photoelectric effect—the discovery, associated with Einstein's 1905 Nobel Prize winning paper, that light has properties of both a wave and of a particle. Additionally, the students are learning about the de Broglie hypothesis—the discovery that won Louis de Broglie the Nobel Prize in 1929, which stated that moving matter (for example an electron) has an associated wavelength. The concepts associated with these two ideas play an important role in students developing an understanding that light and matter display both wave and particle characteristics, and hence, in beginning to understand the foundations of quantum mechanics.

In both PLTL groups, the peer leaders have one group of students work on I(*A*) and the other work on I(*B*). After students have finished working on either I(*A*) or I(*B*) in small groups, they discuss Part II, "Explain the trend," as a whole group. The peer leaders were instructed to present the problem in this way in the weekly meeting that they have with the course instructor.

I. Compute the de Broglie wavelength of an electron ejected from manganese (work function $= 6.6 \times 10^{-19}$ J) by one photon at each of the following wavelengths

(*A*) 3.0×10^{-7} m
(*B*) 2.5×10^{-7} m

II. Explain the trend.

I(*A*) and I(*B*) can be solved in two different ways: either by the rote application of formulas, or by engaging with the deeper concepts that underlie the formulas. Because it is not possible to interpret the two transcripts without some understanding of these concepts, we provide a brief overview of the concepts and equations underlying the problem.

In the early 1800s, a series of interference experiments was performed that showed fairly conclusively that light was a wave and hence had a wavelength. However in the early 1900s, the results of the photoelectric effect experiment were contradictory to light behaving only as a wave. Albert Einstein resolved this conflict by developing the concept of the particle nature of light (a light particle was called a photon) to describe the results of the photoelectric effect experiment.

In the photoelectric effect experiment a beam of light—a stream of photons—strikes the surface of a metal, and electrons from the metal are ejected if the energy of each photon is high enough. Each metal has a distinct *work function*, which is the minimum energy required to eject an electron; if the photon has an energy greater than the work function, one electron will be ejected with a kinetic energy that corresponds to the photon's energy minus the work function. A longer wavelength photon (corresponding to lower frequency) has less energy; a shorter wavelength photon (corresponding to higher frequency) has higher energy.

Hence in this problem, there are three key concepts:

The dual wave and particle nature of light (i.e., light has both wave and particle characteristics).

Light interacts with matter as a photon (particle). The photon transfers energy to the metal to eject one electron if the photon is energetic enough. The amount of energy required to eject one electron is a characteristic of each metal and is called the work function.

Matter (for example an electron) has both wave and particle characteristics (where matter refers to anything that has mass).

The students have been taught these key concepts and the related formulas listed below in lecture and from the textbook.

The light equation (9.1) is related to the first concept: the relationship between the wavelength (λ) of a photon and the energy (E_p) of the photon:

$$E_p = \frac{hc}{\lambda_{(photon)}} \; or \; E_p = hv \tag{9.1}$$

Where h = Planck's constant; c = the speed of light; ν = frequency

The photoelectric effect equation (9.2) represents the second concept: the relationship between the energy of the incoming photon (E_p), the work function (Φ) (the amount of energy required to release the electron from its atom), and the kinetic energy of the electron (E_k) as it leaves the atom:

$$E_p = E_k + \Phi \; or \; E_p - \Phi = E_k \tag{9.2}$$

The de Broglie equation (9.3) represents the third concept: that all matter has both particle and wave characteristics. Hence, (9.3) captures the relationship between wavelength (λ), mass (m), and velocity (v) of all matter, including the electron:

$$\lambda_{(matter)} = \frac{h}{mv} \tag{9.3}$$

The kinetic energy of the ejected electron is found using equation

$$E_k = \frac{1}{2}mv^2 \tag{9.4}$$

It is possible to solve the above problem purely by algebraic symbol manipulation. Approaching the problem in this manner, a student would note the given information—the wavelength of the photon and the work function of the metal. Then, (9.1) through (9.4) can be applied as follows:

$E_p = \frac{hc}{\lambda_{(photon)}}$	h, c, and lambda (λ) are known; solve for E_p
$E_p - \Phi = E_k$	E_p and phi (Φ) are known; solve for E_k
$E_k = \frac{1}{2}mv^2s$	E_k and m are known; solve for v
$\lambda_{(matter)} = \frac{h}{mv}$	h, m, and v are known; solve for lambda (λ)

However, the goal of PLTL is for students to do more than simply solve the problem algebraically; the goal is to encourage them to engage in knowledge-building discourse about the deeper concepts underlying the equations. After all, the students are already given homework problems that require application of the formulas, and they have opportunities to check their work in help sessions, office hours with the instructor, and in recitations with a Teaching Assistant. The purpose of PLTL is to provide an opportunity for students to go beyond this relatively superficial level of understanding, and to help them master the deeper concepts underlying the equations.

Summary

There is a great deal of research evidence that students who participate in PLTL acquire higher levels of chemistry understanding than students who learn individually and alone. However, no studies have looked inside the "black box" of the PLTL session to examine exactly how peer discourse contributes to chemistry understanding. We offer this dataset for analysis in this volume to help explain how PLTL contributes to improved learning outcomes.

References

AAAS. (1989). *Science for all Americans*. Washington, DC: AAAS.

Bargh, J. A., & Schul, Y. (1980). On the cognitive benefits of teaching. *Journal of Educational Psychology, 72*(5), 593–604.

Basili, P. A., & Sanford, J. P. (1991). Conceptual change strategies and cooperative group work in chemistry. *Journal of Research in Science Teaching, 28*, 293–304.

Bereiter, C. (2002). *Education and mind in the knowledge age*. Mahwah, NJ: Erlbaum.

Blosser, P. E. (1992). Using cooperative learning in science education. Retrieved December 20, 2008, from *Clearinghouse for Science, Mathematics and Environmental Education* web site: http://www.ericse.org/Bulletins/SEB92-1.html

Bossert, S. T. (1988–1989). Cooperative activities in the classroom. *Review of Research in Education, 15*, 225–252.

Bowen, C. W. (2000). A quantitative literature review of cooperative learning effects on high school and college chemistry achievement. *Journal of Chemical Education, 77*, 116–119.

Brown, P., Sawyer, R. K., & Frey, R. (2010, June 29–July 2). *What are they talking about? Findings from an analysis of the discourse in peer-led team learning in general chemistry*. Paper presented at the International Conference of the Learning Sciences, Chicago, IL.

Cooper, M. M. (1995). Cooperative chemistry laboratories. *Journal of Chemical Education, 71*(4), 307.

Engle, R. A., & Conant, F. R. (2002). Guiding principles for fostering productive disciplinary engagement: Explaining an emergent argument in a community of learners classroom. *Cognition & Instruction, 20*(4), 399.

Fuchs, L. S., Fuchs, D., Hamlett, C. L., Phillips, N. B., Karns, K., & Dutka, S. (1997). Enhancing students' helping behavior during peer-mediated instruction with conceptual mathematical explanations. *The Elementary School Journal, 97*(3), 223–249.

Gafney, L., & Varma-Nelson, P. (2008). *Peer-Led team learning: Evaluation, dissemination, and institutionalization of a college level initiative*. Dordrecht, The Netherlands: Springer.

Gosser, D. K., Cracolice, M., Kampmeier, J., Roth, V., Strozak, V., & Varma-Nelson, P. (2001). *Peer-led team learning: A guidebook*. Upper Saddle River, NJ: Prentice Hall.

Gosser, D. K., & Roth, V. (1998). The workshop chemistry project: Peer-led team learning. *Journal of Chemical Education, 75*(2), 185–187.

Greeno, J. G. (2006). Learning in activity. In R. K. Sawyer (Ed.), *The Cambridge handbook of the learning sciences* (pp. 79–96). New York: Cambridge University Press.

Hockings, S. C., DeAngelis, K. J., & Frey, R. F. (2008). Peer-led team learning in general chemistry: Implementation and evaluation. *Journal of Chemical Education, 85*(7), 990–996.

Johnson, D. W., & Johnson, R. T. (1989). *Cooperation and competition: Theory and research*. Edina, MN: Interaction Book Company.

Johnson, D. W., & Johnson, R. T. (1992). Positive interdependence: Key to effective cooperation. In R. Hertz-Lazarowitz & N. Miller (Eds.), *Interaction in cooperative groups: The theoretical anatomy of group learning* (pp. 174–199). New York, NY: Cambridge University Press.

Johnson, D. W., Johnson, R. T., & Maruyama, G. (1983). Interdependence and interpersonal attraction among heterogeneous and homogeneous individuals: A theoretical formulation and a meta-analysis of the research. *Review of Educational Research, 53*(1), 5–54.

Johnson, D. W., Johnson, R. T., & Smith, K. A. (1998, July/August). Cooperative learning returns to college: What evidence is there that it works? *Change*, 27–35.

Lewis, S. E., & Lewis, J. E. (2005). Departing from lectures: An evaluation of a peer-led guided inquiry alternative. *Journal of Chemical Education, 82*(1), 135–139.

Lewis, S. E., & Lewis, J. E. (2008). Seeking effectiveness and equity in a large college chemistry course: An HLM investigation of peer-led guided inquiry. *Journal of Research in Science Teaching, 45*(7), 794–811.

Lundeberg, M. A. (1990). Supplemental instruction in chemistry. *Journal of Research in Science Teaching, 27*(2), 145–155.

Mullen, B., Johnson, C., & Salas, E. (1991). Productivity loss in brainstorming groups: A meta-analytic integration. *Basic and Applied Social Psychology, 12*(1), 3–23.

National Research Council. (1996). *National science education standards*. Washington, DC: National Academy Press.

Paulson, D. R. (1999). Active learning and cooperative learning in the organic chemistry lecture class. *Journal of Chemical Education, 76*(8), 1136–1140.

Rogoff, B., Matusov, E., & White, C. (1996). Models of teaching and learning: Participating in a community of learners. In D. R. Olson & N. Torrance (Eds.), *Handbook of education and human development* (pp. 338–414). Malden, MA: Blackwell Publishing.

Sarquis, J. L., Dixon, L. J., Gosser, D. K., Kampmeier, J. A., Roth, V., Strozak, V. S., et al. (2001). The workshop project: Peer-led team learning in chemistry. In J. E. Miller, J. E. Groccia, & M. S. Miller (Eds.), *Student-assisted teaching* (pp. 150–156). Boston, MA: Anker.

Sawyer, R. K. (2006). Analyzing collaborative discourse. In K. R. Sawyer (Ed.), *The Cambridge handbook of the learning science* (pp. 187–204). New York, NY: Cambridge University Press.

Scardamalia, M., & Bereiter, C. (2006). Knowledge building: Theory, pedagogy, and technology. In R. K. Sawyer (Ed.), *Cambridge handbook of the learning sciences* (pp. 97–118). New York, NY: Cambridge University Press.

Schray, K., Russo, M. J., Eglof, R., Lademan, W., & Gelormo, D. (2009). Are in-class peer leaders effective in the peer-led team-learning approach? *Journal of College Science Teaching, 38*(4), 62–67.

Seymour, E., & Hewitt, N. (1994). *Talking about leaving: Factors contributing to high attrition rates among science, mathematics, and engineering undergraduate majors*. Final report to the Alfred P. Sloan Foundation. Boulder, CO: University of Colorado, Bureau of Sociological Research.

Siebert, E. D., & McIntosh, W. J. (2001). *College pathways to the science education standards*. Arlington, VA: NSTA Press.

Slavin, R. E. (1995). *Cooperative learning: Theory, research, and practice* (2nd ed.). Boston, MA: Allyn & Bacon.

Tien, L. T., Roth, V., & Kampmeier, J. A. (2002). Implementation of a peer-led team learning instructional approach in an undergraduate organic chemistry course. *Journal of Research in Science Teaching, 39*(7), 606–632.

Webb, N. M., & Palincsar, A. S. (1996). Group processes in the classroom. In D. C. Berliner & R. C. Calfee (Eds.), *Handbook of educational psychology* (pp. 841–873). New York, NY: Simon & Schuster Macmillan.

Chapter 10
Knowledge Building Discourse in Peer-Led Team Learning (PLTL) Groups in First-Year General Chemistry

Keith Sawyer, Regina Frey, and Patrick Brown

Introduction

Although we know that cooperative techniques enhance student learning in PLTL, and more broadly in undergraduate chemistry courses, previous studies (reviewed in the corresponding dataset description chapter) have not investigated the interactional mechanisms that account for students' improved academic performance from cooperative learning in General Chemistry. In (Brown, Sawyer, & Frey, 2010), we examined the influence of peer-leader discourse in PLTL in General Chemistry. We showed that a peer leader's interactional style (whether instructional or facilitative) influenced student discussions. When a peer leader's interactional style was almost entirely facilitative, the students' discourse was characterized by longer chains of student-to-student conversations and more equal student participation. Conversely, when peer leaders used equal amounts of instructional and facilitative discourse, students consistently demonstrated unequal participation and engaged in mostly short chains of interactions. This finding corroborates a number of K-12 studies that have shown that teachers play a pivotal role in both enabling and constraining student discourse (Carlsen, 1993; Crawford, 2005; Hanrahan, 2005; Kelly, Brown, & Crawford, 2000; Klaassen & Lijnse, 1996; van Zee, Iwasyk, Kurose, Simpson, & Wild, 2001; van Zee & Minstrell, 1997).

In this chapter, we build on the above findings by presenting detailed analyses of how the conversation unfolds across two extended problem-solving sessions. One of the sessions is led by a peer leader with a largely facilitative style, and the

K. Sawyer (✉)
University of North Carolina, Chapel Hill, NC, USA
e-mail: rksawyer@email.unc.edu

R. Frey
Washington University in St. Louis, St. Louis, MO, USA

P. Brown
DuBray Middle School, Fort Zumwalt School District, St Peters, MO, USA

D.D. Suthers et al. (eds.), *Productive Multivocality in the Analysis of Group Interactions*, Computer-Supported Collaborative Learning Series 16,
DOI 10.1007/978-1-4614-8960-3_10,

other by a peer leader with a roughly balanced use of facilitative and instructive styles. The major research questions we address in this study are: (1) How are students' contributions responsive to those of other students? (2) What types of collaborative discourse practices are used by students working in small groups that lead to building knowledge of chemistry content? (3) What peer leader actions facilitate student collaborative discourse?

The Five Dimensions Characterizing Our Approach

1. Theoretical assumptions. We take a broadly positivist and realist approach: We maintain that individual phenomena, such as conceptual change, and social phenomena, such as conversation, are real and exist in the world, and can be studied objectively. We maintain that learning occurs at both the individual and the social levels of analysis simultaneously; that learning emerges over time; and that an explanation of these emergent processes at either the individual or the group level cannot be complete without a complementary consideration of those processes at the other level (Sawyer, 2005).
2. Purpose of analysis. Our general goal is a practical one: to make PLTL groups more effective at enhancing individual learning outcomes. We seek data that would provide practical advice on how to improve the organization of PLTL groups, how to better design and present problems to be solved collectively, and how best to train peer leaders. Specifically, in this study we hope to accomplish this practical goal by (1) identifying sequences of dialogue among students that indicate engagement with deep concepts, rather than exchange of superficial information, and (2) identifying the contextual factors correlated with these sequences, including group organization, problem design, and peer leader interactional style.
3. Unit of interaction. We follow a fairly conventional conversation-analytic methodology in which the unit of interaction is, at the lowest level of analysis, the adjacency pair, and at a higher level of analysis, an extended sequence of acts that form a coherent episode.
4. Representations. Our representation is the transcript.
5. Manipulations. In our larger study, we manipulated the transcript representation by applying a coding scheme to categorize individual acts. The categories in the coding scheme emerged from a grounded theory approach, and the reliability of the coding scheme was demonstrated by attaining satisfactory intercoder reliability. In the analysis presented here, we do not use this coding scheme; rather, we present a narrative analysis of how knowledge building unfolds differently in the two groups.

Methodology

Both groups were videotaped and transcribed verbatim. The written transcripts were annotated with relevant gestures to include what students were doing when they were not talking. An utterance was defined as a single phrase or sentence spoken by one participant; utterances were delimited by short pauses for breath. A turn was defined as a continuous segment of talk uttered by the same speaker. A single turn could consist of one or more utterances. Each utterance was assigned a code (see Brown et al., 2010). The codes were developed using the constant comparative method of qualitative data analysis (Glaser & Strauss, 1967).

Cohen's Kappa is an inter-rater reliability measure for qualitative studies (Bakeman & Brownlee, 1980; Lunn, 1998). The Cohen's Kappa was 0.91; meeting the criteria for inter-rater reliability (greater than 0.70). All disagreements were resolved through discussions.

Results

In this section, we present turn-by-turn analyses of the conversations that took place in the two different groups as the students solved this problem. These analyses reveal that the first group engages in collaborative discourse exploring the deeper concepts underlying the equations, and that the second group focuses on algebraic manipulation to solve the equations.

When the groups split into two smaller groups to work in parallel on I(*A*) and I(*B*), we moved the microphones to capture the discourse of the smaller groups solving I(*B*). Although both groups took similar amounts of time working on parts I(*B*) and II (916.1 s and 1045.8 s, respectively), the two groups differed dramatically in the ways that they used the time allotted to solve the problem (see Fig. 10.1).

The most dramatic difference is that students in Gillian's group spent more than twice as much time talking as did Matt's students (675 s versus 320.8 s), and Matt's students' spent almost ten times as much time as Gillian's students working

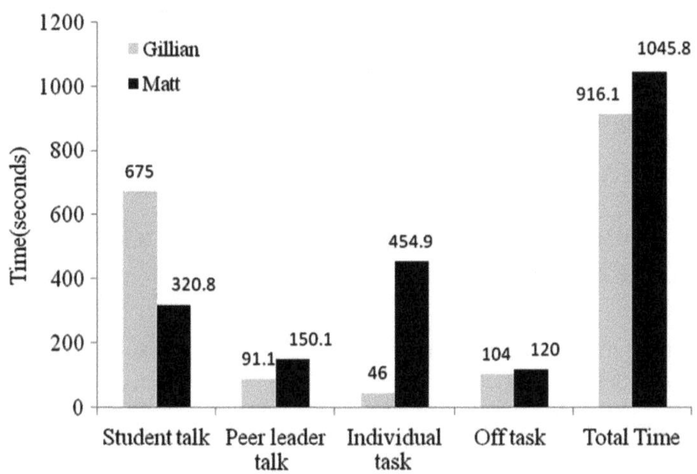

Fig. 10.1 Time of peer leader and student discourse, individual tasks, and off task during Part I(*B*) and II of the de Broglie problem

individually (454.9 s versus 46 s). During their individual task time, Matt's students worked in silence. In contrast, during the 46 s that Gillian's students engaged in individual tasks, they were silent for only 15 of those seconds. Students in both groups spent comparable amounts of time off task (Gillian's students spent 104 s and Matt's students spent 120 s) while their peers spent time writing the group's work on the board. In sum, Matt's students are individually working on the problem and only occasionally interacting with each other; Gillian's students are constantly engaged in conversation.

Extended Analysis of Problem-Solving Discourse

We next explore whether the greater proportion of conversation among Gillian's students might also result in a greater focus on the deeper underlying concepts. Previous research shows that when students engage in learning conversations, they are more likely to address underlying concepts (Hanrahan, 2005; Kelly et al., 2000; van Zee et al., 2001; van Zee & Minstrell, 1997). We also hypothesized that because Matt's students were working predominantly alone, that they might be focused solely on algebraic symbol manipulation (Sfard, 1991) without discussing the underlying concepts associated with the problem.

We begin with an extended analysis of the conversation among Gillian's students, and we find that indeed, their conversations frequently address the underlying concepts. We then turn to an extended analysis of Matt's students; they also solved the problem, but their discourse did not reveal any engagement with the underlying big ideas. Rather, their discourse demonstrated an emphasis on algebraic symbol manipulation. For example, to solve part I(A) or I(B), students can use (9.1) to find the energy of a photon (E_p); (9.2) to find the kinetic energy of the ejected electron (E_k); (9.4) to find to find the velocity of the ejected electron; and (9.3) to solve for the wavelength of the ejected electron, lambda (λ). (The four equations can be found in the dataset description chapter.) In both analyses, we focus on portions of the transcripts that highlight differences in the conversations among students between the two PLTL groups solving the problem.

Gillian's Group: Solving the Problem While Discussing the Underlying Concepts

Gillian's group discusses the equations used to algebraically solve the problem and some of the underlying concepts.[1]

[1] F = Female, M = Male, S = Multiple students in unison, PL = Peer leader.

10 F1: (Time: 32:56.1) So we need the de Broglie wavelength and it's giving us the two wavelengths (referring to the photon) so with the (photon's) wavelength, we can find velocity. And with...

This is correct; the wavelength of the ejected electron is calculated using the de Broglie equation. F1 is correctly noting that the de Broglie equation captures the relationship between wavelength and velocity.

11 F4: And we don't necessarily have to work with the work function right away.
12 F1: Not right away, but we do need it at the end.

Both are correct; the work function is required in the second step of the four steps.

14 F1: Because the important thing is that for the de Broglie wavelength we are finding the wavelength of the electron not of light.
15 F4: Right, exactly. So, first for the electron we use h/mv because we know the mass of...

This is correct; F1 and F4 are demonstrating their agreement with the statement of the problem, and their correct understanding of the de Broglie equation: it captures the relationship between wavelength and velocity for an electron, but not for a photon—that relationship is captured with the light equation. So in lines (14–15), F1 and F4 demonstrate a partial understanding of the big idea that matter has both mass and wave characteristics.

16 F1: We need to find the mass of...
17 F4: No, we know the mass of an electron. It's an electron.
18 F1: Very true, very true.

The mass of an electron is a known number that is readily available in the textbook. In line (16) F1 misspoke, F4 corrects her, and F1 quickly agrees.

19 F4: but we don't know v.

They are on the right track: knowing the velocity of the electron (9.4), the wavelength can then be determined using the de Broglie equation (9.3).

20 F1: With this wavelength (pointing to handout), we would be finding velocity.
21 F2: What did they give us? For the following wavelengths. So, well $\lambda = h/mv$ right?

This is the first time F2 speaks. F2 thinks there is a direct relationship between the wavelengths given and the de Broglie equation (9.3). She seems a bit behind the other two since in line (15) F4 has already mentioned the importance of using the de Broglie equation (9.3) to solve the problem.

22 F4: We can find the energy. I got it.
23 F1: Yes it's telling us to use this, so we can find kinetic energy. (talking over)
24 F4: Of the photon, (pause) right?
25 S: ...(inaudible)
26 F1: No, you can find kinetic energy.

This is the correct next step—(9.4) can be used to find the velocity, if the kinetic energy of the electron can be determined—and, the kinetic energy of the electron (E_k) can be determined using (9.2). However, F4's statement in line 24, "of the photon" is incorrect and (9.2) is used to find the kinetic energy of the ejected electron, not the energy of a photon, which is found using (9.1).

27	F4:	Kinetic energy of a photon.
28	F1:	You can use kinetic energy of the work function; $e_k = h\,\nu$ (nu) minus work function. We can find energy of a photon (pointing to board) (talking over, F2: Ohhh, hc/ν, I mean hc/λ) and using energy of the photon we can find the, energy (catches herself using wrong term), the wavelength of light.

In line 28, F1 is trying to work through the algebraic steps necessary to solve the problem. Although finding the energy of a photon is a correct approach (9.1), there are multiple errors in F1's explanation. First, F1 misspeaks; they are finding the kinetic energy of the ejected electron, not "of the work function." Second, F1 suggests that they need to find the wavelength of light; but they were given the wavelength of light in the problem. Hence, F1 appears to be talking backwards through some of the algebraic steps necessary to solve the problem. Finding the energy of the photon (9.3) would be the last step.

At this point, all three students have slightly wrong conceptions of the experiments used to determine that light and matter display both wave and particle characteristics and the relationship between the experiments and equations used to solve the problem (see lines 21, 24, and 28). However, the preceding analysis illustrates that the students are collaboratively working towards a deeper understanding. For example, students asked questions (lines 21 and 24), provided equations (lines 20, 21, and 28), gave explicit instructions for using (9.1–9.4) to solve the problem (line 28), and provided conceptual explanations (line 14). These students engage in collaborative discourse aimed at increasing the collective knowledge of the group.

The students continue to talk about the problem:

36	F4:	(Time: 34:41.0) So first, kinetic energy of a photon we're finding first? What are we finding first?
37	F1:	We need to find first the velocity. Unless you've found it already.
38	F4:	No I haven't found velocity yet.
39	F1:	Ok, so wavelength equals $h/(mv)$, right?
40	F2:	No, wait that's the de Broglie. That's to find the ejected electron. I'm sure, you actually use, I'm sure you probably use the E_k first.
41	F4:	We have to find hc/λ.
42	F4	The energy of a photon equals hc/λ, which we are given. 2.5×10^{-7} m. So that is…

In turn (36), F4 is still using incorrect terminology (i.e., "kinetic energy of a photon we're finding first" when actually they are finding the kinetic energy of the ejected electron). F1 is a bit confused as well; it's true that the velocity of the electron must be found, but it cannot be found until the kinetic energy of the electron is known. Although up until this point F2 has spoken slightly less than F1 and F4, she is correct that the de Broglie equation comes only after E_k is found (line 40).

43	F2:	Ohhh, they gave us the work function because we have to find out if the kinetic energy is greater than zero so that we can say that 1 electron is equal to 1 proton. Do you remember like before?
44	F1:	I know, I remember that.
45	F4:	No it's ejected.
46	F1:	Ohhh, it's ejected.
47	F4:	We do need the work function to find kinetic energy.

F2 is getting close to the conceptual underpinnings: It is true that the work function is a minimum amount of energy that the photon must transfer, "greater than zero." In the latter portion of F2's statement, she misspeaks and says "proton" instead of "photon." F1 also seems to have this conceptual understanding, and says "I know, I remember that." It is unclear whether F4 has a conceptual understanding or whether she has made an assumption that an electron is ejected and the kinetic energy is greater than the work function from the wording of the problem.

At this point, all three students use the light equation to derive the photon's energy from the photon's wavelength (9.1). While students work individually to do calculations, they talk through the steps and values they find from using each equation. F2 does not finish her calculation; F4 and F1 get the same answers, and then use (9.2), the photoelectric effect, to calculate the kinetic energy of the electron. F4 and F1 again get the same answer.

57	F1:	So now using E_k you can find...
58	F4:	E_k = one-half mass times velocity squared.
59	F1:	You can find velocity
60	F4/1:	and then you can find the (point to board) yeah (clapping) (students excited)
61	F4:	Are you with us? (looking at F2)
62	F2:	What, no, not at all.
63	F4:	Ok, we have E_k. We now know E_k (showing F2 work on her paper)
64	F2:	Yes
65	F4:	That equals (1/2) mv^2 and we can find (pointing towards board)(talking over)
66	F2:	This DOES equal velocity. You are awesome. (Laughing)

At line (60), F1 and F4 know they are nearing the solution. While the students work, they talk about the values they get from their calculations. At line (61), F4 realizes that F2 is a bit behind and checks to make sure she understands what they are doing. She does not, so F4 explains that knowing E_k, and using (9.4), the velocity of the electron can be found, and then the de Broglie equation (9.3) can be used to find the wavelength, knowing the velocity. F2, who had earlier emphasized the de Broglie equation (9.3), in line (66) realizes that what is missing from the de Broglie equation is the velocity of the electron. Lines 57–66 illustrate that the students continue to use collaborative discourse aimed at ensuring that all group members understand the problem.

At this point, the students do basic algebra to solve for velocity, using (9.4). F2 is now on board; all three students do the calculations independently and then compare answers to confirm they have the same answer.

The final step is to apply the de Broglie equation (9.3) to find the wavelength, knowing the velocity of the electron:

95	F1&4:	…then λ equals $h/(mv)$, finally.
99	F2:	105th?
100	F4:	Yep.
101	F2:	And then what?
102	F4&F2:	And then λ equals $6.626\times$…

After further calculations, they again confirm that all three have reached the same answer. While students work on their calculations, they check their answers at every step of the process. At this point, the peer leader asks them to write their work on the board, and then asks the second group to write their work. It turned out there was a minor difference in the equations used.

The above analysis shows that students are not applying the equations in a rote manner, but that they are beginning to develop an understanding of the concepts associated with the equations. Students are engaging in collaborative conversations aimed at increasing the collective knowledge of the group.

Matt's Group: Algebraic Symbol Manipulation

In contrast to the group knowledge building observed in Gillian's group, Matt's group focuses on calculations and equations with little discussion of the underlying concepts. The beginning of the students' conversation does not include any reference to, or explanation of, the underlying structure of the problem or the concepts associated with the equations.

5	M	(Time: 11:34.6) Are you doing e photon?
6	F5:	Yeah, and then figure out that (pointing to work) and put it in the work function equation (referring to the photoelectric effect equation).
7	S:	(Time: 11:46–12:05) Individual Task (students are silent)
8	F5:	I got 7.95×10^{-19}.
9	M:	Yep.
10	S:	(Time: 12:07–12:22) Individual Task (students are silent)

During this excerpt, M and F5 engage in superficial talk focusing on the correct calculation. In line 5, M is correct; the energy of a photon is calculated first; however, both students (M and F5) focus on using the equation to calculate the energy of a photon and carry out individual tasks at line (7). F5 (line 8) and M (line 9) demonstrate their agreement about the value for the energy of a photon. M's and F5's discussion consisted of a short-answer question that required recall of equations (see line 6) and non-elaborate answers (see line 8).

After working individually, students briefly exchange information gained from their calculations.

11	F5:	So we know we are going to eject an electron. (Pause)
12	M:	(Nods his head indicating yes)

F5 further demonstrates her understanding of the relationship between the energy of the photon and the work function as she says in line 11, "so we are going to eject an electron"; this relationship is captured with the photoelectric effect (9.2). F5 provides no indication of the meaning of the numbers from which she drew this conclusion.

Students engage in individual work for approximately 25 s, from 12:33 to 12:57. During this individual task work, the students are silent. Then, they discuss the next step in the problem.

17	F4:	Would you use that as the kinetic energy to get (overlapping speech)
18	F4/F5:	*v*?
19	F5:	I think so, don't you think. (Looking towards male.)
20	M:	I would think so.
21	S	(Time: 13:08–13:16) Individual Task (students are silent)

This is correct; the students are demonstrating their agreement that the next calculation necessary to solve the problem is the kinetic energy equation (9.4). F4 and F5 ask each other algebraic manipulation questions (lines 17 and 18) that do little more than require M to provide non-elaborated feedback (see line 20). Again, the brief conversation leads to students doing calculations individually in silence (see line 21). Students' discourse focuses on procedural knowledge and using equations to calculate the correct answers and they do not discuss the conceptual reasons for carrying out the mathematics in the problem.

Once the students have finished working individually, they discuss their calculations.

26	F5:	(Time: 14:16.2) Did you get $7.14 \times 1{,}012$ for the velocity? (Looking towards F4.)
27	F4:	I got 1.32×106. (Pause) Did you square root it? (Looking towards F5.)
28	F5:	Ohhh, good point. (Looking towards F4.)
29	S:	(Time: 14:28.3–15:41) Individual Task (students are silent)
30	F5:	(Time: 15:41.7) did you get 5.51×10^{-10} (referring to the de Broglie wavelength)?
31	M:	No, I think I got the velocity a little different then you did too; so that's probably the problem.
32	F4:	I got 1.32×106.
33	M:	I got 5.44×105, so that's different. (Pause) All right, let's see. (Male looks over female student's work) I didn't get the same kinetic energy.

Although in lines 17–21 the students agree that they need to calculate velocity, they all arrive at different calculations of velocity (see lines 26–33). For example in line 27, F4 had calculated a value of 1.32×106 because she used the energy of a photon (9.1) as the kinetic energy (9.4) needed to find velocity; this is incorrect. F5 also calculated an incorrect value for velocity (see line 26) and did not manipulate the equation for kinetic energy (9.4) of an electron correctly to find velocity (line 30) (see also line 38). Meanwhile, M has the correct calculation (line 33). M focuses on the errors F4 and F5 (line 33) made with their calculations. The students do not provide each other with explicit details of their procedures or what confusion formed the basis for the incorrect procedure.

Eventually, F4 notices differences in the equations they used and their calculations.

34	F4:	I didn't subtract the work function.
35	F5:	Weren't suppose to, are we?
36	M:	Well E_k is e-photon minus the work function. (Pause). I thought that kinetic energy was, (Pause) like that. (Pause) (F4 and F5 look towards male.)

F4 and F5 are uncertain if they need to subtract the work function to find kinetic energy (34 and 35). In line 36, M tells F4 and F5 the correct equation, without engaging them in a discussion of the underlying concepts.

The students focus on using the correct equation to solve the problem, instead of working to understand the relationship between the energy of a photon and the work function of a metal.

38	F5:	I thought we only used that (referring to (9.1)) to find out whether or not electrons are actually ejected and once we know that they are, wouldn't they (referring to the ejected electron) have the same energy as the photon? (Looking towards male.)
39	M:	I don't think so. What did you guys use as the ...(Talking to other group of students.)
40	F4:	I think you do have to subtract it (referring to Phi).
41	M:	For kinetic energy, did you subtract the work function? (Talking to other group.)
42	F1:	Yes. (F1 is from the other group.)
43	M:	Ok.

In line 38, F5 has an inaccurate understanding and thinks that if the energy of a photon is greater than the work function, then the kinetic energy of the ejected electron is equal to the energy of the photon. Now, she is trying to discuss the concept, but M reverts the group's focus back to the algebraic manipulation and he reconciles the differences in his group members' (F4 and F5) calculations by asking the other small group how they calculated kinetic energy (see lines 39 and 43). Although F4 agrees with M as she says "I think you have to subtract it (referring to the work function)" (see line 40), she does not elaborate, and focuses on the equation rather than the underlying concepts. F4 and F5 do not request a specific explanation; they carry out the setup proposed by M, and verified by F1 from the other group (line 42), to do the algebra necessary to solve the problem.

Now F4 and F5 are on the right track: knowing how to calculate E_k using (9.2).

50	F4:	I got... (Showing her answer to F5.)
51	F5:	Yeah.
52	M:	I multiplied that by 109 in my calculator, so I had 1.34 (referring to de Broglie wavelength). Yes, so that's (referring to de Broglie wavelength) 1.34 nm

After further calculations, all three students in the group (M, F4, and F5) confirm that they have the same answer. For approximately 2 min, F4 writes the problem on the board while students either sit quietly or talk (off task) with the peer leader.

Discussion of Group Contrasts

Gillian's group engaged in extended conversations and exhibited more collective knowledge building. The talk of Gillian's students indicated that they were thinking about the salient features of the problem, and their comments were more often made in coordination with each other, rather than independent of each other (see Gillian, lines10–28; 36–47; 57–66; 95–102). These excerpts show that Gillian's students acknowledged, built upon, and elaborated on each other's ideas when discussing the problem. Extended discourse episodes are associated with the sort of active, participatory activities that learning sciences research shows contributes to deeper conceptual understanding, greater transferability of knowledge, and better retention (Engle & Conant, 2002; Greeno, 2006; Sawyer, 2006; Scardamalia & Bereiter, 2006). Additionally, her students' explanations went beyond algebraic manipulations and began to address the underlying concepts (see Gillian, lines 14; 43; 45–46). A critical component of effective knowledge building is that it supports and facilitates student collaboration as students engage in explaining, clarifying, and debating their ideas (Hiebert et al., 1996; Yackel, Cobb, & Wood, 1991). As a result of their discourse, students collectively improved their ideas through active, intellectual discussions.

Gillian's students used managerial/structure statements and refocusing statements that were directed at collaboration and learning processes. In Gillian's group, students were active participants in making sure everyone understood the process necessary to solve the problem, and all students made intellectual contributions (see Gillian, lines 61–65).

Gillian's group displays the sort of group knowledge building discourse that is currently advocated by the science education research community. Their collaborative problem-solving conversations provided support for, and challenged, individuals' thinking. Over time, the students' ideas became more coherent and elaborated. Two discursive moves made by the students—elaborating on each other's ideas, and self-monitoring the group's understanding of the content—enabled knowledge-building discourse. In these ways, Gillian's students engaged in what (Lave & Wenger, 1991) called a "community of practice," where the goal was to support both the growth of individual cognitive advancement and the collective knowledge of the group. Many education researchers have stressed that once a collaborative group culture has emerged, it can motivate and engage students in knowledge building and in constructing understandings that support integration and application of the content (Blumenfeld, Kempler, & Krajcik, 2006; Engle & Conant, 2002; Greeno, 2006; Sawyer, 2006; Scardamalia & Bereiter, 2006).

Matt's students exhibited shorter discourse episodes, frequently engaged in individual tasks, and mostly provided each other with algebraic manipulations that did not deal with the underlying concepts. Our analyses show that the questions asked by Matt's students were often task-oriented and used to coordinate the group's interactions (see Matt, lines 5–10; 17–21; 26–33; 34–35; 38–43) in preparation for individual tasks (see Matt, lines 7; 10; 21; 29). The lack of explicit focus on the important features of the problem and underlying concepts led students to a false sense of competence. During one episode, although all three students discussed and agreed upon the equations to use, they all arrived at different calculations. Our analysis revealed that the students had, in fact, not all used the same equation (see Matt lines: 26–33). Webb (1995) suggests that asking closed questions and providing non-elaborated help involves less cognitive restructuring or clarifying on the part of help-givers and does not enable help-receivers to correct their misconceptions or lack of understanding.

Not only did Matt's students focus on getting the correct answer, but the participatory structure was unequal. Students in Matt's group held different positions according to their perceived competence. For example, the male consistently initiated ideas and validated his peers' conceptions (see Matt, lines 5–9; 11–12; 17–20; 26–33; 34–36; 38–43; 50–52). It appears from the transcripts that students in Matt's group adopted an unequal participatory structure where the male in the group was viewed as an authority.

In summary, Matt's group was ineffective at promoting group knowledge-building discourse; students focused instead on individually attempting to understand the content. Neither the students nor the peer leader encouraged in-depth conversations of the underlying concepts associated with the problem. The excerpts demonstrate that students in Matt's group showed little evidence of building upon, debating, and elaborating upon each other's ideas.

Conclusion

There is a great deal of research evidence that students who participate in PLTL acquire higher levels of chemistry understanding than students who learn individually and alone. However, no studies have looked inside the "black box" of the PLTL session to examine exactly how peer discourse contributes to chemistry understanding. This study has shown that not all peer group experiences are equally effective at promoting student knowledge building. First, we investigated how students' contributions were responsive to those from other students. We found that in Gillian's group, students engaged in intellectual conversations where they asked each other questions, provided procedural and conceptual explanations, and checked each other's understanding of the problem. Even while Gillian's students worked on individual tasks, they constantly talked about their calculations with each other. In contrast, the discourse of Matt's students rarely included any reference to, or explanation of, the equations or underlying concepts. The conversations in Matt's group were mostly superficial; students provided each other with equations and non-elaborated explanations. In Matt's group students' discourse focused on the algebraic steps necessary to solve the problem.

We found that Gillian's students frequently elaborated on each other's ideas and self-monitored the groups' understanding of the problem in part I(*B*). Even though at times Gillian's students had slightly incorrect conceptions, as a result of their discussions they collectively developed a better understanding of the problem.

Conversely, in Matt's group, there was little evidence that the group jointly developed a more in-depth understanding of the content from their discourse; rather, their conversations focused on rote application of formulas and calculating a "correct" solution.

Implications for Practice and Research

Examining student discourse also has implications for the redesign of PLTL problems. Although the problem was basically a closed question, with a single correct answer, we expected a different type of peer leader and student discourse than we observed. First, we thought that students would discuss the underlying concepts associated with the problem based on their experiences in lecture and recitation sessions. Second, when the students failed to engage in conceptual discussions on their own, we thought that the peer leader would challenge students to verbally explore the concepts involved with the problem. Based on the current study, we plan to redesign many of our PLTL problems. We recommend that chemistry instructors design PLTL problems to begin with guiding questions that allow for students to discuss key concepts and experiments in addition to equations and variables. The purposes of these guiding questions are to unveil students' prior knowledge and review the phenomena discussed in lectures, recitations, and other problems. Once students have identified the important ideas associated with the phenomena, the problem set should provide students with the opportunity to apply equations and concepts to other contexts. Altering existing problem sets to provide explicit questions that have students discuss phenomena before problem solving may engage students in higher-order thinking and alter students' interactions with each other and the peer leader. Redesigning the problem sets could provide an even more active environment for students to engage in science discourse and further improve students' conceptual understanding of the problems they solve in PLTL. Hence, future research is needed that investigates whether revising PLTL problems in this way does in fact foster the type of conversations that lead to deep conceptual understanding. Restructuring the problems could favorably affect student's chemistry understanding, critical thinking, and knowledge building from collaborative discourse.

References

Bakeman, R., & Brownlee, J. R. (1980). The strategic use of parallel play: A sequential analysis. *Child Development, 51*, 873–878.

Blumenfeld, P. C., Kempler, T. M., & Krajcik, J. S. (2006). Motivation and cognitive engagement in learning environments. In R. K. Sawyer (Ed.), *The Cambridge handbook of the learning sciences*. New York, NY: Cambridge University Press.

Brown, P., Sawyer, R. K., & Frey, R. (2010, June 29–July 2). *What are they talking about? Findings from an analysis of the discourse in peer-led team learning in general chemistry*. Paper presented at the International Conference of the Learning Sciences, Chicago, IL.

Carlsen, W. S. (1993). Teacher knowledge and discourse control: Quantitative evidence from novice biology teachers' classrooms. *Journal of Research in Science Teaching, 30*(5), 471–481.

Crawford, T. (2005). What counts as knowing: Constructing a communicative repertoire for student demonstration of knowledge in science. *Journal of Research in Science Teaching, 42*(2), 139–165.

Engle, R. A., & Conant, F. R. (2002). Guiding principles for fostering productive disciplinary engagement: Explaining an emergent argument in a community of learners classroom. *Cognition & Instruction, 20*(4), 399.

Glaser, B. G., & Strauss, A. L. (1967). *A discovery of grounded theory: Strategies for qualitative research*. Chicago, IL: Aldine.

Greeno, J. G. (2006). Learning in Activity. In R. K. Sawyer (Ed.), *The Cambridge handbook of the learning sciences* (pp. 79–96). New York, NY: Cambridge University Press.

Hanrahan, M. U. (2005). Highlighting hybridity: A critical discourse analysis of teacher talk in science classrooms. *Science Education, 90*(1), 8–43.

Hiebert, J., Carpenter, T. P., Fennema, E., Fuson, K., Human, P., Murray, H., et al. (1996). Problem solving as a basis for reform in curriculum and instruction: The case of mathematics. *Educational Researcher, 25*(4), 12–21.

Kelly, G. J., Brown, C., & Crawford, T. (2000). Experiments, contingencies, and curriculum: Providing opportunities for learning through improvisation in science teaching. *Science Education, 84*(5), 624–657.

Klaassen, C. W. J. M., & Lijnse, P. L. (1996). Interpreting students' and teachers' discourse in science classes: An underestimated problem? *Journal of Research in Science Teaching, 33*(2), 115–134.

Lave, J., & Wenger, E. (1991). *Situated learning: Legitimate peripheral participation*. New York, NY: Cambridge University Press.

Lunn, M. (1998). Applying simple k-sample tests to conditional probabilities for competition risks in a clinical trial. *Biometrics, 54*, 1662–1672.

Sawyer, R. K. (2005). *Social emergence: Societies as complex systems*. Cambridge: Cambridge University Press.

Sawyer, R. K. (2006). Analyzing collaborative discourse. In K. R. Sawyer (Ed.), *The Cambridge handbook of the learning science* (pp. 187–204). New York, NY: Cambridge University Press.

Scardamalia, M., & Bereiter, C. (2006). Knowledge building: Theory, pedagogy, and technology. In R. K. Sawyer (Ed.), *Cambridge handbook of the learning sciences* (pp. 97–118). New York, NY: Cambridge University Press.

Sfard, A. (1991). On the dual nature of mathematical conceptions: Reflections on processes and objects as different sides of the same coin. *Educational Studies in Mathematics, 22*, 1–36.

van Zee, E., Iwasyk, M., Kurose, A., Simpson, D., & Wild, J. (2001). Student and teacher questioning during conversations about science. *Journal of Research in Science Teaching, 38*(2), 159–190.

van Zee, E., & Minstrell, J. (1997). Using questioning to guide student thinking. *Journal of the Learning Sciences, 6*(2), 227–269.

Webb, N. M. (1995). Group collaboration in assessment: Multiple objectives, processes, and outcomes. *Educational Evaluation and Policy Analysis, 17*(2), 239–261.

Yackel, E., Cobb, P., & Wood, T. (1991). Small-group interactions as a source of learning opportunities in second-grade mathematics. *Journal of Research in Mathematics Education, 22*, 390–408.

Chapter 11
A Multivocal Process Analysis of Social Positioning in Study Groups

Iris Howley, Elijah Mayfield, Carolyn Penstein Rosé, and Jan-Willem Strijbos

Introduction

One advantage as well as challenge of multivocal approaches to analysis of collaborative learning interactions is that it reveals the ways in which our individual operationalizations of complex constructs are limited. In bringing together analyses from multiple perspectives addressing similar issues with the same dataset, our eyes are opened to the richness and complexity of how these constructs are manifest in language. In this chapter, we compare two multidimensional approaches to assessing collaborative learning processes, which are based on a similar theoretical foundation and sound superficially similar. However, when a line by line comparison is made between the specific codings, we find interesting differences that serve to highlight subtle nuances in the operationalization of these theories. We are left with a deeper appreciation for the difficulty of our task as analysts to capture the intricacies of the ways in which collaborative processes are displayed through the language that we see.

The scope of the analytical work we present in this chapter is defined by our theoretical assumptions regarding formative assessment of collaborative learning interactions (Howley, Mayfield, & Rosé, 2013; Strijbos, 2011). Specifically, we assume that collaborative learning processes are an integration of three orthogonal dimensions, namely, cognitive, relational, and motivational. Furthermore, we assume that each dimension can be operationalized as a set of mutually exclusive codes, each of which is defined at the level of an individual contribution to a conversation. Thus, assessments within each dimension are performed by analyzing sequences or distributions of these codes. In this chapter, we focus specifically on distributional analyses. Overall, the purpose of the analysis could be considered broadly to be that

I. Howley (✉) • E. Mayfield • C.P. Rosé
Carnegie Mellon University, Pittsburgh, PA, USA
e-mail: ihowley@cs.cmu.edu

J.-W. Strijbos
Ludwig-Maximilians-Univerity Munich, Munich, Germany

D.D. Suthers et al. (eds.), *Productive Multivocality in the Analysis of Group Interactions*, Computer-Supported Collaborative Learning Series 16,
DOI 10.1007/978-1-4614-8960-3_11,

of identifying the discourse contributions that support or hamper the unfolding collaborative learning process. Automatic assessment of collaborative learning in real time using such approaches can be used to trigger support for collaborative learning in a context sensitive way, for example conversational agents triggered by detection of an attempt at an explanation that prompts other group members to respond with their evaluation (see (Kumar & Rosé, 2011) for a review of such context sensitive support techniques). Formative assessment of collaborative learning processes can also be used to measure progress within iterative development processes for design of CSCL environments or for supporting the facilitation efforts of instructors who work with collaborative groups.

In the remainder of this chapter we offer an overview of two multidimensional frameworks for assessment of collaborative learning interactions, each of which can be thought of as including separate dimensions for exploring group processes from the perspective of three core dimensions: cognitive, relational, and motivational (Strijbos, 2011). Along with each of those frameworks we offer a high level assessment of both group discussions included in the PTL Chemistry dataset (Sawyer, Frey, and Brown, Chap. 9, this volume).

Despite the similarity in conceptualization of the two multidimensional frameworks, we find that they make different predictions within one of the groups where we explore social positioning as it is negotiated through the style of information presentation. While our codes are assigned at the contribution level, we can consider that our analysis is used to do an assessment for each student in a group within the span of time it takes the group to solve one chemistry problem together. Important differences are those that emerge through comparisons across distributions of codes from the two analysis frameworks per student along single dimensions within problems. Thus the basic unit of interaction is the student within a group per problem, because each group in our data only worked on one problem. The coding schemes we use can be thought of as our analytic representations. They impose a structure on the stream of conversational contributions that we analyze. The subtlety in our analytic work comes in the operationalization of those codes. We do not employ any sophisticated transformations of our codings beyond simple statistical comparisons. Using a distributional approach, we are able to characterize behavior of students within a problem solving session in terms of what was most typical for them during that interaction. Within this approach, we can define a pivotal moment for a student as a moment in which that student diverges from this typical behavior and does something that is atypical.

The Souflé Framework

Howley et al. (2013) first introduced the Souflé framework as a linguistic analysis approach for studying small groups. The intention was to define the codes at the level of basic language processes without reference to theoretical constructs that are specific to a particular theory of learning or collaboration. More specifically, the aim

Table 11.1 Codes, definitions, and examples for the cognitive dimension

Transactivity	Code	Definition	Example
Not reasoning	No	A reasoning display includes a causal mechanism, rationale, interpretation, or abstraction. If the contribution is missing this, then it fits in this category	"You are doing e photon."
Externalization	Ext	A reasoning statement that introduces an novel idea into the conversation that does not build on or comment on an earlier reasoning statement	"Using kinetic energy allows you to compute V."
Transactive	Trans	A reasoning statement that builds on or comments on a previously articulated reasoning statement	"You got that answer because you didn't add correctly."

was to provide a neutral way of describing collaborative processes that might serve as a boundary object for researchers from different theoretical perspectives. Here we define its Cognitive, Relational, and Motivational dimensions in turn.

The Cognitive Dimension

The Cognitive dimension of Souflé is distinct from the other two in that its definition is not strictly linguistic. However, the values underlying the construct of transactivity (Berkowitz & Gibbs, 1979) are not controversial. The simple idea behind the concept of transactivity is a value placed on making reasoning explicit and elaborating expressed reasoning by building on or evaluating instances of expressed reasoning that came earlier in the discussion. In our prior work, we have developed and applied machine learning techniques for automatic analysis of transactivity in discussion forums (Rosé et al., 2008), chat transcripts (Joshi & Rosé, 2007), transcribed group discussions (Ai, Sionti, Wang, & Rosé, 2010), and speech recordings of dyadic discussions (Gweon, Jain, McDonogh, Raj, & Rosé, 2012) (Table 11.1).

The unit of analysis we have adopted in Souflé was first established for analysis of transactivity. In particular, one unit is the minimal amount of text required to express reasoning. Our formulation of what counts as a reasoning display comes from the Weinberger and Fischer (2006) notion of what counts as an "epistemic unit," where what they look for is a connection between some detail from the given task (which in their case is the object of the case study analyses their students are producing in their studies) with a theoretical concept (which comes from the attribution theory framework, which the students are applying to the case studies). When they have seen enough text that they can see in it mention of a case study detail, a theoretical concept, and a connection between the two, they place a segment boundary. Occasionally, a detail from a case study is described, but not in connection with a theoretical concept. Or a theoretical concept may be mentioned, but not tied to a case study detail. In these cases, the units of text are considered degenerate, not quite counting as an epistemic unit. In our coding of the PLTL corpus, degenerate contributions are coded as "no" for "not reasoning." Note that degenerate does not necessarily imply contentless or unimportant. For example, questions such as "Are you doing e photon?" do not count as reasoning displays, but they nevertheless serve an important function within the collaboration.

The simple way of thinking about what constitutes a reasoning display is that it has to communicate an expression of some causal mechanism or express an evaluation or comparison. Often that will come in the form of an explanation, such as X because Y. However, it can be more subtle than that, for example "Increasing the tension makes the spring springier." The basic premise was that a reasoning statement should reflect the process of drawing an inference or conclusion through the use of reason. Note that in the example with the spring, although there is no "because" clause, one could rephrase this in the following way, which does contain a "because" clause: "The spring will be springier because we will increase the tension." Reasoning statements stand in contrast to mere information sharing statements, which can be thought of as sharing of rote knowledge. An example of a reasoning display from the PLTL corpus is "and then you use the kinetic energy to get V." Because face-to-face discussion frequently leaves much implicit, we do not require all portions of the reasoning display to be articulated if the context makes the full articulation of the expressed reasoning clear. Because of this, even some questions can count as reasoning displays. For example, "and then did you subtract the work function?" counts as a reasoning display in that it is an expression of a student checking that her understanding was correct about how another student just derived a recently articulated result.

In our work, we have needed to adjust our specific definition of what counts as a reasoning display each time we have applied our transactivity coding scheme to a new domain (Gweon et al., 2012). In each case, however, the thinking behind the operationalization was the same. In particular, when students are working on a given task or a project in a team, they typically receive a certain amount of information that would help them solve the problem, in the form of a task statement and training materials. In order to solve the given problem, students discuss the materials that are given to them and try to apply them to a potential solution. These shared materials provide a frame of reference for anchoring our definition of a reasoning display. The displayed reasoning that we are interested in capturing is what goes beyond what is given and displays some understanding of a causal mechanism. Typically some causal mechanism would be referenced in a discussion of how something works or why something is the way it is. It is important to note that what we are coding is *attempts* at displayed reasoning. Thus, we need to allow for displays of incorrect, incomplete, and incoherent reasoning to count as reasoning, as long as in our judgment we can believe an attempt at reasoning was made. That will necessarily be quite subjective—especially in the case of incoherent explanations.

Statements that display reasoning can be coded as either Externalizations, which represent a new direction in the conversation, not building on prior contributions, or Transactive contributions, which operate on or build on prior contributions. In our distinction between Externalizations and Transactive contributions, we have attempted to take an intuitive approach by determining whether a contribution refers linguistically in some way to a prior statement, such as through the use of a pronoun or deictic expression. In the PLTL corpus, we defined transactive contributions as those reasoning displays that were positioned as contingent on at least one earlier expression. For example, one student reported, "I didn't get the same kinetic energy." Then another student responded, "I didn't subtract the work function." The response is an explanation for the difference between the answers obtained by the two students. Thus, that expression of reasoning (i.e., explanation for the difference) was also an evaluation of the other student's approach. Its contingent relationship with the prior utterance makes it transactive. The definition of transactive employed in the analysis of the PLTL corpus might be seen as reading a lot into the contributions of students that goes beyond what is literally found in the text. However, the terse nature of the majority of student contributions in the discussions necessitated such an approach in order for the distributions of codes on this dimension to exhibit a nontrivial amount of variance between students.

The Relational Dimension

The Relational dimension in Souflé is meant to capture the level of openness to the ideas of others that is communicated in a student's framing of assertions. Whereas in the Cognitive dimension we adopted an approach in which we read into the text in order to identify expressions of reasoning and transactivity, in the Relational dimension, we base our work on the earlier work of Martin and White (2005), whose theoretical approach explicitly mandates not going beyond the evidence that is explicit in a text.

The important distinction in our application of the Martin and White's Heteroglossia framework is the distinction between a monoglossic assertion that is framed as though it leaves no room for questioning, in contrast to those framed in a heteroglossic manner, where the assumed perspective of others is explicitly acknowledged within the framing. For example, whereas "For e photon it would be 6.6×10^{-19}." is monoglossic, "I would say that for e photon it would be 6.6×10^{-19}." would be heteroglossic.

The specifics of the definition for the heteroglossic versus monoglossic distinction are adapted from Martin and White's (2005) original discussion. First, some propositional content must be being asserted in some form, although it may be done in such a way as to communicate extreme uncertainty. Thus, questions that are framed in such a way as the reader believes the speaker was asking an honest question, for which no specific answer seems to be supposed do not count as heteroglossic. Interjections, like "Yay," that cannot be interpreted as ellipsis, and thus have no propositional content are not considered heteroglossic. However, fixed expressions like "no," and "yes" that implicitly assert the propositional content of the yes/no question they are a response to do count as expressing propositional content. Other forms of ellipsis (e.g., "13.4 angstroms" in response to "What did you get?") and

Table 11.2 Codes, definitions, and examples for the relational dimension

Heteroglossia	Code	Definition	Example
No assertion	NA	A contribution in which no claim is being made	"What is the value of e?" "I don't know" "wow"
Monoglossic	Mon	A bald assertion that is made unequivocally	"Now you multiply by e."
Heteroglossic expand	Het-E	An assertion that is offered as one option, up for discussion	"I think multiplying by e sounds reasonable." "Multiplying by e might work."
Heteroglossic contract	Het-C	An assertion that is offered in such a way that options are eliminated from consideration	"Multiplying by e is the only way it can work." "Multiplying by anything other than e won't work."

do-anaphora (i.e., "I did." In response to "Did you also get 13.4 angstroms?") also count as having propositional content. Second, an awareness must be made visible to the presence of alternative perspectives than that represented by the propositional content of an utterance. Thus, bald claims, even if they are biased, do not acknowledge alternative perspectives. For example, "13.4 angstroms is the obviously the answer." May be subjective, but it is not heteroglossic. It does not show any awareness that someone else might disagree. If a speaker goes on to give reasons to defend the statement, however, then that speaker is showing awareness of other perspectives. These cases will be caught by the third requirement. Third, in order to count as heteroglossic, the acknowledgement of other perspectives must be expressed grammatically (e.g., through a model auxiliary like "might") or paraphrastically (e.g., "I think") within the articulation of that propositional content. If it is implicit or signaled through the discourse structure, then that is not enough to count as heteroglossic in the Martin and White sense as represented by their Engagement system, although they would acknowledge it as heteroglossic "in spirit."

There are two types of contributions we code as Heteroglossic, one type that shows openness to other perspectives, which we refer to as Heteroglossic Expand, and another that explicitly expresses a rejection of some other perspective, which we refer to as Heteroglossic Contract (Table 11.2).

The Motivational Dimension

The Motivational dimension in Souflé is meant to capture conversational behavior that reflects the self-efficacy of students related to their ability to participate meaningfully in the collaborative learning interaction. In our prior work we have seen correlations between self-report measures of collective self-efficacy from collaborative groups and measures of authoritativeness of stance derived from our coding in this dimension (Howley et al., 2012). In short, on this dimension we consider that an authoritative presentation of knowledge is one that is presented without seeking external validation for that knowledge.

This dimension, which we have referred to as the Authoritativeness Framework, is rooted in Martin's Negotiation Framework (Martin & Rose, 2003), from the systemic functional linguistics community. This framework highlights the moves that are made in a dialogue as they reflect the authoritativeness with which those moves were made, and gives structure to exchanges back and forth between participants. Previous work has studied the complexity of, for instance, the difference between authority to alter the direction of a conversation and authority to contribute new information to a conversation. We are interested in this framework because of its descriptiveness for social interactions, and how it boils down the intricacies of power management within an interaction to a few simple codes, making it easy to track shifts in positioning over time. An application of this framework to analysis of social shift in response to bullying in computer supported collaborative learning offers an example of that use (Howley et al., 2013).

While the Negotiation framework as formulated by Martin is highly descriptive for sociolinguists, and has been widely used by Martin himself as well as by other sociolinguistics, it is difficult to replicate reliably from the previously published formulations, as this was not a methodological goal of the original researchers. This makes its immediate use for quantitative analysis difficult without introducing threats to internal validity. To remedy this, we have worked iteratively on a coding manual that incorporates the insights from that framework that are relevant to our task and makes them precise and concrete enough to be reproducible. Our inter-rater agreement for this coding has achieved a Cohen's Kappa of 0.78. A full treatment of the details of our development process is beyond the scope of this chapter. However, we would like to acknowledge that we have developed this Authoritativeness Framework through consultation with experts from a variety of backgrounds (socio-cultural researchers, education researchers, sociolinguists, computational linguists, computer scientists, interaction analysts, learning scientists, etc.). As in our work on transactivity, we have had success with automating our analysis of authoritativeness with high reliability in transcriptions of face to face interactions (Mayfield & Rosé, 2011) as well as chat transcripts (Howley et al., 2012).

Our formulation of the Authoritativeness framework is comprised of two axes with six and three codes, respectively, and incorporates structural and pragmatic knowledge of language. To simplify our analysis for this chapter, we will focus on two moves in particular. The first is K1, or "primary knower," and the second is K2, or "secondary knower." A "primary knower" move includes a statement of fact, an opinion, or an answer to a factual question, such as "yes" or "no." It only counts as "primary knower" if it is not presented in such a way as to elicit an evaluation from another participant in the discussion. Conversely, a "secondary knower" move includes statements where the speaker is not positioned as authoritative on the topic at hand, such as asking a question eliciting information, or presenting information in a context where evaluation is the expected response or formulated to elicit feedback (Table 11.3).

Table 11.3 Codes, definitions, and examples for the motivational dimension

Core moves	*Code*	*Definition*	*Example*
Primary knower	K1	A contribution that provides information to another participant. It shows assertiveness in that it seeks no ratification from another person	"Your answer is wrong." "It's an electron."
Secondary knower	K2	A contribution that indicates a need for someone else to provide information or ratification	"What is the value of e?" "I'm not sure I have the value of e correct." "It's e, right?"
Primary actor	A1	A contribution that marks the speaker as a source of action	"I'm on it!" "I got the answer."
Secondary actor	A2	A contribution that marks the speaker as needing someone else to do some action on that person's behalf	"I need help." "Can you compute the value of e?"
Preparatory and follow-up moves	*Code*	*Definition*	*Example*
Challenge	Ch	A contribution that marks a previous contribution as not licensed in the context	"You're not making any sense." "You're assuming that light is a wave rather than a particle."
Other	O	Any other preparatory or follow up move	"Wow" "Can I tell you something?"

There is no strict form-function relationship between these codes and the text being analyzed. The simplest example of this is a line such as "yeah," which could be authoritative in response to a question or could be non-authoritative response to someone else's evaluation. Additionally, factual statements where the speaker is uncertain of their correctness and is looking for approval from a listener would be coded as a K2 move, even though it is structurally similar to most K1 moves. The roles that speakers take through these codes can shift rapidly within a conversation, and are dynamic, being heavily based on the context of what has happened leading up to an utterance, and how that utterance is responded to by other participants. Figure 11.1 displays an excerpt from the PLTL corpus where three students are discussing an intermediate result within the scope of their problem solving session. Here we see that M's contributions and F5's contributions are both entirely framed as authoritative, but F4 is positioned, both by her contribution, and by M's response, as non-authoritative.

Souflé Analysis

Now we apply the Souflé analysis to the two separate PLTL groups and interpret the distribution of codes, at the group level and at the individual level. These are the Gillian (G-group) and Matt (M-group) groups referred to throughout this section of the book.

Speaker	Contribution	Authority
M	No, I think I got the velocity a little different then you did too, so that's probably the problem.	K1
F5	I got 1.32 x 10^6.	k1
M	I got 5.44 x10^5 so that's different.	K1
M	Alright let's see. I didn't get the same kinetic energy.	k1
F5	I didn't subtract the work function.	k1
F4	Weren't suppose to are we?	k2
M	Well Ek is e-photon minus the work function. I thought that kinetic energy was, like that.	K1

Fig. 11.1 An example analysis using Martin and Rose's (2003) Negotiation system, labeled as Authority

Table 11.4 Frequency and proportion for occurrences of each code in each of the three dimensions for both groups

G-group (*N* segments = 105)									M-group (*N* segments = 49)								
Cognitive			Relational			Motivational			Cognitive			Relational			Motivational		
	f	%		*F*	%		*f*	%		*f*	%		*f*	%		*f*	%
No	80	76	Na	34	32	K1	50	47	No	27	55	Na	13	27	K1	27	55
Ext	15	14	Mon	40	38	K2	16	15	Ext	10	20	Mon	15	30	K2	12	24
Trans	10	10	HetE	21	20	A1	3	3	Trans	12	25	HetE	21	43	A1	2	4
			HetC	10	10	A2	5	5				HetC	0	0	A2	4	8
						Ch	4	4							Ch	0	0
						O	27	26							O	4	8

Group Level Analyses (Table 11.4)

Based on the distribution of codes on each dimension for each group, we are able to see that the two groups behave quite differently. For example, on the cognitive dimension, only 24 % of the contributions in the G-group are some form of reasoning display, whereas in the M-group, 45 % fall into this category. Furthermore, while in both groups the split between Externalizations and Transactive contributions is not very skewed, in the G-group, more than half of the reasoning displays are Externalizations, whereas in the M-group more than half are Transactive. Thus, the M group is not only displaying more reasoning, but is also engaging in building complex reasoning more frequently. In both groups, roughly 70 % of the contributions are some form of assertion on the Relational dimension, but the distribution across the types of assertions was different in the two groups. In the G-group, the majority of assertions were coded as Monoglossic, whereas in the M-group, the majority of assertions were Heteroglossic Expand. And only the G-group had any heteroglossic contract assertions of either type. Thus, we see a more open attitude communicated in the G-group. In both groups, we see a similar distribution of codes on the Motivational dimension.

Individual Level Analysis

For the next phase of the analysis, we examine more closely the inner workings of teams at the student level.

G-Group

The G-group is composed of five women. Two of them each only contribute once in the discussion. So we focus our analysis on the other three participants.

- *Cognitive dimension*: Most of the reasoning contributions come from two of the three participants, which we refer to as F1 and F4. F1 has twice as many Externalizations as Transactive contributions, whereas F4 has equal numbers of both. In contrast, F2 contributes about 25 % as many reasoning contributions as the other two. Thus, we see F2 as less engaged in the active reasoning process. And F1 may be an idea leader.
- *Relational dimension*: On the Relational dimension, we also see a contrast between F1 and F4 on the one hand and F2 on the other. For F1 and F4, the dominant code on this dimension is Monoglossic, even more so for F4 than F1, whereas for F2 it is no assertion. Furthermore, F4 is roughly balanced between Heteroglossic Expand and Contract, whereas the other two have twice as many Heteroglossic Expand as Contract. On this dimension we see F4 start to distinguish herself as a little more of a firm leader than F1, whereas F1 appears to be more of a supportive and open leader. In combination with the findings at the cognitive level, we can see F1 as an idea leader who does not push her own view, but places her ideas on the table for discussion.
- *Motivational dimension*: The Motivational dimension echoes the view of F4 as a more dominant leader in that F4 has the highest ration of K1 to K1 + K2 contributions of the three participants, and F2 has the lowest.

M-Group

The M-group has three members, one of which is male, referred to as M, and two of which are female, referred to as F4 and F5. Roles are much less pronounced in the M-group than in the G-group.

- *Cognitive dimension*: In the Cognitive dimension, the three participants utter close to the same percentage of reasoning display contributions. F5 is the lowest at 41 %, whereas F4 is the highest at 55 %. However, whereas F5 has a lower percentage of reasoning displays overall, her percentage of Transactive contributions is roughly twice that of each of the other two (i.e., 41 % in contrast to 22 % for F4 and 21 % for M). This could be seen as a way in which F5 places herself in less of a leadership position by recognizing the idea leadership of others.

- *Relational dimension*: There is little distinction between participants on the Relational dimension. In all cases, Heteroglossic Expand is the dominant category.
- *Motivational dimension*: There was little distinction between participants on the Motivational dimension, however, complementary to the finding on the Cognitive dimension, we see F4 as slightly more authoritative than the other two, with a K1 to K1 + K2 ratio of 86 % in comparison for 75 % for each of the other two. Thus we see a slight pattern where F4 is taking more leadership, and F5 is taking a little more of a follower role.

When comparing the two groups, we see a stronger leadership pattern in the G-group, with less openness, less reasoning and fewer contributions, whereas we see a richer and longer discussion in the M-group, with a greater level of equality between participants, more reasoning displays, more openness, and more collaborative knowledge building.

CRM Coding Scheme

Reviewing the literature on (CS)CL, it is apparent that cognitive outcomes are central to the assessment of learning in past and present (CS)CL studies, however, cognitive outcomes are not the only outcomes of collaborative learning. Slavin (1996) already identified three major perspectives in cooperative learning research—the motivational, social (cohesion), and cognitive—and stated that they "…may be seen as complementary, not contradictory" (p. 52) and that there are many other outcomes like "…intergroup relations, self-esteem, acceptance of mainstreamed classmates, pro-social norms, and so on" (p. 64). Social (cohesion) aspects, such as intergroup relations, are typically emphasized in the "Learning Together" approach (Johnson & Johnson, 1994) and the "Group Investigation" approach (Sharan & Sharan, 1992). In the context of Group Investigation, there also appear to be positive effects in relation to aspects commonly associated with intrinsic motivation, such as interest, enjoyment, and (mutual) encouragement (Ryan & Deci, 2000).

The CRM coding scheme is a tentative conceptualization of the "Group Experience" (GE) metaphor applied to the analysis of collaborative interaction. The GE metaphor contends that (a) both the individual and group level should be analyzed, (b) that the collaborative interaction cannot solely be reduced to the cognitive plane, and (c) that concurrent strands of experience exist, that is, cognitive, relational and motivational processes are affected differentially within as well as between individuals (Strijbos, 2011). The present coding scheme draws from various coding schemes and coding dimensions in previous and forthcoming publications.

Table 11.5 Codes, definitions and examples for the cognitive dimension

Verification	*Code*	*Definition*	*Example*
Positive	VP	A positive verification of a previous statement or calculation outcome	"Very true, very true."
Negative	VN	A negative verification of a previous statement or calculation outcome	"I got 5.44×10^5 so that's different."
Elaboration	*Code*	*Definition*	*Example*
Question	EQ	Any type of task-related (problem-solving) question	"Wouldn't they have the same energy as the photon?"
Correction	EC	A previous statement by a fellow group member (or one-self) is corrected	"I didn't subtract the work function."
Affirmation	EA	A previous statement by a fellow group member (or one-self) is affirmed	"That's what I got too (laughing)"
Suggestion	ES	A suggestion for an approach to handle or solve the problem at hand	"I think we need to put that in angstroms."
Justification	EJ	An additional argument to support a prior verification or suggestion	"Because the important thing that for the de Broglie wavelength we are finding the wavelength of the electron not of light."

The Cognitive Dimension

The Cognitive dimension focuses on the content plane. Within the context of this chapter, the Cognitive dimension is operationalized in terms of feedback processes. Widely investigated types of feedback are (a) simple feedback types providing outcome-related information, and (b) elaborated feedback types providing additional information besides outcome-related information. Narciss (2008) developed a content-related classification providing a structured overview of simple and elaborated components. Simple feedback components evaluate the performance level achieved—i.e., knowledge of performance, knowledge of result, and knowledge of the correct response (also referred to as the "verification" component; Kulhavy & Stock, 1989). An elaborated feedback component (also referred to as the "informational" component) depends on the elaborated information provided, which might address: (a) knowledge on task constraints, (b) knowledge about concepts, (c) knowledge about mistakes, (d) knowledge on how to proceed, and (e) knowledge on meta-cognition.

The present codes for the Cognitive dimension were developed as part of a study on peer feedback (Strijbos, Van Goozen, & Prins, 2012). The codes are a subset of that coding scheme. More specifically, the Verification and Elaboration codes. Krippendorff's alpha for the entire coding scheme (24 categories) was 0.67 and 0.73 for the Verification and Elaboration distinction. Table 11.5 provides an overview of the Cognitive dimension codes, their definitions and examples from the present dataset.

Table 11.6 Codes, definitions and examples for the relational dimension

Orientation	*Code*	*Definition*	*Example*
Collaborative	CL	A statement reflecting a collaborative orientation to the group process	"That's what I got too (laughing)"
Individual	IN	A statement reflecting an individual orientation to the group process	"I got 1.32×10^6."
Dominance	*Code*	*Definition*	*Example*
Positive	DP	Positive dominance is aimed at including other group member, for example tutoring behavior	"Are you with us?"
Negative	DN	Negative dominance closes the floor to further discussion, for example blocking behavior	"No it's ejected. We do need the work function to find kinetic energy."

The Relational Dimension

The Relational (or social) dimension and its associated outcomes are considered to a certain degree in recent literature, e.g., Gillies (2007), Strijbos and Stahl (2007), and Sarmiento and Shumar (2010). A recent study by Tolmie, Kenneth, Topping, Christie, Donaldson, Howe, Jessiman, Livingston, and Thurston (2010) investigated social effects of collaborative learning among 575 primary schools students (aged 9–12) and revealed that (a) collaborative learning leads to a dual impact in terms of cognitive and social gains, (b) collaborative skills improve alongside understanding and optimal social relations need not be in place prior to collaboration, (c) social context (rural versus urban schools) did not affect cognitive or social gains; rather the engagement in collaborative learning raises both cognitive and social gains counteracting prior social differences, and (d) the convergence over time between transactive dialogue and collaborative skills (in terms of work relations) suggests that "…cognitive and social gains would appear to be interlinked, if distinguishable, outcomes" (p. 188). In the context of (CS)CL, however, social interaction is still often taken for granted, or restricted to cognitive processes (Kreijns, Kirschner, & Jochems, 2003).

The present codes for the relational dimension were inspired from the coding scheme by Kumpulainen and Mutanen (1999) and more particularly the "social processing" dimension, consisting of six codes: collaborative, tutoring, argumentative, individualistic, dominative, conflict, and confusion. The present scheme operationalizes the Relational dimension in terms of "social climate." The codes Collaborative Orientation and Individual Orientation were previously used in the social dimension of the VMT coding scheme (Strijbos & Stahl, 2007). The single dominance code by Kumpulainen and Mutanen (1999) was extended in terms of Positive and Negative

Table 11.7 Codes, definitions and examples for the motivational dimension

Orientation	*Code*	*Definition*	*Example*
Encouragement	ME	A statement aimed to encourage other group members during the task (future orientation)	"Are you with us?"
Performance	MP	A statement on the quality of the performance or problem solving (past orientation)	"You are awesome. (Laughing)"
Dominance	*Code*	*Definition*	*Example*
Interest	MI	A statement expressing interest in the task or in other group members	"Reads (humor about term de Broglie wavelength)"
Enjoyment	MJ	A statement expressing enjoyment in the task or working with other group members	"That's what I got too (laughing)"

Dominance. Table 11.6 provides an overview of the Relational dimension codes, their definitions and examples from the present dataset.

The Motivational Dimension

The Motivational dimension and associated outcomes have also received increased attention in recent (CS)CL literature, see for example Boekaerts and Minnaert (2006), Dillenbourg, Järvelä, and Fischer (2009), and Järvelä, Volet, and Järvenoja (2010). In contrast to an extrinsic operationalization of motivation in early studies on cooperative and collaborative learning (e.g., rewards), present motivational perspectives, such as the "dual processing self-regulation model" (Boekaerts & Niemivirta, 2000), "self-determination theory" (Ryan & Deci, 2000), and "person-object theory" (Krapp, 2005), share the premise that students have multiple goals with their subsequent motivations, actions, and affective responses. Likewise, students have multiple goals and motivations in the context of collaborative learning. Hijzen, Boekaerts, and Vedder (2007) found that mastery goals ("I want to learn new things") and social responsibility goals ("I want help my peers") prevail in effective collaborative learning groups. Furthermore, belongingness goals (e.g., "I want my peers to like me") were more important than mastery goals in ineffective collaborative groups, whereas the opposite was observed for effective groups.

The present scheme operationalizes the Motivational dimension as "motivation/affect." The codes Encouragement and Performance were taken from the peer feedback coding scheme discussed in the section on the Cognitive dimension (Strijbos et al., 2012). The codes Interest and Enjoyment were added based on recent insight in motivation research (Krapp, 2005; Ryan & Deci, 2000). Table 11.7 provides an overview of the Motivational dimension codes, their definitions and examples from the present dataset.

Analyses and Initial Interpretations

Group Level Analyses

In this analysis, a first striking difference between both groups is visible with respect to the proportion of the overall number of statements for the Cognitive dimension (48 % in the G-group versus 65 % in the M-group) and more specifically with respect to Elaboration codes, which are solely of the Suggestion type in the M-group. This is very consistent with the findings from the Souflé analysis. Furthermore, although the groups have an almost equal proportion of statements in the Relational dimension (36 % in the G-group and 42 % in the M-group), more statements with an Individual Orientation are made in the M-group (20 % versus 9 %) and slightly more Collaborative Orientation in the G-group (19 % versus 16 %). This is the opposite of the finding from the Souflé analysis where we see a more open atmosphere in the M-group than in the G-group. Finally, the Motivational dimension is virtually nonexistent in the M-group (2 %), whereas these types of the statements constituted 10 % of all statements in the G-group. This distinction was not evident in the Souflé analysis.

Individual Level Analysis

The individual analysis complements the group level analysis, adopting more of a qualitative flavor and focusing on pivotal moments during the collaborative learning episode (Table 11.8).

G-Group

- *Cognitive dimension*: Consistent with the Souflé analysis, at the cognitive plane there is collaborative dialogue and it is interspersed with short argumentative instances. There is a roughly equal input by F4 and F1, whereas F2 is "on the side" for most of the collaborative episode.
- *Relational dimension*: On the Relational dimension it is evident that F2 is not especially involved and this student's early suggestions ("no wait that's the de

Table 11.8 Frequency and proportion for occurrences of each code in each of the three dimensions for both groups

G-group (*N* segments = 105)									M-group (*N* segments = 49)								
Cognitive			Relational			Motivational			Cognitive			Relational			Motivational		
	f	%		*f*	%		*f*	%		*f*	%		*f*	%		*f*	%
VP	13	12	CL	20	**19**	ME	3	**3**	VP	6	12	CL	8	**16**	ME	0	**0**
VN	9	8	IN	9	**9**	MP	2	2	VN	5	8	IN	10	**20**	MP	1	2
EQ	15	14	DP	5	5	MI	1	1	EQ	10	20	DP	3	6	MI	0	0
EC	2	2	DN	3	**3**	MJ	5	**5**	EC	3	6	DN	0	**0**	MJ	0	**0**
EA	4	**4**							EA	0	**0**						
ES	6	**5**							ES	12	**25**						
EJ	3	**3**							EJ	0	**0**						
CT	52	**48**	ST	37	36	MT	11	**10**	CT	36	**65**	ST	21	42	MT	1	**2**

Broglie. That's to find the ejected electron" and "oh, they gave us the work function because we have to find out if the kinetic energy is greater than zero so that we can say that 1 electron is equal to 1 proton") are not taken up by F1 and F4. In fact a Negative Verification is given by F1 "no it's ejected. We do need the work function to find kinetic energy" to the suggestion by F2. Then F2 replies with "oh ok" and subsequently stays on the side line for quite some time. In this case the reply "no it's ejected" also signals that there is a Negative Dominance by F1, which results in F2 "staying out of the way." Further on there is a Positive Dominance and Encouragement by F4 to involve F2 again in a tutoring episode ("are you with us?", "ok, we have ek", "We know that ek equals 1/2mv^{2}"), and F2 responds with appreciation for F4's competence in solving the task, "you are awesome." A couple of lines later there is again Negative Dominance since the statement "we are so smart" seems to refer to F4 and F1 working on the problem in a dyadic mode with F2 on the side. F4 appears to be the "pack leader" from this analysis, as in the Souflé analysis. However the impression we get about Negative and Positive Dominance is the opposite of what we might have expected given that F4 has many more Heteroglossic Contract and Monoglossic moves than F1 in the Souflé analysis. We will address this in our conclusion.

- *Motivational dimension*: On the Motivational dimension there are expressions of interest and enjoyment during the task, although mostly at the end and by F1 and F4.

M-Group

- *Cognitive dimension*: Within the Cognitive plane there are mostly Suggestions provided, as well as the comparison of calculations. In most cases this is followed by a Positive or Negative Verification. We do not see any of the students distinguish themselves at this level.
- *Relational dimension*: Within the Relational plane none of the students is dominant (if any could be considered dominant, then F5 enacts the most Positive Dominance reflected by the statements "and then you subtract the work function?", "yeah,

and then you use the kinetic energy to get V", "would you use that as the kinetic energy to get V"). This is the opposite impression than we got from the Souflé analysis, but in neither analysis is the distinction very pronounced.
- *Motivational dimension*: Within the Motivational plane there are no expressions of either Interest or Enjoyment.

When comparing both groups from the perspective of this analysis, it appears that the G-group resembles a much more interactive group, whereas the M-group is focused on finding the answer as soon as possible, working individually interspersed with short episodes of sharing and checking answers. This difference is also visible in the lack of expressions of encouragement, interest or enjoyment.

Conclusions

In this chapter we have explored two separate three-dimensional analysis frameworks for formative assessment of collaborative learning processes in the PLTL dataset. Within each framework, the separate dimensions provide distinct lenses through which collaborative processes can be viewed. Students have the opportunity to take leadership along any one of these dimensions independent of the others.

The analysis along the relational dimensions across is most interesting comparison from the perspective of multivocality. Here we see the subtleties behind the idea of dominance and the different ways that positive versus negative may be viewed. In the Souflé framework, contributions are characterized as expanding or contracting the set of ideas that remain up for consideration. In the other framework, contributions are characterized as either enacting a positive or negative polarity. We see that there is a many-to-many correspondence between these distinctions. The question is where this leaves us in terms of defining the Relational dimension. Within both frameworks, one is viewed as more imposing (contracting, negative) and the other less imposition (expanding, positive). We leave it to future work to determine how these differing conversational constructs can be validated through correlation with external measures of power relations, leadership, and social roles, etc. that are sensitive enough to measure the impact of conversational positioning in collaborative groups.

Acknowledgments This work was supported in part by NSF grant SBE 0836012 to the Pittsburgh Science of Learning Center.

References

Ai, H., Sionti, M., Wang, Y. C., & Rosé, C. P. (2010). Finding transactive contributions in whole group classroom discussions. In K. Gomez, L. Lyons, & J. Radinsky (Eds.), *Proceedings of the*

International Conference of the Learning Sciences (pp. 976–983). Chicago, IL: International Society of the Learning Sciences.

Berkowitz, M.W., & Gibbs, J.C., (1979). A Preliminary Manual for Coding Transactive Features of Dyadic Discussion. Unpublished Manuscript. Marquette University, Milwaukee, WI.

Boekaerts, M., & Minnaert, A. (2006). Affective and motivational outcomes of working in collaborative groups. *Educational Psychology, 26*(2), 187–208.

Boekaerts, M., & Niemivirta, M. (2000). Self-regulated learning: Finding a balance between learning goals and ego protective goals. In M. Boekaerts, P. R. Pintrich, & M. Zeidner (Eds.), *Handbook of self-regulation* (pp. 417–450). New York, NY: Academic.

Dillenbourg, P., Järvelä, S., & Fischer, F. (2009). The evolution of research on computer-supported collaborative leaning: From design to orchestration. In N. Balacheff, S. Ludvigsen, T. de Jong, A. Lazonder, & S. Barnes (Eds.), *Technology-enhanced learning: Principles and products* (pp. 3–20). New York, NY: Springer.

Gillies, R. M. (2007). *Cooperative learning: Integrating theory and practice*. Thousand Oaks, CA: Sage.

Gweon, G., Jain, M., McDonogh, J., Raj, B., & Rosé, C. P. (2012). Predicting idea co-construction in speech data using insights from sociolinguistics. In M. Jacobson & P. Reimann (Eds.), *Proceedings of the International Conference of the Learning Sciences*. International Society of the Learning Sciences: Sydney, Australia.

Hijzen, D., Boekaerts, M., & Vedder, P. (2007). Exploring the links between students' engagement in cooperative learning, their goal preferences and appraisals of instruction conditions in the classroom. *Learning and Instruction, 17*(6), 673–687.

Howley, I., Adamson, D., Dyke, G., Mayfield, E., Beuth, J., & Rosé, C. P. (2012). Group composition and intelligent dialogue tutors for impacting students' self-efficacy. In S. Cerri, W. Clancey, G. Papadourakis, & K.-K. Panourgia (Eds.), *Intelligent tutoring systems: 11th International Conference, ITS 2012*. Crete, Greece: IOS Publisher.

Howley, I., Mayfield, E., & Rosé, C. P. (2013). Linguistic analysis methods for studying small groups. In C. Hmelo-Silver, A. O'Donnell, C. Chan, & C. Chin (Eds.), *International handbook of collaborative learning*. New York, NY: Taylor and Francis.

Järvelä, S., Volet, S., & Järvenoja, H. (2010). Research on motivation in collaborative learning: Moving beyond the cognitive-situative divide and combining individual and social processes. *Educational Psychologist, 45*(1), 15–27.

Johnson, D. W., & Johnson, R. T. (1994). *Learning together and alone: Cooperative, competitive and individualistic learning* (4th ed.). Boston, MA: Allyn & Bacon.

Joshi, M., & Rosé, C. P. (2007). *Using transactivity in conversation summarization in educational dialog* (Proceedings of the SLaTE Workshop on Speech and Language Technology in Education). Farmington, PA: ISCA.

Krapp, A. (2005). Basic needs and the development of interest and intrinsic motivational orientations. *Learning and Instruction, 15*(5), 381–395.

Kreijns, K., Kirschner, P. A., & Jochems, W. M. G. (2003). Identifying the pitfalls for social interaction in computer-supported collaborative learning: A review of the research. *Computers in Human Behavior, 19*(3), 335–353.

Kulhavy, R. W., & Stock, W. A. (1989). Feedback in written instruction: The place of response certitude. *Educational Psychology Review, 1*(4), 279–308.

Kumar, R., & Rosé, C. P. (2011). Architecture for building conversational agents that support collaborative learning. *IEEE Transactions on Learning Technologies, 4*(1), 21–34.

Kumpulainen, K., & Mutanen, M. (1999). The situated dynamics of peer group interaction: An introduction to an analytic framework. *Learning and Instruction, 9*(5), 449–473.

Martin, J. R., & Rose, D. (2003). *Working with discourse: Meaning beyond the clause*. London: Continuum.

Martin, J. R., & White, P. R. R. (2005). *The language of evaluation: Appraisal in English*. London: Palgrave.

Mayfield, E., & Rosé, C. P. (2011). Recognizing authority in dialogue with an integer linear programming constrained model. In Y. Matsumoto & R. Mihalcea (Eds.), *Proceedings of the 49th Annual Meeting of the Association for Computational Linguistics: Human Language Technologies*. Portland, OR: Association for Computational Linguistics.

Narciss, S. (2008). Feedback strategies for interactive learning tasks. In J. M. Spector, M. D. Merrill, J. J. G. Van Merriënboer, & M. P. Driscoll (Eds.), *Handbook of research on educational communications and technology* (3rd ed., pp. 125–143). Mahwah, NJ: Erlbaum.

Rosé, C. P., Wang, Y. C., Cui, Y., Arguello, J., Stegmann, K., Weinberger, A., et al. (2008). Analyzing collaborative learning processes automatically: Exploiting the advances of computational linguistics in computer-supported collaborative learning. *International Journal of Computer-Supported Collaborative Learning, 3*(3), 237–271.

Ryan, R. M., & Deci, E. L. (2000). Self-determination theory and the facilitation of intrinsic motivation, social development, and well-being. *American Psychologist, 55*(1), 68–78.

Sarmiento, J. W., & Shumar, W. (2010). Boundaries and roles: Positioning and social location in the Virtual Math Teams (VMT) online community. *Computers in Human Behavior, 16*(4), 524–532.

Sawyer, Frey, & Brown. (this volume). Peer led team learning in general chemistry. In Suthers, D., Lund, K., Rosé, C. P., Teplovs, C., Law, N. (Eds.), *Productive multivocality in the analysis of group interactions*, Chapter 9. New York, NY: Springer.

Sharan, Y., & Sharan, S. (1992). *Expanding cooperative learning through group investigation*. New York, NY: Teachers College Press.

Slavin, R. E. (1996). Research on cooperative learning and achievement: What we know, what we need to know. *Contemporary Educational Psychology, 21*(1), 43–69.

Strijbos, J. W. (2011). Assessment of (computer-supported) collaborative learning. *IEEE Transactions on Learning Technologies, 4*(1), 59–73.

Strijbos, J. W., & Stahl, G. (2007). Methodological issues in developing a multi-dimensional coding procedure for small group chat communication. *Learning and Instruction, 17*(4), 394–404.

Strijbos, J. W., Van Goozen, B., & Prins, F. (2012, August). *Developing a coding scheme for analysing peer feedback messages*. Paper presented at the EARLI-SIG 1 Assessment and Evaluation Conference, Brussels, Belgium.

Tolmie, Andrew Kenneth, Keith J. Topping, Donald Christie, Caroline Donaldson, Christine Howe, Emma Jessiman, Kay Livingston, and Allen Thurston. (2010). Social effects of collaborative learning in primary schools. *Learning and Instruction, 20*(3), 177–191.

Weinberger, A., & Fischer, F. (2006). A framework to analyze argumentative knowledge construction in computer-supported collaborative learning. *Computers and Education, 46*(1), 71–95.

Chapter 12
Application of Social Network Analysis to Collaborative Problem Solving Discourse: An Attempt to Capture Dynamics of Collective Knowledge Advancement

Jun Oshima, Yoshiaki Matsuzawa, Ritsuko Oshima, and Yusuke Niihara

Introduction

Theoretical Assumptions

The theoretical assumption that we base our analysis on is knowledge building. Scardamalia (2002) discusses that the main aim in a knowledge building community is collective knowledge advancement, and that group members should take up their personal cognitive responsibility to contribute to that collective knowledge advancement.

Purpose of the Analysis

We are interested in the analysis of collective knowledge advancement; however, our conjecture is that none of the existing methodologies are fully successful in capturing collective knowledge advancement. First, no existing methodology has been capable of representing dynamic change in collective knowledge advancement as it unfolds over time. Knowledge building is a process in which multiple participants are engaged in building knowledge collaboratively, mainly through their social discourse (Bereiter, 2002). We need an approach for capturing such dynamics in collective knowledge construction. At the same time we are also concerned with individual participants who are involved in collective knowledge advancement and

J. Oshima (✉) • Y. Matsuzawa • R. Oshima
Shizuoka University, Shizuoka, Japan
e-mail: joshima@inf.shizuoka.ac.jp

Y. Niihara
InfoCom, Tokyo, Japan

D.D. Suthers et al. (eds.), *Productive Multivocality in the Analysis of Group Interactions*, Computer-Supported Collaborative Learning Series 16,
DOI 10.1007/978-1-4614-8960-3_12,

contribution that occurs at that level. In knowledge building practices, participants with different knowledge resources collaborate with one another to build new knowledge objects. At the individual level, the focus of analysis should be on how differently or uniquely each individual contributes rather than on how correctly individuals can develop their own knowledge. In this paper, we propose a social network analysis (SNA) of discourse as an alternative approach to representing both collective knowledge advancement and identification of each individual's contribution to that advancement.

Unit of Interaction

In this chapter, we look at how participants contribute to their collective knowledge advancement through their discourse in a face-to-face context. Therefore, the unit of interaction for us is the exchange of ideas between conversation turns. We do not analyze specifically the exchange between contiguous turns. Rather, we examine how each conversation turn contributes to participants' collective knowledge built through the preceding turns.

The unit of observation we are dealing with in the analysis is transcribed data from oral discourse. The observation is examined at two different levels of analysis. At the collective level (in other words, a group as a whole), we analyze how collective knowledge develops through participants' discourse in solving a chemistry problem. At the individual level, the same representation is more finely segmented into each individual's contribution. Finally, we integrate the two levels of analyses for answering our research questions: (1) how collective knowledge develops through interaction and (2) how each individual contributes to it.

Representations

We apply Social Network Analysis (SNA) to transcribed data from participants' conversation in solving a chemistry problem. In our SNA, the original data representation is a bipartite graph of words selected by analysts and conversational turns. By projecting that bipartite graph into a unimodal projection three different ways in our analysis, we are able to use these disparate representations of interactions to examine the same interactions using three very distinct lenses that bring out different insights. First, we create a network of words. The word network may provide us with insights about how participants' ideas would contribute to the collective knowledge advancement. Second, we create a network of participants (a typical social network). The participants network may inform us how different participants' ideas are related to each other. Finally, we create a network of conversation turns. The turns network shows us how different turns are related to one another on the basis of our selected words with links representing participants' ideas.

Manipulations

As we discussed in the previous section, the networks created based on the original bipartite graph from transcribed data are the first type of analytic representation. We can visually inspect how participants contribute to their collective knowledge advancement through their discourse by examining how the networks are structured turn by turn. In addition, we used several indices for analysis of collective knowledge advancement that can be captured through traditional measures such as network centrality coefficients used in SNA studies. The quantitative analysis using our numerical measures is conducted both at the collective level and at the individual level.

Social Network Analysis Approach to Collective Knowledge Advancement in the Knowledge Creation Metaphor

Recent studies in the learning sciences have discussed a new approach that integrates two prevailing metaphors of learning: acquisition and participation (Paavola, Lipponen, & Hakkarainen, 2004; Sfard, 1998). However, current assessment techniques do not act in concert with the development of such a theoretical approach to learning. Hence, to address this deficiency, social network analysis (SNA) is introduced as a novel assessment approach for learning interactions inspired by the knowledge-creation metaphor. In this chapter, we propose the social network analysis (SNA) of discourse as an alternative approach to analyzing collective knowledge advancement. In the following, we briefly review SNA research in CSCL and how the approach could be applied to learner discourse in knowledge building environments. We then describe our SNA of discourse data by two groups of university students that showed different dynamics in their collective knowledge advancement.

In CSCL research, there have been discussions on the advantages of using SNA to investigate community knowledge advancement and individual learners' engagement in this advancement from the perspective of the knowledge-creation metaphor (e.g., Martinez, Dimitriadis, Rubia, Gomez, & de la Fuente, 2003; Reffay, Teplovs, & Blondel, 2011; Reuven, Zippy, Gilad, & Aviva, 2003). de Laat, Lally, Lipponen, and Simons (2007) considered the application of SNA in CSCL research. They outlined an approach to synthesize and extend the understanding of CSCL teaching and learning processes so as to balance SNA, content analysis and critical event recall. In this complementary approach, SNA was used to study interaction patterns within a networked learning community, as well as to study how learners share and construct knowledge. de Laat et al. concluded that SNA would be advantageous to include in any multi-method approach because of the following advantages: (a) researchers and learners are provided with a tool that is capable of illustrating mutual understanding and cohesion with group activities, and (b) a method is made available to researchers for selecting appropriate groups to study.

A limited number of studies have used SNA, especially those espousing the knowledge-creation metaphor in their work. Over a period of 3 years, Zhang,

Scardamalia, Reeve, and Messina (2009) implemented a complementary approach that used SNA to visualize and compare classroom collaboration among fourth grade elementary school students through a CSCL environment designed to support them in knowledge building. An analysis of the students' online participatory patterns and knowledge advancement indicated that this learning process facilitated students' knowledge advancement effectively, and that this was the case through critical changes in organizations within the classroom: from fixed small groups in the first year of the study to appropriate collaboration through dynamic formation of small teams based on emergent goals.

In previous work (Oshima, Oshima, & Knowledge Forum® Japan Research Group, 2007), we further extended the potential of SNA as a core assessment technique by describing a different type of social network. Ordinary SNA illustrates the social patterns of learners, namely, the learners' social network. As de Laat et al. (2007) suggested, this approach is thus informative when examining developments or changes in the participatory structure of a community. However, we argued that existing social network models are unable to examine how community knowledge advances through learners' collaboration (Oshima et al., 2007). Instead, we used a procedure similar to ordinary SNA, but proposed a different type of social network, one based on the words learners use in their discourse in a CSCL environment. We compared this social network, in which words were selected as nodes representing learners' knowledge or ideas during discourse on a study topic, with a network of words from the discourse of a group of experts on the same topic. The results showed that there were remarkable differences in the community knowledge of elementary school students and of experts that can be revealed in terms of the words centered within the networks. We concluded that SNA can provide a new type of representation of community knowledge building by learners, enabling researchers to adopt a new complementary assessment technique for investigating knowledge building community models.

Although studies have proposed the application of SNA to learning analysis as a new assessment technique combining word level analysis and the knowledge-creation metaphor, an exact methodology has yet to be established. The purpose of this study is to propose an SNA approach to analyzing students' discourse that is consistent with the knowledge creation metaphor. Using the data provided by Sawyer, Frey, and Brown (this volume a, Chap. 9), we demonstrate how SNA is useful for us to analyze the collective knowledge advancement in either a qualitative or quantitative manner.

SNA of Discourse from the Perspective of Knowledge Creation Metaphor

We analyzed two groups of university students, one from what is called the Gillian class and the other from what is called the Matt class. As Sawyer, Frey, and Brown (Chap. 10, this volume b) discussed, these two groups were quite different in their strategic approaches to solving a chemistry problem, i.e., calculating the wavelength of an electron discharged from an object by utilizing formulas related to the photoelectric effect and the de Broglie equation. In their own conversation analysis, Sawyer and colleagues described the distinct profiles of the two groups as follows: the Gillian group went beyond pure calculation by discussing conceptual ideas about what they had learned and engaged in collaborative knowledge construction through mutual reflection of ideas. The Matt group, on the other hand, was involved in calculation activities without deep, reflective articulation of what they had learned.

With the same data set, we conducted our SNA for how each group of students was engaged in their collective knowledge advancement. In addition, we further analyzed how students in the Gillian class constructed their conceptual understanding after solving the given problem. In the first phase when students solved a problem in the Gillian and Matt classes, our analysis was focused specifically on the collective knowledge advancement. In the second phase when students discussed the trend related to the photoelectric effect and de Broglie wave after solving problems in Gillian class, our analysis was focused on how a peer leader supported students' collective knowledge advancement.

For the analysis of collective knowledge advancement, we referred to Scardamalia (2002) as our theoretical framework. Scardamalia proposed 12 socio-cognitive determinants of a knowledge building community. She discusses that the main aim in such a community is collective knowledge advancement, and that any members should take responsibility to contribute to the collective knowledge advancement.

Indicators for Collective Knowledge Advancement

Our effort in this study is focused on the establishment of indicators for collective knowledge advancement. Referring to Scardamalia's socio-cognitive determinants of a knowledge building community, we selected three aspects for our network analysis: (1) the continuous improvement of ideas, (2) learners' collective responsibility for community knowledge, and (3) their cognitive effort to rise above their previous ideas. What we attempt to do is to computationally measure the three aspects of collective knowledge advancement by the target groups. The basic assumption behind this analysis is that learners' ideas are represented as clusters of nodes, i.e., sets of words as nodes with links among them.

Based on our assumption, the improvement of ideas is captured by measuring degree centrality coefficients of nodes in a network of words. Degree centrality is a straightforward concept that indicates cumulative path lengths by which each node is linked to other nodes in the network. High degree centrality means that the node is at the center of the network as a whole, or near the center of a local cluster in the network. We are interested in the sum of degree centrality coefficients of all nodes as an indicator for the continuous improvement of ideas (i.e., increase in nodes and links). The learners' collective responsibility for collective knowledge is examined by calculating displacements of three centrality coefficients of all nodes by using a stepwise technique (Oshima et al., 2007). By comparing the displacements by three students in each group, we evaluate how each student individually contributes to

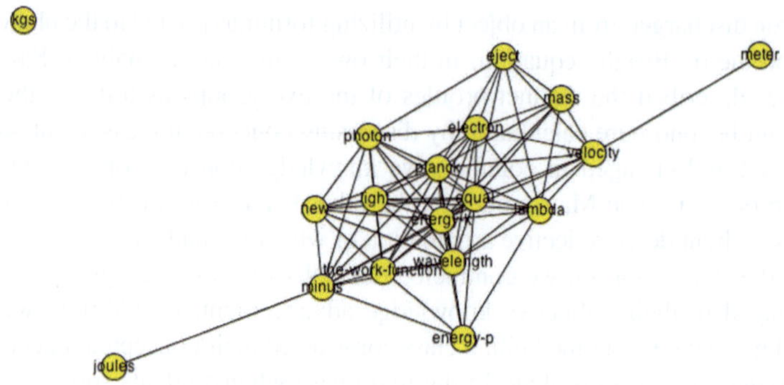

Fig. 12.1 The network structure of conceptual words in Gillian class

collective knowledge advancement. Finally, learners' efforts to rise above their own specific ideas is captured by displacements of closeness and degree centrality coefficients in our stepwise analysis. Closeness centrality is a measure of how close the node is to other nodes in a network, based on the geodesic distance. When a conversation turn works to integrate previous ideas, the turn is considered to contribute to the increase in closeness and degree centrality coefficients more than other conversation turns.

Visual Inspection of Network Structures by Two Groups

We selected words for the analysis that we believe to be representative of student explanations about their problem solving at the conceptual level and calculation level. Words selected to represent the conceptual level are nouns and verbs by which students engaged in planning and motoring their problem solving by referring to related formulas. Words selected at the calculation level are numbers they produced as they worked towards reaching their final answers. The agreement of word selection between two independent researchers was over 90 %. Disagreements were resolved through discussion. There were 18 conceptual words and 14 calculation words selected for the Gillian class. For the Matt class, 14 conceptual and 12 calculation words were selected for the analysis. Using an SNA application we developed in our earlier work (Matsuzawa, Oshima, Oshima, Niihara, & Sakai, 2010), we visually inspected the progressive turn-by-turn development of network structures and found critical differences between the two groups as well as one particular pivotal conversation turn in the discourse.

One critical difference between the two groups was in the cohesiveness of network structure. While the two groups were solving exactly the same problem, their usage of conceptual words in their contributions was quite different from each other. The network structure of conceptual words in the Gillian class was more cohesive. Although one word was isolated, other words were gathered into one big cluster (Fig. 12.1). This suggests that students related these conceptual words in their discussion. On the other hand, the network structure of words in the Matt class was segmented: the network consisted of two completely separate clusters of words

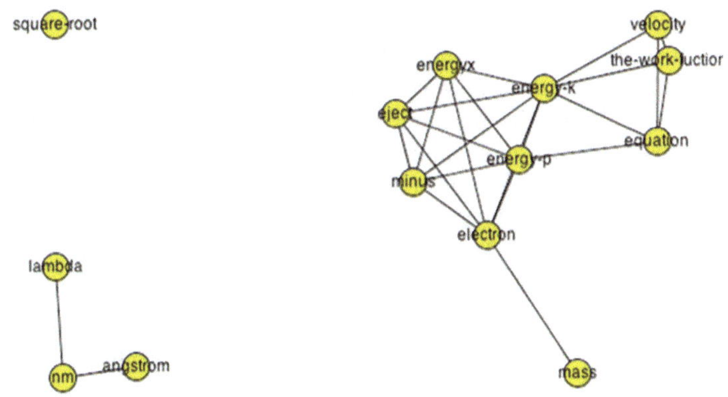

Fig. 12.2 The network structure of conceptual words in Matt class

Table 12.1 An excerpt of discourse by Gillian Group with a pivotal turn found in the network analysis

F1	so we need lambda and its give us the *** so with the wavelength we can find velocity. And with…
F4	And we don't have to works with the-work-function right away.
F1	not right away but we do need the-work-function at the end.
F4	Ok to find the…
F1	Because the important thing that for the lambda we are find the wavelength of the electron not of light.
F4	Right, exactly. So, first for the electron we use Planck/mass velocity because we know the mass of an electron
F1	We need to find the mass of an electron
F4	no, we know the mass of an electron. It's an electron.
F1	very true, very true.
F4	but we don't know…
F1	with this wavelength we would be find velocity
F2	what did they give us. For the following wavelength. So, well lambda equal Planck/mass velocity right?
F1	Yes its tell us to use lambda equal Planck/mass velocity
F1	You can find energy-k.
F4	energy-k of a photon.
F1	use energy-k of the-work-function equal energy-k equal Planck new minus the-work-function equal energy-k of a photon and use energy-k of a photon to find the wavelength of light.

*** inaudible portion

(Fig. 12.2). This result suggests that students in the Matt class were not having a cohesive discussion at the conceptual level of problem solving.

We also found a conversation turn that was pivotal for the cohesiveness of network structure in the Gillian class. By observing changes in the network structure turn by turn, we found a pivotal point at which segmented clusters of conceptual words merged into one cohesive cluster. In this pivotal conversational turn, a student (F1) offered the idea of relating different formulas to one another during their planning before actually calculating the answer using the formulas. Table 12.1

shows their discourse from the beginning until the pivotal conversation turn by F1. Their discourse started with sharing their ideas of formulas for solving a problem. Their problem-solving strategy was a backward-chaining approach setting their final goal of calculating the wavelength of an electron ejected from manganese. They first considered the application of the equation to calculate the wavelength of a matter by finding what was still unknown, the velocity. Then they further explicated their inference to use the equation of the relationship between kinetic energy of matter, mass and velocity. Their focus was mainly discussing de Broglie hypothesis. The last conversation turn by F1 gave a new idea to go back to the correct equation for the photoelectric effect. After this conversation turn, the Gillian group was able to establish their solution and began their calculations in earnest.

Network Analysis with Indicators for Collective Knowledge Advancement

The Continuous Improvement of Ideas

For the analysis of idea improvement, we calculated the sum of degree centrality coefficients of nodes in the network of conceptual words and examined its time-series change turn by turn (Fig. 12.3). The network structure of conceptual words in the Gillian class was more inter-connected than that in the Matt class. We also examined which students had conversation turns that steeply increased the sum of degree centrality coefficients. We referred to these as "jumping turns." There were ten such turns found to increase the degree centrality in this way in the Gillian class and four in the Matt class. In the Gillian class, contribution by each student was quite even. F1, F2 and F4 were involved in the jumping turns by four, three and

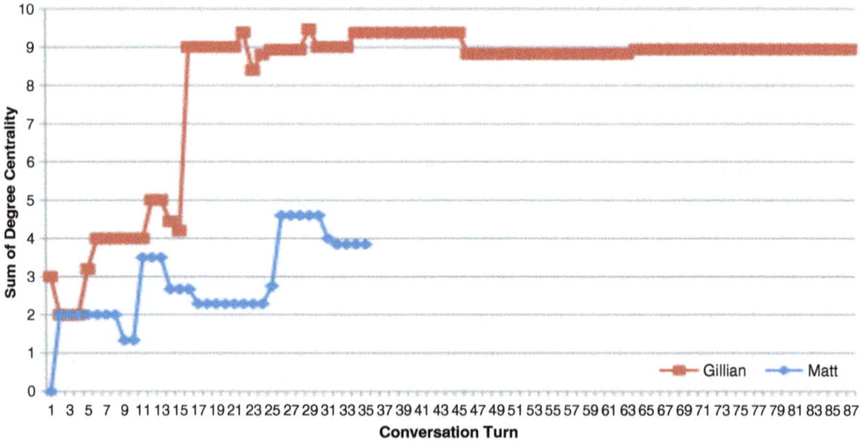

Fig. 12.3 Transition of the sum of degree centrality coefficients by Gillian and Matt groups

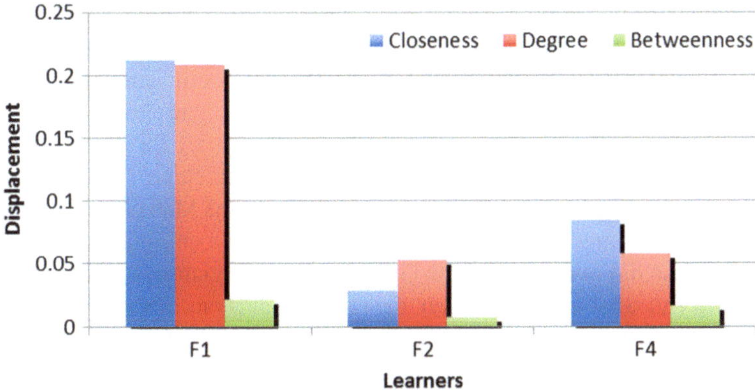

Fig. 12.4 Means of displacements of centralities by learners in Gillian

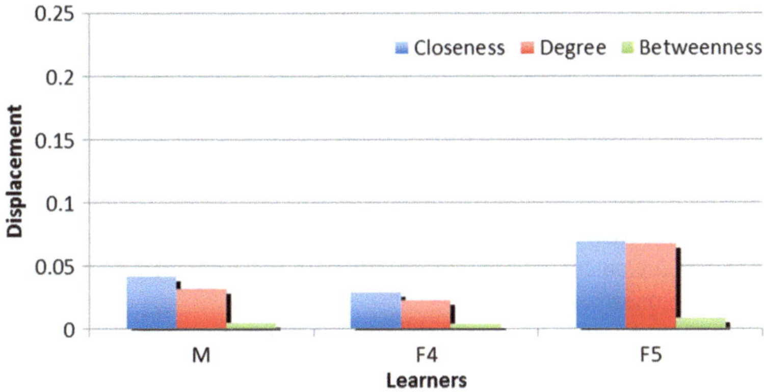

Fig. 12.5 Mean displacements of centralities by learners in Matt

three times, respectively. In the Matt group, F5 was involved in jumping turns three times. M was once. F4 was never. The results suggested that the Gillian class students were more oriented towards continuous idea improvement than were the Matt class students. In addition, the contributions in the Gillian class were more distributed than in the Matt class.

Collective Responsibility for Community Knowledge, and Effort to Rise Above

Learners' collective responsibility was evaluated by the displacements of three centrality coefficients of nodes in their network of conceptual words when their discourse contributions are excluded. Figures 12.4 and 12.5 show mean displacements of nodes by learners. 3 (Learners)×3 (Centralities) ANOVAs with mean displacement as a dependent variable for the two groups demonstrated that (1) mean displacements of closeness and degree centrality coefficients by F1 were significantly higher than those by two other students in the Gillian class, $F(2, 51)=7.52$, $p<0.01$ for closeness centrality, and $F(2,51)=8.01$, $p<0.01$ for degree centrality; and that no significant differences were found in the Matt class.

These results suggest that collective responsibility in the Matt class was relatively equal among the three students whereas F1 had more contribution to collective knowledge than did two others in the Gillian class. At a glance, the results here are contradictory to what we discussed in the previous analysis of the continuous idea improvement. In the idea improvement, we found that the Gillian students were more equally engaged in the improvement of collective knowledge than were Matt students. We have to be mindful that the analysis of idea improvement is focused on time-series change in the network structure whereas the analysis here is focused on comparison of each learner's contribution to the final state of the network structure. In other words, the stepwise analysis by excluding each learner's discourse represents how much "unique" contribution (i.e., links or nodes) each learner has in the network structure. In taking the differences in focus of the analyses into consideration, we discuss the differences in closeness and degree centrality coefficients between F1 and others in the Gillian group. One possible interpretation of the differences may be that F1 took a unique role to integrate others' ideas in some way. Closeness and Degree centrality coefficients are considered to be indicators of learners' effort to rise above ideas. Results here suggest that F1 contributed turns that enabled previously disconnected portions of the network to get linked or to move closer to one another.

How a Peer Leader Facilitated Students' Collective Knowledge Advancement

After solving problems, the two sub-groups in the Gillian class were merged as one group of seven students. They discussed the trend from their answers to the two problems under the supervision of the peer leader. This phase was an opportunity for students to make use of their conceptual understanding by explaining their process and what they could find as principles. The peer leader took her role to support students in making progress in constructing their shared meaning. She provided students with five key prompts during their discussion (see Fig. 12.6). We analyzed students' discourse data with two different purposes. First, we were concerned with the contribution by the peer leader to students' collective knowledge advancement. Since we considered that the leader's contribution would affect students' discourse following it, we segmented students' conversation turns into five parts following each key prompt by the peer leader. Second, we conducted stepwise analysis for identifying each student's contribution to each part of the discourse. For creating

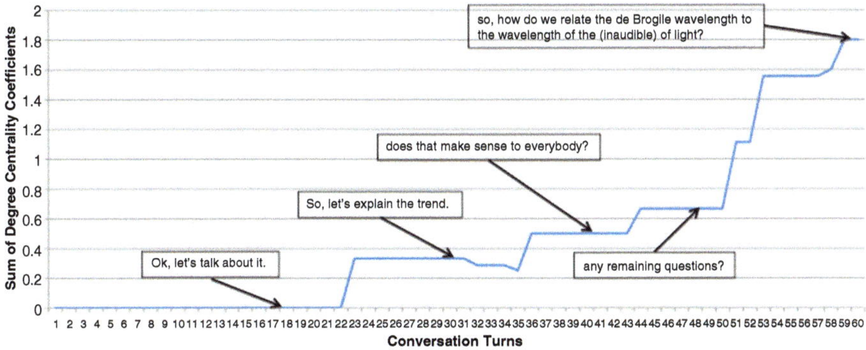

Fig. 12.6 Transition of the sum of degree centrality coefficients during students' discourse following the peer leader's key prompt

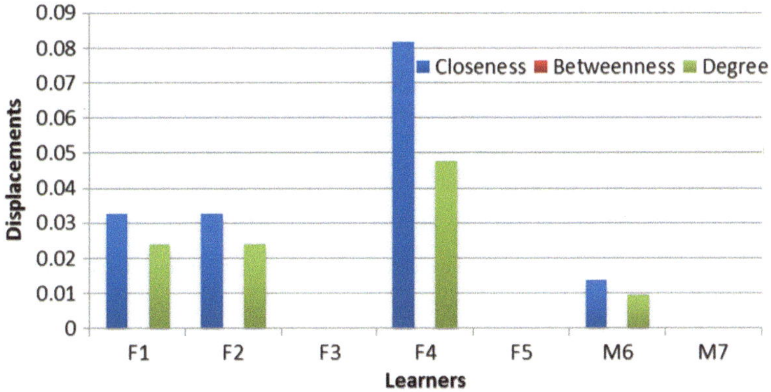

Fig. 12.7 Individual contributions in discourse following the first prompt

network structures, we used a subset of conceptual words that we used in the previous analysis because students did not use all the words we had previously selected.

Figure 12.6 shows the transition of the sum of degree centrality coefficients and conversation turns that the peer leader spoke to students. The first key prompt was "OK, let's talk about it." Here, the peer leader encouraged students to argue about what they found in solving the two problems. In between this first prompt and the second prompt, students "talked about" their answers, e.g., the unit of energy "joules." At this stage, we could not find any links in networks of learners and conversation turns. There was just one link between words, which means that a student used the two words in the same conversation turn. An individual students' contribution to the structure of network of words can be seen in the closeness and degree centrality but not in betweenness centrality (Fig. 12.7).

In the second prompt, the peer leader further directed students' discussion toward what trends they could find, i.e., "So, let's explain the trend." Following this second

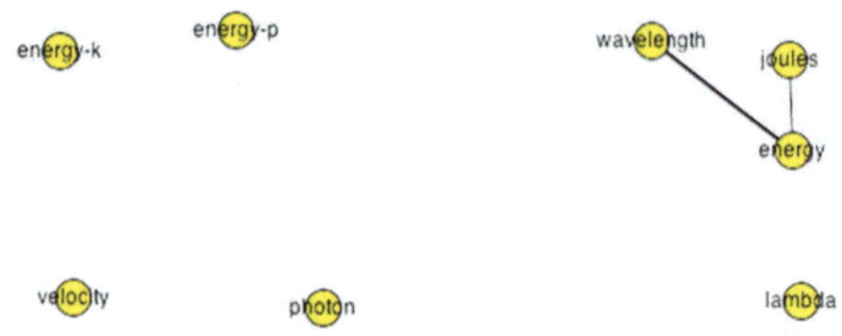

Fig. 12.8 The network structure of words in discourse following the second prompt

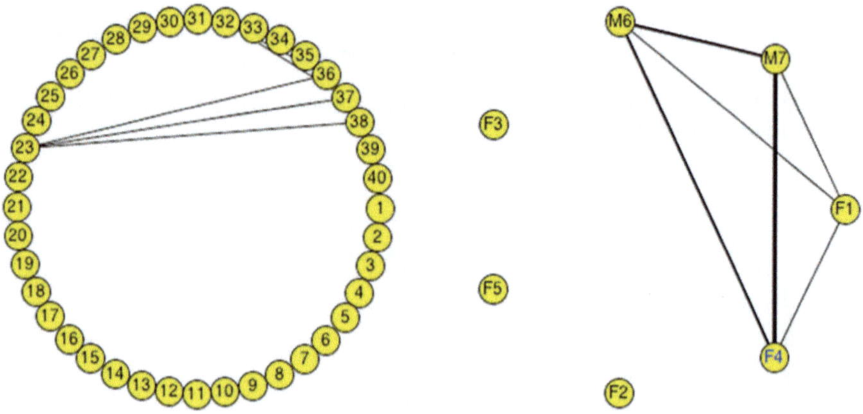

Fig. 12.9 The network structures of conversation turns (*left*) and students (*right*) following the second prompt

prompt, students discussed the relationship between the wavelength and energy (Fig. 12.8). Student F1 and M6 had conversation turns about this issue, e.g., "larger wavelength, less energy." From this second stage, students' conversation turns came to be linked to one another, and a network structure of students appeared (Fig. 12.9). This suggests that their ideas came to be linked to one another as the conversation proceeded. At the end of this stage, each individual student's contribution to the network of words was equally unique except for student F5 who did not use any of the selected words (Fig. 12.10). Students F2 and F3 had unique contributions although they were not linked to anybody in the network of students because there were several unique words used only by them in the conversation.

After encouraging students to explain the trends they found, the peer leader further encouraged their discussion by her third and fourth prompts. As seen in Fig. 12.11, these two prompts were found to stimulate students' further construction of their shared meaning of the trends. In her third prompt ("does that make sense to everybody?"), the peer leader attempted to confirm students' conceptual understanding of the de Broglie wavelength by facilitating reflection on what they had just discussed. With this third prompt, the students' conversation was more focused on what they found as the trend (Fig. 12.11). More conversation turns and students became linked in the networks (Fig. 12.12). After the third stage, however, a

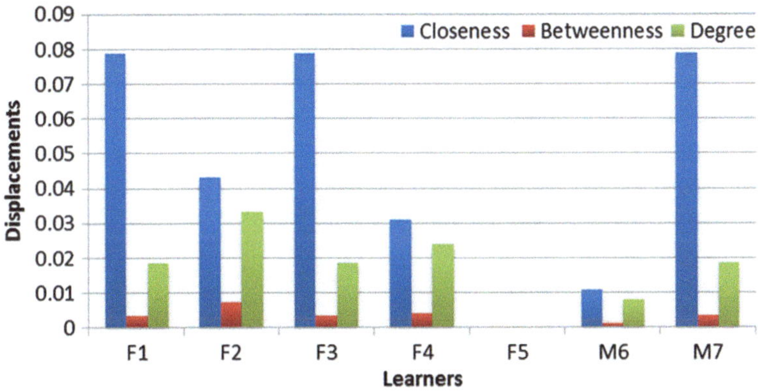

Fig. 12.10 Individual contributions in discourse following the second prompt

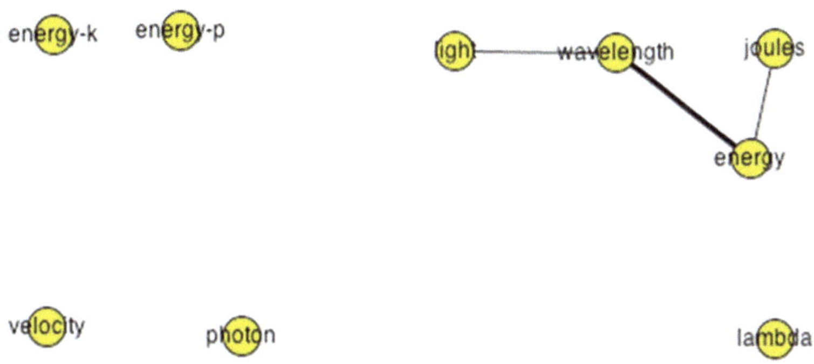

Fig. 12.11 The network structure of words in discourse following the third prompt

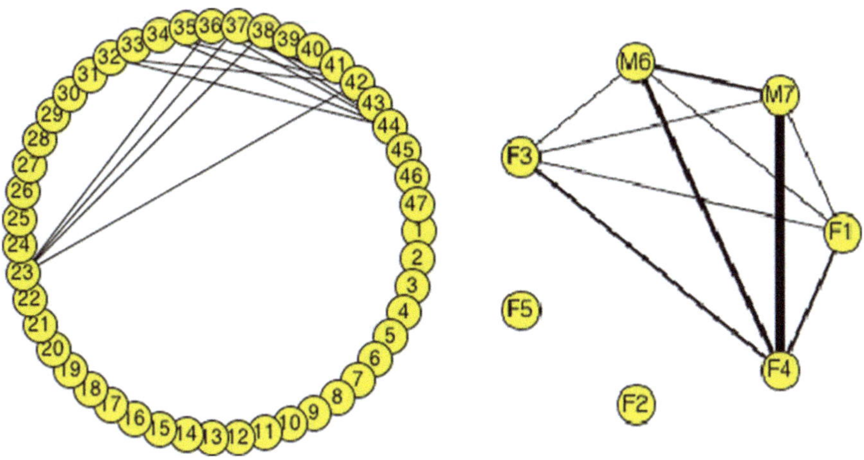

Fig. 12.12 The network structures of conversation turns (*left*) and students (*right*) following the third prompt

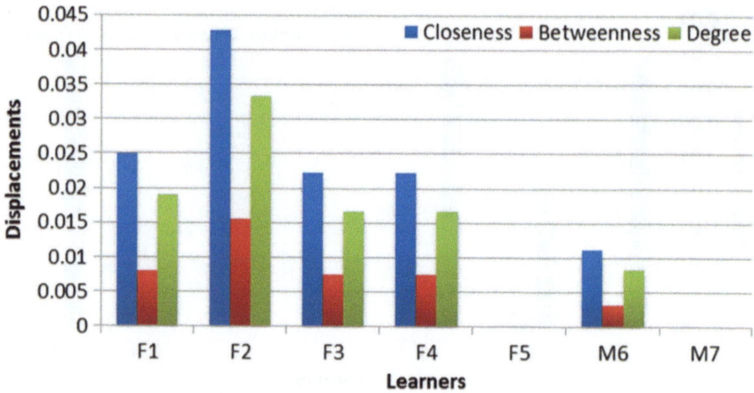

Fig. 12.13 Individual contributions in discourse following the third prompt

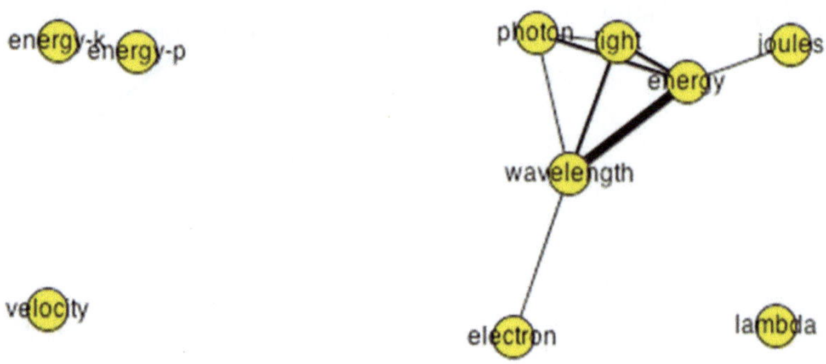

Fig. 12.14 The network structure of words in discourse following the fourth prompt

contribution by one student (M7) disappeared from the evolving representation because other students generated the same links of words in other conversation turns (Fig. 12.13).

The fourth prompt by the peer leader, "any remaining questions?", was found to be so critical that all the students were involved in the construction of shared meaning that followed. We could not identify what prompted the peer leader to utter this turn based on what is visible in the transcription only. But, there must have been some reason for her to prompt her students for further discussion after they had already made sense of the trend. This fourth prompt led students to be deeply involved in a more complete explanation of the trend. Networks of words, students and conversation turns became even more robust in their structure through this process. One of the most remarkable findings here was that all the students finally became linked in their social network at this stage. Based on the more robust network structure of words (Fig. 12.14) with the social network structure of students (right side in Fig. 12.15), we can claim that the fourth stage was a very important discussion process by which every student offered a meaningful contribution to the

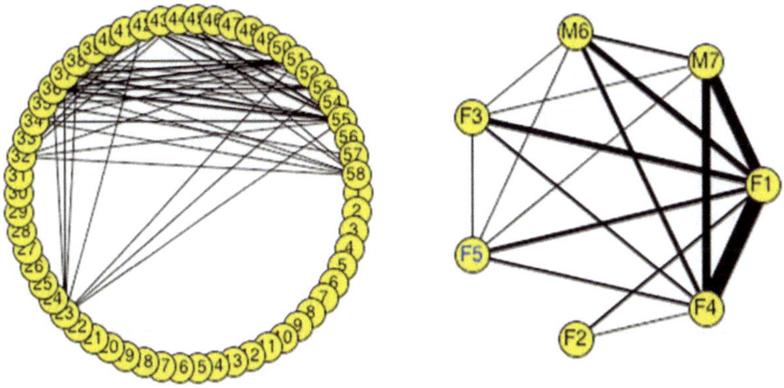

Fig. 12.15 The network structures of conversation turns (*left*) and students (*right*) following the fourth prompt

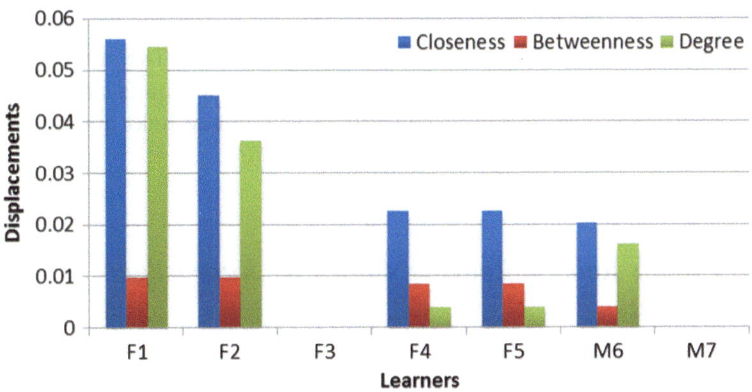

Fig. 12.16 Individual contributions in discourse following the fourth prompt

collective knowledge advancement, although some students did not contribute a unique contribution to it (Fig. 12.16). The fifth prompt "so, how do we relate the de Broglie wavelength to the wavelength of the [inaudible] of light?" was followed by only one conversation turn that did not make a big change in the network structures, we therefore omit description here.

Summary

We have presented two analyses in this chapter: (1) comparison between small group problem solving supervised by two peer leaders (i.e., Gillian and Matt), and (2) student discussion for constructing shared understanding of the de Broglie wavelength after their problem solving activities in the Gillian class. The first analysis was aimed at describing how different the two groups were in solving the same problem from the perspective of the knowledge creation metaphor. The second section was more directed at how we can identify the peer leader's contribution to the students' discussion for the purpose of constructing shared understanding. Based on our visual inspection and network analyses of indicators for collective knowledge advancement in the first section, we developed profiles of the two groups as follows:

Gillian Class. In solving the problem, the Gillian students devoted much effort to conceptual idea improvement. Only after exploring the problem space did they start their calculations. Each learner made a significant contribution to the group idea improvement, but one of them (F1) was found to have a more unique contribution to their collective knowledge advancement. Her contribution was unique in the sense that she attempted to rise above the previously expressed ideas.

Matt Class. The Matt group was calculation-centered. They did not devote much effort to exploration of the problem space. One student (F5) was somewhat involved in conceptual idea improvement. However, the contributions contributed by the three students were not significantly different. The non-significance suggests that the three students used conceptual words in a quite similar way and their conversation turns did not frequently create unique links among nodes in the network.

Peer Leader's Role in Students' Collective Knowledge Advancement

As the first section of analysis suggested, the peer leader in the Gillian class was more concerned with students' intentional involvement in constructing conceptual understanding, and she seemed to have an intention to support her students' engagement in such an activity. We, therefore, further analyzed how the peer leader attempted to be involved in the students' discussion after solving problems. During students' discussion in figuring out the trend, she gave students five key prompts. In early stages, her intention was to direct students' attention to the issue of discussion, i.e., "OK, let's talk about it," and "So, let's explain the trend." After successfully involving students in discussing the trend, she further asked students to reflect on what they found twice, i.e., "does that make sense to everybody?" and "any remaining questions?" We found that these two prompts activated students' deep involvement in conceptual understanding. The fourth prompt, in particular, elicited student discourse that created network structures of words, students and conversation turns and increased robustness in the structure. After the third prompt, students demonstrated their understanding quite visibly. Nevertheless, for some reason the peer leader offered them her fourth prompt. This remains a mystery that should be further examined by conducting more micro-level of analysis or the ethnographic studies in the classroom.

Final Remarks

In this study, we analyzed collaborative problem solving discourse from the perspective of the knowledge creation metaphor. As a methodological tool, we selected the SNA approach by which we can visually and computationally investigate the dynamics of collective knowledge advancement. For computationally analyzing discourse, we referred to the theoretical framework of knowledge building (Scardamalia, 2002) to create indicators for collective knowledge advancement.

Our challenge might be evaluated with the following criteria: (1) whether our findings match those from the original study by Sawyer et al., and (2) whether we can propose new insight beyond their original analysis. Regarding the first criterion, results of analysis mostly match what Sawyer et al. discussed in their original analysis. However, we further found a possibility that one student in the Gillian class, namely, F1, also identified as a leader in the Souflé analysis by Howley and colleagues, might be a key player in their collaborative problem solving. Conversation turns by F1 had significantly higher effect in increasing the extent to which conceptual words became linked and closer to one another within the evolving network structure. Regarding the second criterion, we described our first step to establish the methodological approach by using SNA for interaction analysis with discourse as data. The second section of our analysis might provide readers with a new perspective on the computational analysis of discourse and how instruction (appropriate prompts by the peer leader in our case) can affect students' discourse.

Our future effort will be focused in two directions. One direction will be the development of application software for educational researchers to easily engage in SNA of discourse. Our tools under development are still in a progressive refinement stage. The other direction will be the establishment of indicators for collective knowledge advancement by using SNA. Knowledge building is one possible methodological framework for us to use in creating indicators. However, other possibilities should also be explored.

References

Bereiter, C. (2002). *Education and mind in the knowledge age*. Hillsdale, NJ: Lawrence Erlbaum.

de Laat, M., Lally, V., Lipponen, L., & Simons, R.-J. (2007). Investigating patterns of interaction in networked learning and computer-supported collaborative learning: A role for social network analysis. *International Journal of Computer-Supported Collaborative Learning, 2*(1), 87–103.

Martinez, A., Dimitriadis, Y., Rubia, B., Gomez, E., & de la Fuente, P. (2003). Combining qualitative evaluation and social network analysis for the study of classroom social interactions. *Computers & Education, 41*(4), 353–368.

Matsuzawa, Y., Oshima, J., Oshima, R., Niihara, Y., & Sakai, S. (2010). KBDeX: A platform for exploring discourse in collaborative learning. *Proceedings of COINs2010: Collaborative Innovation Networks Conference*.

Oshima, J., Oshima, R., & Knowledge Forum® Japan Research Group. (2007). Complex network theory approach to the assessment on collective knowledge advancement through scientific discourse in CSCL. *Proceedings of CSCL1997* (pp. 563–565).

Paavola, S., Lipponen, L., & Hakkarainen, K. (2004). Models of innovative knowledge communities and three metaphors of learning. *Review of Educational Research, 74*(4), 557–576.

Reffay, C., Teplovs, C., & Blondel, F.-M. (2011). Productive re-use of CSCL data and analytic tools to provide a new perspective on group cohesion. In H. Spada, G. Stahl, N. Miyake, & N. Law (Eds.), *Proceedings of CSCL 2011* (pp. 846–850). Mahwah, NJ: Lawrence Erlbaum.

Reuven, A., Zippy, E., Gilad, R., & Aviva, G. (2003). Network analysis of knowledge construction in asynchronous learning networks. *Journal of Asynchronous Learning Networks, 7*(3), 1–23.

Sawyer, K., Frey, R., & Brown, P. (this volume a). Peer-led team learning in general chemistry. In D. Suthers, K. Lund, C. P. Rosé, C. Teplovs, & N. Law (Eds.), *Productive multivocality in the analysis of group interactions*, Chapter 9. New York, NY: Springer.

Sawyer, K., Frey, R., & Brown, P. (this volume b). Knowledge Building Discourse in Peer–Led Team Learning (PLTL) Groups in First–year General Chemistry. In D. Suthers, K. Lund, C. P. Rosé, C. Teplovs, & N. Law (Eds.), *Productive multivocality in the analysis of group interactions*, Chapter 10. New York, NY: Springer.

Scardamalia, M. (2002). Collective cognitive responsibility for the advancement of knowledge. In B. Jones (Ed.), *Liberal education in a knowledge society*. Chicago, IL: Open Court.

Sfard, A. (1998). On two metaphors for learning and the dangers of choosing just one. *Educational Researcher, 27*(2), 4–13.

Zhang, J., Scardamalia, M., Reeve, R., & Messina, R. (2009). Designs for collective cognitive responsibility in knowledge-building communities. *Journal of the Learning Sciences, 18*, 7–44.

Chapter 13
A Multivocal Analysis of the Emergence of Leadership in Chemistry Study Groups

Carolyn Penstein Rosé

Introduction

One advantage as well as challenge of multivocal approaches to analysis of collaborative learning interactions is that it reveals the ways in which our individual operationalizations of complex constructs such as social positioning, idea development, or leadership are limited. In bringing together analyses from multiple perspectives addressing similar issues with the same dataset, our eyes our opened to the richness and complexity of how these constructs are embodied in language. Here we bring together three distinct analytical approaches as they are applied to the same dataset of Peer-Led Team Learning (PLTL) groups in an undergraduate chemistry class (Sawyer, Frey, & Brown, Chap. 9, this volume a).

The three analysis chapters in this section fall on three distinct places on the continuum between the highly quantitative and the highly qualitative. At the qualitative end, Sawyer, Frey, and Brown (Chap. 10, this volume b) approached the conversations in a situated, blow-by-blow fashion. Many excerpts from the corpus are included in the text in raw form along with commentary and reflection on the substance of the interactions, including an assessment of the knowledge that was communicated and the manner in which it was communicated. At the opposite end of the spectrum, Oshima, Matsuzawa, Oshima, and Niihara (Chap. 12, this volume) applied a social network analysis approach to the same data and viewed the data through that lens. Connections are made between the topology of the network representation and how it evolved over time and a theoretical framework from the Knowledge Building community. In between these two end points stands the coding and counting approach of the chapter by Howley, Mayfield, Rosé, and Strijbos (Chap. 11, this volume), which contains analyses from two different multidimensional coding schemes, including the Souflé analysis lead by Rosé and the CSM

C.P. Rosé (✉)
Carnegie Mellon University, Pittsburgh, PA, USA

D.D. Suthers et al. (eds.), *Productive Multivocality in the Analysis of Group Interactions*, Computer-Supported Collaborative Learning Series 16,
DOI 10.1007/978-1-4614-8960-3_13, © Springer Science+Business Media New York 2013

analysis lead by Strijbos. Both of these coding schemes consist of the same three dimensions, namely, Cognitive, Relational, and Motivational, and are thus expected to reveal similar insights about the data. What is most interesting in that chapter is the ways in which the frameworks actually brought out different insights from the data. Like the Sawyer and colleagues chapter, the Howley chapter contains a number of example excerpts from the corpus. However, like the Oshima chapter, the inferences are made primarily based on quantitative measures as viewed through the lens of the theoretical frameworks that underlie the coding schemes. Only the Sawyer analysis includes an extended, blow-by-blow thick description of the two problem solving experiences.

While the three analyses are different in many respects, the questions the analysts sought to answer were highly overlapping. All were interested in the articulation of complex reasoning, social interconnectedness, and hierarchy. These are made explicit in the two three dimensional coding schemes used in the Howley chapter, and less explicit in the Sawyer and Oshima chapters. Nevertheless, these three distinct aspects of collaborative learning fit naturally together and can be thought of as three facets of leadership taking.

The multivocal analysis process for the PLTL dataset proceeded in two major phases, each discussed separately in this chapter. The first stage was presented at the ICLS 2010 workshop, which featured the Oshima social network analysis approach and the two coding and counting approaches (one presented there by Rosé and the other by Strijbos). Prior to the workshop, the analysts worked independently. It was during the discussion of these three analyses that the idea that all three approaches somehow captured the idea of leadership emerged. Of particular note was the structural similarity between the Rosé and Strijbos analyses, both including a Cognitive, Relational, and Motivational dimension. However, while there was a striking commonality across analyses, an interesting contrast between the Rosé and Strijbos coding schemes on the relational dimension was noted with respect to the idea of dominance. While on the surface the idea of dominance, as encoded in the Relational dimension within the two coding schemes, sounded similar, subtle differences in operationalization lead to different findings with respect to openness to ideas in the contrasts between groups and social positioning of individuals within groups. Further contrasts between all three of these analyses came out over iterations of these initial analyses as the chapters for this section were in preparation. The detailed comparison of the final analyses using the Rosé and Strijbos approaches are covered in the Howley et al. chapter included earlier within this PLTL Chemistry section and recounted as part of the analysis of the first iteration of multivocal analysis of the PLTL dataset discussed below.

In part as a way of gaining greater insight into the data for addressing the differences in findings from the earlier iteration, Sawyer and colleagues kicked off the second iteration of analysis by conducting a qualitative analysis of the PLTL corpus for this extended investigation. The rich contextualized blow-by-blow analysis of Sawyer and colleagues serves as a counter-point to the other analyses, which are quantitative in nature and attempt to draw conclusions from patterns found within a structure imposed on the data as an analytic lens. Because of its importance in contrasting with the other approaches, and because of the detailed view of the data it presents, that chapter is placed first among the analysis chapters within this section, although it came later in time in terms of the multivocal process. Within this section, the qualitative and the quantitative challenge one another's interpretations. Whereas the qualitative analysis has the benefit of contextual knowledge and human insight, it is limited with respect to the ability to distinguish between the typical and the idiosyncratic. On the other hand, the quantitative approaches, while they provide the machinery to not only make but quantify this important distinction, they are prone to misinterpretation caused by over-generalization between instances of behaviors that are treated as the same type. For purposes of this meta-analysis, the role of the qualitative analysis will be to challenge the treatment of the significance of individual events within the three quantitative analyses. The corresponding role for the quantitative approaches will be to challenge summative conclusions drawn within the Sawyer et al. analysis.

We begin with a discussion of the first iteration of multivocal analysis, which is a comparison across analyses using the concept of leadership as a lens as portrayed in the Rosé, Strijbos, and Oshima analyses. We will then discuss Iteration 2, which focuses more on knowledge building and idea improvement, including all four analyses. Finally, reflections on these two separate perspectives on the collection of analyses will be offered in conclusion.

Iteration 1: Reflections on Leadership Within Groups

The Rosé, Strijbos, and Oshima analyses all agreed that the Gillian group had more of a clear leader, whereas leadership within the Matt group was more diffuse. However, at a finer grained level, the analyses revealed important distinctions that bring out ways in which each of these operationalizations is limited, and how we can deepen our understanding of complex constructs such as leadership through multivocal analysis. Broadly, we consider the idea of leadership from two directions, first in terms of how authoritative a speaker presents him or herself as being, and second, in terms of how receptive other group members are to a leader's positioning of him or herself as a leader, which is based on the nature of their conversational responses.

Let us first consider the issue of authoritativeness in presentation of self. One of Jun Oshima's innovative word level network analysis was accomplished by quantifying and comparing "social relationships" between words within discourses. To this end, a social network analysis methodology was applied, whereby an edge represented a co-occurrence relationship between words within contributions to a discourse. And influential students identified in this analysis were those who used words that appeared in central locations within the graph. One student emerged as a central player within the Gillian group, but not within the Matt group. The same student was identified as authoritative within the Strijbos analysis. In this case, authoritativeness was measured by patterns of occurrence of turns labeled as

Positive Dominance and Negative Dominance, where positive dominance statements show leadership through positive polarity statements, such as declaring an idea as correct, while negative dominance statements show leadership through negative polarity statements, such as providing corrections or challenges.

The Rosé analysis approached the idea of presentation of self in terms of leadership on the Relational and Motivational dimensions, through two constructs from the field of systemic functional linguistics, namely, Martin and Rose's negotiation framework (Martin & Rose, 2003) and Martin and White's operationalization of heterogolossia (Martin & White, 2005). In the negotiation framework, authoritativeness is demonstrated by making a contribution to a discourse that is not offered as an invitation for validation from another group member. For example, an assertion that is made in response to a question that is framed as a hint rather than a serious question, and then followed by an evaluation, is not coded as an authoritative assertion. Based on this analysis, the Motivational dimension of the Rosé framework identified the same student as authoritative that the Oshima and Strijbos analysis did.

However, the heteroglossia framework that forms the Relational dimension of the Rosé analysis painted an alternative picture. Within that framework, assertions framed in such a way as to acknowledge that others may or may not agree, are identified as heteroglossic. Such assertions can be either expanding, in other words adding to the set of items up for negotiation, or contracting, in other words eliminating items from consideration. This could potentially be seen as similar to the notion of positive dominance and negative dominance within the Strijbos framework, and frequently there is a correlation between these two constructs. However, as noted at the first workshop where both the Rosé and Strijbos analyses were first presented and as highlighted earlier, it is not always the case that heteroglossic assertions that are framed as negative polarity statements perform the function of contracting the set of options under negotiation. For example, if it is a constraint that is eliminated, then more items are made negotiable since fewer constraints need to be satisfied. This subtle distinction between the Strijbos and Rosé approaches to measurement of leadership lead to differences in how students were ranked in the less clear cases. Furthermore, while the student identified as authoritative by all three analysts was identified as having more negative dominance statements in the Strijbos analysis, one might expect that student to have more contracting than expanding heteroglossic statements, it ended up being the case that the student had more expanding statements. Thus, we see how these similarly sounding operationalizations actually bring out different details in the analyses, and therefore lead to different conclusions.

Receptivity of leadership as evidenced by the response of other group members is another dimension along which interesting differences emerge in the investigation of leadership within the Chemistry discussions. Within the Oshima analysis, a much higher level of interconnectivity was evidenced within the Gillian group. High levels of vocabulary sharing could indicate higher levels of receptivity between students. In the Strijbos analysis, receptivity was indicated though Collaborative Orientation vs. Individual Orientation codes. Similar to the Oshima analysis, Strijbos identified the Gillian group as having more Collaborative than Individual utterances, whereas the Matt group was the opposite. In the Rosé analysis, however, receptivity is analyzed on the Cognitive dimensions through identification of Transactive contributions, which operate on reasoning displayed in a prior contribution (Berkowitz & Gibbs, 1979). If the Rosé analysis was consistent with the Oshima and Strijbos analyses, we would see more transactivity in the Gillian discussion, but interestingly, the opposite turns out to be the case. This is because although the discussion was focused on moving step by step through the problem rather than discussion concepts at length, the group reasoned out the problem solving steps together, checking each other's work. Thus, we see that there is a way of viewing the Matt group as having a more meaningful interaction than the Sawyer and Oshima analyses give them credit for.

This multivocal analysis of receptivity to leadership reveals how quantitative approaches to analysis of such constructs may inadvertently adhere too tightly to shallow features of interactions, such as the way highly transactive exchanges appear in conceptually oriented discussions, and thus miss valuable interactions occurring where the focus is more procedurally focused. In looking at these four different analyses, a multifaceted image of ideal leadership emerges that would not be visible in any single one of the frameworks investigated. For example, our multivocal separation between different leadership constructs allows us to view how it is possible to present ones views as standing on their own without denying others the right to have their own voice. With such a multivocal analysis, we can view how few leaders have mastered both aspects. Furthermore, a person who presents him or herself as authoritative might be more likely to elicit receptivity from his or her collaborators, but that result may not always occur for various reasons. Thus, this multivocal analysis has a greater potential as an assessment framework for identifying where strong but not stellar leaders may need improvement.

The concept of leadership provided a convenient theoretical lens through which to view the differing patterns identified within the three analyses. In treating the complex construct of leadership in a flexible, multifaceted way, we are able to view the differences that emerged as pointing to separate facets of leadership in such a way that it was not necessary to view the findings as contradictory. Instead, the differences challenged the initial simplicity with which we viewed the construct.

Iteration 2: Knowledge Building and Idea Improvement

The second iteration of analysis on the PLTL corpus began as the analysts began to write their chapters. While the concept of leadership proved to be a useful tool for integrating perspectives across analyses in the first iteration, the second iteration shifted focus to a combination of social positioning and idea improvement as discussion and comparison focused more in detail on the patterns found within the corpus using analytic tools, with interpretation playing a slightly less prominent role up front. Here we will compare across analyses at the group and individual levels as well as exploring the concept of pivotal moments.

Group Level Analysis

Most of the analysts agreed that the Gillian group was more conceptually oriented and interactive whereas the Matt group was more narrowly focused on problem solving. A major thrust of the Sawyer et al. analysis is making this contrast and showing many examples where the students in the Gillian group made reference to mathematical concepts and ideas, whereas the Matt group referred to formulas and problem solving steps. In concert with this view, the Oshima et al. analysis, wherein the shape of the network structure says something about the nature of the group dynamics, shows much more interconnectivity in the graphs associated with the Gillian group with a sparser, less well connected graph for the Matt group. The Rosé and Strijbos analyses specifically targeted individual behavior, and thus group level analyses are simply comparisons of group averages of individual behavioral tendencies. The Cognitive dimension of both of these analyses relate most directly to the conceptual versus procedural contrast. The Strijbos analysis echoes these earlier analyses, displaying argumentation in the Gillian group and suggestions paired with comparisons between computations in the Matt group.

The Rosé analysis presents a slightly different view, however. In both the Gillian and Matt groups the same number of reasoning statements are uttered, however, this number is a smaller percentage of moves in the Gillian group than in the Matt group. Thus, while the articulation of reasoning might be framed differently in the two groups, both groups are equally open with one another about their reasoning, and the Matt group is more singularly focused on reasoning. In the Gillian group there is more packaging around the reasoning. This packaging might be what makes the conversation hang together better and appear more highly inter-connected. And yet, the fact that the same number of reasoning statements are uttered raises questions about what the conceptual versus procedural contrast signifies. When we probe more deeply, looking at transactive contributions rather than simply reasoning statements, we see that a slightly higher proportion of reasoning statements in the Matt group are transactive. These mainly took the form of comparisons between problem solving approaches. While these comparisons had been written off as simply procedural by many of the analysts, the Rosé analysis calls these out as places where the students are considering each other's approaches and comparing them with their own in order to determine how best to solve the problem. These places where students are checking each other's reasoning are not called out as argumentation in the Strijbos coding scheme, although transactive utterances have been treated as argumentative knowledge construction in earlier work (Weinberger & Fischer, 2006).

Despite the differences highlighted above, the Strijbos and Rosé analysis do make similar observations about the Gillian group being argumentative with one another. On the Relational dimension, the Rosé analysis displays the Gillian group has having a high concentration of Monoglossic statements and Heteroglossic contract statements, both of which show a stronger stand in favor of one's own view and less openness to an alternative view. In the Matt group, in contrast, we see more Heteroglossic Expand statements, which explicitly express openness to other views. It may be that the high level of interconnectivity displayed within the Oshima analysis is an indicator of this more intensive interchange within the Gillian group rather than a reflection of deeper reasoning and knowledge construction per se. Consistent with this view, the Strijbos analysis displays more Negative Dominance statements in the Gillian group, and less dominance of any kind within the Matt group.

The differences in what these analyses bring out about the group discussions raise questions about what is desirable in PLTL groups. All of the analysts agreed with value placed on students sharing their reasoning and working together to refine that reasoning. Not all of the analysts agreed with what that should look like on the surface. Thus, there was not a consensus on where we see a higher quality discussion.

Individual Level Analysis

While the Oshima analysis offers a unique perspective at the group level using social network analysis techniques, with relatively few observations about individuals, the Sawyer, Rosé, and Strijbos analyses spend more time talking about each student's individual role within the discussions. Some of the most striking differences between approaches come out when examining the transcripts at the individual level.

The Gillian Group

Most of the analysts found the Gillian group more interesting and thus spent more time looking in detail at the interactions at the individual level within that group. The Social Network Analysis approach of Oshima enabled a unique perspective on responsibility for idea improvement, which was operationalized for each student as displacement of closeness and degree centrality coefficients within the original graph as compared to a version with that student's contributions not included. Using this metric, F1 stands out with respect to responsibility taking in the Gillian group. This is consistent with the finding in the Rosé analysis where F1 stands out as an idea leader, with twice as many Externalizations as Transactive statements within the Cognitive dimension. This might also be related to the identification of F1 as taking up a dominant role on the Relational dimension in the Strijbos analysis. However, in neither case do the codes really capture the same notion of responsibility that can be captured by Oshima's network analysis, which displays not the nature of the contributions, but their effect on the discussion. The statistical orientation of the Rosé and Oshima analyses reveal a significance to the nature of contributions that come from F1. However, this significance does not surface in the qualitative analysis of Sawyer and colleagues. There, the main thing that stands out about F1 is that she shows more and more signs of confusion as the discussion progresses. It could be that in focusing deeply and in detail at the interaction at the blow-by-blow level, it is not as natural to take a step back and see a big picture view. As we consider these alternative perspectives, they are not so difficult to reconcile. It is very possible for the questions raised by someone who is confused to trigger reflection on the part of group mates, and thus end up having an influential effect on the problem solving session.

Detailed comparison across the Rosé, Strijbos, and Sawyer analyses was possible at the contribution level, since all three analysts conducted their analysis at this level. Within the Cognitive dimension, the Howley chapter points out that both Rosé and Strijbos see relatively equal contribution from F1 and F4, but Rosé points to F1 as more of an idea leader than F4. In a line by line comparison, the finding was that there was very little overlap between the operationalization of codes on the two Cognitive dimensions. A chi-squared test failed to find any statistically significant relationship between the codes. Only 10 out of 52 of the codes on the Strijbos Cognitive dimension occurred on contributions coded as Reasoning in the Rosé analysis. So then in retrospect, it is more the commonality in interpretation than the distinction between the two analyses that is the surprise, since a similar conclusion is drawn from different codes bringing out details from a different portion of the discussion.

On the Relational dimension, again we find no statistically significant relationship between codes from the two coding schemes. Monoglossic contributions are frequent within all of the Strijbos codes. Heteroglossic Expand contributions are frequent within the Positive Dominance category from the Strijbos analysis, but Heteroglossic Contract contributions are equally represented within the Individual orientation and the Negative Dominance category. And all of the codes from this dimension in the Rosé analysis occur frequently in the uncoded contributions of the Strijbos analysis. The lack of relationship between Heteroglossic Expand/Contract and Negative/Positive Dominance explains the difference in conclusion pointed out in the Howley et al. chapter.

While each of the coding schemes includes a Motivation dimension, they focus on different aspects of Motivation. Whereas the Rosé analysis focuses exclusively on a connection with Self-Efficacy, the Strijbos analysis contains many more codes related to engagement and affect. Thus, the lack of a statistically significant relationship between codes on these two Motivation related dimensions is not surprising. Over the three dimensions, the story is the same. Whereas the two three dimensional coding schemes sounded very similar from a high level, when we look closely at the transcripts, we see how distinct they are in practice. Thus, again, the surprise is more in how similar the conclusions were between the two coding schemes rather than in the small distinctions that were raised. Nevertheless, the differences raise important questions for design of assessment frameworks for collaborative problem solving in terms of what we should consider desirable patterns of interaction in high functioning collaborative groups.

The Matt Group

Despite striking differences in approach and operationalizations, all three quantitative approaches saw the members of the Matt group as less distinct from one another than the Gillian group members. One reason, brought out by the qualitative analysis of Sawyer et al., is that the members of the Matt group appear to have been more equal in their ability level. Nevertheless, while the quantitative approaches see less distinction between the group members, the Sawyer analysis alludes to potential gender differences within the Matt group between lines 26 and 52. Specifically, in the Sawyer et al. analysis, the one male student is cast as playing a supervisory role, overseeing the calculations of the two female students, and sometimes offering feedback, corrections, or suggestions along the way. This might have been hinted at in the relatively high proportion of reasoning statements that are Externalizations by M, but that percentage is roughly the same as that of F4, who does not really stand out as distinct either in the Sawyer analysis or in the Strijbos analysis. This calls into question the interpretation of Sawyer that there was a gender effect in the style of interaction and social positioning within groups.

Pivotal Moments

All of the quantitative approaches pointed out places within the interactions they identified as pivotal. However, although all of the quantitative analysts adopted a notion of pivotal moment that was related to the idea of something unusual happening at a certain place, the authors were different with respect to which moments they picked out as pivotal.

The Oshima analysis pointed to one event in particular within the Gillian analysis that they identified as pivotal. That moment was identified as pivotal because the shape of the network structure representing the interconnectedness of articulated concepts dramatically changed with the introduction of this turn. This was a turn where student F1 said "you can find energy-k to find the wavelength of light." It is interesting that this turn would end up being pivotal from the network perspective. In form it does not look any different from many turns in the Matt group that were written off as uninteresting. The Oshima analysis did not mark this turn as pivotal based at all on their own interpretation of its form, content, or assumed intent. Instead, their analysis was based on its effect within their network representation.

Neither the Rosé analysis nor the Strijbos analysis specifically call out particular moments as pivotal, but they define a pivotal moment as one in which a student behaves in a way that is not in keeping with his or her tendency within the interaction. For example, both Rosé and Strijbos mention F2 as less active and somewhat marginalized in the Gillian group, but line 40, displayed within the Sawyer chapter, is a place where F2 behaves in an uncharacteristic way. First she offers a challenge and then two primary knower moves. In concert with this view of this contribution as pivotal for F2, we see an even greater significance of this move in the qualitative analysis presented in the Sawyer chapter. Sawyer notes that in this instance, F2 is correctly objecting to some incorrect assertions that F1 and F4 have made, and thus plays an important role in bringing the conversation back on track. The Rosé and Strijbos chapters present F2 as the less capable peer, and as less engaged in the interaction. But the Sawyer analysis offers a different interpretation of the pattern of behavior displayed in the Rosé analysis. The qualitative analysis shows F2 as a little bit slower in getting to the conclusion than F1 and F4, but she is actively trying to catch up, and eventually shows that she has more knowledge than perhaps she gives herself credit for knowing, and perhaps F1 and F4 are overly confident in their abilities. Thus, we see the importance of the interplay between a qualitative and quantitative analysis.

Integration

While the concept of leadership and the lack of qualitative analysis to ground the interpretation of patterns, it was possible to construct an interpretation in the first iteration that combined the separate analyses without finding them contradictory. That is more difficult to do from the perspective of contrasts like conceptual versus procedural, idea leader versus follower, more or less skilled. If leadership means pointing the way to a correct solution, then we might be forced to conclude that the Sawyer analysis casts doubt on the interpretations discussed in the first iteration. Nevertheless, the contrasting analyses provide some clarifications and refinements to constructs. For example, in contrasting the Strijbos and Rosé analyses, we see that what is revealed as higher involvement on the Cognitive, Relational, and Motivational dimensions does not necessarily correspond to ability level. We also see that what counts as a cognitively relevant contribution is not necessarily an articulation of reasoning, and vice versa. We also see that openness is not necessarily expressed through positive polarity expressions, and negative polarity expressions do not necessarily display a closed attitude. We are reminded that motivation is a multifaceted construct.

In contrasting these analyses with the Oshima analysis, we are faced with the question of whether one's role is more accurately characterized by the manner in which one presents oneself, or in the effect ones contributions have on the conversation. The question is raised as to what it really means to play a dominant role in a conversation. Is it how much of one's view is expressed? Is it in making a unique contribution? Is it in being correct? Is it in effecting change in the direction or nature of the discussion?

Reflections

This chapter recounts two iterations of analysis of the PLTL Chemistry corpus. In both cases, we find alternative interpretations of the inner workings of the two groups whose discussion we analyze. In the first iteration, the comparison was done in a highly theory focused way. The focus of the integration is on refinement of constructs rather than grappling with what really happened in the interactions themselves. In the second iteration, a different theoretical lens is chosen, and the comparison remains closer to the data and less focused on construct development. As we grapple with the blow-by-blow insights offered by the Sawyer et al. analysis, we are faced much more directly with questions about ground truth—what really happened in these analyses? Was F2 disengaged? Or was she a reflective participant who waited for a strategic time to step up and play her influential role? Was the Matt group just trying to get the problem solved, or were they checking each other's reasoning to make sure they were all on the same page? Was the Matt group approach an equal partnership? Or was M playing a chauvinistic managerial role over his two female group mates? In both iterations, we are left with challenges to the constructs that provide a way of layering meaning on the analysis.

As analysts part of a multivocal process, we are faced with the question of what is different between multivocal analysis and mixed methods research. My own answer to that question is that whereas both leverage multiple methodologies to improve the quality of the research, multivocality also transforms the researcher. We see that the analytic tools are malleable, lending themselves equally well to the two different theoretical framings for the integration. And yet we are forced to see them as brittle and imperfect as we are forced to realize that the tools of others are equally valid and equally brittle, and that only in grappling together can we move past our own limitations and move towards greater clarity. The alternative analytic lenses provided boundary objects that stimulated conversation between analysts. For me as analyst and discussant, this experience provided the first opportunity of this kind to explore interpretation of the same corpus from so many different angles. It was a pleasure to work side by side with my partner analysts, many of whom are leaders in the field of learning sciences, and to learn from considering their work at such a detailed level.

Acknowledgments This work was supported in part by NSF grant SBE 0836012 to the Pittsburgh Science of Learning Center.

References

Berkowitz, M., & Gibbs, J. (1979). *Unpublished manual for coding transactivity.*

Howley, I. K., Mayfield, E., Rosé, C. P., & Strijbos, J.-W. (this volume). A multivocal process analysis of social positioning in study groups. In D. D. Suthers, K. Lund, C. P. Rosé, C. Teplovs, & N. Law (Eds.), *Productive multivocality in the analysis of group interactions*, Chapter 11. New York, NY: Springer.

Martin, J. R., & Rose, D. (2003). *Working with discourse: Meaning beyond the clause*. New York, NY: Continuum.

Martin, J. R., & White, P. R. R. (2005). *The language of evaluation: Appraisal in English*. Palgrave: Macmillan.

Oshima, J., Matsuzawa, Y., Oshima, R., & Niihara, Y. (this volume). Application of network analysis to collaborative problem solving discourse: An attempt to capture dynamics of collective knowledge advancement. In D. D. Suthers, K. Lund, C. P. Rosé, C. Teplovs, & N. Law (Eds.), *Productive multivocality in the analysis of group interactions*, Chapter 12. New York, NY: Springer.

Sawyer, R. K., Frey, R., & Brown, P. (this volume). Peer-Led team learning in general chemistry. In D. D. Suthers, K. Lund, C. P. Rosé, C. Teplovs, & N. Law (Eds.), *Productive multivocality in the analysis of group interactions*, Chapter 9. New York, NY: Springer.

Sawyer, R. K., Frey, R., & Brown, P. (this volume). Knowledge building discourse in peer-led team learning (PLTL) groups in first-year general chemistry. In D. D. Suthers, K. Lund, C. P. Rosé, C. Teplovs, & N. Law (Eds.), *Productive multivocality in the analysis of group interactions*, Chapter 10. New York, NY: Springer.

Weinberger, A., & Fischer, F. (2006). A framework to analyze argumentative knowledge construction in computer-supported collaborative learning. *Computers and Education, 46*(1), 71–95.

Part IV
Case Study 3: Multimodality in Learning About Electricity with Diagrammatic and Manipulative Resources

Section Editor: Dan Suthers,
University of Hawai'i

The data for this section is from an innovative primary school science classroom in Singapore that used Group Scribbles collaborative sketching software in conjunction with physical manipulatives (batteries, light bulbs, and wires) in an exercise to understand how basic electric circuits work. Wenli Chen and Chee Kit Looi describe the setting and data collection in Chap. 14, "Group Scribbles-Supported Collaborative Learning in Primary Grade 5 Science Class." Four teams analyzed this corpus; the chapters are organized by certain similarities discussed below.

As reported in their chapter, "Identifying Pivotal Contributions for Group Progressive Inquiry in a Multimodal Interaction Environment" (Chap. 15, this volume), Chee Kit Looi, Yanjie Song, Yun Wen, and Wenli Chen analyzed their data using uptake and content analysis guided by a theory of progressive inquiry. Pivotal moments were identified along the uptake graph according to changes in content focus of activity.

Richard Medina also applied uptake analysis, but with an ethnomethodological orientation. His chapter "Cascading Inscriptions and Practices: Diagramming and Experimentation in the Group Scribbles Classroom" (Chap. 16, this volume), showed how group accomplishments are contingent on their material and interactional settings over time and across media and modes of interacting.

Kristine Lund, CNRS, and Karine Bécu-Robinault's chapter, "Conceptual Change and Sustainable Coherency of Concepts Across Modes of Interaction" also examines cross-media and cross-modal interaction, but with a focus on coherence and conceptual change. Motivated by a theory of semiotic bundles, they examine how translations between media and modes can be taken as evidence for individual students' evolving understanding.

Heisawn Jeong was also interesting in assessing understanding, but at the group rather than individual level. Her chapter, "Development of Group Understanding via the Construction of Physical and Technological Artifacts," (Chap. 18, this volume), assesses "group understanding" as evidenced by the group's collective diagrams and circuits, much as an archeologist assesses culture through its artifacts.

The discussion chapter by Suthers, "Issues in Comparing Analyses of Uptake, Agency, and Activity in a Multimodal Setting" (Chap. 19, this volume), identifies two major themes across the analyses: what evidences understanding, and practices of multimodal interaction across various media. He provides a little history with a related Group Scribbles case study that preceded the present one, discusses pragmatic issues concerning transcript sharing, discusses two points where the interpretations of the analyses differed in theoretically or epistemologically interesting ways, and then characterizes their different conceptions of individual and group agency and the distribution of activity in multimodal settings.

Chapter 14
Group Scribbles-Supported Collaborative Learning in a Primary Grade 5 Science Class

Wenli Chen and Chee-Kit Looi

Introduction

Computer technologies can play an important role in supporting students' collaborative learning. Various research efforts have examined the effectiveness of technologies that support collaboration among learners, by providing rich opportunities for students to engage in group work and to share group artifacts. There is CSCL research centrally concerned with investigating group interaction processes in virtual online environments (Cakir, Zemel, & Stahl, 2009; Stahl & Hesse, 2010). There is relatively less research on how group interactions take place across multiple media in a networked environment, where face-to-face (F2F) and online interaction spaces are intertwined, even though their respective affordances have long been studied (Dillenbourg & Traum, 2006; Suthers & Hundhausen, 2003).

The data reported in this chapter is from a large-scale 3-year research project investigating how to design and support students' collaborative learning using a networked technology called Group Scribbles (GS) in a F2F classroom. A very common pattern in classroom talk is IRE: a teacher initiation (I) is followed by a student reply (R), followed by an evaluation of this reply (E) by the teacher (Mehan, 1979). IRE has been observed to account for up to 70 % of teacher–student classroom interactions in the classroom (Nassaji & Wells, 2000; Wells, 1999) and is continuously reproduced as part of institutionalized schooling. IRE has been criticized for leading to unrewarding and boring classroom discussions. Changing such deep-seated traditional patterns of classroom discourse poses a considerable degree of challenge for classroom reform. The aim of the project is to transform the traditional IRE patterns of classroom talk into more student-centered ones by connecting students together by GS. There are three actors in a GS classroom: the

W. Chen(✉) • C.-K. Looi
National Institute of Education, Nanyang Technological University, Singapore, Singapore

D.D. Suthers et al. (eds.), *Productive Multivocality in the Analysis of Group Interactions*, Computer-Supported Collaborative Learning Series 16,
DOI 10.1007/978-1-4614-8960-3_14, © Springer Science+Business Media New York 2013

teacher as the facilitator, the student as the task performer/problem solver, and the different communication modes (GS and F2F interaction) as the mediator of the collaboration process. Introducing GS in the F2F classroom provides different kinds of scaffolding and support for the cognitive and social interactions between the participants involved.

Networked Technology: Group Scribbles

The CSCL technology used in classroom is GS 2.0, which was co-developed by SRI international and National Institute of Education Singapore. The GS user interface presents each user with a two-paned window (Fig. 14.1). The lower pane is the user's personal work area, or "private board," with a virtual pad of fresh "scribble sheets" on which the user can draw or type. A scribble can be shared by being dragged and dropped on the public board in the upper pane, which is synchronized across all devices. The essential feature of GS is the combination of the private board where students can work individually, engaging in the sense-making processes with the materials without being influenced by others (Vahey, Tatar, & Roschelle, 2004), and the public boards where students engage in group- or class-level interactions as they post and position their work relative to others, view others' posts, initiate discussion and critique ideas generated, and take items back to the private board for further elaboration. It is evident that GS technology scaffolds the process of different levels of interactions and the seamless switch between them, private interaction–group interaction–class interaction–group/private interaction, enabling a synergy between autonomy and collaboration by combining both private and collaborative learning. The F2F GS environment leverages resources such as shared screen, gestures, and conversation norms to help students jointly construct meaning, become more proficient in participating in representation-based interactions, and build a common understanding of the subject matter (Chen, Looi, & Tan, 2010; Vahey, Tatar, & Roschelle, 2004).

GS is a general-purpose collaboration tool in the sense that it does not assume a predefined topic or task but rather is intended to be appropriated for different tasks. GS enhances the characteristics of sticky paper notes and student response systems (SRS) by providing their key features for supporting brainstorming, idea response aggregation, and collaborative decision making while avoiding some of their physical constraints. It enables collaborative generation, pooling, and improvement of ideas through a synchronized public virtual space, eschewing the substantial manual work needed when paper sticky notes are used in classrooms (in terms of supplying, distributing, duplicating, moving, collecting, archiving, publishing, and sharing the notes). It complements other SRS technologies (e.g., Clickers and Classroom Presenter) in supporting coordinated use of the technology among students, liberating teachers from explicitly coordinating all classroom interactions.

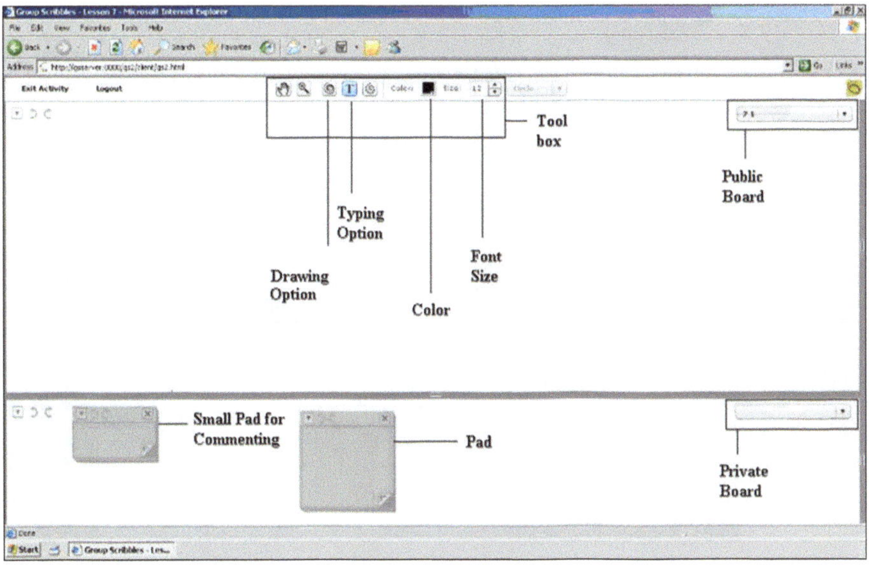

Fig. 14.1 The user interface of GS with a two-paned window

Context and Participants

In a 3-year school-based research project, we have worked with one primary school and three secondary schools in Singapore, systematically designing and implementing collaborative learning supported by GS for mathematics, science, and English and Chinese language learning (Chen & Looi, 2011; Looi & Chen, 2010). A design-research approach is adopted to address complex problems in real-classroom contexts in collaboration with practitioners and to integrate design principles with technological affordances to render plausible solutions (Brown, 1992; Collins, 1992). The GS lessons are integrated tightly with the national-mandated curriculum and co-designed by the researchers and the teachers.

In our work with a primary (elementary) school in Singapore, students from two primary grade 5 classes (one high-ability class and one mixed-ability class, each class having 40 students) have used GS technology in learning science, mathematics, and the Chinese language for 1.5 years (Chen, Looi, & Chen, 2009; Looi, Chen, & Ng, 2010) at the time of data collection. During each week, they had 1–2 sessions (1 h per session) of GS-based science lessons in the computer laboratory (Fig. 14.2).

In a 1-h GS-based lesson, about half of the time was devoted to students using GS to do collaborative learning tasks with the facilitation of the teacher. One session can have 1–2 collaborative activities, depending on the complexity of the tasks. When doing a collaborative task, students worked in groups of four. In the computer lab, there was an interactive whiteboard in the front so that the teacher and the students could write or draw on the large screen directly. Each student was equipped with a Tablet PC with GS

Fig. 14.2 GS classroom

Fig. 14.3 Seating arrangement of the target group

Agnes	Serena
Bruno	Joel

client software installed. As the students had been using the Tablet PCs as a learning tool for more than a year, they were used to them as part of the repertoire in class lessons.

The data reported in this chapter and analyzed in subsequent chapters is from a primary grade 5 science class. The data is from one target group (two target groups were chosen randomly from the ten groups formed in the high-ability GS class) and comprises four students: Agnes, Serena, Bruno, and Joel (all pseudonyms). They sat together at a separate desk with Agnes and Serena facing Bruno and Joel. The seating arrangement is as shown in Fig. 14.3. This is a group in which the abilities of the members were considered diversified, as it consisted of two students (Joel and Bruno) who had high scores of 81–90 % from previous science tests, one student (Serena) with a score of 71–80 %, and one student (Agnes) with a lower score of 61–70 %.

The Learning Task

The weekly GS lessons covered topics in line with the Science Syllabus Primary 2008. The topics include Cycles in Plants and Animals, Cycles in Water and Matter, Plant System, Human System, Electricity System, Interaction of Forces, Interactions within the Environment, Energy Forms and Uses, and Energy Conversion (Ministry of Education, Singapore, 2007). The reported lesson was the first GS lesson on the

Table 14.1 GS lesson flow: Electricity system lesson 1

	"Electricity System" GS activity	Time
1	Teacher divides the pupils into groups of four. Teacher gives the instruction of the learning activity: students need to find for themselves the different arrangements possible to light a bulb.	1 min
2	Students draw as many possible arrangements of bulb, wires, and battery in 5 min on GS group board.	10 min
3	Teacher distributes the materials (each group gets four wires, two bulbs, and two batteries). Students construct the circuits according to the diagrams created on GS. If the bulb lights up, they put a check next to the diagrams (successful diagrams).	10 min
4	Students browse through their classmates' group boards and view the diagrams created, endorsing the correct ones and commenting on the incorrect ones.	10 min
5	Students comment on and discuss the correct arrangements and give reasons why the incorrect arrangements failed to light up. Teacher randomly selects two groups to come up to the front of the class to present their circuits.	14 min

topic of "Electricity System." It was also the first GS lesson of the term (lessons, both with and without GS, were on different topics in the previous term). The lesson was intended to achieve two main objectives: (1) students would understand the internal arrangements of wires in a bulb and (2) students would construct a circuit to light a bulb. The collaborative learning activity was called "Bulbs and Circuits." It required the students to discuss configurations of connecting a light bulb with batteries in a circuit so that the bulb would light up (GS lesson plan, see Table 14.1). Before the lesson, students had learnt in previous lessons or in grade 4 that (1) electrical current can only flow through a closed circuit; (2) an electric circuit is an unbroken chain of conductors; and (3) an electric circuit consisting of an energy source (battery or batteries with positive and negative poles) and other circuit components (wire, bulb, and switch) forms an electrical system. The teacher had not taught the students that they need to connect the wire to the metal casing and to the metal tip of the light bulb. Therefore, the teacher expected that the students would learn through a process of trial and error to find the correct way of connecting the light bulb.

The activity started by getting the students to individually sketch out their initial impressions of how to connect closed circuits with a light bulb in their GS private board. They contributed their scribble sheets to their own GS group board and then discussed as a group. This task of consolidating the ideas on the same platform was intended to help them to infer the key similarities in constructing a working closed circuit from the various contributions posted in their group board. The students were also provided with some electrical components (batteries, light bulb, and wires) to physically connect the circuits following the manner they had sketched earlier in GS and to test if they would work. In testing the circuits, they could work freely as individuals, in pairs, or even as a group. Later, they had opportunities to look at other GS boards to be exposed to the different ideas contributed by the other groups. They could also comment on other GS posts if they desired to do so. This would reinforce their newly learnt concept of a closed circuit with a light bulb. The teacher followed the GS lesson plan closely when enacting it in the classroom.

Data Collection

When collecting data on the collaborative learning of this group, one video camera was set behind the classroom to record the whole classroom, while another camera recorded the interactions in the target group. The group video captured students' interactions and activities outside the GS boards. The screen capturing software Morae 2.0 was installed on the Tablet PCs of all the four group members to capture the process of each student's work on the Tablet PC and their verbal talk, facial expressions, and nonverbal behaviors via each laptop's webcam. Six videotapes were available in the data set: one for whole-class interaction, one for target group interaction, and one for each member of the group. The classroom interaction video was the longest with about 35 min in length, but the rest of the videos were all around 28 min long.

The data and materials provided to our colleagues in the Productive Multivocality project include the group video, student Morae videos, screenshots, GS inscriptions, GS lesson plan, and four Morae transcripts. The transcriptions were done by one of the researchers in the team. The researcher watched the Morae videos of each group member carefully and transcribed the actions captured (four Morae Transcripts). Then, she checked the accuracy of the transcripts by comparing the group video, Morae videos, and transcripts. In the transcripts provided, there are incidences of utterances in Chinese. The researcher has translated them literally into English.

We were aware that the requirements of researchers for the transcription would be different due to their different methodological approaches. The transcripts provided were not intended to serve their needs for analysis but to serve as a resource for obtaining a better understanding of the data (e.g., the students' language spoken in a colloquial manner is simplified). Researchers may need to generate their own transcripts, which serve their own different purposes from the raw data provided.

Acknowledgements This research was funded by the National Research Foundation (Singapore) under Grant NRF2007-IDM003-MOE-001. We are grateful to Mayflower Primary School for collaborating with us on this research.

References

Brown, A. L. (1992). Design experiments: Theoretical and methodological challenges in creating complex interventions in classroom settings. *The Journal of the Learning Sciences, 2*(2), 141–178.

Cakir, M. P., Zemel, A., & Stahl, G. (2009). The joint organization of interaction within a multimodal CSCL medium. *International Journal of Computer-Supported Collaborative Learning, 4*(2), 155–190.

Chen, W., & Looi, C. K. (2011). Active classroom participation in a group scribbles primary science classroom. *British Journal of Educational Technology, 42*(4), 676–686.

Chen, F. H. J., Looi, C. K., & Chen, W. (2009). Integrating technology in the classroom: A visual conceptualization of teachers' knowledge, goals and beliefs. *Journal of Computer Assisted Learning, 25*(5), 1–19.

Chen, W., Looi, C. K., & Tan, S. (2010). What do students do in a F2F CSCL classroom? The optimization of multiple communications modes. *Computers and Education, 55*(3), 1159–1170.

Collins, A. (1992). Towards a design science of education. In E. Scanlon & T. O'Shea (Eds.), *New directions in educational technology* (pp. 15–22). Berlin: Springer.

Dillenbourg, P., & Traum, D. (2006). Sharing solutions: Persistence and grounding in multimodal collaborative problem solving. *Journal of the Learning sciences, 15*(1), 121–151.

Looi, C. K., & Chen, W. (2010). Community-based individual knowledge construction in the classroom: A process-oriented account. *Journal of Computer Assisted Learning, 26*(3), 202–213.

Looi, C. K., Chen, W., & Ng, F. K. (2010). Collaborative activities enabled by Group Scribbles (GS): An exploratory study of learning effectiveness. *Computers and Education, 54*(1), 14–26.

Mehan, H. (1979). *Learning lessons.* Cambridge, MA: Harvard University Press.

Ministry of Education, Singapore. (2007). Science syllabus primary 2008. Retrieved on Mar 10, 2012, from http://www.moe.gov.sg/education/syllabuses/sciences/files/science-primary-2008.pdf.

Nassaji, N., & Wells, G. (2000). What's the use of 'triadic dialogue'? An investigation of teacher-student interaction. *Applied Linguistics, 21*(3), 376–406.

Stahl, G., & Hesse, F. W. (2009). Classical dialogs in CSCL. *International Journal of Computer-Supported Collaborative Learning, 4*(3), 233–237.

Suthers, D., & Hundhausen, C. (2003). An empirical study of the effects of representational guidance on collaborative learning. *Journal of the Learning Sciences, 12*(2), 183–219.

Vahey, P., Tatar, D., & Roschelle, J. (2004). *Leveraging handhelds to increase student learning: Engaging middle school students with the mathematics of change. Proceedings of the sixth international conference of the learning sciences* (pp. 553–560). Hillsdale, NJ: Lawrence Erlbaum Associates.

Wells, G. (1999). *Dialogic enquiry: Towards a sociocultural practice and theory of education.* Cambridge, MA: Cambridge University Press.

Chapter 15
Identifying Pivotal Contributions for Group Progressive Inquiry in a Multimodal Interaction Environment

Chee-Kit Looi, Yanjie Song, Yun Wen, and Wenli Chen

Introduction

Inquiry learning has its origins in the practices of scientific inquiry and focuses on posing questions, gathering and analyzing data, and constructing evidence-based explanations and arguments. Hakkarainen (2003) proposes an inquiry pedagogical approach termed as "progressive inquiry" for young learners in learning science. Progressive inquiry is a progressive and cyclic process consisting of six interacting elements that guide learners to (a) systematically generate their own research questions, (b) construct their own intuitive working theories, (c) critically evaluate various intuitive conceptions generated, (d) search for new scientific information, (e) engage in progressive generation of subordinate questions, and (f) build new working theories as the inquiry process continues. The process is not linear and may not involve all the components in each learning cycle.

From the knowledge advancement and creativity perspective, Hakkarainen and Paavola (2009) posit that in progressive inquiry, a focus should be placed on how students collaboratively organize their activities for developing something new. Rather than focusing solely on idea improvement, they propose using a trialogical approach to guide the examination of learning. The trialogical approach emphasizes the collaborative development of mediating objects or artifacts rather than monologues within a mind (the acquisition view) or dialogues between minds (the participation view). The aim of the inquiry is to progressively refine concrete knowledge artifacts or to further elaborate upon a shared object.

C.-K. Looi (✉) • Y. Wen • W. Chen
National Institute of Education, Singapore, Singapore
e-mail: cheekit.looi@nie.edu.sg

Y. Song
Hong Kong Institute of Education, New Territories, Hong Kong

D.D. Suthers et al. (eds.), *Productive Multivocality in the Analysis of Group Interactions*, Computer-Supported Collaborative Learning Series 16,
DOI 10.1007/978-1-4614-8960-3_15, © Springer Science+Business Media New York 2013

There are various ways to examine collaborative knowledge advancement in inquiry learning in the existing literature (Stahl, Koschmann, & Suthers, 2006). Some researchers examined collaborative process at both the individual and group levels (e.g., Arvaja, Salovaara, Häkkinen, & Järvelä, 2007). Other studies have examined collaboration as knowledge convergence, which focuses on individual contributions independently of each other (e.g., Weinberger, Stegmann, & Fischer, 2007). However, Stahl (2002) posits that to understand collaborative learning, it is important to understand how groups work together to make sense of the problem inquiry situations.

In this paper, we explore a trialogical approach that focuses on concrete objects in mediating collaboration to examine how progressive inquiry in science learning happened in one target group of a grade 5 class supported by Group Scribbles (GS)—a collaborative technology (Chen, Looi, & Tan, 2010). We are interested in investigating how a group of students working as a group co-constructed shared artifacts in a science lesson on the topic of electricity via different modes of interactions. To achieve these aims, uptake analysis (Suthers, Dwyer, Medina, & Vatrapu, 2010) is adopted to discover the meaning-making process in the group cognition, leading to progressive inquiry in science learning. We extend uptake analysis to explore the notions of different types of uptakes in a multimodal environment setup comprising verbal, online, and experimental science practice interactions. We also further explicate the notion of pivotal contributions (a contribution that plays the role of shifting the direction of the subsequent events seamlessly or abruptly) by identifying them in the dataset and explore how these contributions shape the direction of the group inquiry productively. This chapter constitutes our contribution to productive multivocality in the analysis of collaborative learning (Suthers et al., 2011).

The research context, the participants, and the data sources are covered in the data chapter (Chen & Looi, Chap. 14, this volume). The organization of the chapter is as follows: we first discuss the analytic approach and how each contribution is coded and categorized and thereafter present the transcript of the dataset. We next identify pivotal contributions via transitional stages of interactions that influence the course of inquiry as well as uptake paths for group progressive inquiry. The chapter ends with a discussion of the results.

Research Methods for Analyzing Progressive Inquiry in GS

Data Sources

To understand the group progressive inquiry process, we focused on three data sources: speech and gesture interactions, artifacts in GS, and captures of experimental practices, as described in Table 15.1.

Table 15.1 Descriptions of data sources

No.	Data sources	Description
1	Speech and gesture interactions	Verbal interactions consisting of conversations and gestures between students and between the student(s) and the teacher in the course of collaborative progressive inquiry
2	Artifacts in GS	Data related to the posting or the editing of a GS note on the private board of individual students or on their group board in GS
3	Captures of experimental practices	Data related to the captures of hands-on experiment of connecting the circuit and the outcome of experiment (trial–error actions)

Uptake Analysis

One approach to analyzing collaborative interactions is sequential analysis deployed to examine the meaning of an act or an utterance as a function of its context of the prior sequence of acts and utterances. To understand collaborative interactions, conversation analysis (Sacks, 1992) and video analysis (Koschmann, Stahl, & Zemel, 2005) are adopted, focusing on turns or adjacency pairs to produce interpretive results (Stahl et al., 2006). However, analysis based on turn-taking or adjacency pairs is appropriate neither for synchronous nor for non-synchronous online communications in which contributions may be produced in parallel and are persistent due to the nature of the collaborative technology. We cannot reduce the complexity of the analysis by shrinking the time window to search for relevance relations only in adjacent contributions. Any previous contribution produced on the technology may be taken up again in later nonadjacent contributions (Suthers, 2006; Wee & Looi, 2009).

In addition, in many cases, multiple media are involved and the data are collected in a variety of formats; we need to transcribe and examine the data for collaborative interactions, which may not be apparent upon inspection, being distributed across these different media (Suthers et al., 2010). To address such issues, Suthers (2006) proposes the concept of "information uptake" which refers to "the event of a participant doing something with previously expressed information"; and this previously expressed information is termed as a previous contribution. In later definition of uptake, Suthers et al. (2010) posit, "Uptake is present when a participant takes aspects of prior events as having relevance for ongoing activity" (p. 5). Uptake can add to or modify the prior contributions or relate it to new contributions. It can take up a participant's own prior contribution as well as those of others. By identifying both types of uptakes, it is possible for researchers to characterize the mixture of intrasubjective and intersubjective knowledge constructions (Suthers, 2006). It is noted that a prior contribution refers not only to a relevant immediately adjacent event but also to relevant nonadjacent events that have "logical adjacency" (Stahl, 2006, p. 91).

In our study, we attempt to examine student collaborative meaning making through the evolvement of artifacts created in GS in multimodal interactions among speech, artifacts created in GS, and captures of experimental practices, using a trialogical approach. Due to the complexity of the interactions, we adopt uptake analysis method to discover patterns of the collaborative activities. Building on the concept of "uptake," we define an "uptake" as the event of a participant doing something with previous contribution(s) by the student, a group, or the teacher. Doing something in our study means the process of verbal communication, artifact creation in GS, and experiments in lighting up bulbs in the course of progressive inquiry. The uptake analysis does not provide explanations or make predictions of the relationships. Uptake is treated as "a fundamental building block of interaction, and the basis for construing interaction as an object of study" (Suthers et al., 2010, p. 7).

We attempted to address three questions in the analysis:

1. What are the types of uptake in the interactions between speech interactions, artifacts, and experimental practices?
2. How are pivotal contributions identified across the interactions between speech interactions, artifacts, and experimental practices in the uptake paths?
3. How do the uptake paths support progressive inquiry adopting a trialogical approach?

We next summarize our theoretical framework and methodology through a discussion of five specific dimensions (cf. Chap. 2).

Theoretical Assumptions

In the collaborative learning environment supported by GS, there are multiple media that mediate students' progressive inquiry, which is represented at two levels:

1. The first level (uptake analysis): It is assumed that uptake happens in the learning environment when a participant takes aspects of prior events as having relevance for ongoing activity. Progressive inquiry can be examined through pivotal contributions marked out at different stages in the uptake paths.
2. The second level (a trialogical approach): Tracing the development of the students' shared artifacts in GS helps make the progressive inquiry "materialized."

Purpose of the Analysis

The purposes of using uptake analysis are:

1. To interpret pivotal contributions in the context of the uptake diagram for progressive inquiry.
2. To interpret how the uptake paths support group progressive inquiry by focusing on visualizing the evolving process of the shared group artifacts.

Unit of Interaction

The data are sequential traces of contributions (data belonging to the same event was grouped together and coded as a contribution if they manifest a single interactional move or behavior). The unit of interaction is "uptake" which presents when a participant takes aspects of prior events as having relevance for ongoing activity.

Representations of Data and Analyses: Coding Contributions and Uptakes

For representing the data and the analysis, we first studied video recordings (all around 30 min) in the electricity lesson (one target group video recording, four individual Morae video recordings, and one class video recording). Secondly, we discussed segmentation of the video data for the uptake analysis. Thirdly, we chose the group video clip between the timing 00:12:15 and 00:16:16 as our object of analysis because we concurred that interesting interactions happened in this segment. We then synchronized what happened during this time period with the other video clips accordingly.

Fourthly, to do the data analysis, we adopted the family of methods loosely classified as exploratory sequential data analysis (Sanderson & Fisher, 1994), especially interaction analysis (Jordan & Henderson, 1995), to identify ethnographic chunks (easily identifiable behavioral units) first and then to transform the data into representations that are more suitable for analytic interpretation (Suthers et al., 2010). One of the authors transcribed the video clips based on the chunking units of: speech communication artifacts created in GS and captures of experimental practices that were represented using screenshots and were logged together with the transcribed verbal text in a sequential order. Another author read the data logs, identified obscure transcriptions, and suggested going back to the raw data for re-transcription of those parts.

Fifthly, the transcribed data were coded and analyzed line by line along with the screenshots of the artifacts in GS and experiments. Data belonging to the same observable action was grouped together and coded as a contribution with an assigned contribution number. Each contribution was numbered in a sequential order chronologically. The numbered contributions can be (a) an individual utterance; (b) an act of artifact creation in GS and experiments; or (c) sets of sequential utterances or acts that form a single interactional move by one participant or the group.

Next, in order to distinguish the three forms of contributions by different participants, we coded the contributions in four ways: (a) To represent a specific participant's utterance, we chose to use the first capitalized letter of the contributing participant's name in front of the numbered contribution. For example, J2 represents the contribution from the participant Joel, which is the second coded contribution in sequence. (b) To represent an act of artifact creation in GS, we chose to use the first capitalized letter of the contributing participant's name, followed by a small letter "g" in front of the numbered contribution. For example, Bg17 represents the artifact created in GS from the participant Bruno that is the 17th coded contribution in sequence. (c) To represent an act of experimental practice, we chose to use capitalized "G" to represent the group doing the experiment, followed by a small letter "e" to represent experimental practice. For example, Ge22 represents the experiment conducted by the group that is the 22nd coded contribution in sequence. (d) To represent special coded contributions such as when two participants were involved in creating the artifacts, we used artifact coding method and used a plus to combine the two in front of the numbered contribution. For example, "Ag+Bg26" represents artifact creation by Agnes and Bruno, which is the 26th coded contribution in sequence. A similar coding method was applied to the coding of an experimental practice. Finally we triangulated the coding and presented the contributions chronologically (see Appendix I). Referring to the coded contributions and their related transcripts, we identified uptake relationships. We identified the contributions that added to or modified the previous contributions to a new form. We then identified whether the interaction relationships between contributions was intrasubjective—a participant did an uptake on his/her own contribution(s), or intersubjective—a participant did an uptake on others'/group's prior contribution(s). These contributions had "logical adjacency" rather than ordinary adjacent pairs. We generated uptake diagram using arrows to demonstrate uptake relationships (Suthers, 2006). The results of the uptake analysis were triangulated between two researchers.

Manipulations

After the phase of coding to identify contributions in sequence and uptakes, two other phases were involved in the data analysis: identifying pivotal contributions via transitional stages of the interactions and identifying uptake paths for group progressive inquiry.

Identifying Pivotal Contributions via Transitional Stages of the Interactions

We define a pivotal contribution as a contribution that plays the role of shifting the direction of the subsequent events (contributions) seamlessly or abruptly through uptake between the subsequent event (contribution) and the transitional stage. Identifying transitional stages goes hand in hand with identifying pivotal contributions.

Jordan and Henderson (1995) posit that events of any duration are always *segmented* in some way. They stated that analysts are interested in the ways in which participants make the internal structure of the events visible to themselves and to each other and are interested in how they can present in some sense that they have reached a segment boundary in the work and that the next stretch of interaction will be of a different character. Thus, a segment boundary is the place where a transition occurs from one segment of an event to another indicated by a shift in activity. In some cases, the students are aware of the sequence of the learning activities and the problems they need to solve in general, although how they are involved in the activities and how they solve the problems in real learning situations may vary. When the students finish doing something and something new is starting, it is considered a smooth transition from one stage to another. Such a transition was termed "a seamless transition" (Jordan & Henderson, 1995). In some cases, the transition from one stage to another is not smooth. It is stopped abruptly and shifts to a new stage, which is termed as "abrupt transition" by Wee and Looi (2009).

In our analysis, we adopted the methods of segmenting a series of events in the form of coded contributions (the utterances, artifact creation, and experimental practices) into different stages based on the segmented transitional boundaries to trace the progressive inquiry. Both seamless and abrupt transitions were identified in the uptake graph. For example, contributions O1, J2, and B3 (see Appendix) were a series of events about proposing, praising, and accepting the idea of using two batteries to light a bulb. S4 proposes doing an experiment on the idea, and from Je5, activity shifted to doing the experiment. Thus the transitional boundary is between S4 and Je5, S4 is a pivotal contribution, and the transition is smooth and seamless. From contributions Je5 to J9, the events were concerned with students doing an experiment successfully with verbal utterances. S10 raised a new question of trying a new experiment, so here is a transitional boundary between J9 and S10 and S10 is another pivotal contribution. But this approach suggested by S10 failed to be further explored, as at this moment, the teacher (T12) facilitated the students to represent their understandings gained from speech interactions and experimental practices onto the GS space. T12's utterance of "No draft, no draft [on the group board]. Then people will look at a blank board" caused the abrupt shift of the event from doing and discussing about the experiments to B13's working on drawing the artifact in GS. Thus one more transitional boundary lies between T12 and B13; and the transition is abrupt. T12 is the real subsequent pivotal contribution after S4. To distinguish between successfully and unsuccessfully explored pivotal contributions, we named them manifested (the former) and latent (the latter) pivotal contribution, respectively, in this paper. S4 and T12 are *manifested pivotal contributions*, and S10 is a *latent pivotal contribution.*

By identifying the uptakes, transitional stages, and pivotal contributions of uptake in a graph representing interaction process, we are able to discern uptake paths. These paths are helpful for us to make interpretations of students' group meaning-making process of progressive inquiry illustrated in the next section.

Identifying Uptake Paths for Group Progressive Inquiry

This phase of uptake analysis focused on (a) connecting interactional relationships to identify uptake paths for the group progressive inquiry and (b) providing evidence that supported group knowledge co-construction in the inquiry process.

We did coding of video data and produced the transcript in Appendix I. We distinguished the different modalities of the contributions comprising conversations and gestures; GS artifacts being created, edited, or moved about; and the state of the students' trial–error experiments as represented by the still shots extracted from the videos. For some of the students' trial–error experiments, because there were contributions in which the view of the experimental circuits was blocked or occluded, we were not able to observe what they were working on. In such situations, we relied on verbal utterances and the GS artifacts to infer what might be going on.

Results

Types of Uptake

We used symbols to visualize the flow of uptake in the progressive inquiry (see Fig. 15.1). Different modalities of contributions are represented in different shapes: a square represents a contribution of a verbal utterance; a circle represents an artifact creation in GS; and a triangle represents a contribution of an experiment. The dotted line refers to intrasubjective uptake, and the solid line refers intersubjective uptake.

Table 15.2 shows two dimensions of contributions in terms of uptakes that are intrasubjective or intersubjective (SS stands for student–student, and ST stands for student–teacher) and the modalities. According to the table, there are a total of 27 ($3 \times 9 = 27$) types of uptake. Among them, the percentage of speech-initiated uptakes (42.3 %) is higher than that of the other two modalities initiated uptakes (30.8 and 26.9 %, respectively). Nevertheless, the difference among different modalities is not huge. This means that every modality of interaction is important for the emergence of uptake.

We also counted the number of the various uptakes between the same participant, and between different participants. It was noted that the majority of the interactions ($61.5 + 23.1 = 84.6$ %) happened between different participants, and the interactions between the same participants (15.4 %) were identified mainly from Bruno and Agnes. This indicates that the two students were more engaged in reflection and evaluation of their working theories. The majority of intrasubjective uptakes happened via GS artifacts. That means Bruno and Agnes were leading the role of externalizing their understandings by means of working artifacts in GS.

Pivotal Contributions in the Progressive Inquiry

We seek to identify the pivotal contributions via examining transitional stages of the interactions. In our study, we identify seven pivotal contributions in the

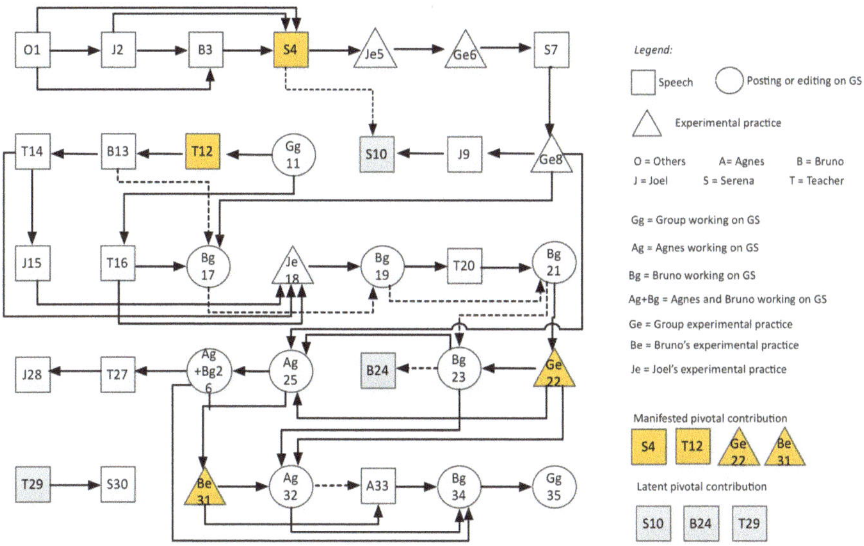

Fig. 15.1 The uptake diagram with pivotal contributions

uptake diagram (see Fig. 15.1) in different transitional stages. Among the pivotal contributions, four are manifested pivotal contributions S4, T12, Ge22, and Be31, and three are latent pivotal contributions S10, B24, and T29. Linking to the structural features of the graph, we describe how each of the pivotal contributions led to significant shifts of the progressive inquiry in understanding the ways to light one or more bulbs using one or more batteries and how to conceptualize a closed electric circuit comprising batteries and bulbs.

Manifested pivotal contribution S4: From inspecting Je5, Ge6, and Ge8 as a path of experimental practice and the preceding event S4 as a high-degree node (that is, one with a good number of incoming and outgoing uptakes), S4 is a candidate for a pivotal contribution. In the uptake path (O1–S4), although a student from another group (O1) suggested that the group tried to light a bulb using two batteries, which was appreciated by Joel (J2) and Bruno (B3), it was Serena (S4) who set up the question for the group to connect a circuit to light a bulb using two batteries for the group using both verbal language and gestures. This sparked the group's (J2, B3, and S4) interest in exploring new scientific information through a new experiment. Thus, Serena (S4) contributed to the group inquiry path shift from the previous experiment of using one battery to light a bulb to a new experiment of using two batteries to light a bulb.

Latent pivotal contribution S10: After completing the experiment of lighting a bulb using two batteries, instead of reflecting and externalizing their intuitive working theories by drawing their understandings of the closed electrical circuit in GS, Serena (S10) raised a new question of trying to understand how to light two bulbs

Table 15.2 Types of uptake and its percentages

		Speech initiated			Artifacts initiated			Hands-on initiated			
		to F2F	to Artifact	to Hands-on	to Artifact	to F2F	to Hands-on	to Hands-on	to F2F	to Artifact	Sum (%)
Intrasubjective		1	1	1	3	2	0	0	0	0	8 (15.4)
Intersubjective	SS	8	1	2	6	0	3	1	3	8	32 (61.5)
	ST	5	2	1	0	2	0	0	2	0	12 (23.1)
Sum (%)		14 (26.9)	4 (7.7)	4 (7.7)	9 (17.3)	4 (7.7)	3 (5.8)	1 (1.9)	5 (9.6)	8 (15.4)	52 (100)
Total (%)		22 (42.3)			16 (30.8)			14 (26.9)			

using two batteries. At this critical moment, the teacher (T12) facilitated the students to represent their understandings gained from speech interaction and experimental practice onto the GS space. In the uptake diagram, we can see a breakpoint between S10 and subsequent contribution. Thus here S10 is a latent pivotal contribution which fails to be further explored.

Manifested pivotal contribution T12: In the uptake path (Gg11–Bg21), the teacher's facilitation and orchestration (T12) was picked up by Bruno (B13) and was crucial to change the direction of the group inquiry path from trying to do a new experiment to reflecting and conceptualizing their working theories of how to light a bulb and the mechanism of the circuit. So Bruno (B13) began to draw the circuit in GS. While Bruno (B13) was drawing, the teacher (T14) reminded him, "Where are the two batteries?" which indicated that Bruno should show the circuit of connecting two batteries to light the bulb. This further fostered the group to do the experiment and draw their working theories by Bruno (B17) gained from the experimental practice.

Manifested pivotal contribution Ge22: In the uptake path (Ge22–J29), before posting their conceptualized working theories of the mechanism of the closed electrical circuit in a graph on the private board to the group board in GS, the group (Ge22) did the hands-on experiment again to evaluate their intuitive understanding. After the evaluation, Bruno seemed confident that his drawing of the circuit (Bg21) was "correct" after further checking the experiment results by the group (Ge22) and posted it to the group board (Bg23). Ge22 is pivotal as it helped Bruno confirm his conceptualization of the circuit before initiating further inquiries. Then, Bruno (B24) proposed to try to light three bulbs. At this moment, it was Agnes (A25) who identified the problem in the drawing by uptaking the information from previous contributions Ge8, Ge22, and Bg23 and began to externalize her understanding of the working theories by starting to draw another circuit in GS. Bruno and Agnes (Ag + Bg26) began to work together on the evaluating, reflecting, and revising their group working theory.

Latent pivotal contribution B24: When Bruno (B24) proposed to try to use three bulbs, his proposal was not answered. Instead, students zeroed in on evaluating and reflecting their temporal inscriptional artifact.

Latent pivotal contribution T29: While Agnes and Bruno (Ag + Bg26) were working in GS to conceptualize their new understanding of drawing the circuit, the teacher (T29) asked the students to stop working on their own group board and to visit other group boards. This was only picked by Serena (S30) but failed to be continued.

Manifested pivotal contribution Be31: In the uptake path (Be31–Bg35), while referring to Agnes' (Ag25) drawing on his private board, Bruno (Be31) began to explore new information further through experiment in order to externalize his understanding. The new experimental practice transformed Bruno's understanding. So Be31 played a pivotal role in making Bruno accept Agnes' (A32 and A33) advice on deleting the wrong drawing and completing a correct one (Bg34) and posting it (Bg35) to the group board. This contributed to Bruno's success in externalizing the working theories of the closed electric circuit to light a bulb using two batteries.

Agnes and Bruno manifested group cognition in this process that was made possible because of the mediation of F2F, online, and experimental practices.

Uptake Paths for the Progressive Inquiry

The pivotal contributions marked out different stages of progressive inquiry as distilled out in the uptake paths described below:

Stage 1: Generate students' own research questions (from O1 to S4): A student from another group (O1) suggested to Serena (S4) to try to light the bulb using two batteries (after the group successfully lighted a bulb using one battery). This information was overheard by Joel (J2) and Bruno (B3). Serena (S4) had an uptake on the information from O1, J2, and B3 and set up the question of trying two batteries to light a bulb based on their intuitive working theories of how to connect the light bulb using one battery. S4, the manifested pivotal contribution, helped the group to shift from proposing the inquiry problem to the problem experimentation seamlessly.

Stage 2: Search for scientific information through experimentation (from Je5 to S10): The group, initiated by Joel (Je5), began to search new ways through experimentation to connect the circuit with the bulb and two batteries. They connected the bulb but failed in lighting the bulb in their first group attempt (Ge6). It was Serena (S7) who found the problem ("wrong side" was taken up by the group). The group quickly succeeded in lighting the bulb in their second attempt (G8). Serena then (S10) had a new proposal but was not adopted by the group.

Stage 3: Construct their own intuitive working theories of how to light the bulb and draw the bulb circuit (from Gg11 to Bg21): After successfully connecting the bulb with the batteries, the group did not externalize their working theories onto the group board (Gg11). It was the teacher (T12) who advised the group in time to show their co-constructed understanding of the bulb circuit through experiment on the group board. This information enabled an uptake by Bruno (B13) which led to the group's later work on externalizing their understanding of how to connect two batteries to light a bulb. In this stage of progressive inquiry, the manifested pivotal contribution T12 helped facilitate students' inquiry to a new height—to conceptualize their working theories in lighting up the bulbs. T12 introduced an abrupt transition of the group's inquiry process from doing further experimentation to conceptualizing the group working theories.

Stage 4: Evaluate different intuitive understandings of the bulb circuit (from Ge22 to T29): Before posting the drawing of the bulb circuit to the group board, the group (Ge22) connected the light bulb again for the evaluation of the group's intuitive working theories on bulb circuit. Bruno (Bg23) checked how the bulb was lighted using two batteries again and then posted the drawing of the circuit from his private board to the group board. Bruno's proposal (B24) of testing three bulbs was not picked up. Instead, Agnes (Ag25) began to draw the circuit on Bruno's private board again to explicate whether Bg23's bulb circuit was correct. This triggered Bruno (Ag + Bg26) to observe and think whether his previous drawing was wrong. In this stage of inquiry, the pivotal contribution Ge22 played a crucial role in leading the process of revising the group's working theories seamlessly. On the contrary, the latent pivotal contribution T29 of the teacher's instruction of asking the students to stop working on their own group work was not followed by the group and hence did not influence the group's progressive process.

Stage 5: Build new working theories of the bulb circuit and share the co-constructed artifacts on the group board through experimentation and collaboration (Be31–Bg35): Enlightened by Agnes, Bruno (Be31) began to explore new information further through experiment in order to externalize his understanding. Agnes (A33) deemed that Bruno was able to understand how to build the new working theories of the bulb circuit and asked Bruno to continue Agnes' (Ag32) uncompleted drawing. Bruno (Bg34) completed the drawing of the bulb circuit with one bulb and two batteries. Agnes and Bruno exhibited mutual understanding at this stage, and the manifested pivotal contribution Be31 helped the group advance their working theories posted on the group board seamlessly.

At each stage of the group progressive inquiry, we also examined the uptake paths to find out the types of uptake (see Table 15.2).

Table 15.3 shows that Stage 1 had the least types of uptake, and Stages 3, 4, and 5 had the most types of uptake. This indicates that in the course of the progressive inquiry, more varieties of the types of uptake might be involved. In addition, Stages 3, 4, and 5 accounted for more percentage of uptake (35.42, 20.83, and 20.83 %, respectively) than Stages 1 and 2 (12.50 and 10.42 %). This indicates that students were more engaged in the progressive inquiry over time.

Uptake Paths for the Progressive Inquiry at a Theoretical Level

In this section, we provide evidence of how the uptake paths supported group progressive inquiry by focusing on visualizing of the evolving process of the shared artifacts using the trialogical approach in GS. The dataset was chosen from Stages 4 and 5 (from Ge22 to Gg35) shown in Fig. 15.2.

The essence of trialogical approach is to examine student inquiry learning through the collaborative development of mediating shared knowledge artifacts or objects instead of focusing on monologues in mind or dialogues between minds (Hakkarainen & Paavola, 2009). The analysis of the two stages was to elaborate how the students progressively revised concrete shared knowledge artifacts to construct knowledge. For the convenience of elaborating the approach, we put the development of shared knowledge artifacts in GS in the upper row and the other modalities (the verbal interactions and experimental practices in the two progressive inquiry stages) in the lower row as shown in Fig. 15.3. In the figure, the arrows indicate the uptakes of information from previous contributions.

Table 15.3 Types of uptake at each stage of group progressive inquiry

Stage (contributions)	Stage 1 (O1–S4)	Stage 2 (Je5–S10)	Stage 3 (Gg11–Bg21)	Stage 4 (Ge22–T29)	Stage 5 (Be31–Gg35)
Types of uptake (number)	1	3	9	7	7
Total number of uptakes (%)	6 (12.50)	5 (10.42)	17 (35.42)	10 (20.83)	10 (20.83)

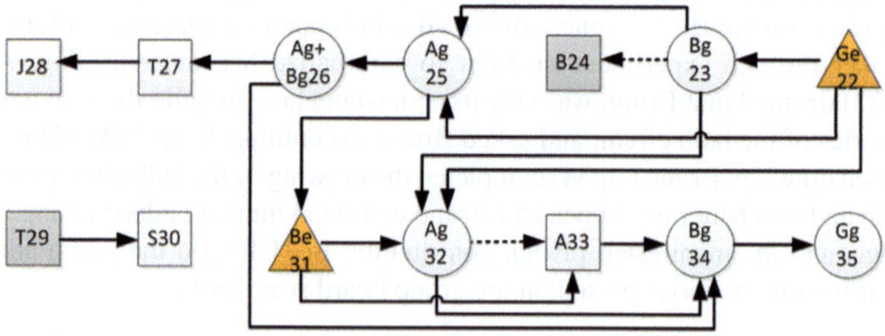

Fig. 15.2 Evolving process of shared artifacts

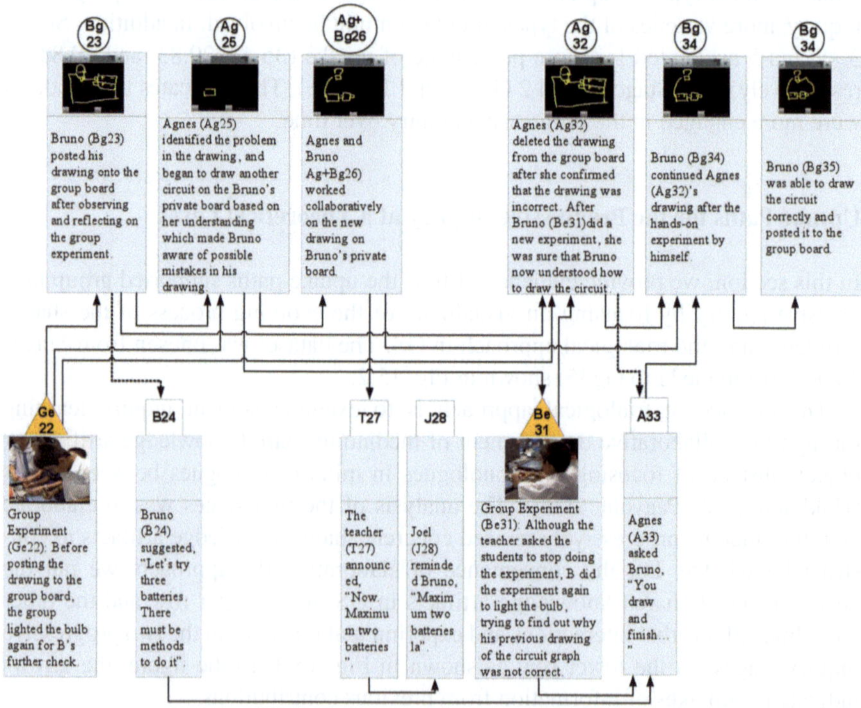

Fig. 15.3 Co-construction of shared knowledge artifacts in GS in the progressive inquiry

Figure 15.3 shows that the progress of co-construction of the shared knowledge artifacts (Bg23, Ag25, Ag+Bg26, Ag32, Bg34, and Bg35) involved the multimodal interactions and mediations between experimental practices, speech, and online communication in GS. First, Bruno (Bg23) posted the artifact of his conceptualized working theories of lighting up the bulbs onto the group board after further checking the group experimental practice (Ge22). The group shared artifact helped Agnes to visualize the working theory, which encouraged her to work on a new artifact to externalize her understanding. The artifact (Ag25) created by Agnes could be traced back to the group experimental practice (Ge22). The shared experimental practice shown visibly to Agnes mediated her to re-work on the artifact (Ag25) in GS, which, in turn, mediated Bruno to reflect on the artifact that Agnes was working on and work together with Agnes on collaboratively constructing the artifact (Ag+Bg26). While working on the shared artifact, Bruno did the experiment again (Be31), which mediated him to work out the knowledge artifact (Bg34) and post it to the group board as group shared knowledge artifact (Gg35). We have to point out that the process of the evolvement of the student knowledge artifacts was contributed by the interactions and mediations between different artifacts, experimental practices, and online F2F communications. The pivotal contributions from the group (Ge22) and Bruno (Be31) played significant roles in the evolvement of artifact construction.

Discussion

In this chapter, using an interaction analysis method we identified pivotal contributions in different transitional stages and five stages for progressive inquiry in the uptake paths. The method provides a lens to help us understand the group progressive inquiry into how to connect a closed electric circuit that can light up a bulb or bulbs and how to conceptualize the mechanism of the closed electrical circuit. In addition, adopting the trialogical approach to analyze the development of shared knowledge artifacts created in GS, we are able to visualize how the group of students constructed knowledge progressively through multiple mediations and interactions between different modalities.

Regarding pivotal contribution/moments, we extracted and analyzed a short dataset and identified seven pivotal contributions (four manifested and three latent) using an interaction analysis method. A pivotal contribution is defined as a "contribution that shifts the direction of subsequent events, whether seamlessly or abruptly, through uptake between the pivotal and subsequent contributions." Direction refers to a transition from one segment of an event to another indicated by a shift in an activity occurring at a segment boundary. Further, we classified the pivotal contributions into two types: manifested and latent pivotal contributions. It was noted that only the four manifested pivotal contributions were taken up and shape the course of the group inquiry. The three latent pivotal contributions were unrealized potentials. The pivotal contributions demonstrated characteristics of group processes of seamless transition to a new stage/direction of inquiry or an abrupt transition to a new line of inquiry. The pivotal contributions were identified not only by looking into the uptake structure but also by referring to the transcripts. However, the uptake structure helped us locate some possible pivotal moments and transitional boundaries in the uptake diagram for students' progressive inquiry.

For the purpose of multivocality, the other papers (e.g., Medina, this volume, Chap. 16; Lund & Bécu-Robinault, this volume, Chap. 17; and Jeong, this volume, Chap. 18) also analyzed the same dataset with different time stamps in this volume. Due to the different referential frames and analytical methods/approaches adopted, each paper interpreted the data and presented the research results in terms of pivotal contributions/moments and group meaning-making process differently. As we have different understanding of pivotal contributions and adopted different analysis approaches, the focus of the data interpretation was different, which influenced the results of the analysis. In our analysis, an uptake analysis method was adopted to identify pivotal contributions and transitional stages for progressive inquiry at a surface level first, and then we focused intensively on the development of shared knowledge artifacts created in GS using the trialogical approach to analyze it at a theoretical level. The progressive development of shared knowledge artifacts is indispensable to the interaction and mediation (coordination in Medina's case) of the multimodalities (media in Medina's case). The uptake analysis presented in our paper shows that it was not only the GS technology that mediated learning, but also the student, teacher, GS artifacts, and experimental practice all became mediating means to make the progressive inquiry happen and at different points in time. In addition, adopting the trialogical approach to examining the group's evolvement of shared knowledge artifacts helped us understand the process of group cognition as students developed and explored working theories of electrical circuit visible or "materialized" (Hakkarainen & Paavola, 2009) in GS progressively.

Medina's analysis (Chap. 16) adopts a sequential analysis approach in which pivotal moments are related to pivotal sequence of interactions, in which "the group members develop an innovation for lighting two bulbs with one circuit." In his analysis, the focus is on pivotal moments where "uptakes from multiple media [verbal, nonverbal, textual, and visual-spatial] converge to the identification of a discrepancy and the need to correlate them." It seems that the verbal media that originated from the teacher is considered "a third party" (cited from Medina's e-mail dated on 24 March 2012) to the convergence. The analysis approach is to perform "iterative readings to build a tractable structure for understanding how persistent media is appropriated in contexts of interaction" and to then construct an innovation as a joint activity. Although the dataset in our analysis correlates with Medina's dataset of experiments 2 and 3, our interpretation of the teacher's intervention (T12) to ask the group to draw their electrical circuit of lighting one bulb using two batteries in GS was considered pivotal to shape the students' inquiry to a higher level for conceptualization, while Medina's interpretation is that the teacher's intervention was abrupt and "splintered, temporarily, a coherent element of the building of intersubjectivity" (cited from Medina's e-mail dated on 24 March 2012).

In Lund and Bécu-Robinault's analysis (Chap. 17), a given mode/medium couplet is defined as "a potential pivotal moment that may be important for conceptual change." In their analysis, pivotal moments resulted from different types of talk—"both for reformulation across semiotic systems, and for profiting from reformulation in order to extend or change theories and ways of linking objects and events to move towards talk and actions compatible with canonical physics theories." Lund and Bécu-Robinault's interpretation of group understanding focused on the potential of pivotal moments—a reformulation across modes/media. A reformulation is pivotal when it contributes to the students' conceptual change of physics theories. The authors postulate that "particular types of talk—in their interactional context, coupled with the use of external representations and gestures in their immediate interactional vicinity—will play a role in instigating such theory change." This appears to be at a higher level of analysis to identity transformation to demonstrate understanding. The identified pivotal moment in the dataset that overlapped with ours is the third instance of reformulation provided in Lund and Bécu-Robinault's writing-up, which started by Serena, "who, inspired by the collective experiment, begins a drawing in Bruno's GS space, which he then finishes." Their interpretation was that Serena performed "types of reformulation, one that is a hetero reformulation of Bruno's incorrect drawing, but also one that is reformulating the (manipulating, object-events) collective experiment to the (drawing, GS) couplet." The conceptual change is revealed through Bruno correcting his drawing and thus "potentially experienced conceptual change."

Jeong's analysis (Chap. 18) focuses on one modality—artifacts—to investigate group understanding, which was revealed progressively via the construction of physical (experiment) and digital (GS) artifacts. The analysis "examined [students'] circuit understandings reflected in either GS or physical artifacts only." The dataset in our analysis corresponds to the dataset in Jeong's analysis of group co-construction of Circuit B (both B1 and B2). Circuit B1 refers to the bulb connected to the batteries with two pieces of wire, while Circuit B2 refers to the bulb connected to the batteries directly on the top. Based on the analysis of group co-constructed physical (experiment) and digital (GS) artifacts separately, Jeong posits that each understanding of circuit was a close extension of the preceding circuit.

We also note similarities in our analyses. For example, in Medina's analysis of Experiment 4 of lighting a bulb using two batteries, it states, "when Serena notes a discrepancy between the manipulated circuit and the diagram, and the group adjusts to bring the two into alignment. This is pivotal as there are uptakes from multiple media that converge to the identification of a discrepancy and the need to correlate them." Although we did not distinguish "media" with "modalities" in our writing, we agree that the convergence of discrepancy between the diagram (in our case the GS artifacts) and the manipulated circuit (in our case, experimental practice) was the product of interactions and mediations between different multiple modalities (media in Medina's case).

Acknowledgments This material is based on the work supported by the National Research Foundation (Singapore) under Grant NRF2007-IDM003-MOE-001. We are grateful to Mayflower Primary School for collaborating with us on this research.

Appendix

Transcripts of data on electricity for the uptake analysis (from 00:12:15 to 00:16:16)

Time	Label	Name	Observable action, discourse (including graphs), or content: Face-to-face conversations and gestures	GS artifacts	Screenshots of student trial–error experiments
12:15	O1	Others	*(To S)* You can try to use two batteries (*using a gesture with his fingers*).		
12:18	J2	Joel	*(J overheard what O said and understood his gestures.)* Bright idea.		
12:19	B3	Bruno	*(To the group members)* Yeah.		
12:21	S4	Serena	*(To the group members)* Let's try two batteries. We connect all of them? *(Using her fingers to make a gesture of two. All of them got excited.)*		
12:38	Je5	Joel starting the experiment	*(To B)* Wait! Wait! I know! Quickly, connect the wire! B, hold here! Hold it hold it. Hold it properly. Ok, hold it.		
12:47	Ge6	Group experiment	*(B connected the wires with the two batteries to light the bulb, but the bulb was not lighted.)*		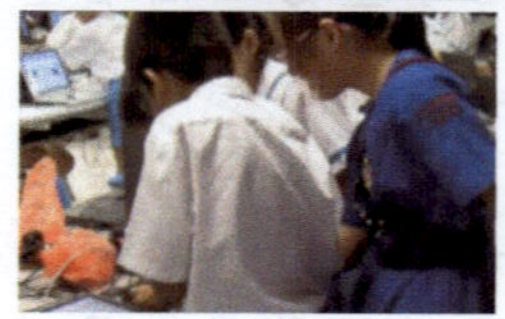
12:49	S7	Serena	*(To B)* Hey! Wrong side *(The bulb was not lighted.)*		

12:52	Ge8	Group experiment	*(The group tried again, and this time the bulb was lighted by B's final touch.)*	
12:53	J9	Joel	Hey, very bright.	
12:54	S10	Serena	*(To the group members)* How about we try two batteries, two bulbs? Two bulbs?	
13:05	Gg11	Group in GS	*(The group board is empty. T looked at the group board.)*	
13:06	T12	Teacher	*(To the group)* No draft, no draft [*on the group board*]. Then people will look at a blank board.	
13:07	B13	Bruno	*(To T)* I'll draw, I'll draw.	
13:10	T14	Teacher	*(To the group)* Where are the two batteries?	
13:13	J15	Joel	*(To T)* We can do with two batteries.	
13:15	T16	Teacher	*(To the group)* One draws, one member draws ... *(Stopped for a few seconds)* One member draws. The other three do the fix and circuit.	
13:25	Bg17	Bruno in GS	*(To T)* I'm drawing, I'm drawing.	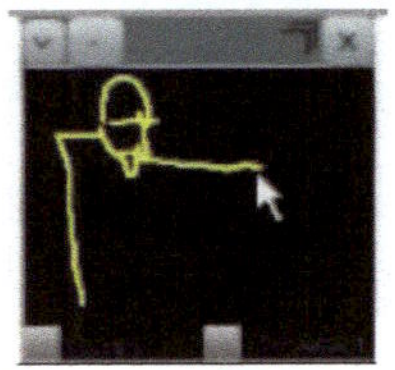

(continued)

Appendix (continued)

13:27	Je18	Joel starting the experiment	*(To S)* Hey, *S*. Hold this. Hold this. You hold here. Hold here. Hold here. *(S and J begin to connect the wires with the two batteries to light the bulb again to make B understand how to draw the circuit.)*		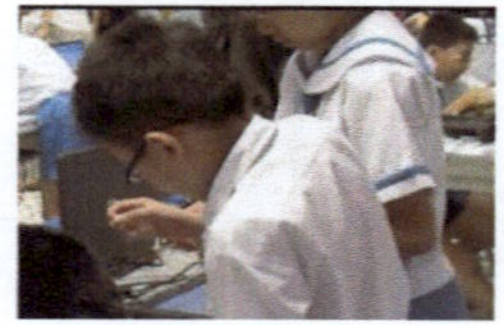
13:38	Bg19	Bruno in GS	*(B is drawing and did not look up)* Hey, don't knock my hand.	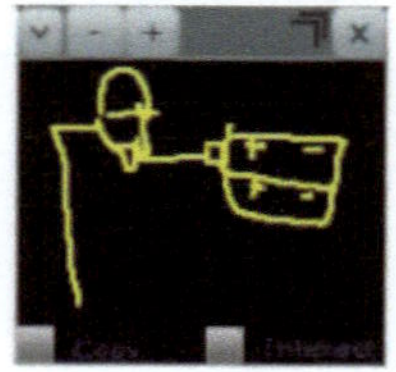	
13:47	T20	Teacher	*(To the class)* Try various ways. Use two wires. *(Stopped for a few seconds.)* Now children, 5E *(this class number)* gave me more varieties you know? 5E, I want team work.		
14:16	Bg21	Bruno in GS	*(B finished the drawing of the graph of circuit with two wires and two batteries to light the bulb they experimented just now on his private board.)*	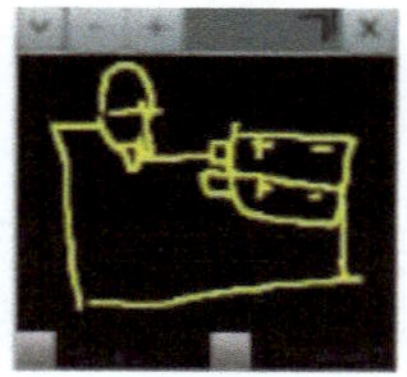	
14:28	Ge22	Group experiment	*(Before posting the drawing to the group board, S and A lighted the bulb again for B's further check.)*		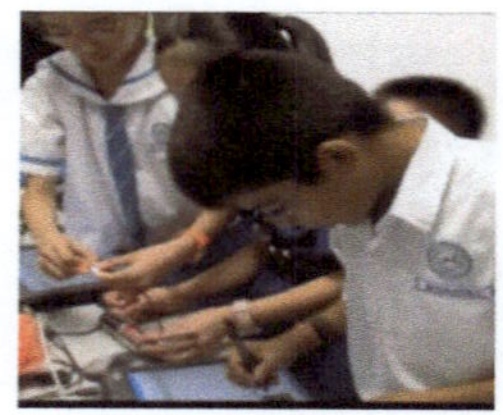

14:29	Bg23	Bruno in GS	*(B dragged the drawing of the circuit with two wires and two batteries on the left together with his previous drawing of the circuit with two wires and one battery to light the bulb on the right from the private board to the group board.)*	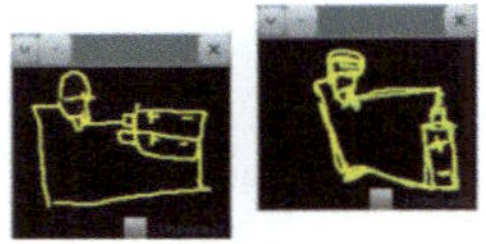 B posted his drawing (*on the left*) together with his previous drawing (*on the right*) onto the group board.
14:30	B24	Bruno	*(To group members)* Let's try three batteries. There must be methods to do it.	
15:03	Ag25	Agnes in GS	*(After observing the hands-on activity about using two batteries to light one bulb, A began to draw on B's private board according to her understanding.)*	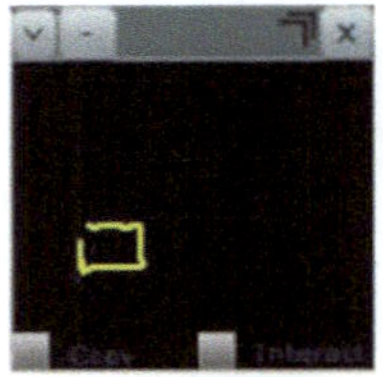
15:10	Ag+Bg26	Bruno in GS	*(While A is drawing, B is looking at and pointing to the drawing. He realized that his previous circuit graph was wrong. Ag had no time to finish the drawing; she asked Bruno to finish the rest.)*	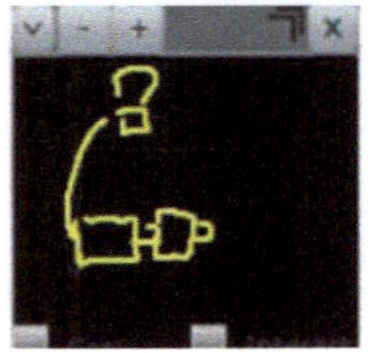
15:11	T27	Teacher	*(To class)* Now. Maximum two batteries.	
15:13	J28	Joel	*(To B)* Maximum two batteries la.	

(continued)

Appendix (continued)

15:19	T29	Teacher	*(To class)* All right, children, I think it's enough now. Stop what you're doing. Now, I want you to visit other people's board and see whether there are new ideas. If you got different ideas, you can clone your friends' board and paste it into your own group board. And circle it to show me that you borrow the ideas. Now you can visit your neighbor's board.		
15:50	S30	Serena	*(To group members)* Teacher said we cannot do la.		
15:52	Be31	Bruno's experiment	*(Although the teacher asked the students to stop doing the experiment, B did the experiment again to light the bulb, trying to find out why his previous drawing of the circuit graph was not correct.)*		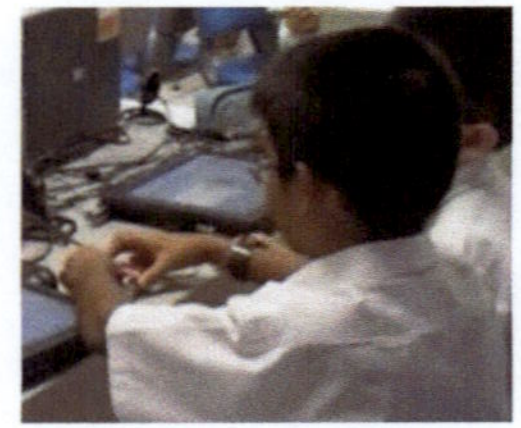
15:53	Ag32	Agnes in GS	*(To B, pointing to the drawing of the circuit with two batteries to light the bulb on the group board that B drew previously.)* It's not like that. Not like that. Cancel this one la.	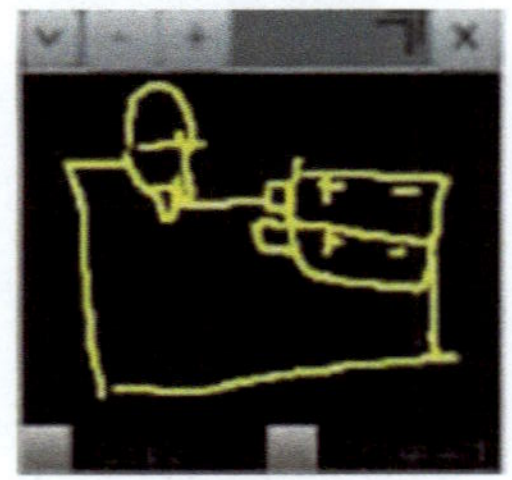 A deleted the drawing from the group board	
16:03	A33	Agnes	*(To B, pointing to the drawing)* B, you draw and finish. Draw and finish *(A is observing B's drawing)*.		

16:10	Bg34	Bruno in GS	*(B is drawing.)*	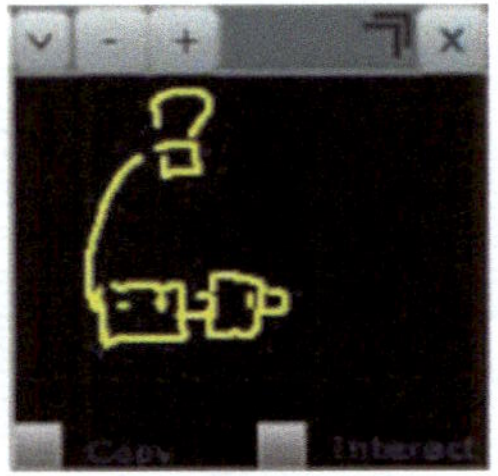 B continued A's drawing (16:03)
16:16	Gg35	Bruno in GS	*(B removed his previously published drawing and posted the revised drawing of the circuit with two batteries to light the bulb as the group work to the group board.)*	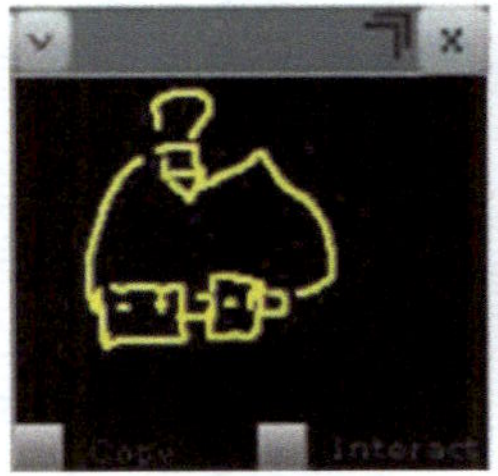 

O=Others, A=Agnes, B=Bruno, J=Joel, S=Serena, T=Teacher, G=Group
Gg=Group working in GS (creating, editing, or moving a GS post)
Ag=Agnes working in GS, Bg=Bruno working in GS, Ag+Bg=Agnes and Bruno both working in GS
Ge=Group experimental practice, Be=Bruno's experimental practice, Je=Joel's experimental practice

References

Arvaja, M., Salovaara, H., Häkkinen, P., & Järvelä, S. (2007). Combining individual and group-level perspectives for studying collaborative knowledge construction in context. *Learning and Instruction, 17*(4), 448–459.

Chen, W., & Looi, C.-K. (this volume). Group Scribbles-supported collaborative learning in a primary grade 5 science class. In D. D. Suthers, K. Lund, C. P. Rose, C. Teplovs, & N. Law (Eds.), *Productive multivocality in the analysis of group interactions*, Chapter 14. New York, NY: Springer.

Chen, W., Looi, C. K., & Tan, S. (2010). What do students do in a F2F CSCL classroom? The optimization of multiple communications modes. *Computers & Education, 55*(3), 1159–1170.

Hakkarainen, K. (2003). Progressive Inquiry in a computer-supported biology class. *Journal of Research in Science Teaching, 40*(10), 1072–1088.

Hakkarainen, K. (2009). A knowledge-practice perspective on technology-mediated learning. *International Journal of Computer-Supported Collaborative Learning, 4*(2), 213–231.

Hakkarainen, K., & Paavola, S. (2009). Toward a trialogical approach to learning. In B. Schwarz, T. Dreyfus, & R. Hershkowitz (Eds.), *Transformation of knowledge through classroom interaction* (pp. 65–80). New York, NY: Routledge.

Jeong, H. (this volume). Development of group understanding via the construction of physical and technological artifacts. In D. D. Suthers, K. Lund, C. P. Rose, C. Teplovs, & N. Law (Eds.), *Productive multivocality in the analysis of group interactions*, Chapter 18. New York, NY: Springer.

Jordan, B., & Henderson, A. (1995). Interaction analysis: Foundations and practice. *The Journal of the Learning Sciences, 4*(1), 39–103.

Koschmann, T., Stahl, G., & Zemel, A. (2005). The video analyst's manifesto (or the implications of Garfinkel's policies for the development of a program of video analytic research within the learning sciences). In R. Goldman, R. Pea, B. Barron, & S. Derry (Eds.), *Video research in the learning science*. New York, NY: Routledge.

Lund, K., & Bécu-Robinault, K. (this volume). Conceptual change and sustainable coherency of concepts across modes of interaction. In D. D. Suthers, K. Lund, C. P. Rose, C. Teplovs, & N. Law (Eds.), *Productive multivocality in the analysis of group interactions*, Chapter 17. New York, NY: Springer.

Medina, R. (this volume). Cascading inscriptions and practices: Diagramming and experimentation in the Group Scribbles classroom. In D. D. Suthers, K. Lund, C. P. Rose, C. Teplovs, & N. Law (Eds.), *Productive multivocality in the analysis of group interactions*, Chapter 16. New York, NY: Springer.

Paavola, S., & Hakkarainen, K. (2009). From meaning making to joint construction of knowledge practices and artefacts - A trialogical approach to CSCL. In C. Malley, D. Suthers, P. Reimann, & A. Dimitracopoulou (Eds.), *Computer supported collaborative learning practices: CSCL2009 conference proceedings* (pp. 83–92). International Society of the Learning Sciences (ISLS): Rhodes, Creek.

Sacks, H. (1992). Lectures on conversation. 2 vols. Edited by Gail Jefferson with introductions by Emanuel A. Schegloff. Oxford: Basil Blackwell.

Sanderson, P., & Fisher, C. (1994). Exploratory sequential data analysis: Foundations. *Human-Computer Interaction, 9*, 251–317.

Stahl, G. (Ed.). (2002). *Computer support for collaborative learning: Foundations for a CSCL community*. Lawrence Erlbaum Associates: Mahwah, NJ.

Stahl, G. (2006). Sustaining group cognition in a math chat environment. *Research and Practice in Technology Enhanced Learning (RPTEL), 1*(2), 85–113.

Stahl, G., Koschmann, T., & Suthers, D. D. (2006). Computer-supported collaborative learning: A historical perspective. In R. K. Sawyer (Ed.), *Cambridge handbook of the learning sciences* (pp. 406–427). New York, NY: Cambridge University Press.

Suthers, D. D. (2006). A qualitative analysis of collaborative knowledge construction through shared representations. *Research and Practice in Technology Enhanced Learning, 1*(2), 1–28.

Suthers, D. D., Dwyer, N., Medina, R., & Vatrapu, R. (2010). A framework for conceptualizing, representing, and analyzing distributed interaction. *International Journal of Computer Supported Collaborative Learning, 5*(1), 5–42.

Suthers, D. D., Lund, K., Rosé, C., Dyke, G., Law, N., Teplovs, C., et al. (2011). Towards productive multivocality in the analysis of collaborative learning. In H. Spada, G. Stahl, N. Miyake & N. Law (Eds.), *Connecting computer-supported collaborative learning to policy and practice: Proceedings of the 9th international conference on computer-supported collaborative learning (CSCL 2011)* (Vol. III, pp. 1015–1022). Hong Kong: University of Hong Kong.

Wee, J. D., & Looi, C. K. (2009). A model for analyzing math knowledge building in VMT. In G. Stahl (Ed.), *Studying virtual math teams*. New York, NY: Springer.

Weinberger, A., Stegmann, K., & Fischer, F. (2007). Knowledge convergence in collaborative learning: Concepts and assessment. *Learning & Instruction, 17*(4), 416–426.

Chapter 16
Cascading Inscriptions and Practices: Diagramming and Experimentation in the Group Scribbles Classroom

Richard Medina

The analysis discussed in this chapter draws attention to the interactional and inscriptional practices observed in the Group Scribbles data set. The critical finding is the identification of a pivotal sequence of interaction occurring in the later half of the activity in which one member of the group proposes an innovation for illuminating two light bulbs in a single circuit. The proposal and its subsequent endorsement by the other members are contingent on an immediately prior interaction in which the group appropriates another group's circuit diagram. Together, this pair of adjacent sequential structures exposes multiple instances of uptake (Suthers, Dwyer, Medina, & Vatrapu, 2010) between participants. These uptake relations are realized through an ensemble of contingencies consisting of persistent drawings, tabletop materials, and a locally situated interactional practice. It will also be shown how the group's history of action is a prerequisite for these uptake relations.

Analytic Approach

Conducting interaction analysis from what is a rich matrix of multimodal interaction, such as that in our current study, poses a number of analytic puzzles. A central problem confronted here is how to carry out an analysis of interactional practice in which resources for action are distributed across time, space, and media. How do the situated actions of the participants unfold in relation to prior shared contexts in analytically observable ways? To explore this phenomenon, a novel approach to sequential analysis is applied to the investigation of four students' interaction during a classroom exercise. The aim of the analysis is to reveal how sequential structures of

R. Medina (✉)
University of Hawai'i at Mānoa, Honolulu, HI, USA

D.D. Suthers et al. (eds.), *Productive Multivocality in the Analysis of Group Interactions*, Computer-Supported Collaborative Learning Series 16,
DOI 10.1007/978-1-4614-8960-3_16,

multimodal interaction and the availability of persistent artifacts permeate successive situated activity (Suthers & Medina, 2010) and influence the meaning-making processes demonstrated by the participants (Koschmann et al., 2005). This chapter presents this analytic approach and discusses its application to the Group Scribbles data set (Chen & Looi, this volume, Chap. 14). The analytic approach is summarized below along the five dimensions of Chap. 2 (Lund & Suthers, this volume): theoretical assumptions, purpose of analysis, unit of interaction, representations of data and analyses, and analytic manipulations of the representations.

Theoretical Assumptions

Investigating interaction within a setting is a matter of careful consideration of the connection between the sequential organization of participants' actions and the semiotic, material, and embodied aspects of the setting. Situated interaction has a mutual elaborating relationship with the environment (Suchman, 1987). The environment offers an array of external resources for action-relevant appropriation. Simultaneously, situated action modulates, redefines, or otherwise reconfigures the resources within the environment, thereby enabling and constraining subsequent acts. This dynamic ebb and flow of appropriation and modulation are sequentially organized (Garfinkel, 1994; Goodwin & Heritage, 1990). With respect to multimodal and embodied interaction Goodwin (2003) writes,

> The issues posed for the analysis of action in such a setting involve not simply the resources provided by different semiotic systems as self-contained wholes, but also the interactive practices required to juxtapose them so that they mutually elaborate each other in a way relevant to the accomplishment of the actions that make up the setting.

The study of multimodal interaction requires an understanding of both the properties of semiotic systems *and* how those systems are coordinated and appropriated in joint activity. This suggests two overlapping planes of study: recognizing the external properties of the setting in one and meaning-making practices in the other.

Multimodality as Semiotic System

Modes of communication have a materiality that enable and constrain what kinds of meanings can be expressed through their appropriation (Jewitt, 2008). Baldry and Thibault (2006) refer to the coupling between meaning and modality as defining a semiotic system. Verbal and nonverbal modes and their aural and visual realizations can project very different meanings in isolation and in coordination with each other. The relation between media, its mode, and meaning potential is not absolute nor is it deterministic. A semiotic system certainly delimits meaning-making potential; however, as the quote by Goodwin alludes, these meditational constraints are discernable only with respect to how they are appropriated and thereby exposed in joint activity.

Multimodal Interaction as Sequential Organization of Joint Meaning Making

Understanding interaction with multimodal resources entails careful consideration of how participants' actions in the environment are made relevant for emergent, sequentially organized, and shared structures of joint activity (Sacks, Schegloff, & Jefferson, 1974). Sequential analysis techniques based on ethnomethodology and conversation analysis (EMCA) demonstrate, in a very detailed manner, the contingent nature of human interaction. These techniques expose how the resources available in the setting of interaction are integrated in the very structures of communication that emerge in the activity. Taking the turn-by-turn pattern of interaction in any social activity as an analytic starting point has yielded valuable insights into the in situ emergence of meaning-making practices (Koschmann et al., 2005).

Although conversational analysis was originally proposed to handle speech exchanges, numerous scholars have taken an EMCA approach, in principle, as an inroad to understanding interaction mediated by more semiotic-rich settings such as online environments, scientific field work, and classrooms (Çakır, Zemel, & Stahl, 2009; Goodwin, 1995; Roth, 2001). Streeck and Kallmeyer's (2001) analysis of a rather mundane two-party business negotiation offers an example that suggests that graphic inscriptions can be taken as a form of interaction that offers a different set of opportunities for meaning making altogether. The act of inscribing during interaction carries with it not only that which is being represented, its instrumental purpose, but also perhaps less obviously its discursive function. Inscriptions, once recorded in a medium (paper, whiteboard, computer screen, etc.), offer structures for making arguments, substantiating claims, and indexing a range of situation relevant and epistemologically consistent communicative action. The sequential organization of inscriptional activity carries structural (e.g., rhetorical, canonical, or discursive) information that embodies taken as shared conceptions, concerns, and meanings that are relevant to the situation at hand.

Inscriptional action draws upon an extended vocabulary from the visual field. They can embody forms of action such as a line-intimating gesture (e.g., a line drawn around a figure may be a deictic reference to an aspect of the figure of concern in the interaction). Gesture is highly coupled with talk; however, inscriptions and instrumental acts occur independently of talk yet articulate it. This has not been studied at length especially in regard to how inscriptional action is sequentially organized. Streeck and Kallmeyer write,

> Actions that can occur independently of talk, however - instrumental acts, inscriptions, and so on - have so far only rarely been studied for their possible participation in the construction of 'projectable' turns-at-talk. (p. 469)

Inscriptions that once served as a field of calculation and measurement can be reinstated in rhetorical contexts to persuade, compare, and express ideas. Further, persistent inscriptions enable variable, situation-relevant courses of action over time and setting (Latour, 1990).

Still, while arguing for analytic accountability of inscriptions and nonverbal modalities in the setting, Streeck and Kallmeyer warn against oversimplifying or fragmenting components of interaction across modal and material properties. Rather, they suggest that ongoing interaction draws upon multiple vocabularies in the making of meaning. Thibault (2011) goes further in advising against the rush to discover and extrapolate upon regularities of symbolic systems. The prudent starting place is the distributional character of communication across the senses, materiality, and symbols.

The analytic approach considered in this chapter takes interaction as fundamentally multimodal and sequentially organized. Underlying theoretical assumptions are based on the notion that participants build their interaction through the moment-by-moment or otherwise sequential exchange of actions. These actions are potentially distributed across all available aspects of the setting. As Goodwin's quote makes clear, the study of interaction practices exposes the relationship between the setting and the joint activity of the participants.

Purpose of Analysis

Through analysis of practice we gain a rich understanding of the distributional character of action and its implications for computer-supported collaborative learning and teaching in classrooms, online settings, and instances of both. Sequential multimodal interaction analysis can be used to uncover the relationship between the properties of the environment and the interactional practices that make those properties relevant and consequential for joint meaning making (Schegloff, 1991). More specifically, analysis of interactional practices is useful for understanding how inscriptional devices (verbal and nonverbal) are integrated in joint meaning-making structures.

Unit of Interaction

Uptake: A Relational Unit of Interaction

Making sense of the sequential structure of multimodal interaction presents a degree of complexity for analysis. Participant actions may be distributed across a diverse range of media. A useful strategy to begin with might be the recognition of how any participant actions are evidenced to be relevant and consequential for the activity. How and where are actions *positioned* in the sequential unfolding of the activity and how and through what means do those actions relate to prior actions? The notion of uptake has been proposed as a useful concept for investigating precisely these questions.

Suthers, Dwyer, Medina, and Vatrapu (2010) describe uptake as a relational construct that identifies a participant action as appropriating aspects of a prior or an ongoing setting as relevant for ongoing interaction. This definition is deliberately abstract, enabling it to be purposed in a wide range of interactional settings. It is also intended to support a diverse range of theoretic and methodological approaches. Uptake provides an interpretive heuristic rather than a specific method of analysis (the authors describe it as a proto-analytic). The potential gain by interpreting interaction as uptake is that uptake does not privilege one particular communicative modality or granularity over another. A warranted interpretation of uptake only specifies that one human action is appropriating aspects of a prior or an ongoing element of the setting while also transforming that setting. The value of uptake for the analysis of multimodal interaction is its provision for a more flexible consideration of sociological and environmental contingencies.

Representations of Data and Analyses

In addition to formulating uptake as unit of interaction, Suthers et al. (2010) also detail how the construct can be organized as an analytic framework. The following discussion about analytic data representations adapts components of this framework to qualitative microanalysis.

Data Elements

A representation of action at the lowest granularity is a participant action. Actions are represented as data elements. Data elements are not limited to single-participant actions such as posting a message or making a single utterance. A data element may also encompass a set of actions.

Contingencies

A contingency specifies a connection between two sequentially ordered data elements. A criteria for this connection is that the first element provides a portion of the setting in which the second element, an action, took place. This connection must be manifest (empirically observable), e.g., temporal or spatial proximity, involve the same actor, or overlap in content or form. Collections of contingencies provide the basic empirical foundation for the identification of traces of interaction (i.e., contingencies are interpreted as uptake).

Traces of Sequential Interaction

A trace is a sequence of data elements that together form a sequential structure. These sequential structures are emic (derived directly from participant actions). They may range from short exchanges such as a reply to a question or may extend into wider structures concerned with, for example, specific topics or problems that are of concern to the participants. Despite an apparent etic (derived from an analyst's observation) attribution of structure, a trace is not intended to obfuscate the continuous nature of interaction in its setting. Instead, it is a construct that encompasses a sequence of data elements cohered by a relational structure specified through contingencies.

Manipulations of the Representations

Interpreting Uptake

An uptake relation specifies a relationship between two sequentially ordered data elements. The uptake construct constitutes the relation between two elements (elements being a representation of a participant action or actions) in which the second action appropriates the first as relevant for ongoing interaction. As noted above, uptake is an interpretive heuristic that is applied where the empirical evidence (collections of contingencies) warrants its identification. Here, the relational evidence is a foundation for making interpretations about the interaction and its structure.

Segmentation

Traces are organized into segments. The natural coherence of participants' interaction and the discontinuities within this coherence lead to the identification of segments (Jordan & Henderson, 1995): for example, opening and closing remarks or other statements that indicate a beginning and end of a course of action-bound sequences of talk. Each segment, as a representation of a trace, can be defined based on different analytical objectives. For example, in the current analysis traces are analyzed with respect to two concerns—inscriptional and ideational interaction (these are discussed further below). Figure 16.1 below visualizes this idea. The figure indicates two different traces of the same data set. Each trace is segmented (Fig. 16.1 has seven such segments). Each segment encapsulates a sequentially ordered structure represented as a collection of data elements (participant actions) connected by contingencies.

Comparisons/Connections Between Traces

The relations between situated interactions over time can be specified by drawing on the trace analysis (segments) as a resource for comparison. Contingencies, for example, may extend from local interaction (i.e., a segmented trace in Fig. 16.1) back into prior situated settings (other segmented traces). Persistent aspects of the setting such as inscriptions or other material and symbolic artifacts may mediate the

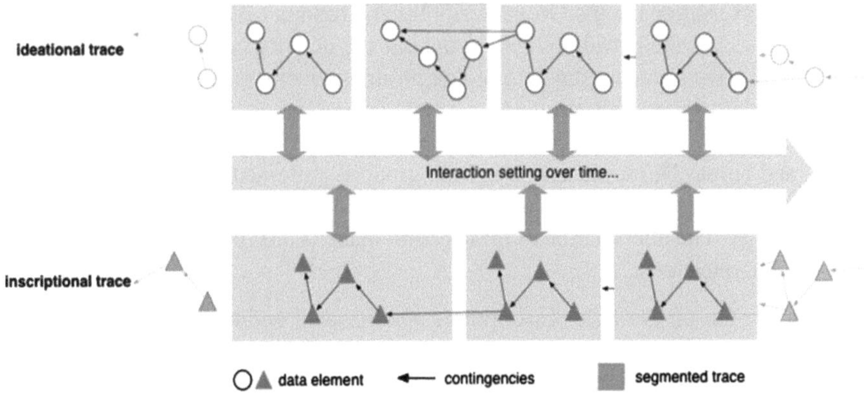

Fig. 16.1 Analytic representation scheme

temporal extension of contingencies across situated interactions. Other forms of connections based on patterned interactions can also be traced across segments to expose the development of interactional competencies and meaning-making practices (Medina, Suthers, & Vatrapu, 2009). Zooming “in” and “out” (Roth, 2001) between micro and macro traces is ultimately achievable by maintaining a tractable record of the analytic representations at multiple granularities. The following sections describe the application of this analytic approach to the Group Scribbles classroom.

Group Scribbles Analysis

The analysis presented in the following sections takes as a topic the multimodal interactional mechanisms demonstrated by the participants in the Group Scribbles classroom exercise (see Chen & Looi, this volume, Chap. 14, for details of the setting). These mechanisms are investigated using sequential microanalysis techniques for examining the moment-by-moment flow of action among the participants. Uptake is appropriated here because it offers a general heuristic at a suitable granularity for describing the empirical evidence. This is not to suggest that uptake is only a macro-level construct. It can be utilized at any descriptive level appropriate for the analytic evidence on which it is grounded. In the present analysis, uptake is used to explicate the critical relations between participant actions that are distributed across time, media, participants, and material (classroom) artifacts.

The Group Scribbles classroom is organized like many similar data sets involving learning environments that are computer supported, networked, and embodied in classroom situations. In these contexts interaction is distributed across modalities (verbal, nonverbal, textual, and visual-spatial). Three aspects of the current data set were identified at the onset of the analysis. First, the students constructed a series of persistent inscriptions in the Group Scribbles environment. The production of these inscriptions, and their availability throughout the activity, suggested that inscriptions might have a role to play in ongoing group interactions. Second, the centrally located tabletop materials provided a visual-spatial modality with respect to the spatial arrangement and placement of the various circuit parts such as batteries, wires, and bulbs. Third, the activity of the group is patterned. That is, the students' work occurs in a series of sequences, each oriented to a unique problem or concern in the activity. The interrelation of these phenomena formed the basis for the following analytic questions.

1. How are Group Scribbles inscriptions appropriated and/or coordinated in joint action involving two- and three-dimensional visual and material resources?
2. How are the elements of the (classroom) environment modulated by the situated actions of the participants, and what are the consequences of this for their meaning-making practice?

This analysis relied on three iterations (readings) over a 20-min segment of the data record. The first two iterations were conducted with a specific focus on the students' drawings and situated interaction, respectively (inscriptional and ideational interactional traces). Figure 16.1 illustrates these traces conceptually. A third iteration integrated the understandings gained from the two prior trace analyses. The rationale for performing iterative readings was to build a tractable structure for examining the interactional and media-based contingencies of the classroom activity.

Segmenting Inscriptional Interaction Traces

The objective of this trace analysis was to segment the video data into discrete units corresponding to the construction of electrical circuit diagrams made using the Group Scribbles software. Time segments were recorded in the transcript denoting sequences of activity in which drawings were constructed. Determining the opening and closing of such segments required microanalysis in which the students' inscriptional processes were studied to distinguish one inscription from another. This segmentation relied on temporal, ideational, or interpersonal contingencies. For example (respectively), many drawings can be deemed relatively complete with respect to the time frame in which they are produced (temporal), "made complete" by a participant's description or representation of a contextually relevant idea or concept (ideational), or attributable to an individual or a group of individuals (interpersonal). In essence, a segment of inscriptional activity is an assembly of finer grained levels of drawing actions. Each analyzed segment was by no means complete with respect to its contextual complexities. These trace analyses provided a stratum for the next trace analysis (described in the next section).

The transcript is annotated to indicate the onset and completion of each segment of inscriptional activity. Additionally, screen captures were made showing the final states of the drawings identified in each segment.

Segmenting Ideational Interaction

The objective of the ideational trace analysis was to mark off identifiable chunks of activity related to specific topics or objectives of concern to the participants. This involved a reading of the video data to set off the opening and closing of sequences of activity. Each segment was distinguished by the degree to which participants made particular concerns relevant for the immediate situation at hand (Jordan & Henderson, 1995). Temporality is a primary indicator. The moment-by-moment organization of interaction is, for its members, a resource that is both context defining and context shaping (Schegloff, 1991). Entwined with this ordering are particular concerns, questions, ideas, goals, complaints, and so on.

As with the trace of inscriptions, the ideational interaction traces involved a microanalytic approach but examined the verbal and nonverbal actions of the participants. It involved a turn-by-turn, moment-to-moment reading of the video record in order to delineate one set of interactions from another. Analytic representations of segments were recorded as transcriptional annotations and in narrative descriptions.

Interaction Tracing

Interaction tracing involved making observations about the relationship between the two prior traces (inscriptional and ideational) as overlapping and co-articulating analyses. This pass was a (re)reading of both the organization of context and the organization of inscriptions. Recorded evidence from inscription and ideational segmentation analyses became an analytic resource for reasoning about and examining the relevancy of prior inscriptions and interaction. These resources remain indexed by time to the data source (video and transcript) and provide analytic parameters for reexamination of micro-interactions or contrastively enable the identification of patterned interaction across situated contexts.

Results

During the 30-min activity the students experiment with five different electricity circuit configurations. Figure 16.2 maps the ideational and inscriptional traces as a series of segmentations. The segments above the timeline in Fig. 16.2 were determined from the ideational trace analysis described above (each trace is numerically ordered; E0–E6). The segments below the timeline were determined from the inscriptional analysis (letters in each segment indicate which participant constructed the inscription, and the subscript identifies a specific inscription). The results from the inscriptional trace include a series of ten drawings made by the students using the Group Scribbles environment.

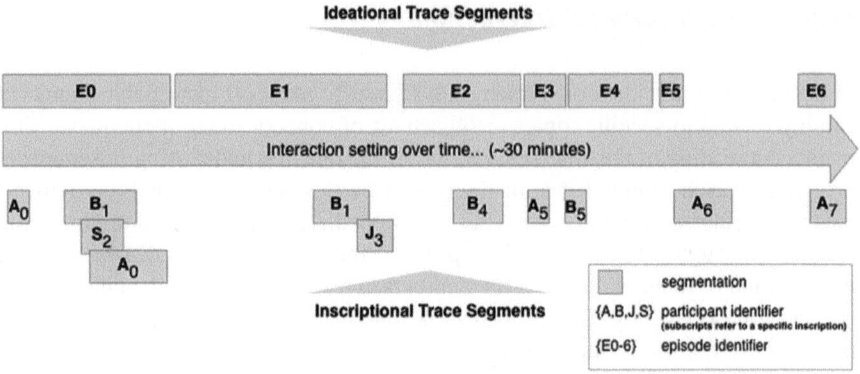

Fig. 16.2 Overview of the segmented traces of the activity

The following sections describe the interactions observed in the activity. These descriptions are organized around the identified segments referenced in Fig. 16.2. In the ensuing descriptions each student is referred to by his or her pseudonym (Bruno, Serena, Agnes, and Joel). Time stamps refer to the video log time relative to the beginning of the group's work in the classroom. Overall, the analysis reveals a sequence of interaction during the fourth and fifth episodes (*E4* and *E5* in Fig. 16.2), in which the group members develop an innovation for lighting two bulbs with one circuit. The episodes prior to these two are first provided in the ensuing description as a way to adequately background the crucial findings.

Initial Diagrams (E0)

At the onset of the activity (01:57) the students are instructed to begin composing a drawing of their circuit diagrams. The drawing exercise precedes any hands-on work with the actual materials. Three of the members (Agnes, Serena, and Bruno) individually construct a diagram consisting of one battery, two wires, and a light bulb. The three circuit diagrams appear to be identical with respect to their organization of wires, battery, and bulb with subtle differences in the spatial arrangement of the circuit components (see Figs. 16.3, 16.4, and 16.5). This short episode concludes as the teacher instructs the students to begin testing their ideas with the circuit materials (06:43).

Episode 1: One-Battery-One-Bulb (E1)

The group attempts two experiments using the tabletop materials provided by the teacher. The constructed circuits are consistent with the diagrams in Figs. 16.3, 16.4, and 16.5. Agnes and Serena pair up to attempt a circuit. Bruno and Joel pair up to

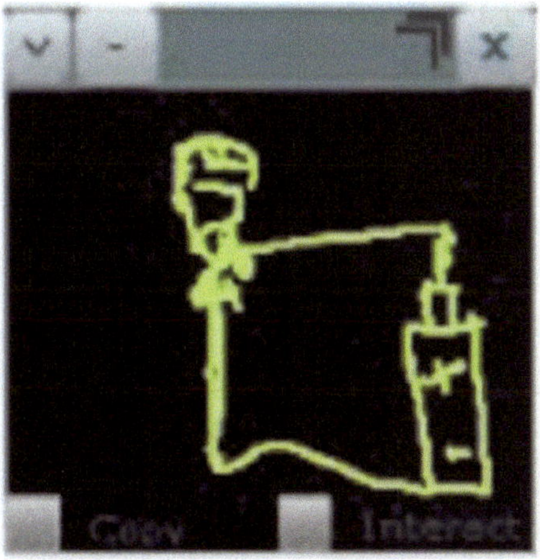

Fig. 16.3 Bruno's (B_1)

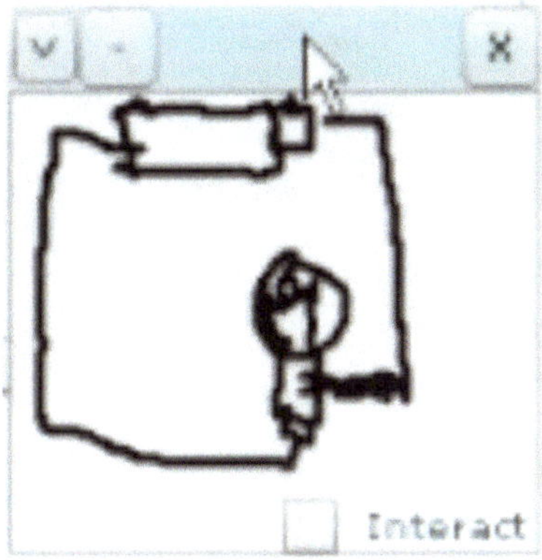

Fig. 16.4 Serena's (S_2)

Fig. 16.5 Agnes' (A_0)

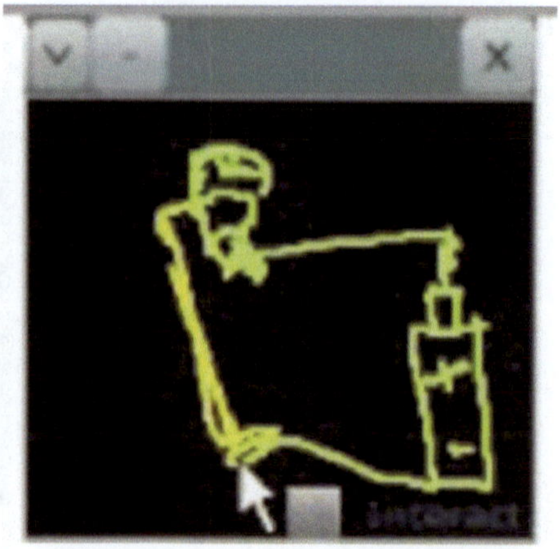

Fig. 16.6 Bruno's adjustment (06:43 through 12:02) (B_1)

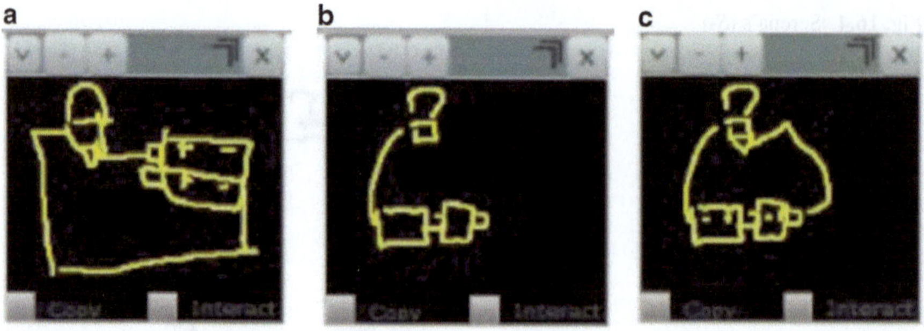

Fig. 16.7 (**a**) Bruno's two-battery circuit. (**b**) Agnes begins diagram. (**c**) Bruno completes

complete their circuit with success. Bruno and Joel demonstrate their successful circuit to Agnes and Serena. The initial diagrams are validated. After discussion with Joel, Bruno adjusts his diagram to reflect that the negative contact must be placed above the contact point of the positive wire (see Figs. 16.3 and 16.6).

Episode 2: Two-Battery-One-Bulb in Horizontal Series (E2)

As the group wraps up the previous experiment Serena suggests to the group that they attempt to use two batteries to light the bulb. Joel, Serena, and Agnes begin by laying the batteries in a series on the table. During this time Bruno sketches an alternative design (Fig. 16.7a and B_4 in Fig. 16.2) to the one being attempted by the others. This experiment concludes as the teacher approaches the group directing them to draw diagrams associated with their tests.

Episode 3: Two-Battery-One-Bulb in Parallel (E3)

At the teacher's prompting Agnes partially diagrams (Fig. 16.7b; A_5 in Fig. 16.2) the circuit the group worked on in the previous episode (E2). She appropriates Bruno's tablet to begin drawing the diagram. Concurrently, Bruno attempts to implement the circuit he sketched in Fig. 16.7a without success. He returns to his tablet and completes the diagram begun by Agnes (Fig. 16.7c; B_5 in Fig. 16.2).

Episode 4: Two-Battery-One-Bulb in Vertical Series (E4)

Serena has scanned other groups' public screens. She finds a diagram layout from another group in the class named "SF_2" that has two batteries stacked upright one above the other with the light bulb contact directly on the positive end of the top battery. Two wires are arranged from the bottom of the lower battery up to the bulb shielding (see Fig. 16.8a). Serena shares this diagram with the others by swiveling her screen around so that it faces Bruno and Joel. All four students orient to the diagram being referenced by Serena and agree to test its arrangement. The subsequent experimentation moves through two phases. In the first, Bruno and Joel have made an interpretation of the stacked diagram by proceeding to use a wire to complete the circuit connection between the bulb and the positive post of the battery. Serena publicly notes this inconsistency with respect to the diagram posted by group "SF_2." A subtle departure in the stacked diagram from the other diagrams the group has been working with is that the bulb appears to make direct contact with the positive battery post, bypassing the need for a wire (see Fig. 16.8b). Serena points this out and grasps the bulb and places it directly on the positive battery post. The group then proceeds to successfully construct a working circuit using the stacked arrangement (diagrammed in Fig. 16.9). The subtle yet critical diagrammatic placement of the bulb and its successful implementation appear to set up the group's immediate next experiment with two bulbs.

Episode 5: Two-Battery-Two-Bulb in Vertical Series (E5)

The last experiment has the group attempting to light two bulbs. They are building on the "stacked" arrangement of the circuit they successfully tested in the immediately prior experiment. Joel initiates this experiment by spontaneously picking up two bulbs and placing them both directly on the positive post of the top battery. Joel performs this act while Bruno remains holding the batteries in the position they were in during the prior experiment. At this moment there is no direct evidence from the video record that the bulbs are actually lit. At the very least Joel's act is presenting an idea to the group. He appears to position the bulbs and then looks up to the group as if requesting their noticing (this moment is captured in the frame presented in Fig. 16.10, inset 3 below). The group takes notice and excitedly proceeds to successfully construct the circuit with two bulbs.

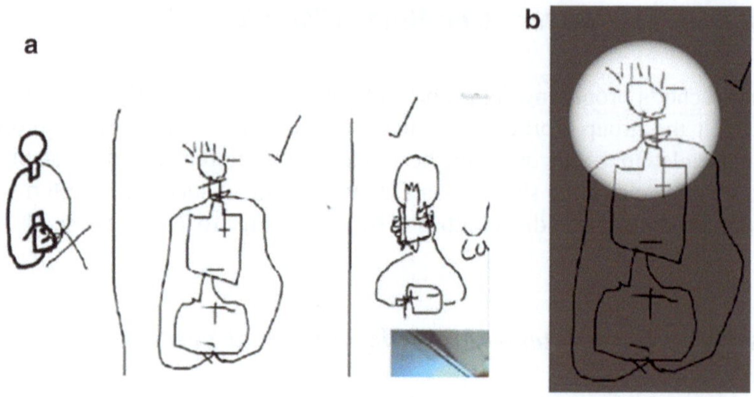

Fig. 16.8 (a) Group SF_2's diagrams. (b) Bulb contact (*highlighted* by the author)

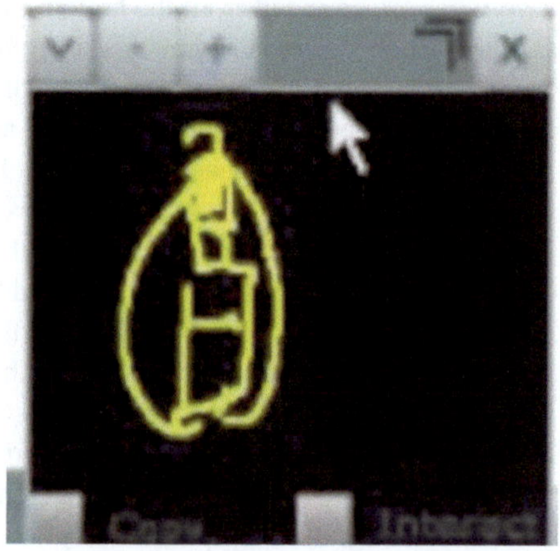

Fig. 16.9 Agnes draws configuration

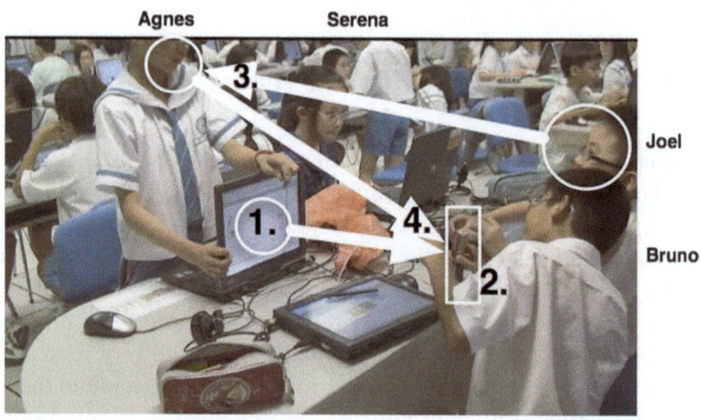

Fig. 16.10 Proposing an innovation

As noted above the group's approach does not require a wire for the positive contact (i.e., the bulb is placed directly on the positive post of the top battery). This innovation opens up the possibility of adding a second bulb to the positive post of the top battery. It is this opportunity that Joel seizes and demonstrates and that the others recognize. Subsequently, their competence at constructing a working circuit is evidenced in the efficiency at which they move from Joel's initial idea to a working two-bulb circuit.

Discussion

Representational Competence: Modalities Appropriated as Resource for Interactional Practices

The sequence of interaction in episodes 4 and 5 described above demonstrates the group members' appropriation of the visual-spatial modality for conducting their problem-solving activity. In episode four the group members have integrated another group's diagram into their interaction. The diagram depicts a vertical arrangement of the circuit in which the bulb makes direct contact with the battery post (no wire), which the group had not previously considered. As Bruno and Joel work at reconstructing the circuit as an experiment, Serena notes that they have misinterpreted the diagram. She observes and demonstrates to the others a novel detail in the placement of the bulb with respect to the top battery (see Fig. 16.8b). The group subsequently arranges their circuit to maintain its consistency with the diagram.

Here we can observe a series of uptake relations. Bruno and Joel's reconstruction (using the tabletop materials) of the diagram displayed on Agnes' screen is the first in the series. Next, Serena takes up the result of this reconstruction when she observes how their arrangement is inconsistent with the diagram. Bruno and Joel rearrange the tabletop materials accordingly as they take up Serena's point. By definition, there are multiple uptake relations in which a participant action takes up aspects of a prior resource as relevant for the situation at hand and by so doing transforms those resources (Suthers et al., 2010). From an interactional exchange perspective, experiment 4 might appear rather straightforward and mundane. Interpreted as uptake, however, the brief exchange sheds a different light on the significance of interactions with respect to how the participants are cooperatively and opportunistically appropriating a range of resources in the environment well beyond verbalization. Uptake relations show how the actions configure persistent artifacts, visual-spatial properties of inscriptions, and prior practices for handling the tabletop materials. Further, the very notion that Bruno and Joel can be referred to as jointly performing an action is a notable example of how tightly coordinated the participants' actions are with their local experimental practice.

Episode four is crucial as it demonstrates a level of representational competence with respect to the group's ability to correlate the phenomena (electrical circuits) with its associated diagrammatic inscription (Kozma, 2003). In this group's work, there is a clear, non-abstract, relation between diagram and phenomena (bulbs look like bulbs in the diagram). The critical point, however, is that it is the group's orientation to Serena's indexical act that reveals that the others clearly understand the misinterpretation and immediately rectify the relations between diagram and phenomena. The visual-spatial modality is appropriated for representation and as a resource for negotiated meanings.

Joel's Proposal: Configuring Semiotic Fields Across Modalities

A second critical moment comes immediately after the group has completed episode four. At this moment (Fig. 16.10) Joel makes his proposal for lighting two bulbs using the circuit configuration constructed in experiment 4. There are a number of points to note here. (1) The prior configuration, originally taken up via a diagram displayed in the Group Scribbles software, remains intact on the table. Further, Bruno's hand remains on the batteries to keep them in the stacked position. These are physical contingencies for the formulation of ideas. The arrangement of batteries, wires, and bulbs requires bodily coordination to be distributed across individuals, and this is taken up by Joel to make his proposal. (2) Joel places the two bulbs on the post and looks up towards Agnes as if to request a response. Agnes and Bruno take up Joel's act as demonstrating a new arrangement (two bulbs). After some excitement Serena, who has directed her attention at Bruno and Joel's attempt to implement the proposal, reaches over the table to position the necessary wire to complete the circuit. In this instance, Serena is taking up the now established practice that requires coordinated arrangement of the components to complete an electric circuit. This practice became necessary over the course of the entire activity and is taken up once again here.

The proposed innovation offered by Joel and the subsequent uptake and implementation performed by the group demonstrate the coordination of multiple visual-spatial modalities. The proposal is initiated when the group reconstructs a circuit diagram depicted on Agnes' computer screen, which she has faced towards the group (Fig. 16.10, *inset 1*). Joel's proposal and its recognition by the group members are achieved by a configuration of an indexical field consisting of the batteries held in place by Bruno along with Joel's placement of two bulbs (Fig. 16.10, *inset 2*). Here, Joel is taking up a prior tabletop configuration and transforming it into another. This is followed by Joel's upward gaze towards the shared space above the table (Fig. 16.10, *inset 3*) and Agnes' uptake of the gaze and position of the two batteries as communicating an idea (Fig. 16.10, *inset 4*). Bruno's endorsement may be implicit as he maintains the physical apparatus to sustain Joel's communicative action. The group members are configuring the environment through multiple modalities to accomplish their meaning making.

Mutability as Achievement

The final observation of these results rests on the notion of mutability. At the onset of experiment 4, the group has been shown a diagram constructed by another group in the classroom. This diagram, as a visual-spatial and nonverbal media, provides an immutable mobile depiction (Latour, 1990). Bruno and Joel's uptake of the diagram is to *reconstruct* its features using the tabletop materials. As detailed above, the next series of actions project through experiments 4 and 5. Here we also see demonstrated the transposition between modes such that the immutable object displayed on the screen is *made mutable* through its reconstitution in the tabletop materials. Realization of the affordance of mutability establishes the necessary relationship between the actors and the setting making it *procedurally consequential* (Schegloff, 1991). This observation suggests that uptake relations that take the form of modal transposition (Peeters, 2010) may evidence and contribute to interactional practice (Alac & Hutchins, 2004). The development of interactional practices directly implicates the multimodal resources of the environment and the interactional construction of its meanings (Kupetz, 2011).

Uptake of Multimodal Contingencies in the Development of Group Practices

There are multiple instances of uptake identified in experiments 4 and 5. The first instance is Joel and Bruno's initial reconstruction of the "stacked" electrical circuit, depicted in another group's workspace, into the tabletop materials. This act of appropriation sets up a two-phase *modulation* of the depicted inscription in which Serena and Agnes point out a discrepancy that is subsequently resolved. The second instance of uptake involves the group operating on the now *transformed* battery, wire, and bulb arrangement to propose and negotiate a circuit with two bulbs.

The purposing of the tabletop materials in experiments 4 and 5 was to first appropriate and then construct an innovation. This uptake relation between the activity of the group and the inscribed diagram on the Group Scribbles screen is dependent on a second uptake relation. In this case the members of the group build on their prior locally available interactional practices for constructing experiments with electrical circuit materials. Evidence for this association is the placement of Bruno's right hand (Fig. 16.10), as Joel enacts his proposal for two bulbs. The configuration of body parts and orientation of the learners with respect to the table are entwined with the group's interactional context. Thus, the uptake of the diagram takes place through the physical and embodied arrangement of materials and hands. The diagram is brought into coordination with the setting. The participants effectively take up the idea represented in the diagram by leveraging prior arrangements of bodies and hands. Bruno's maintenance of the battery arrangement is required for Joel's proposal. This arrangement is established over the course of entire activity and constitutes important part of the group's interactional practice in an ongoing and changing semiotic context (Goodwin, 2007).

Conclusions

The directive given to the participants was to use diagrams to reason about and explore the concept of electricity flow. This analysis showed how the use of diagrams provided a resource on which the participants *juxtaposed* and configured their local interactional practices and developed competencies for experimentation and intragroup communication. Uptake was used to describe the interaction contingencies that were observed and to demonstrate how contingencies scale across the embodied, physical, semiotic, and temporal facets of the setting.

References

Alac, M., & Hutchins, E. (2004). I see what you are saying: Action as cognition in fMRI brain mapping practice. *Journal of Cognition and Culture, 4*(3–4), 629–661.

Baldry, A., & Thibault, P. J. (2006). *Multimodal transcription and text analysis*. London: Equinox.

Çakır, M. P., Zemel, A., & Stahl, G. (2009). The joint organization of interaction within a multimodal CSCL medium. *International Journal of Computer-Supported Collaborative Learning, 4*, 115–149.

Chen, W., & Looi, C.-K. (this volume). Group Scribbles-supported collaborative learning in a primary grade 5 science class. In D. D. Suthers, K. Lund, C. P. Rose, C. Teplovs, & N. Law (Eds.), *Productive multivocality in the analysis of group interactions*, Chapter 14. New York, NY: Springer.

Garfinkel, H. (1994). *Studies in ethnomethodology*. Malden, MA: Polity press.

Goodwin, C., & Heritage, J. (1990). Conversation analysis. *Annual Review of Anthropology, 19*(1), 283–307.

Goodwin, C. (1995). Seeing in depth. *Social Studies of Science, 25*, 237–274.

Goodwin, C. (2003). *Pointing as situated practice. Pointing: Where language culture and cognition meet* (pp. 217–241). Mahwah, NJ: Lawrence Erlbaum.

Goodwin, C. (2007). Participation, stance and affect in the organization of activities. *Discourse and Society, 18*(1), 53.

Jewitt, C. (2008). Multimodality and literacy in school classrooms. *Review of Research in Education, 32*(1), 241–267.

Jordan, B., & Henderson, A. (1995). Interaction analysis: Foundations and practice. *Journal of the Learning Sciences, 4*(1), 39–103.

Koschmann, T., Zemel, A., Conlee-Stevens, M., Young, N. P., Robbs, J. E., & Barnhart, A. (2005). *How do people learn? In R. Bromme, F. W. Hesse, & H. Spada (Eds.), Barriers and biases in computer-mediated knowledge communication: And how they may be overcome (pp. 265–294).* Amsterdam: Kluwer Academic.

Kozma, R. (2003). The material features of multiple representations and their cognitive and social affordances for science understanding. *Learning and Instruction, 13*(2), 205–226.

Kupetz, M. (2011). Multimodal resources in students' explanations in CLIL Interaction. *Novitas-ROYAL, 5*(1), 121–142.

Latour, B. (1990). Drawing things together. In M. Lynch & S. Woolgar (Eds.), *Representation in scientific practice*. Cambridge, MA: MIT Press.

Lund, K., & Suthers, D. D. (this volume). Methodological dimensions. In D. D. Suthers, K. Lund, C. P. Rosé, C. Teplovs, & N. Law (Eds.), *Productive multivocality in the analysis of group interactions*, Chapter 2. New York, NY: Springer.

Medina, R., Suthers, D., & Vatrapu, R. (2009). Inscriptions becoming representations. In C. O'Malley, P. Reimann, D. Suthers, & A. Dimitracopoulou (Eds.), *Computer supported collaborative learning practices: CSCL 2009 conference proceedings* (pp. 18–27). Rhodes, Greece: International Society of the Learning Sciences.

Peeters, H. (2010). Multimodality and its modes in novelizations. *Image and Narrative, 11*(1), 118–129.

Roth, W. M. (2001). Situating cognition. *Journal of the Learning Sciences, 27–61.*

Sacks, H., Schegloff, E. A., & Jefferson, G. (1974). A simplest systematics for the organization of turn-taking for conversation. *Language, 50*, 696–735.

Schegloff, E. A. (1991). *Reflections on talk and social structure. Talk and social structure: Studies in ethnomethodology and conversation analysis* (pp. 44–70). Cambridge, MA: Polity Press.

Streeck, J., & Kallmeyer, W. (2001). Interaction by inscription. *Journal of Pragmatics, 33*(4), 465–490.

Suchman, L. A. (1987). *Plans and situated actions*. Cambridge: Cambridge University Press.

Suthers, D. D., Dwyer, N., Medina, R., & Vatrapu, R. (2010). A framework for conceptualizing, representing, and analyzing distributed interaction. *International Journal of Computer-Supported Collaborative Learning, 5*, 5–42.

Suthers, D. D., & Medina, R. (2010). *The temporal development of representational practices: Implications for theory and analysis of situated learning. Proceedings of the 2010 43rd Hawaii international conference on system sciences* (pp. 1–10). Washington, DC: IEEE Computer Society Press.

Thibault, P. J. (2011). First-order languaging dynamics and second-order language: The distributed language view. *Ecological Psychology, 23*(3), 210–245.

Chapter 17
Conceptual Change and Sustainable Coherency of Concepts Across Modes of Interaction

Kristine Lund and Karine Bécu-Robinault

Introduction

Physics teachers habitually organize teaching sequences in order to favor students' co-construction of physics concepts across different ways of communicating and interacting. Such an organization is meant to provoke conceptual change. In this chapter, we present two types of conceptual change. The first is changing one's conception from an intuitive everyday view on physics to a canonical view of physics. The second is maintaining a canonical view of physics (hence, the sustainability in our title) but while also integrating more complexity—either more objects from the experimental field into one's understanding (e.g., adding a second bulb to a simple battery–bulb circuit), by appropriating a new external representation (e.g., a new Group Scribbles drawing of a more complex circuit), or by integrating into one's talk references to new objects, new external representations, or new concepts.

In this chapter we address students' conceptual difficulties in learning electricity, through the study of multimodal and multimedial reformulation (MMR) as a tool to co-construct discourse (Apothéloz, 2001; De Gaulmyn, 1987; Lund, 2007), thus taking into account all the semiotic resources that co-participants do when they work together to solve physics problems. We first present our theoretical framework and methodology through a discussion of five specific dimensions (cf. Lund & Suthers, Chap. 2, this volume), our results, and finally our conclusions.

K. Lund (✉)
ICAR Research Lab, CNRS—University of Lyon, 15 parvis René Descartes, 69007 Lyon, France
e-mail: kristine.lund@ens-lyon.fr

K. Bécu-Robinault
Institut Français de l'Education, ENS of Lyon—University of Lyon, 15 parvis René Descartes, 69007 Lyon, France
e-mail: karine.robinault@ens-lyon.fr

D.D. Suthers et al. (eds.), *Productive Multivocality in the Analysis of Group Interactions*, Computer-Supported Collaborative Learning Series 16,
DOI 10.1007/978-1-4614-8960-3_17,

Dimensions of Our Analysis

Below, each dimension is defined according to our analyses on that dataset with the view of facilitating comparisons with other analysts.

Theoretical Assumptions

We use five theoretical positions and/or constructs, construed as a series of steps. Each step is relevant for our approach to understanding difficulties students have in co-constructing and sustaining coherency of physics concepts across modes of communication. We begin with a view on describing physics learning as acquiring the competence to connect theories and models with objects and events in an experimental field (e.g., Bécu-Robinault, 2007). These modeling activities allow the prediction or the interpretation of phenomena on the basis of theories. These theories can be intuitive and based on everyday experience or canonical or some combination of both. In this case study, we assume that students have manipulated flashlights during their everyday experience and that this has led them to build intuitive knowledge about ways to light a bulb with batteries. This intuitive knowledge, essentially based on causal relations, is not necessarily compatible with canonical knowledge concerning electricity flow and the ways to connect electric dipoles. We note that conceptual change—some of which is the transfer of knowledge from one situation to another (Schwartz, Varma, & Martin, 2008)—involves progressively changing how knowledge is organized through the use of talk and both internal and external representations (Ainsworth, 2006). Concerning the former, knowledge is organized in domain-specific theory-like structures that students can change by both bottom-up and top-down learning mechanisms. Neither the situative nor the cognitive perspective on learning can alone account for knowledge transfer and so we propose—along with Vosniadou (2007)—that Hutchin's distributed cognition (1995) be adopted as it allows for describing change in both internal representations (some form of knowledge represented in human memory) and knowledge represented in external representations.

Our second step concerns the theoretical assumptions underlying multimodal discourse analysis: Which elements are involved in the learners' communication and interaction? We note that language is a form of action and interpreted as a system of meanings, accompanied by forms through which meanings can be realized (Halliday, 1994), rather than statements corresponding to phenomena with an independent existence. We choose to study what we call the ⟨mode, medium⟩ couplet where modes are the abstract, nonmaterial resources of meaning-making and media are the specific material forms in which modes are carried out (Kress and Van Leeuwen, 2001). For example, each mode of communication in this corpus is coupled with a particular medium that allows for its expression: ⟨talk/speech⟩, ⟨drawing/GS⟩, and ⟨manipulation/battery-bulb-wires⟩. Such couplets allow us to distinguish where meaning-making is going on in terms of physics theories, models, and experimental manipulations that consist of using objects to construct events. Although ⟨gesture/body⟩ is also a crucial couplet in meaning-making, this dataset was not always filmed so that they were visible.

In our third step, we build on literature on internal and external representations in order to relate these theoretical constructs with modes expressed by particular media and conceptual change in terms of physics theories, models, and objects/ events. We maintain that it is possible to infer particular characteristics of internal representations of learners by their performance patterns and that there is a specific relationship between internal and external representations. However, we do not propose a specific mental structure for the internal representation, nor do we describe the steps a learner goes through in creating a mental model. Rather, like Zhang (1997) we explore how "cognitive activity is distributed across internal human minds, external cognitive artifacts, and groups of people, and across space and time" (p. 180). We agree with Vosniadou (2007) in that it is not possible to conceptualize learning if not in terms of some change in what is already known, but we add that these changes occur in both internal and external representations, both while taking the individual and the group as the cognitive unit; pinpointing learning becomes a matter of perspective and granularity.

In our fourth step, we consider the manner in which the kind of conceptual change we describe takes place using the notion of MMR. In this view, thought is constructed by and within language and interaction across talk, images, drawings, gestures, body movements, and manipulation of artifacts. We introduce three types of talk that will help us navigate through our tracking of conceptual change: object talk, tool talk, and conceptual talk. The first is talk centered on objects in the experimental environment, such as batteries, bulbs, and wires. The second is talk centered on the GS tool, and the third is talk accounting for concepts found in the model of electricity.

Finally, in our fifth step, we propose using a particular lens—the semiotic bundle (Arzarello, 2004)—for viewing as a conceptual unit the reformulations occurring between modes, media, and amongst learners. A semiotic bundle is a collection of semiotic sets and a set of relationships between the sets of the bundle. A semiotic set is composed of a set of signs produced with different intentional actions, a set of modes for producing signs and possibly transforming them, and a set of relationships among these signs and their meanings embodied in an underlying meaning structure. A semiotic bundle is a dynamic structure changing over time due to the semiotic activities of the participants and can be "owned" individually or by a group. Considering the studied situation, different sets of signs can be a priori identified, such as drawing on GS, words, or gestures. In our context, a mode never changes its medium. Talk is expressed through human speech, drawings are expressed through GS, and manipulations are expressed through handling batteries, bulbs, and wires. This leads us to speak of ⟨mode, medium⟩ couplets. The semiotic bundle we infer from the data can be attributed either to the group or to the students, depending on the degree of collaboration among students. We note that although Arzarello's semiotic bundles are made up of an ensemble of semiotic sets (or systems), like Vosniadou, he also proposes the work of Vygotsky as a link between these sets and the human mind. Arzarello (2004) argues that "the notion of semiotic bundle properly frames the most important point in Vygotsky's analysis, namely, the semiotic transformations that support the transformation from outer to inner speech (internalization)" (p. 283).

Although the literature reviewed here is not typically seen together, we argue that the theoretical assumptions underlying each of our five theoretical constructs are compatible with one another and thus provide for a stable foundation for analyses.

Purpose of Analysis

The purpose of our analysis is to track relations between and transformations of conceptual content that four students in a group express from the domain of physics as they communicate by talk, gestures, drawings in the GS interface (external representations), and manipulations involving experimental apparatus. In other words, our goal is to track how cognitive activity moves across space and time and is distributed across the internal representations of each of our four students as we can glean them from their interactions and the external representations they work with. Our account of internal representations will be limited to their reflection through learners' actions, both talk and activity. We wish to illustrate the difficulties and competencies students have in either showing or not sustainable coherency of physics concepts while reformulating these concepts between all of these media.

Finding Pivotal Moments as a Goal for Analysis

Each time a given mode/medium couplet is auto- or hetero-reformulated (meaning either self- or other-reformulated) into another mode/medium couplet, it is a potential pivotal moment that may be important for conceptual change that progresses towards canonical physics because content is being transformed in one way or another. We argue for two types of pivotal moments. The first is when the semiotic bundle of either an individual or a group evolves towards a structure that is progressing in its conforming to canonical physics and thus exhibiting conceptual change. The second is when the semiotic bundle reflects more complexity in terms of canonical physics while at the same time sustaining canonical coherency (no conceptual change per se, but still sustaining coherency while augmenting complexity).

Research Questions

Given all of the above, we can state three research questions. First, we wish to characterize when reformulation of conceptual content takes place across the mode/medium couplet, regardless of whether it is auto- or hetero-reformulation. Second, we wish to describe how naïve theories of physics are changed to canonical theories of physics, not only through punctual reformulation but also through the building-up of the semiotic bundle as the interaction proceeds. Finally, we expect to illustrate that some learners are able to maintain sustainable coherency of physics concepts over time while moving across mode/medium couplets whereas others have difficulties.

Basic Units of Analysis Leading to Units of Interaction and to Semiotic Bundles

We analyze both single utterances or actions for their conceptual physics content as well as how the content of pairs of utterances and/or actions are reformulated, either by the same person (auto-reformulation) or by another (hetero-reformulation). Each utterance or action is described in the form of a particular mode and the medium used to express it. We also group together these reformulations (together with utterances or actions that have not been reformulated) across a session of work in order to form a semiotic bundle (made up of different semiotic sets or systems). The semiotic bundle allows us to determine the state of individual student's and group's conceptualizations of physics and how they progress over time and space. Semiotic bundles are always in flux, as new sets of signs are continually being used to reflect upon content to be learned.

Representations of Data and Representations of Analytical Interpretations

Considering the theoretical perspectives we presented earlier in this section, we choose to represent our analytical work through annotated sequences of mode and medium couplets gleaned from both videos and transcriptions (e.g., talk through human voice, drawing through the GS interface, and artifact manipulation through physics experiments with bulbs, batteries, and wires). Annotations illustrate in what ways one mode/medium couplet is transformed into another mode/medium couplet and if talk is involved, what type of talk it is (tool talk, object talk, or concept talk). Finally, we group mode and medium couplets into semiotic bundles in order to get a more global picture of individual learner's and the group's conceptualizations. Based then on what learners say and do over time and on what we know about learners' conceptions about physics, we ascertain to what extent their theories about electricity are canonical or based on everyday experience and pinpoint pivotal moments preceding conceptual change or illustrating the sustaining of coherency with canonical physics.

Manipulations on Data Representations and Analytical Approach

As our research questions focused in part on actual talk, we could not credibly rely on the data providers' synthesis of the interaction because it summarized what the speaker said or did in the transcriber's own words, instead of the actual words of the speaker. Our analytical approach thus first consisted in obtaining more complete transcriptions from the data providers (Chen & Looi, Chap. 14, this volume), synchronizing them within Tatiana[1] (Dyke, Lund, & Girardot, 2009) with the synthesis of the interaction given by the data providers as well as with the videos and the GS interface in order to replay the interaction at will, thus creating an analytical artifact. The road to a complete transcription was rocky. Chen and Looi provided two competing transcriptions that had been transcribed from two different microphones so that they differed in their accuracy, and then Jeong (this section) provided a merged version of this transcription. The synchronization of this merged version and its replay was necessary in order to understand the dynamics of the interaction. However, at times the speaker was not specified in the merged version and so we needed to refer to the video to decide. There were also problems of repetition as the merged version often contained two versions of a speaker's talk (one recorded from each microphone but apparently considered to be different utterances by the person who merged the transcriptions). In order to correctly categorize the interventions (see below), at times we needed to cut them into distinct propositional units, each of which reflected a single propositional phrase during talk or a single action.

In order to understand and decrypt the interaction we performed the following steps:

1. Using the merged and corrected transcription as our data representation and replaying it with the synchronized video, we found where in the interaction the beginning and end of each drawing occurred.
2. We evaluated to what extent the drawing was coherent with canonical physics.
3. We divided the video into three episodes where different experiments of increasing complexity were performed (one battery and one bulb; two batteries and one bulb; two batteries and two bulbs). We could not always evaluate the experiments from the physics point of view because they were not always visible in the video.
4. When pertinent and possible (due to video quality), we looked at how gestures were used in the interaction in general, not the technical gestures used for manipulating objects and events, but gestures used in communicating.

Performing these four steps allowed us to understand the interaction and to get to a point where we felt we could analyze it. We first developed our analysis of the dataset on paper and in PowerPoint before coding it in Tatiana, as we felt that

[1] Trace Analysis Tool for Interaction Analysts: http://code.google.com/p/tatiana/

Tatiana was better suited to be used with a stabilized coding system than for supporting its development, although we used Tatiana in this step as a replayer.

Each intervention that we coded is defined as either (1) a propositional phrase in talk, (2) a single action in GS, or (3) a moment in time during an experimental action. We speak of ⟨mode, medium⟩ couplets and in our context a mode never changes its medium: ⟨talk, speech⟩; ⟨drawing, GS⟩; and ⟨manipulating/batteries-bulbs-wires⟩. Each ⟨mode, medium⟩ couplet is coded in the following way:

1. Is the physics naïve or canonical? If an intervention (GS drawing or experimental action) was coherent in terms of canonical physics, it was colored white. If it was not completely coherent, it was colored grey.
2. A type of talk was chosen for each ⟨talk, speech⟩ couplet: tool talk (signified by horizontal lines), object talk (signified by dots), and concept talk (signified by a checkered motif—but in fact never appears). For each intervention, when there was talk about either GS (circle) or experiments (triangle), we distinguished whether the speaker was talking about his or her own action (auto or self) or an action of another (hetero or other). In the former case, we put the circle or triangle in the square (talk). In the latter, we put the circle or triangle outside of the square.

Figure 17.1 shows a hypothetical sequence of talk. The first utterance is talk about tools (a square containing horizontal lines), followed by a first GS drawing without talk and not compatible with canonical physics (grey circle: utterance 2). The second GS drawing is accompanied by talk about tools and not compatible with canonical physics (grey circle within a square containing horizontal lines: utterance 3).

Next, there is an experimental manipulation coherent with canonical physics (as evidenced by the bulb lighting up) where the manipulator speaks about her own manipulations of experimental objects (a white triangle inside a square containing dots: utterance 4), followed by another correct experimental manipulation where the manipulator speaks about her partner's manipulations of experimental objects (a white triangle outside a square containing dots: utterance 5). Rows indicate the learners where each one has his or her mode/medium couplets shown in a temporal line.

Such temporal visualizations (Fig. 17.1) illustrate three phenomena: (1) whether or not and how reformulation of conceptual content takes place across the mode/medium couplet, (2) how naïve theories of physics progress towards canonical theories, and (3) to what extent learners are able to reach or maintain sustainable coherency of physics concepts over time while moving across mode/medium couplets. The MMR present in this diagram is between utterances 3 and 4 where Learner 1 performs a transformation from a first mode/medium couplet (GS/drawing) to a second mode/medium couplet (manipulating/batteries–bulbs–wires). We argue that each MMR forms the frontier between two semiotic bundles. The grey circle that turns into a white triangle illustrates that the physics knowledge becomes canonical during the experimental manipulation. Finally, because the student in utterance 4 has reached a canonical understanding, the next MMRs (not shown in this short temporal sequence) should show us how the student is able to sustain this understanding, even while facing more complexity. In the next section, we illustrate how we applied this procedure to the dataset and we discuss our results.

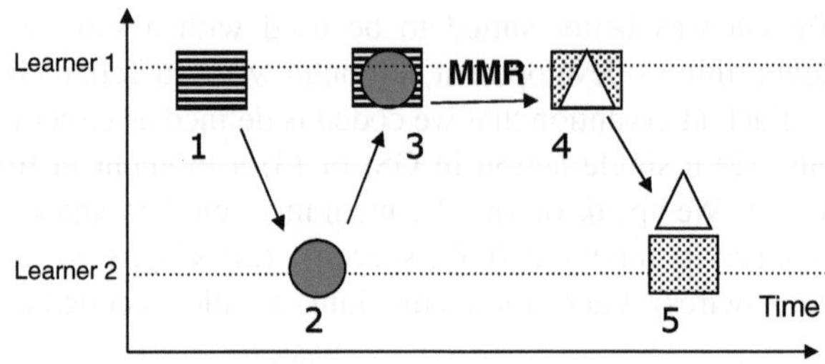

Fig. 17.1 Example visualization based on our approach

Results

We defined three episodes corresponding to the three experiments the learners performed in which we located nine MMRs that lead to either coherent or non-coherent additions to the semiotic bundle of each student or to the semiotic bundle of different incarnations of the group, depending on who was involved in the interaction at that point. For each episode, we (1) illustrate the students' GS drawings at that point in their discussion, (2) comment a coded sequence of student and teacher talk and actions while showing where the drawing occurred, and (3) provide a commented temporal visualization of our analysis. After discussion of the nine MMRs and the corresponding semiotic bundles within the three episodes, we summarize results.

Connecting a Battery and a Bulb with Two Wires: Episode One

Figure 17.2 shows drawings done by Agnes (A1), Serena (S1), and Bruno (B1) before a particular sequence in the first episode (Table 17.1). Figure 17.3 shows Bruno's (B2) and Joel's (J1) drawing after the sequence. Agnes' and Serena's drawings are correct canonical physics drawings. Bruno's first drawing (B1) is incorrect: one wire should be connected to the casing of the light bulb and the other to the end of the light bulb. Bruno corrects this in B2. However, he does not clearly show that the wire going to the minus pole of the battery actually touches the pole and does not just end at the bottom left of the battery, so we still consider it as incorrect. Joel's drawing illustrates the wire going to the minus pole a bit better.

Each turn in the transcription in Table 17.1 is coded with a type of talk and is meant to be read in conjunction with Fig. 17.3. Turns are labeled as object talk when they deal with objects and events in the experimental field. Turns are labeled as tool talk when they refer to GS. It may seem elementary, but students talk object talk when they manipulate objects and talk tool talk when they manipulate tools. They do not speak about concepts and we argue that this will prove to be important.

In this sequence, there are two simultaneous discussions taking place: one between Agnes and Serena (in bold) and the other between Bruno and Joel. Figure 17.3 shows a representation of these discussions that can be linked back with

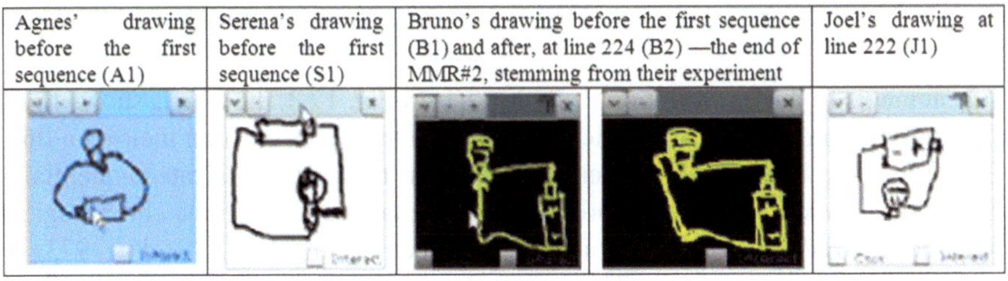

Fig. 17.2 Agnes', Serena's, and Bruno's GS drawings before the sequence shown in Table 17.1 and Joel's and Bruno's drawings after

Table 17.1 Our first sequence shows two dyads working in parallel: Serena and Agnes and Bruno and Joel; at this point, drawings A1, S1, and B2 have already been published on GS (cf. Fig. 17.2). Parentheses signal short pauses; Singlish (e.g., leh, mah) interpreted according to http://en.wikipedia.org/wiki/Singlish; overlap not transcribed. This part of the transcript is not included in the dataset given by Chee Kit and team

N°	Partic.	Talk	Type of talk
202	Bruno	ok () ok you put it at the bottom cause just now you were on the table	Object talk
204	Bruno	ok then put it on top put it on top	Object talk
205	Bruno	ok then let's write the answer la	Tool talk
206	**Agnes**	aiyer cannot connect that one cannot play leh no fair	Object talk
207	Bruno-	so there's the minus	Tool talk
208	Joel	no cannot write the answer first do the draft first	Tool talk
209	Bruno	where's the pen where's the pen	Tool talk
210	Joel	we're doing the draft first mah	Tool talk
212	**Serena**	one at the bottom one at the side	Object talk
213	**Agnes**	I know there	Object talk
214	**Serena**	ei got got light already got light already	Object talk
219	Bruno	object which one is the one for drawing	Tool talk
222	Joel	you haven't changed the color	Tool talk
224	Bruno	oh no wonder it's black and black	Tool talk

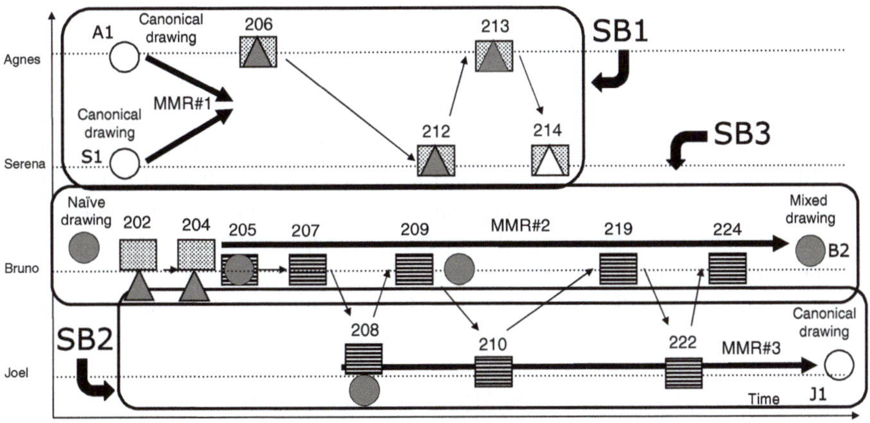

Fig. 17.3 A temporal visualization of a sequence from episode one

line numbers to the transcription in Table 17.1. In Fig. 17.3, the *X*-axis shows time, although not proportionally. Participants are on the *Y*-axis, each on their own line. Thin arrows going from one utterance or action to another indicate sequentially occurring talk/action. But sequentially occurring talk/actions from a chronological standpoint do not necessarily mean adjacency pairs (Schegloff & Sacks, 1973) or sequences of meaningful talk. The sequences and figures we show will not always have all turns noted because some turns were duplicates for the reasons explained previously.

During the sequence in Fig. 17.3, Agnes and Serena are connecting a bulb to a battery, based on their canonical GS drawings (white circles). Squares signify talk, and triangles signify experiments. The experiment succeeds (the bulb glows)—i.e., represents canonical physics—and so the triangle is white (214). The triangles are within dotted squares where the dots represent talk about objects in the experimental field (206, 212, 213, 214). Triangles or circles within squares mean that the talk is a reference to one's own talk or actions; this is often only verifiable in the video. For example, in 206, Agnes is talking about the wires she is connecting herself.

It seems evident from the transcription and from viewing the video that each understands the other's actions. We argue that what we call Agnes' and Serena's coordinated four-handed experiment is a reformulation from ⟨drawing/GS⟩ to ⟨manipulating/batteries–bulbs–wires⟩. Let's call this MMR#1 (cf. Fig. 17.3). MMRs are illustrated by thick arrows. The construction of this MMR leads to the first semiotic bundle (labeled SB1 in Fig. 17.3) that integrates talk about drawing with GS, drawings A1 and S1, as well as the objects that Agnes and Serena manipulate together. Agnes and Serena share this semiotic bundle. The two girls sustain coherency with canonical physics during the construction of their semiotic bundle despite performing a complex transformation between mode/medium couplets. Students do not always succeed at such transformations, and indeed, success is rather the mark of a competent student or expert (Kozma, 2003; Lund & Bécu-Robinault, 2010).

In the second parallel interaction thread, Bruno and Joel are talking together and trying to connect the circuit with a battery, bulb, and wires. Bruno has already drawn his circuit on GS, but Joel has not. Unfortunately, we do not see how they attempt to connect the circuit because they are filmed from behind without a view on the experiment. As before, squares signify talk, and the dots signify that talk is about objects in the experimental field. The triangle signifies objects or actions taken in the experimental field. Bruno is speaking about Joel's actions in the first two utterances in regard to the experimental field (202, 204), so the triangles are outside of the squares. Next, Bruno proposes to redo the drawing in GS and searches for how to erase part of his drawing (205). This part of the interaction is represented by a square because it's talk, and the square contains horizontal lines for the purposes of representing Bruno talking about GS tools (tool talk). The circle is inside the square in order to represent him talking about his own action and it is grey because he is referring to changing his drawing and his drawing is only partly coherent (see below).

After more talk (207, 209, then 219, 224), Bruno redraws the wire (grey circle) that leaves the minus pole so that it connects to the bulb's casing, instead of to the bulb's endpoint (as in his previous drawing). However, it still is not clear that Bruno has identified the minus pole of the battery because the wire could be hooked only to the corner end of the battery and not to the pole on the middle of the battery's end. The drawing is grey (B2 in Figs. 17.2 and 17.3) because part of it respects canonical physics and part does not. On the other hand, Joel does a GS drawing (J1 in Figs. 17.2 and 17.3) that is clear and completely coherent (white circle), both in terms of battery pole and bulb poles. Note the difference between Bruno's second drawing and Joel's first drawing in terms of how the wire leaves the minus pole of the battery (compare them in Fig. 17.2). Both boys reformulate from ⟨manipulating/batteries–bulbs–wires⟩ to ⟨drawing/GS⟩. Let's call these, respectively, MMR#2 (Bruno) and MMR#3 (Joel)—cf. Fig. 17.3. The construction of these MMRs leads to the two next semiotic bundles: SB2: talk during manipulation of objects between Joel and Bruno, J1 drawing, attempts to connect wires and battery and bulb; and SB3: talk about objects with Joel after posting B1 and with the girls after Bruno posts B2, B2 drawing, successful connection of one battery, one bulb, and one wire. SB2 is Joel's semiotic bundle and SB3 is Bruno's. Both are labeled in Fig. 17.3.

At this point in the interaction, three out of the four students have completed a coherent drawing. Only Bruno does not exhibit sustainable coherency during semiotic bundle construction while reformulating between mode and medium couplets. However, none of the students make reference to physics concepts throughout this sequence or at any other time in the corpus. Thus, even if on the surface they succeed, we may question their ability to understand the physics concepts behind their drawings and their experimental manipulations.

Connecting Two Batteries and a Bulb: Episode Two

In a second sequence from episode two, all four students are interacting, mostly trying to connect two batteries together with wires to a bulb. Before this sequence, Bruno posted two drawings (B2 and B3) into the public space (cf. Fig 17.4). In the drawing representing the experiment with two batteries (B3), the minus signs are drawn, but he shows the same ambiguity as he did for B2 in regard to connecting the wire to the actual minus pole in the middle of the battery and not just to the bottom corner of the battery. In addition, his possible attempt at modeling a parallel circuit with the batteries (B3) is not successful; the batteries are not connected together so that current flows. Bruno's model of the new experiment (B3) is thus not coherent with canonical physics, but he has nevertheless performed a second MMR (MMR#4—cf. Fig. 17.4). He performs a transformation from ⟨manipulating/batteries–bulbs–wires⟩ to ⟨drawing/GS⟩. The construction of MMR#4 leads to the construction of Bruno's SB5 (done in parallel to SB4—see below). SB5 consists of talk (sequence not shown; occurs before that in Table 17.2) with others during the manipulation phase, drawing B3 and a successful connection of one battery, one bulb, and one wire. We note that although Bruno's drawing B3 represents two

Bruno's second drawing (B2)	Bruno's third drawing (B3) — end of MMR#4 leading to SB5	Agnes begins drawing in Bruno's space (A2) — end of MMR#5	Serena shows drawing from another group (G s)	Bruno finishes the drawing Agnes began (B_{A2}) — end of MMR#6

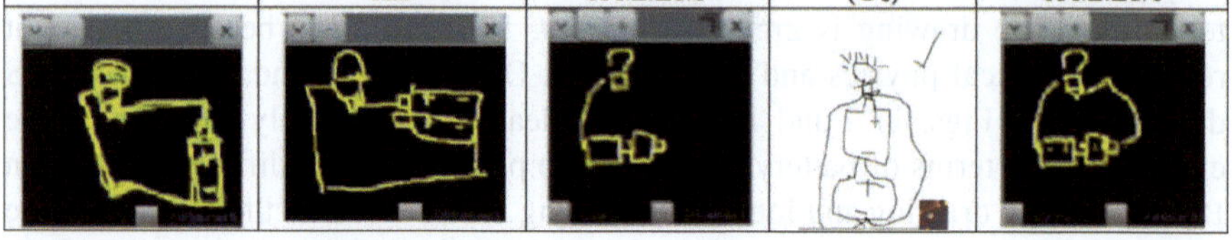

Fig. 17.4 Bruno's publically posted drawings, going into the second sequence of episode two; the drawing Agnes begins in Bruno's space; the drawing that Serena notices in another group, and how Bruno finishes the drawing Agnes begins in his space

Table 17.2 Sequence showing the students first working together to hook two batteries to one bulb, then Agnes and Bruno exchange about Bruno's GS drawing, and finally Serena brings everyone's attention to a drawing she found on another group board. A2 was posted at 15:03 in the dataset provided by Chee Kit and team

N°	Partic.	Talk	Type of talk	GS drawings
380	Bruno	worked out already heh heh posted already	Tool talk	B2, B3 been posted
381	Serena	we got () we got four wires leh	Object talk	
382	Bruno	never mind just cancel oh I know put here easier	Tool talk	
385	Agnes	just do another one la	Tool talk	
386	Teacher	now () maximum two batteries	Object talk	A2 posted here
387	Serena	maximum two batteries leh Bruno	Object talk	
388	Friend	no I know put this here	Object talk	
394	Joel	then then one () one here one here	Object talk	
396	Agnes	bruno you draw the rest	Tool talk	
408	Joel	it's not connected to the light bulb	Tool talk	
409	Agnes	Bruno Bruno you draw finish () draw finish	Tool talk	
413	Bruno	( ) cancel this one	Tool talk	
415	Joel	( ) already finish drawing	Tool talk	
416	Agnes	Bruno, you draw finish ah	Tool talk	
420	Agnes	and this one.	Tool talk	
421	Serena	[to the rest] we all try this one. () we all try this one	Tool talk	Shows other group's GS

batteries, he is only experimenting with one, which may illustrate his current incapacity to sustain coherency with canonical physics. SB4—shared by Agnes, Serena, and Joel—is a reconfiguration of SB1 and SB2 while also integrating a second battery. SB4 is the only example of a semiotic bundle that is not produced by an MMR. Rather it contains all the elements of the previous semiotic bundles (SB1 and SB2) but also adds an extra battery.

In this second sequence (cf. Table 17.2), not all turns are represented, as explained previously. Students talk object or tool talk with no discussion about concepts such as poles or current. We will return to this observation in the conclusions.

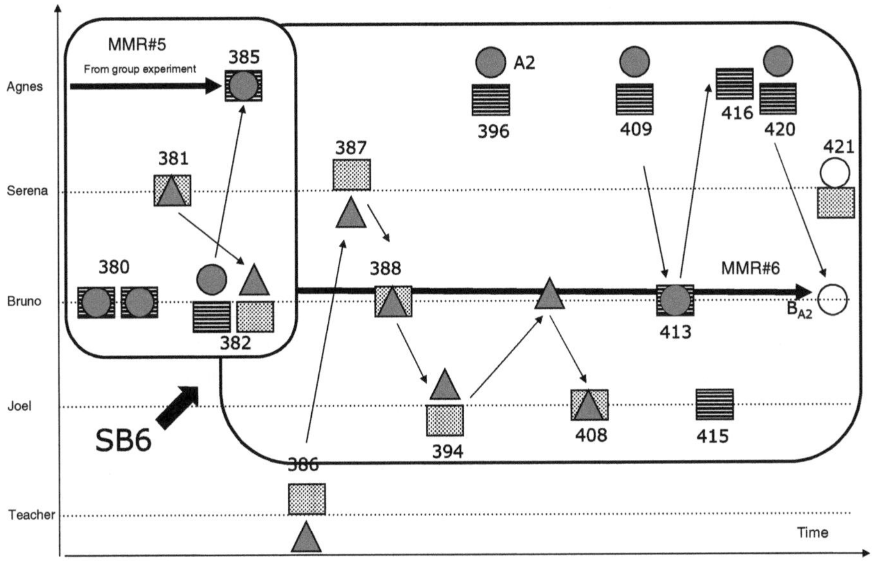

Fig. 17.5 A temporal visualization of a sequence from episode 2

After having noticed that Bruno's drawing (B3) is incorrect, Agnes begins a correct one in his GS space and tells him to finish it ("Bruno, you draw finish draw finish": turns 409, 416). This is MMR#5, represented by A2 in Fig. 17.4 and shown in Fig. 17.5. Agnes' talk is about tools all along the sequence (squares containing horizontal lines in 396, 409, 416, and 420); the circle represents GS; the ones in 396, 409, and 420 are outside the square because Agnes is taking about Bruno's drawing, and they are grey because the drawing is not finished so we can't decide if it respects canonical physics (white) or not.

Bruno finally agrees to follow Agnes' directions (no corresponding talk), finishes the drawing correctly, and posts it. Let's call this MMR#6 (labeled in Figs. 17.4 and 17.5; it occurs after the sequence in Table 17.2 ends). It is distinguished from the others in that it is cooperative: begun by one person (Agnes) and finished by another (Bruno). In other words, in this sequence (Table 17.2 and Fig. 17.5), Agnes reformulates from ⟨manipulating/batteries–bulbs–wires⟩ to ⟨drawing/GS⟩, a drawing that Bruno later finishes, although we consider that his MMR reformulates from ⟨manipulation/batteries–bulb–wires⟩ to ⟨drawing/GS⟩: it's the shared experiment that allows Bruno to complete Agnes' drawing. Figure 17.4 (far right, B_{A2}) shows that Bruno added labels to the poles of both batteries (– and +) and then drew a wire from the plus side of the second battery to the endpoint of the bulb. Note that the batteries are now clearly connected serially, contrary to his drawing (B3). At this point, both MMR#5 and MMR#6 help Bruno to build semiotic bundle SB6 (labeled in Fig. 17.5): talk about ways to handle the objects (turns 386–394), drawing BA2 (turns 385, 396, 409, 416), and successful connection of two batteries, two wires, and one bulb (previous turns not shown). Like the first sequence, students only talk about tools and objects, not concepts. However, contrary to the first sequence (two parallel discussions), here the students are exchanging more and moving a bit more between GS tool and objects.

Connecting Two Batteries, Two Bulbs, and Two Wires: Episode Three

In the continuing interaction—still shown at the very end of turn 421 and Fig. 17.5—as Bruno finishes up the drawing Agnes began, Serena turns around her pivot screen to show her GS public space to the others in her group (see also Fig. 17.6, left).

This moment is interesting because of Agnes' drawings (she draws B_{A3} (cf. Fig. 17.7) during the talk in Table 17.3). She enlarges her semiotic bundle while sustaining coherency with canonical physics. Indeed, when the environment changes, Agnes is able to adapt her understanding of how a circuit functions and integrate this into her semiotic bundle. She appropriates as her own and integrates two mode/medium couplets: (1) a new GS drawing (the one Serena shows to the group: G_S) and (2) corresponding experimental actions, both being compatible with previous elements she has already integrated. Her performing this integration does not seem to modify the coherency she has already built and enables her to reformulate the experiment done by the others with her drawing A4.

MMR#7 is *collective* in that it starts with the drawing/GS couplet (turn 421, Table 17.2) that Serena shows the group (we argue that it is *appropriated* by the group members) and it ends with ⟨manipulating/batteries–bulb–wires⟩, carried out by Joel, Bruno, Agnes, and Serena (cf. Fig. 17.6, right). It is not Serena's drawing, but the MMR is still performed from this other group's drawing. The construction of MMR#7 leads to the seventh semiotic bundle SB7, shared for the first time by all participants: Agnes, Serena, Joel, and Bruno. SB7 is constructed simultaneously with SB8 and consists of talk concerning comparison of events with one or two batteries, the GS drawing from another group that Serena shows to the group, and the two batteries connected to one bulb (sequence not shown). MMR#8 occurs when Agnes transforms the group's ⟨manipulating/batteries–bulbs–wires⟩ to ⟨drawing/GS⟩ on Bruno's GS (B_{A3} in Figs. 17.7 and 17.8). Figure 17.8 shows the talk around the MMR that the group performs (turns 536–554) and the two MMRs that Agnes performs (MMR#8 and MMR#9). SB8 is also shared by all participants and consists of talk about handling devices (turn 537), drawing BA3 (turn 536), and the manipulation of two batteries connected to one bulb (turns 553–554). MMR#9 occurs when Agnes begins with the group's manipulating/batteries–bulbs–wires couplet and ends with a drawing/GS couplet (A4 in Figs. 17.7 and 17.8). SB9 is also shared by all participants and consists of talk (sequence not shown) about ways to handle batteries, bulbs, and wires; the drawing A4; and the manipulation of two batteries, two wires, and two bulbs. Yet again, in this third sequence, students only speak about tools and objects and not concepts.

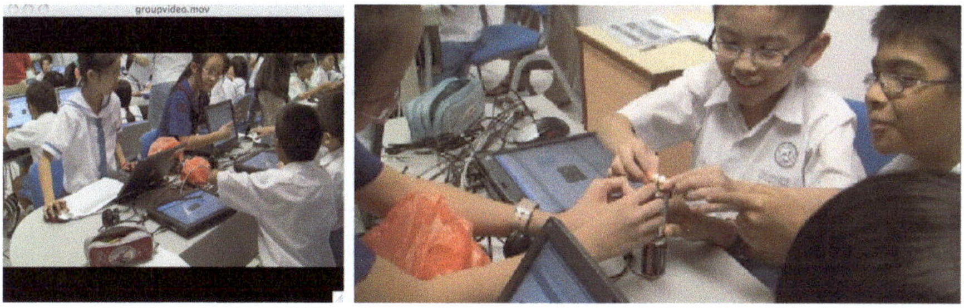

Fig. 17.6 Serena shows the members of her group another group's drawing, and her group decides to perform a new experiment. This episode ends with MMR#7, represented above *right*, when the students light the two bulbs with two batteries (and this results in SB7)

Table 17.3 Sequence where Agnes draws in GS (cf. Fig. 17.7, on left, published at the time of line 554), the others look at other students' productions and manipulate the batteries, bulb, and wires. B_{A3} is not referred to in the common transcript provided in the dataset chapter, as it occurs later

N°	Partic.	Talk	Type of talk	GS drawings
536	Agnes	I draw the first method ah	Tool talk	
537	Joel	we can test we can test out more	Object talk	
538	Joel	( ) like that can ( ) no	Tool talk	
543	Serena	you all () you all try other ways	Tool talk	
544	Friend	cannot this one lighted	Tool talk	
546	Serena	you know why it's the same battery	Tool talk	
553	Friend	you want to take another wire	Object talk	
554	Joel	( ) arrow () take two ( ) where's the other black wire	Object talk	B_{A3} posted

Agnes does the drawing to the right on Bruno's GS after the experiment (B_{A3}),	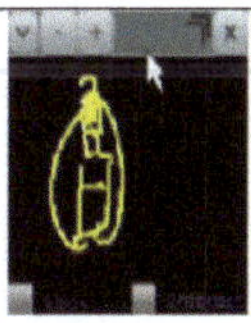	Serena notices the drawing on the right from another group (G_S)	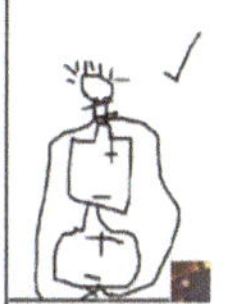	At the end of the sequence, Agnes draws what looks like two batteries and two bulbs (on right), but her wire connections are not clear (A4)	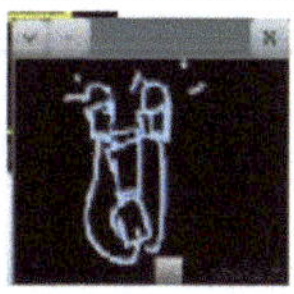

Fig. 17.7 Agnes draws on Bruno's GS, Serena notices another group's drawing, and Agnes draws her group's experiment

Synthesis of Results

Table 17.4 describes our synthesis of MMRs and semiotic bundles (SBs); it begins with the MMR#, then specifies the drawing as referenced in this chapter, describes the beginning and ending MMR couplet, gives the number of the SB, evaluates its state (coherent, incoherent in terms of whether or not the physics is correct), and

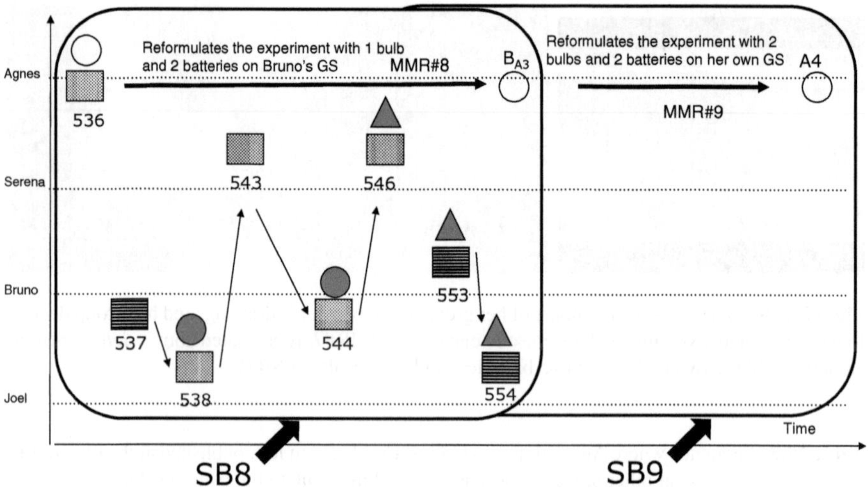

Fig. 17.8 A temporal visualization of a third sequence from episode three

Table 17.4 Synthesis of MMRs and the semiotic bundles they participate in constructing; MMRs 4, 5, and 6 appear in the shared transcript, and the corresponding time stamps are shown in the far left column.

MMR#	GS#	Departing MM couplet	Arriving MM couplet	Semiotic bundle#, state of and owner
1	A1 & S1	Drawing/GS	Manipulating/ batteries–bulb–wires (BBW)	SB1, coherent, shared by Agnes and Serena
2	B2	Manipulating/BBW	Drawing/GS	SB3, probably incoherent, Bruno
3	J1	Manipulating/BBW	Drawing/GS	SB2, coherent, Joel
4/14:16	**B3**	**Manipulating/BBW**	**Drawing/GS**	**SB5, incoherent, Bruno** **SB4 (contains elements of SB1 and SB2 + an extra battery), Agnes, Serena, Joel**
5/15:27	**A2**	**Manipulating/BBW**	**Drawing/GS**	**Coherent as MMR#5, but not finished**
6/16:16	**B_{A2}**	**Manipulating/BBW**	**Drawing/GS**	**SB6, probably coherent, Bruno**
7	G_S	Drawing/GS	Manipulating/BBW	SB7, coherent, all participants
8	B_{A3}	Manipulating/BBW	Drawing/GS	SB8, probably coherent, all participants
9	A4	Manipulating/BBW	Drawing/GS	SB9, coherent, all participants

The lines in bold show time stamps that enable comparison with other analysts in Suthers (Chap. 19, this volume)

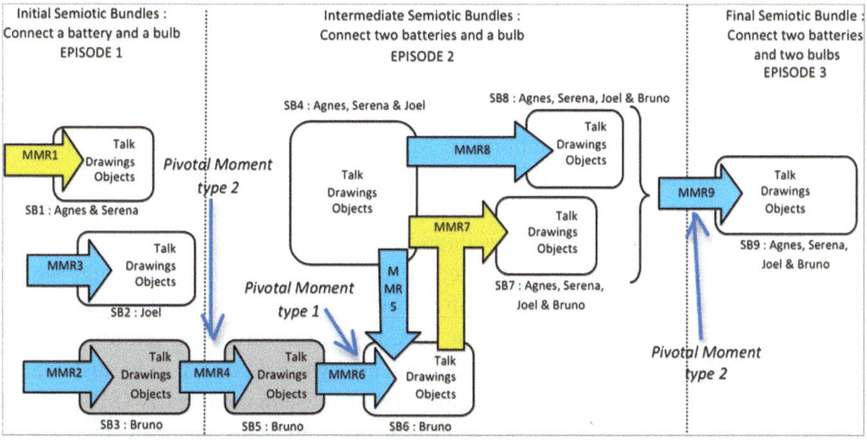

Fig. 17.9 The *boxes* represent the semiotic bundles built by the students. *Boxes* are *grey* if the semiotic bundle is not totally coherent with physics. *Boxes* are *white* if the semiotic bundle is coherent with physics. Using a lighter *grey arrow* (MMR1 and MMR7) we represent multimodal and multimedial reformulations from drawing/GS to manipulating/batteries–bulbs–wires. Using a darker *grey arrow*, we represent multimodal reformulations from manipulating/batteries–bulbs–wires to drawing/GS

finally lists the SB "owners." There are four individual SBs (2,3,5,6) and six collective (1,4,6,7,8,9).

Figure 17.9 synthesizes the three episodes referred to in our results. Each square is an SB incorporating talk, drawings, and experimental objects. Analyses show two MMR types: from drawing/GS to manipulating/batteries–bulbs–wires and the reverse.

Figure 17.9 represents the evolution of the individual student's and group SBs. There are nine SBs. A snapshot of our analysis of the SB is taken at each MMR, shown here in succession. The students who participated in the interaction at that point are listed as "owning" that particular SB at that moment. We argue that the MMRs are partly responsible for the semiotic bundles' evolution as they can illustrate moments of conceptual change (pivotal moment type 1) but also increasing levels of complexity (pivotal moment type 2) without giving up coherency with canonical physics.

At the top left of Fig. 17.9, we argue that Agnes and Serena use MMR#1 to establish a semiotic bundle concerning the way to connect a bulb to a battery (SB1). Joel does not reach a semiotic bundle coherent with canonical physics until after MMR#3 (SB2), whereas Bruno's production, even after MMR#2, is still not coherent with canonical physics (SB3). We have seen that Bruno's misunderstanding disrupts him as he tries to connect two batteries to a single bulb (i.e., he has still not identified that minus poles and plus poles should be connected so that current flows). He is thus not able with MMR#4 to build a semiotic bundle coherent with canonical physics (SB5). Bruno is only able to build a coherent semiotic bundle with the help of Agnes' MMR#5 (SB4), when he performs MMR#6 (SB6), continuing the MMR initiated by Agnes (ending with B_{A2}).

From this point, the group seems to share the same semiotic bundles (cf. Fig. 17.9). As Serena shows a new drawing on her GS, the whole group starts a collaborative experiment producing the collaborative MMR#7 (SB7). This reformulation from drawing to manipulating enables the whole group to integrate in their shared semiotic bundle the connection of two bulbs to two batteries, respecting canonical physics. MMR#5 compared to MMR#7 is important from a learning perspective. Indeed, the students navigate between the electric devices to their GS drawing, which implies modeling activities including physics knowledge. These modeling activities reveal relationships among the semiotic sets, which reinforces the semiotic bundle coherence. This coherence, respecting canonical physics, is shown to be sustainable with MMR#9 as coherence remains despite the increase in complexity (SB9). Right before this, Agnes takes the responsibility of drawing, with MMR#8, the experiment performed by the group. No one questions this reformulation, which thus seems to obtain the tacit agreement of the group. The semiotic bundle of the group is enlarged to become SB8 due to integrating a new way to connect the battery and bulb and because of its drawing.

The group then starts a new experiment with two bulb and two batteries. As soon as the bulbs light, Agnes reformulates this experiment (MMR#9) with a new drawing. This leads the group to build the last common semiotic bundle SB9, which integrates all the previous elements as well as the method to light simultaneously two bulbs with two batteries.

Pivotal Moments and Sustainable Coherency

Earlier in this chapter, we postulated that each time a given mode/medium couplet is auto- or hetero-reformulated into another mode/medium couplet; it is a potential pivotal moment that may be important for conceptual change because content is being transformed. We argued for two types of pivotal moments. The first was when the semiotic bundle of either an individual or a group evolves towards a structure that is progressing in its conforming to canonical physics and thus exhibiting conceptual change (i.e., Bruno's MMR#6). The second was when either an individual's or a group's semiotic bundle reflects more complexity in terms of canonical physics while at the same time either sustaining canonical coherency or progressing towards canonical coherency. The MMRs that satisfy this definition of being pivotal are MMR#4 and MMR#9. MMR#4 is found at the frontier of episode 1 and 2 (cf. Fig. 17.9) where complexity is increased by connecting two batteries instead of one to the bulb. MMR#9 is found at the frontier of episode 2 and 3 (cf. Fig. 17.9) where complexity is again increased by connecting together two batteries and two bulbs. MMR#4 maintains its coherency with canonical physics, whereas MMR#9 progresses partially towards canonical physics (i.e., Bruno correctly draws the batteries but not how the poles should be connected).

Conclusions and Perspectives

Our first research question sought to characterize when reformulation of conceptual content took place across the mode/medium couplet, regardless of whether it is auto- or hetero-reformulation. This distinction was important for deciding whether or not an MMR was individual or collective and thus to see when semiotic bundles were shared. We characterized nine instances of MMR in the three episodes that we studied. Second, we sought to describe through pivotal moments how naïve theories of physics are changed to canonical theories of physics, not only through punctual reformulation but also through the building-up of the semiotic bundle as the interaction proceeds. We identified three pivotal moments that either illustrated conceptual change (MMR#6) or showed how students' physics knowledge remained canonical (#MMR4) or progressed in respect to canonical physics (MMR#9) despite an increase in complexity (e.g., moving from episode 2 to episode 3). Third, our results show that some students can increase complexity while maintaining coherency (Agnes, Serena) while others have difficulties (Bruno).

In conclusion, although students were mostly likely not conscious of this fact, the pedagogical situation that gave rise to the dataset studied for this chapter gave them the opportunity to test the robustness of their conceptions about electricity. If they were able to perform successful MMRs across couplets of ⟨talk, speech⟩, ⟨drawing, GS⟩, and ⟨manipulations, batteries, bulbs, wires⟩ and thus build coherent semiotic bundles, they therefore illustrated their robust canonical knowledge of physics. If they had difficulties, then perhaps the reformulations they were required to do helped them progress from naïve physics to canonical physics, as in the case of Bruno correcting his drawing after having done the experiment with Joel. A next step could be to reflect more generally on how methods and tools in pedagogical situations could help learners to develop such sustainable coherency, especially if the pedagogical situation could be structured so that student talk centered more on concepts rather than only on objects and tools.

References

Ainsworth, S. E. (2006). DeFT: A conceptual framework for considering learning with multiple representations. *Learning and Instruction, 16*(3), 183–198.

Apothéloz, D. (2001). Les formulations collaboratives du texte dans une rédaction conversationnelle : modes d'expansion syntaxique, techniques métalangagières, grandeurs discursives manipulées, etc. In M.-M. de Gaulmyn, R. Bouchard, & A. Rabatel (Eds.), *Le processus rédactionnel. Ecrire à plusieurs voix* (pp. 49–66). Paris: L'Harmattan.

Arzarello, F. (2004). Semiosis as a multimodal process. *Relime, Numero Especial*, 267–299.

Bécu-Robinault, K. (2007). Modélisation et investigation autour d'une séquence d'électrocinétique: Introduction d'une analogie pour expliquer. In L. Morge & J.-M. Boilevin (Eds.), *Séquences d'investigation en physique-chimie* (pp. 117–129). Scéren: CRDP Auvergne.

Chen, W., & Looi, C. -K. (this volume). Group Scribbles-supported collaborative learning in a primary grade 5 science class. In D. D. Suthers, K. Lund, C. P. Rose, C. Teplovs, & N. Law (Eds.), *Productive multivocality in the analysis of group interactions* (Chap. 14). New York, NY: Springer.

De Gaulmyn, M. -M. (1987). Actes de reformulation et processus de reformulation. In P. Bange (Ed.), *La dame de Caluire* (pp. 83–98). Berne: Peter Lang.
Dyke, G., Lund, K., & Girardot, J. -J. (2009, June 13–18). Tatiana: An environment to support the CSCL analysis process. *Proceedings of the International Conference on Computer Supported Collaborative Learning (CSCL) 2009 on CD-ROM*. University of the Aegan: Rhodes, Greece.
Halliday, M. A. K. (1994). *An introduction to functional grammar* (2nd ed.). London: Edward Arnold.
Hutchins, E. (1995). *Cognition in the wild*. Cambridge Press, MA: MIT Press.
Kozma, R. (2003). The material features of multiple representations and their cognitive and social affordances for science understanding. *Learning and Instruction, 13*(2003), 205–226.
Kress, G., & Van Leeuwen, T. (2001). *Multi-modal discourse*. London: Arnold.
Lund, K. (2007). The importance of gaze and gesture in interactive multimodal explanation. *International Journal of Language Resources and Evaluation, 41*(3/4), 289–303.
Lund, K., & Bécu-Robinault, K. (2010). Learning physics as coherently packaging multiple sets of signs. In K. Gomez, L. Lyons, & J. Radinsky (Eds.), *Learning in the Disciplines: Proceedings of the 9th International Conference of the Learning Sciences (ICLS 2010)—Volume 1, Full Papers* (pp. 404–411). Chicago IL: International Society of the Learning Sciences.
Lund, K., & Suthers, D. D. (this volume). Methodological dimensions. In D. D. Suthers, K. Lund, C. P. Rosé, C. Teplovs & N. Law (Eds.), Productive *multivocality in the analysis of group interactions* (Chap. 2). New York, NY: Springer.
Schegloff, E. A., & Sacks, H. (1973). Opening up closings. *Semiotica, VIII*(4), 289–327.
Schwartz, D. L., Varma, S., & Martin, L. (2008). Dynamic transfer and innovation. In S. Vosniadou (Ed.), *International handbook of research on conceptual change* (pp. 479–506). New York, NY: Taylor & Francis.
Vosniadou, S. (2007). The cognitive-situative divide and the problem of conceptual change. *Educational Psychologist, 42*(1), 55–66.
Zhang, J. (1997). The nature of external representations in problem solving. *Cognitive Science, 21*(2), 179–217.

Chapter 18
Development of Group Understanding via the Construction of Physical and Technological Artifacts

Heisawn Jeong

Introduction

In real life, small group activity is not an option, but a way of life. People are often asked to learn, solve problems, or carry out tasks in groups or teams. Many of those tasks are intensely cognitive in nature, requiring lots of knowledge and problem-solving skills. Often, it is the groups, not its individual members, who are asked to learn, solve problems, or make decisions (e.g., navigation team, Supreme Court). In such cases, we may be interested in how individual members think and respond, but it is the group's collective decision and action that are of interest to us. In this chapter, I introduce the notion of *group understanding* to refer to a cognitive state or understanding that lie behind the group's collective action and attempt to capture it based on the physical and technological artifacts a group constructs during learning.

Group Understanding

There have been increasing attempts to understand how groups function cognitively at the group level (Börner et al., 2010; Hutchins, 1995; Stahl, 2006). Notable in learning sciences community is the notion of *group cognition* by Stahl (2006). The notion of group cognition refers to a form of cognition that arises during group interaction. It refers to the shared meaning and knowledge that is collaboratively constructed through group discourse. As such, group cognition is not an automatic outcome of a group membership but arises only when meaning is mutually created and shared through the group discourse. The notion of *group understanding* that I

H. Jeong (✉)
Hallym University, Chuncheon, South Korea
e-mail: heis@hallym.ac.kr

D.D. Suthers et al. (eds.), *Productive Multivocality in the Analysis of Group Interactions*, Computer-Supported Collaborative Learning Series 16,
DOI 10.1007/978-1-4614-8960-3_18,

propose in this chapter differs from Stahl's notion of group cognition (the term "understanding" was specifically introduced to emphasize the difference), in that it refers to ALL cognitive states that groups can manifest. It includes cognitive states of the group that may be only partially co-constructed and shared among its members as well as the state of group cognition proposed by Stahl.

One may wonder why such a notion of group understanding is needed, that is, why we should study the group as a group when the group members have not constructed a mutual understanding and thus have not become a group in a genuine sense. One might argue that if a group behaves merely as a collection of individuals, then we should just study them as individuals instead of as groups. The rationale that I offer in this chapter is that groups are often required to act as a unit of action and decision making even when they have not fully engaged in the group meaning-making process and/or achieved mutual understanding. Constructing and maintaining mutual understanding can be challenging, especially in a large group with little communication and discussion. In addition, even when mutual understanding is achieved, it is difficult to maintain, constantly being challenged by poor communication practices and a host of cognitive, social, and/or technological factors. Cognitive science research has shown that individual understandings are not always a single coherent entity. Students are often confused about what they know and hold more than one, often conflicting conceptions (Chinn & Brewer, 2001; Vosniadou & Brewer, 1992). And yet, they still function as a unit, giving answers and making decisions. Likewise, a group can act and thus be studied as a unit in spite of some disagreeing voices or lack of mutuality or intersubjectivity in its members' understandings. Understanding the full range of cognitive states that groups can manifest would help us to better understand how group understanding develops and why some groups succeed and/or fail to achieve mutual and/or intersubjective understanding.

Like individual understanding, group understanding undergoes a constant change and reconstruction during its lifetime. It may begin with a poorly developed notion about a certain topic. The understanding may not be shared either at the beginning. As groups process new information and engage in active interaction among its members, however, their understanding changes in response. I propose that there are at least two dimensions along which group understanding develops. *First*, group understandings can differ in the amount and complexity of the domain knowledge they display. When groups first start interacting about a certain topic or a problem, they might begin with a naïve conception. As children's understanding develops from naïve to more complex ones (Vosniadou & Brewer, 1992), however, group understanding would also develop in a similar fashion, progressing from primitive to more coherent and sophisticated conceptions. *Second*, group understanding can differ in intersubjectivity. While some groups would achieve a singular state of mutual understanding, other groups would display a range of conceptions. Some group understanding may be a mere collection of disjointed individual understandings. Other groups may display an understanding that may not be mutual but still be intersubjective so that group members understand each other although they do not yet fully agree with each other yet (Matusov, 1996).

Artifacts and Assessment of Group Understanding

In order to understand and document the developmental trajectory of group understanding, we need to find reliable and systematic ways to assess it. The most common analysis method to assess group cognition and other group cognitive processes has been to analyze verbal interaction of the group (Jeong, 2013). In this chapter, I add artifacts as another data source from which we can trace the development of group understanding. Small group work often involves manipulation and creation of artifacts. Artifacts reveal the nature of the meaning-making practices of the group (Hollan, Hutchins, & Kirsh, 2000). They can reveal aspects of group understanding not easily captured in verbal dialogues. By examining artifacts, its contents, and the process of its construction, we can better understand how groups co-construct and develop their understandings.

Jeong, Chen, and Looi (2011) carried out an analysis of group understanding based on digital artifacts, using methods similar to the ones reported in this chapter, and revealed some of the characteristics of group understanding development. They examined a group of four students who collaboratively solved how to divide two pizzas equally among three students (Looi & Chen, 2010). Students interacted using Group Scribbles (GS), sharing drawings and notes with each other in the GS group board. The analyses explored how group understanding, in the form of *contributions* to the group board, evolved over time. The group board started out with zero contributions but evolved into a state with six contributions (i.e., a set of related postings) as students added, revised, or removed postings from the group board. In other words, the group collectively considered a number of different ideas but settled on a subset of them through interaction. Group understanding, as reflected in the evolution of the group board, was quite dynamic and was under steady improvement and revisions.

In spite of all the changes and activities, the group board in Jeong et al. (2011) study remained fragmented till the end of the episode. Students responded to others' postings by putting comments such as "nice" (prompted by the teachers), but not much interaction occurred about the substance of the drawings. The ideas in the group board were not organized in any meaningful manner except for physical rearrangements. Students just published their postings without regard to what was already in the group board. As a result, a lot of the postings were redundant and poorly integrated. The group appeared to have achieved some degree of intersubjectivity, acknowledging each other's contribution with evaluative comments, but the group understanding was largely unintegrated both in terms of idea development and intersubjectivity.

Analysis Objectives

The primary aim of the current analysis is to examine the two different dimensions of group understanding development discussed earlier, the development of domain understanding and the development of intersubjectivity, based on the artifacts that students construct during learning. In addition, in the current data set, students constructed two different types of artifacts in the process of learning about the electrical circuits: GS-based digital artifacts (e.g., drawings) and physical electrical circuits. The analysis thus additionally examines how group understanding develops in two different kinds of artifacts and what kinds of roles different artifacts play in mediating the development of group understanding. The current analysis extends Jeong et al. (2011) in two important ways. First, it examines the development of group understanding in two separate dimensions. Second, it extends the previously used analytic framework of artifact analysis to different learning domains and artifact types.

The Five Dimensions Characterizing the Approach

Theoretical and Methodological Assumptions

The analysis reported in this chapter is based on the assumption that knowledge is constructed through social interaction as well as individual construction. Although the analytic approach is strongly rooted in the cognitive framework, the analysis does not privilege individual over group, nor treat social interaction as a mere constraint or stimulant of individual cognition. The individual dimension is considered, but the goal is on assessing and characterizing group understandings in its diversity.

Methodologically, the analysis is based on the assumption that cognitive states of both individuals and groups can be inferred from public manifestations of their cognition, such as verbalizations/dialogues, artifacts, and/or other responses and activities that can be empirically observed in naturalistic as well as in more controlled settings such as a laboratory. The analysis also assumes that individuals' and groups' understandings can be studied and analyzed in more or less objective fashion, that is, an outside account of cognition and learning other than the learners' own account is useful and possible to an extent. This also means that an understanding can be judged to be *incorrect* or *misconceived* against the canonical views of science, even though it may be perfectly coherent and appropriate to the person or the group who constructed it.

Purpose of the Analysis

The goal of the analysis is to examine how group understanding develops using the group artifacts constructed during collaborative learning. The analysis will attempt to capture how group understanding develops from less to more sophisticated ones and how intersubjective it is. Pivotal moments are defined as moments where a change in group understanding occurs in terms of the development of both domain understanding and intersubjectivity.

Unit of Interaction

Group understanding refers to ideas, conceptions, or models that the group constructs about a topic or a problem (e.g., electrical circuits in this study). It is assessed using group artifacts in the current analysis, much in the same way that various analytic concepts are operationalized and assessed using interaction dialogues. Group's domain understanding was assessed based on the display of understanding in artifacts group constructed, initially using separate units for GS-based artifacts and physical artifacts (e.g., postings vs. segments) but later combined.

The analysis then examines the interaction that occurred around each domain understanding displayed in the group artifacts. Acts or activities (e.g., confirmation, building on) related to the display or the transformation of the same domain understanding are considered to be the unit of interaction. The unit of agency is by definition the group in this analysis. However, note that according to the definition of group understanding proposed earlier, even when groups act as an agent, they may display a varying degree of intersubjectivity. Group agency is assumed even in cases where the changes in the artifacts involve an individual action (e.g., individual student publishing in the group board). Even though it is created by a single individual, publication in the group space makes it a group-level act, to which other members can respond with a range of actions from active build-on, explicit confirmation, and acknowledgement to ignoring or rebuking. When all members of the group participate in the construction of the group artifacts, the understanding is assumed to be mutual unless there is a disagreement or a discontinuity in their activity coordination.

Representations

Analysis is predominantly based on the group and individual videos. Verbal transcripts are used as a context to interpret and verify the activities captured in the videos. For representing different understandings that groups constructed, I mainly use tables with pictures (e.g., GS drawings and video screen captures) along with verbal descriptions. Additionally, I use a CORDTRA graph (Hmelo-Silver & Chernobilsky, 2007) to represent the time sequence of the group understanding development.

Manipulations

A number of manipulations are carried out: (1) video segmentation and coding, (2) coding of GS artifact construction in terms of postings and state changes in the group board, (3) analysis of physical artifact construction in terms of episodes, and

(4) characterization of different domain understandings and their emergence during group work based on (a) conceptual analysis of the domain along the comparative analysis of the domain understandings manifested in GS and physical circuits and (b) descriptive accounts of their emergence during interaction. Each manipulation is described in more detail below. Note that although tables and figures corresponding to each manipulation are mentioned in this section, the actual tables and figures are presented in the result section.

Video Segmentation and Coding

In order to get a sense of how students distribute their activities across the two types of artifacts, the group video is segmented at 30-s intervals. Segmentation is a way to get a handle on the stream of video data as it is not easy to tell when an activity begins and ends in the video. I first experimented with 10-s intervals but settled with 30-s intervals, because this granularity produced a more manageable number of total segments with a reasonable length to make sense of the activities within the segments. The 29-min-long group video produces 58 segments. Each segment is coded in terms of whether students engaged in *GS-related*, *circuit-related*, or *other* activities. A given segment often contains both GS- and physical artifact-related activities, as students could engage in multiple acts in one segment. Thus, *GS-related* segments mean segments where students' activities are *predominantly* GS related such as login, drawing, publishing, accessing, and examining GS diagrams. *Circuit-related* segments refer to segments in which student activities are predominantly about the construction of physical artifacts such as arranging and holding bulbs, wires, and/or batteries as well as discussing ideas of arrangements. *Other* segments refer to segments that could not be clearly coded, either GS or physical, or segments where students engage in miscellaneous activities such as talking to students in another group, or listening to the teacher instructions, and answering the teacher's questions. Descriptive statistics will be presented as an outcome.

Analysis of GS Artifacts: Coding of Postings and State Changes

Analyses of GS artifacts are mostly based on the individual videos that recorded the contents of the GS board and examine (a) individual postings in the private GS boards and (b) state changes of the GS group board. All postings created in the GS environment are identified and then examined in terms of who created it, what it was about, and whether it was shared in the group board. When students published something in the group board, the group board changed its content. Whenever there was a change in the group board, it is noted as a unique state. A state change can involve an addition or a removal of a single or multiple postings if more than one student published at the same time (Table 18.1).

Analysis of Physical Artifacts: Identifying and Coding Episodes

Video segments, a purely time-based unit, are further grouped into meaning-based units called episodes where students worked on a distinct physical circuit (Table 18.2). Reconstruction of the same circuit is counted as a separate episode. Incomplete constructions are noted. Episodes consist of circuit-related segments where students predominantly worked on physical circuits but may include GS-related segments if the circuit building activity began in the segment and/or if relevant GS activity was carried out (e.g., the division of labor among students). Each episode, once identified, is further examined in terms of what kinds of circuits were constructed and who was involved in circuit building. Unfortunately, students' bodies sometimes blocked the camera view, and it is not always possible to identify the circuits being built. In such situations, transcripts and GS artifacts are used to infer what students were doing.

Documentation of Domain Understandings and Their Emergence in Group Interaction

What kinds of domain understandings did the group construct as students engaged in the construction of GS and physical circuits? Based on the circuit understandings manifested in GS and physical artifacts, four circuit understandings are identified: Circuit A, B1 and B2, and C understandings (Table 18.3). Circuit understandings are identified only when they were implemented in either GS or physical artifacts. Unimplemented circuit understandings are not coded as a separate domain understanding.

Lastly, descriptive accounts of how the four circuit understandings emerged in the process of group interaction are provided. The account examines how the idea for the new domain understanding came about and how it was co-constructed and mutually shared by its members along with the role/relationships of different artifacts (Fig. 18.4).

Results

Artifact Construction

Throughout the class, students actively constructed GS and physical artifacts. Students spent about an equal amount of time with each artifact: 42 % of the segments were coded as GS related, and 41 % were coded as physical circuit related, with 17 % being coded as other activities. Due to the teacher's instruction to draw GS sketches at the outset of the class, GS artifact construction was more dominant at the beginning, while physical artifact construction became more predominant later in the class.

Table 18.1 State changes in Group 11's group board

State	Contents of the group board	Description of change
1	[blank state]	
2		Bruno publishes two drawings
3		Bruno deleted one of the drawings
4	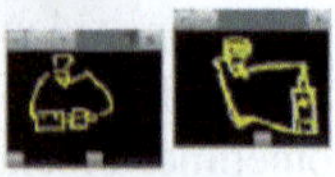	Bruno adds a drawing that Agnes started and he later finished
5		Agnes publishes her drawing she made a while ago
6		Someone put "nice" on top of the two drawings
7		Agnes publishes another drawing
8	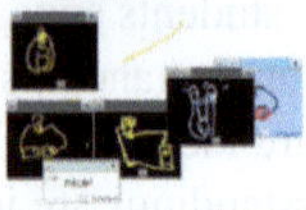	Agnes publishes yet another drawing

GS Artifacts

Students created nine GS postings. Six of them were shared in the group board, and three remained in the private board. The group board also contained a text note saying "nice" by an unknown student from another group. In terms of the changes in the group space, the group board experienced eight different states (see Table 18.1). It evolved from a blank space to a space with five circuit sketches and one text note at the end. Reflecting the group's growing understandings of the electrical circuits, the contents of the group board became richer and more accurate as students shared their initial sketch of the circuits, removed incorrect ones, and/or drew sketches of the circuits that the group built together.

Physical Artifacts

Students constructed a number of physical circuits. Fourteen different episodes were identified in which students worked on distinct circuits. Each episode lasted from 1 to 9 segments (i.e., 30 s to 4.5 min). Unlike GS drawings that were mostly created by individual students with an exception of one drawing by Agnes and Bruno, circuit construction was mostly collaborative because it often requires more than one pair of hands to hold the wires at two different points of the bulb and the battery. Students initially constructed circuits in pairs, with Agnes and Serena

Table 18.2 Episodes (E) of circuit construction

E	Segments	Who	Activity description	Circuit
1	4–6 (1:30–3:00)[a]	Bruno, Joel	Bruno and Joel worked on a circuit. Joel appeared to say something to Bruno. When Bruno was not responsive, Joel went back to work on the GS board. Bruno continued alone for a while but was not successful in lighting the bulb 	A
2	14–22 (6:30–11:00)	Agnes, Serena	Agnes and Serena constructed a circuit. They got light 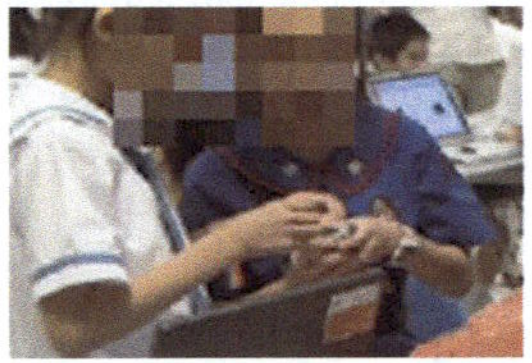	A?
3	17–20 (8:00–10:00)	Bruno, Joel	Bruno and Joel constructed a circuit. The camera view was blocked, but the transcripts indicated that Bruno was holding the wires for the battery while Joel connects the bulb to the wires	A?
4	23–24 (11:00–12:00)	Bruno	Bruno lit a bulb again this time by himself	A
5	25 (12:00–12:30)	Bruno, Joel	Bruno and Joel began a circuit. The camera view was blocked, but their body language and transcripts suggest that they successfully lit a bulb	A?
6	25–26 (12:00–13:00)	All	Upon Serena's suggestion, all four students started working together on a circuit with two batteries. Their bodies blocked the camera view, but their transcript indicated that they lit the bulb. They noted that it is very bright	B1

(continued)

Table 18.2 (continued)

E	Segments	Who	Activity description	Circuit
7	27–30 (13:00–15:00)	All[b]	Agnes, Serena, and Joel worked on a circuit and lit the bulb. Initially, Serena held the wires to the batteries, while Joel worked on connecting the wires to the bulb. He was unsuccessful. Agnes took over and lit the bulb Bruno drew a diagram in GS while the others were working on the circuit. This was in response to the teacher's request that "one member draws." 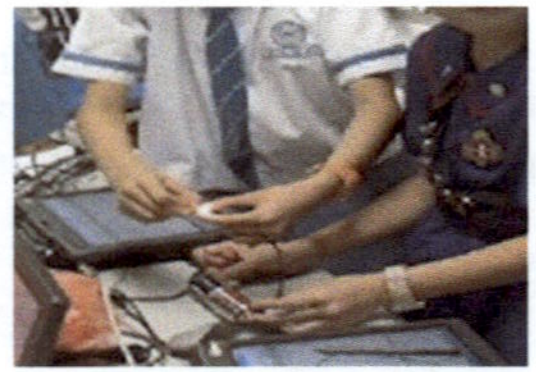	B1
8	31–32 (15:00–16:00)	All[b] → Bruno, Joel	Students attempted another circuit, while Agnes started another GS drawing of the circuit that they constructed in Episode 7. Serena went back to her seat at the request of the teacher, but Bruno and Joel continued to work on the circuit for a while. Based on the arrangement of the batteries on the table it appears that they were testing Bruno's incorrect drawing that he just published in Episode 7	B1?
9	33–36 (16:00–18:00)	All[b]	At the request from the teacher to visit and test neighbor's idea, Serena showed Group 2's drawing from her screen. Bruno and Joel first built a circuit. It did not quite match Group 2's diagram. Upon the feedback from Agnes and Serena, they redid the circuit. The group noted that the circuit was "very bright" 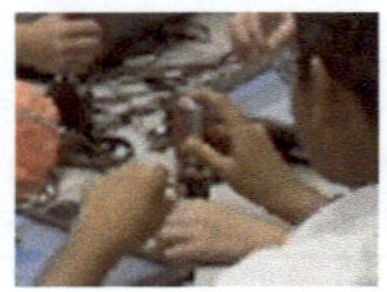	B2

10	37–38 (18:00–19:00)	All[b]	At Joel's initiation, they constructed another circuit, this time one with two bulbs. Initially all four were engaged, but Agnes broke away and started a drawing in Bruno's GS board 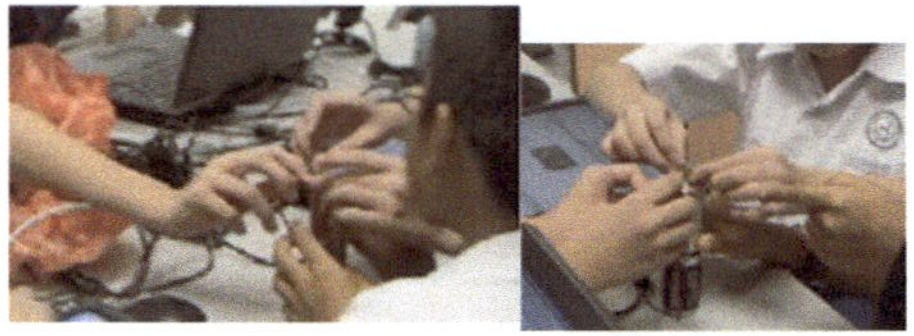	C
11	39–42 (19:00–21:00)	Serena, Bruno, Joel	Bruno, Joel, and Serena began another circuit idea with two batteries apart but did not finish. Agnes kept working on the GS board, initially on Bruno's computer and later on her computer	?
12	46–48 (22:30–24:00)	Bruno, Joel	Bruno and Joel tried another circuit. They successfully lit two bulbs	C
13	53–55 (26:00–27:30)	All	At the end of their session in front of the class, the teacher suggested that group try whether the two-bulb circuit was indeed equally bright as the one-bulb circuit. They came back to their seat and compared the two circuits	B2, C
14	55–58 (27:00–29:00)	Bruno, Joel → All	Joel said he got a new method and was trying something, initially with Bruno and later briefly with Agnes and Serena, although it was not clear what they constructed or whether they even completed it	?

[a]Group video time stamp
[b]Role division between GS drawing and circuit construction

teaming as a pair and Bruno and Joel forming another pair. The four students began to work as a team in Episode 6. A role division emerged later in response to the teacher's request that one person should draw while the rest build the circuits. Even with the role division, the group was still considered to work as a group since they worked on the same circuit. Table 18.2 describes each episode in detail.

Group's Domain Understandings

Based on the circuit understandings manifested in GS and physical artifacts, four circuit understandings are identified: Circuit A, B1 and B2, and C understandings (Table 18.3). They differ in terms of the number of components in the circuits and also whether they are series or parallel circuits (Hallyday, Resnick, & Walker, 2007). Each circuit understanding was a close extension of the preceding understanding. Circuit A is the simplest circuit with one battery and one bulb. Both Circuit B1 and B2 consist of two batteries and a bulb, but they were arranged differently. With an additional battery in series, the bulb in Circuit B1 and B2 is brighter than that in Circuit A. Circuit C is the most complex one with two batteries and two bulbs. It is also a parallel circuit where the two bulbs are connected in parallel and the batteries in series. With an additional bulb, the lights are dimmer in Circuit C compared to Circuit B1 or B2. Note that circuit understandings are coded only when the idea was implemented into artifacts. There are a few incidences where students only talked about circuit ideas. For example, Bruno suggested a circuit with three batteries right after they built Circuit B1, saying things like "Let's try three batteries." His idea, however, was neither materialized into physical nor GS drawings. Table 18.3 presents how the four circuit understandings were displayed in both of the artifacts. Circuit understandings displayed in the GS and physical artifacts were closely aligned with each other.

Emergence of the Group Understandings

How did each circuit understanding of the group emerge? As for the Circuit A understanding, all Circuit A sketches were first drawn individually in the GS private board. Students might have seen each other's drawing while talking to each other, but the drawing was done individually in their private board. Although all four students drew at least one sketch of Circuit A, only Bruno and later Agnes contributed their individual drawings to the group board. Why did the other two students, Serena and Joel, not contribute their sketches to the group board even though they already have drawn their own sketches (see Fig. 18.1)? We do not know for certain, but once Bruno first published Circuit A drawing, other students, especially Joel and Serena, might have felt that it would be redundant to publish their drawings to the group

Table 18.3 Artifact construction associated with each circuit understanding

Circuit	GS artifacts	Physical artifacts
A	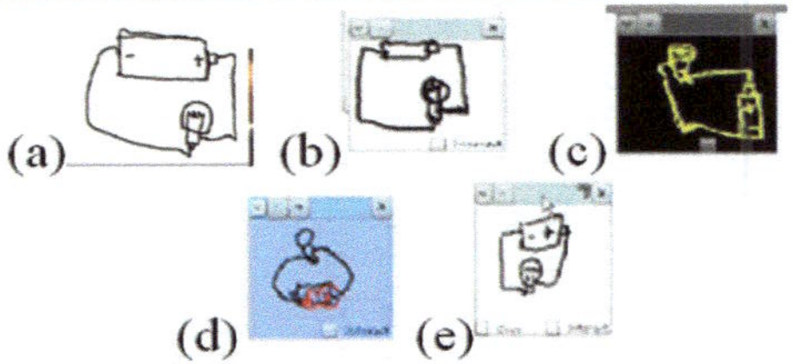 (a) (b) (c) (d) (e) The students separately drew five sketches of Circuit A. Two of them (c and d) were shared in the group space. The initial version of (c) was incorrect but later revised.	Before GS drawing, Bruno and Joel attempted to construct a circuit but were unsuccessful (Episode 1). After finishing GS drawings, students constructed circuits in pairs, with Agnes and Serena in one pair (Episode 2) and Bruno and Joel in another pair (Episode 3). Bruno constructed the circuit alone (Episode 4) after he revised his GS drawing.
B1	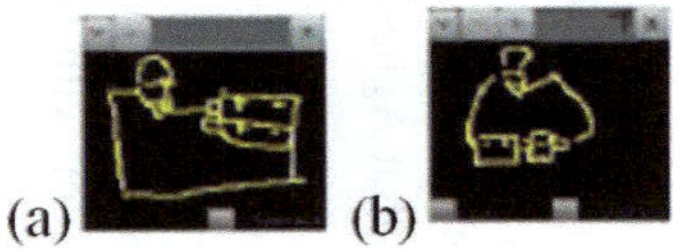 (a) (b) Bruno drew and published a sketch (a) after constructing the circuit but later removed from the group space. Agnes initiated another sketch of the circuit (b) that was later finished by Bruno.	The group built the circuit after Serena's interaction with another group (Episode 6). The group constructed it one more time (Episode 7), while Bruno is drawing on the GS board (a). Bruno initiated a test of his initial drawing (a) but did not finish it (Episode 8).
B2	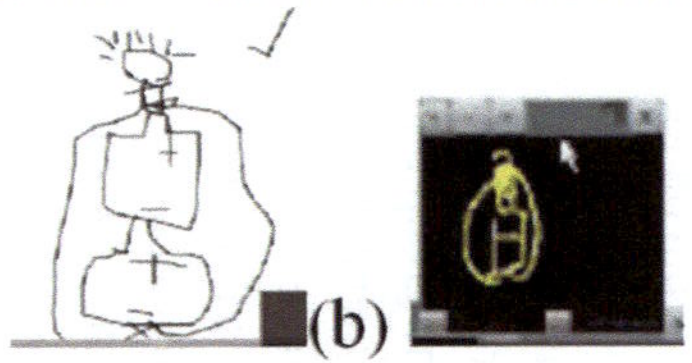 (a) (b) Serena found (a) in Group 2's group board and showed it to the rest of the group. Agnes drew a sketch of the circuit (b).	After seeing Group 2's drawing, Joel and Bruno constructed it. Upon the feedback from Agnes and Serena, they rebuilt the circuit this time correctly (Episode 9). The group constructed the circuit one more time at the teacher's request after the question answering session with the teacher (Episode 13).
C	 Agnes published a sketch of Circuit C right after the group constructed it.	The group constructed the circuit at Joel's initiation (Episode 10). Bruno and Joel constructed the circuit one more time (Episode 12). The group constructed the circuit one last time after the question answering session with the teacher (Episode 13).

Fig. 18.1 Unshared GS drawings by Serena (**a**) and Joel (**b**)

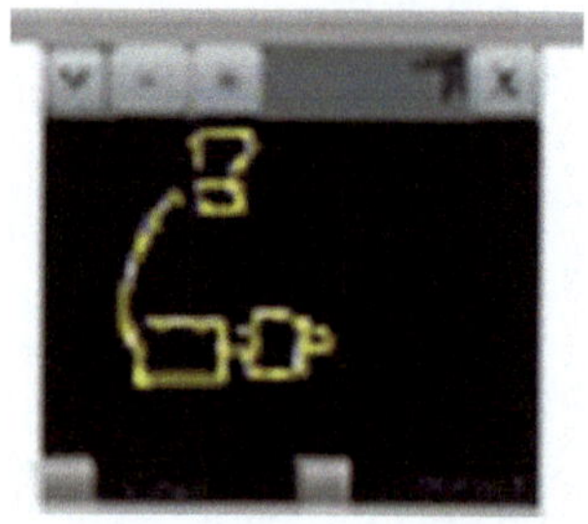

Fig. 18.2 Unshared GS drawing by Agnes

board since they were all about the same circuit. This means that they considered what the group already has in the group board in deciding whether to share theirs.

Note, however, that unlike Joel and Serena, Agnes published her drawing of Circuit A after Bruno (State change #5 in Table 18.1). It might be the case that Agnes, unlike Serena and Joel, did not care about making a redundant contribution, but it might also be the case that she saw a difference in her drawing. Her drawing also consisted of one battery and one bulb like Bruno's but had a different layout with the battery being on the bottom. The difference does not have any significance from the physics standpoint, but she might still have felt that it was different enough to warrant publishing in the group board. The latter possibility is more likely, since she did not share her drawings indiscriminately. Although she was the most active contributor to the group board, there was one drawing she did not share with other students. It was a drawing of an incomplete circuit (Fig. 18.2). By avoiding publication of an incomplete drawing, it appears that Agnes was also trying to be informative with her contributions.

All the initial sketches of Circuit A were correct except for Bruno's. In his first drawing, he connected both wires to the bottom tip of the bulb, although he later corrected it. How did this revision come about? After Bruno finished his initial drawing, he went on to build a physical circuit with Joel (Episode 3). The transcripts indicated that Bruno disagreed with Joel on where to connect the wire, indicating that Bruno still held the incorrect view about how the wires should be connected to the bulb. However, Bruno was just holding the wires to the battery, and it was Joel

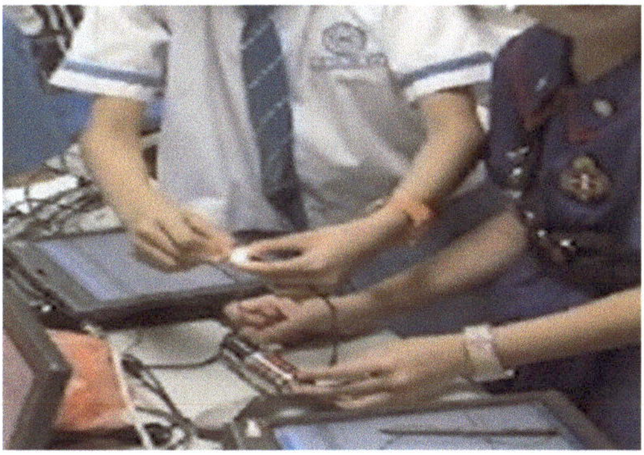

Fig. 18.3 Construction of circuit B1

who connected the wires to the bulb. The camera view was blocked, but their body language suggested that they eventually succeeded in lighting the bulb. This experience must have made Bruno realize his faulty understanding. Immediately after Episode 3, he came back to his computer and started revising his earlier drawing. Upon finishing the revision, he then picked up the wires again, constructed another Circuit A, and lit the bulb this time all by himself (Episode 4). His revised understanding was evident in his re-drawing and also in his subsequent circuit construction. Constructing a physical circuit with a partner provided Bruno an opportunity to realize his misconception and prompted a change in his individual understanding. The interaction also prompted a change in the intersubjectivity of the group understanding as Bruno's individual understanding becomes aligned with Joel's.

The group's understanding of Circuit A did not result from a joint construction. It emerged largely as a result of pooling individual member's prior knowledge. One might say that it is just a collection of individual understandings. And yet, it is more than a collection of disjointed individual understandings. As described earlier, implicit regulation seems to be going on with respect to the pooling of individual understandings. Students were trying to be informative with their individual contributions by avoiding redundant or incomplete contributions. It appears that students were following something akin to conversational maxim or cooperative principles (Grice, 1975) in publishing their private drawings in the public space. In addition, although the group's Circuit A understanding was not a mutually constructed one, by sharing individual ideas and engaging in activities that translate GS circuits into physical circuits, it became mutually shared by the group, from which subsequent understandings developed.

The idea for Circuit B1 came about when Serena interacted with a student from another group (Episode 6). The nature of their interaction was not clear, but it appeared to give an idea to Serena. After she finished talking to this student, she suggested to her group "How about we do two batteries? We all connect ours." The group got all excited, stood up, and gathered on one side of the table. As pointed out earlier, up until this point, their circuit construction activities were mostly pair-wise, but all four students more or less worked as a team from this episode on. At Serena's suggestion, they first assembled and pooled together the materials that they had in two separate bags. The student bodies blocked the camera view, but the transcript showed that they were asking each other to hold wires and eventually were able to light the bulb. They noted that it is "very bright" compared to Circuit A that they just constructed moments ago. Thus, although the initial idea of Circuit B1 appears to have been seeded by an interaction with another group, it was elaborated and implemented collaboratively by the group (Fig. 18.3).

After they successfully constructed Circuit B1 for the first time, students started discussing various ideas for a new circuit (e.g., Serena suggested adding another bulb, while Bruno suggested adding one more battery). At this point, the teacher came over and asked the group whether they had a draft of the two-battery circuit that they just built. When it turned out that they did not have a draft, she told the group that one member should draw while the others fix the circuit. In response to this teacher prompt, Bruno started drawing, and the rest of the students successfully constructed Circuit B1 the second time. This is the first time that a role division between the drawer and builder appeared in the group, which was repeated in later episodes. The drawing that Bruno did in response to the teacher's prompt in Episode 7, however, turned out to be an incorrect one, and he later removed it in Episode 8. Agnes, who assumed the role of a scribe in Episode 9 and 10, began drawing another sketch of Circuit B1. She could not finish it because the teacher called students to their seat. Bruno finished the drawing later. Thus, unlike Circuit A understanding that was first constructed individually and later shared, the Circuit B1 understanding was mutually constructed by the group. In addition, unlike Circuit A understanding that appeared in GS artifact first, Circuit B understanding appeared in physical circuit construction first and then later transferred into GS artifacts. The process of converting an understanding from one artifact format to another was not a straightforward matter. Even though they just constructed the circuit, the group was able to succeed in it only after repeated attempts by two of its members.

The Circuit B2 understanding first appeared as a result of an *indirect* interaction with another group (Episode 9). The teacher asked students to visit other people's board and test their neighbor's ideas. Serena visited Group 2's board and found a new circuit diagram. She showed it to the rest of the group by turning her screen around. Bruno and Joel built it, but Agnes and Serena pointed out that the circuit did not match Group 2's diagram ("It's not like that, not like that"). With their feedback, Bruno and Joel realized that the bulb was touching the metal casing of the battery in Group 2's diagram and rebuilt the circuit. Again this episode shows that transforming an understanding displayed in one artifact to another is not a straightforward matter. In the process of figuring out the mapping from the drawing to the physical circuits, Bruno's and Joel's incomplete understanding of Circuit B2 was revealed. Aided by the feedback from Serena and Agnes, they were able to match their construction more closely to the drawing in the Group 2's board and also to Serena's and Agnes' conceptions. Although the initial ideas for Circuit B2 came from another group, the group was able to construct and share their own Circuit B2 understanding through interaction around the artifacts. Agnes later codified it into their own Circuit B2 drawing.

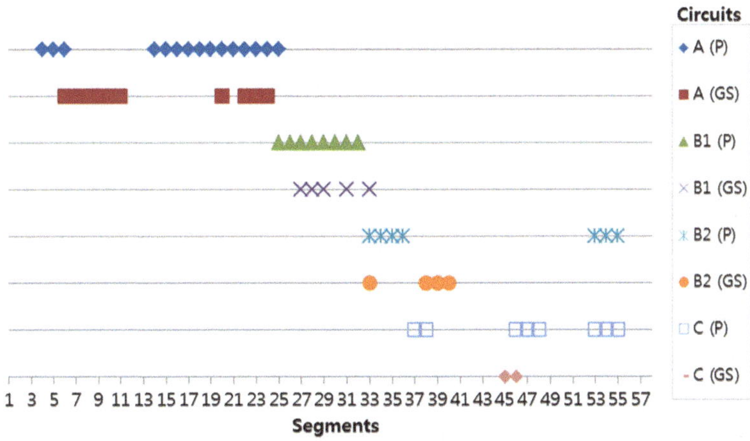

Fig. 18.4 Emergence of the group circuit understandings in physical (P) and GS artifacts (GS)

Serena first suggested the idea for Circuit C right after the group first constructed Circuit B1. She said, "How about we all try two batteries, two bulbs?" Joel initially rejected her idea by saying, "Don't want. Two bulbs cannot. There are only two wires." However, it was Joel who later reintroduced the idea after they finished Circuit B2 (Episode 10). With two bulbs in each of his hands, it was not much of a leap from a one-bulb circuit to a two-bulb circuit. He put the two bulbs on top of the stacked batteries and said "How about this?" At seeing this, the rest of the group got quite excited again and went on constructing Circuit C. After they were done, Agnes drew a diagram of Circuit C and posted it in the group board. Bruno and Joel later constructed the circuit one more time (Episode 12), consolidating their understanding.

Later during the question answering session in front of the class, the teacher asked the group whether the bulbs were brighter or equally bright in this circuit compared to the circuit with one bulb. After hearing students' answers that Circuit C seems to be brighter than the circuit with one bulb, the teacher told them to go and try. Students came back to their seats and compared the circuits (Episode 13), but they still could not draw a correct conclusion about the bulb intensity, with some saying "weaker" while others saying "brighter." It was unclear why they were unable to make a correct and stable observation (it might have been the case that the change in light intensity is not clear enough without the use of measurement devices), but such a failure suggests that successful construction of working circuits and direct observation were not enough to deduce or understand underlying mechanisms. It also appears that students did not yet understand concepts such as current, voltage, and resistance. The group may have mutually constructed a working circuit, but their domain understanding was not fully developed yet, suggesting that the two dimensions of group understanding develop more or less independently.

As can be seen from Fig. 18.4, the development of group understanding was an incremental process as the group developed new ideas by extending what they did before, that is, by incorporating a new component in a piecemeal fashion. At least in the current data set, there was no noticeable conceptual leap where students suddenly came up with an understanding that involves changes in more than one aspect of the circuits. Rather, the understandings progressed in a step-by-step fashion with occasional errors and subsequent revision along the way. In addition, the development of the four circuit understandings was closely paralleled in the digital and physical artifacts. A new understanding typically appeared first in the form of physical circuits, followed by the construction of GS drawing.

Discussion

How can we capture group understanding, the various cognitive states that are behind groups' activities and decisions? How does group understanding develop during small-group interaction? The analyses reported in this chapter used group artifacts as a way to capture group understanding and analyzed the development of group understanding along two dimensions. In terms of the domain knowledge development, the analyses identified four circuit understandings. The group's understanding of electrical circuits became more sophisticated over time as it expanded to include additional ways to light a bulb. The development process was piecemeal but not always that of a linear improvement. A number of incorrect ideas were developed due to initial misconceptions and/or incorrect interpretation or rejection of others' contributions. They were resolved as the group shared and co-constructed artifacts, challenged, and built on each other's ideas and construction of artifacts.

Another dimension of group understanding development is its intersubjectivity, that is, the extent to which it is co-constructed and shared by the group. The four group understandings varied in terms of their intersubjectivity. The first group understanding was more or less the results of pooling individual ideas, although it appears that an implicit group process regulated the pooling process. The rest of the circuit understandings were more or less co-constructed by the group but to a different degree. The Circuit B1 and B2 understandings were influenced by interaction with another group, while the Circuit C understanding was generated from the internal activities of the group. The ideas were not adopted or processed accurately by all its members initially but eventually implemented and shared by the group. Artifacts played an important role helping the group to achieve intersubjective understanding, as it was in the process of the building of physical circuits or drawing GS sketches that the misalignment in their understanding was detected and revised.

The pivotal moments were defined in this analysis as moments when changes in group understandings occurred both in terms of the development of the domain understanding and intersubjectivity. As described in the result section, these processes were incremental. A number of different events or moments contributed to the development of group understanding such as when new ideas are first suggested and accepted, when an incorrect idea is challenged and corrected, or when an idea is made public, shared, and/or acted upon. There were thus many different pivotal moments rather than *the* pivotal moment(s). Instead of listing them all, I propose three underlying mechanisms of group understanding development. *First*, group understanding developed as a result of pooling and sharing individual resources. This was especially the case for the group's understandings of Circuit A. Circuit A understanding emerged and shared as students pooled their individual ideas. It does not involve co-construction, but it is nonetheless a mechanism that group develops its understanding. Sharing allows groups to learn from what others have done. When individuals build upon it and share it back with the rest of the group, the group advances its understanding. Note that pooling and sharing does not always contribute to the development of group understanding. If individual members do not have anything to contribute or do not contribute what they know, it would not be possible for the group to advance its understanding this way. *Second*, group understanding developed as a result of interacting with another group. As we have seen in Circuit B2 and to a lesser extent in Circuit B1, the group understanding was prompted by interaction with other groups. The Circuit B2 understanding initially came about as the group interacted indirectly with another group through the GS board, which was later elaborated and implemented in subsequent interaction. Circuit B1 was also prompted by a face-to-face interaction with another group. The interaction gave the group an idea to use two batteries in their circuit. *Lastly*, group understanding developed through group interaction around the artifacts. The idea for Circuit C came about as the group interacted around the physical circuits. The idea to connect two bulbs was suggested twice, as an obvious extension of the circuits that they have just completed. This idea was not accepted at first. It was accepted the second time and implemented by none other than the student who rejected the idea in the first place. It appears that although the idea seemed untenable at first, playing and experimenting with physical artifacts made this student see the feasibility of the idea. Interaction around the artifacts also gave a chance for the group to find points where they lacked mutuality. Bruno's initial faulty understanding of Circuit A was revised after he constructed physical artifacts with Joel. Likewise, Bruno's and Joel's misinterpretation of Group 2's drawing was spotted and corrected by Agnes and Serena after they built the circuit. Interacting around the artifacts was pivotal both to the development of domain understanding and intersubjectivity.

The two types of artifacts both played important yet different roles in the development of group understandings. The GS artifacts served as a device to pool individual ideas from which subsequent understanding developed. In addition, GS artifacts helped the group elaborate and codify what they co-constructed in physical circuit construction. Drawing a GS sketch of the circuits that they constructed requires them to translate from one form of understanding to another, a kind of representational competence (Kozma & Russell, 1997), and helped the group to elaborate and codify what it learned from building the physical artifacts. Also, after students put their understanding into more permanent form and shared it with others, it served as a source of new ideas for other students and groups. As for the physical circuits, they helped the group to develop ideas by making it easier to take the next step in the progression of ideas. Although students spent about equal amount of time working on GS and physical circuits, new ideas often emerged first in the process of constructing and experimenting with physical artifacts. Unlike Circuit A and B2 that most of the students either knew at the beginning or copied from another group, the Circuit C and, to a lesser extent, Circuit B1 understandings emerged first in the process of experimenting with physical objects. It appears that constructing physical artifacts prompts students to think in concrete terms and stimulates them to examine and manipulate ideas more easily (Kafai, 2006). In addition, constructing physical circuits also provided an opportunity for the group to test and detect problems with their ideas. It was by testing their diagrams that they learned whether their ideas worked.

This chapter reported on an analysis of group understandings based on group artifacts. In spite of much interest in and emphasis on group-level cognitive phenomena, we do not understand fully how groups develop cognitively as a unit. This chapter argued for the need to understand the full trajectory of group's cognitive states and proposed the notion of group understanding. Group understanding represents a diverse range of cognition that groups can display and can differ in terms of domain knowledge and intersubjectivity. The analysis reported in this chapter attempted to capture the group understanding(s) based on the artifacts a group constructed. More research is needed with respect to the notion of group understanding and the analysis method, but I hope that the analysis reported in this chapter is a small step toward characterizing cognitive states at the group level. It is also hoped that the current analysis provided a useful technique in carrying out artifact-based assessment, complementing existing discourse-based approaches. The multivocal efforts reported in this book have contributed a great deal to refining the notion of group understanding and analytic approach reported in this chapter. The analyses reported in this chapter addressed different questions and did not give rise to conflicting interpretations with the rest of the analytic approaches in this section. Still, the different analytic approaches helped to broaden the contextual perspective and sharpen the notion of the group understanding. I hope to see such efforts in other areas of research as well.

Acknowledgement This work was presented in a workshop that was partially financed by the European STELLAR Network of Excellence (PF7). The analysis and writing of this chapter were supported by the grant from the National Research Foundation of Korea (NRF) (Grant No. 2009-0068919). Any opinions, findings, and conclusions or recommendations expressed in this chapter are those of the author and do not necessarily reflect the views of the funding agency.

References

Börner, K., Contractor, N., Falk-Krzesinski, H. J., Fiore, S. M., Hall, K. L., Keyton, J., et al. (2010). A multi-level systems perspective for the science of team science. *Commentary, 2*(49), 1–5.

Chinn, C. A., & Brewer, W. F. (2001). Models of data: A theory of how people evaluate data. *Cognition and Instruction, 19*(3), 323–393.

Grice, H. P. (1975). Logic and conversation. In P. Cole & J. L. Morgan (Eds.), *Syntax and semantics* (Vol. 3, pp. 59–83). New York, NY: Academic.

Hallyday, D., Resnick, R., & Walker, J. (2007). *Fundamentals of physics*. New York, NY: Wiley.
Hmelo-Silver, C. E., & Chernobilsky, E. (2007). *CORDTRA: A representational lens for analyzing tool-mediated collaborative learning*. Paper presented at the AERA.
Hollan, J., Hutchins, E., & Kirsh, D. (2000). Distributed cognition: Toward a new foundation for human-computer interaction research. *ACM Transactions on Computer-Human Interaction, 7*(2), 174–196.
Hutchins, E. (1995). *Cognition in the wild*. Cambridge, MA: The MIT Press.
Jeong, H. (2013). Verbal data analysis for understanding interactions. In C. Hmelo-Silver, A. M. O'Donnell, C. Chan, & C. Chinn (Eds.), *The international handbook of collaborative learning*. London: Taylor and Francis.
Jeong, H., Chen, W., & Looi, C. K. (2011). *Analysis of group understanding in artifact-mediated discourse*. In H. Spada, G. Stahl, N. Miyake & N. Law (Eds.), Connecting Computer-Supported Collaborative Learning to Policy and Practice: Proceedings of the 9th International Conference on Computer-Supported Collaborative Learning (CSCL 2011) (Vol. II, pp. 786-790). Hong Kong: International Society of the Learning Sciences.
Kafai, Y. B. (2006). Constructionism. In R. K. Sawyer (Ed.), *The Cambridge handbook of the learning sciences*. Cambridge: Cambridge University Press.
Kozma, R. B., & Russell, J. (1997). Multimedia and understanding: Expert and novice responses to different representations of chemical phenomena. *Journal of Research in Science Teaching, 34*(9), 949–968.
Looi, C. K., & Chen, W. (2010). Community-based individual knowledge construction in the classroom: A process-orientation account. *Journal of Computer Assisted Learning, 26*(3), 202–213.
Matusov, E. (1996). Intersubjectivity without agreement. *Mind, Culture, and Activity, 3*(1), 25–45.
Stahl, G. (2006). *Group cognition: Computer support for building collaborative knowledge*. Cambridge, MA: The MIT Press.
Vosniadou, S., & Brewer, W. F. (1992). Mental models of the earth: A study of conceptual change in childhood. *Cognitive Psychology, 24*, 535–585.

Chapter 19
Issues in Comparing Analyses of Uptake, Agency, and Activity in a Multimodal Setting

Daniel D. Suthers

Introduction

The task of this chapter is to bring forth and summarize the lessons that we learned from analyzing a shared data corpus and what we learned about how to make such research collaborations productive. This task differs from that of a typical edited volume, in which a discussion chapter *for the first time* makes sense of, compares, and finds relationships between independently produced chapters. Here, it was also the analysts' responsibility to engage in such a comparative discussion (with my assistance as coordinator of this case study) while conducting their analyses, and it is my responsibility to report on this process as well as on our conclusions.

The data corpus was provided by Chen and Looi (Chap. 14, this volume) and derives from a primary grade classroom in Singapore using the Group Scribbles collaborative whiteboard (Roschelle et al., 2007) and electrical components to learn about electric circuits. The Group Scribbles dataset is unique among those in this volume in that it mixes face-to-face interaction, collaborative physical manipulation of objects, and computer-mediated interaction (both face-to-face and to a lesser degree asynchronous-distributed). These attributes motivated bringing this dataset into the productive multivocality collaboration. I had previously visited a Group Scribbles implementation in Singapore, examined data from another class using Group Scribbles, and mentored Richard Medina on analyzing an earlier Group Scribbles corpus. The richness of this multimodal setting was obvious from these experiences. Including Group Scribbles in this volume improves the diversity of our case studies, not only by its multimodality but also in being a different science topic and a school setting in a country renowned for its performance in science and mathematics education (Gonzales et al., 2007).

D.D. Suthers (✉)
Department of Information and Computer Sciences, University of Hawai'i at Manoa,
1680 East West Road, POST 309, Honolulu, HI 96822, USA
e-mail: suthers@hawaii.edu

D.D. Suthers et al. (eds.), *Productive Multivocality in the Analysis of Group Interactions*, Computer-Supported Collaborative Learning Series 16,
DOI 10.1007/978-1-4614-8960-3_19,

The corpus was analyzed by Looi, Song, Wen, and Chen (Chap. 15, this volume) using uptake and content analysis guided by a theory of progressive inquiry; Medina (Chap. 16) using uptake analysis with an ethnomethodological orientation towards unpacking group accomplishments; Lund and Bécu-Robinault (Chap. 17) focusing on coherence and conceptual change in translations between media and modes motivated by a theory of semiotic bundles; and Jeong (Chap. 18) using content analysis under her conception of "group understanding." The analyses focused on two major themes: what evidences understanding, and practices of multimodal interaction across various media. Understanding was analyzed via uptake structures, the coordination of multimodal acts in multiple media, and/or the contents of resulting artifacts in relation to canonical physics. Multimodality was understood in three ways: in terms of the unique affordances of each medium, conceptual coherence sustained through translations across media, and group accomplishments through simultaneous coordinated use of media. My discussion of these themes and approaches is organized around three topics: a pragmatic issue concerning sharing transcripts that also has theoretical overtones; comparisons of the results obtained (including but not limited to pivotal moments), with ensuing discussion that worked out some differences; and comparisons of the approaches on their conceptions of individual and group agency and the distribution of activity in multimodal settings. But before discussing the present analyses, I will place this effort in context of a related effort involving many of the same researchers and a similar data corpus.

Experiences with an Earlier Group Scribbles Corpus

Before settling on the present data corpus, the Singapore team offered us a different corpus also involving Group Scribbles in a primary school classroom but on the topic of fractions. This earlier corpus was analyzed in the activity surrounding the fourth workshop in our series, as summarized in Suthers et al. (2011). (Data was shared and analyzed in advance of the ICLS 2010 workshop; analyses were presented discussed in the workshop; and revisions were conducted afterwards in preparing the CSCL 2011 paper.) In this corpus, four students in a class that was studying mathematical fractions interacted verbally and via the Group Scribbles software (Roschelle et al., 2007) to solve the problem of dividing two pizzas equally among three children, comparing their own solutions. They were then required to view and comment on solutions provided by other groups.

Four individuals or teams of researchers analyzed this data: three from the present section (Jeong; Looi et al.; and Medina) and Jan van Aalst, who examined the data from a knowledge building perspective (Scardamalia & Bereiter, 2003). Analyses by Jeong, Medina, and van Aalst were presented at the ICLS 2010 workshop, and analyses were also reported in (Jeong, Chen, & Looi, 2011; Looi & Chen, 2010; Suthers et al., 2011). Initially, the analyses appeared to show wide discrepancies in the accounts of what took place and their interpretations. Closer examination revealed that some of these discrepancies were gratuitous differences resulting from conflicting choices in the time span considered, the granularity of the unit of analysis, whether verbal and/or nonverbal acts were included in the analysis, and whether or not activity in private spaces was considered (in Group Scribbles, one can construct notes in one's private space before copying to the group space).

Once we identified that these were nonessential differences, we agreed to examine both verbal and nonverbal events in the public space, and further analyses were conducted. Gregory Dyke and Kristine Lund provided a meta-analysis by representing and aligning the various analyses in Tatiana (Dyke, Lund, & Girardot, 2009). This meta-analysis, reported in Dyke et al. (2011), supported our further progress. It became apparent that all analysts agreed on their accounts of the empirical facts of the interaction: Two students formulated similar graphical solutions and agreed on their equivalence; a third student had difficulty at first but after having seen the graphical solution wrote an equivalent textual solution; and a fourth student developed a symbolic solution without interacting with others.

Yet, in spite of this agreement, analysts differed in their assessments of whether the interaction was interesting from a learning perspective, with the greatest contrast between van Aalst and Medina. Examining verbal data in both private and public spaces from a knowledge building perspective, van Aalst found only minimal discussion about the graphical solutions, no coordination with the symbolic solution, and in general that the students had not taken control of the learning opportunities offered to them (as in the "intentional learning" paradigm, Scardamalia & Bereiter, 1991). His frank assessment was that the data was uninteresting. On the other hand, Medina examined the sequential structure of uptake (Suthers, Dwyer, Medina, & Vatrapu, 2010) across graphical as well as verbal acts and found a pivotal moment in which a graphically expressed proposal was contingent on its social and material setting in a way that brought two lines of activity together and evidenced the development of a representational practice.

From this experience we noted the importance of eliminating gratuitous differences in the scope of the data to be analyzed and the utility of aligning different analyses against a common transcript (or at least a shared timeline) so that essential differences can be foregrounded. Productive multivocality does not mean coming to agreement on everything, but it does require sufficient congruence to highlight essential differences such as the question of what analysts see as evidence for learning.

We also debated about the role of "unexceptional" data in the literature. Researchers understandably prefer to analyze episodes that show interesting and positively valued forms of interaction, and these are more likely to be published than uneventful episodes, yet the educational reality is that these group accomplishments are rare and teachers need to be informed about how to transform less exceptional but more typical situations into learning opportunities. Perhaps we should deliberately focus on case studies where the interaction fell short of its potential? However, our own reality is that we cannot ask authors to put substantial time into analyzing and writing about situations they do not find to be interesting, nor publishers to publish or readers to read about such cases.

In the end, we requested a different Group Scribbles corpus from the Singapore team, and they graciously provided us with the present data corpus on electricity, which not only evidenced richer dialogue and closer collaboration between participants but also added the dimension of physical manipulatives (batteries, wires, and light bulbs). This corpus was analyzed in activity surrounding the fifth workshop in our series. At this time, van Aalst left the effort, Song and Wen joined the Singapore team, and Lund and Bécu-Robinault took on this corpus due to their interest in multimodal interaction.

Transcripts and Other Analytic Representations as Boundary Objects

Participants in the fourth workshop were by this time well aware of the importance of establishing shared empirical foundations and aligning analytic representations (beginning with segmented transcripts). However, not all analysts of the present corpus were present at that workshop, and other practical realities brought these issues again to the fore when we moved to a new corpus in the fifth workshop. Transcription is a time-consuming process, especially when trying to coordinate multiple recordings from a noisy classroom. Research teams are generally not funded to provide data to other teams, and the present situation was no exception: the Singapore team was only able to provide us with documents they had generated for their own purposes (from which the transcript in Chap. 15 is derived). These included partial transcripts from each of the four students' microphones that did not include everything that was said. Events relevant to the Singapore team's analysis were included but not all events. These documents were useful in that they translated occasional Chinese utterances and made the "Singlish" dialect easier to understand. However, the documents did not include events or data needed by other analysts (e.g., some verbal events, the use of gesture, direction of gaze, and precise timing of manipulations of objects). Therefore the other analytic teams undertook the process of augmenting or completing the transcripts as needed for their own purposes.

Jeong's team attempted to combine the four individual student transcripts that were provided to us. Sometimes utterances differed slightly across the transcripts, and the original audio had to be reviewed to see whether they were two utterances or two different transcriptions of one utterance. In addition to this combined transcript, Gregory Dyke's work importing the videos into Tatiana (Dyke et al., 2009) was a significant aid to the analyses undertaken by Medina and by Lund and Bécu-Robinault (as well as my own review), as analyses could be done by replaying synchronized videos and transcripts. Subsequently, some effort was made to align the segmented and annotated transcripts along with other analytic representations generated or used by each of the four analyst teams so that we could compare analytic results. The representations differed in basic matters such as the reference point for time stamps, the units of analysis (including what events were included), and how events were labeled.

The discrepancies between the analytic teams' requirements for transcripts highlighted for us the purpose-driven and hence theoretical nature of transcripts (Duranti, 2006; Ochs, 1979). What is suitable as a transcript for one analytic purpose is inadequate for another. For example, analyses concerned with the conceptual content of students' utterances have different requirements than those concerned with the coordination of gestures or of cross-model translations. For some analyses the interleaved timing of utterances is crucial and for others less so. It is also difficult to align transcript representations designed for different units of analysis: Jeong used 30-s intervals to enable her to systematically code the progression of artifact configurations over the session; Looi et al. worked at the granularity of acts (utterances, drawings, or experiments) between which uptake relations were asserted; Medina worked with uptake between acts at a similar granularity but also divided the activity into artifact-construction and interactional episodes; and Lund and Bécu-Robinault organized acts into episodes of multimedia/multimodal reformulations (see Table 19.1 for a partial alignment).

These different conceptions raise fundamental questions concerning the role shared transcripts play in a multivocal analytic collaboration, beginning with whether shared transcripts are even possible. With adequate resources, it would be possible to construct a transcript that contains the union of all analysts' requirements (of all features required by each analytic approach). One might object that a "super-transcript" might not make salient the more specific features a given team wishes to attend to—the desired features could be obscured in all the detail required by others. However, in a computational age, we no longer need work with static (e.g., paper) representations: this objection could be overcome with adequate interface tools for selecting appropriate views of a multidimensional digital representation. Then each team could work with their preferred transcript representations upon which they layer further analytic representations, but all representations would be aligned in the super-transcript and hence could be juxtaposed across analysts as needed. This requires the development of more powerful analytic tools, and indeed such a vision motivated the first workshop in our series and the development of Tatiana.

However, a shared transcript is a means to an end, not the end in itself. We should consider the process that would support productive multivocality. It may not be wise to contract out the task of producing a super-transcript to one team. Productive dialogue may arise when analytic teams collaborate on producing the transcript, each providing what they see as essential and negotiating their merger. This process would help make other analysts aware of different dimensions of the data they might not otherwise have considered and form a common boundary object for discussions concerning the constitution of the phenomenon of interest as an object of study. Some of the same benefits might arise even from trying to align analysis-specific transcripts without producing a super-transcript, and even when the attempt is unsuccessful. To the extent that differences between transcript representations are motivated by different theoretical and analytic points of view (rather than being

Table 19.1 Partial alignment of units of analysis by group video time

Group video time	Looi label (event)	Medina ideational seg.	Medina inscript. seg.	Lund label[a]	Lund MMR	Lund semiotic bundle	Jeong 30-s seg.	Jeong episode	Jeong circuit
12:15	O1						25	5 & 6	A/B1
12:16		E2							
12:18	J2								
12:19	B3								
12:21	S4								
12:38	Je5					SB4, SB5	26	6	B1
12:47	Ge6								
12:49	S7								
12:52	Ge8				MMR4				
12:53	J9								
12:54	S10								
13:05	Gg11						27	7	
13:06	T12								
13:07	B13								
13:10	T14								
13:13	J15		B4						
13:15	T16								
13:25	Bg17								
13:27	Je18								
13:38	Bg19						28		
13:47	T20								
14:16	Bg21			B3			29		
14:22					MMR5, MMR6				
14:28	Ge22								
14:29	Bg23								
14:30	B24						30		
15:02			A5						
15:03	Ag25						31	8	
15:10	Ag+Bg26								
15:11	T27								
15:12		E3							
15:13	J28								
15:19	T29								
15:27				A2					
15:50	S30				MMR6 Cont.	SB6	32		
15:52	Be31								
15:53	Ag32								
15:56			B5						
16:03	A33						33	9	B2
16:10	Bg34								
16:15									
16:16	Gg35	E4		BA2					
16:18									

[a]The analyst did not provide a mapping of line numbers to group video time

gratuitous), productive multivocality is advanced if these essential differences are exposed by representational mismatches.

But these thoughts are about the possible; we return now to what actually happened. There was one positive effect of the unavailability of a transcript produced by data providers. This situation led most analysts to examine the six video records directly (synchronous viewing was made possible by Tatiana) and arguably gave the analysts a closer relationship to the primary data than would have been the case if working from a textual transcript, no matter how detailed. Yet this is a benefit for individuals. We only began to realize the benefits expected from having analysts try to align their transcripts and derived analytic representations late in our process. Once each analyst had done their work in producing their chapter, it was left to the present writer to attempt to bring analyses into alignment for comparison. Individual analysts each added their analytic summaries to tables intended to align the analyses, but this was done primarily in responses to my requests rather than appropriated by the group as a means towards a shared objective. Therefore we missed earlier opportunities for valuable discussions afforded by the problem of comparing each other's data and analytic representations and understanding what the differences mean conceptually.

However, some progress was made late in our collaboration in my conversations with analysts, reported later in this chapter. The lesson for future multivocal analysis efforts includes our prior advice: try to align analytic representations, eliminate gratuitous differences, discuss essential differences, and repeat. The additional lessons are that this process must be owned by analysts to enable them to confront their conceptual differences and be done early to allow for iteration. In the following sections I discuss what surfaced in my own comparison of analyses and ensuing discussions with analysts. I begin with a detailed comparison of analyst's identification of pivotal moments and then raise the level of discussion to underlying conceptions of agency and distributed activity implied by the approaches taken.

Comparing the Analyses

We examine how the analyses compare in the portion that is analyzed by all four analyst teams. Looi et al.'s analysis covers the shortest time span, approximately 12:00–16:10 in the timeline defined by the video of the whole group (Table 19.1). In this time period, Looi et al. identify uptake relations between individual acts (contributions) labeled O1 to A33: the letter indicates the actor and the number the sequential position of the act. Medina discusses two episodes (E2 and E3); Jeong sees a transition from circuit A to B1 across four episodes (5–8); and Lund and Bécu-Robinault identify three multimodal and multimedial reformulations (MMRs) and three semiotic bundles, all within their episode 2. MMR #4 leads to semiotic bundle 5 (SB5), and both MMR #5 and #6 lead to semiotic bundle 6 (SB6). Table 19.1 summarizes the mapping. The discussion below will use both Looi et al.'s labeling and group video times for reference (e.g., O1, 12:15).

Comparing Pivotal Moments

As discussed in Chap. 1, we learned in our earlier workshops that analyzing the same data is not enough to generate productive dialogue between multiple analysts, as they could "talk past" each other. Our solution was to ask analysts to identify the "pivotal moments" of the interaction. The definition was left open so that analysts could choose definitions that match their traditions and interests, and a comparison of the definitions chosen is as much a part of the desired dialogue as a comparison of the actual "moments" chosen. As might be expected given that analysts relied on different theories and segmented the data differently, the definitions of "pivotal" differed across teams.

For Looi, Song, Wen, and Chen (Chap. 15), a contribution is pivotal if it shifts the direction of subsequent events, whether seamlessly or abruptly. Looi et al. distinguished *manifest* pivotal moments: those that were actually taken up; and *latent* pivotal moments: those that had the potential to shift the direction but were not taken up (e.g., S10 and B24; compare to the "missed opportunities" of Baker & Bielaczyc, 1995). Their manifest pivotal moments included the following:

S4 (12:21), where Serena suggests, "Let's try two batteries. We connect all of them?"

T12 (13:06), where the Teacher reminds students that they should diagram their circuit ("No draft, no draft. Then people will look at a blank board.")

Ge22 (14:28), the conclusion of the group experiment in which Serena and Agnes light a bulb with two batteries for the second time for the benefit of Bruno, who is drawing the circuit.

Be31 (15:52), where Bruno repeats his attempt to light the bulb, trying to find out why his previous drawing of the circuit graph was not correct.

Looi et al. (personal communication) identified pivotal moments by looking at changes in the content of the acts (compare to Chiu's breakpoints in Part II, the case study of a Japanese mathematics class), rather than in terms of uptake structure. This approach may be biased towards attributing change to immediately prior acts, but the act just before a change in focus of activity is not necessarily the act that initiated that change. Alternatively, pivotal moments could be defined in uptake-graphical terms, e.g., by finding nodes with high in-degree or out-degree (for two kinds of "pivotality"). If the uptake graph is constructed with a larger horizon than immediately prior acts, such an approach might identify acts that initiated changes further "downstream" in the activity. The act immediately prior to a content change could be simply wrapping up an activity, with a previous act initiating the change. In my judgment, T12 is an initiator of change and Ge22 is wrapping up an activity. Bruno's posting of his incorrect diagram in Bg23 actually initiated the subsequent attempts to correct his error.

Medina (Chap. 16) worked at the granularity of "pivotal sequences of interaction." Although he did not publish an explicit uptake graph, Medina's analysis implies that a pivotal sequence is convergence of uptake in enacting an innovation.

Those he identified (Episodes 4 and 5) fall out of the time range across which the four analyses overlap. However, his analysis shows how the previous episodes (including those in the present overlapping segment) contribute towards the multiple contingencies of action that converge to enable group enactment of Episodes 4 and 5.

Lund and Bécu-Robinault (Chap. 17) work with MMRs as the units to be evaluated. An MMR is pivotal when it evidences (a) conceptual change towards canonical physics or (b) progressions towards more complexity while at least sustaining the same level of coherency. Within the segment under consideration, they consider MMR#4 (about 12:52–14:16, ending at Bg21) to meet their definition of (b) progression towards greater complexity (by involving two batteries rather than one). The later MMR#6 (Be31/14:22-Gg35/16:18), initiated by Agnes and completed by Bruno, is pivotal in sense (a), conceptual change progressing towards canonical physics.

Jeong (Chap. 18) defined pivotal moments as "moments when changes in group understandings occurred both in terms of the development of the domain understanding and intersubjectivity." Rather than identifying pivotal moments, Jeong discusses mechanisms behind these changes, so we do not have specific commitments to compare to the others' analyses.

Comparing the various conceptions of pivotal moments, Looi, Song, Wen, and Chen emphasize changes in the direction of events; Medina seeks an enactment of an innovation; and Lund and Bécu-Robinault require evidence of conceptual change. All of these concepts of "pivotality" identify change in some aspect of cognition or behavior that evidence change (change being a fundamental aspect of learning). The innovation aspect of Medina's approach has similarities to Lund and Bécu-Robinault's progression towards greater complexity in semiotic bundles. Medina and Looi et al. both analyze uptake, implying the importance of interaction structure, but Looi et al. rely on change in content rather than uptake per se. These analyses all examine multiple acts to identify pivotal entities. Interaction analysis is a matter of examining relations between acts, not properties of acts in isolation (Jordan & Henderson, 1995; Sacks, Schegloff, & Jefferson, 1974; Suthers et al., 2010). We now focus our attention on two issues surfaced by discrepancies that I identified and posed to analysts.

The Role of a Teacher Intervention

In Medina's Episode 2, during which what Jeong calls circuit understanding B1 was displayed, the group successfully lit the bulb with two batteries. This was approximately from 12:21 to 12:52 or S4-Ge8 in Looi et al.'s notation. Serena suggests trying two batteries and two bulbs (S10, 12:54), but the teacher interrupted the group experiment and pointed out that there is "no draft" in the Group Scribbles board (T12, 13:06). (Recall that the suggested procedure was to draw the circuits first and then experiment to see whether they worked.) In the ensuing interaction the teacher implied that the two-battery configuration should be diagrammed (T14, 13:10) and that one person should draw while the others work on the circuit (T16, 13:15).

Medina (in e-mail discussion) initially saw the teacher's intervention as a disruptive move that "splintered the group's intersubjectivity and re-prioritized how they proceeded to manage their interaction." For Medina, T12 was pivotal in a negative sense of having broken up the group's enthusiastic and coordinated work towards Serena's suggestion (S10). Looi, Song, Wen, and Chen also mark T12 as pivotal according to their definition of acts that change the direction of student activity, but they interpret T12 in a more positive light: Bruno's uptake of T12 (verbally in B13 at 13:07 "I'll draw" and in Bg17 at 13:25, where he is drawing) "changed the direction of the group inquiry path from trying to do a new experiment to reflecting and conceptualizing their working theories of how to light a bulb and the mechanism of the circuit."

When I brought this discrepancy to the attention of analysts, Medina came to *simultaneously* sustain his position and agree with Looi's more positive interpretation. T12 led Bruno to become disengaged from the group's work, and his diagram is problematic from a canonical physics point of view (Lund & Bécu-Robinault also point this out): the batteries are not placed in series as they were in the experiment and the diagrammed configuration would not work. But then other group members noticed this inconsistency (Ag32, 15:53: "not like that. Cancel this one"), leading ultimately to repair of Bruno's understanding of how to diagram the circuit (Agnes' initiation and Bruno's completion of the corrected drawing in 16:03–16:16, ending in Gg35).

Lund and Bécu-Robinault do not define a single act such as T12 as a pivotal moment, because they were concerned specifically with MMRs as the unit to be judged as pivotal or not, and MMRs comprise more than one act. Recall that to be pivotal an MMR must evidence (a) conceptual change or (b) increase in complexity. But (entering our discussion of this issue), they agreed that T12 contributes to (initiates) Bruno's attempt to perform an MMR (#4) from the group's manipulating/batteries–bulbs–wires to drawing/GS—that is, from the group's experiments at the beginning of the shared segment (Ge6 at 12:47 and Ge8 at 12:52) to the drawing ending in Bg21 at 14:16. Although Bruno's MMR#4 does not succeed in doing a drawing that respects canonical physics, MMR#4 displays (b) progression towards greater complexity. Also, the end product ultimately leads, through a chain of uptakes, to MMR#6, which is pivotal to Lund and Bécu-Robinault in sense (a), conceptual change progressing towards canonical physics.

Through our efforts to make sense of each others' interpretations, we all came to agree that T12 was a trade-off. The teacher disrupted the group's indigenous activity (suggested by S10) for exogenous objectives. The group did not get to explore "two batteries, two bulbs" until later, but by imposing the activity structure of the original lesson plans, the teacher led the participants to represent their understanding in a different medium that exposed an individual's conceptual weak point, prompting the help of other students. Medina reports that this exchange led to a breakthrough in his thinking regarding "the pragmatic dynamic between individual and group." A disruption to group activity led an individual (Bruno) to expose his issues, and this exposure led others (group) to respond in a manner that helped the individual. The resulting dynamic benefited other individuals and the group: it surfaced (made explicit) features that everyone should attend to and helped make their agreement more explicit.

In a sense, the same thing happened in our analyses. My identification of discrepancies between analyses (the identification of pivotal moments and interpretation of their value) led Medina to think about the basis for his interpretation and see the value of Looi et al.'s alternative interpretation. The fact that both interpretations were well grounded led us to a more dimensional understanding of trade-offs and individual/group dynamics.

Misconception or Innovation?

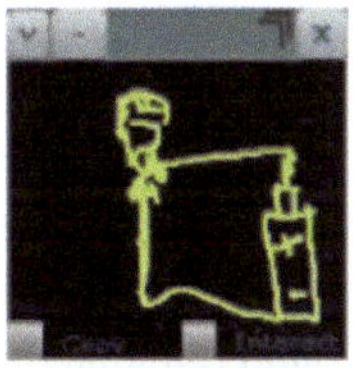

One point of contention emerged, first in this editor's commentary on Lund and Bécu-Robinault and then in discussion between the analysts, concerning whether Bruno demonstrated understanding (had a "canonical conception") of how to connect a wire to the negative end of a battery. Based on Bruno's series of diagrams and Agnes' attempts to correct them in MMR#2 and MMR#4, Lund and Bécu-Robinault concluded that Bruno did not have a complete understanding in terms of being able to clearly represent his knowledge in each mode/medium couplet. They wrote, "he does not clearly show that the wire going to the minus pole of the battery actually touches the pole and does not just end at the bottom left of the battery." For example, see the inset of the diagram that Lund and Bécu-Robinault label B2 in their Figure 4, Medina labels B_1 in his Figure 3, and Jeong identifies as Circuit A, instance (c) in her Table 3 (this diagram was made prior to the scope of the Looi et al. transcript, but appears in Bg23 and Gg35).

I found the diagram to be inconclusive, because an effective strategy for attaching a wire to a battery at its flat negative end is to press the wire flat against the battery. (One would actually get a more tenuous connection the way Agnes was drawing the circuit.) If one drew a wire going flat against the "–" end of a battery it would look like it ended at the edge of the battery, as the battery itself would overlap with the line representing the wire. To test my interpretation, I returned to the video record, looking both at Bruno's diagramming in Group Scribbles and his manipulation of the physical batteries and wires in his individual screen capture video. In drawing B1 (as it is labeled in both Lund & Bécu-Robinault's and Medina's chapters), Bruno clearly clicked on the *middle* of the bottom of the battery when he started the line for the wire. Also, in drawing B3 (Lund & Bécu-Robinault) or B4 (Medina) (also a manifestation of Jeong's B1, completed at Looi et al.'s Bg21/14:16),[1] Bruno clearly drew the wire into the middle of the first rectangular end of the first

[1] If the reader finds this concatenation of references in five coordinate systems confusing, then the reader is experiencing a hint of the challenges I faced in comparing the analyses.

battery: he did not stop at the corner. He has a misconception at this point about how to arrange two batteries, but this is different than a misconception about how to connect a wire to the negative end of a battery.

The evidence is stronger when examining video of Bruno's manipulation of physical batteries. Three times, Bruno clearly placed the wire flat against the flat end of the battery, establishing a solid connection. In fact, at 16:56–58 (in Medina's Episode 4), Bruno (possibly in collaboration with Joel, not visible in the video) introduced an innovation that relies on this placement: he has just arranged the physical batteries in series, and clearly lifts the batteries up, puts a wire under them on the table, and presses the batteries down with the bottom end flat on the wire on the table, thereby accomplishing with the table what would otherwise need a third hand. (Prior to this innovation, we can see Agnes and Serena struggling to keep all the wires in place using only their hands.) Subsequently their experiment concludes successfully. Bruno maintained this configuration through subsequent work towards constructing the two-wire, two-battery configuration. Medina discusses how participants' actions in Episode 5 take up aspects of the situation relevant for ongoing activity. He notes: "The prior configuration, originally taken up via a diagram displayed in the Group Scribbles software, remains intact on the table. Further, Bruno's hand remains on the batteries to keep them in the stacked position. These are physical contingencies for the formulation of ideas." We can add to this set of contingencies the innovation that the stacked batteries are on top of the wire flat on the table. This eliminates the need for a hand to hold the wire against the first battery and makes the entire battery configuration more stable (sitting on the table) and remaining in place, ready to be pressed into service. Thus, when one examines the actual video record rather than just his static diagrams, Bruno arguably had a better understanding of how to connect the wire than the others, leading to an innovation upon which a subsequent group accomplishment was contingent.

However, in discussion it became clear that Lund and Bécu-Robinault were not making a claim about whether the drawing correctly showed the actual situation in the physical environment. Rather, Lund (personal communication) is concerned with whether Bruno has shown representational competence in correct use of the semiotic resources of Group Scribbles as a physics student trained in circuit modeling would. For example, using diagrammatic conventions for modeling circuits that are commonly taught in school (in France in particular), circuit diagrams conventionally show the wire drawn to the middle of the battery pole. We were not able to find out whether students were taught to draw battery bulb diagrams in any kind of formal way in the class studied in this section. If Bruno was not taught to draw a diagram in this particular way, then we cannot expect it of him. However, other students did draw the diagram in this way. Yet, even if diagrammatic conventions were taught, our dispute comes down to whether it is appropriate to evaluate Bruno's understanding based on etic standards for abstract representations or rather to accept his emic display of understanding by successfully lighting the bulb in a complex physical configuration that led to a group innovation as well.

There are implications for the other two analyses. Looi, Song, Wen, and Chen install uptake relations that show how each act builds on prior acts. Uptakes of the type discussed above, where Bruno's innovation in laying the battery flat against the wire is taken up in later experimental efforts by the group, could be added to Looi et al.'s graph. We might then ask whether the moment in which Bruno invents this innovation should be considered pivotal for subsequent group interaction. This episode presents challenges for Jeong's analysis of "group understandings" as displayed in artifact configurations, particularly whole-circuit configurations. The matters discussed above cannot be addressed in terms of whole-circuit configurations as the granularity of analysis: one must examine nuances of how wires are connected in both inscriptions and physical circuits as evidence of (individual) understanding. Therefore, a classification of circuits into A, B1, and B2 is too coarse to capture what Bruno has accomplished. (To be fair, Jeong is concerned with group rather than individual understanding.) Also, "understanding" (whether group or individual) is not displayed merely by static configurations but by sequential situated action (Koschmann et al., 2005). My analysis showed that the temporal aspect, such as whether a line is drawn to the center of the battery even if it is not visible in the final product, is critical.

I now raise the level at which the analyses are compared, stepping away from the interpretation of specific events and instead examining theoretical issues indicated by how the analyses construe the objects of study.

Agency and the Distribution of Activity Across Modalities

Learning has been theorized as taking place at different granularities of agency, including individuals, small groups, or larger networks such as communities as the agent and locus of learning (Greeno, Collins, & Resnick, 1996; Suthers, 2006b). Thus, it is not only interesting but also relevant to compare the chapters on their units of *agency*. Is the unit of agency individuals, the group acting or analyzed as a unit, or something in between, such as intertwined individual agencies from which group agencies emerge? An analyst's conception of agency is evident in the primary activity or process described by his or her analysis (e.g., problem solving, practices, cognition, or understanding) and how this process distributes across candidate agents. We can treat individuals as separate actors or view their intertwined voices as part of a collective phenomenon.

A related question is how *activity* distributes across the available mode/media couplets (Lund & Bécu-Robinault, this volume). Acts in different media can be treated as separate realms between which participants translate or as simultaneously coordinated in a unified activity. There is an implicit third issue: To what extent and in what ways do analysts treat artifacts or media as having agency, whether passively through their affordances and constraints (Norman, 1999), or actively as in Actor-Network Theory's mediators (Latour, 2005)? Some of the analyses hint at the role that media and artifacts play in influencing activity, but the agency of artifacts is not strongly developed in the present case study and will not be discussed further.

For Looi et al. (Chap. 15, this volume), individuals are the primary agents. They identify uptake between acts of individuals, with each act taking place in one of multiple modalities: in Group Scribbles, in "face-to-face" (verbal) interaction, and in the "experimental practice" of manipulating the devices. However, their interest is in transitions in the collective activity. Activity begins with individual acts in specific modalities, but compositions of such individual acts into uptake sequences constitute threads of progressive inquiry. Individuals not only influence each other to progress in their understanding, but also the network of uptake across the group, mediated by two kinds of artifacts, forms a collective progressive inquiry.

For Jeong (Chap. 18, this volume), the agent of interest is the group, but not necessarily in the sense of there being a singular group cognition or shared understanding and purpose. Rather, in her analysis "group understanding" is defined to be the understandings captured via the artifacts students construct. Visible artifacts are taken as proxies for cognitive states of understanding, and the group's understanding is the totality of what has been expressed in shared artifacts through individual acts. The collective display of understanding is distributed across diagrammatic and manipulative modalities: each of these modalities is independently interpreted in terms of how it displays the capability of constructing particular circuit configurations and then later integrated. (Jeong excluded the verbal modality due to the incomplete transcript.) There is a tension in her approach: Jeong analyzes individual acts but makes claims about the group. An analogy offered in personal communication between Jeong and Gerry Stahl helps resolve this tension: Jeong's approach is similar to how archeology characterizes a culture through examples of artifacts produced by individuals.

Medina's discussion (Chap. 16, this volume) implicitly postulates a dynamism between individual and collective agency. Individuals bring issues, ideas, and resources to the attention of the group. Sometimes the group acts in a coordinated collective way that not only happens to involve collective agency but also could not have been accomplished by an individual. The necessity of the group is not merely due to the need to have more than two hands to assemble a complex circuit: the collectivity of Medina's analysis is intrinsic to his way of looking at activity. Rather than individuals acting in ways that change the focus of the group (Looi et al.), for Medina, agency exists across the simultaneous coordinated actions of individuals.

The contrast between Medina's analysis and those of Looi et al. and Jeong raises the related question of how *activity* distributes across modalities. The other analyses treat activities in each modality as separate from each other: proposals are made in a given medium and taken up by others in this or another medium (Looi et al.) or understandings are displayed in one modality (Jeong) and translated to another (Lund & Bécu-Robinault). Medina's discussion of his Figure 10 shows how activity is distributed simultaneously across modalities in an essential way: activity is not analyzed as different acts in each modality being coordinated but one activity distributed across coordinated use of the modalities.

Returning to individual and collective agency, Lund and Bécu-Robinault's analysis offered in Chap. 17 (this volume) tends more towards Looi et al.'s conception of interacting individual agents rather than either Jeong's or Medina's version of group agency, as they are concerned with individuals' conceptual coherence (sometimes as influenced by other individuals). Lund and Bécu-Robinault's conception of the distribution of activity occupies an interesting in-between place: the coordination of activity across modalities is clearly central to their analysis, but it is posed as a reformulation: there are two separate acts, in one medium and then another, and the analysis is concerned with coherence between the two, or how the reformulation is accompanied with evidence of conceptual progress.

The difference is reflected in traditions cited. Lund and Bécu-Robinault cite distributed cognition (Hutchins, 1995), a tradition that examines transformations across different representations. A representation constructed in one medium is subsequently transformed into another medium: this is a sequence of acts, one in one medium and then another in the other. Medina cites Goodwin (I would add Goodwin, 2000, 2003 to his citations), who emphasizes the *simultaneous* use of multiple modalities, in a manner such that it is not possible to understand what is being conveyed without taking multiple modalities into account at once. Therefore, to the extent to which it relies on distributed cognition, Lund and Bécu-Robinault's analysis implies a conception of activity that is distributed across modalities/media in a sequential rather than simultaneous manner.

However, Lund and Bécu-Robinault cite another theory that includes an account of simultaneous distribution of activity: semiotic bundles (Arzarello, 2004). A *semiotic set* (or *semiotic system* if it is more structured) is a set of signs with means of production in a mode and associated meanings. Each set has its own logic, but sometimes sets are associated with each other. Language and gesture are a prime example: the two have their independent signs and modes of production, but they are also used in a highly coordinated way to constitute something more than their mere sum or concatenation. This coordinated pair is the *semiotic bundle*, and it includes ways of coordinating across the semiotic sets (i.e., across modalities/media). Also the sets and their coordination in a bundle can change over time. In education we are particularly interested in this change. The coordination of semiotic systems across modalities can be *diachronic* (modalities are invoked sequentially over time), as we see in the Lund and Bécu-Robinault analysis, or *synchronic* (invoked at the same time), as in Medina's analysis. Lund and Bécu-Robinault have focused on the diachronic not because they would downplay the synchronic, but rather because (1) in this analysis they are concerned with the learning-relevant question of conceptual change, which will be seen over time (from one act to another), and (2) in this domain, as well as other physical sciences and mathematics in general, the ability to translate between representations is a key skill and indicator of competence in the domain (c.f. representational competence, Kozma & Russell, 2005).

It is interesting to compare the Medina and Looi, Song, Wen, and Chen analyses because both analyses are explicitly based on uptake between media-specific acts. Uptake is a proto-analytic concept that makes a commitment to the idea that interaction (and hence *any* analysis of interaction) is constructed out of how agents take up aspects of prior acts or the situation as relevant for their ongoing activity (Suthers et al., 2010). The concept of uptake was originally motivated by the need for a cross-modal, and hence media-independent, unit of interaction, making it a suitable concept for multimodal data such as in this case study (Suthers, 2006a). Uptake is proto-analytic because it can (and must) be specialized and operationalized for specific analytic purposes. In our case examples, the difference in how uptake is used reflects the theoretical orientations just discussed. Looi et al. make a graph of acts in the various media (Group Scribbles; face-to-face speech, gesture and gaze; and with the physical apparatus) and identify moments that were followed by significant changes in the direction of activity. Acts are taken individually and are related by various types of uptake relations (not made explicit in their chapter, but see Looi & Chen, 2010). Even if Medina agreed on the same uptake relations, his analysis would still differ: he takes the unit of activity of interest to be the convergence of uptake from multiple prior relevant acts and environmental situations. The activity is constituted of a network of modality-coordinating uptakes rather than a sequence translating from one to another. So his theoretical position (that activity is simultaneously distributed across modalities) is reflected in how he makes use of uptake structures: as simultaneous uptake from multiple modalities. Lund and Bécu-Robinault use a more specific concept of MMRs, the translation of "conceptual content" across media/modes. Uptake is clearly present, as the act of creating the reformulation is uptake of the acts creating the original formulations, but in Lund and Bécu-Robinault's analysis, uptake takes a background role. Their interest is in the conceptual content that transits the uptake, evidencing a "semiotic bundle" that includes this content and how it is manifested in the two (or more) media/modes. Like Jeong, they are looking for objects that evidence conceptual understanding and are more interested in the contents of these objects than in sequential interaction, but while Jeong defines an understanding to be evidenced by a single formulation (and hence is not analyzing interaction per se), Lund and Bécu-Robinault take as their object one that requires an uptake structure: the reformulation.

In summary, for Jeong, the different modalities and media serve as multiple sources of evidence in reference to a more abstract conceptual realm of "understandings"; for Looi et al., the modes and media are the loci of individual acts that are in relation to each other in ways that can be analyzed for group focus shifts over time that evidence progressive inquiry; and the other two analyses place the modes in more essential relations. Although Lund and Bécu-Robinault draw on a theory that emphasizes synchronic coordination of multiple modes/media, their analysis is diachronic, examining how translations across modes/media evidence canonical understanding ("coherence"). Lund and Bécu-Robinault are like Jeong in that they analyze modalities as evidence for conceptual understanding but are like Looi et al. in that they attend to sequential transitions between acts (not an emphasis of Jeong). Medina attends to modalities/media as closely as Lund and Bécu-Robinault but in a more nearly synchronic manner, examining how simultaneous (or at least converging) acts in multiple modalities/media are brought together to accomplish the group's activity as a whole.

Lessons Learned

Our experience has been that interaction between multiple analysts with different viewpoints can drive advances in both analysts' individual and collective understanding. Multivocal analysis differs from the use of mixed methods by a single investigator because the involvement of independent agents advocating for competing and sometimes conflicting different points of view initiates processes of intersubjective meaning-making not found in the single-investigator case. For example, Medina may not have had the insight concerning the pragmatic dynamic between individual and group activity if he had not been confronted with an alternative interpretation of T12, and although Lund and I did not come to agreement concerning Bruno's understanding, the discussion surfaced assumptions behind the analyses. The value of collaborative data analysis has long been established (e.g., Jordan & Henderson, 1995 for collaborative video review), but there are additional opportunities as well as challenges for multivocality in multimethodological settings. The following lessons for achieving productive multivocality can be drawn from this experience.

The first challenge we encountered was that different analytic approaches make different demands on transcripts, so a transcript produced for one group's needs is not likely to be appropriate for others. There are *potential opportunities in the negotiation of shared transcripts as boundary objects*, although these opportunities were not realized in this collaboration.

A related point is that (unlike collaborative video review) analysts in a multimethodological collaboration may be using different representations and tools that are integral to their ways of viewing the world. Therefore, if we are to achieve productive multivocality in such a collaboration by comparing analyses, it is essential to *attempt to map between analytic representations or learn from the intrinsic incommensurabilities that prevent such a mapping*.

While abstractions such as transcripts, snapshots, and analytic structures play important roles in each analytic tradition, *it may be necessary to go back to the original data record to resolve disputes*, as was illustrated by my observations concerning Bruno's production of circuit diagrams. The sequential and situated dynamic progression of action offers important information to analysts, as it did to participants. Nuances of how things are drawn and how things are held can change interpretation.

In order to be able to do both of the above (mapping, and returning to the original data), it is essential that the abstractions used by analysts *index back to the data record in some shared coordinate system*. Typically the shared coordinate is time, but we saw that analysts are likely to parse time into different kinds of units (e.g., utterances, episodes of inscription or interaction, 30-s intervals) and even to label their units using different naming conventions. In the present case, I had some trouble aligning the analyses for these reasons, and there were different time lines based on the six videos in the corpus.

But we also saw that not all of the benefits are found in the attempt to align and compare analyses. Some of the productivity of multivocality is found by *comparing how analyses constitute the object of study, thereby making alternative theoretical conceptions explicit*, such as in our discussion of the distribution of agency and activity across persons and media.

Finally, *a third party tasked with moderating multivocal dialogue plays an important role* in achieving the above. Some of the issues that turned out to be productive to address arose through my persistent questioning of authors in a public forum. We are accustomed to going our own ways, writing papers that are contributed as independent units, and avoiding conflict. Badgering by a provocateur may be needed to get analysts to look at each others' work, identify differences, and work them through. Having served as a provocateur on this project, I can attest to both the frustrations and fruitfulness of multivocal analysis.

Acknowledgements Many thanks to the Singapore team for their efforts in providing us with rich data and to the analysts for contributing their analyses and tolerating my queries and requests. Portions of this work were supported by NSF Award 0943147. The views expressed herein do not necessarily represent the views of NSF.

References

Arzarello, F. (2004). Semiosis as a multimodal process. *Relime, Numero Especial*, 267–299.

Baker, M. J., & Bielaczyc, K. (1995, August). Missed opportunities for learning in collaborative problem-solving interactions. In J. Greer (Ed.), *Proceedings of AI-ED '95: World Conference on Artificial Intelligence in Education* (pp. 210–217). Washington, DC: AACE.

Chen, W., & Looi, C. -K. (this volume). Group Scribbles-supported collaborative learning in a primary grade 5 science class. In D. D. Suthers, K. Lund, C. P. Rose, C. Teplovs, & N. Law (Eds.), *Productive multivocality in the analysis of group interactions*, Chapter 14. New York, NY: Springer.

Duranti, A. (2006). Transcripts, like shadows on a wall. *Mind, Culture & Activity, 13*(4), 301–310.

Dyke, G., Lund, K., & Girardot, J.-J. (2009). Tatiana: An environment to support the CSCL analysis process. In C. O'Malley, P. Reimann, D. Suthers, & A. Dimitracopoulou (Eds.), *Computer supported collaborative learning practices: CSCL 2009 Conference Proceedings* (pp. 58–67). Rhodes, Greece: International Society of the Learning Sciences.

Dyke, G., Lund, K., Jeong, H., Medina, R., Suthers, D. D., van Aalst, J., et al. (2011). Technological affordances for productive multivocality in analysis. In H. Spada, G. Stahl, N. Miyake & N. Law (Eds.), *Connecting computer-supported collaborative learning to policy and practice: Proceedings of the 9th International Conference on Computer-Supported Collaborative Learning (CSCL 2011)* (Vol. I, pp. 454–461). Hong Kong: International Society of the Learning Sciences.

Gonzales, P., Williams, T., Jocelyn, L., Roey, S., Kastberg, D., & Brenwald, S. (2007). *Highlights from TIMSS 2007: Mathematics and Science Achievement of U.S. Fourth- and Eighth-Grade Students in an International Context. NCES 2009-001*. Washington, DC: National Center for Education Statistics.

Goodwin, C. (2000). Action and embodiment within situated human interaction. *Journal of Pragmatics, 32*, 1489–1522.

Goodwin, C. (2003). The semiotic body in its environment. In J. Coupland & R. Gwyn (Eds.), *Discourses of the body* (pp. 19–42). New York, NY: Palgrave/Macmillan.

Greeno, J. G., Collins, A. M., & Resnick, L. B. (1996). Cognition and learning. In D. C. Berliner & R. C. Calfee (Eds.), *Handbook of educational psychology* (pp. 15–46). New York, NY: Simon & Schuster Macmillan.

Hutchins, E. (1995). *Cognition in the wild.* Cambridge, MA: The MIT Press.

Jeong, H. (this volume). Development of group understanding via the construction of physical and technological artifacts. In D. D. Suthers, K. Lund, C. P. Rose, C. Teplovs, & N. Law (Eds.), *Productive multivocality in the analysis of group interactions*, Chapter 18. New York, NY: Springer.

Jeong, H., Chen, W., & Looi, C.-K. (2011). Analysis of group understanding in artifact-mediated discourse. In H. Spada, G. Stahl, N. Miyake & N. Law (Eds.), *Connecting computer-supported collaborative learning to policy and practice: Proceedings of the 9th International Conference on Computer-Supported Collaborative Learning (CSCL 2011)* (Vol. II, pp. 786–790). Hong Kong: International Society of the Learning Sciences.

Jordan, B., & Henderson, A. (1995). Interaction analysis: Foundations and practice. *Journal of the Learning Sciences, 4*(1), 39–103.

Koschmann, T., Zemel, A., Conlee-Stevens, M., Young, N., Robbs, J., & Barnhart, A. (2005). How *do* people learn: Member's methods and communicative mediation. In R. Bromme, F. W. Hesse, & H. Spada (Eds.), *Barriers and biases in computer-mediated knowledge communication (and how they may be overcome)* (pp. 265–294). Amsterdam: Kluwer Academic Press.

Kozma, R. B., & Russell, J. (2005). Students becoming chemists: Developing representational competence. In J. Gilbert (Ed.), *Visualization in science education.* London: Kluwer.

Latour, B. (2005). *Reassembing the social: An introduction to Actor-Network-Theory.* New York, NY: Oxford University Press.

Looi, C. -K., Song, Y., Wen, Y., & Chen, W. (this volume). Identifying pivotal contributions for group progressive inquiry in a multi-modal interaction environment. In D. D. Suthers, K. Lund, C. P. Rose, C. Teplovs, & N. Law (Eds.), *Productive multivocality in the analysis of group interactions*, Chapter 15. New York, NY: Springer.

Looi, C.-K., & Chen, W. (2010). Community-based individual knowledge construction in the classroom: A process-oriented account. *Journal of Computer Assisted Learning, 26*(3), 202–213.

Lund, K., & Bécu-Robinault, K. (this volume). Conceptual change and sustainable coherency of concepts across modes of interaction. In D. D. Suthers, K. Lund, C. P. Rose, C. Teplovs, & N. Law (Eds.), *Productive multivocality in the analysis of group interactions*, Chapter 17. New York, NY: Springer.

Medina, R. (this volume). Cascading inscriptions and practices: Diagramming and experimentation in the Group Scribbles classroom. In D. D. Suthers, K. Lund, C. P. Rose, C. Teplovs, & N. Law (Eds.), *Productive multivocality in the analysis of group interactions*, Chapter 16. New York, NY: Springer.

Norman, D. A. (1999, May–June). Affordance, conventions, and design. *ACM Interactions, 6*, 38–42.

Ochs, E. (1979). Transcription as theory. In E. Ochs & B. B. Schieffelin (Eds.), *Developmental pragmatics* (pp. 43–72). New York, NY: Academic.

Roschelle, J., Tatar, D., Chaudhury, S. R., Dimitriadis, Y., Patton, C., & DiGiano, C. (2007). Ink, improvisation, and interactive engagement: Learning with tablets. *IEEE Computer, 40*(9), 38–44.

Sacks, H., Schegloff, E. A., & Jefferson, G. (1974). A simplest systematics for the organization of turn-taking for conversation. *Language, 50*(4), 696–735.

Scardamalia, M., & Bereiter, C. (1991). Higher levels of agency for children in knowledge building: A challenge for the design of new knowledge media. *Journal of the Learning Sciences, 1*(1), 37–68.

Scardamalia, M., & Bereiter, C. (2003). Knowledge building. In James W. Guthrie (Ed.), *Encyclopedia of education* (pp. 1370–1373). New York: Macmillan Reference.

Suthers, D. D. (2006a). A qualitative analysis of collaborative knowledge construction through shared representations. *Research and Practice in Technology Enhanced Learning, 1*(2), 1–28.
Suthers, D. D. (2006b). Technology affordances for intersubjective meaning-making: A research agenda for CSCL. *International Journal of Computer-Supported Collaborative Learning, 1*(3), 315–337.
Suthers, D. D., Dwyer, N., Medina, R., & Vatrapu, R. (2010). A framework for conceptualizing, representing, and analyzing distributed interaction. *International Journal of Computer-Supported Collaborative Learning, 5*(1), 5–42.
Suthers, D. D., Lund, K., Rosé, C. P., Dyke, G., Law, N., Teplovs, C., et al. (2011). Towards productive multivocality in the analysis of collaborative learning. In H. Spada, G. Stahl, N. Miyake & N. Law (Eds.), *Connecting computer-supported collaborative learning to policy and practice: Proceedings of the 9th International Conference on Computer-Supported Collaborative Learning (CSCL 2011)* (Vol. III, pp. 1015–1022). Hong Kong: International Society of the Learning Sciences.

Part V
Case Study 4: Knowledge Building Through Asynchronous Online Discourse

Section Editor: Chris Teplovs, Problemshift, Inc.

In this section we investigate data from an online graduate-level course in education that used Knowledge Forum as its principal communication medium. Nobuko Fujita uses the concept of progressive discourse to frame the presentation of her data set. She describes how the data set was collected as part of a design-based research study that examined pedagogical and technological supports for knowledge building.

The three analysis chapters highlight different approaches to the analysis of asynchronous discourse data. In the first analysis chapter, Nobuko Fujita is joined by Chris Teplovs to present an analytic technique that seeks to associate patterns of social interaction with semantic analysis. They describe a framework for learner modeling that uses visualizations of semantic and social networks among learners to highlight opportunities to optimize learning.

Nancy Law and On-Wing Wong also use knowledge building theory to develop analytics that use discourse markers and patterns of participation to identify pivotal moments in collective knowledge advancement. Their goal is to develop a methodology that can potentially be automated with the hopes of providing a pedagogical aid to teachers in addition to the more traditional goal of developing a researcher's analytic tool.

The next chapter, by Ming Ming Chiu, differs from the previous chapters in that it is not couched in terms of knowledge building theory and approaches. Chiu demonstrates how statistical discourse analysis (SDA) can be adapted for use with online discussions. He uses the revised SDA to examine cognitive and social metacognitive relationships among the asynchronous messages.

Nobuko Fujita closes the section with a critical reflection on some of the advantages and problems of multivocal analyses and presents a model of iterative design-based research that capitalizes on some of the unique affordances of multivocality.

Chapter 20
Online Graduate Education Course Using Knowledge Forum

Nobuko Fujita

Introduction

Progressive discourse is a collaborative process through which participants share, question, and revise their ideas to develop "a new understanding that everyone involved agrees is superior to their own previous understanding" (Bereiter, 1994, p. 6). While norms for quality, quantity, relevance, and manner (Grice, 1975) may apply to all kinds of discourse, what distinguishes progressive discourse are a set of commitments that amount to devotion to advancement in the state of knowledge (Bereiter, 1994, 2002; Scardamalia & Bereiter, 2006). The number of commitments for progressive discourse varies across different publications authored by Carl Bereiter and Marlene Scardamalia, but commitments to seek common understanding and to expand the base of accepted facts appear in all versions. Upholding these commitments represent what it means to engage in knowledge building discourse (Scardamalia & Bereiter, 2006). However, fostering progressive discourse in online graduate education is a challenge as many students are practicing teachers who find it difficult to engage in such collaborative dialogue. Previous research has focused on examining the epistemic nature of students' written contributions in elementary science classrooms (Bereiter, Scardamalia, Cassells, & Hewitt, 1997; Hakkarainen & Palonen, 2003). These show that CSILE/Knowledge Forum can support improved learning and knowledge building over time (Scardamalia & Bereiter, 1994; Zhang, Scardamalia, Reeve, & Messina, 2009). Few studies have examined how to detect the emergence of progressive discourse in the participation patterns in asynchronous conferencing environments or what kinds of instructional interventions are most effective to support its development in online graduate education settings.

This chapter describes the context of the discourse that unfolds during an online graduate education course using Knowledge Forum at University of Toronto in fall

N. Fujita (✉)
University of Windsor, Windsor, ON, Canada, N9B 3P4
email: nfujita@uwindsor.ca

D.D. Suthers et al. (eds.), *Productive Multivocality in the Analysis of Group Interactions*, Computer-Supported Collaborative Learning Series 16,
DOI 10.1007/978-1-4614-8960-3_20,

2004 (September 13–December 13). Educational Applications of Computer Mediated Communication (CMC) is a master's level course exploring applications and issues of teaching and learning in the online environment related to all levels of education. The course was taught exclusively online using Web Knowledge Forum (version 4.5.3), without any face-to-face meetings.

The instructor organized the weekly course discussions into 13 separate folders or "views" in the Knowledge Forum database. There were separate views for Class Biographies, Course Café, Course Administration, Course Assignments, and Practice (to practice technical features of Knowledge Forum). This serves to keep off-task posts to a minimum in the course discussion views, to enable online students to get to know each other, and to deal with procedural or technical questions elsewhere.

For the first 2 weeks, the instructor led the course discussion in Knowledge Forum to model online teaching practices; thereafter, a pair of students led discussions on the course readings. Participation in the course discussion views including discussion leadership accounted for 20 % of the student's grade. The instructor did not specify a minimum number of contributions per week but from the beginning of the course made explicit the expectations for students to engage in progressive discourse for deepening understanding. In addition, students were also responsible for pre- and post-course assignments on conceptions of collaborative discourse (5 % each); an online learning journal that took the form of a single note in Knowledge Forum that students added to weekly starting in week 3 (20 %, self-assessed); a group assignment on a relevant educational application issue in CMC (15 %); and a final assignment (35 %).

Literature Review

Researchers in various disciplines including communications, distance education, and educational technology have studied issues of interactivity and learning in asynchronous CMC environments. More recently, the computer-supported collaborative learning (CSCL) research community has turned its attention from studying face-to-face settings in which the computer is one element of a collaborative learning environment to also studying exclusively online teaching and learning contexts (Wallace, 2003). CSCL studies place an emphasis on examining how participants use technology tools to communicate as well as exploring the effectiveness of innovations to support the needs of local learners (Lipponen, Rahikainen, Lallimo, & Hakkarainen, 2003; Orrill, Hannafin, & Glazer, 2003).

Concurrently, CSCL research has built on the theoretical foundations of the field to understand the processes of social knowledge construction in both face-to-face and online CMC environments (Orrill et al., 2003). Sociocultural learning theories emphasize analyses of discourse in order to understand learning and stress the importance of tools in mediating knowledge construction (Cole & Engeström, 1993; Hmelo-Silver, 2003; Palincsar, 1998; Pea, 1993). From this perspective, instructional interventions focus on redistributing the responsibility for generating questions and evaluating explanations from teachers to students, highlighting the importance of student-centered discourse compared to instructor-led classroom discourse in traditional classrooms (Greeno, Collins, & Resnick, 1996).

Promisingly, CSCL research suggests that specially designed tools like CSILE/Knowledge Forum can support student-centered, collaborative discourse to improve learning and enhance knowledge building in classroom settings (Scardamalia & Bereiter, 1994; Zhang et al., 2009). Researchers question, however, whether technology tools alone can mediate online discourse aimed at higher order goals (e.g., Hewitt, 2001; Lai & Law, 2006). Online discussions typically feature a minimal number of interpersonal or conversational turns of talk compared to face-to-face discussions (Guzdial & Carroll, 2002). Online discussions also involve low levels of participation and lack continuity without considerable direction from the instructor (Guzdial, 1997; Hewitt, 2001). Moreover, patterns of online engagement or disengagement are established quickly, are persistent, and show stability and robustness over time (Brett, 2002). These patterns may be seen even among "expert" learners like graduate students, who rarely engage in convergent processes (synthesizing or summarizing ideas) in threaded discussion (Hewitt, 2001). Indeed, content analysis studies show that students' online discourse mostly consists of information sharing and exploration, rather than negotiation of meaning and integration to co-construct knowledge (Gunawardena, Lowe, & Anderson, 1997; Kanuka & Anderson, 1998).

Whereas content analysis is useful for answering certain kinds of research questions, it is a reductive analysis (Stahl, 2002). Mixed methods may be required to understand the multi-faceted phenomena of collaborative knowledge construction (Hmelo-Silver, 2003). Documenting the complex interplay between individual and collaborative understanding is not only crucial in CSCL research but also a major challenge in knowledge building (Lee, Chan, & van Aalst, 2006). Progressive discourse is a focal point in knowledge-building communities, but little is known whether it is possible to detect its emergence in participation patterns or how to support its development instructionally.

Methods

This dataset forms the second iteration of a larger design-based research study (Fujita, 2009). Data sources in the larger study included questionnaire responses, learning journals, and discourse in Knowledge Forum, but only the student discourse data, with pseudonyms replacing participant names in compliance with Research Ethics Board approval, were shared with the analysts. The discourse data consisted of 1,330 asynchronous online discussion messages or "notes" contributed by 17 graduate student participants, the researcher (author), and the instructor in a Web-based Knowledge Forum database.

The author participated in the course both as a design researcher collaborating closely with the instructor and as a teaching assistant interacting in course discussions with students. The goals for this study were twofold: to improve first hand the quality of online graduate education in this particular instance and to contribute to the theoretical understanding of how students collaborate to learn deeply and build knowledge through progressive discourse.

To orient the students to the course theme of progressive discourse, three intervention activities were used: a reading by Bereiter (2002), classroom materials called Discourse for Inquiry (DFI) cards, and the scaffold supports feature built into Knowledge Forum. In the first week of the course, the students read a chapter by Bereiter (2002) that presented rationales for engaging in progressive discourse for knowledge building. This reading was more theoretical and challenging than the other course readings, which helped set high expectations for online discussion.

The DFI cards were adapted from classroom materials originally developed by Woodruff and Brett (1999) to help elementary students and preservice education students take a more advanced approach to face-to-face collaborative discussion. These activities were adapted with the aim to help online graduate students engage in progressive discourse by modeling thinking processes and discourse structures that could be possible in the online Knowledge Forum environment. There were three DFI cards: Managing Problem Solving outlined commitments to progressive discourse (Bereiter, 2002); Managing Group Discourse suggested guidelines for voicing a supporting view or an opposing view; and Managing Meetings provided two strategies to help students with dealing with anxiety. The cards were available as a portable document file (.pdf) in the database for students to download, print out, and refer to as they worked online.

Knowledge Forum, an extension of the Computer-Supported Intentional Learning Environment (CSILE), is specially designed to support knowledge building. Students work in virtual spaces or "views" to develop their ideas, represented as "notes." It offers sophisticated features not available in other conferencing technologies, such as "scaffold supports" (labels of thinking types), "rise-above" (summary note), and a capacity to connect ideas through links between notes in different views. Scaffold supports are typically used as sentence openers that students use while composing notes in the database. They place yellow highlights of thinking types in the text that bracket segments of body text in the notes. Knowledge Forum comes with Theory Building and Opinion scaffold types, but scaffold supports are customizable by instructors and, in some cases, students. At the beginning of the course, only the Theory Building and Opinion scaffolds were available. Later, in week 9, two students designed the "Idea Improvement" scaffolds as part of their discussion leadership (see Table 20.1). The Idea Improvement scaffolds emphasize the socio-cognitive dynamics of "improvable ideas," 1 of the 12 knowledge building principles (Scardamalia, 2002) relevant for progressive discourse.

Participants

Participants were 17 students (12 females, 5 males) (see Table 20.2). They ranged in age from mid-twenties to mid-forties. Five were students in academic programs (4 M.A., 1 Ph.D.); 12 were students in professional programs (9 M.Ed., 3 Ed.D.). All participant names provided in this chapter have been replaced with pseudonyms to ensure confidentiality.

Table 20.1 Knowledge Forum scaffolds and scaffold supports used in iteration 2

	Scaffolds		
Scaffold supports	Theory building	Opinion	Idea improvement
	My theory	Opinion	Idea
	I need to understand	Different opinion	Idea advancement
	New information	Reason	What do we need this idea for
	This theory cannot explain	Elaboration	Problem/question
	A better theory	Evidence	
	Putting our knowledge together	Example	
		Conclusion	

Table 20.2 Participant demographics for iteration 2

Name	Gender	Age	Degree	Reg	Occupation	Residence	Familiar with KF	Previous online courses
Adam	M	26–35	M.A.	P/T	Graduate student	Toronto, ON	Yes	5
Anne	F		M.Ed.	P/T	Secondary teacher	Newmarket, ON	No	0
Belinda	F	36–45	M.Ed.	P/T	Postsecondary instructor	Kelowna, BC	Yes	5
Chloe	F	36–45	M.Ed.	P/T	Software trainer	Toronto, ON	Yes	4
Christine	F	26–35	M.A.	P/T	Postsecondary instructor	Taiwan	Yes	5
Dylan	M	36–45	M.Ed.	P/T	Postsecondary instructor	Toronto, ON	Yes	0
Evelyn	F	<25	M.Ed.	F/T	Graduate student	Toronto, ON	Yes	0
Gail	F	36–45	Ed.D.	P/T	Elementary school principal	Simcoe County, ON	Yes	3
Ian	M	26–35	M.Ed.	P/T	Secondary teacher	Toronto, ON	Yes	0
Jeff	M	26–35	M.Ed.	P/T	Elementary teacher	Toronto, ON	Yes	3
Kelly	F	26–35	M.Ed.	P/T	Teacher	Toronto, ON	No	0
Laurel	F	26–35	Ed.D.	P/T	Educational technology consultant	Markham, ON	Yes	6
Maria	F		MA	F/T	Postsecondary instructor	Toronto, ON	No	0
Megan	F	36–45	Ph.D.	F/T	Postsecondary instructor	Millbrook, ON	Yes	8
Sharon	F	26–35	M.A.	P/T	Elementary teacher	Toronto, ON	Yes	2
Paul	M	36–45	Ed.D.	F/T	Information and planning analyst	Peterborough, ON	Yes	7
Yvonne	F	36–45	M.Ed.	P/T	Educational technology consultant	Ottawa, ON	Yes	4

Summary

Research suggests that computer supports can improve learning and enhance knowledge building, but little is known about what progressive discourse looks like as it unfolds in online graduate education courses or how to support its development. I have analyzed this dataset previously using both manifest and latent types of analyses aligned with the Knowledge Building theoretical framework (Fujita, 2009; Fujita & Brett, 2007, 2008). However, I offered this dataset for analysis in this volume to discover how diverse methods of analysis and theoretical assumptions underlying them can offer insights into the emergence of progressive discourse in participation patterns of asynchronous conferencing transcripts and the design of instructional interventions.

References

Bereiter, C. (1994). Implications of postmodernism for science, or science as progressive discourse. *Educational Psychologist, 29*(1), 3–12.

Bereiter, C. (2002). *Education and mind in the knowledge age*. Mahwah, NJ: Lawrence Erlbaum Associates.

Bereiter, C., Scardamalia, M., Cassells, C., & Hewitt, J. G. (1997). Postmodernism, knowledge building, and elementary science. *Elementary School Journal, 97*(4), 329–340.

Brett, C. (2002). *Investigating online engagement and community among preservice mathematics teachers* (Unpublished doctoral thesis). Ontario Institute for Studies in Education of the University of Toronto, Toronto, ON.

Cole, M., & Engeström, Y. (1993). A cultural-historical approach to distributed cognition. In G. Salomon (Ed.), *Distributed cognitions. Psychological and educational considerations* (pp. 1–46). New York, NY: Cambridge University Press.

Fujita, N. (2009). *Group processes supporting the development of progressive discourse in online graduate courses* (Unpublished Doctoral Dissertation). University of Toronto, Toronto, ON.

Fujita, N., & Brett, C. (2007). *Progressive online discourse: Identifying characteristics in graduate in-service teacher education courses*. Paper presented at the Annual Meeting of the American Educational Research Association, Chicago, IL.

Fujita, N., & Brett, C. (2008). Rethinking analysis of progressive discourse in online learning: An activity theory perspective. In G. Kanselaar, V. Jonker, P. A. Kirschner, & F. Prins (Eds.), *Proceedings of the International Conference of the Learning Sciences 2008: Cre8 a learning world*. Utrecht, NL: International Society of the Learning Sciences.

Greeno, J. G., Collins, A. M., & Resnick, L. B. (1996). Cognition and learning. In D. C. Berliner & R. C. Calfee (Eds.), *Handbook of educational psychology*. New York, NY: Macmillan.

Grice, H. P. (1975). Logic and conversation. In J. J. Morgan (Ed.), *Syntax and semantics* (Vol. 3, pp. 41–58). New York, NY: Academic.

Gunawardena, C. N., Lowe, C. A., & Anderson, T. (1997). Analysis of a global online debate and the development of an interaction analysis model for examining social construction of knowledge in computer conferencing. *Journal of Educational Computing Research, 17*(4), 397–431.

Guzdial, M. (1997, April 7, 2004). *Information ecology of collaborations in educational settings: Influence of tool*. Paper presented at the CSCL '97, Toronto, ON.

Guzdial, M., & Carroll, K. (2002). Exploring the lack of dialogue in computer-supported collaborative learning. In G. Stahl (Ed.), *Computer support for collaborative learning: Foundations for a CSCL community* (pp. 418–434). Mahwah, NJ: Lawrence Erlbaum.

Hakkarainen, K., & Palonen, T. (2003). Patterns of female and male students' participation in peer interaction in computer-supported learning. *Computers & Education, 40*(4), 327–342.

Hewitt, J. G. (2001). Beyond threaded discourse. *Journal of Educational Telecommunications, 7*(3), 207–221.

Hmelo-Silver, C. E. (2003). Analyzing collaborative knowledge construction: Multiple methods for integrated understanding. *Computers & Education, 41*(4), 397–420.

Kanuka, H., & Anderson, T. (1998). Online social interchange, discord, and knowledge construction. *Journal of Distance Education, 13*(1), 57–74.

Lai, M., & Law, N. (2006). Peer scaffolding of knowledge building through collaborative groups with differential learning experiences. *Journal of Educational Computing Research, 35*(2), 123–144.

Lee, E. Y. C., Chan, C. K. K., & van Aalst, J. (2006). Students assessing their own collaborative knowledge building. *International Journal of Computer-Supported Collaborative Learning, 1*, 57–87.

Lipponen, L., Rahikainen, M., Lallimo, J., & Hakkarainen, K. (2003). Patterns of participation and discourse in elementary students' computer-supported collaborative learning. *Learning and Instruction, 13*(5), 487–509.

Orrill, C. H., Hannafin, M. J., & Glazer, E. M. (2003). Disciplined inquiry and the study of emerging technology. In D. H. Jonassen (Ed.), *Handbook of research on educational communications and technology* (2nd ed., pp. 335–353). New York, NY: Macmillan.

Palincsar, A. S. (1998). Social constructivist perspectives on teaching and learning. *Annual Review of Psychology, 49*, 345–375.

Pea, R. D. (1993). Practices of distributed intelligence and designs for education. In G. Salomon (Ed.), *Distributed cognitions. Psychological and educational considerations* (pp. 47–87). New York, NY: Cambridge University Press.

Scardamalia, M. (2002). Collective cognitive responsibility for the advancement of knowledge. In B. Smith (Ed.), *Liberal education in a knowledge society* (pp. 67–98). Chicago, IL: Open Court.

Scardamalia, M., & Bereiter, C. (1994). Computer support for knowledge-building communities. *The Journal of the Learning Sciences, 3*(3), 265–283.

Scardamalia, M., & Bereiter, C. (2006). Knowledge building: Theory, pedagogy, and technology. In K. Sawyer (Ed.), *Cambridge handbook of the learning sciences*. New York, NY: Cambridge University Press.

Stahl, G. (2002). Rediscovering CSCL. In T. Koschmann, R. Hall, & N. Miyake (Eds.), *CSCL2: Carrying forward the conversation* (pp. 169–181). Mahwah, NJ: Lawrence Erlbaum.

Wallace, R. M. (2003). Online learning in higher education: A review of the research on interactions among teachers and students. *Education, Communication and Information, 3*(2), 241–280.

Woodruff, E., & Brett, C. (1999). Collaborative knowledge building: Preservice teachers and elementary students talking to learn. *Language and Education, 13*(4), 280–302.

Zhang, J., Scardamalia, M., Reeve, R., & Messina, R. (2009). Designs for collective cognitive responsibility in knowledge building communities. *The Journal of the Learning Sciences, 18*(1), 7–44.

Chapter 21
Socio-Dynamic Latent Semantic Learner Models

Chris Teplovs and Nobuko Fujita

Introduction

The nascent field of learning analytics focuses on "the measurement, collection, analysis and reporting of data about learners and their contexts, for purposes of understanding and optimising learning and the environments in which it occurs".[1] One approach to learning analytics is social network analysis, which examines the patterns of interaction among learners. Social network analysis of e-learning in particular is facilitated by the availability of digital data that are amenable to such analysis. Considerably less attention has been paid to the content of the artefacts around which the learners are interacting. Content analysis is a time-consuming, painstaking detailed work. Without content analysis, however, claims about the nature of the dynamics among learners are left wanting. Understanding learning, it seems, requires digging deeply into the data that are available.

This chapter begins with a brief introduction and survey of relevant literature using social network and latent semantic analysis (LSA) to analyse online discourse. Next, a description of the prototypic software environment (the Knowledge Space Visualiser or KSV) is presented. The new Knowledge, Interaction and Semantic Student Model Explorer (KISSME software) is an extension of the KSV. The use of LSA in the generation of learner models suitable for studies of collaborative learning is then proposed.

[1] https://tekri.athabascau.ca/analytics/call-papers

C. Teplovs (✉)
Problemshift, Inc., Windsor, ON, Canada
e-mail: dr.chris@problemshift.com

N. Fujita
University of Windsor, Windsor, ON, Canada

D.D. Suthers et al. (eds.), *Productive Multivocality in the Analysis of Group Interactions*, Computer-Supported Collaborative Learning Series 16,
DOI 10.1007/978-1-4614-8960-3_21, © Springer Science+Business Media New York 2013

Background

Theoretical Assumptions

We frame our approach as the nexus of knowledge building (Scardamalia & Bereiter, 2006) and networked learning (de Laat, Lally, Lipponen, & Simons, 2007). In this chapter, we introduce a framework that interweaves social network analysis, semi-automated content analysis and information visualisation to help us understand and optimise learning.

Purpose of the Analysis

We are interested in examining the relationship between social interactions and the semantics of the written products from online learner environments. We are also interested in examining the degree to which automated content analysis, using techniques such as LSA, can be used for analysis with the hope that successful application of such techniques can be used for real-time educational interventions.

Units of Interaction

The units of interaction for our analyses are documents called "notes". Patterns of interaction amongst participants are mediated by these documents.

Data Representations and Analytic Manipulations

The content of the notes is used to create high-dimensional vector representations via LSA. Relationships between these vectors, as well as relationships amongst participants, are represented using network graphs and highlighted adjacency matrices.

Wasserman and Faust (1997) describe social network analysis (SNA) as a methodology that focuses on relationships and patterns of relationships. As such it "requires a set of methods and analytic concepts that are distinct from the methods of traditional statistics and data analysis" (p. 3). They cast SNA in the broader list of topics that have been studied using network analytic methods, including communities (Wellman, 1979), group problem solving (Bavelas, 1950; Bavelas & Barrett, 1951; Leavitt, 1951), diffusion and adoption of innovations (Coleman, Katz, & Menzel, 1957, 1966; Rogers, 1979) and cognition (Freeman, Romney, & Freeman, 1987; Krackhardt, 1987). Irrespective of the objective of the study, network analytic methods focus on the relations between units.

Studies have explored the use of a variety of analytic methods to explore learning and knowledge construction in Networked Learning/Computer-Supported Collaborative Learning (NL/CSCL) environments. However, researchers have yet to achieve consensus on what methods to use to study patterns of interaction. For example, de Laat, Lally, Lipponen, & Simons (2007) used content analysis, critical event recall and SNA to study interaction patterns. They suggest that SNA can be used to complement content analysis (Hara, Bonk, & Angeli, 2000; Henri, 1992) to describe and understand patterns of interaction in NL/CSCL. Of the various network metrics that are available, Wasserman and Faust (1997) focus on density and centrality. In contrast, Reffay and Chanier (2003) applied SNA to determine the cohesion of groups engaged in CSCL. They argue that embedding tools that perform such analyses in the design of the learning environment itself may be more effective than conducting time-consuming content analysis to support teaching and learning.

The importance of time-based analyses has also been noted (Haythornthwaite, 2001; Martínez, Dimitriadis, Rubia, Gomez, & de la Fuente, 2003). The study by de Laat et al. (2007) was the first application of using SNA to illustrate how patterns change over time and the relationship of those patterns to teaching and learning. An important generalization from the literature is that the essential goal in conducting SNA is to elucidate the relationship between two or more units, usually learners. However, there is another equally important type of network analysis to be considered in learning analytics and knowledge work: the network of ideas. Ideas, unfortunately, are difficult to delineate.

Latent Semantic Analysis

LSA represents both a statistical technique and a model of human knowledge acquisition. Landauer and Dumais (1997) propose LSA as a model that could answer the following question: How do individuals know so much given as little information as they get? This problem is variously known as Plato's Problem, the "Problem of Induction", the "poverty of the stimulus", or "the problem of the expert" (Plato's solution was that individuals possess innate knowledge and only need some stimulation to reveal it).

LSA provides a high-dimensional representation of the associations between words and the documents containing those words. The final output from LSA is a series of measures that describe the relationships between units such as words, documents or words-and-documents. In LSA, each document or word is represented by a vector in high-dimensional latent semantic space. The vector is calculated by examining patterns of co-occurrence of words in a term-by-document matrix, which is subsequently simplified using singular value decomposition (SVD). Thus, each document is represented by a vector of numbers, typically numbering between 100 and 300 elements. Whereas dimensions resulting from the application of SVD to data can typically be interpreted (e.g. the dimensions from principal components analysis), the dimensions resulting from LSA are not typically interpretable. This limitation has made the interpretability of LSA-based analyses difficult in the past.

Information visualisation techniques seem to be a natural next step in interpreting LSA and can be used to create meaningful representations of ongoing learning processes. Visualisation of LSA-derived similarities may be problematic, though, due to an unacceptable reduction of dimensionality to two or three dimensions suitable for visualisation from that which is optimal for LSA (typically around 300) (Landauer, Laham & Derr, 2004).

Software

In this section we describe software designed to support the visualisation of learner models based on social and semantic networks. We present a description of the KSV, a prototypic software system on which our new software, KISSME, is based.

The Knowledge Space Visualiser

KISSME extends the KSV, which was developed by the first author for his doctoral dissertation. The KSV was designed to allow researchers to use computer-assisted two-dimensional visualisation of learner-generated contributions to an online discourse space. In its simplest form this generates a graph in which nodes are contributions and links are relationships between those contributions such as "reply", "reference" and "annotate" (see Fig. 21.1).

These explicit relationships between contributions are based on the behaviours of the contributors. A learner, for example, can intentionally choose to make a contribution that is a reply to another learner's contribution. In the resulting graph the links are based on these behavioural relationships. Content is not considered.

In addition to the explicit linkages defined by behaviours such as replying, referencing and annotating there exist implicit linkages between contributions to the discourse space. These implicit linkages concentrate on the similarity of the content of the contributions. Whereas human raters can evaluate the similarity between documents reliably and with good validity, it is a very tedious and time-consuming work. There are a variety of automated and semi-automated techniques that can be used to determine the similarity of text-based contributions. One powerful technique is LSA, described above.

Other representations can be used to highlight different properties of the data. For example, Reimann (2009) highlights the importance of chronology when studying the dynamics of learning communities. The KSV supports this sort of inquiry by facilitating the positioning of notes chronologically. More generally, the KSV supports the use of any categorical, ordinal or continuous variable from the data set to define either of the axes for the display. So in addition to the use of a continuous chronological scale to define the horizontal axis, authorship can be used to define the vertical axis.

Fig. 21.1 Structural links between contributions. *Blue lines* indicate "build-on" or "reply-to" relationships. *Magenta lines* indicate "reference" links

Once contributions are positioned on whatever set of operationally defined axes the analyst has chosen, links between nodes can be overlaid without affecting the positioning of the nodes. For example, the links based on actions such as "building on" and "replying" (hereafter referred to as "structural links") can be overlaid on the learner-time display to show how patterns of interaction change over time.

In a similar way, links between contributions based on latent semantic analysis can be overlaid on the same learner-time display to show the degree to which contributions are similar over time and authorship. More computationally intensive measures can also be visualised. For example, one can determine which contributions were opened (and likely read) by a learner within some specified time interval before that contributor added a new contribution to the discourse space.

Perhaps some of the most interesting diagrams that can be produced using the KSV are based on the superposition of different link types on the same layout. An example of a representation that combines chronology, authorship, structural links and recently opened documents is shown in Fig. 21.2.

The KSV also allows the user to constrain the analysis by specifying beginning and end dates for the analysis. Rather than specifying the dates a priori, the user can manipulate the beginning and end dates with a specially designed slider. In addition to being able to manipulate the beginning and end dates independently of one another, the user can manipulate both dates simultaneously, effectively providing time slices of the network graph.

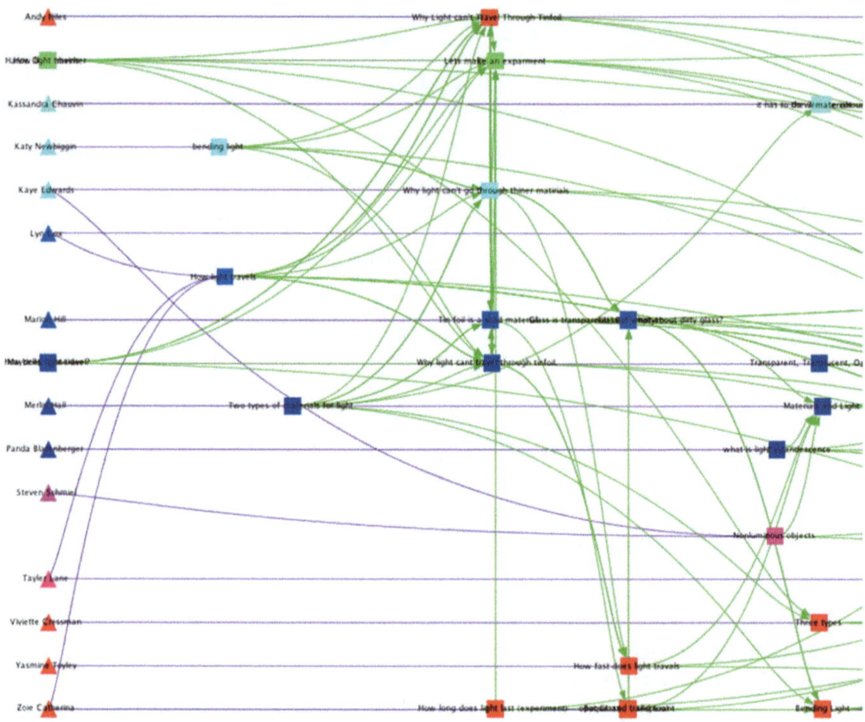

Fig. 21.2 Chronological (time along *x*-axis) and authorial (authors along *y*-axis) layout overlaid with structural links (*magenta lines*) and recently opened notes (*green lines*)

One of the key innovations of the KSV was the use of flexible thresholds in the creation of network representations. This is what allowed us to create visualisations of LSA-based representations of texts. Rather than attempting to provide a two-dimensional layout based on the first few dimensions resulting from the matrix decomposition used in LSA, our approach has been to determine the similarities between documents based on the cosines between the vectors representing documents. A graph is then created in which the nodes correspond to the documents and the edges correspond to the LSA-based similarities between them. A force-directed layout algorithm is then applied to the graph such that the positions of nodes in the two-dimensional representation minimize the distortion of the (very low dimensional) representation. This representation of a maximally connected graph typically lacks clarity, and in typical cases where there are tens or hundreds of nodes the graph is essentially unintelligible due to the large number of edges.

This problem of overly connected graphs also presents a conceptual problem: Does it make sense to connect two document nodes if their LSA-based similarity is very low? Other researchers (Penumatsa et al., 2006) have attempted to address the "threshold problem", but their research suggests that no typical value of cosine threshold for determining document similarity exists. Our approach to tackle this problem is to provide the end user with control over the choice of threshold to use. We do so by providing a slider control in the software that allows the user to specify the cosine value below which edges are not drawn between document nodes. The dynamic nature of this control allows the user, for example, to examine patterns of cluster formation as the similarity threshold is varied.

This provides an example of how visual approaches to learning analytics can provide solutions to previously intractable problems. The answer to the question of "when are two documents (or ideas) different?" is typically "it depends on what you're looking for". Consider a collection of documents generated by students on, for example, the physics of light. At the most permissive level of similarity threshold, all documents are related by virtue of being in the same language. This corresponds to a similarity threshold of zero. At a value slightly higher than zero, one could imagine the documents cluster into two groups: one about colours of light and one about reflection. As one raised the threshold higher yet one could imagine the colours cluster fragmenting into smaller clusters of related notes about topics such as rainbows, wavelength and so on. The interactive nature of being able to manipulate the threshold supports this broad range of possibilities for determining the diversity of ideas that are present in discourse space.

The KSV, while providing powerful visualisations of multidimensional networks, has several limitations. First, it relies on the end user having a functional installation of a recent version of Java. Recent advances in browser-based technology, specifically the widespread adoption of HTML5, have enabled the production of highly interactive browser-based visualisations. Perhaps more significantly, the KSV was limited by its focus on document-based networks. The KSV enables the visualisation of relationships between documents, based on both explicit and implicit linkages, but other than examining patterns of authorship and co-authorship it was not particularly good at generating visualisations of author-based networks. We are working on creating next-generation software that will facilitate the examination of networks of authors. In its earliest versions, the KSV was highly tuned to data from Knowledge Forum. The KSV was recently enhanced to allow the importation of data from almost any data source that provided indications of authorship, chronology and content.

Visualising Student Models: The Knowledge, Interaction and Semantic Student Model Explorer

Recent work has led to the implementation of a learner model based on interactions with other learners. The functionality of the KSV, in terms of being able to manipulate the threshold at which two nodes are considered similar enough to be joined by visible edges, was extended from document nodes to learner nodes. That is, the KSV allowed researchers to investigate patterns of interaction via SNA. The KSV allowed the analyst to exercise considerable control over various parameters such as the intensity of interaction necessary to establish a social link between participants as well as the date at which the social network was analysed. The ability of the analyst to vary these parameters allowed the detection of patterns of interaction that were previously obscured (Reffay, Teplovs & Blondel, 2011). However, the network between authors was based solely on their patterns of interaction. No information about the content of their contributions was used in the generation of the graphs.

The ability to model students or other participants and then to visualise those models in an interactive visualisation environment offers the potential to gain insights into the nature and outcomes of interactions between learners. In the work with the STEF lab we constrained our analyses to focus on the social networks that formed among learners. While this approach revealed interesting patterns of interaction, we felt the results were incomplete because no attention was paid to the content of the learners' contributions to the online discourse space.

Other researchers have conducted studies that meld automated interaction analysis with manual content analysis (de Laat et al., 2007; Martínez et al., 2003). However, manual content analysis represents the rate-limiting step in this sort of analysis. Because manual content analysis takes so long it is incommensurable with real-time analysis, which is one of our goals. Therefore, we are interested in using some sort of automated or semi-automated content analysis. For reasons specified earlier we have chosen to use LSA to help us conduct automated content analysis. For our purposes, all that we are using LSA for is to generate mathematical representations of the participants' contributions to the discourse space. We can then use those mathematical representations in a variety of ways. LSA uses a vector representation of text. One characteristic of these vectors is that they are additive: the vectors of two documents can be added together to get the vector of the combined documents. We can extend this property to generate latent semantic models of participants by adding together the vector representations of all their contributions to the discourse space.

This is not the first application of LSA to student modelling. Other researchers (Dessus, 2009; Dessus, Mandin & Zampa, 2008; Zampa & Lemaire, 2002) have used LSA in student modelling, but they have not focused on the collaborative nature of learning. Still others have extended techniques from earlier research on LSA to apply to e-learning contexts (Kintsch, Caccmise, Franzke, Johnson & Dooley, 2007; Rehder et al., 1998; Wolfe et al., 1998). Zampa and Lamaire's (2002) recent work builds on the notion of matching students to text based on Vygotsky's (1978) Zone of Proximal Development. However, theirs is an individualistic model: the selection of "stimuli" is meant to effect individualised optimisation of learning.

Our approach is somewhat different: we are interested in combining information about patterns of interaction among participants with information about the content of those contributions. We too take a Vygotskian approach: that optimal learning will take place when interactions occur between individuals who are neither too similar nor too dissimilar from each other, based on the semantics of what they have written. This approach of combining SNA and latent semantic network analysis is an example of the sort of "multidimensional" network championed by Contractor (2009).

Our current work includes the implementation of software that will allow us as researchers to examine the interplay of interactions between learners and the latent semantic models of those learners. We are interested in testing the Vygotskian hypothesis that uptake (Suthers, Dwyer, Medina & Vatrapu, 2010) is most likely to occur when the semantic relatedness of the corresponding contributor models is neither too high nor too low. We refer to this intermediate similarity of latent semantic learner models (LSLM) as "compatibility" in the next section where we apply this framework to a case study.

Case Study

Here we present the application of this analysis framework to the data described by Fujita (Chap. 20, this volume). Our approach was to generate network diagrams based on interaction patterns and the similarity of LSLM. Such an approach yielded diagrams such as those shown in Fig. 21.3. We show the progression of the semantic network over the 109 days of the course in Figure. For the LSLM networks we set the threshold of similarity to 0.4 (that is, the LSA-based cosine between learner models had to be at least 0.4 to have a visible tie). In the case of the social interaction network, there had to be at least 50 read events shared by two users for a tie to be shown in the network diagram.

A common problem with network graphs is that they are often complex "hairballs" and unless the presentation can be simplified they do not aid in the interpretability of the underlying data. The network graphs from the Fujita data comprise only 22 people, but they border on being too complex to aid in interpretability. In network diagrams it is generally accepted that the strength of the ties is proportional to the intensity of the visual representation. In our case, however, we are not as interested in the most similar learner models as we are in those models that are of intermediate similarity, based on the Vygotskian notion of a zone of proximal development.

We could represent intermediate-intensity ties with strong visual links, but a more promising approach might be to use an alternative representation. We present in Figs. 21.4 and 21.5 representations of the semantic adjacency matrix that corresponds to the data in Fig. 21.3. In Fig. 21.4 high values (those greater than 26) are shaded. In contrast, intermediate values (those between 14 and 26) are shaded in Fig. 21.5. A complementary representation of the adjacency matrix based on the social interaction of reading each other's notes is shown in Fig. 21.6. In this representation the most intense interactions are shown with the darkest shading.

We were interested in examining the relationship between the intensity of social interaction (reading each others' notes) and the semantic similarity of what each participant had written. Examining the highlighted values in Figs. 21.4 and 21.5 and looking for correlations with highlighted values in Fig. 21.6 provide some glimpses into these sorts of relationships.

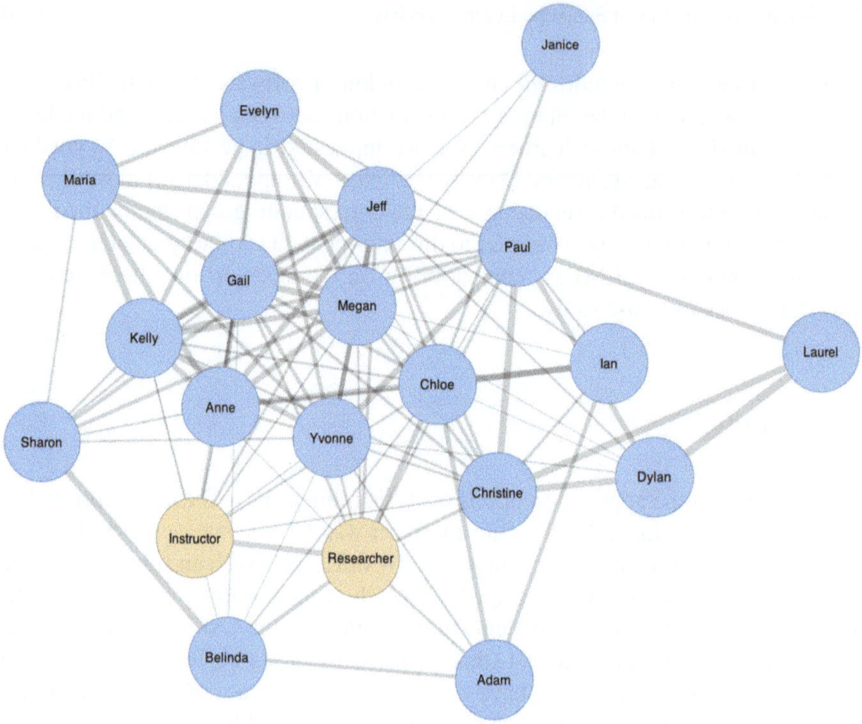

Fig. 21.3 Network graph of participants as determined by similarities of latent semantic learner models

	Ian	Jeff	Evelyn	Sharon	Adam	Belinda	Dylan	Yvonne	Gail	Marla	Megan	Kelly	Laurel	Chloe	Janice
Ian		4			14						1			26	
Jeff	4		35	12			4	15	22	21	33	19		7	1
Evelyn		35						13	6	13	21	19		2	
Sharon		12				34		4	5	4	12	5			
Adam	14					11		4						18	
Belinda				34	11			1	1			1		3	
Dylan		4						2			1		52	1	
Yvonne		15	13	4	4	1	2		10	12	17	21			
Gail		22	6	5		1		10		15	32	28		11	
Marla		21	13	4				12	15		28	34			
Megan	1	33	21	12			1	17	32	28		26		7	2
Kelly		19	19	5		1		21	28	34	26			5	
Laurel							52								
Chloe	26	7	2		18	3	1		11		7	5			8
Janice		1									3			9	
Anne	29	26	8	4				7	17	27	25	15		6	
Paul	4	7	4				22	30	7		11	11	21	2	
Christine	6	9					32	5	8		8	5	34	11	
Researcher		8			6	13		5	7		3	1		22	
Instructor		5						2	7		3	2		4	

Fig. 21.4 Adjacency matrix based on similarities of latent semantic learner models. Highest values are *shaded*

	Ian	Jeff	Evelyn	Sharon	Adam	Belinda	Dylan	Yvonne	Gail	Marla	Megan	Kelly	Laurel	Chloe	Janice
Ian		4			14						1			26	
Jeff	4		35	12			4	15	22	21	33	19		7	1
Evelyn		35						13	6	13	21	19		2	
Sharon		12				34		4	5	4	12	5			
Adam	14					11		4						18	
Belinda				34	11			1	1			1		3	
Dylan		4						2			1		52	1	
Yvonne		15	13	4	4	1	2		10	12	17	21			
Gail		22	6	5		1		10		15	32	28		11	
Marla		21	13	4				12	15		28	34			
Megan	1	33	21	12			1	17	32	28		26		7	2
Kelly		19	19	5		1		21	28	34	26			5	
Laurel							52								
Chloe	26	7	2		18	3	1		11		7	5			8
Janice		1									3			9	
Anne	29	26	8	4				7	17	27	25	15		6	
Paul	4	7	4				22	30	7		11	11	21	2	
Christine	6	9					32	5	8		8	5	34	11	
Researcher		8			6	13		5	7		3	1		22	
Instructor		5						2	7		3	2		4	

Fig. 21.5 Adjacency matrix based on latent semantic learner model similarities. Intermediate values (between 14 and 26) are *shaded*

	Ian	Jeff	Evelyn	Sharon	Adam	Belinda	Dylan	Yvonne	Gail	Marla	Megan	Kelly	Laurel	Chloe	Janice	Anne	Paul	Christine	Researcher	Instructor
Ian	0	18	7	16	226	34	16	23	10	24	27	4	27	97	1	35	52	4	16	41
Jeff	18	0	87	93	177	150	31	56	28	65	72	18	44	60	3	45	163	24	89	176
Evelyn	7	87	0	13	26	26	4	19	2	7	10	3	8	3	0	4	60	12	19	18
Sharon	16	93	13	0	67	200	28	56	45	120	68	14	41	53	2	37	154	31	57	193
Adam	226	177	26	67	0	196	15	104	14	37	27	22	34	120	4	32	211	29	76	162
Belinda	34	150	26	200	196	0	53	153	68	58	86	49	54	107	6	72	252	44	115	183
Dylan	16	31	4	28	15	53	0	6	13	25	25	9	50	21	2	5	108	9	33	107
Yvonne	23	56	19	56	104	153	6	0	27	51	55	52	36	46	0	28	187	32	59	166
Gail	10	29	2	45	14	68	13	27	0	24	28	9	9	20	0	18	23	27	18	60
Marla	26	65	7	120	37	58	25	51	24	0	51	45	19	40	2	33	64	23	48	134
Megan	27	72	10	68	27	86	25	55	28	51	0	19	21	46	1	17	145	29	44	79
Kelly	4	18	3	14	22	49	9	52	9	45	19	0	15	14	1	7	42	6	35	65
Laurel	27	44	8	41	34	54	50	36	9	19	21	15	0	20	1	11	94	15	16	75
Chloe	97	60	3	53	120	107	21	46	20	40	46	14	20	0	4	16	99	26	74	130
Janice	1	3	0	2	4	6	2	0	0	2	1	1	1	4	0	1	10	0	2	10
Anne	35	45	4	37	32	72	5	28	18	33	17	7	11	16	1	0	56	9	14	38
Paul	52	163	60	154	211	252	108	187	23	64	145	42	94	99	10	56	0	57	96	293
Christine	4	24	12	31	29	44	9	32	27	23	29	6	15	26	0	9	57	0	89	88
Researcher	16	89	19	57	76	115	33	59	18	48	44	35	16	74	2	14	96	89	0	101
Instructor	41	176	18	193	162	183	107	166	60	134	79	65	75	130	10	38	293	88	101	0

Fig. 21.6 Adjacency matrix based on frequency of reading events. *Darker shading* corresponds to higher frequency

A final step in identifying points of interest in the data set requires us to consider those points where there was intermediate overlap in terms of the semantic similarity and determine the degree to which there was the opportunity for the exchange of ideas. In this case that opportunity was provided by the presence of reading events. If there were abundant reading events shared between two learners and their LSLM suggested they were compatible then we would predict that there might have been "pivotal moments" or a point in time at which the semantic structure of the community changed in an important way. In the case that their latent semantic learner models suggest they're compatible but there were insufficient interactions between them we can postulate that there may have been unrealized potential for productive interaction or pivotal moments. The consideration of other cases (e.g. highly similar or dissimilar LSLM) is beyond the scope of this chapter.

Consider, for example, the highlighted values for Ian. Ian is semantically most similar to Evelyn and Megan but has the most interactions with Adam and Chloe. In terms of semantic similarity, Adam and Chloe are of intermediate similarity to Ian, which provides some support for the hypothesis that there is a relationship between intermediate semantic similarity and intensity of interaction. Similar patterns exist for Adam and Chloe.

Different patterns exist for Evelyn, who shared the most reading events with Jeff and to whom she was semantically most similar. Other participants were showed little evidence of a discernable relationship between their reading events and semantically high or moderate similarity.

This is a retrospective analysis, and therefore we cannot intervene in an attempt to engineer change. However, there is some preliminary evidence that there is a relationship between the intensity of interactions and learners who are semantically similar, but not too similar, to each other based on what they have written in the online discourse space. We propose that the bringing together of individuals who are neither too similar nor too dissimilar can be considered a "pivotal moment", inasmuch as these can lead to higher levels of interaction. We concede that these preliminary results are not indicative of any causal mechanism and are merely correlational in nature. Future design-based research could see the use of learner models based on latent semantic analysis to inform the opportunistic formation of groups in a way that might enhance group processes.

Summary

We have proposed a framework that combines SNA and latent semantic analysis of online discourse. The proposal is speculative: previous work with latent semantic analysis has yielded promising results that may help us understand the nature of interactions among learners. Examining those interactions may allow us to gain insight into the nature of community dynamics and in particular identify pivotal moments—either manifest or potential—that can occur when individuals whose similarity based on semantic analysis of their writing are given the opportunity to interact.

References

Bavelas, A. (1950). Communication patterns in task-oriented groups. *Journal of the Acoustical Society of America, 22*, 271–282.

Bavelas, A., & Barrett, D. (1951). An experimental approach to organizational communication. *Personnel, 27*, 366–371.

Coleman, J. S., Katz, E., & Menzel, H. (1957). The diffusion of an innovation among physicians. *Sociometry, 20*, 253–270.

Coleman, J. S., Katz, E., & Menzel, H. (1966). *Medical innovation: A diffusion study*. Indianapolis, IN: Bobbs-Merrill.

Contractor, N. (2009). The emergence of multidimensional networks. *Journal of Computer-Mediated Communication, 14*, 743–747.

de Laat, M., Lally, V., Lipponen, L., & Simons, R.-J. (2007). Investigating patterns of interaction in networked learning and computer-supported collaborative learning: A role for Social Network Analysis. *International Journal of Computer-Supported Collaborative Learning, 2*, 87–103.

Dessus, P. (2009). An overview of LSA-based systems for supporting learning and teaching. In *Proceeding of the 2009 conference on Artificial Intelligence in Education: Building Learning Systems that Care: From Knowledge Representation to Affective Modelling*. IOS Press: Amsterdam, Netherlands.

Dessus, P., Mandin, S., & Zampa, V. (2008). What is teaching? Cognitive-based tutoring principles for the design of a learning environment. In S. Tazi & K. Zreik (Eds.), *Common innovation in e-learning, machine learning and humanoid (ICHSL.6)* (pp. 49–55). Paris, France: Europa/IEEE.

Freeman, L. C., Romney, A. K., & Freeman, S. C. (1987). Cognitive structure and informant accuracy. *American Anthropologist, 89*, 310–325.

Fujita, N. (this volume). Online graduate education course using knowledge forum. In D. D. Suthers, K. Lund, C. P. Rosé, C. Teplovs, & N. Law (Eds.), *Productive multivocality in the analysis of group interactions*, Chapter 20. New York, NY: Springer.

Hara, N., Bonk, C. J., & Angeli, C. (2000). Content analyses of on-line discussion in an applied educational psychology course. *Instructional Science, 28*, 115–152.

Haythornthwaite, C. (2001). Exploring multiplexity: Social network structure in a computer-supported distance learning class. *The Information Society, 17*, 211–226.

Henri, F. (1992). Computer conferencing and content analysis. In A. R. Kaye (Ed.), *Collaborative learning through computer conferencing*. London, UK: Springer.

Kintsch, E., Caccmise, D., Franzke, M., Johnson, N., & Dooley, S. (2007). Summary street®: Computer-guided summary writing. In T. K. Landauer, D. S. McNamara, S. Dennis, & W. Kintsch (Eds.), *Handbook of latent semantic analysis*. Mahwah, NJ: Lawrence Erlbaum Associates.

Krackhardt, D. (1987). Cognitive social structures. *Social Networks, 9*, 109–134.

Landauer, T. K., & Dumais, S. T. (1997). A solution to Plato's problem: The latent semantic analysis theory of the acquisition, induction, and representation of knowledge. *Psychological Review, 104*, 211–240.

Landauer, T. K., Laham, D., & Derr, M. (2004). From paragraph to graph: Latent semantic analysis for information visualization. *Proceedings of the National Academy of Sciences of the United States of America, 101*, 5214–5219.

Leavitt, H. J. (1951). Some effects of communication patterns on group performance. *Journal of Abnormal and Social Psychology, 46*, 38–50.

Martínez, A., Dimitriadis, Y., Rubia, B., Gomez, E., & de la Fuente, P. (2003). Combining qualitative evaluation and social network analysis for the study of classroom social interactions. *Computers & Education, 41*, 353–368.

Penumatsa, P., Ventura, M., Graesser, A. C., Louwerse, M. M., Hu, X., Cai, Z., et al. (2006). The right threshold value: What is the right threshold of cosine measure when using latent semantic analysis for evaluating student answers? *International Journal on Artificial Intelligence Tools, 15*, 767–778.

Reffay, C., & Chanier, T. (2003). How social network analysis can help to measure cohesion in collaborative distance-learning. In *Designing for change in networked learning. Proceedings of the international conference on Computer Supported Collaborative Learning 2003* (pp. 343–352). Kluwer Academic Publishers: Bergen, Norway.

Reffay, C., Teplovs, C., & Blondel, F.-M. (2011). Productive re-use of CSCL data and analytic tools to provide a new perspective on group cohesion. In *Proceedings of the 10th International Conference on Computer Supported Collaborative Learning, 2011*, Hong Kong.

Rehder, B., Schreiner, M. E., Wolfe, M. B., Laham, D., Landauer, T. K., & Kintsch, W. (1998). Using latent semantic analysis to assess knowledge: Some technical considerations. *Discourse Processes, 25*, 337–354.

Reimann, P. (2009). Time is precious: Variable- and event-centred approaches to process analysis in CSCL research. *International Journal of Computer-Supported Collaborative Learning, 4*, 239–257.

Rogers, E. M. (1979). Network analysis of the diffusion of innovations. In P. W. Holland & S. Leinhardt (Eds.), *Perspectives on social network research* (pp. 137–164). New York, NY: Academic.

Scardamalia, M., & Bereiter, C. (2006). Knowledge building: theory, pedagogy, and technology. In K. Sawyer (Ed.), *Cambridge Handbook of the Learning Sciences* (pp. 97–118). New York: Cambridge University Press.

Suthers, D., Dwyer, N., Medina, R., & Vatrapu, R. (2010). A framework for conceptualizing, representing, and analyzing distributed interaction. *International Journal of Computer-Supported Collaborative Learning, 5*, 5–42.

Vygotsky, L. S. (1978). *Mind in society: The development of higher psychological processes.* Cambridge, MA: Harvard University Press.

Wasserman, S., & Faust, K. (1997). *Social network analysis: Methods and applications.* Cambridge, UK: Cambridge University Press.

Wellman, B. (1979). The community question: The intimate networks of East Yorkers. *American Journal of Sociology, 84*, 1201–1231.

Wolfe, M. B., Schreiner, M. E., Rehder, B., Laham, D., Foltz, P. W., Kintsch, W., et al. (1998). Learning from text: Matching readers and text by latent semantic analysis. *Discourse Processes, 25*, 309–336.

Zampa, V., & Lemaire, B. (2002). Latent semantic analysis for user modeling. *Journal of Intelligent Information Systems, 18*(1), 15–30.

Chapter 22
Exploring Pivotal Moments in Students' Knowledge Building Progress Using Participation and Discourse Marker Indicators as Heuristic Guides

Nancy Law and On-Wing Wong

Introduction

This paper sets out to identify pivotal moments in students' knowledge building progress for an online asynchronous corpus generated by a class of master's level students in the context of a totally online course. The main motivation for this study is to develop a methodology that can be effectively automated to aid teachers and/or researchers to quickly gain a good overview of students' progress in understanding at an overall class level from a very large, semantically rich and complex discourse corpus. The methodology incorporates the use of participation and discourse marker indicators to provide an overview of the nature and depth of students' engagement in relation to key concepts targeted for student learning, and to support the heuristic selection of a small sample of notes for use by the teacher and/or researcher for further in-depth qualitative analysis. This methodology has the potential of being developed into a teacher's pedagogical aid to more effectively facilitate students' collaborative inquiry and knowledge building. As a researcher's productivity tool in understanding students' developmental trajectory in learning through discourse, it offers a distinct possibility for developing and validating knowledge building theory on the basis of empirical discourse analysis of large sets of corpus.

Theoretical and Methodological Overview

This study is underpinned by the theory of knowledge building, which can be taken both as an epistemology and a pedagogy. For the latter, Scardamalia and Bereiter has pointed out that knowledge building is not something that someone does naturally.

N. Law (✉) • O.-W. Wong
University of Hong Kong, Hong Kong, Hong Kong
e-mail: nlaw@hku.hk

D.D. Suthers et al. (eds.), *Productive Multivocality in the Analysis of Group Interactions*, Computer-Supported Collaborative Learning Series 16,
DOI 10.1007/978-1-4614-8960-3_22, © Springer Science+Business Media New York 2013

It requires collective intentional efforts of the community involved. The 12 knowledge building (KB) principles (Scardamalia, 2002) describes socio-metacognitive characteristics that can be observed in a KB community. When a group of people is said to be involved in knowledge building on some topics or problems, it is meaningful to identify the trajectory of KB development of this community from two perspectives—(1) progress in the level at which this group of people exhibits characteristics of the 12 KB principles, and (2) advances in understanding of the core topics/themes identified. In this paper, we are exploring students' KB trajectory from both perspectives.

If we take pivotal moments as indicative of moving from one stage of the trajectory into another, we may consider this from the perspective of both the community as a whole and that of an individual, which could be different. In this case, the assumption is that an individual would not be able to advance to another stage of the trajectory unless the community as a whole has been able to do so. On the other hand, improved ideas or emerging insights from a productive discourse cannot be attributed to any individual or subgroup of individuals' contributions (Bereiter, 2002). Furthermore, it is not necessary that individuals within the community are able to achieve the advanced understanding made by the collective.

Purpose of the Analysis

There is a strong pedagogical motivation underpinning the present study: Is it possible to design a kind of dashboard of indicators derived from automated analysis that can help teachers to identify roughly the stage of developmental trajectory the students' discourse is in, the key problems of understanding being explored, and the "at-risk" students, if any?

Unit of Interaction

As the outcome of this study is to provide some form of learning analytics to help teachers to understand their students' KB engagement and advancement, we have used different units of analysis and representations to provide different layers of insight to teachers, from the overall quantitative statistics of participation, questioning and keywords used on a week-to-week basis, to qualitative content analysis of the discourse involving two units of analyses: a note (i.e., a message) and a sentence.

Representations

Befitting the purpose of providing helpful tools to support teachers' pedagogical decision-making, simple, easy to understand graphical displays accessible to novices are used.

Manipulations

The tools used in this study currently only provide static displays. It is hoped that in future, users can traverse across different layers of analyses.

Knowledge Building Principles and Trajectory of Knowledge Building Advancement

The 12 KB Principles describe the characteristics of a knowledge building community's behavior (including their discourse), and serves as a theoretical framework for the research community to identify indicators for KB advancement (e.g., Law & Wong, 2003; Lee, Chan, & van Aalst, 2006). From a pedagogical perspective, the strengths to which these characteristics are exhibited serve as indicators of students' depth of KB engagement. There are reports of a developmental trajectory in students' KB engagement (Law, 2005; Law & Wong, 2003), based on the observation of co-presence clustering for the 12 KB principles, and the hierarchy of accessibility for these clusters. The four dimensions of KB community development reported, together with their associated cluster of KB principles are as follows:

1. Establishment of a social dynamic conducive to sharing and open exploration of ideas (principles involved: Community knowledge collective responsibility, Democratizing knowledge and Idea diversity)
2. Establishment of a progressive inquiry orientation (principles involved: Epistemic agency, Knowledge building discourse, Improvable ideas, Constructive use of authoritative sources)
3. Establishment of a socio-metacognitive orientation for KB (principles involved: Real ideas authentic problems, Rise above, Embedded and transformative assessment)
4. Establishment of a KB habit of mind (principles involved: Pervasive knowledge building, Symmetric knowledge advancement)

The above description of KB progress refers to a community of learners taken as a whole rather than as individual learners. This does not mean that everyone in the same class is making the same progress. At the same time one cannot draw conclusions on an individual learner's progress in understanding based only on what the individual autonomously expresses in a discourse. In this paper, the only source of data upon which analysis is made is students' asynchronous online data posted on the Knowledge Forum®. While individual student's written contributions to the discourse to some extent reflect his/her related understanding and learning outcomes, one cannot draw any conclusion based on what has not been expressed explicitly. Gaining an understanding of individual students' progressive understanding is hence much more complex, and would need other data collection mechanisms, such as individual learning journals (e.g., Lee et al., 2006) and student achievement data in addition to the collective discourse data. In this paper, we do not examine students' contributions individually, but focus on understanding the progress of the class as a totality.

The Curriculum Design Context That Generated the Discourse Data

Students' behaviors are very much influenced by the curriculum design and facilitation of the teacher. The title of the course was Educational Applications of Computer Mediated Communications, which was conducted entirely online and learning was orchestrated through asynchronous online discussions on Knowledge Forum® around topics and readings specified for each of the 13 weeks in the course outline. Knowledge Forum® provides many features for structuring discussions, one of which is the provision of views for users to enter their discussions on different topics. The instructor created 13 views, one for each of the weekly discussion topics.

One important aspect of the course design was that starting from Week 3, pairs of students would take turns to lead the weekly discussions on the readings. The discussion leaders should identify critical questions from assigned readings and facilitate the class discussion. The class originally had 18 students enrolled but one dropped out of the course shortly after the course started. Hence the researcher who was studying the class interactions (the data presenter for this book section) took on the discussion co-leader role for week 10 as a replacement. The nature of this course and the important role of the asynchronous discussions in this course ensured that much of the learning interactions are captured through the discourse data. Weeks 1 and 2 were designed as warming up sessions, the purpose of which was to introduce the course outline, course expectation and to get the class started. Two of the weeks towards the end of the course, weeks 11 and 13, were designed as wrapping up sessions where students can work with groupmates on their assignment projects, and for final synthesis and evaluation respectively.

It has to be noted that week 5 was somewhat anomalous in two respects. First of all, the two student facilitators wanted to help the class to explore the role of a teacher in the context of an online course by being a "bad teacher," i.e., by not being a responsive facilitator. A second feature of this week was that the students experimented with the "rise above" function of Knowledge Forum®. This is a function that allows users to pull together a number of notes into a "rise-above note" so that the key ideas of those notes can be summarized in this one note, which acts as a container. This feature allows users to organize and reorganize the discussion as the discourse progresses. However, it poses serious difficulties to the analysis process as the notes placed inside these "rise-above" containers are no longer visible at the first level and it is difficult to interpret the relationship during the actual discourse process among these "invisible" notes and between these "invisible" notes and the other notes. Fortunately (from the authors' point of view), rise-above notes were mainly found in week 5. Hence, the data for week 5 is incomplete. In this paper, while we do not have data contained in the "rise-above" notes from week 5, we do not make any interpretation of the quantitative data from week 5 for the above reasons.

Pivotal Weeks in Terms of Knowledge Building Social Dynamic

The three KB principles associated with the presence of a social dynamic conducive to KB are community knowledge collective responsibility, democratizing knowledge and idea diversity. A prerequisite for achieving such a condition is wide participation from all students. For a teacher, it is thus important to know the level and parity of participation from the students. As the course was organized in the form of 13 weekly "sessions," as a first pass, we use the number of notes contributed by each student each week as an indicator of student engagement. Figure 22.1 is a boxplot of the number of notes each of the students wrote in the 13 weekly views of this course. It can be seen that other than for the first week and the last 3 weeks, the median of the number of notes written are rather similar at about four notes per week per student, with the exception of weeks 3 (5.5) and 9 (1.5). Can we take these 2 weeks as pivotal points of some sort, and inquire what might have led to these deviations?

Week 3 was when student facilitated discussions first started. There could be many reasons for the much higher median number of notes posted per person this week. May be the students were excited to launch into the student-facilitated stage of discussion. May be they feel less intimidated when their peers were facilitating. Or it could just be that they wanted to show support for their peers leading in this first week of student-facilitated discussion. On the other hand, it is not immediately apparent what was "pivotal" about week 9.

In addition to central tendency, the boxplot also shows the dispersion (or spread) in the number of notes written by the students. It can be seen that in each of the 13 weeks, there were always students who had not written a single note for that week. On the other hand, the maximum number of notes written by a single student increased from 5 in the first week, to a maximum of 18 in week 6. This maximum then tailed off, except for week 10 when it jumped to 13 from a maximum of 6 in week 9. Hence again, it appears that the motivation to contribute notes to the discussion was lowest even among the most engaged students during week 9, making this week another candidate for consideration as a pivotal, low engagement week in this course. If we focus on the dispersion in the number of notes written by students, then week 6 and week 9 are again the obvious "pivotal" points, the former showing the largest dispersion while the latter the smallest. However, a high dispersion is a negative indicator from the perspective of KB as it reveals a lack of collective responsibility, which is counter to the KB principle of democratizing knowledge.

While the number of notes written is an indicator of engagement and contribution to the knowledge building process, it cannot be equated with students' interest or commitment to learning. It is well acknowledged in CSCL literature that "lurkers" who read and reflect on posts without posting may still benefit greatly from this "passive" form of participation (Dennen, 2008). Figure 22.2 shows the boxplot of the percentage of the total notes posted in the weekly views read by each of the students. It can be seen from the statistics presented here that indeed many of the students took the online discussions very seriously. For each of the 13 weeks, at least 25 % of all students read 100 % of all the notes posted. On the other hand, there were also always students who had not read any of the notes in the view, showing that even at the postgraduate student level, getting all of the students' attention and commitment to learning is still a challenge. The dispersion statistics reveals that week 9 is again a very special week (pivotal?) in not only showing the smallest dispersion, but that it has the highest percentage of notes read by students, with half of the students having read 90 % or more of all the notes posted during that week and only four students reading fewer than 75 %. So if the percentage of notes read can be taken as an indication of students' interest and engagement in learning, then week 9 should be taken as a "positive" pivotal week based on the KB principles associated with the establishment of a social dynamic conducive to knowledge building.

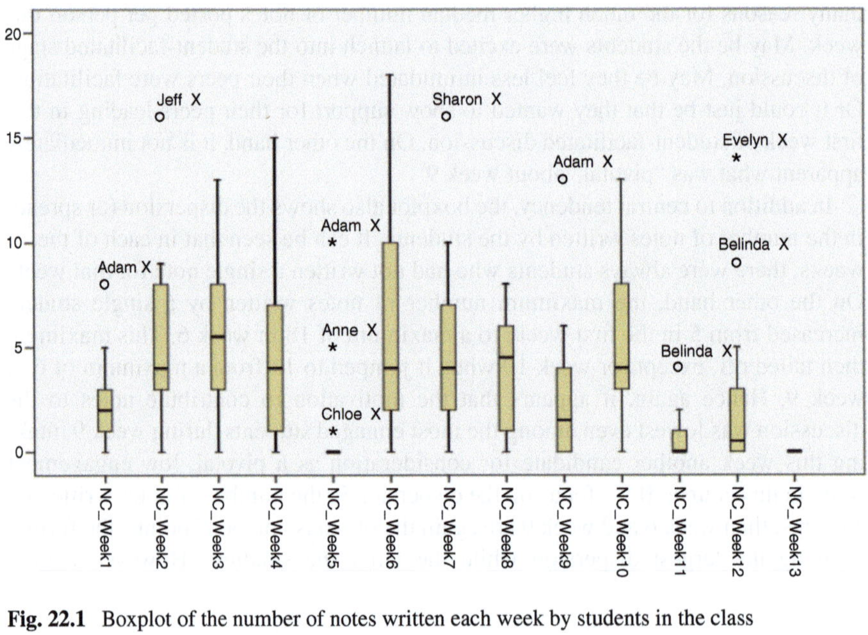

Fig. 22.1 Boxplot of the number of notes written each week by students in the class

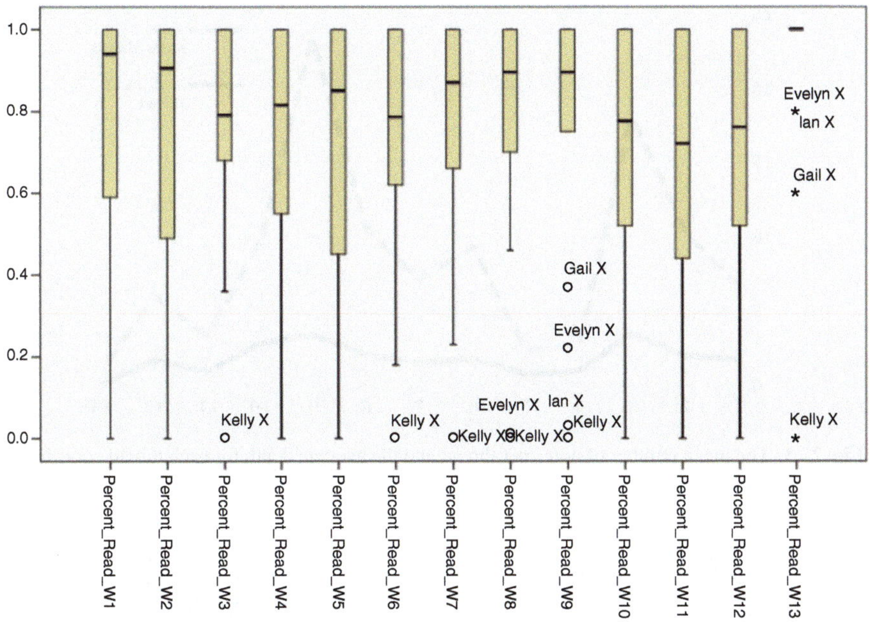

Fig. 22.2 Boxplot of the proportion of notes read each week by students in the class

Pivotal Weeks in the Establishment of a Progressive Inquiry Orientation

Using Thread-Level Indicators

In order that online discourse contributes to knowledge building around problems of understanding and inquiry, sustained exploration of ideas is needed. Unconnected notes or very short threads with one or two levels of build-on are thus unlikely to contribute much to knowledge advancement. Hence the number of notes per thread and the deepest level of build-on achieved can be taken as indicators of depth of inquiry. Figure 22.3 is a graph showing the mean number of notes per thread and the average thread depth for each of the weeks.

The graphs in Fig. 22.3 show that week 9 has the highest mean thread size and deepest average thread depths, followed by week 3. These statistics pose a conundrum to our interpretation of these figures. In particular, week 9 had the lowest number of notes written per week if we exclude the last 3 weeks when the students were too busy with completing their project assignments. Was this week particularly productive in terms of contributing to knowledge advancement since it has large and deep threads? How productive is this week's inquiry compared to the other weeks when students wrote many more notes overall?

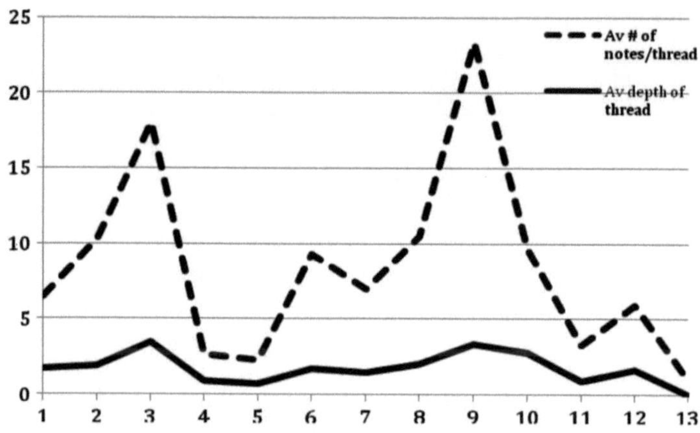

Fig. 22.3 The mean number of notes per thread and the average depth for each thread contributed in each of the weekly views

Using Inquiry-Related Discourse Markers and Knowledge Forum® Built-in Functions to Gauge Students' Depth of Engagement

There are features and functions embedded in Knowledge Forum® specifically designed to support knowledge building. One such feature is the provision of metacognitive scaffolds such as "I need to understand" and "my theory" which are in some ways similar to sentence openers, and users are free to select as appropriate to identify the nature of their contribution in the note. The presence of such scaffolds indicates more conscious efforts and intentionality in their engagement in the discourse as a knowledge building effort, and can serve as an indicator for the KB principle "Epistemic Agency." Another important feature of a KB discourse is the presence of questions, particularly explanation seeking ones, which focus on identifying problems of understanding and on identifying inconsistencies and suggesting better explanations rather than content topics (Hakkarainen, Lipponen, & Järvelä, 2002). The autonomous raising of questions by students is also an indicator of epistemic agency.

Machine identification of discourse markers was used in this study to track the likely presence of factual, explanatory and elaboration questions in the discourse. Table 22.1 lists the discourse markers used in this study. It has to be cautioned here that the presence of markers cannot be equated with the presence of these types of questions. It would be necessary to study the reliability of this methodology in identifying the three types of communication functions in the present discourse data if we are actually using this process to code the presence of different types of questions. However, as we are only using the discourse markers to identify text with a high probability of belonging to these three types of communication functions as a heuristic method, reliability statistics is deemed unnecessary. Figure 22.4 shows the total number of notes posted in each of the

Table 22.1 Discourse markers used to identify communication functions in the discourse data

Discourse type	Communication function	Discourse markers
Fact questions	What questions	What's, What're, What is, What are, What do, What does, What did, What kind, What about, What can, What could, What will, What would, What shall, What should, What has, What have, What had
	Confirmatory questions	Is there, Are there, Is this, Are these, Do you think, Do you agree, Don't you think, Don't you agree
	Quantitative	How many, How much, How often
Explanatory questions	How question	How is, How are, How's, How're, How do, How does, How did, How will, How would, How can, How could, How to
	Why question	Why's, Why is, Why are, Why do, Why did, Why does, Why it is
	Conditional	What if
Elaboration questions	Who question	Who, Whom, Whose
	When question	When
	Where question	Where's, Where're, Where is, Where are, Where do, Where did, Where does, Where can, Where could, Where will, Where would, Where shall, Where should, Where has, Where have, Where had

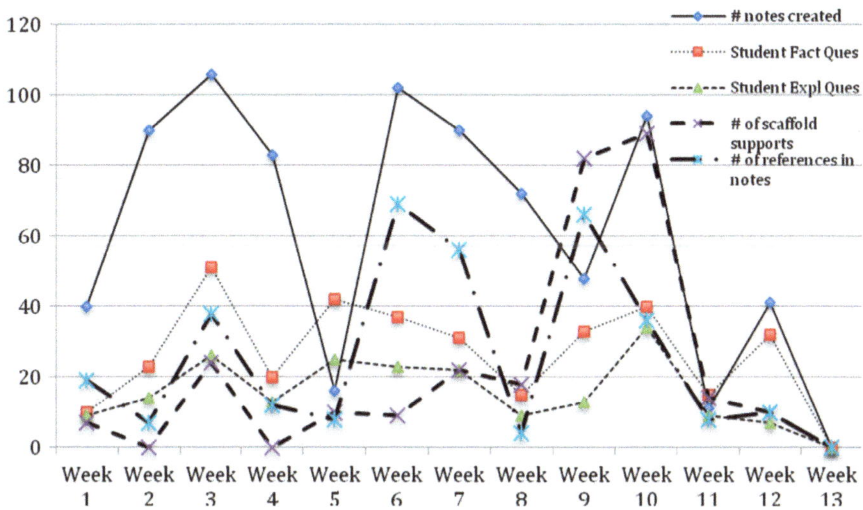

Fig. 22.4 The total number of question markers, scaffolds, and references in notes posted in each of the weekly views

weekly views as well as the total weekly counts for four of the quantitative indicators described: the numbers of fact(-seeking) question markers, explanatory (i.e., explanation-seeking) question markers, scaffolds, and references in the notes posted on each of the weekly views.

If the students were as likely to ask questions or to make use of the scaffold and referencing functions in their postings throughout the 13 weeks, the numbers of each type of question markers and the numbers of scaffolds and references used should follow the same trend as the total number of notes written over this period. The results presented in Fig. 22.4 are consistent with this expected trend for most, but not all of the 13 weeks. Week 9 again stood out as pivotal (or anomalous). From week 6–9, the total number of notes written by students was dropping steadily from 102 to 48. The total numbers of fact and explanatory question markers were also dropping from week 6 to 8. However, when the total number of notes further dropped from 72 in week 8 to 48 in week 9, the number of fact question markers rose from15 to 33 and explanatory question markers from 9 to 13. This anomaly in week 9 is even more pronounced in terms of the number of scaffolds and references found. The ratio of scaffold to notes rose from 0.25 in week 8 to 1.71 in week 9. A similar pattern is also observed in the number of references present in the notes across the weeks. These all point to some very in-depth engagements in the knowledge building process in week 9. Hence, even without reading the note contents, the quantitative indicators show week 9 to be pivotal in having a relatively low total in the number of notes written, but high in all the quality indicators related to equitable collective engagement and progressive discourse.

Pivotal Weeks from the Perspective of Students' Engagement with Core Concepts Targeted in the Curriculum

The course outline identified *progressive discourse*, defined as "the kind of interactive discussion that deepens over time and that is potentially enhanced through an online, asynchronous context," to be an important theme for the course. The course outline and readings indicate that the central theme was knowledge building. The core reading for week 1 was a chapter from Bereiter's (2002) book *Education and Mind in the Knowledge Age*. In order to track students' progress in exploring and understanding the concepts associated with knowledge building, we have identified eight keywords from the required reading to trace: idea, knowledge building (collocated as one noun-phrase, KB for short), knowledge (and not collocated with the word "building"), discourse, conceptual artifact (CA for short), belief mode, design mode, and world (as in world 1, world 2, and world 3 discussed in Bereiter's chapter). Figure 22.5 shows the total number of notes written and the frequency count of all eight keywords in the 13 weekly views.

It can be seen from Fig. 22.5 that variations in keyword frequency count were far larger than variations in the numbers of notes written per week. Furthermore, the highest frequencies of occurrence of these eight keywords were found in weeks 1 and 10. Since the eight keywords were the focal ideas introduced in the week 1 reading and the discussion was facilitated by the instructor, this high observed frequency in week 1 is not surprising. Week 3 was the first week to be facilitated by the students. This week registered the highest number of notes found in the weekly

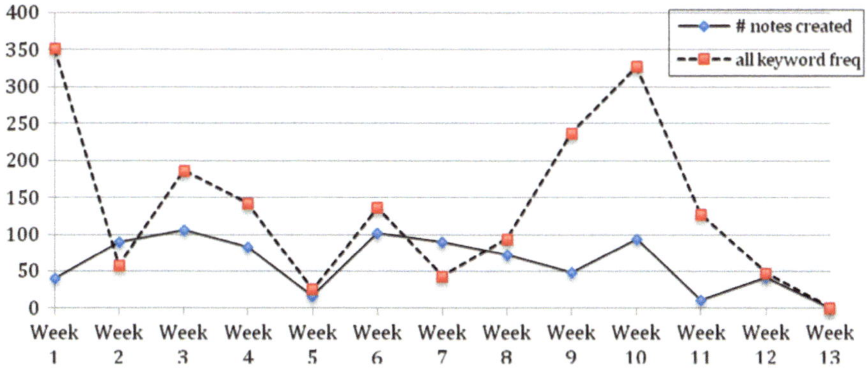

Fig. 22.5 Total number of notes written and frequency count of all eight selected keywords in the weekly views

views and the number of keywords found in this view was also the third highest, indicating that the students were wanting to clarify these key concepts. The keywords occurrence frequency gradually subsided until the ninth week, when it suddenly shot up. So again here, the ninth week was pivotal in that the students' notes reflected a strong focus on the core concepts introduced in the first session, even though the number of notes written during this week was relatively low. This strong presence of identified keywords in the ninth week triangulates well with the quantitative indicators of progressive inquiry such as thread size, thread depth, number of question markers, scaffolds, and reference notes.

Assuming that the frequency counts of the keywords reflect the intensity of students' engagement with the concepts, the frequency distribution for the eight keywords presented in Table 22.2 sheds light on the evolving engagement (and understanding) of the students with these core concepts. The data show that among the eight keywords, four were more frequently used: idea, knowledge, knowledge building and discourse. Furthermore, the frequency distribution profiles over time of the keywords were different. In the first week, six of the eight keywords were present and the frequencies for most of these were relatively high. We focus our attention only on keywords use from week 3 onwards, when the discussion was led by the students. The focal concept in week 3 was idea, followed by knowledge in week 4. As mentioned earlier, because of the missing notes placed into rise-above notes in week 5, the data for that week is not accurate. The engagement in both concepts then dropped, but the keyword *idea* picked up again prominently in week 9 together with another keyword, *discourse*. The keyword *discourse* has so far not drawn much attention until week 9, and then dropped off again after this week. However, *progressive discourse* was designated the key focus for this course, and is an important concept in the theory of knowledge building (Bereiter, Scardamalia, Cassells, & Hewitt, 1997). The change in week 10 is even more notable—the frequencies for five important keywords peaked during this week: idea, knowledge, knowledge building, design mode, and belief mode. It is important to note that

Table 22.2 Frequency of presence of the eight selected keywords in the 13 weekly views

	Week number													
Keyword	1	2	3	4	5	6	7	8	9	10	11	12	13	Total[a]
Idea	57	28	83	32	15	56	8	39	89	108	60	15	0	505
Knowledge	160	12	42	61	6	40	13	32	31	77	26	5	0	333
KB	31	2	25	24	3	16	2	2	19	55	32	6	0	184
Discourse	16	0	12	6	1	12	13	2	88	13	1	6	0	154
CA	31	0	11	0	0	3	0	0	2	2	1	0	0	19
Belief mode	0	0	0	0	0	0	0	0	0	25	1	4	0	30
Design mode	0	0	0	0	0	0	0	0	0	37	4	6	0	47
World	56	16	13	19	0	9	7	19	8	10	2	5	0	92

[a]The total listed only includes keywords for weeks 3–13 when the discussion was led by students

while knowledge was discussed quite a bit prior to this week, knowledge building was given comparatively much lower attention until week 9. Also, while the terms design mode and belief mode were introduced in the reading for week 1 (Bereiter, 2002), both did not appear in any of the discussions in the previous weeks until week 10. So from the perspective of succeeding in getting students to really engage in discussing and understanding the theory of knowledge building, week 10 is pivotal. However, just from the analyses so far, it is not clear what might have contributed to such changes and whether the success in week 10 was connected with the pivotal characteristics in week 9 as described earlier.

Tracing Advances in Conceptual Understanding: Pivotal Sentences

In order to gain a more in-depth understanding of students' cognitive and metacognitive engagement in the online discourse, it is necessary to read and understand the contents of the discourse. However, this is far easier said than done. There were about 1,000 notes found in the 13 weekly views, about 800 of which were written by students. As these notes were written by postgraduate students as their main means of learning interactions, many of the notes were very lengthy. The total word count contained in these notes amounted to more than 180K words. Hence conducting systematic, detailed, manual qualitative coding of the entire content to track conceptual growth and development as is reported in the knowledge building literature (e.g., Zhang, Scardamalia, Lamon, Messina, & Reeve, 2007) is a great challenge even for the seasoned researcher and not really feasible in the context of a practitioner. In this part of the study, we attempt to devise a heuristic to select a small number of sentences from the entire set of discourse data for in-depth reading and interpretation that will afford a reasonable understanding of the foci and pivotal conceptual development of the students throughout the course. The heuristic consisted of a two-stage process. First, we selected all the sentences from the discourse data that contained the eight KB-related keywords reported earlier.

This was still a very large dataset. As we were more interested in identifying instances where the postings involved exploration of ideas rather than information giving, we then chose only sentences which contained text markers for fact questions, explanatory questions, or elaborations. This resulted in a total selection of 353 sentences, and a total word count that is under 10K, which is about 5 % of the total size of the discourse.

With this selected set of text, we read through them to identify the sentences that contain *new* ideas that are *relevant* to the theme. Sometimes, the keywords were used as common terms and not as a technical conceptual term as in the readings and for which these keywords were selected. This was found to be particularly the case with the keyword "world," which was selected because an important theoretical underpinning of Bereiter's chapter was Popper's (1972) theory of the three worlds. However, the sentences selected with the keyword "world" were mostly about virtual world, worldwide web, the real world, etc. These sentences were irrelevant to our further analysis and were removed. We then went through the remaining sentences in chronological order to identify ideas that were new to the discussion. So if a relevant idea had been expressed earlier in this selected set of discourse and was revisited at a later point, it is identified as not new. Using these criteria, we identified 48 sentences with new, relevant ideas.

Of the 48 sentences so identified, 13 were from week 1, indicating the importance of this week in introducing key concepts into the discourse. Of these 13, 4 were written by the instructor or the researcher, and 9 by the students. An inspection of these new ideas reveal that the notes written by the "teachers" (i.e., the instructor or researcher) were drawing attention to key aspects of the knowledge building theory, for example discourse as objects for discussion and improvement, the need for knowledge sharing for knowledge building to happen. On the other hand, the students' ideas indicated attempts to make sense of the concepts gleaned from the reading and discussion, expressed in the form of questions. Some questions were asking for clarification, such as "What is progressive discourse" and whether knowledge sharing should be equated with KB. Most of the students' questions were trying to link the reading with their prior knowledge. For example, a student tried to make sense of conceptual artifact (as equated with objectified knowledge) with the explicit curriculum, and another asked about the "merit" of one's own thinking in comparison with KB as a collaborative process.

During weeks 2–4, the number of relevant new ideas found in the discourse was only 10 (1, 5 and 4 respectively for each of these 3 weeks). Further, all these ten ideas were raised by the students themselves. In week 5, four new ideas were introduced, two of which were by the instructor. This was followed by a relatively "quiet period with only 2, 1 and 1 new ideas during weeks 6–8. Of these four again two were from the "teachers." Then came another "pivotal week" in terms of productivity of new ideas. Altogether nine new ideas were introduced in this ninth week, and similar to the first week, four of these were from the teachers. There were seven new ideas emerging during the tenth week, only one of which was from the teacher. So, in terms of new ideas being put forward by students, the tenth week was even more productive and "pivotal." Subsequent to this, only one further new idea was found in the selected sentences, which belonged to the 11th week.

Table 22.3 A chronological listing of the relevant key ideas related to the theme of idea improvement in the sentences selected on the basis of keywords and discourse markers

	Author[a]				
Week	Teacher	Student	Note id	Keyword category	Key ideas in related note
1	Res.		244	Idea	How do we know if an idea is improved?
3		Belinda	615	Idea	How do you define idea improvement?
3		Adam	667	CA	CA [conceptual artifact] appears when we say this is not my/his/her idea, let's now try to make it [idea] work better
6		Adam	1306	Discourse + knowledge	How to create a progressively deepening discourse that participants will recognize the communal knowledge improvement and advancement?
9		Belinda	1678	Knowledge	How do you evaluate collective knowledge advancement? What does it look like?
9		Paul	1688	CA + idea + discourse	Conceptual artifacts are created intentionally and purposefully through discourse for idea improvement—what do we need this idea for?
9	Instr.		1746	Idea	How to evaluate the collective advancement of an idea? How do we implement idea improvement or collective cognitive responsibility?
10		Adam	2064	KB + idea + knowledge	Can KB be more holistic so as to result in building different kinds of knowledge, i.e., why is holistic KB more effective for idea improvement and knowledge advancement
10		Belinda	2224	Design mode + idea	The essence of the design mode is idea improvement—what is this idea good for?

[a]The students' names used are fictitious

Since we are only examining new ideas found in the notes selected on the basis that these contain the identified keywords and question/elaboration markers, this crude selection and analysis process does not provide us with information about whether any of these ideas were followed up or further developed outside of this selected set of text. However, if we examine the contents of the 48 sentences, a prominent theme emerged around the concept of idea improvement in knowledge building. The development of this theme can be seen through 10 of the 48 sentences, which are listed in Table 22.3.

The theme began with a note written by the Researcher (note id 244) asking, "How do we know if an idea is improved?" This theme was picked up in week 3 by Belinda (note id 615) who asked, "How do you define idea improvement?" and Adam (note id 667) who explained that an idea becomes a conceptual artifact when people try to make it work better. So while the researcher's question was about how one knows that there is idea improvement, Belinda problematizes the concept itself and ask for the definition. Adam took a different approach and provided a pragmatic operational definition of idea improvement. So here we can see a chronological progression in the exploration of the concept of idea improvement, which is in fact the basis for their selection. Within this sequence of ten questions involving nine different notes, we can identify three points to be pivotal progressions in contributing to more significant advances in the discourse (and understanding) through connecting key concepts.

Pivotal Point 1

The fourth note in this sequence was posted in week 6, again by Adam (note id 1306). Two important new concepts were introduced: *communal* knowledge improvement and *progressively deepening discourse*. He was explicitly referring to knowledge advancement as a communal event, that this was to be achieved through a progressive discourse, and his problem was how people engaged in the discourse would recognize the improvement. This is a significant advancement in understanding the theory of knowledge building and can be considered a pivotal point in the students' learning. This "breakthrough" provided a conceptual platform for the students' further progressive discourse on the theme of idea improvement. The next note on the list was posted by Belinda in week 9 (note id 1678), in which she raised the question of how people evaluate collective knowledge advancement. This is an extension of Adam's problem of recognition (meaning to identify whether there is), and evaluation is more refined and formal than recognition. Her use of the term "collective" shows that she was distinguishing it from individual knowledge advancement. Her second question indicates that she was still puzzled and unclear about what collective knowledge advancement looks like.

Pivotal Point 2

The next pivotal point was a remarkable note by Paul (note id 1688), which brought together three of the key concepts in one statement: *Conceptual artifacts are created intentionally and purposefully through discourse for idea improvement.* He further elaborated that to achieve idea improvement one has to focus clearly on what the idea is needed for. Note here that these three concepts were introduced in week 1, and were revisited here with the improved understanding. However, this statement did not make reference to the collective aspect of idea improvement.

Pivotal Point 3

At this juncture, the Instructor posted a note (note id 1746) that introduced two new ideas through two questions: how is collective advancement of an idea to be evaluated and how idea improvement and collective cognitive responsibility are to be implemented? The first question connects the advancements made by the two previous pivotal points, connecting *idea improvement* with *collective knowledge advancement*. It echoes the question raised by the Researcher in note 244, but now explicitly articulating that idea improvement is to be achieved through the "collective." The second question introduced the concept of collective cognitive responsibility, and identifying this as an implementation condition for idea improvement. Bringing these concepts into focus apparently led to further enhancement in understanding as reflected in Belinda's note (note id 2224) posted in week 10 in which she categorically declared that the essence of working in the design mode is idea improvement.

Note 2064 posted by Adam in week 10 raised a question about a new idea introduced in earlier discussions—"holistic" knowledge building, meaning more varied ways of knowledge building beyond carefully orchestrated rational discussions and considerations as described by Bereiter. It is difficult to say whether this note should be considered pivotal as we have not yet identified further new ideas that emerged to further extend this concept.

Collective Inquiry and Individual Knowledge Advancement

In the previous section, we reported on students' evolving understanding about knowledge building as reflected by the notes selected from the entire discourse. However, it is also noticeable that the seven student-contributed new ideas were from three students only. This raises the question of whether the progressive understanding pertains only to these three students, and whether the pivotal points were only pivotal for these three students. We do not have definitive data about individual students' understanding and can only interpret what the students have written. On the other hand, we have evidence that these pivotal points in students' understanding about knowledge building coincided with periods of high activity and engagement from a high proportion of students. Figure 22.6 displays three line graphs representing the extent to which students were engaged with the exploration of new ideas.

The upper line shows for each week the number of students who wrote notes that contained at least one of the eight identified keywords in a sentence that had question or elaboration markers. This line hence reflects the proportion of students in the class who were engaged in exploring ideas related to the eight keywords during each week. A second line plots the number of authors who contributed the new ideas listed in Table 22.2. Comparing these two lines, we see that in general, more new ideas were generated when more students were engaged in asking questions. The third line plots the number of knowledge building (KB)-related new ideas reported in Table 22.2 per week. Again, there is a close match of the profile of this line with

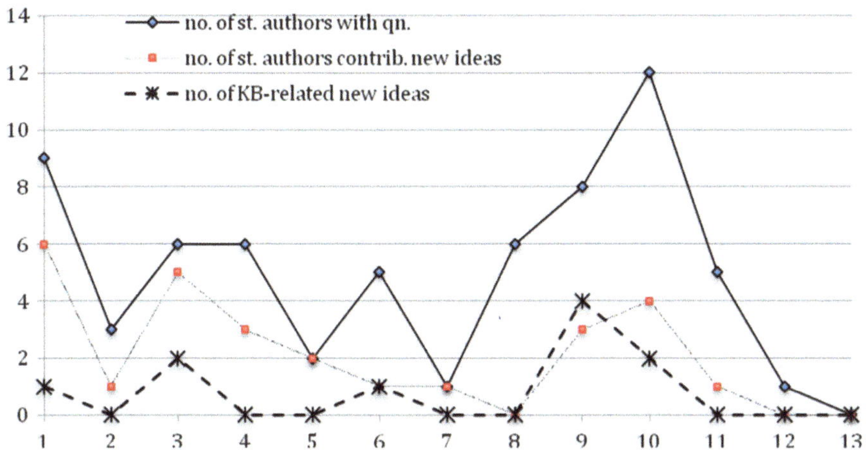

Fig. 22.6 A comparison of indicators of student author engagement with the emergence of KB-related new ideas over the 13 weeks of the course

the other two lines, indicating that even though the KB-related new ideas picked out using the current methodology were contributed by three of the students, the emergence of these ideas benefitted from the active engagement of the wider group of students.

Discussion: Pivotal Moments in Collective Knowledge Advancement

This paper sets out to explore pivotal moments in collective knowledge advancement in a class of master's students studying a totally online course on Educational Applications of Computer Mediated Communications. The choice of examining collective knowledge advancement as the focus for identifying pivotal moments is underpinned by a strong theoretical alignment with and pedagogical commitment to knowledge building as proposed by Scardamalia and Bereiter (2003a). Our approach to the identification of "pivotal moments" is driven by a strong pedagogical motivation—what kind of information provided to teachers "on the fly" could possibly help them in understanding students' overall engagement and specific conceptual advancement when a knowledge building approach is adopted. In our analysis, we have adopted two different definitions for the concept of pivotal. The first refers to time periods (matched to the curriculum design) in the discourse that may be particularly productive or otherwise. As the course discussion data was organized around weekly themes, our unit of analysis was a week. Weekly statistics on participation (medians and dispersions in the number of notes written and notes read) and on discourse structure (thread size, thread depth, the number of scaffolds and references used) can help teachers in monitoring the overall levels of engagement in the discussion, and to decide whether intervention and/or changes in the curriculum design is necessary. Multiple triangulations of the pivotal weeks using different statistics consistently demonstrate that read statistics and statistics related to discourse structure are more indicative of productive knowledge building discourse.

The second meaning of pivotal refers to breakthroughs in students' understanding of the key curriculum concepts, which requires semantic analysis of the discourse content. A major goal of the course was to help students understand knowledge building as collective idea improvement through progressive discourse, and is the focus of our semantic analysis. The unit of analysis used is a sentence within a posting (or a note). Hence there can be more than one "pivotal moment" within the same note if there are more than one conceptual advancement articulated within the same note. This part of the analysis began with selecting sentences that contained keywords core to the theory of knowledge building. Eight important keywords were identified: idea, knowledge, knowledge building, discourse, conceptual artifact, belief mode, design made and world. A further heuristic was employed to further narrow down the number of sentences selected for semantic analysis by choosing only those containing discourse markers indicative of fact, explanatory or elaboration questions. This resulted in 353 sentences containing fewer than 10,000 words, which is about 5 % of the entire corpus of more than 180K words.

The final semantic content analysis was a two-step process. First, the sentences were read carefully in chronological order to identify all instances where the semantic meaning of the keywords were relevant to knowledge building and had not been expressed earlier in this set of selected corpus. This resulted in 48 sentences, 10 of which were related to the concept of idea improvement in knowledge building. In one sense, these ten sentences can be taken as ten pivotal points in students' progressive understanding of this theme. Upon closer examination of these ten sentences, we further identified three pivotal points that made important connections among related concepts in this chain of progressive discourse. The first pivotal point problematized the *collective perspective* of understanding idea improvement, and connecting it with communal knowledge advancement. The second pivotal point connected idea improvement with *intentional improvement of conceptual artifacts through discourse*. The third pivotal point connected idea improvement and collective knowledge advancement with the concept of collective cognitive responsibility. The first pivotal point was located in week 6 and the other two were in week 9. These 2 weeks were identified as pivotal in terms of quantitative indicators of productive discourse. These findings indicate that the methods used in this study for identification of pivotal weeks and pivotal advances in conceptual understanding triangulates consistently and so can be used as a starting point for the construction of a teacher "dashboard" to support the understanding, monitoring and facilitation of students' knowledge building engagement and learning through CSCL.

References

Bereiter, C. (2002). *Education and mind in the knowledge age*. Hillsdale, NJ: Lawrence Erlbaum.

Bereiter, C., Scardamalia, M., Cassells, C., & Hewitt, J. (1997). Postmodernism, knowledge building, and elementary science. *Elementary School Journal, 97*(4), 329–340.

Dennen, V. P. (2008). Pedagogical lurking: Student engagement in non-posting discussion behavior. *Computers in Human Behavior, 24*(4), 1624–1633.

Hakkarainen, K., Lipponen, L., & Järvelä, S. (2002). Epistemology of inquiry and computer-supported collaborative learning. In T. Koschmann, R. Hall, & N. Miyake (Eds.), *CSCL 2: Carrying forward the conversation* (pp. 129–156). Hillsdale, NJ: Lawrence Erlbaum Associates.

Law, N. (2005). Assessing learning outcomes in CSCL settings. In T.-W. Chan, T. Koschmann, & D. Suthers (Eds.), *Proceedings of the Computer Supported Collaborative Learning Conference (CSCL) 2005* (pp. 373–377). Taipei, Taiwan: Lawrence Erlbaum Associates.

Law, N., & Wong, E. (2003). Developmental trajectory in knowledge building: An investigation. In B. Wasson, S. Ludvigsen, & U. Hoppe (Eds.), *Designing for change in networked learning environments* (pp. 57–66). Dordrecht, The Netherlands: Kluwer.

Lee, E. Y. C., Chan, C. K. K., & van Aalst, J. (2006). Students assessing their own collaborative knowledge building. *International Journal of Computer-Supported Collaborative Learning, 1*, 277–307.

Popper, K. R. (1972). *Objective knowledge: An evolutionary approach*. New York, NY: Oxford University Press.

Scardamalia, M. (2002). Collective cognitive responsibility for the advancement of knowledge. In B. Smith (Ed.), *Liberal education in a knowledge society* (pp. 67–98). Chicago, IL: Open Court.

Scardamalia, M., & Bereiter, C. (2003). Knowledge building. In J. W. Guthrie (Ed.), *Encyclopedia of education* (2nd ed.). New York, NY: Macmillan Reference.

Zhang, J., Scardamalia, M., Lamon, M., Messina, R., & Reeve, R. (2007). Socio-cognitive dynamics of knowledge building in the work of 9- and 10-year-olds. *Educational Technology Research and Development, 55*, 117–145.

Chapter 23
Statistical Discourse Analysis of an Online Discussion: Cognition and Social Metacognition

Ming Ming Chiu

Introduction

This study considers how to revise a statistical method designed for face-to-face talk, statistical discourse analysis (SDA), to apply it to participant-coded online discussions (Fujita, Chap. 20, this volume). Unlike the linear sequence of turns of talk however, asynchronous online messages often branch out into separate threads. Applying a successful, revised SDA to online discussion can capitalize on participants' self-coding of messages to enable analyses of large databases and extend online discussion research beyond *messages' aggregate attributes* (e.g., Gress, Fior, Hadwin, & Winne, 2010) to *relationships among messages*. As earlier turns of talk affect later turns of talk, earlier online messages might influence later messages (Chiu, 2000a; Chiu, 2001; Jeong, 2006). Specifically, I examine how cognitive and social metacognitive aspects of earlier messages affect ideas and explanations in later messages. Whereas individual metacognition is monitoring and control of one's own knowledge, emotions, and actions (Hacker & Bol, 2004), *social metacognition* is defined as group members' monitoring and control of one another's knowledge, emotions, and actions (Chiu & Kuo, 2009). By understanding how cognitive and social metacognitive components of recent online messages create a micro-time context that aid or hinder students' ideas and explanations, educators can help students engage in beneficial online processes to learn more.

This study contributes to the research literature in two ways. First, I introduce a new method to model branches of online messages across multiple topics. Second, this method tests how explanatory variables at multiple levels (individual characteristics, cognitive and social metacognitive aspects of messages) influenced 1,330

I appreciate the generous sharing of data by Nobuku Fujita, the research assistance of Choi Yik Ting, and comments from participants in the Alpine Rendez-Vous workshop.

M.M. Chiu (✉)
University at Buffalo—State University of New York, Buffalo, NY, USA

D.D. Suthers et al. (eds.), *Productive Multivocality in the Analysis of Group Interactions*, Computer-Supported Collaborative Learning Series 16,
DOI 10.1007/978-1-4614-8960-3_23, © Springer Science+Business Media New York 2013

asynchronous online messages during a 13-week educational technology course. By examining students' asynchronous online messages, researchers can build a more comprehensive understanding of students' online processes and their influences to develop appropriate teacher interventions and computer environments.

Theoretical Framework

Unlike students talking face-to-face, those in asynchronous online discussions can participate at different places and times, a valuable resource for improving their learning (Dubrovsky, Kiesler, & Sethna, 1991; Harasim, 1993; Tallent-Runnels et al., 2006). As students writing asynchronous, online messages have more time than those in face-to-face conversations to gather information, contemplate ideas, and evaluate claims before responding, they often display higher levels of decision making, problem solving and writing (Hara, Bonk, & Angeli, 2000; Luppicini, 2007; Tallent-Runnels et al., 2006). During higher quality discussions, students explain and synthesize ideas more often, so they typically learn more (Clark & Sampson, 2008; Glassner, Weinstoc, & Neuman, 2005).

A natural follow-up question is whether students' sequences of online messages affect their content. Researchers have shown that online discussions can begin with students sharing ideas, recognizing conflicts, and then resolving them by synthesizing ideas (Gunawardena, Lowe, & Anderson, 1997; Howe, 2009). In addition to expressing ideas (cognition), students also monitor and control one another's ideas and actions through questions, evaluations (agree vs. disagree), and summaries (social metacognition).

Many researchers advocate using clear, formal concepts rather than imprecise, informal concepts (also known as preconceptions or intuitions; e.g., Piaget, 1985; Vygotsky, 1986). However, informal concepts may not necessarily compete with formal concepts; instead, students might initially activate familiar, informal concepts before activating less familiar, formal concepts (Chiu, 1996). During a discussion, a student's comments (e.g., a key word) might spark another student to activate related concepts in his or her semantic network and propose a new idea (Nijstad, Diehl, & Stroebe, 2003). Consider the following example. Ada and Bill are posting messages about whether teachers should allow students to use the Internet during class lessons.

Ada: I think students can use the Internet to access useful pages, such as …
Bill: Yes, they can use the mathematical tools on these pages to solve problems.

When students share ideas, they implicitly recognize and agree with one another's ideas. When other students disagree or do not understand these ideas, they can ask questions to get facts, explanations, or examples of how to use these ideas (Hakkarainen, 2003). Such questions can also serve as polite disagreements.

Ada: I'm not clear on what you mean by Internet tools? How could you use them?

Students can respond with facts, explanations or uses (Lu, Chiu, & Law, 2011). Ideally, the explanations incorporate facts into theoretical models with specific applications.

Bill: Internet tools are computer programs on a webpage that everyone can access. For example, anyone can graph a line by typing its equation at this website …

Even in the absence of questions, people often support their ideas with explanations, especially when they anticipate disagreements (Chiu & Khoo, 2003; Clark & Sampson, 2008). Explanations also often foster further explanation by others (Chiu, 2008b).

As students share more ideas, they are more likely to disagree with at least one of their groupmate's ideas (Jeong, 2003). Disagreements can include identifying areas of disagreement, their sources, bases, or their extents.

Dan: While Internet tools can be useful, they can also be a crutch …

In response, other students might ask questions (as above) or propose different opinions along with facts, anecdotes, and explanations (Clark & Sampson, 2008).

Ada: That's possible, but it needn't be a crutch if students have to plot points …

In an advanced discussion, students try to reconcile different views into an integrated summary by identifying areas of agreement, clarifying meanings, proposing and negotiating compromises and syntheses (Wise & Chiu, 2011).

Fay: I think we can all agree that Internet tools can be useful in these six ways: … However, we need to be careful to …

Students summarizing ideas often show higher levels of cognition, and these summaries often elevate the levels of cognition in the subsequent time period, suggesting that summaries are *pivotal messages* that radically change the interaction (Wise & Chiu, 2011).

Table 23.1 summarizes the hypotheses. To reduce omitted variable bias, the explanatory model controls for several individual variables (such as gender; for a full list of control variables, see analysis section below). For example, earlier studies showed that male students were more likely than female students to make claims, argue, elaborate, explain, and critique others (Lu et al., 2011).

Method

In this study, I examine relationships among asynchronous discussion messages posted by students in a 13-week online graduate educational technology course delivered using Web-Knowledge Forum. For a description of the data, see Fujita (this volume, Chap. 20).

Table 23.1 Hypotheses 1–4 regarding the effects of classroom problem solving process on the outcome variables correct contributions (symbols in parentheses indicate expected relationship with the outcome variables: positive and supported [+], hypothesized but not supported [+])

Explanatory variable	→ Dependent variable			
Cognition	H-1 New fact	H-2 Ask for Explanation	H-3 Theorize	H-4 Summarize
Opinion	+	+	+	+
Anecdote	+	+	+	+
Elaboration	+	+	+	+
New fact	+	+	+	+
Theorize			+	+
Social metacognition				
Ask about use	+	+	+	+
Ask for explanation	+	+	+	+
Different opinion	+	+	+	+
Summarize				
Any of the above variables (vs. none)			+	+

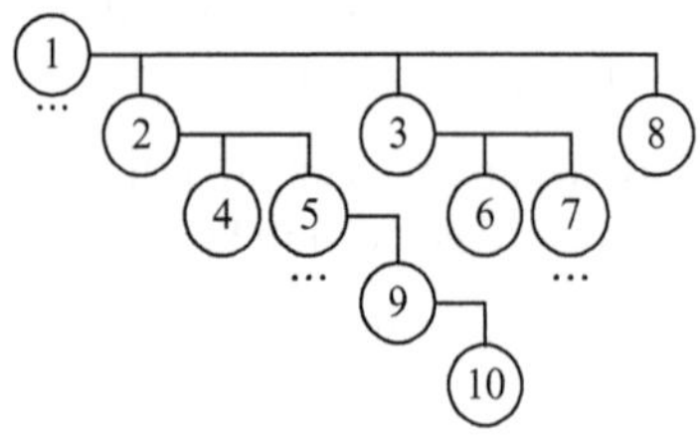

Fig. 23.1 The tree structure of relationships between a problem and its reply messages

Data

As SDA was designed for turns of talk, it required revision to analyze branches of messages.

Unlike a linear, face-to-face conversation in which one conversation turn typically follows the one before it, an asynchronous message in an online discussion might follow a message written much earlier (*branches of messages*), forming a message tree. See Fig. 23.1 for an example of relationships among 10 messages. The number "1" denotes the initial message; "2" through "10" indicate 9 reply messages in the order of time.

The messages occurred along five discussion threads: (a) 1 → 2 → 4, (b) 1 → 2 → 5 → 9 → 10, (c) 1 → 3 → 6, (d) 1 → 3 → 7, and (e) 1 → 8. Messages in each thread were ordered by time, but they were not necessarily consecutive. In thread (c) for example, message #3 followed message #1 (not #2) and message #6 followed message #3 (not #5). By storing each message's previous message on its thread in a variable, I can capture the structure of the tree of messages. Then, I change my application of SDA to examine the previous message on a thread, not the most recent message.

Analysis

This section specifies the assumptions underlying the analysis, its purpose, units of interaction, representations of the data, and the analytic manipulations.

Assumptions Underlying the Analysis

Theoretical assumptions. Statistical discourse analysis (SDA, Chiu & Khoo, 2005) has several theoretical assumptions. First, as with any statistics (e.g., count, mean, standard deviation), SDA assumes that instances of a category (e.g., summarize) with the same value (e.g., is vs. is not [coded as 1 vs. 0]) are sufficiently similar to be treated as equivalent for the purpose of this analysis.

This specific study has at least four additional theoretical assumptions. Second, *participant-coded* message characteristics are sufficiently similar to be treated as equivalent for the purpose of this analysis. Third, aspects of recent messages, participating individuals and time constitute a micro-context in which future messages emerge. Fourth, aspects of recent messages, their authors and the time period can influence later messages. Fifth, residuals reflect attributes related to the dependent variables that are not specified in the theoretical model and not correlated with the explanatory variables.

Methodological assumptions. Like other regressions, SDA assumes a linear combination of explanatory variables (Nonlinear aspects can be modeled as non-linear functions of variables [e.g., age squared] or interactions among variables [new fact x opinion].) SDA also requires independent residuals and a modest, minimum sample size.

Purpose of Analysis

This analysis has two purposes. First, the revised SDA shows how to model trees of messages rather than linear turns. Second, the revised SDA tests whether variables are linked to greater or reduced likelihoods of cognitive (new information, theory) and social metacognitive (ask for explanation, summary) characteristics of each message.

Units of Interaction That are Taken as Basic in the Analysis

While the unit of analysis is a message, the unit of interaction is a sequence of one type of message following another. The interaction as a whole is characterized by the probabilities of these sequences, which is modeled with SDA.

Table 23.2 Statistical Discourse Analysis strategies to address each analytic difficulty

Analytic difficulty	Statistical Discourse Analysis strategy
Data set	
Missing data (0110??10)	Markov Chain Monte Carlo multiple imputation, Peugh and Enders (2004)
Branches of messages (Λ)	Identify preceding message to replicate tree structure
Topics differ ($T_1 \neq T_2$)	Multilevel analysis (aka Hierarchical linear modeling, Bryk and Raudenbush (1992), Goldstein (1995))
Serial correlation ($t_6 \sim t_7$)	I^2 index of Q-statistics, Huedo-Medina, Sanchez-Meca, Marin-Martinez and Botella (2006)
Dependent variables	
Binary (yes/no)	Logit, Kennedy (2008)
Infrequent	Logit bias estimator, King and Zeng (2001)
Multiple (Y_1, $\mathbf{Y}_{2,\ldots}$)	Multivariate outcome models, Goldstein (1995)
Explanatory variables	
Sub-threads of messages (X_{t-2}, $X_{t-1} \rightarrow Y_0$)	Vector Auto-Regression VAR, Kennedy (2008)
Interactions across levels (e.g., Topic X Message)	Random effects model, Goldstein (1995)
False positives (Type I errors)	Two-stage linear step-up procedure, Benjamini, Krieger, and Yekutieli (2006)
Robustness	Single outcome, multilevel models for each outcome variable
	Testing on subsets of the data
	Testing on unimputed data

Representations of Data and Analytic Interpretations

I used the standard representations of a database table, a summary statistics table, and a path diagram. The database table initially had one message per row. Next, I added columns (variables) for coding whether each attribute occurred in each message. Then, I performed statistical analyses to test relationships across this table of vectors, resulting in a summary statistics table and a table of results of regression models (via SDA). To aid reader comprehension, I capitalize on readers' understanding of spatial relationships to convert the regression results into a path diagram.

Analytic Manipulations

Testing the above hypotheses requires addressing analytic difficulties involving the data set (missing data, branches of messages, topic differences, serial correlation), dependent variables (binary, infrequent, multiple), and explanatory variables (sub-threads of messages, cross-level interactions, indirect effects, false positives) see Table 23.2.

To address these difficulties, a simplified version of statistical discourse analysis (SDA) is used (Chiu, 2008a; Chiu & Khoo, 2005). First, missing data can reduce estimation efficiency, complicate data analyses, and bias results. Computer simulations showed that estimating the missing data with Markov Chain Monte Carlo multiple imputation (MCMC-MI) addressed these missing data issues more effectively than deletion, mean substitution, or simple imputation (Peugh & Enders, 2004). Second, to capture the tree structure of branches of messages, a variable identifies and stores the message to which the current message responds along a thread (in Fig. 23.1 for example, message 4 responds to message 2 [not message 3]), thereby enabling identification of any ordinal predecessor of any message along a thread. Third, messages within the same topic (especially those near one another) likely resemble one another more than messages across topics, so they are likely not independent. Modeling messages across topics requires multilevel analysis (Goldstein, 1995; also known as hierarchical linear modeling, Bryk & Raudenbush, 1992). Fourth, resemblances among adjacent messages can result in serial correlation of errors if not modeled properly (Kennedy, 2008). An I^2 index of Q-statistics can test messages across many topics simultaneously for serial correlation, which can be modeled if needed (Goldstein, Healy, & Rasbash, 1994; Huedo-Medina et al., 2006; Ljung & Box, 1979).

The four dependent variables were binary and infrequent (*new fact*, *ask for explanation*, *theory*, and *summarize*). To model a binary dependent variable, Logit or Probit is used. When dependent variables occur far less than 50 % of the time, standard regressions will yield biased results. To remove this bias, King and Zeng's (2001) bias estimator is used. Multiple outcomes can have correlated residuals that underestimate standard errors. To model several dependent variables properly, a multivariate outcome analysis is needed (Goldstein, 1995).

The explanatory variables can include sub-threads of messages, have interactions across levels, yield indirect effects, show false positives, or yield different results during robustness tests. Sub-threads of explanatory variables are modeled with vector auto-regression (VAR, Kennedy, 2008). To model interactions across levels, multilevel random effects are used (Goldstein, 1995). As single-level mediation tests on nested data can bias results downward, multilevel M-tests test for indirect, multilevel mediation effects, in this case, messages within topics (MacKinnon, Lockwood & Williams, 2004). Testing many hypotheses of potential explanatory variables increases the likelihood of a false positive (Type I error). To control for the false discovery rate (FDR), the two-stage linear step-up procedure was used, as it outperformed 13 other methods in computer simulations (Benjamini et al., 2006). To test the robustness of the results, three variations of the core model can be used. First, a single outcome, multilevel model can be run for each dependent variable. Second, subsets of the data (e.g., halves) can be run separately to test the consistency of the results for each subset. Third, the analyses can be repeated with the original data set.

Analysis Procedure

After MCMC-MI estimation of the missing data to yield a complete data set (Peugh & Enders, 2004), the message to which each message responded was identified to store the data set's tree structure. Then, four process variables in students' messages (*new fact*, *ask for explanation*, *theory*, and *summarize*) were simultaneously modeled as follows (Chiu & Khoo, 2005).

$$\mathbf{Process}_{ynt} = \beta_y + \mathbf{e}_{ynt} + \mathbf{f}_{yt} \tag{23.1}$$

For $\mathbf{Process}_{ynt}$ (the process variable *y* [e.g., *theorize*] for message *n* in topic *t*), β_y is the grand mean intercept. The unexplained message-level component (or *residual*) is $\mathbf{e}_{nt}$, and the unexplained topic-level component is $\mathbf{f}_t$. As analyzing rare events (these processes occurred in less than 10 % of all messages) with Logit/Probit regressions can bias regression coefficient estimates, King and Zeng's (2001) bias estimator was used to adjust them.

First, a vector of student demographic variables was entered: *male* and *young* (**Demographics**). Each set of predictors was tested for significance with a nested hypothesis test (χ^2 log likelihood, Kennedy, 2008).

$$\begin{aligned}\mathbf{Process}_{ynt} = {} & \beta_y + \mathbf{e}_{ynt} + \mathbf{f}_{yt} + \beta_{ydt}\mathbf{Demographics}_{ynt} + \beta_{yst}\mathbf{Schooling}_{ynt} \\ & + \beta_{yjt}\mathbf{Job}_{ynt} + \beta_{yxt}\mathbf{Experience}_{ynt} + \beta_{ypt}\mathbf{Previous\_One}_{ynt} \\ & + \beta_{ypt}\mathbf{Previous\_Two}_{ynt} \ldots\end{aligned} \tag{23.2}$$

Next, schooling variables were entered: *doctoral* student, *Master's of Education* student, *Master's of Arts* student, and *part-time* student (**Schooling**). Then, students' job variables were entered: *teacher*, *post-secondary teacher*, and *technology* (**Job**). Next, students' experience variables were entered: *Knowledge Forum experience* and *number of past online courses* (**Experience**).

Then, aspects of the previous message were entered: *ask for explanation* (-1), *ask about use* (-1), *new fact* (-1), *theory* (-1), *summarize* (-1), *different opinion* (-1), *elaboration* (-1), *anecdote* (-1), *opinion* (-1), and *any of these supportive* processes (-1) (**Previous_One**). Next, the above aspects of the message two responses ago along the same thread (-2) were entered (**Previous_Two**). Then, those of the message three responses ago along the same thread (-3) were entered, and so on until none of the explanatory variables in a message along a thread were significant.

Structural variables (**Demographics**, **Schooling**, **Job**, **Experience**) might show moderation effects, so a random effects model was used. If the regression coefficients of an explanatory variable in the **Previous** message (e.g., evidence; $\beta_{ypt} = \beta_{y0t} + f_{y0j}$) differed significantly ($f_{y0j} \neq 0$?), then an interaction effect across levels might occur and tested accordingly with multilevel random effects cross-level interaction variables (Goldstein, 1995).

The multilevel M-test captures indirect multilevel, mediation effects (within and across levels, MacKinnon, Lockwood & Williams, 2004). For significant mediators, the percentage change is $1-(b'/b)$, where b' and b are the regression coefficients of the explanatory variable, with and without the mediator in the model, respectively. The odds ratio of each variable's total effect (E = direct effect plus indirect effect) was reported as the increase or decrease (+E% or –E%) in the dependent variable

(Kennedy, 2008). As percent increase is not linearly related to standard deviation, scaling is not warranted.

An alpha level of .05 was used. To control for the false discovery rate, the two-stage linear step-up procedure was used (Benjamini et al., 2006). An I^2 index of Q-statistics tested messages across all topics simultaneously for serial correlation, which was modeled if needed (Goldstein et al., 1994; Huedo-Medina et al., 2006; Ljung & Box, 1979).

Sample Size

SDA has modest sample size requirements. Green (1991) proposed the following heuristic sample size, N, for a multiple regression with M explanatory variables and an expected explained variance R^2 of the outcome variable:

$$N > \left(\left\{8 \times \left[\left(1 - R^2\right) / R^2\right]\right\} + M\right) - 1 \tag{23.3}$$

For a large model of 20 explanatory variables with a small expected R^2 of 0.10, the required sample size is 91 messages: $= 8 \times (1 - 0.10)/0.10 + 20 - 1$. Less data are needed for a larger expected R^2 or for smaller models. Note that statistical power must be computed at each level of analysis (message, topic, group, class, school … country). With 1,330 messages, statistical power exceeded 0.95 for an effect size of 0.1 at the message level. At the individual level, the sample size (17) is very small, so any individual results must be interpreted cautiously.

Results

Summary Statistics

There were 1,330 messages by 17 students on 13 topics in the study. Students who were older, enrolled in master's of arts programs, were part-time students, were not teachers, worked in technology fields, or had Knowledge Forum (KF) experience posted more messages on average than other students (*older*: $m = 47$ vs. other $m = 37$ messages; *master's of arts*: 64 vs. 36; *part-time*: 47 vs. 27; *not teachers*: 55 vs. 36; *technology*: 54 vs. 39; *KF*: 44 vs. 32). Students posted few messages with the following attributes (see Table 23.3, panel B): *summarize* (3 %), *theory* (4 %), *ask for explanation* (9 %), *new fact* (1 %), *ask about use* (2 %), *different opinion* (1 %), *elaboration* (2 %), *opinion* (5 %), *example* (1 %). Indeed, most messages (83 %) lacked any of these attributes. As some messages included more than one of these attributes, these percentages do not sum up to 100 %.

Table 23.3 Summary statistics at the individual level (panel A) and message level (panel B)

(A) Individual Variables ($N=17$)	Mean	Description
Man	0.28	28 % of students were men. 72 % were women
Young (under 35 years of age)	0.50	Half of the students were under 35 years of age
Doctorate	0.22	22 % had either a PhD or an EdD
Master's Art	0.22	22 % had a Master's of Art (MA) degree
Master's Education	0.50	50 % had a Master's in Education degree
Part-time Student	0.78	78 % were part-time students. 22 % were full-time
Teacher	0.67	67 % worked as teachers
Post-Secondary Teacher	0.28	28 % taught at the post-secondary level
Technology	0.22	22 % worked in the technology industry
Knowledge Forum (KF)	0.83	83 % had used Knowledge Forum before
Past Online Courses	2.89	Students took an average of 2.89 online courses. SD=2.74; Min=0; Max=8
(B) Message Variable ($N=1,330$)	**Mean**	**Description**
Man	0.26	Men posted 26 % of all messages. Women posted 74 %
Young (under 35)	0.44	Young students posted 44 % of all messages
Doctorate	0.20	Those with doctorates posted 20 % of all messages
Master's Art	0.33	MAs posted 33 % of all messages
Master's Education	0.47	EdMs posted 47 % of all messages
Full time Student	0.14	Full-time students posted 14 % of all messages
Teacher	0.57	Teachers posted 57 % of all messages
Post-Secondary Teacher	0.23	Post-Secondary Teacher posted 23 % of all messages
Technology	0.28	Those working in technology posted 28 % of all messages
Knowledge Forum (KF)	0.87	Those who used KF before posted 87 % of all messages
Past online courses	3.35	SD=2.21; Min=0; Max=8. The average number of author's online courses, weighted by number of messages
Summarize	0.03	3 % of the messages had summaries. 97 % did not
Ask for explanation	0.09	9 % of the messages had a request for explanation
Ask about use	0.02	2 % of the messages had a request for a use
New fact	0.01	1 % of the messages had at least one new fact
Theorize	0.04	4 % of the messages had theorizing
Different opinion	0.01	1 % of the messages had a different opinion than others
Elaboration	0.02	2 % of the messages had an elaboration of another's idea
Reason	0.01	1 % of the messages gave a reason to support an idea
Anecdote	0.01	1 % of the messages gave evidence to support an idea
Opinion	0.05	5 % of the messages gave a new opinion
Example	0.01	1 % of the messages gave an example of an idea
Any above discussion process	0.17	17 % of the messages had at least one of the above features

MESSAGE: Except for past online courses, all variables have possible values of 0 or 1

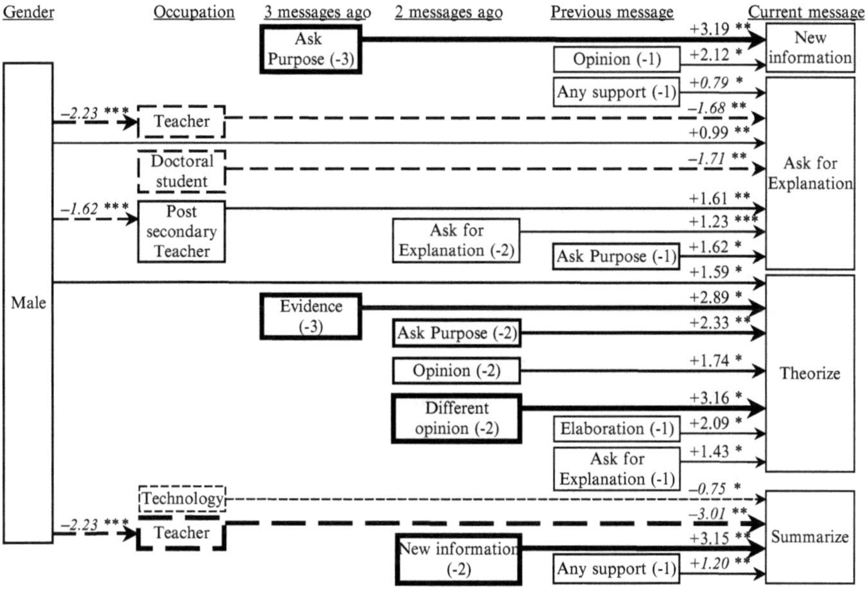

Fig. 23.2 Path diagram of *Ask for explanation*, *Theorize*, *New information*, and *Summarize*. *Solid lines* indicate positive links. *Dashed lines* indicate negative links. *Thicker lines* indicate larger links. *$p<0.05$, **$p<0.01$, ***$p<0.001$

Explanatory Model

As none of the topic-level (level 2) variance components were significant, a single-level (message level) analysis was sufficient. All results discussed below describe first entry into the regression, controlling for all previously included variables. Ancillary regressions and statistical tests are available upon request.

H-1: New information. The attributes of previous messages were linked to a new fact in current message. After an *opinion*, new information was 7 % more likely in the next message. After a *question about use* (-3) three messages ago, new information was 10 % more likely. Together, these explanatory variables accounted for about 26 % of the variance of new information (see Fig. 23.2).

H-2: Ask for explanation. Students' gender, educational study and occupation, and discussion process were all significantly linked to asking for an explanation. *Men* were 24 % more likely than *women* to ask for an explanation. Meanwhile, students in *doctoral* programs were 19 % less likely to ask for an explanation. *Post secondary teacher* and *non-post secondary teachers* were 1 % and 22 % less likely to ask for an explanation respectively. Controlling for teacher occupation, the gender effect was reduced by 21 %. Demographic and occupation variables accounted for 11 % of the variance in explanation requests.

Attributes of earlier messages were linked to explanation requests. After a *question about use*, an explanation request was 14 % more likely. After *any discussion process*, an explanation request was 9 % more likely. After an *explanation request* (-2) two messages ago, another explanation request was 8 % more likely. Together, these explanatory variables accounted for about 22 % of the variance of an explanation request.

H-3: Theorize. Gender and attributes of previous messages were significantly linked to theorizing. *Men* were 21 % more likely than women to theorize. Demographics accounted for 5 % of the variance in theorizing.

Attributes of earlier messages up to three messages ago were linked to theorizing. After an *explanation* (-1) or an *elaboration* (-1), theorizing was 21 % or 38 % more likely, respectively. If someone *asked about the use of an idea* (-2), gave an *opinion* (-2) or gave a *different opinion* (-2) two messages ago, theorizing was 21 %, 56 %, or 12 % more likely, respectively. After *anecdotal evidence* (-3) three messages ago, theorizing was 33 % more likely. Altogether, these explanatory variables accounted for 38 % of the variance of theorizing.

H-4: Summarize. Gender, occupation, and attributes of previous messages were linked to summary. *Men* were 22 % more likely to summarize than women. Meanwhile, *teachers* or *technology workers* were 14 % or 1 % less likely to summarize respectively. Controlling for teacher, the link between gender and summary was no longer significant. Demographics accounted for 15 % of the variance in summary.

After any discussion process, a summary was only 1 % more likely. After a *new fact* (-2) two messages ago however, a summary was 10 % more likely. Together, these explanatory variables accounted for about 22 % of the variance of summaries.

Other variables were not significant and the results did not differ significantly across topics. The I^2 index of Q-statistics for each dependent variable was not significant, indicating no serial correlation. Robustness tests showed similar results.

Discussion

To analyze relationships among asynchronous online messages, I revised SDA to apply to branches of messages. As a result, researchers can use this revised SDA to analyze large data sets of participants' self-coded online messages, with the potential for semiautomatic analyses through integrated computer programs. Specifically, this analysis showed that both individual characteristics and recent messages' cognitive and social metacognitive aspects affected the likelihoods of new information, explanation requests, theories, and summaries.

Extending SDA to Online Data

A large data set of 1,330 participant-coded online messages that branch off into multiple threads offers opportunities for multivocality advances in analytic methods in two ways: extending SDA to analyze relationships among messages and taking steps toward semiautomatic analyses. Unlike the linear sequence of turns of talk, Fujita's data set of online messages often branch out into separate sub-threads. To capture this branching structure, I store each message's previous message along its thread in a variable. Tracing messages backwards along this variable, I can identify any ordinal predecessor of any message along each thread. Then, I change my application of SDA to examine the previous message on a thread, not the most recent message (according to time). Hence, one benefit of multivocality is improving statistical methods (e.g., SDA) in response to challenging data structures (e.g., nonlinear branches of messages).

As the large data set includes participant-coding of their messages, it offers the potential for semiautomatic analyses that integrate multiple analyses encoded into computer programs. Unlike transcripts of audiotapes or videotapes that must be coded afterwards, the participant coding occurs during the writing of the message and reflects the author's intention (Fujita, this section). Whether participant coding yields sufficiently similar categories of codes is an open question and a valuable research area. If participant coding is viable in some cases, the codes can be entered into specific computer programs to yield descriptive and temporal analyses, as shown by the other authors in this section (Law & Wong, Chap. 22, this volume; Teplovs & Fujita, Chap. 21, this volume). As the revised SDA algorithm can be encoded into a computer program, it can be integrated with other software [e.g., Teplovs & Fujita's KISSME in this section; Dyke, Lund, and Girardot's (2009) TATIANA]. Guided by descriptive statistics and extended social network analyses (KISSME) from this potential integrated software, users can select participant-coded explanatory variables and dependent variables in the SDA portion of the software, which can test the model to show all results and only significant results. As SDA identifies both typical results and exceptions to the model, both types of sub-threads of messages can be further examined (e.g., via TATIANA). Thus, two additional potential benefits of multivocality are (a) understanding and appropriating other analysts' user interfaces and (b) integration of multiple analyses into a computer program capable of semiautomatic analyses.

Demographics and Occupation

In this specific analysis, the results show the need to examine explanatory variables at the individual level as well as the message level. Past studies of students had shown that individual differences in gender, past achievement and status accounted for little of the variance in discussion behaviors (e.g., Chen & Chiu, 2008; Chiu, 2008b; Lu et al., 2011), but this study showed that individual differences in adults, specifically gender and occupation, accounted for a mean of 10 % of the variance in explanation, theories and summaries. Compared to women, men were more likely to ask for explanations, theorize and summarize. These results are consistent with the research that men are more active than women during online discussions (e.g., Lu, Chiu, & Law, 2011). Compared to gender, job accounted for much more of the differences in explanation requests and summaries. Doctoral students and teachers (especially primary and secondary teachers) were less likely to ask for explanations. Cumulatively, job had the largest effects on explanation requests. Meanwhile, teachers and technology workers were less likely than other students to summarize. Further research can examine the origins of these substantial job differences in online behaviors and on larger data sets.

Micro-time Context of Recent Messages

Beyond the effects of individual characteristics, these results showed that asynchronous messages are more than simply lists of individual cognition (Thomas, 2002); instead, these messages both influence and respond to one another. Specifically, both cognitive and social metacognitive aspects of recent messages showed micro-time context effects.

Informal and formal cognition do not compete; instead, informal cognition preceded formal cognition. Opinions, anecdotes, elaborations and information increased the likelihoods of subsequent information, theories and summaries. After an opinion, new information or theorizing was more likely to follow. Anecdotes and elaborations were also more likely to be followed by theorizing. Together, the last three results are consistent with the view that familiar, informal cognition is activated faster than formal cognition (Chiu, 1996), and that the former can facilitate the latter through spreading activation of related semantic networks both in the individual and among group members (Nijstad et al., 2003).

Social metacognition, in the form of questions and different opinions, affected the likelihood of new information, explanation requests and theories. Questions about use had the largest effect on inducing more information, showing the power of questions to influence other's behaviors, consistent with earlier research (e.g., Chen, Chiu, & Wang, 2010). Furthermore, both types of questions elicited more explanation requests and theories; the latter is consistent with earlier studies (e.g., Lu et al., 2011). Lastly, a different opinion had the largest effect on a subsequent theory, consistent with face-to-face research showing that disagreements provoke explanations (e.g., Chiu, 2008a).

Conclusion

Showing several benefits of multivocality, this study revised a statistical method designed for linear sequences of turns of talk to apply to branches of messages in asynchronous online discussions, in this case to test for cognitive and social metacognitive relationships among messages. To capture the branching structures of messages, each message's previous message on along its thread was stored in a variable. Then, changing SDA to examine the previous message on a thread expanded SDA's scope to analyses of messages in asynchronous online discussions as well as face-to-face talk. Exposure to other authors' computer programs and displays also suggest opportunities to improve the SDA user-interface and integration of multiple analyses into a computer program capable of semiautomatic analyses.

The results showed that both individual characteristics and the micro-time context of recent messages' cognition and social metacognition affected the likelihoods of subsequent new facts, explanation requests, theories and summaries. Unlike past studies of students, this study showed that gender and occupation differences in adults account for substantial differences in online behaviors. Specifically, men were more likely than women to ask for explanations, theorize and summarize. Doctoral students and teachers were less likely to ask for explanations, and teachers and technology specialists were less likely to summarize.

Rather than simply being lists of individual cognition, asynchronous messages create a micro-time context that affects subsequent messages. Informal cognition (opinions, anecdotes, elaborations) facilitates formal cognition (facts and theories). Meanwhile, social metacognition, in the form of questions and different opinions, had the strongest effects on subsequent facts and theories. Together, revised SDA and its results offer opportunities to improve understanding of the relationships among online messages, which can help educators and students to improve online discussion processes.

References

Benjamini, Y., Krieger, A. M., & Yekutieli, D. (2006). Adaptive linear step-up procedures that control the false discovery rate. *Biometrika, 93*, 491–507.

Bryk, A. S., & Raudenbush, S. W. (1992). *Hierarchical linear models*. London: Sage.

Chen, G., & Chiu, M. M. (2008). Online discussion processes. *Computers & Education, 50*(3), 678–692.

Chen, G., Chiu, M. M., & Wang, Z. (2010). Group micro-creativity in online discussions. In K. Gomez, L. Lyons, & J. Radinsky (Eds.), *Proceedings of the 9th International Conference of the Learning Sciences* (Vol. 1, pp. 357–364). Chicago, IL: International Society of the Learning Sciences.

Chiu, M. M. (1996). Exploring the origins, uses and interactions of student intuitions. *Journal for Research in Mathematics Education, 27*(4), 478–504.

Chiu, M. M. (2000a). Group problem solving processes: Social interactions and individual actions. *Journal for the Theory of Social Behavior, 30*(1), 27–50.

Chiu, M. M. (2000b). Status effects on solutions, leadership, and evaluations during group problem solving. *Sociology of Education, 73*(3), 175–195.

Chiu, M. M. (2001). Analyzing group work processes. In F. Columbus (Ed.), *Advances in psychology research* (Vol. 4, pp. 193–222). Huntington, NY: Nova Science.

Chiu, M. M. (2008a). Effects of argumentation on group micro-creativity. *Contemporary Educational Psychology, 33*, 382–402.

Chiu, M. M. (2008b). Flowing toward correct contributions during group problem solving: A statistical discourse analysis. *Journal of the Learning Sciences, 17*(3), 415–463.

Chiu, M. M., & Khoo, L. (2003). Rudeness and status effects during group problem solving. *Journal of Educational Psychology, 95*, 506–523.

Chiu, M. M., & Khoo, L. (2005). A new method for analyzing sequential processes. *Small Group Research, 36*, 1–32.

Chiu, M. M., & Kuo, S. W. (2009). From metacognition to social metacognition. *Journal of Education Research, 3*(4), 1–19.

Clark, D. B., & Sampson, V. (2008). Assessing dialogic argumentation in online environments to relate structure, grounds, and conceptual quality. *Journal of Research in Science Teaching, 45*(3), 293–321.

Dubrovsky, V. J., Kiesler, S. B., & Sethna, B. N. (1991). The equalization phenomenon. *Human-Computer Interaction, 6*, 119–146.

Dyke, G., Lund, K., Girardot, J.-J. (2009). Tatiana: an environment to support the CSCL analysis process. *CSCL 2009*, Rhodes, Greece.

Fujita, N. (this volume). Online graduate education course using knowledge forum. In D. D. Suthers, K. Lund, C. P. Rosé, C. Teplovs & N. Law (Eds.), *Productive multivocality in the analysis of group interactions, Chapter 20*. New York, NY: Springer.

Glassner, A., Weinstoc, M., & Neuman, Y. (2005). Pupils' evaluation and generation of evidence and explanation in argumentation. *British Journal of Educational Psychology, 75*, 105–118.

Goldstein, H. (1995). *Multilevel statistical models*. Sydney: Edward Arnold.

Goldstein, H., Healy, M., & Rasbash, J. (1994). Multilevel models with applications to repeated measures data. *Statistics in Medicine, 13*, 1643–1655.

Gress, C. L. Z., Fior, M., Hadwin, A. F., & Winne, P. H. (2010). Measurement and assessment in computer-supported collaborative learning. *Computers in Human Behavior, 26*, 806–814.

Gunawardena, C. N., Lowe, C. A., & Anderson, T. (1997). Analysis of a global online debate and the development of an interaction analysis model for examining social construction of knowledge in computer conferencing. *Journal of Educational Computing Research, 17*(4), 397–431.

Hacker, D. J., & Bol, L. (2004). Metacognitive theory. In D. M. McInerney & S. Van Etten (Eds.), *Big theories revisited* (Vol. 4, pp. 275–297). Greenwich, CO: Information Age.

Hakkarainen, K. (2003). Emergence of progressive-inquiry culture in computer-supported collaborative learning. *Learning Environments Research, 6*(2), 199–220.

Hara, N., Bonk, C. J., & Angeli, C. (2000). Content analysis of online discussion in an applied educational psychology course. *Instructional Science, 28*, 115–152.

Harasim, L. M. (1993). *Global networks*. Cambridge, MA: MIT Press.

Howe, C. (2009). Collaborative group work in middle childhood. *Human Development, 52*(4), 215–239.

Huedo-Medina, T. B., Sanchez-Meca, J., Marin-Martinez, F., & Botella, J. (2006). Assessing heterogeneity in meta-analysis. *Psychological Methods, 11*, 193–206.

Jeong, A. (2003). The sequential analysis of group interaction and critical thinking in online threaded discussions. *American Journal of Distance Education, 17*, 25–43.

Jeong, A. (2006). The effects of conversational language on group interaction and group performance in computer-supported collaborative argumentation. *Instructional Science, 34*(5), 367–397.

Kennedy, P. (2008). *Guide to econometrics*. Cambridge: Wiley-Blackwell.

King, G., & Zeng, L. (2001). Logistic regression in rare events data. *Political Analysis, 9*, 137–163.

Law, N., & Wong, O.-W. (this volume). Exploring pivotal moments in students' knowledge building progress using participation and discourse marker indicators as heuristic guides. In D. D. Suthers, K. Lund, C. P. Rosé, C. Teplovs & N. Law (Eds.), *Productive Multivocality in the Analysis of Group Interactions*, Chapter 22. New York, NY: Springer.

Ljung, G., & Box, G. (1979). On a measure of lack of fit in time series models. *Biometrika, 66*, 265–270.

Lu, J., Chiu, M., & Law, N. (2011). Collaborative argumentation and justifications: A statistical discourse analysis of online discussions. *Computers in Human Behavior, 27*, 946–955.

Luppicini, R. (2007). Review of computer mediated communication research for education. *Instructional Science, 35*(2), 141–185.

MacKinnon, D. P., Lockwood, C. M., & Williams, J. (2004). Confidence limits for the indirect effect: Distribution of the product and resampling methods. *Multivariate Behavioral Research, 39*, 99–128.

Nijstad, B. A., Diehl, M., & Stroebe, W. (2003). Cognitive stimulation and interference in idea generating groups. In P. B. Paulus & B. A. Nijstad (Eds.), *Group creativity: Innovation through collaboration* (pp. 137–159). New York, NY: Oxford University Press.

Peugh, J. L., & Enders, C. K. (2004). Missing data in educational research. *Review of Educational Research, 74*, 525–556.

Piaget, J. (1985). *Equilibration of cognitive structures*. Chicago: University of Chicago Press.

Tallent-Runnels, M. K., Thomas, J. A., Lan, W. Y., Cooper, S., Ahern, T. C., Shaw, S. M., et al. (2006). Teaching courses online. *Review of Educational Research, 76*(1), 93–135.

Teplovs, C., & Fujita, N. (this volume). Socio-dynamic latent semantic learner models. In D. D. Suthers, K. Lund, C. P. Rosé, C. Teplovs & N. Law (Eds.), *Productive multivocality in the analysis of group interactions*, Chapter 21. New York, NY: Springer.

Thomas, M. J. W. (2002). Learning within incoherent structures: The space of online discussion forums. *Journal of Computer Assisted Learning, 18*, 351–366.

Vygotsky, L. (1986). *Thought and language*. (A. Kozulin, Trans.). Cambridge, MA: MIT Press. (original work published 1934).

Wise, A., & Chiu, M. M. (2011). Analyzing temporal patterns of knowledge construction in a role-based online discussion. *International Journal of Computer-Supported Collaborative Learning, 6*, 445–470.

Chapter 24
Critical Reflections on Multivocal Analysis and Implications for Design-Based Research

Nobuko Fujita

Introduction

This chapter presents critical reflections on the multivocal analyses presented by Chris Teplovs and Nobuko Fujita (2014, Chap. 21), Nancy Law and On-Wing Wong (Chap. 22), and Ming Ming Chiu (Chap. 23) on asynchronous online discussion data that was collected in an online graduate education course using Knowledge Forum (Fujita, 2014, Chap. 20). These analyses work towards identifying and exploring collaborative interactions and "pivotal moments" in dynamic group processes that support the progress towards knowledge building over 13 weeks of the course.

This data forms the second iteration of a larger design-based research study (Fujita, 2009) and differs from the other datasets in this book by featuring asynchronous, text-based discourse that unfolded in a higher education online learning context. It is also a large dataset that focuses on the 1,330 notes contributed by 17 graduate student participants. A tenure-stream faculty instructor and a researcher closely collaborated to design instructional interventions and participated in the forum to foster progressive discourse for knowledge building.

A common teaching problem in online courses is moving students beyond expressions of social connection and opinion exchange. Earlier research indicates that desirable educational outcomes such as critical thinking, knowledge construction, and critical discourse rarely occur in these settings (Garrison, Anderson, & Archer, 2001; Gunawardena, Lowe, & Anderson, 1997; Rourke & Kanuka, 2007). My study departed from previous studies by focusing on higher goals for collaborative interaction. It refined designs of instructional interventions that support group processes towards knowledge building in online graduate education courses and offered a unique perspective to identifying characteristics of resulting high quality online

N. Fujita (✉)
University of Windsor, Windsor, ON, Canada
e-mail: nfujita@uwindsor.ca

D.D. Suthers et al. (eds.), *Productive Multivocality in the Analysis of Group Interactions*, Computer-Supported Collaborative Learning Series 16,
DOI 10.1007/978-1-4614-8960-3_24, © Springer Science+Business Media New York 2013

discourse. I combined multiple levels of analyses, including an in-depth analysis of group discourse to explore ways to assess individual and group level learning and knowledge building in online courses.

Findings from my various analyses converged to suggest that peer scaffolding that made norms for progressive discourse for knowledge building was most effective at the beginning of the course for newer online learners and newer graduate students, and least effective for students who were practicing K-12 teachers. A significant barrier to knowledge building discourse was the tendency for teachers to reject these norms and revert to "belief-mode thinking" (Bereiter & Scardamalia, 2003) and "devotional discourse" typical of traditional schooling (Woodruff & Brett, 1999). Additionally, findings suggested that software-based scaffolding (as found in Knowledge Forum's scaffold support feature) is a most promising avenue for future innovations to promote knowledge building discourse.

However, identifying and describing patterns of collaborative interaction in large textual data sets is a daunting task for even seasoned researchers. The quantitative and qualitative analyses yielded deep insights into online learning and teaching processes, but were time-consuming and laborious. A significant issue arising from the study was the need for future research to draw a more complete picture of the complex learning unfolding online in meaningful, timely, and actionable ways. I sought to extend the suggestive findings of the characteristics of knowledge building discourse with visualizations and advanced quantitative analyses.

Therefore, to gain new perspectives of collaborative learning through alternate analytic approaches, this dataset was contributed for the 2009 Alpine Rendez-Vous, along with the Shirouzu dataset. In that workshop, the analyses on this data were presented by Teplovs and Fujita (2009) and Tscholl and Dowell (2009). Discussion of the analyses were provided by Rosé; meta-discussion was given by Law. Teplovs and Fujita used latent semantic analysis (Landauer, Laham, & Derr, 2004; LSA) and visualizations generated by the Knowledge Space Visualizer (Teplovs, 2008) to pinpoint moments in the online discussion that showed promise for being a "rise above" or synthesis moment. While Rosé pointed to the difficulty in interpreting the visualizations, the limitations of LSA, and the possibility of using alternative approaches such as latent dirichilet allocation (Blei, Ng, & Jordan, 2003), this automated analysis quickly and accurately identified a few interesting, possibly "pivotal" moments in the data. It thus offered the potential to make analyses of large textual corpora more practicable during design-based research iterations and to inform teachers' pedagogical decision making while the course was still in session to enhance the quality of student learning.

In contrast, Tscholl and Dowell (2009) took a more traditional qualitative analysis approach. Their analysis traced "individualistic appropriations of words, propositions and objects (e.g. symbols) in a collaborative learning situation, and to show that often these constitute pivotal moments in collaboration." Beginning with the notion of "uptake" (Suthers, Dwyer, Medina, & Vatrapu, 2010), they identified myriad instances of uptakes and shifting problem frames in online exchanges. While highly desirable, instances of knowledge building are very difficult to foster in online discourse and happen infrequently even in graduate education.

The meta-discussion by Law and comments with other workshop participants concurred with some reservations I had about the findings from Tscholl and Dowell's analysis. In addition, Law's meta-discussion presented another promising automated method that incorporates participation patterns and discourse markers to provide an overview of the nature and depth of students' engagement with course concepts. Later, Law's meta-discussion evolved into an analysis presented at the 2011 Alpine Rendez-Vous along with analyses by Teplovs and Fujita, and Chiu.

In the sections that follow, I critically reflect on the analyses presented by Teplovs and Fujita, Law and Wong, and Chiu that were introduced at the 2011 Alpine Rendez-Vous workshop along the five dimensions for productive multivocality. Then, I turn to discuss the implications of multivocal analysis for design-based research with recommendations for future research.

Five Dimensions for Reflecting on Productive Multivocality

This section delves into the multivocal analysis by Teplovs and Fujita, Law and Wong, and Chiu along the following five dimensions exhorted at the Alpine Rendez-Vous workshops: theoretical assumptions, purpose of analysis, unit of analysis/unit of interaction, data representations, and manipulations on data representations.

Theoretical Assumptions

The theoretical assumptions of knowledge building underpin the original study for which the data were collected (Fujita, 2009). These assumptions also drive the analyses by Teplovs and Fujita (Chap. 21) and Law and Wong (2014, Chap. 22). In contrast, Chiu's analysis (Chap. 23) explicates a method that has relatively few theoretical assumptions and *may* be compatible with diverse theoretical lenses. Knowledge building is defined as the production and continual improvement of ideas of value to a community, through means that increase the likelihood that what the community accomplishes will be greater than the sum of individual contributions (Scardamalia & Bereiter, 2003, p. 1370).

Knowledge building may be considered a theory, pedagogy, and technology (Scardamalia & Bereiter, 2006). As a theory, it places an overt focus on improving ideas. In knowledge building, ideas are considered "conceptual artifacts" and knowledge work is defined as "work that creates or adds value to conceptual artifacts" (Bereiter, 2002b, p. 69). Broadly, conceptual artifacts are cultural artifacts, as in communities of practice. Of these, some cultural artifacts are abstract rather than concrete. Conceptual artifacts are abstract cultural artifacts (theories, abstract models) that "can be distinguished by the logical relations that exist between them" (Bereiter, 2002b, p. 76).

As a pedagogy, it is an attempt to reform education in a fundamental way to enculturate students into the culture of knowledge creation (Scardamalia & Bereiter, 2006). From this view, being able to advance the state of community knowledge is not a social process exclusive to experts, but rather one in which students can and should engage in if they are to progress along a developmental trajectory from childhood inquisitiveness to mature, disciplined creativity. Knowledge building differs from other learning community models by putting ideas at the center and focusing on idea improvement rather than on collaborative learning activities such as in "communities of learners" (Brown & Campione, 1990, 1994). It is also guided by a set of 12 principles (Scardamalia, 2002) that characterize the complex socio-cognitive and technological dynamics it involves. Although "collective cognitive responsibility" (Scardamalia, 2002) seems to be an overarching principle for knowledge building, all of the 12 principles work in concert with each other, not separately, to drive the knowledge building process.

Knowledge Forum, a second generation of CSILE (Computer Supported Intentional Learning Environment) software, is a technology especially designed to support knowledge building. Students work in virtual spaces or "views" to develop their ideas, represented as "notes." Knowledge Forum offers sophisticated features not available in other conferencing technologies including "scaffold supports" (labels of thinking types), "rise above" (summary note), and a capacity to connect ideas through links between notes in different views. These features provide means to overcome the chronological sequence of threaded discussion, in which important ideas may be lost. In addition, Knowledge Forum facilitates the collection of data that are amenable to analysis with a variety of assessment tools. These include behavioral and interaction analyses (Burtis, 1998), traces of vocabulary development (Hewitt, 1999), social network analysis (SNA; Teplovs, Donoahue, Scardamalia, & Philip, 2007), and semantic analysis (Fujita & Teplovs, 2009).

In addition to the theoretical assumptions of knowledge building, Teplovs and Fujita's current analysis is informed by those of networked learning and computer-supported collaborative learning (de Laat, Lally, Lipponen, & Simons, 2007). Fundamental to all these approaches is the notion of community dynamics, in which patterns of interaction develop over time and may be investigated using social or semantic network analyses. Assessment tools built in to an online environment can provide participants with formative feedback on the progress towards advancing the community's emergent discourse and embodies the knowledge building principle of concurrent, embedded assessment in knowledge building. Further, their approach assumes that a semantic model based on student contributions (you are what you write) is a reasonable one, and that LSA is valid on such short texts.

Likewise, Law and Wong's analysis is strongly committed to knowledge building. They seek to identify the trajectory of knowledge building from the two perspectives which they deem most meaningful: (1) the extent to which a group of students exhibits characteristics of the 12 principles; and (2) the advances in the emergent understanding of the key ideas. Law and Wong suggest that pivotal moments indicate movement from one stage of the trajectory to another, which may be considered from the perspective of both the community and the individual. Following Bereiter (2002b), they accept that the improved ideas or emerging insights from the community cannot be attributed to any one individual or subgroup of individuals. In addition, they assume that an individual would not be able to advance to another stage of the trajectory unless the community as a whole has been able to do so. Furthermore, they acknowledge that not all individuals within the community may achieve the understanding made by the collective.

Chiu's chapter showcases Statistical Discourse Analysis (SDA). Chiu explains that SDA has few theoretical assumptions or commitments and suggests that it is possible to use SDA with many theoretical frameworks. Nonetheless, Chiu points to at least three assumptions underpinning his analysis of this particular dataset. First, the analysis assumes that participant-selected scaffolds in notes (participants labeled the characteristics of their own notes by inserting a Knowledge Forum scaffold support when they composed a note), individual differences in participants, and time period (weekly discussions were organized around a topic and led by student discussion leaders) are sufficiently similar to be treated as equivalent for the purpose of this analysis. Second, it takes as given that notes containing scaffolds, participating individuals, and time together constitute a "micro-context" in which future notes emerge. Lastly, the analysis supposes that the characteristics of recent notes, their authors, and time period can influence the characteristics of later notes.

Thus theoretical assumptions of social metacognition (Chiu & Kuo, 2009; Chiu & Pawlikowski, 2013) also play a key role in Chiu's analysis. Social metacognition goes beyond individual metacognition that involves monitoring and control of one's own knowledge, emotions, and actions (Hacker & Bol, 2004) to consider group members' monitoring and control of one another's knowledge, emotions, and actions (Chiu & Kuo, 2009). Social metacognition can enhance "micro-creativity," or the creation of new, useful ideas (Chiu & Pawlikowski, 2013). In Chiu's analysis, social metacognition refers to how students monitor and control one another's ideas and actions through questions, evaluations (agree vs. disagree), and summaries. Social metacognition may be likened to the knowledge building principle of collective cognitive responsibility, in which "the responsibility for the success of a group effort is distributed across all the members rather than being concentrated in the leader" (Scardamalia, 2002, p. 68). However, these differ because social metacognition attends to the emotions, public self-image (face), and social rapport building in thinking, whereas collective cognitive responsibility emphasizes the "cognitive" dimension over other aspects. This is not to say that knowledge building pedagogical designs do not pay heed to social aspects of student and teacher interactions. On the contrary, knowledge building teachers take great pains to establish a culture of safety in the database to enable students to take risks in voicing nascent ideas (Zhang, Scardamalia, Reeve, & Messina, 2009). In the original study (Fujita, 2009), students were encouraged to use specially designed materials (Discourse for Inquiry cards) to help them structure their discourse for problem solving in polite and supportive ways. Additionally, opportunities to engage in metacognitive reflection have been found to enhance knowledge building in online courses (Brett, Forrester, & Fujita, 2009; Cacciamani, Cesareni, Martini, Ferrini, & Fujita, 2012).

In all three of the analyses, the theoretical assumptions drive the methodology to go beyond the analysis of an individual student's behavior or the content of a single note. They all trace the development of collaborative interactions involving more than one student over time. Although there are some similarities between the theoretical underpinnings that inform Chiu's work with knowledge building, it is likely that researchers from the knowledge building community will point to the fundamental incompatibility of diverse theoretical assumptions in this application of SDA. They may experience tension between how the theoretical assumptions influenced the resultant methodological choices.

Purpose of Analysis

The purpose of Teplovs and Fujita's analysis is to examine the relationship between social interactions and the semantics of the written contributions of students participating in an online graduate course. To do so, they introduce a framework and software for learner modeling that interweaves social network analysis and latent semantic network analysis of online discourse called the Knowledge, Interaction, and Semantic Student Model Explorer (KISSME). KISSME uses highly interactive visualizations of semantic and social interactions among learners. It enables researchers to examine the interplay of students' social interactions and the latent semantic models of those students. They attempted to test the hypothesis that uptake (Suthers et al., 2010) is most likely to occur when the semantic relatedness of the corresponding student models is neither too high nor too low, but at the optimal level of compatibility for collaboration.

Law and Wong's analysis is driven by a strong pedagogical motivation to investigate the possibility of designing a dashboard of indicators derived from automated analysis for teachers to help them identify the state of students' progress, the key problems of understanding that they are exploring, and to identify any "at-risk" students. Thus, the authors seek to establish a form of learning analytics for teachers to access information "on the fly" to help them in understanding students' overall engagement and conceptual advancement in knowledge building.

The purpose of Chiu's analysis is to use SDA to (1) identify pivotal moments along specific dimensions that divide the data into distinct time periods; and (2) examine variables that significantly increase or decrease the likelihoods of dependent variables of interest. The dependent variables of interest were as follows:

H-1. Online discussions have proportionately more ideas, facts and explanations than face-to-face discussions
H-2. New fact
H-3. Ask for explanation
H-4. Theorize
H-5. Summarize

All three analyses seek to identify and explore "pivotal moments" and are compared to broader analysis of dynamics of group processes that support knowledge building.

Two of these analyses (Teplovs & Fujita; Law & Wong) investigate the potential of automated analyses for use by students, teachers, and researchers. The third analysis (Chiu) extends SDA to understand the probabilities of one kind of note following another in a sequence in asynchronous online discussions, which would yield useful insights for researchers.

Unit of Analysis/Unit of Interaction

The units of analysis in Teplovs and Fujita's analysis are documents—online discussion messages called "notes." It follows that the smallest unit of interaction would be two notes. This unit focuses on the interpersonal system and the patterns of interaction between students mediated by notes and goes beyond individual contribution to knowledge building. They attempted to show points in the data over time, 109 days of the course, where there was progression in (1) the latent semantic learner model (LSLM) networks; and (2) the social interaction network determined through the intensity of shared reading events.

Law and Wong's analysis combines quantitative and qualitative methods and sundry units of analysis. First, to identify pivotal weeks, they compute median and dispersion statistics of individual students' writing and reading behavior on a week-to-week basis using the note as the unit of analysis. Second, Law and Wong use threads (discussion threads or tree structures of at least two notes) to examine pivotal weeks. Third, they utilize keywords (nouns or noun phrases collocated as one word as units of analysis) to examine students' engagement with core concepts on a week-by-week basis. The keywords they used were based on a chapter from Bereiter (2002b): idea, knowledge building (collocated as one noun phrase; KB), discourse, conceptual artifact (CA), belief mode, design mode, and world (as in Popper's (1972) world 1, world 2, and world 3). Fourth, Law and Wong used machine identification of discourse markers to track the presence of question markers that might indicate the presence of factual, explanatory, and elaboration questions. This analysis, like the thread-level indicators, was used to delve into the establishment of a progressive inquiry orientation. The discourse markers comprise words and phrases to identify various question types. Fifth, Law and Wong trace advances in students' conceptual understanding by conducting qualitative content analysis of a subset of data using the sentence as a unit of analysis. Finally, although Law and Wong concede that this "crude selection and analysis process" does not indicate whether any of these ideas were developed outside of this subset of data, they found a prominent theme around the concept of idea improvement emerging through qualitative analysis of 10 of the 48 sentences.

Chiu's unit of analysis is the sequence of one type of note following another and how this affects their content (Chiu, 2000a). At a minimum, this involves two notes. His analysis examines sequences among a subset of data (306 student notes) that contain scaffold supports. Students can label a particular "thinking type" by inserting one or more Knowledge Forum scaffold support(s) while composing a note. Chiu's SDA models the probabilities of these sequences.

In short, the three analyses examine manifold units of analysis or interaction over time. The analyses by Teplovs and Fujita and by Chiu concentrate on interactions, or relationships among students, rather than focus on the properties of individual notes. In Teplovs and Fujita's analysis, the relationship assessed is between documents or notes written by students, where they assume each student author is represented by notes that he or she writes. In Chiu's analysis, the relationship examined is the sequence of at least two notes containing particular scaffolds that are labeled with a thinking type or discourse process. One vulnerability in Chiu's analysis is that it assumes that the scaffold supports accurately reflect the discourse processes in the text and is susceptible to critique unless a neutral observer can predict the scaffold supports that the students used to label or self-code their own note in the database. In Fujita's (2009) study, however, this problem was addressed through randomly selecting 56 segments of student discourse containing a scaffold support from the sample (scaffold supports either bracketed or preceded segments of text, setting it apart from the rest of the note). Then, the scaffold support that the student participants actually used were omitted from the text and another graduate student was asked to guess correctly the appropriate scaffold based on the discourse processes reflected in the text. Next, percentage agreement was calculated. This found that 79 % of the time a graduate student can predict the scaffold support that another graduate student would use.

Law and Wong's analysis employs several units of analysis at varying levels of granularity: a thread, a note, a sentence, and discourse markers (word or phrase). The analytic toolkit (ATK; Burtis, 1998) built-in to Knowledge Forum facilitates some of these analyses to investigate knowledge building dynamics. Previous researchers have reported findings correlating such quantitative indicators of participation to portfolio scores and conceptual understanding (e.g., Lee, Chan, & van Aalst, 2006). Relationship between extensive writing, reading, and use of features such as build-on notes, rise-above notes (summaries and higher-order syntheses), referencing, and scaffold use have also been identified with knowledge building Zhang et al. (2009). Law and Wong's week-by-week analysis differs from the summary analysis that researchers often compute to get an overview of a Knowledge Forum database. Their week-by-week approach is likely to be useful for researchers and teachers to identify and describe changes in participation patterns and engagement in knowledge building. The authors' strong background in knowledge building gives purchase to their final qualitative analysis, particularly as they can discern relevant and irrelevant concepts that students discuss in the data. For example, in tracing the development of concepts, they eliminated sentences containing keywords that were not unique conceptual terms as others such as "world" when they were not in reference to Popper's (1972) theory of the three worlds vs. commonplace usage such as "virtual world" (c.f. Tscholl & Dowell, 2009).

Data Representations

While much attention has been paid to the development of graphical representations of quantitative data, less attention has been paid to the graphical displays of qualitative and mixed methods data (Onwuegbuzie & Dickinson, 2008). Visual techniques can assist with data reduction and conclusion drawing/verification in qualitative and mixed methods research (Leech & Onwuegbuzie, 2007; Miles & Huberman, 1994). Thus, information visualizations may reveal "pivotal" moments unfolding online; concurrently, attention must be paid to the crucial information they may conceal.

Visualizations play a central role in Teplovs and Fujita's analysis. First, Teplovs and Fujita generated high-dimensional vector representations of the content of notes via LSA. That is, the words in the notes are turned into a vector of numbers. Examining the co-occurrence of words in a term-by-document matrix followed by singular value decomposition of that matrix reveals the semantic similarity between notes. LSA enhances the structural (build-on or reply to notes) and the social (read notes) relationships that may exist among students over time. Second, network and adjacency matrixes representations of productive collaborative interactions among students were explored to test predictive models of similar semantic contributions and shared reading behavior over the 109 days of the course data. One bias in their approach is that they assume that students' cumulative written notes or artifacts in Knowledge Forum can be considered learner models. Another bias may be that they apply the Vygotskian notion of scaffolding in the zone of proximal development to predict that the individuals most likely to benefit from collaborative learning situations are those who are semantically not too close or not too far, but just right (c.f., Zampa & Lemaire, 2002). These assumptions conceal cognition and metacognition not written in Knowledge Forum notes but perhaps communicated among students via other modes available in the course (synchronous chats, telephone conversations, videoconferences, and online learning journals/blogs). The interpretability of the network diagrams and adjacency matrices is also a concern, but consistent with Tufte's (2001) six principles of analytical design, they attempt to compare and explain the evidence from social interaction and semantic content recorded in the database. Their approach is promising as it advances current visual techniques for quantitative data representation of online discussion data and identifies compatible students based on LSLMs (for instance, students who coauthored notes or led discussions together).

Law and Wong's analysis aims to use "simple, easy to understand graphical displays accessible to teachers." They propose different representations to reveal different layers of insight from both quantitative analyses (participation statistics, week-by-week questions and keywords) and qualitative analyses (content analysis at the note and sentence level). For example, they utilize a boxplot graph to represent the number of notes created and the percentage of notes read by students each week. This representation differs from the tabular format that the ATK (Burtis, 1998) generates of these metrics. Alternatively, they employ line graphs to show the various discourse markers for various communication functions. These sorts of data representations offer learning analytics or "teaching analytics" (Vatrapu, Teplovs, Fujita, & Bull, 2011) that may be meaningful and actionable to those teachers who are able to decipher them. However, Bachelor of Education programs do not prepare teachers to be researchers (Donald, 2002; Labaree, 2003). Teachers' comprehension of statistical data displays are limited through the lack of exposure in teacher preparation programs (Jacobbe & Horton, 2010). Thus, professional development is necessary to enable teachers to make sense of the graphical displays so that they can understand their students' engagement in knowledge building and enact timely decisions to foster epistemological growth.

Chiu's analysis employs standard representations of quantitative information such as a database table, a summary statistics table, a breakpoints table, a time series graph, and a path diagram to convey the nonlinear sequence of notes with scaffolds (self-coded notes) and the probabilities of group problem solving outcomes. These tables and figures are conventional ones following the APA style guide (American Psychological Association, 2009). They offer clear ways to make the complexity of the SDA more accessible to the reader by organizing the sequences of words (cognition and social metacognition) and numbers (probabilities) together in a diagram.

Analytic Manipulations on Data Representations

Teplovs and Fujita's analysis first generated network diagrams, which are dynamic and illustrate changes in interaction patterns over time. To increase interpretability, they manipulated the data representation into an adjacency matrix based on LSLMs in which intermediate similarity values are indicated by intensity of color. This predicts pairs of students who should interact productively. Next, they generated an adjacency matrix based on intensity of shared reading events of students who indeed interacted by reading each other's notes in Iteration 1. Comparing the pairs of students who should have interacted based on semantic similarity and actual reading interactions, they found that such automated analyses can accurately identify productive collaborative relationships (e.g., pairs of students who led weekly discussions, authors of cowritten notes). However, as these analyses were conducted post hoc as a summary analysis, it was not possible for the researcher or teacher to explore the effect that formative feedback might have had in cases where pairs of students who should interact productively based on intermediate levels of semantic similarity but who did not interact through reading as might be predicted.

Law and Wong's analysis currently offers only static displays of quantitative and qualitative information. The authors would like to work towards more dynamic, open displays that enable users to access different layers of the analyses.

Chiu's SDA incorporates sophisticated and comprehensive analytic manipulations to address the analytic difficulties in modeling (1) sequences of notes within a tree structure that differs across topics; (2) four infrequent, dependent variables; (3) many explanatory variables that might yield mediation effects or false positives; and (4) general issues of missing data and robustness procedures (see Table 2 in Chiu's chapter).

Pivotal Moments

Teplovs and Fujita defined a pivotal moment as "a rise-above or synthesis moment" at the Alpine Rendez-Vous 2009. In the current paper, they consider a pivotal moment as "a point in time at which the semantic structure of the community changed in an important way." If we take intermediate semantic similarity and intensity of reading behavior of pairs of students as the measure of optimal collaborative interactions, the point in time when such interactions happened could be considered pivotal moment. The primary goal of Teplovs and Fujita's analysis, however, was not to pinpoint pivotal moments, but rather to examine the relationships between social interactions and semantic relationships between notes in an online learning environment. They attempted to extend the state-of-the-art in graphical representation of such collaborative interactions in asynchronous discussion data. Accordingly, they present a summary analysis from the 100th day of the course (no notes were posted on the final 9 days of the course), showing pairs of students who collaborated or have the potential to collaborate effectively. Generating visualizations at earlier points in time would be helpful to trace the development of concepts over time and people over time.

Law and Wong adopt two different definitions for the notion of pivotal. The first definition refers to time periods—pivotal weeks—that may be particularly productive. This matches the curriculum design for the online course data, in which discussions were organized around weekly themes. To this end, weekly statistics on participation (medians and dispersions in the number of notes written and read) and on discourse structure (thread size, thread depth, number of references used) were calculated using the ATK. The second definition refers to breakthroughs in students' understanding of key concepts—pivotal moments—found through qualitative content analysis of the students' discourse. Analyses dovetailed to identify week 9 as a pivotal week in terms of: (1) group dynamics conducive to knowledge building (smallest dispersion in writing and reading, but highest percentage of notes read); (2) establishment of a progressive inquiry orientation (smallest number of notes but high in indicators related to collective cognitive responsibility such as question markers, scaffolds, and references); and (3) engagement with targeted curricular concepts (high frequency of eight keywords). Tracing advances in conceptual understanding through qualitative analysis, Law and Wong found that week 10 was pivotal in terms of new ideas introduced. They suggest that the previous week may have had a positive effect, since week 9 was also productive. In week 9, four of the nine new ideas introduced were from the "teachers" (instructor and researcher) rather than from the students. Perhaps modeling in week 9 helped students to exercise higher levels of agency to build knowledge in week 10.

In Chiu's chapter, summaries are seen as "pivotal messages that radically change the interaction" (Wise & Chiu, 2011). Summaries of discussion have been shown to enhance knowledge construction both in the note and in subsequent notes (Wise & Chiu, 2011). Chiu's analysis did not find pivotal moments thus defined among sequences of notes containing certain scaffold supports (*My Theory, New Information*). However, Chiu's analysis did identify six notes containing scaffold supports recoded into "summarize" (e.g., *Putting our knowledge together*) and "ask for explanation" (e.g., *I need to understand*) variables. Chiu asserts that certain individual characteristics like gender and occupation as well as the presence of new information predicts summaries. Omitting one moment that occurred in a private view, these notes containing summaries that might be pivotal occurred in course weeks 6, 8, 9, and 10. These correspond with some of the same weeks that Law and Wong (weeks 6, 9, 10) also identify as being unusual in some way.

In summary, all three analysts may have found pivotal moments in the data, but their definitions of pivotal moments are diverse and their findings appear to have few commonalities. Teplovs and Fujita and Law and Wong suggest that there are pivotal moments that changed discussion on a larger scale, whereas Chiu does not. Chiu's pivotal moments are found on a finer time scale, or "micro-time contexts." The lack of shared pivotal moments may be influenced by the disparate theoretical and methodological assumptions undertaken. Both Teplovs and Fujita's and Law and Wong's analyses are faithful to knowledge building theory and empirically driven, while Chiu's analysis employs SDA to analyze knowledge creation in asynchronous discussion data and is methodologically driven.

Teplovs and Fujita's exploratory analysis attempted to identify points in time at which the semantic structure of the community changed in an important way, but presents a summary analysis from the end of the course. It introduces a vision for designing a leading-edge software system, KISSME, that can be used for future intervention studies, but more analysis is needed in order to optimize its potential and further development.

Law and Wong's analysis identify multiple pivotal weeks and moments through an array of quantitative and qualitative methods that is somewhat messy but nonetheless compelling. In my original study, I too chose Week 9 to begin qualitative discourse analysis because it appeared most promising for discovering instantiations of progressive discourse. I later abandoned it because it featured relatively small number of notes written by mostly four doctoral students, the Instructor, and the Researcher. Additionally, the use of scaffold supports, especially "Idea Improvement" scaffolds introduced for this week, was made mandatory by the student discussion leaders. Although "disciplined creativity" is characteristic of knowledge building (Scardamalia & Bereiter, 2003), students vehemently complained of the structure the scaffold supports imposed on their thinking. To compare the discourse from the beginning and end of the course, I eventually selected Week 3 and Week 10 for in-depth analysis. Yet, Law and Wong's analysis has renewed my interest in revisiting Week 9 as a pivotal week for knowledge building in this dataset. Refinement of Law and Wong's methodological design and the further development of Knowledge Forum assessment tools offer much promise.

Even expert students like graduate education students rarely engage in convergent processes such as writing synthesis or summary notes without considerable direction from the instructor (Hewitt, 2001, 2005). Chiu's analysis showed that demographics and occupation can account for differences in discussion behaviors. The small number of participants may limit generalizability of the statistical inferences made here, but the large number of notes in the dataset lend credibility.

Interestingly, Chiu found that informal cognition in the form of opinions, anecdotes, elaboration and facts increased the likelihoods of subsequent formal cognition in the form of more facts, theories, and summaries. Chiu defines pivotal moments as summaries and found only one instance of such a note in Week 6. One explanation may be that Chiu's analysis is restricted to a subset of sequences of notes that contain scaffold supports. Aside from Week 9, scaffold use was optional and idiosyncratic. A few students used them prolifically, and others just once. It is possible to compose summary notes without inserting a scaffold support, but Chiu's current analysis would conceal such pivotal moments.

Chiu reveals that social metacognition in the form of questions and different opinions, affected the likelihood of new facts, explanation requests, and theories. This is consistent with students posing wonderment questions to investigate a problem of understanding through knowledge building discourse (Scardamalia & Bereiter, 1991, 1994), but a knowledge building researcher would not consider questioning to be social metacognition. Chiu found that a new fact has the largest effect on a subsequent summary. Knowledge building involves an emergent process of explanatory coherence (Thagard, 1989), where groups of students contribute ideas and advance theories that best explain facts. Students can use a *New Information* scaffold to contribute new facts gleaned from authoritative sources and use a *Putting our Knowledge Together* scaffold to label better theory, but they may not. Further SDA using micro-context codes applied by researchers, rather than by students, would increase methodological rigor and perhaps convince knowledge building researchers who are familiar with the particular difficulties of analyzing group processes supporting knowledge building in large textual corpora within the Knowledge Forum platform.

Implications for Design-Based Research

I was given the particular distinction of serving as a discussant for multivocal analyses on data that I collected for a larger design-based research study (Fujita, 2009). In the original study, I investigated how instructors could foster progressive discourse for knowledge building in three online graduate education courses. Productive multivocality is more than just data sharing. I shared the data from an online graduate education course using Knowledge Forum (Fujita, 2014) to seek new perspectives from alternate analyses and the effect that this would have on my own insights. Seeing, reading, and being exposed to other researchers' analyses of my data influenced my research by encouraging me to collaborate with colleagues to examine: (1) leading-edge automated approaches such as visualizations and network analyses that make assessments for learning more applicable to educational practice (Teplovs, Fujita, & Vatrapu, 2011; Vatrapu et al., 2011); and (2) sophisticated statistical modeling methods for investigating knowledge creation processes in education (Chiu & Fujita, accepted).

These forays into automated and quantitative approaches to analyzing asynchronous discourse data urged me to reexamine my epistemological beliefs and practices as a design-based researcher. In the sections that follow, I outline in brief characteristics of design-based research relevant for this discussion, note advantages and challenges of the multivocal analyses for design-based research, and summarize the implications for future design-based research studies.

Characteristics of Design-Based Research

Responding to major changes in the focus of learning theory from the study of individual behavior and cognition to larger interactive systems, Ann Brown (1992) introduced the term "design experiments" to label a new methodology for carrying out studies of educational interventions (Collins, Joseph, & Bielaczyc, 2004). Design experiments are iterative, situated and theory-based attempts to understand and improve educational processes (Brown, 1992; Cobb, Confrey, diSessa, Lehrer, & Schauble, 2003; Collins, 1992, 1999; diSessa & Cobb, 2004; Edelson, 2002). Design-based research allows researchers to study complex learning where it is difficult to test the causal impact of particular variables with experimental designs (Barab, 2006). It deals with complexity by iteratively changing the design of the environment over time, collecting evidence of its effects, and recursively refining successive designs. Quantitative and qualitative methods may be combined, but similar to qualitative research, it uses criteria to ensure rigor such as *trustworthiness* and *credibility* akin to reliability and validity, and *usefulness*, analogous to generalizability or external validity (Barab & Squire, 2004). Like participatory action research, design-based research also involves participants to bring their different expertise into producing and analyzing designs. However, design-based research is distinguished by its goal to advance new theories and practices that can be generalized to other educational settings. Following Hoadley (2002), I use the term "design-based research" rather than "design experiments" or "design research" to avoid mistaken identification with experimental design, studies of designers, or trial teaching methods.

Design-based research and traditional psychological experiments differ on paradigmatic issues such as ontology, epistemology, methodology, and axiology. Design-based researchers assume a participative reality instead of positing that the knower has an independent existence from the subject (Barab & Kirschner, 2001). Design-based researchers' epistemological stances also vary along a continuum (see Fig. 24.1):

As Dede (2004) notes, some researchers (e.g., diSessa & Cobb, 2004) are on the objectivist end of the epistemological continuum, but suggests that most are in the middle, with cognitivists closer to the objectivist stance and the situated learning theorists on the subjectivist side. In terms of methodology, design-based researchers typically use mixed methods to describe the complex phenomena over time. For example, traditional pretest and posttest data may be combined with a few in-depth analyses of some students (A. L. Brown, 1992). Additionally, values play a large role in interpreting results. Bereiter (2002a) argues that design-based research is not

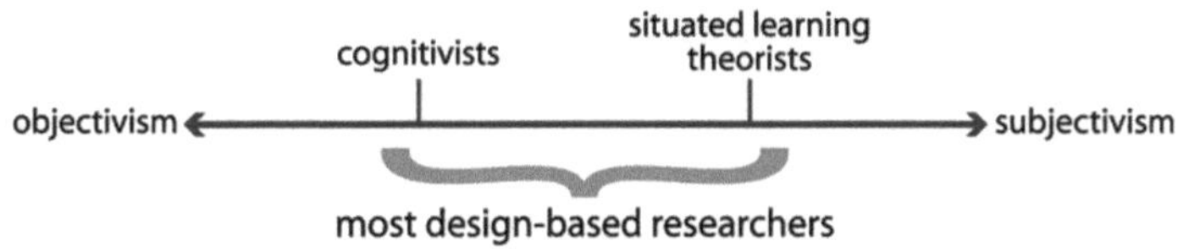

Fig. 24.1 Epistemological stances among design-based researchers adapted from Dede (2004)

defined by its methods, but the goals for sustained innovation of education. Likewise, diSessa and Cobb (2004) suggest that the goal of design-based research should be ontological innovations. Finally, design-based research shares philosophical characteristics of pragmatism with mixed-method research (Tashakkori & Teddlie, 2003), but differs in that one of its goals is to advance theory (Barab & Squire, 2004; Cobb et al., 2003; diSessa & Cobb, 2004).

The multivocal analyses by Teplovs and Fujita, Law and Wong, and Chiu seek to identify and explore "pivotal moments" in relation to the broader analysis of group processes that support knowledge building. As I participated in the data collection and the analyses, I accept a participative reality. Previously, my epistemological stance leaned towards the situated learning end, but after embracing multivocality, I seem to have moved a little closer to the objectivist side of the epistemological continuum. Being a pragmatist, I have always deployed mixed methods. As a member of the knowledge building community, I aimed to advance understanding of how knowledge building discourse can be fostered among students in online graduate courses.

Advantages and Challenges of the Multivocal Analyses for Design-Based Research

Multivocal analyses offer new perspectives for design-based researchers open to critically reflect on their own theoretical and methodological contributions. The divergent voices along the five facets of multivocality offer some advantages as well as disadvantages to reconsidering the findings from analyses of the shared data, which came from the second of three iterations of a larger design-based research study.

One significant advantage of multivocal analyses is the sharing of data. Design-based research projects collect large amounts of data over several iterations. Inevitably, some of this data is left unanalyzed. Collins et al. (2004) recommended that the design-based research community "establish an infrastructure that would allow researchers at other institutions to analyze the data collected in design studies, in order to address their own questions about learning and teaching" (p. 40). A recent analysis by Anderson and Shattuck (2012) of the five most-cited design-based research articles from each year in the past decade revealed that there was no evidence of data sharing among diverse research teams in their sample of 47 articles. The multivocal analyses in this section embrace data sharing among researchers at three different geographic locations (Canada, United States, and China) and encourage a complex social construction of meaning around the asynchronous online discussion data.

Another advantage is that the automated and quantitative analyses presented in this section offer researchers the possibility of conducting just-in-time assessments of student learning within and between iterations of the design-based research study. Modeling student learning during the redesign cycles define this iterative research approach (Kelly, Baek, Lesh, & Bannan-Ritland, 2008). Increasingly, it is possible to model student learning and more complex twenty-first century skills such as collaboration, problem solving, and learning to learn in real-time and generate usable visualizations of this activity (Johnson, Bull, Reimann, & Fujita, 2011). Cutting-edge assessment tools may enable design-based researchers to collaborate with teachers to modify and improve interventions, whether they are instructional design or technological design interventions, in real-time instead of in retrospect. This may also open the possibility for other stakeholders such as students, parents, and policy makers to participate in the design process and voice their needs in their local contexts of use. While the particularity of the intervention means that the impact of design-based research on practice is on a small scale, the design principles that can emerge out of these rich conversations hold great promise.

Finally, multivocal analyses may resolve some concerns about the question of causality in design-based research, which often employs mixed methods. As Reimann (2011) argues, causality in design-based research is a "particular causation" (Miles & Huberman, 1994, p. 147) or "action causality" (Abell, 2004) that pertains to the local needs of the particular participants involved, similar to qualitative research (Maxwell, 2004). As design-based research is distinguished by its goal to advance new theories, however, explaining how theoretical conjectures will function in the designed features of the environment, mediate learning and produce intended outcomes is an important concern (Sandoval, 2013).

Multivocal analyses were instrumental for me in reflecting on the "conjecture map" or argumentative grammar of my design-based research study and articulating the "mediating processes" or "design conjectures" (Sandoval, 2013). From Knowledge Building theory, the intended outcomes of the study were to foster students' understanding of the commitments to progressive discourse (seek common understanding, to expand the base of accepted facts) and to produce high-quality discourse among online graduate students. To do so, the "embodiment" included three intervention designs (tools and materials, activity structures, discursive practices): a reading by Bereiter (2002b), Discourse for Inquiry cards, and Knowledge Forum scaffold supports. The mediating processes, theorizing (explanations supported by authoritative sources or new information) and summaries (rise above or convergent processes), became more explicit for me though the multivocal analyses as I collaborated with the analysts. While the process of explanatory coherence (Thagard, 1989) is crucial to Knowledge Building and theorizing, it was not made so explicit in my own work at the beginning of the study. Moreover, since previous related studies had examined younger students, the emphasis had been to encourage students to provide explanations rather than facts, but my study found that the graduate student participants actually provided little evidence to back their claims in online discussions. Students did contribute summaries, but it would have been useful for instruction if summaries were framed as potential pivotal moments for shared conceptual understanding within and across weekly views in the Knowledge Forum database. Future research would benefit from such reflections on the mediating processes for supporting progressive discourse for Knowledge Building.

Yet, challenges remain in productive multivocality in design-based research. Chief among them is the lack of convergence on the definition and findings of pivotal moments, perhaps as a result of divergent theoretical assumptions underpinning the analyses. For example, while Chiu claims his SDA method may have few theoretical assumptions or commitments and may be compatible with many theoretical frameworks, researchers in the knowledge building community will have difficulty reconciling the modeling of social metacognition as knowledge building. Even when the theoretical assumptions underlying the analyses are the same, as in Teplovs and Fujita's and Law and Wong's analyses, inconsistencies in the units of analysis used make the accumulated findings difficult to interpret and apply in practice. Future research is needed to refine the design of methodologies and assessment tools presented. For example, Chiu's analysis would benefit from micro-context coding of the larger dataset to showcase his SDA on the asynchronous discussion dataset, which should reveal more pivotal moments. Teplovs and Fujita's analysis would capture pivotal moments more effectively if the analyses could be conducted in earlier weeks of the course, perhaps as a pretest and posttest to assess the effectiveness of a particular intervention (e.g., introduction of new scaffold supports) or for formative rather than summative assessments. It would also be useful to see what is happening in between iterations of the larger study. Finally, Law and Wong's analysis would benefit from streamlining to harness the most important aspect of the multifaceted analyses. This would also make the assessments more usable for researchers and teachers who must collaborate closely in design-based research to optimize the learning outcomes for students.

Acknowledgements I would like to thank Chris Teplovs, Nancy Law, Wing Wong, and Ming Ming Chiu for their insightful analyses of this dataset. This data was originally collected with funding from the Social Sciences Research Council of Canada and the Canadian Foundation for Innovation awarded to Clare Brett at the University of Toronto.

References

Abell, P. (2004). Narrative explanations: An alternative to variable-centered explanation? *Annual Review of Sociology, 30*, 287–310.

American Psychological Association. (2009). *Publication manual of the American Psychological Association* (6th ed.). Washington, DC: Author.

Anderson, T., & Shattuck, J. (2012). Design-based research: A decade of progress in education research. *Educational Researcher, 41*(2), 16–25.

Barab, S. A. (2006). Design-based research: A methodological toolkit for the learning scientist. In R. K. Sawyer (Ed.), *The Cambridge handbook of the learning sciences* (pp. 153–169). New York, NY: Cambridge University Press.

Barab, S. A., & Kirschner, D. (2001). Guest editor's introduction: Rethinking methodology in the learning sciences. *The Journal of the Learning Sciences, 10*(1&2), 5–12.

Barab, S. A., & Squire, K. (2004). Design-based research: Putting a stake in the ground. *The Journal of the Learning Sciences, 13*(1), 1–14.

Bereiter, C. (2002a). Design research for sustained innovation. *Cognitive Studies, Bulletin of the Japanese Cognitive Science Society, 9*(3), 321–327.

Bereiter, C. (2002b). *Education and mind in the knowledge age*. Mahwah, NJ: Lawrence Erlbaum Associates.

Bereiter, C., & Scardamalia, M. (2003). Learning to work creatively with knowledge. In E. De Corte, L. Verschaffel, N. Entwistle, & J. van Merrienboer (Eds.), *Powerful learning environments: Unravelling basic components and dimensions* (pp. 55–68). Oxford: Elsevier Science.

Blei, D., Ng, A. Y., & Jordan, M. I. (2003). *Journal of Machine Learning Research, 3*, 993–1022.

Brett, C., Forrester, B. C., & Fujita, N. (2009). Online learning journals as an instructional and self-assessment tool for epistemological growth. *Canadian Journal of Learning and Technology, 35*(1).

Brown, A. L. (1992). Design experiments: Theoretical and methodological challenges in creating complex interventions in classroom settings. *The Journal of the Learning Sciences, 2*(2), 141–178.

Brown, A. L., & Campione, J. C. (1990). Communities of learning and thinking, or a context by any other name. *Contributions to Human Development, 21*, 108–126.

Brown, A. L., & Campione, J. C. (1994). Guided discovery in a community of learners. In K. McGilly (Ed.), *Classroom lessons: Integrating cognitive theory and classroom practice* (pp. 229–270). Cambridge, MA: Bradford Books/MIT Press.

Burtis, J. (1998). Analytic Toolkit for Knowledge Forum. Centre for Applied Cognitive Science, The Ontario Institute for Studies in Education/University of Toronto.

Cacciamani, S., Cesareni, D., Martini, F., Ferrini, T., & Fujita, N. (2012). Influence of participation, facilitator styles, and metacognitive reflection on knowledge building in online university courses. *Computers & Education, 58*(3), 874–884.

Chiu, M. M. (2000). Group problem solving processes: Social interactions and individual actions. *Journal for the Theory of Social Behavior, 30*(1), 27–50.

Chiu, M. M. & Fujita, N. (in press). Statistical discourse analysis of online discussions: Informal cognition, social metacognition and knowledge creation. In S. C. Tan, H. J. So, and J. Yeo (Eds.) *Knowledge creation in education*. New York: Springer.

Chiu, M. M., & Kuo, S. W. (2009). From metacognition to social metacognition. *Journal of Education Research, 3*(4), 1–19.

Chiu, M. M. (2014). Cognition and social metacognition: A statistical discourse analysis of an online discussion. In D. D. Suthers et al. (Eds.) Productive Multivocality in the Analysis of Group Interactions, Computer-Supported Collaborative Learning Series 15, DOI 10.1007/978-1-4614-8960-3_23, ©Springer Science+Business Media New York 2013.

Chiu, M. M., & Pawlikowski, M. (2013). Social metacognition and micro-creativity. In E. G. Carayannis (Ed.), *Encyclopedia of creativity, invention, innovation, and entrepreneurship*. New York, NY: Springer.

Cobb, P., Confrey, J., diSessa, A., Lehrer, R., & Schauble, L. (2003). Design experiments in educational research. *Educational Researcher, 32*(1), 9–13.

Collins, A. (1992). Toward a design science of education. In E. Scanlon & T. O'Shea (Eds.), *New directions in educational technology* (pp. 15–20). Berlin: Springer.

Collins, A. (1999). The changing infrastructure of education research. In E. C. Lagemann & L. S. Shulman (Eds.), *Issues in education research: Problems and possibilities* (pp. 289–298). San Francisco, CA: Jossey-Bass.

Collins, A., Joseph, D., & Bielaczyc, K. (2004). Design research: Theoretical and methodological issues. *The Journal of the Learning Sciences, 13*(1), 15–42.

de Laat, M., Lally, V., Lipponen, L., & Simons, R.-J. (2007). Investigating patterns of interaction in networked learning and computer-supported collaborative learning: A role for Social Network Analysis. *International Journal of Computer-Supported Collaborative Learning, 2*, 87–103.

Dede, C. (2004). If design-based research is the answer, what is the question? *The Journal of the Learning Sciences, 13*(1), 105–114.
diSessa, A., & Cobb, P. (2004). Ontological innovation and the role of theory in design experiments. *The Journal of the Learning Sciences, 13*(1), 77–103.
Donald, J. (2002). *Learning to think: Disciplinary perspectives*. San Francisco, CA: Jossey-Bass.
Edelson, D. C. (2002). Design research: What we learn when we engage in design. *The Journal of the Learning Sciences, 11*(1), 105–121.
Fujita, N. (2009). Group processes supporting the development of progressive discourse in online graduate courses. Ph.D. Doctoral Dissertation, University of Toronto, Toronto, ON.
Fujita, N. (2014). Online graduate education course using Knowledge Forum. In D. D. Suthers et al. (Eds.) *Productive Multivocality in the Analysis of Group Interactions, Computer-Supported Collaborative Learning Series 15*, DOI 10.1007/978-1-4614-8960-3_20, ©Springer Science+Business Media New York 2013.
Fujita, N., & Teplovs, C. (2009). Automating analysis of collaborative discourse: Identifying idea clusters. In C. O'Malley, D. Suthers, P. Reimann, & A. Dimitracopoulou (Eds.), *Computer supported collaborative learning practices* (Vol. 2, pp. 162–169). Rhodes, Greece: International Society of the Learning Sciences.
Garrison, D. R., Anderson, T., & Archer, W. (2001). Critical thinking, cognitive presence and computer conferencing in distance education. *The American Journal of Distance Education, 15*(1), 7–23.
Gunawardena, C. N., Lowe, C. A., & Anderson, T. (1997). Analysis of a global online debate and the development of an interaction analysis model for examining social construction of knowledge in computer conferencing. *Journal of Educational Computing Research, 17*(4), 397–431.
Hacker, D. J., & Bol, L. (2004). Metacognitive theory. In D. M. McInerney & S. van Etten (Eds.), *Big theories revisited* (Vol. 4, pp. 275–297). Greenwich, CT: Information Age.
Hewitt, J. (1999). *The Vocabulary Analyser for Knowledge Forum.* Centre for Applied Cognitive Science, The Ontario Institute for Studies in Education/University of Toronto.
Hewitt, J. G. (2001). Beyond threaded discourse. *Journal of Educational Telecommunications, 7*(3), 207–221.
Hewitt, J. G. (2005). Towards an understanding of how threads die in asynchronous computer conferences. *The Journal of the Learning Sciences, 14*(4), 567–589.
Hoadley, C. (2002). Creating context: Design-based research in creating and understanding CSCL. In G. Stahl (Ed.), *Computer support for collaborative learning 2002* (pp. 453–462). Mahwah, NJ: Erlbaum.
Jacobbe, T., & Horton, R. M. (2010). Elementary school teachers' comprehension of data displays. *Statistics Education Research Journal, 9*(1), 27–45.
Johnson, M., Bull, S., Reimann, P., & Fujita, N. (2011). Epistemic beliefs and open learner models. In D. Pérez-Marin, M. Kravcik, & O. C. Santos (Eds.), *Proceedings of the International Workshop on Personalization Approaches in Learning Environments (PALE 2011)* (pp. 19–24). Girona, Spain: Springer.
Kelly, A. E., Baek, J. Y., Lesh, R. A., & Bannan-Ritland, B. (2008). Enabling innovations in education and systematizing the impact. In A. E. Kelly, R. A. Lesh, & J. Y. Baek (Eds.), *Handbook of Design Research Methods in Education*. New York, NY: Routledge.
Labaree, D. F. (2003). The peculiar problems of preparing educational researchers. *Educational Researcher, 32*(4), 13–22.
Landauer, T. K., Laham, D., & Derr, M. (2004). From paragraph to graph: Latent semantic analysis for information visualization. *PNAS, 101*, 5214–5219.
Law, N. & Wong, O.-W. (2014). Exploring pivotal moments in students' knowledge building progress using participation and discourse marker indicators as heuristic guides. In D. D. Suthers et al. (Eds.) Productive Multivocality in the Analysis of Group Interactions, Computer-Supported Collaborative Learning Series 15, DOI 10.1007/978-1-4614-8960-3_22, ©Springer Science+Business Media New York 2014.
Lee, E., Chan, C., & van Aalst, J. (2006). Students assessing their own collaborative knowledge building. *International Journal of Computer-Supported Collaborative Learning, 1*(1), 57–87.

Leech, N. L., & Onwuegbuzie, A. J. (2007). An array of qualitative data analysis tools: A call for data analysis triangulation. *School Psychology Quarterly, 22*(4), 557–584.

Maxwell, J. A. (2004). Causal explanation, qualitative research, and scientific inquiry in education. *Educational Researcher, 33*(2), 3–11.

Miles, M. B., & Huberman, A. M. (1994). *Qualitative data analysis: A sourcebook of new methods* (2nd ed.). Thousand Oaks, CA: Sage.

Onwuegbuzie, A. J., & Dickinson, W. B. (2008). Mixed methods analysis and information visualization: Graphical display for effective communication of research results. *The Qualitative Report, 13*(2), 204–225.

Reimann, P. (2011). Design-based research. In L. Markauskaite, P. Freebody, & J. Irwin (Eds.), *Methodological choices and research designs for educational and social change: Linking scholarship, policy and practice* (pp. 37–50). New York, NY: Springer.

Rourke, L., & Kanuka, H. (2007). Barriers to online critical discourse. *International Journal of Computer-Supported Collaborative Learning, 2*(1), 105–126.

Sandoval, W. (2013). Conjecture mapping: An approach to systematic educational design research. *Journal of the Learning Sciences, 22*(1), 1–19.

Scardamalia, M., & Bereiter, C. (1991). Higher levels of agency in knowledge building: A challenge for the design of new knowledge media. *The Journal of the Learning Sciences, 1*(1), 37–68.

Scardamalia, M., & Bereiter, C. (1994). Computer support for knowledge-building communities. *The Journal of the Learning Sciences, 3*(3), 265–283.

Scardamalia, M. (2002). Collective cognitive responsibility for the advancement of knowledge. In B. Smith (Ed.), *Liberal education in a knowledge society* (pp. 67–98). Chicago: Open Court.

Scardamalia, M., & Bereiter, C. (2003). *Knowledge Building. Encyclopedia of Education* (2nd ed.). New York, NY: Macmillan Reference.

Scardamalia, M., & Bereiter, C. (2006). Knowledge building: Theory, pedagogy, and technology. In K. Sawyer (Ed.), *Cambridge Handbook of the Learning Sciences*. New York, NY: Cambridge University Press.

Suthers, D., Dwyer, N., Medina, R., & Vatrapu, R. (2010). A framework for conceptualizing, representing, and analyzing distributed interaction. *International Journal of Computer-Supported Collaborative Learning, 5*(1), 5–42.

Tashakkori, A., & Teddlie, C. (2003). *Handbook of mixed methods in social & Behavioral research*. Thousand Oaks: Sage.

Teplovs, C., & Fujita, N. (2009). Asynchronous Knowledge Forum postings from a graduate course in education. Paper presented at the Pinpointing Pivotal Moments in Collaboration Workshop, STELLAR Alpine Rendez-Vous 2009, Garmisch-Partenkirchen, Germany.

Teplovs, C. (2008). The Knowledge Space Visualizer: A tool for visualizing online discourse. In G. Kanselaar, V. Jonker, P. A. Kirschner, & F. Prins (Eds.), *Proceedings of the International Conference of the Learning Sciences 2008: Cre8 a learning world*. Utrecht, NL: International Society of the Learning Sciences.

Teplovs, C., Fujita, N., & Vatrapu, R. (2011). Generating predictive models of learner community dynamics through visualization of networks based on social network analysis and latent semantic analysis. In P. Long, G. Siemens, G. Conole, & D. Gasevic (Eds.), *Proceedings of the First International Conference on Learning Analytics & Knowledge 2011* (pp. 147–152). New York, NY: ACM.

Teplovs, C. & Fujita, N. (2013). Socio-dynamic Latent Semantic Learner Models. In D. D. Suthers et al. (Eds.) *Productive Multivocality in the Analysis of Group Interactions, Computer-Supported Collaborative Learning Series 15*, DOI 10.1007/978-1-4614-8960-3_21, ©Springer Science+Business Media New York 2013.

Teplovs, C., Donoahue, Z., Scardamalia, M., & Philip, D. (2007). Tools for concurrent, embedded, and transformative assessment of knowledge building processes and progress. In C. A. Chinn, G. Erkens, & S. Puntambekar (Eds.), *Proceedings of the 8th International Conference on Computer Supported Collaborative Learning* (pp. 721–723). New Brunswick, NJ: International Society of the Learning Sciences.

Thagard, P. (1989). Explanatory coherence. *Behavioral and Brain Sciences, 12*, 435–502.

Tscholl, M., & Dowell, J. (2009). Analysis of uptakes and individualistic appropriation in Knowledge Forum: Theoretical perspectives and an example. Paper presented at the Pinpointing Pivotal Moments in Collaboration Workshop, STELLAR Alpine Rendez-Vous 2009, Garmisch-Partenkirchen, Germany.

Vatrapu, R., Teplovs, C., Fujita, N., & Bull, S. (2011). Towards visual analytics for teachers' dynamic diagnostic pedagogical decision-making. In P. Long, G. Siemens, G. Conole, & D. Gasevic (Eds.), *Proceedings of the First International Conference on Learning Analytics & Knowledge 2011* (pp. 93–98). New York, NY: ACM.

Wise, A., & Chiu, M. M. (2011). Analyzing temporal patterns of knowledge construction in a role-based online discussion. *International Journal of Computer-Supported Collaborative Learning, 6*, 445–470.

Woodruff, E., & Brett, C. (1999). Collaborative knowledge building: Preservice teachers and elementary students talking to learn. *Language and Education, 13*(4), 280–302.

Zampa, V., & Lemaire, B. (2002). Latent Semantic Analysis for User Modeling. *Journal of Intelligent Information Systems, 19*, 15–30.

Zhang, J., Scardamalia, M., Reeve, R., & Messina, R. (2009). Designs for collective cognitive responsibility in knowledge building communities. *Journal of the Learning Sciences, 18*(1), 7–44.

Part VI
Case Study 5: A Data-Driven Design Cycle for 9th Grade Biology

Section Editor: Carolyn P. Rosé,
Carnegie Mellon University

The unique focus of this section is on using multivocality to enhance a data-driven design process by offering a multifaceted understanding of how interventions under development interact with group functioning. Four analysts offer their interpretation of what went right and what went wrong in a pilot evaluation of a new form of software-agent-based support for scientific discovery learning in 9th grade biology. The four distinct analytic approaches include linguistic analysis from a systemic functional linguistic perspective, ethnographic analysis, ethnomethodological interaction analysis, and network analysis. Each methodological lens identifies unique opportunities to refine and improve the intervention, which illustrates how a multivocal iterative development process enables each iteration to suggest a wider breadth of opportunities for improvement.

The goal of the design process that is the subject of Part 6 is to develop a form of dynamic support for collaborative learning that draws from the classroom discourse research on Academically Productive Talk (APT). In Chap. 25, Dyke, Howley, Adamson, Kumar, and Rosé introduce the study and data set in their chapter entitled "Towards Academically Productive Talk Supported by Conversational Agents." This chapter situates the study in the context of prior work using other forms of conversational agent-based support for collaborative learning. It describes in detail the design of the instructional materials that provided the context in which student groups interacted with the APT agent.

Howley, Kumar, Mayfield, Dyke, and Rosé kick off the multivocal analysis process in Chap. 26 with a chapter entitled "Gaining Insights from Sociolinguistic Style Analysis for Redesign of Conversational Agent Based Support for Collaborative Learning." Using the Tatiana visualization tool to examine patterns of codes in the three-dimensional Souflé analysis framework, this chapter illustrates how linguistic evidence of social positioning within groups pinpoints the experiences that students responded negatively toward, and how that negative response extended over time and interfered with positive experiences that the activity was meant to provide. Cress and Kimmerle follow with an ethnographic study from the perspective of Group Awareness in their chapter entitled "Successful Knowledge

Building Needs Group Awareness—Interaction Analysis of a 9th Grade CSCL Biology Lesson." Rather than focus on the details of the experimental manipulation, the Cress & Kimmerle analysis views the collaborative setting holistically in terms of the desired affordances not provided for meaningful engagement because of deficiencies in group awareness.

The next two analyses, namely Stahl's "Interaction Analysis of a Biology Chat" and Goggins & Dyke's "Network Analytic Techniques for Online Chat" focused on roles within the interaction, and specifically how the role taken by the agent may have limited the opportunities for leadership role taking of students within the group discussions. While Stahl's analysis was more ethnomethodological in nature, the Goggins and Dyke analysis draws from and integrates ethnographic analysis methods and social network analysis methods.

Hmelo-Silver wraps up Sect. 6 with her chapter "Multivocality as a Tool for Design-Based Research," which is an insightful commentary on the ups and downs of the multivocal analysis process and what lessons can be learned about use of multivocality in iterative design.

Chapter 25
Towards Academically Productive Talk Supported by Conversational Agents

Gregory Dyke, Iris Howley, David Adamson, Rohit Kumar, and Carolyn Penstein Rosé

Introduction

The frequent occurrence of over-full classroom settings particularly in urban schools has lead the classroom discourse community to question how discussions in such classrooms can be academically productive, particularly if we wish to use such situations to develop reasoning skills. A large body of work has shown that certain forms of classroom interaction, termed Academically Productive Talk (APT) are beneficial for learning with understanding (Michaels, O'Connor, & Resnick, 2008; Resnick, Asterhan, & Clarke, in press; Resnick, Bill, & Lesgold, 1992; Resnick, O'Connor, & Michaels, 2007). This work has also shown the crucial role of the teacher in facilitating these discussions. The academically productive talk form of classroom interaction is one in which a facilitator (or an agent) poses a question that calls for a relatively elaborated response (e.g., both a solution and a reason for the solution) and then presses the group to build on or challenge these ideas, with the purpose of keeping student reasoning at center stage and increasing student ownership of ideas.

The study described in this chapter is part of a collaborative effort Lauren Resnick of the Learning Research and Development Center and Carolyn Rosé of Carnegie Mellon University's School of Computer Science are leading under the umbrella of the Pittsburgh Science of Learning Center in which they are partnering with the ninth grade biology team in a nearby urban school district. As part of that partnership, they are working with administrators and teachers across the district to introduce APT practices into classrooms. Early studies of APT in classrooms showed success with highly skilled teachers or privileged student populations.

G. Dyke (✉)
University of Lyon, Lyon, France
e-mail: gregdyke@gmail.com

I. Howley • D. Adamson • R. Kumar • C.P. Rosé
Carnegie Mellon University, Pittsburgh, PA, USA

D.D. Suthers et al. (eds.), *Productive Multivocality in the Analysis of Group Interactions*, Computer-Supported Collaborative Learning Series 16,
DOI 10.1007/978-1-4614-8960-3_25, © Springer Science+Business Media New York 2013

The goal we are working towards is increasing the extent to which these APT based practices are used within typical urban classrooms, using professional development with teachers and Computer Supported Collaborative Learning (CSCL) experiences for small groups of students as tools for reshaping the classroom culture (Rosé & Tovares, in press).

This effort builds on an earlier history of successful deployment of intelligent conversational agents for support of small group learning (Kumar & Rosé, 2011; Kumar, Rosé, Wang, Joshi, & Robinson, 2007). In the past 6 years, technology for dynamic support for collaborative learning has matured both in terms of its ability to monitor collaboration through automatic collaborative learning process analysis (Gweon, Jain, McDonough, Raj, & Rosé, 2013; Rosé et al., 2008) as well as to offer context appropriate support for effective participation in groups (Kumar & Rosé, in press), such as using conversational agent technology. In early studies, these computer agents served the purpose of elevating the conceptual depth of collaborative discussions by leading students in groups through directed lines of reasoning, referred to as knowledge construction dialogues, which were meant to scaffold the process of groups constructing conceptually rich explanations together (Chaudhuri, Kumar, Howley, & Rosé, 2009; Kumar, Gweon, Joshi, Cui, & Rosé, 2007). In this study, we have begun exploring how we might design conversational agents that employ APT practices as scaffolding for online collaborative learning discussions, which were successful at leading to increases in learning in the second year of the data collection effort (Adamson, Ashe, Jang, Yaron, & Rosé, 2013; Adamson, Dyke, Jang, & Rosé, accepted; Dyke, Adamson, Howley, & Rosé, 2013).

This study reports on an iterative design process extending over 2 years during which we have conducted two complete cycles of design development, deployment, and analysis. The second year design draws on and benefits from lessons learnt from the multivocal analyses presented in the chapters within this section, which were conducted after the first year study, which we refer to as the Cell Model Study. Within the context of this district wide design study on the introduction of such practices both to teachers and students, we initially focused mainly on teacher training. Our initial observations, showing a relatively slow rate of adoption by teachers, led us to consider alternative means to accustom students to APT in order that they might be better responsive in the future to the teacher's classroom scaffolding. One way we have chosen to work towards this goal is to introduce APT practices to students in small group activities facilitated by conversational agents. Looking back on the 2 year effort, we find evidence that the computer supported collaborative learning activities played an important enabling role in the professional development effort (Clarke et al., 2013). Students came to whole class discussions better prepared and able to engage in intensive discussion after the CSCL activities, which then elevated the teacher's adoption of APT by 1.7 standard deviations.

While this partnership offers a tremendous opportunity both to pilot a new form of support for collaborative learning and to reach into real classrooms with state-of-the-art teaching practices and CSCL technology, this effort imposes a number of real world constraints that have challenged the ability to foster rich group knowledge building experiences. Nevertheless, this foray into translational research at the high school level provides a realistic picture of what must be overcome in order to achieve an impact in an urban school district with CSCL technology.

In the remainder of this chapter, we first describe the scientific questions that prompted this research both from the discourse for learning perspective and the technology perspective. We then describe the educational setting and the design of the year 1 study. We describe the details of our data collection effort and what data was passed on to analysts. Finally, we describe how our second iteration was different from the first, drawing the lessons learnt from the multivocal analysis of the first year's data. We conclude with a short discussion about lessons learned from the multivocal process.

Scientific Objectives

Theoretical Roots in the Classroom Discourse and Collaborative Learning Literature

The notion of Academically Productive Talk (APT) stems from frameworks that emphasize the importance of social interaction in the development of mental processes, and has developed in parallel to similar ideas from the computer-supported collaborative learning community. Michaels, O'Connor, and Resnick (2008) describe some of the core dialogic practices of Accountable Talk along three broad dimensions:

1. Students should be accountable to the learning community, listening to the contributions of others and building on them to form their own.
2. Students should be accountable to accepted standards of reasoning, emphasizing logical connections and drawing reasonable conclusions.
3. Students should be accountable to knowledge, making arguments that are based explicitly on facts, written texts or other public information.

Such practices are often unfamiliar in the classroom. Not only must they be introduced to students but it is necessary to provide teachers with the means to scaffold these interaction forms. Drawing on over 15 years of observation and study, Michaels, O'Connor, and Resnick propose a number of core "moves" that teachers can draw upon in order to encourage the development of academically productive classroom discussion, among which are:

1. Revoicing: "So let me see if I've got your thinking right. You're saying XXX?" (with time for students to accept or reject the teacher's formulation);
2. Asking students to restate someone else's reasoning: "Can you repeat what he just said in your own words?";
3. Asking students to apply their own reasoning to someone else's reasoning: "Do you agree or disagree and why?";

4. Prompting students for participation: "Would someone like to add on?";
5. Asking students to explicate their reasoning: "Why do you think that?" or "How did you arrive APT that answer?" or "Say more about that".

This work is related to work in collaborative learning process analysis (Berkowitz & Gibbs, 1983; Weinberger & Fischer, 2006). In our bringing together of these two bodies of work, we began by investigating the connection between facilitation moves under the heading of "Academically Productive Talk", and what has been called "transactivity" in the collaborative learning community (Berkowitz & Gibbs, 1983) or Social Modes of Co-Construction (Weinberger & Fischer, 2006). Transactive contributions are arguments constructed in such a way as to reference (sometimes described as "operating on") the previously expressed reasoning of self or others. For example, consider the following dialogue excerpt from Chapin, O'Connor, and Anderson (2009):

S1: Well, I don't think it matters what order the numbers are in. You still get the same answer. But three times four and four times three seem like they could be talking about different things.
Teacher: Rebecca, do you agree or disagree with what Eddie is saying?
S2: Well, I agree that it doesn't matter which number is first, because they both give you twelve. But I don't get what Eddie means about them saying different things.

Notice how the first student starts out with an attempt at expressing his reasoning about a mathematical idea. The teacher then enters to encourage another student to attend to and address his reasoning attempt. The second student then responds, articulating not only her own reasoning, but also how it relates to the reasoning already expressed by the first student. In so doing, she has met the two basic criteria for a transactive utterance. She made her reasoning explicit in her articulation, and she made a connection between that reasoning and some reasoning made explicit in a previously articulated utterance. The teacher's facilitation has played a key role in encouraging this transactive conversational behavior and is one of the facilitation moves which has been described for Academically Productive Talk.

A body of work in the collaborative learning community supports the value of this kind of transaction as a property of discussions for learning (Azmitia & Montgomery, 1993). Within the problem based learning community, where discussion groups are smaller, but lead by skilled facilitators, similar ideas have emerged (Hmelo-Silver & Barrows, 2006).

Technology Roots: Conversational Agents for Regulating Social Interactions

Conversational agents have a long history of successful support for individual learning with technology (e.g., Rosé et al., 2001; Rosé & VanLehn, 2005). However, whether because of limitations in agent capabilities or expectations of agent abilities that lead students to interact with them in less rich ways than they would with other humans, conversational interactions with computer agents tend to lack the depth and richness that is desirable from a learning perspective. On the other hand, interactions between human students may or may not exhibit the richness that we hope for, and an active area of research within the Computer Supported Collaborative Learning community continues to be development of a variety of forms of support for collaborative learning.

Beginning with a wizard of oz study with college Freshman learning calculus, we began to experiment with the idea of conversational agents supporting collaborative learning (Gweon, Rosé, Zaiss, & Carey, 2006). Kumar, Rosé, Wang, Joshi, & Robinson, (2007) conducted the first study of fully automatic conversational agent based support for collaborative learning. In that study, college sophomore thermodynamics students worked either individually or in pairs on an engineering design task, with or without the support of a conversational agent. In all conditions, students had access to the same supporting information on the same task and in the same online environment. Students took a pretest before the design task and a posttest afterwards. Students who worked both with a partner student and a supportive agent learned 1.24 standard deviations more than students who worked individually without the support of a conversational agents. Students who worked either with a partner student (but without a support agent) or with a support agent (but without a partner student) learned about 1 standard deviation more than students who worked alone and without a support agent. In this study, the support agents were the same conversational agents that lead individual students through directed lines of reasoning related to domain content that had been evaluated in an earlier study of individual learning with conversational agents.

In the years that followed, we continued to refine the design of these conversational agents who served to inject conceptually deep domain content into collaborative discussions (Ai et al., 2010; Chaudhuri et al., 2009; Howley, Chaudhuri, Kumar, & Rosé, 2009; Howley, Mayfield, & Rosé, 2011; Kumar & Rosé, 2011, in press; Kumar, Rosé, et al., 2007, 2010, 2011). Important innovations include offering students control over timing of feedback (Chaudhuri et al., 2009), using social strategies motivated by the field of collaborative group work (Kumar & Rosé, in press; Kumar et al., 2010, 2011), and showing alignment with student goals (Ai et al., 2010). A key aspect of this iterative development process has been a partnership between our team and Gerry Stahl's Virtual Math Teams group (Stahl & Rosé, 2011; Stahl, Rosé, & Goggins, 2010; Stahl, Rosé, O'Hara, & Powell, 2010). In was in part as a result of the intellectual exchange that has been part of that collaboration that we became dissatisfied with the idea of injecting conceptual content into collaborative discussions and instead turned our attention to stimulating group knowledge construction more directly with the behavior of the agents.

Instructional Context

As part of this design study, ninth grade biology teachers across the district participated in a series of professional development sessions in which they were instructed in using APT techniques in their classroom discussions. The students were also aware of the goal of the district to encourage active participation in group discussions. Posters encouraging APT in the classroom were regularly observed in classrooms across the district. Nevertheless, bringing change to classroom discussion practices is not an easy task in an urban school district. Thus, the data we provide here is not idealized in any sense.

We were requested to keep the materials of the study as close as possible to the regular BSCS (Biological Sciences Curriculum Study) curriculum the students were using in their classes. The instructional activity used in the study consisted of a homework assignment as preparation, a collaborative exercise conducted online in groups of three students as practice and reinforcement, and a whole group teacher led discussion as wrap up to promote reflection and deepen understanding. We chose the "Build a Cell Model" activity that took place during Unit 4 (Homeostasis) of the ninth grade biology curriculum because it was associated with small group and whole group discussions related to the concept of diffusion, so it seemed to have enough conceptual content to provide fodder for discussion. The purpose of the online CSCL activity was to give the students practice using APT moves with each other in small groups in order to prime them to be active and responsive when their teacher attempted to use those classroom discussion facilitation moves in the whole group discussion that followed in the subsequent class period.

This year Cell Model study was conducted in 4 classrooms with the same teacher, for a total of 50 consenting students (out of 78). The teacher has each classroom at a fixed "period" of the day, every day, hence the appellations: period 1, 3, 6 and 9. Class periods were only 43 min long. For several reasons, such as innate ability, time of day, number of students, streaming, and teacher expectations, the different periods had distinct characters. Period 1 is the best performing and has around 20 students. Period 3 is quite small, with 8 students (several of whom were absent for the pullout study), leading to a more conversational habitual type of classroom discussion, but which is more prone to stagnation when a few of the leading student voices happen to lack motivation on a particular day. Period 6 is the largest (29 students) and occurs just before lunch. It is quite noisy but with students who are frequently motivated to work. Period 9 is slightly smaller (26 students) but has several students who are disruptive and very hard to motivate and give attention to within such a large classroom.

Study 1

Content Learning Objective: Understanding Diffusion in Cell Models

The content learning objective of the "Build a Cell Model" activity was to offer students the opportunity to investigate the role of semipermeable membranes in homeostasis. The readings provided to the students ahead of the lab exposed them to the basic vocabulary and concepts, such as permeability, membrane, diffusion, glucose, etc. The lab was designed as an inquiry activity where students actively engaged in making predictions, observing effects, and explaining observations. In being exposed to a series of contrasting cases, the students were meant to actively construct an understanding of how diffusion works. The whole class teacher led discussion offered them the opportunity to discuss their findings with the whole class, to further reflect on the concepts, and deepen their understanding.

It is apparently common practice for this lab not to be performed by students but demonstrated by the teacher instead. However, in the original lab as it was designed within the BSCS curriculum, students actually built their own cell models using dialysis tubing, beakers, water, glucose, starch, and iodine. The students first generated hypotheses, then created cell models to test those hypotheses, observed the cell at 1, 5, and 24 h. And then discussed whether their observations matched their predictions or not, and if not, what that must mean for their concept of diffusion. In the original design of the lab, although the lab was always a group activity, students recorded their predictions, observations, and explanations on their own worksheets rather than jointly create one with their partners. We chose to keep this arrangement so that we would have a record of each student's thinking apart from the group they were participating in.

In the absence of a simulator that would allow students to build a cell model online, we adapted this lab as an online group activity by offering three contrasting cell designs for students, which we illustrated through videos, and then asking students to make predictions about what they would see after 1, 5, and 24 h. While this doesn't offer them as much practice with the scientific method as forming their own hypotheses and basing their cell model designs on that, we were able to identify three strategic configurations that would allow them to construct an understanding of the concept of a semipermeable membrane. The three cell models we picked out are as follows:

(a) Glucose solution inside, water outside
(b) Water inside, glucose solution outside
(c) Starch suspension inside, water + iodine outside

In the first two models, both glucose and water diffuse through the membrane to achieve equilibrium. This molecular movement can be measured through weight changes to the cell model (indicating water movement) and glucose test strips measuring the inside and outside concentration of glucose. In the third model, starch is prevented from diffusing outside by the membrane, iodine and water diffuse in, affecting the weight of the cell model and changing the inside color to dark blue (iodine is used as an indicator of the presence of starch—the outside color on the other hand does not change).

From observing the three cell models, it can be concluded that dialysis tubing is semipermeable, letting glucose, water, and iodine diffuse through freely and preventing the diffusion of starch. Provided reading materials allow students to formulate the hypothesis that this is due to the relative size of the different molecules (starch being bigger).

Process Learning Objective: Academically Productive Talk

The process oriented learning objective for the study was to increase student skill at engaging in APT discussions. During the preparation phase of the discussion, students were provided with written materials in the form of a cartoon to offer examples of the kinds of conversational behavior we expected during the collaborative activity (cf. Fig. 25.1). During the collaborative activity, students continued to have access to these materials. Additionally, this instruction was reinforced through interactive support that was administered during the collaborative activity. Finally, students had the opportunity to gain further practice with engaging in APT discussions with a whole class teacher led discussion. The teacher's own facilitation behavior provided support for their continued engagement in articulation of reasoning and building on each other's reasoning during that phase of the activity.

Training phase. Because the goal was to prepare students for an Academically Productive Talk discussion, we offered them training materials prior to the collaborative activity to illustrate the kind of conversational behavior we were trying to elicit. Students had been encouraged by their instructor to engage in APT discussions. The instructions for the lab referred to these interactions with the instructor along with a reification of those discussion facilitation strategies. A key feature of the instructional cartoon is that it illustrates distinct roles that can be taken within the collaborative discussion that each emphasize one particular APT move and its target response

Online collaborative phase. In the instructions for the online collaborative activity, students were told that each of them would participate both as the instructor and the student in this collaborative activity. In particular, each of them would be responsible for looking for opportunities for a specific APT discussion strategy to be used in the conversation. They were directed to notice that these strategies were illustrated in the cartoon in their materials, which is reproduced here for reference. The specific roles were Revoicer, Challenger, and Explainer labeled on the side of the cartoon. They were told that each of them would be assigned one of these roles. For example, they were told that if they were assigned the Revoicer role, they would be responsible for looking for opportunities where they or another team member could revoice what another student has said. It would be their responsibility to request other students to perform the revoicing move, as they saw in the cartoon below, if they saw such opportunities. And similarly for the other two roles. During the collaborative activity, they continued to have access to the training cartoon illustrating the roles that they were given as preparation. Within each group of three students, the three APT roles were randomly assigned to the students so that each one was represented within each group. In this way, the students were "playing teacher" in addition to being students in the discussion.

The collaborative activity took place online in the ConcertChat environment, where intelligent conversational agents were used to provide task structuring, feedback, and in some conditions support for APT.

Whole class discussion. The whole class teacher led discussion that took place the class period after the online activity offered the students the opportunity to

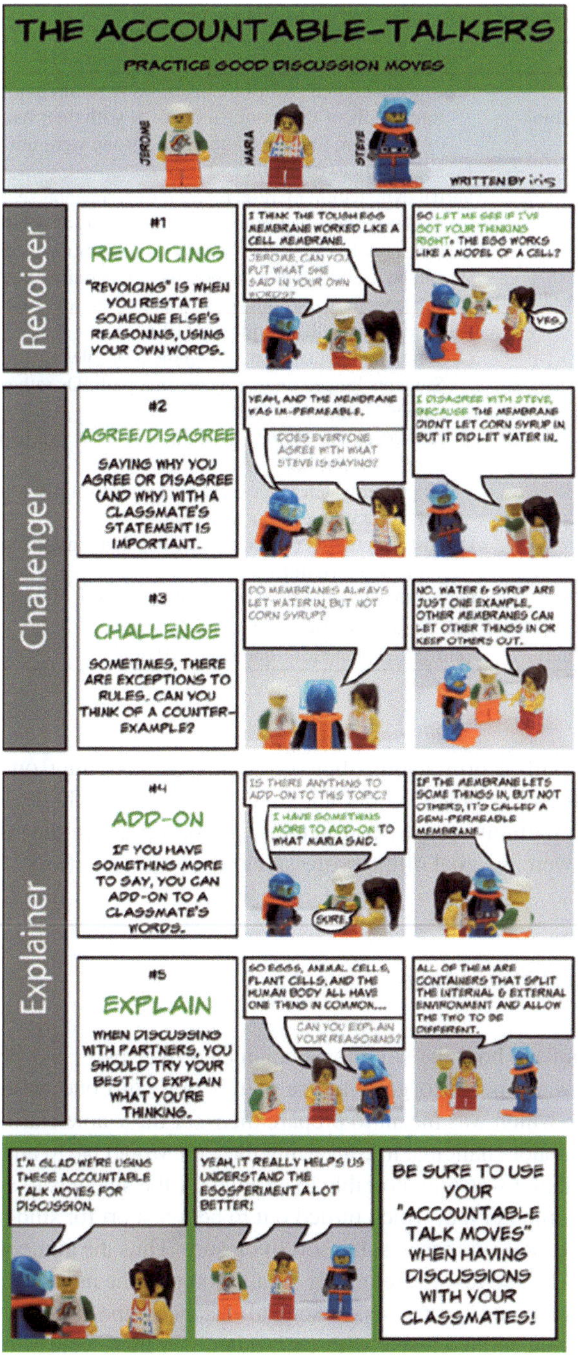

Fig. 25.1 Academically productive talk training materials

Table 25.1 Overview of logistics for the Cell Model Study

Before activity	Students assigned reading materials as homework
Day 1 (1 period duration - 43 min)	Seating assignment—the computers were set up in a U shape around three walls of the room; students sit with their backs to the center of the room. Students in the same group were not placed adjacently to each other
	Verbal explanation by researcher of experimental setup (of the cell model biology experiment, i.e., why dialysis tubing works as a cell model, what the experimental conditions were, how indicators can be used to detect the presence of certain molecules), of the subsequent classroom activity and of the role Academically Productive Talk should play (10 min)
	Log on to concert chat and assignment of roles by tutoring agent. Students logged on to concert chat with IDs rather than their names (2 min)
	Discussion phase 1 (chat): prediction of what will happen in each condition (5 min)
	Video phase: watch videos and make observations (each student on his or her own computer) (6 min)
	Discussion phase 2 (chat): discuss observations and come up with explanation (6 min)
	Quiz 1 (MCQ and Fill in the blank) (10 min)
Day 2 (1 period duration 43 min)	Full class APT discussion (20–30 min)
	Quiz 2 ("explanation" question) (10 min)

display their skill at articulating their reasoning and building on one another's reasoning as they had practiced in their small groups. This whole class discussion was also an opportunity for the instructor to put into practice the APT facilitation practices that were the goal of the professional development effort.

Experimental Procedure

Table 25.1 displays how the study was conducted over two class periods in total. The students were meant to get practice using APT with one another during the online activity, which was meant to prepare them for the whole class discussion the next day. One major challenge in converting the Cell Model lab to an online activity, however, was that it was only possible to arrange for the students to spend one class period in the computer lab, which turned out to be too short for students to get their ideas out on the table and reach some sort of closure. Thus the number of opportunities for using APT moves was relatively small. Some of the most positive outcomes were that in practicing the moves, in several instances, the students realizing they were not communicating clearly with each other, and these confusions and misunderstandings that might have slipped past unnoticed were at least voiced, leading to a very productive classroom discussion the following day.

The purpose of the computer agents was to support students in practicing the prescribed discussion behaviors. We were interested to see whether we would observe any positive effects of the new Academically Productive Talk agents we were piloting for the first time in this study, and whether those effects would also be visible within the whole group discussion that followed the collaborative activities in the next class period. In contrast to previous studies, in this study, the agents provided no conceptual content whatsoever, and only provided instructions prior to the collaborative activity, and process oriented support during the activity.

Measurement

The following data was collected and shared with the analysts:

- Transcripts of chat discussions
- Worksheets
- Test scores for the three tests
- Class discussion audio recording (not shared, as includes some non-consenting students)
- Class discussion transcripts

Domain knowledge was measured at three time points using a paper based test that included both multiple choice questions and open ended, explanation type questions. In order to capture how well the students grasped the material based on their test answers, we differentiated ability to remember what was observed (weight changes, indicator color changes), to understand what the observation meant (weight change means water movement, indicator color change means glucose and starch presence), and to explain and generalize (glucose always moving down the concentration gradient towards equilibrium, starch unable to move through the semipermeable membrane). As the questions asked did not specifically request students to remember or understand, but rather to explain, the students with the better understanding tended to obtain low scores on the first two measurements and better on the final one. As a holistic measure of performance, we also ranked the explanation answers from "best" to "worst", using pairwise comparison, rather than a scoring rubric.

Experimental Design

Students were assigned to discuss in groups of three with an APT tutoring agent, using Concert Chat. Each student was assigned one of three roles: revoicer (prompt for revoicing); explainer (prompt for adequate justifications); challenger (prompt for alternative ideas and agreement/disagreement). The APT tutoring agent was set up to accomplish two roles:

6. Guide them through the phases of the activity

7. Provide various levels of support to the assigned roles

The levels of support formed the 3 experimental conditions of our study:

None	provide no APT support (only guiding through phases of activity)
Direct	directly prompt students using APT moves ("John, could you say what Ann said in your own words")
Indirect	prompt students to fulfill their assigned role ("Susan, could you ask John to say in his own words what Ann said").

Here is an example of an interaction with the conversational agent in the Direct condition:

Student1: I think it's going to get heavier.
Alex (tutor): Student2, do you agree with what Student1 just said?
Student2: Wait I'm confused, please explain this again.
Student1: The egg will get bigger… heavier
Alex (tutor): Student3, do you agree with what Student1 just said?
Student3: I can't understand.
Student2: Why are we talking about eggs?
Student3: oh, ok, I get it.

In this example, when the agent detects that a student has made a prediction, it tries to get the other students to challenge the prediction. In this case, the response is that both of the other students admit that they are confused. This is actually a productive response since voicing confusion can be a precursor to a useful clarification dialogue. If students don't voice their confusion, they are less likely to achieve clarity within the conversation.

Here is an example of an interaction with the conversational agent in the Indirect condition:

Alex (tutor): You should now move on to discussing what will happen in Condition C and your explanation for this change.
Student1: When you put sure and water it's going to get smaller.
Alex (tutor): Student2, check with Student3 if he/she agrees with Student 1.
Student2: Student3, do you agree?
Alex (tutor): When you are in agreement, write down your predictions and explanations for Conditions A, B, and C on your worksheet.
Student3: yea

In this case, rather than ask if a student agrees or disagrees, the tutor directs the student in the Challenger role to play his/her role as Challenger. Here we see the Challenger following the tutor agent's prompting and playing his/her role.

Reflections and Study 2

Several aspects of the Cell Model study diverged from the planned ideal, for practical reasons. Here we outline some of those challenges that emerged.

For consent and anonymity reasons, we chose to use student IDs, rather than their names in the chat. Many students were frustrated at not knowing exactly who they were talking to (and having to do so by chatting in spite of being in the same classroom) and spent considerable energy discussing this.

Despite the experimenters' attempts to keep the students productively engaged, students often lagged behind where they were instructed to be and frequently did not respond as intended to instructions. Evidence of students being out of synch with each other and with the instructions can be seen in many transcripts.

One challenge was to direct the students' attention to the right application on their computer at the right time, since the video that they watched was not embedded in the VMT environment. Furthermore most students engaged conscientiously in filling in their worksheets, occasionally to the detriment of the initial conversation about predictions, which was often slow to get started. As a result, students often didn't watch the videos until 5–10 min after they were instructed to do so by the tutor—experimenters specifically instructed them to ignore the tutor and gave verbal instructions regarding watching the video. Students often took longer to watch the video than planned (watching it a first time through and revisiting it a second time to take notes), leaving very little time for discussion phase 2 (at least 7 min were reserved for the quiz and the period was restricted to a maximum of 43 min).

Modifications to Activity

In the second study, data was collected over seven ninth grade biology classes. The classes were distributed across two teachers (with respectively three and four classes) for a total of 78 consenting students. The material was better inserted into the teacher's planning. The students went through a 3-day activity with four cell models (instead of three). The first cell model (a) starch inside, iodine outside was studied as a physical experiment in the first day, with their teacher. The conditions examined during the computer activity were (b) iodine inside, starch outside, (c) glucose inside, water outside, (d) water inside, glucose outside. Students were not assigned roles, as this appeared to be too much of a cognitive load on top of the science itself.

Students were given the worksheet with a description of all the cell models, but, in order to make the activity more collaborative, the only part of the worksheet they had to fill out was a prediction and explanation for model D. They were instructed to agree as a team on this single pair of answers. They were also told that prizes would be awarded to the group with the best discussion and explanation. Students used their real names with only the data collected on the server being anonymized.

One major difference from the lab design used in the year 1 Cell Model study was in the ordering of activities. Instead of going through a whole round of predictions (all models at once, with no information on which to base the predictions), each model was treated on its own. A recap was given on model A during the introduction by the experimenter. In the computer activity, for models B and C, students were in turn shown the initial setup, then the observations and were *only* asked to come up with an explanation for their observation. They were then shown model D and were asked to come up with a prediction and an explanation for this prediction. Only this last prediction and explanation needed to be written on the worksheet. This allowed for fewer, but longer periods of time intended for discussion among students. Instead of being asked to watch videos to see the outcomes, which caused a number of attention focus problems, students were shown sets of illustrations showing the initial setup and the observation after 1 and 24 h.

Modifications to the Tests

Each of the three tests (pre-test, post-activity test, post-discussion-test) followed a similar format: a multiple choice question, a fill in the blank question and what we refer to as a concept cartoon. Each test was different and was specifically targeted to the kinds of knowledge the students might be expected to have at that associated phase of the activity. The idea of the concept cartoon is to present a contextualized situation with three statements which can all be true given certain assumptions. Respondents are asked to pick the statement they are most in agreement with and to explain why they agree. Each of the concept cartoon explanations were graded in a similar way along two dimensions: the number of science terms (e.g., "diffuse through the membrane" as opposed to "went through the bag") and the degree of understanding exhibited in the explanation provided. A similar rubric was also used to evaluate the predictions and explanations that students wrote at the end of the collaborative learning session prior to the post-Activity test, namely, the Co-constructed explanation.

Experimental Design

This study was run as a 2×2 factorial design in which the APT agents provided different behaviors, namely, Revoicing and Feedback. Across all conditions, the agent provided the same macro level support by guiding the students through the activity in the phases described below.

The conditions of the study were defined based on the microscripting behaviors of the agent. The first variable for manipulation was the presence or absence of the *Revoicing behavior*. The second variable was the presence or absence of the *APT Feedback behavior*. In addition, in each class, a group was provided with "wizard of oz" support in which a human experimenter performed both revoicing and feedback. We did this in order to assess whether any deficiency in positive effect of either factor might be due to technical failure rather than poor design. Results in the Wizard conditions on all measures were always within the same range as in the fully automatic support conditions.

Table 25.2 Sample revoicings by the agent for the second study

Student contribution	Revoicing agent response
Basically the glucose will get inside	Maybe you could state that as "the cell membrane is permeable to glucose."
It changed becouse the tube obsorbed the iodine	So are you saying "the molecules diffused through the membrane?"
I predict that if the holes in the plastic are large enough, the glucose will go into the water solution.	Maybe you could state that as "both water and glucose molecules are able to move between the two environments."

The major design change was a better interaction between the macro-scripting (guiding through the phases of the activity) and the micro-scripting. The improved conversational agent allowed the macro-script to cancel or postpone the occurrence of micro-script interactions to prevent either from undermining the other. Furthermore, we focused on one core APT move: Revoicing. This was selected as the most promising to implement in an effective way. Where the agent would previously have asked for explanations or challenges, we chose to add a different, less disruptive behavior, offering positive feedback for the types of APT the students were doing.

In our implementation of the Revoicing Agent, the agent compared student input (during the discussion of each cell-model experiment) against a list of correct statements drawn from the data collected the previous year's study. If an entry in this list could be interpreted as a paraphrase of the student's input, it was offered by the agent as a "revoicing" to the students. Some examples are given in Table 25.2 below. The same statement was never offered more than once in the same session as a revoicing. When student statements were not close enough to match the revoicing list but contained the first mention of important lesson concepts (like "indicator" or "molecule size"), the agent would nudge the student or a peer to expand or restate their contribution.

The Feedback Agent was the other manipulation implemented using Bazaar and evaluated in this study. The purpose of this agent was to provide positive feedback for APT. Student input was matched against a list of patterns indicating APT moves, including explanation, challenge, revoicing, and requests for others to provide each of the same. If a student statement matched, the agent publicly praised the student's move, and (when appropriate) encouraged the other students to respond. This feedback behavior was meant to indirectly support the prevalence of APT in the discussions by encouraging students to take this facilitation role.

Results

Students showed significant learning gains in all conditions, and there was a significant main effect of Revoicing such that students in the Revoicing condition learned significantly more between Pretest and Postactivity test, with an effect size of 0.34 standard deviations. The difference between conditions was neutralized by the Postdiscussion test, however, demonstrating that the other students benefited from their interactions with students in the Revoicing condition during the whole class discussions. There was no significant main effect of Feedback although there was a trend for it to have a negative effect. And there was no significant interaction between the two factors.

Conclusion

In this chapter, we presented the first two iterations of a design investigating the use of conversational agents as facilitators for academically productive discourse. Our first study compares three conditions of facilitation and is designed to examine the impact both on the discussion and on the following day's discussion and post-test. Our second study is a redesign based on the multivocal analyses of the first study. It investigates a finer distinction in APT conversational agent behaviors and aims to reduce the frustration the students experienced in the first study, and provide better opportunities for productive discussion. We report on the research and educational contexts which motivated us to design the study, and on the issues encountered during data collection. The second study was more successful than the first, with the APT agents producing significant learning gains. These datasets provides a very realistic image of what researchers can expect when piloting CSCL interventions in real classrooms.

Acknowledgments This work was supported by the Pittsburgh Science of Learning Center and Graduate Training Grant from the Department of Education (#R305B040063).

References

Adamson, D., Ashe, C., Jang, H., Yaron, D., & Rosé, C. (2013). Intensification of group knowledge exchange with academically productive talk agents. *Proceedings of the 10th International Conference on Computer Supported Collaborative Learning,* Madison, WI, July 2013.

Adamson, D., Dyke, G., Jang, H., & Rosé, C. P. (accepted). Towards adapting dynamic collaboration support to student ability level. Accepted to the *International Journal of AI in Education.*

Ai, H., Kumar, R., Nguyen, D., Nagasunder, A., Rosé, C. P. (2010). Exploring the Effectiveness of Social Capabilities and Goal Alignment in Computer Supported Collaborative Learning, in Proceedings of Intelligent Tutoring Systems, *Lecture Notes in Computer Science, 6095,* 134–143.

Azmitia, M., & Montgomery, R. (1993). Friendship, transactive dialogues, and the development of scientific reasoning. *Social Development, 2,* 202–221.

Berkowitz, M., & Gibbs, J. (1983). Measuring the developmental features of moral discussion. *Merrill-Palmer Quarterly, 29,* 399–410.

Chapin, S. H., O'Connor, C., & Anderson, N.C. (2009). *Classroom discussions: Using math talk to help students learn, grades K-6* (2nd ed.) (Math Solutions, 2009).

Chaudhuri, S., Kumar, R., Howley, I., Rosé, C. P. (2009). Engaging Collaborative Learners with Helping Agents, Proceedings of the 2009 conference on Artificial Intelligence in Education:

Building Learning Systems that Care: From Knowledge Repres entation to Affective Modeling, pp 365–372.

Clarke, S., Chen, G., Stainton, K., Katz, S., Greeno, J., Resnick, L., Dyke, G., Howley, H., Adamson, D., & Ros., C. P. (2013). The impact of CSCL beyond the online environment. *Proceedings of Computer Supported Collaborative Learning* (pp.105–112), Madison, Wi.

Dyke, G., Adamson, D., Howley, I., & Rosé, C. P. (2013). Enhancing scientific reasoning and explanation skills with conversational agents. *IEEE Transactions on Learning Technologies, 6*(3), 240–247.

Gweon, G., Jain, M., Mc Donough, J., Raj, B., Rosé, C. P. (2013). Measuring prevalence of other-oriented transactive contributions using an automated measure of speech style accommodation. *International Journal of Computer-Supported Collaborative Learning, 8*(2), 245–265.

Gweon, G., Rosé, C. P., Zaiss, Z., & Carey, R. (2006). Providing support for adaptive scripting in an on-line collaborative learning environment. *CHI 06: ACM conference on human factors in computer systems.*

Hmelo-Silver, C. E., & Barrows, H. S. (2006). Goals and strategies of a problem-based learning facilitator. *Interdisciplinary Journal of Problem-based Learning, 1*, 21–39.

Howley, I., Chaudhuri, S., Kumar, R., Rosé, C. P. (2009). Motivation and Collaboration On- Line, Proceedings of the 2009 conference on Artificial Intelligence in Education: Building Learning Systems that Care: From Knowledge Representation to Affective Modeling, pp 743–754.

Howley, I., Mayfield, E., Rosé, C. P. (2011). Missing Something? Authority in Collaborative Learning, in Proceedings of the 9 th International Computer Supported Collaborative Learning Conference, Volume 1: Long Papers, pp 336–373.

Kumar, R., Ai, H., Beuth, J., Rosé, C. P. (2010). Socially-capable Conversational Tutors can be Effective in Collaborative Learning Situations, in Proceedings of Intelligent Tutoring Systems, Lecture Notes in Computer Science Volume 6095, pp 156–164.

Kumar, R., Beuth, J., Rosé, C. P. (2011). Conversational Strategies that Support Idea Generation Productivity in Groups, in Proceedings of the 9th International Computer Supported Collaborative Learning Conference, Volume 1: Long Papers, pp 398–405.

Kumar, R., Gweon, G., Joshi, M., Cui, Y., & Rosé, C. P. (2007). Supporting students working together on math with social dialogue. *Proceedings of the SLaTE Workshop on Speech and Language Technology in Education.*

Kumar, R. & Rosé, C. P. (2011). Architecture for building conversational agents that support collaborative learning. *IEEE Transactions on Learning Technologies special issue on Intelligent and Innovative Support Systems for Computer Supported Collaborative Learning.*

Kumar, R. & Rosé, C. P. (in press). Triggering effective social support for online groups. *ACM Transactions on Interactive Intelligent Systems.*

Kumar, R., Rosé., C. P., Wang, Y. C., Joshi, M., Robinson, A. (2007). Tutorial Dialogue as Adaptive Collaborative Learning Support, Proceedings of the 2007 conference on Artificial Intelligence in Education: Building Technology Rich Learning Contexts That Work, pp 383–390.

Michaels, S., O'Connor, C., & Resnick, L. B. (2008). Deliberative discourse idealized and realized: Accountable talk in the classroom and in civic life. *Studies in Philosophy and Education, 27*(4), 283–297.

Resnick, L., Asterhan, C., & Clarke, S. (in press). *Socializing intelligence through academic talk and dialogue*. Washington, DC: American Educational Research Association.

Resnick, L. B., Bill, V., & Lesgold, S. (1992). Developing thinking abilities in arithmetic class. In A. Demetriou, M. Shayer, & A. Efklides (Eds.), *Neo-Piagetian theories of cognitive development: Implications and applications for education* (pp. 210–230). London: Routledge.

Resnick, L., O'Connor, C., & Michaels, S. (2007). *Classroom discourse, mathematical rigor, and student reasoning: an academically productive talk literature review*, unpublished manuscript.

Rosé, C. P., Jordan, P., Ringenberg, M., Siler, S., VanLehn, K., Weinstein, A. (2001). Interactive Conceptual Tutoring in Atlas-Andes, Proceedings of the 10th International Conference on AI in Education, pp 256–266, San Antonio, Texas.

Rosé, C. P., & Tovares, A. (in press). What sociolinguistics and machine learning have to say to one another about interaction analysis. In L. Resnick, C. Asterhan, & S. Clarke (Eds.), *Socializing intelligence through academic talk and dialogue*. Washington, DC: American Educational Research Association.

Rosé, C. P., & VanLehn, K. (2005). An evaluation of a hybrid language understanding approach for robust selection of tutoring goals. *International Journal of AI in Education, 15*(4).

Rosé, C. P., Wang, Y. C., Cui, Y., Arguello, J., Stegmann, K., Weinberger, A., et al. (2008). Analyzing collaborative learning processes automatically: Exploiting the advances of computational linguistics in computer-supported collaborative learning. *The International Journal of Computer Supported Collaborative Learning, 3*(3), 237–271.

Stahl, G., & Rosé, C. P. (2011). Group cognition in online teams. In E. Salas, S. Fiore, & M. Letsky (Eds.), *Theories of team cognition: Cross-disciplinary perspectives, section V: Social psychology and communication perspectives*. USA: American Psychological Society.

Stahl, G., Rosé, C. P., Goggins, S. (2010). Analyzing the discourse of GeoGebra collaborations. *Proceedings of the GeoGebra NA 2010 Conference.*

Stahl, G., Rosé, C. P., O'Hara, K., & Powell, A. (2010). Supporting group math cognition in virtual GeoGebra teams with software conversational agents. *Proceedings of the GeoGebra NA 2010 Conference.*

Weinberger, A., & Fischer, F. (2006). A framework to analyze argumentative knowledge construction in computer-supported collaborative learning. *Computers & Education, 46*(1), 71–95.

Chapter 26
Gaining Insights from Sociolinguistic Style Analysis for Redesign of Conversational Agent Based Support for Collaborative Learning

Iris Howley, Rohit Kumar, Elijah Mayfield, Gregory Dyke, and Carolyn Penstein Rosé

Introduction

This chapter presents an analysis of collaborative learning interactions that uses the same approach to that employed in one of the analysis chapters of in the PLTL data section of this volume (Howley, Mayfield, Rosé, & Strijbos, Chap. 11, this volume). As such, the relationship of this analysis and the dimensions of multivocality are similar as described in that chapter, although here we take a distinctly troubleshooting orientation. In particular, we similarly assume that collaborative learning processes are an integration of three orthogonal dimensions, namely, cognitive, relational, and motivational. However, we adopt the cognitively focused dimension as our success metric, and use the other two more socially oriented dimensions to examine how collaborative processes that support the cognitive dimension went awry and explain a lack of success. Similar to the earlier analysis in the PLTL Chemistry section, we assume that each dimension can be operationalized as a set of mutually exclusive codes, each of which are defined at the level of an individual contribution to a conversation. Overall, the purpose of the analysis here is more specifically to reveal insights leading to suggestions for redesign of an intervention. Thus, in this analysis, pivotal moments are those key events in the interaction that lead to negative impact on collaborative processes and therefore should be avoided in a redesign of the intervention.

While theories of computer supported collaborative learning are many, there is a considerable consensus about what types of group interactions are desirable (Berkowitz & Gibbs, 1983; Suthers, 2006; Teasley, 1997; Weinberger & Fischer, 2006).

I. Howley (✉) • R. Kumar • E. Mayfield • C.P. Rosé
Carnegie Mellon University, Pittsburgh, PA, USA
e-mail: ihowley@cs.cmu.edu

G. Dyke
University of Lyon, Lyon, France

D.D. Suthers et al. (eds.), *Productive Multivocality in the Analysis of Group Interactions*, Computer-Supported Collaborative Learning Series 16,
DOI 10.1007/978-1-4614-8960-3_26, © Springer Science+Business Media New York 2013

It is widely acknowledged that groups do not operate at an ideal level without support. Thus, researchers in the area of scripted collaboration have worked to develop design principles to guide development of support that elicits the kind of group behaviors that are valued within the CSCL community (Dillenbourg, 2002; Kobbe et al., 2007; Kollar, Fischer, & Hesse, 2006). Through design-based research studies and experimental studies, theories of scripted collaboration have developed, which grapple with issues such as how much constraint or structuring should be applied (Dillenbourg, 2002), how support might be faded over time (Wecker & Fischer, 2007) or offered in a just-in-time fashion (Gweon, Rosé, Zaiss, & Carey, 2006), and what modality should be used for training prior to interaction, prompting during interaction, or eliciting reflection after interaction (Rummel, Spada, & Hauser, 2006). Use of process analysis for evaluation of supportive interventions is well established in the CSCL literature, see for example the extensive work by Weinberger and Fischer (Weinberger & Fischer, 2006). Process analysis approaches that are motivated by theories from Sociolinguistics can be argued to provide certain advantages in terms of providing insights that are cross-cutting with respect to theories of collaborative learning or learning more generally (Howley, Mayfield, & Rosé, 2013). In this chapter we employ a quantitative discourse analysis methodology informed by the Sociolinguistics literature called the Souflé framework (Howley & Rosé, 2011; Howley et al., 2013) paired with time series visualization using the Tatiana tool (Dyke, Kumar, Ai, & Rosé, 2012) to pinpoint specific problems with a conversational agent based intervention at an early stage of development and inform its iterative development for effective support of collaborative interactions.

Conversational agents have a long history of successful support for individual learning with technology (Rosé et al., 2001; Rosé & Van Lehn, 2005). A series of results offer hope that they can be used productively to offer support for collaborative learning, especially in chat environments (Kumar & Rosé, 2011; Kumar & Rosé, in press). The Cell Model study dataset we analyze in this chapter represents an early stage of an effort to develop a new style of Conversational agent-based support for collaborative learning, which can be viewed as structuring interaction both at the macro level (in all conditions), and at the micro level (in the Direct Agent and Indirect Agent support conditions).

In the remainder of the chapter, we first present an analysis of the main outcomes of the study in a style that is typical of quantitative analyses of experimental studies, which motivates a redesign, but does not offer insights into how the redesign should be done. This includes one cognitively focused dimension of the Souflé framework referred to as Transactivity, which we use as our main success metric at the level of collaborative process analysis. Next we present the two Sociolinguistic oriented dimensions of the Souflé analysis scheme, illustrated with examples from the data, in order to show what kinds of insights into social positioning within interaction can be revealed through this style of analysis. Finally, we present a Souflé style analysis of the full dataset as viewed through time series visualizations and draw hypothesized directions from it for redesign.

Main Analysis Motivating Redesign

Effect of Condition on Chat Discussion

In the Cell Model study, the goal of the experimental manipulation was to identify what type of support would most effectively elicit Academically Productive Talk (APT) from students, and subsequently, transactive contributions from their partner students, and finally, more learning. We begin by examining the conversation logs for evidence that we successfully manipulated prevalence of APT. Table 26.1 provides a summary.

The definition of APT moves is presented in the Cell Model study data chapter (Dyke, Howley, Adamson, Kumar, & Rosé, Chap. 25, this volume), where we have listed five separate moves, namely, Revoicing, Asking for a Rephrase, Asking for Agreement or Disagreement, Prompting for Elaboration, or Prompting for Justification. Transactivity is defined in two steps within the Souflé framework. First, a trasactive utterance must be an explicit display of reasoning, where a display of reasoning is one in which some evidence of a causal connection or compare/contrast relation is articulated. This display of reasoning is transactive if it evaluates or builds on a display of reasoning expressed earlier in the discussion. Here we adopt this one dimension of Souflé as a success metric for the manipulation.

The hypothesis that motivated the Cell Model study was that the agent based support would increase the prevalence of APT facilitation moves, which would then elicit more displays of reasoning and more transactivity. Thus, we begin our analysis by investigating the effect of condition on prevalence of APT, which may come either from the agent or from the students. There was no significant effect of condition on total number of APT moves contributed by students. However, there was a significantly higher total number of APT moves in both the supported conditions than the Unsupported condition when we count tutor contributions $F(2,42)=5.5$, $p<0.01$. And when we consider percentage of total contributions that are APT moves, we find a significantly higher percentage in the Direct agent condition than either of the other two conditions $F(2,42)=13.9$, $p<0.0001$. Thus, the first part of the hypothesis was partly supported.

The biggest difference between conditions shows up in terms of explicit displays of reasoning. Here there is a marginal main effect of condition on total number of

Table 26.1 Average number of contributions (and percentage of contributions) of specific types across the three conditions of the study

Condition	Academically productive talk moves	Academically productive talk moves from any source	Reasoning moves	Transactive moves
Unsupported	0.56 (2.7 %)	1.6 (1.8 %)	1.6 (11 %)	0.55 (2.7 %)
Indirect agent	1.2 (4.9 %)	3.8 (3.6 %)	0.53 (3.8 %)	0.13 (1.1 %)
Direct agent	0.67 (6.4 %)	4.25 (7 %)	2 (17 %)	0.92 (5.1 %)

reasoning moves per class period $F(2,42)=2.46, p<0.1$, whereby a student-t posthoc analysis demonstrates that students in the Direct condition produce a significantly greater number of reasoning moves than students in the Indirect condition, with the Unsupported condition not being significantly different from either. There was also a significant effect on percentage of reasoning moves $F(2,42)=4.47, p<0.05$, again where students in the Direct condition produce a significantly greater number of reasoning moves than students in the Indirect condition, with the Unsupported condition not being significantly different from either. Articulation of reasoning is a precondition for transactivity, so the second part of the hypothesis is also partly supported.

Nevertheless, while the effect on displays of reasoning is promising, the hypothesis begins to break down at the second step, in that we did not find evidence of any statistical relationship between the number or percentage of APT moves from the tutor and either student reasoning displays or transactive moves. We did, however, see a significant but weak correlation between total percentage of APT moves in a chat transcript from any source and the percentage of student contributions that were explicit displays of reasoning $R^2=0.11, p<0.05$. This analysis shows that students in the Direct Agent condition were exposed to a greater percentage of contributions that were APT moves and that this higher percentage was associated with a greater percentage of explicit displays of reasoning during the chat. From this we have some weak indication that within the Direct Agent condition specifically, the manipulation was partly working as expected. The story was less positive in the Indirect condition. While the trend was for students in the Indirect Agent condition to employ a greater number of APT moves, the difference was not significant, and when we considered the trend in terms of percentage of contributions, it was not even the highest average. Furthermore, there was no significant main effect of condition on transactivity either within the Direct Agent condition or the Indirect Agent condition.

Overall, we find evidence suggesting some positive effect of condition on prevalence of APT, which is associated with increases in displays of reasoning. However, the goal of increasing transactivity was not reached. The big question here is why there are significantly fewer explicit displays of reasoning in the Indirect Agent condition when across conditions there is a significant positive correlation between APT moves and reasoning moves, and the trend is for there to be more APT moves in the Indirect Agent condition and the other conditions. This pattern suggests that something problematic was occurring, and thus this question will drive much of our process analysis in the subsequent sections of the chapter. In finding an explanation for this pattern, we arrive at insights that ultimately informed our redesign and lead to much more successful implementations of APT agents in subsequent studies (Adamson, Ashe, Jang, Yaron, & Rosé, 2013; Adamson, Dyke, Jang, & Rosé, in press; Dyke, Adamson, Howley, & Rosé, in press).

Effect of Condition on the Whole Class Discussion

In addition to having an effect on behavior within the chat discussions, we also hoped that the collaborative exercise would serve to prepare students for the whole class discussion that took place on the day after the collaborative exercise and before the posttest.

Table 26.2 Average number of contributions to the whole class discussions for students within class periods from each condition

Period	No support	Indirect agent	Direct agent
Period 1	4.2 (3.7)	8.0 (5.9)	3.7 (2.1)
Period 3	N/A	19 (8.5)	60 (49.5)
Period 6	1 (0)	3.2 (2.1)	5.8 (5.3)
Period 9	1 (0)	20 (0)	7 (0)

A summary of student contribution to the whole class discussion that followed in the next class period after the online activity is displayed in Table 26.2.

Analysis of 2 years of recorded classroom discussions from the whole study demonstrates a pattern whereby teacher adoption of APT practices showed a gentle increase over time that was punctuated by local increases in the discussions that immediately followed online CSCL activities (Clarke et al., 2013). This provides more substantial evidence of a connection between the online activities and the whole class discussions that followed them than what we could hope to see from simply examining the one discussion that was part of the Cell Model study. Thus, here we limit ourselves to an informal analysis.

For this chapter, we simply did a cursory analysis of differences in student behavior between conditions in the context of the whole group class discussion that was part of this Cell Model study only. The goal was to investigate the extent to which students in the supported conditions behaved differently than students in the unsupported condition. In this cursory analysis, we simply measured how active the students were in the discussion in terms of number of contributions to the discussion. Because the data were far from normally distributed, we first did a log transformation on the counts of contributions. We then performed an ANOVA analysis to determine whether there was a significant effect of condition. Since there was also a big difference in participation across class periods, we retained class period as an additional factor in the ANOVA analysis. Both class period ($F(3,21)=7.0$, $p<0.005$) and condition ($F(2,26)=4.2$, $p<0.05$) were statistically significant. A student-*t* posthoc analysis demonstrated that students in both the Direct Agent and Indirect Agent condition contributed to the whole group discussion significantly more frequently than students who had been in the Unsupported condition. In both cases the effect size was about 0.75 standard deviations. Thus, we have some evidence that the supported conditions better prepared students for active participation in the group discussion.

In addition to the effect on student activity during the whole class discussion, there was also a significant positive correlation between percentage of APT moves from the chat and the log of number of contributions to the whole class discussion. However, as a caveat we must mention that there were many student entries in the whole class discussion where we were not able to uniquely identify the student from the recordings, transcriptions, and notes. We do not have a reason to suspect that our ability to identify which students were contributing to the whole group discussion was biased by condition. However, we cannot completely eliminate the possibility that our analysis doesn't accurately reflect the effect of condition on participation. Overall, we consider the evidence of a connection between the experimental

Table 26.3 Average posttest score per assessment category in the supported versus unsupported conditions in period 1 only, which was the only period that showed any significant pretest to posttest learning gains

	Supported (period 1)	Unsupported (period 1)
Remember (high is better)	0.6 (0.54)	0.66 (1)
Understand (high is better)	0.4 (0.54)	0.89 (1.1)
Explain (high is better)	2 (0.7)	1.1 (0.9)
Experiment (high is better)	2.2 (1.3)	1.6 (1.4)
Ranking (low is better)	2.4 (1.1)	4.8 (2.6)

manipulation in the CSCL activity and behavior in the whole group discussion in the Cell Model study specifically to be merely suggestive. Despite this small glimmer of hope, based on this analysis, the main conclusion is that manipulation was not successful overall in eliciting the kind of behaviors we wanted from students. This was even clearer when we considered the pattern of results on learning gains.

Effect of Condition on Learning

Perhaps the most disappointing summative result from the Cell Model study was the lack of effect on learning overall. As mentioned in the data chapter (Dyke et al., Chap. 25, this volume), domain knowledge was measured at three time points using a paper based test that included both multiple choice questions and open ended, explanation type questions. In order to capture how well the students grasped the material based on their test answers, we differentiated ability to remember what was observed (weight changes, indicator color changes), to understand what the observation meant (weight change means water movement, indicator color change means glucose and starch presence), and to explain and generalize (glucose always moving down the concentration gradient towards equilibrium, starch unable to move through the semipermeable membrane). As the questions asked did not specifically request students to remember or understand, but rather to explain, the students with the better understanding tended to obtain low scores on the first two measurements and better on the final one. As a holistic measure of performance, we also ranked the explanation answers from "best to worst" using pairwise comparison, rather than a scoring rubric. Table 26.3 contains a summary of post-discussion test scores per assessment category across the Supported and Unsupported conditions in Period 1 only, which is the only class period where we observed significant pretest to posttest learning gains.

To investigate the effect of condition on learning between the pretest and the postdiscussion test, we first verified that the random assignment of students to conditions was successful. This was confirmed in that there was no significant difference between conditions on the pretest $F(2,46)=0.39$, $p=0.69$. Next we checked whether there was evidence of learning. Overall, there was no general effect of learning across conditions. There was a significant effect of Class Period on all posttest measures, which showed that only students in the first class period showed significant pretest to posttest learning gains. Thus, we include an indicator of Class period as a fixed effect in our learning analysis. Due to low statistical power, we conduct a simplified analysis on learning measures, dividing students into Supported (which includes both the Direct and Indirect conditions) versus Unsupported, as well as Class Period 1 versus Other Class Periods.

The only significant effect on learning was on the Explain portion of the test, where there is a significant interaction between the two factors: $F(1,46)=4.3$, $p<0.05$, whereby within the first class period, students scored significantly higher in the Supported condition than the Unsupported condition with an effect size of 1.1 standard deviations. For the other assessments, the trend is in favor of unsupported conditions for recalling details of the experiment and demonstrating understanding of how indicators work in the experiments. However, since students tended to either focus on explanations or these more shallow measures, we consider that lower scores here might actually be preferred. Furthermore, for all other assessments, particularly those that demonstrate understanding of the domain concepts relating to diffusion and explanation for the results observed, the trend is in favor of the Supported conditions. Overall, the trend is weakly in favor of the Supported conditions over the Unsupported condition, although the effect is not strong nor clear.

Considering the whole set of results, the Direct agent condition shows the most promise, whereas the Indirect condition shows the least promise. Overall we must conclude that a redesign is in order. Consistent with the result on the online chat, the Indirect Agent condition shows up as the most problematic. What this analysis does not tell us is what change in design of agent behavior is likely to be more successful at achieving our goals. For this, we turn to the Souflé analysis scheme.

Souflé Analysis Scheme

In order to gain more insights into the inner workings of the chat that occurred during the online activity, we will use the Souflé analysis scheme (Howley et al., 2013; Howley & Rosé, 2011) to obtain a linguistic analysis of the discussion data from the Cell Model study. Souflé has three primary dimensions, one of which is cognitively focused, which we refer to as Transactivity, and the other two of which are more socially orientated, which we refer to as Authoritativeness (Martin & Rose, 2003, Chap. 7) and Heteroglossia (Martin & White, 2005, Chap. 3). Together these dimensions can be considered a multidimensional way of viewing leadership taking in conversation. While the cognitive dimension of that analysis scheme, which codes contributions according to whether they explicitly display reasoning, and whether that reasoning represents a new direction in reasoning within the conversation or builds on previously displayed reasoning (Gweon, Jain, Mc Donough, Raj, & Rosé, 2013; Sionti, Ai, Rosé, & Resnick, 2011), other dimensions of that scheme are drawn from the broad subfield of linguistics referred to as Systemic Functional

Linguistics (SFL). We discussed the transactivity dimension earlier in the chapter. Thus, we focus here on the two more socially oriented dimensions. An alternative presentation of the Souflé framework is presented in the PLTL Chemistry section of this volume (Howley et al., Chap. 11, this volume).

The Authoritativeness and Heteroglossia dimensions of Souflé were inspired by work in the field of Systemic Functional Linguistics, which is a largely descriptive linguistic tradition that provides a firm foundation in analyses of genres of writing or text-based interaction (Martin & Rose, 2003; Martin & White, 2005), as well as face-to-face interaction (Veel, 1999).

Let us first consider the issue of authoritativeness in presentation of self. This analysis is derived from an operationalization of Martin and Rose's negotiation framework (Martin & Rose, 2003). In the negotiation framework, authoritativeness is demonstrated by making a contribution to a discourse that is not offered as an invitation for validation from another group member. For example, an assertion that is made in response to a question that is framed as a hint rather than a serious question, and then followed by an evaluation, is not coded as an authoritative assertion. The Negotiation Framework is a measure of authoritativeness where authority is demonstrated by making a contribution to a conversation that is not meant to be validated by another group member. In our operationalization of the Negotiation framework, there are four core moves, and two secondary moves:

- K1 (Primary Knower), in which the speaker considers herself to be the primary authority on a given (expressed) piece of knowledge
- K2 (Secondary Knower), when the contributor asks for knowledge from someone of higher authority
- A1 (Primary Actor), for contributions that display that one can perform a particular action.
- A2 (Secondary Actor), when instructing someone else to do an action, allowing the other person to either perform the action or reject the request.
- ch (Challenge), in which a speaker rejects the authority of the previous speaker to make the previous move
- o (Other), which encapsulates all other moves that do not fit in the five described above

For our purposes, "Primary Knower" and "Secondary Actor" moves are considered more authoritative (with respect to social relationships), while "Secondary Knower" and "Primary Actor" moves display less authoritativeness. In this chapter we are primarily concerned with authoritativeness over knowledge. As such, to compute a meaningful ratio for the authoritative moves, the formula would be: $K1/(K1+K2)$.

What we can see here is that we can use the Authoritativeness ratio to pick out students that when we look more closely at the content of their interactions with others either appear to be having trouble (in the case of low authoritativeness) or look like leaders (in the case of high authoritativeness). This affirms the face validity of this coding. In other work, we have seen positive correlations between Authoritativeness and self-efficacy and between Authoritativeness and learning (Howley, Mayfield, & Rosé, 2011) or task engagement (Howley et al., 2013).

Speaker	Contribution	Negotiation
Alex(Tutor)	Now discuss what you observed in the video about conditions A and B. Compare what you observed to your predictions.	a2
s008	does ne one know wat they r doing	k2
	(1 minute pause)	
Alex(Tutor)	You should now move on to discussing what you observed in the video about condition C. Compare what you observed to your predictions. How is this different from what happened in condition A?	a2
	(2 minute pause)	
Alex(Tutor)	Ok, I gotta go.	o
Alex(Tutor)	It was nice talking to you all. :-)	o
	(3 minute pause)	
s008	ok this video made no sense at all soes any one know what they are doing	o
s012	maybeee	o
	(1 minute pause)	
s008	well then tell me	k2
s008	well then tell me	o
s012	well A is retarded	k1
s012	it gained twice then lost weight	k1
s008	well i only got the 1 hour thing nd the 5 hour one i dint watch the rest of the video	o

Fig. 26.1 S008 is a student with low authoritativeness

Speaker	Contribution	Negotiation
s006	is anyone else seriously confused	k2
s010	{s006}, did you get any other observation other than the weight	k2
s010	yes	k1
s006	no i didnt write anything down for the observations	k1
s010	nice	o
s010	....	o
Alex(Tutor)	Ok, I gotta go.	o
Alex(Tutor)	It was nice talking to you all. :-)	o
s010	um bye?	o
s002	for 1 hour on a it was .620 they were both clear tube and beaker and	k1
s002	same for the B but the weight was .540	k1

Fig. 26.2 S002 has high authoritativeness

We also see that we can discern something about the dynamic within the group when we look at a student's Authoritativeness ratio in relation to that of the other group members.

In the first interaction displayed in Fig. 26.1 above, s008 has low authoritativeness in comparison to the partner student that we see. In the second interaction displayed in Fig. 26.2 above we see that s002 has relatively high authoritativeness in relationship to the two partner students. In this Cell Model study dataset, we find that both Authoritativeness and difference between an individual's Authoritativeness and the average Authoritativeness of the other two students in the group both individually significantly correlate with the amount of reasoning a student contributes. This suggests that students with high Authoritativeness are students who treat themselves as sources of reasoning in the interaction. In order to facilitate transactive knowledge integration, we want all students within a group to treat themselves this way. Thus, we have reason to hope for groups where there is relative equality of Authoritativeness between students. However, as a caveat we must note that whereas an individual's Authoritativeness explains 11 % of the variance, the difference in Authoritativeness between the individual and the group explains 15 %. This suggests that there might be something more complex going on in the interplay between the students in terms of their Authoritativeness, and in understanding the impact of an intervention on an individual's behavior, we must first factor out the effect of the group dynamic, which also has a significant effect.

The distinction between the core moves and other moves within the Negotiation framework allows us to find the more meaty portions of the interactions. The codes also allow us to see more when we examine them within different spans of the interaction than the statistics computed over the entire interaction. If we compute the statistics within segments of the conversation, we can identify particular segments where a speaker is behaving in a way that is uncharacteristic, for example, a relatively non-authoritative student taking an authoritative stance on some issue. We might refer to these as pivotal moments in the collaboration. We can also investigate whether that uncharacteristic behavior is related to some agent behavior, which may have a different effect locally than we see overall.

The Heteroglossia framework offers a complementary picture to the one just explored in relation to Authoritativeness. Within a heteroglossia analysis, assertions framed in such a way as to acknowledge that others may or may not agree, are identified as heteroglossic. The Heteroglossia framework is operationalized from Martin and White's theory of engagement (Martin & White, 2005), and here we describe it as identifying word choice that allows or restricts other possibilities and opinions. In other words, it captures the extent to which the students show openness to one another's views. This creates a rather simple divide in possible coding terms for contributions:

- Heteroglossic-Expand (HE) phrases tend to make allowances for alternative views and opinions (such as "She *claimed* that glucose will move through the semi-permeable membrane.")
- Heteroglossic-Contract (HC) phrases attempt to thwart other positions (such as "The experiment *demonstrated* that glucose will move through the semi-permeable membrane.")
- Monoglossic (M) phrases make no mention of other views and viewpoints (such as "Glucose will move through the semi-permeable membrane.")
- No Assertion (NA) expressions are ones that do not make an assertion, and therefore cannot be either monoglossic or heteroglossic.

We also code commands expressed as suggestions as Heteroglossic Expand, whereas ones that are stated simply as commands are coded as Monoglossic. Note that it is not always the case that heteroglossic assertions framed as negative polarity statements perform the function of contracting the set of options under negotiation.

Speaker	Contribution	Heteroglossia
Alex(Tutor)	You are now going to watch a video showing the cell in Conditions A, B and C.	m
Alex(Tutor)	As you watch the video, write down your observations on your worksheet.	m
s059	umm a you can say that when you mix them together the colors will change and something may happen..okay	he
Alex(Tutor)	Go to the Videos folder on the Desktop, and watch the video which is there.	m
s056	maybe	na
s062	wacth the video	m
s059	okay so do you want to put that down	he
s056	who is this anyway	na
	(3 minutes pause)	
Alex(Tutor)	Is everyone back?	na
s059	{s059}...and we can say when you mix them together the color changes and they have different weight	he
Alex(Tutor)	Now discuss what you observed in the video about conditions A and B. Compare what you observed to your predictions.	m
s062	glucose will dissolve in the distilled water	m

Fig. 26.3 This chat segment displays a group with a highly monoglossic style

For example, if it is a constraint that is eliminated, then more items are made negotiable since fewer constraints need to be satisfied.

In looking at this analysis of contributions in terms of Heteroglossia, we see how a different attitude is communicated by use of these different styles. See Fig. 26.3 for an example of a group with a highly monoglossic style.

As we see in Fig. 26.3, the monoglossic style does not come across as welcoming or open to discussion. When we consider that, it is not so surprising that when our research group examined giving dialog agents heteroglossic, monoglossic, and neutral language in an idea generation task, we found that dialog agents with heteroglossic language result in the greatest idea generation productivity in a group task (Kumar, Beuth, & Rosé, 2011). Results from that controlled comparison raise questions about whether students may respond to differences in Heteroglossia in tutor agent prompts that arise inadvertently when an intentional effort to control that aspect of style has not been made. We found, for example, in a post hoc analysis of tutor agent prompts across conditions that the proportion of Heteroglossic Expand contributions in the Indirect agent condition was significantly higher than that of the other two conditions $F(2,42)=4.03$, $p<0.05$. This appears to have been caused simply because the majority of tutor prompts were Monoglossic in style, but the prompts that were triggered to support Academically Productive Talk in the Indirect Agent condition were in the style of Heteroglossic Expand.

The final dimension in the Souflé framework is transactivity. It can be viewed as a means for displaying receptivity to the idea leadership of other students in the group. We have already discussed Transactivity in a previous section where we offer the summative analysis of study outcomes.

Together, the multidimensional Souflé framework allow us to view how idea leadership plays out in group knowledge construction and how it is possible to present one's views as standing on their own without denying others the right to have their own voice. But beyond this it might allow us to capture something important in the interplay between this leadership taking and an experimental manipulation. If the manipulation affects how students position themselves socially, it may also affect how students engage in idea contribution and idea integration within the interaction. It is precisely this sort of interplay that we begin to see when we apply the Souflé analysis to the whole corpus and view how it plays out over time using time series visualizations.

Souflé Style Analysis

Table 26.4 displays a summary of the Souflé style analysis, which includes average Authoritativeness ratio per condition as well as the average percentage of Heteroglossic Expand moves and explicit display of reasoning moves. Additionally, we added two informal diagnostic indicators, namely, percentage of contributions that were off-task, which included places where students were joking around or insulting one another, and percentage of cheating moves, where students were engaged in passing answers back and forth rather than discussing the material. We do not find any significant effect of condition on percentage of cheating contributions, although there is trend for there to be the least cheating within the Direct Agent condition and most within the Indirect Agent condition. Similarly, there is significantly more off-task behavior in the two supported conditions than in the No Support condition $F(2,41)=7.7, p<0.005$, although the trend is for there to be more in the Indirect Agent condition than the Direct Agent condition. This adds further support for the conclusion that the Indirect Agent condition was problematic. But we are still left wondering why and what to do about it.

We already mentioned that there was a significant effect of condition on Percent Student Reasoning. As mentioned in the previous section, the big question here is why there are significantly fewer explicit displays of reasoning in the Indirect Agent condition when across conditions there is a significant positive correlation between Academically Productive Talk moves and Reasoning Moves, and the trend is for there to be more Academically Productive Talk moves in the Indirect Agent condition than in the other conditions. If we do a multiple regression with Total

Table 26.4 Overview of distribution of Souflé moves as well as cheating and off-task across conditions

Condition	Authoritativeness ratio	Percent student heteroglossia	Percent student reasoning	Percent cheating	Percent off-task
Indirect agent	0.54 (0.21)	0.08 (0.06)	0.03 (0.1)	0.18 (0.15)	0.35 (0.15)
Direct agent	0.6 (0.21)	0.24 (0.14)	0.17 (0.14)	0.05 (0.09)	0.27 (0.16)
No support	0.62 (0.27)	0.2 (0.16)	0.11 (0.1)	0.12 (0.2)	0.13 (0.17)

percentage of Academically Productive Talk moves, student Authoritativeness, and the average Percent Heteroglossia of the other two students in a group, all of these factors have a positive significant correlation with Percent Student Reasoning, and together they explain 44 % of the variance in Percent Student Reasoning. Without total percentage of Academically Productive Talk moves, we can only explain 38 %. Thus, we have evidence that Academically Productive Talk moves positively contribute to the conversation. However, the condition that was meant to elicit Academically Productive Talk moves did not. We see from this that we have more evidence that we need to find a different way to elicit this behavior other than that the goal itself was not appropriate.

One clue is in connection with the Heteroglossia dimension. Overall, we find a positive and strong correlation between the average percentage of Heteroglossic Expand contributions in a discussion and the percentage of a student's contributions that are explicit reasoning displays, $R^2=0.5$, $p<0.0001$. We see a corresponding negative correlation between percentage of Heteroglossic Expand contributions and percentage of Off-task contributions, $R^2=12$, $p<0.05$ and a similar trend for less cheating when the percentage of Heteroglossic Expand utterances are high. What we notice when we consider all of the dimensions of Souflé is that in addition to a significant drop in percentage of contributions that contain explicit reasoning displays, we also see a significantly smaller percentage of student contributions that are Heteroglossic Expand $F(2,41)=6.79$, $p<0.005$. A student-t posthoc analysis confirms that this percentage is significantly smaller in the Indirect Agent condition than either of the other two conditions. This is a surprise considering that a greater proportion of tutor agent moves are Heteroglossic Expand in the Indirect Agent condition, as discussed in the previous section. We mentioned in the previous section that a higher proportion of Monoglossic contributions is associated with a somewhat negative attitude. All of this connection between the Heteroglossia dimension and other variables supports this interpretation. Clearly something was happening in that condition that was disenfranchising students, but it wasn't simply that they did not want to respond to the Indirect Agent prompts, because if that were true, we would expect to see significantly fewer APT moves from students rather than a trend for more APT moves.

To get a better idea of what was happening, we constructed a visualization using Tatiana (Dyke et al., 2012) where we see the average distribution of codes on the Heteroglossia dimension across time within each of the three conditions. Since some off-task talk nevertheless received codes on the Heteroglossia dimension, but did not contribute towards the substantive part of the conversation, we marked these as off-task rather than any of the four heteroglossia codes. Tutor prompts are indicated with vertical bars. The black bars are the task-related microscripting prompts that were the same in all conditions. The first black bar is where the tutor asks the group to start making predictions. The fourth black bar is where the students are instructed to watch the video. The next black bar after that is when the tutor asks if the students returned. The last black bar is where the tutor says goodbye and leaves the group to interact on their own. The green bars represent places where the tutor intervened with a condition specific prompt in at least one of the conversations (Fig. 26.4).

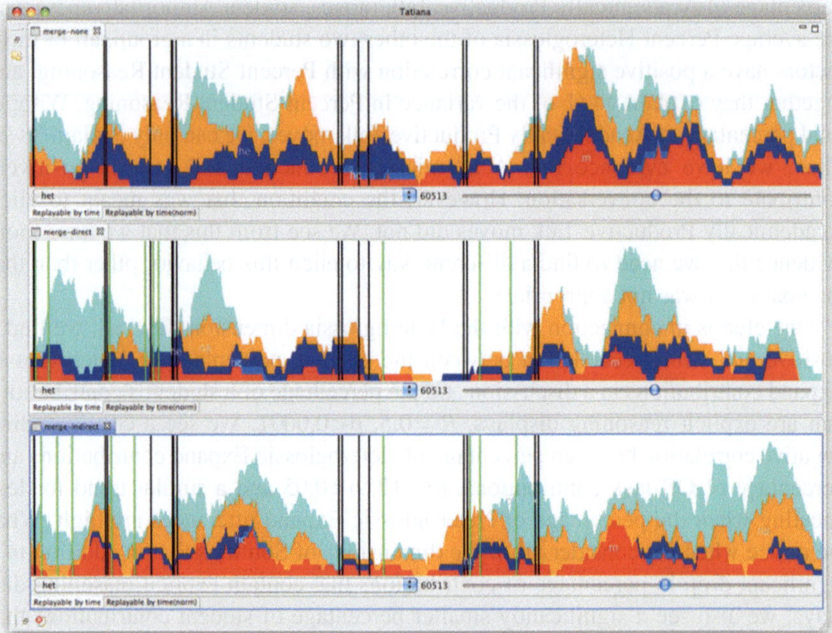

Fig. 26.4 Tatiana visualization of Heteroglossia and Off-task codes over time

The visualization tells an interesting story that cannot be seen using average counts across whole discussions. First we notice that in all conditions, there is some off-task talk early and late in the period. However, in the Indirect condition, the initial segment of the conversation where there is intermittent off-task talk extends throughout the time when students were supposed to be watching the video, and starts up again before the tutor has left the conversation. In the No support condition, the initial off-task talk only extends halfway through the video, and doesn't start up again until after the tutor has left. In the Direct agent condition, students stop their off-task talk early in the period of time when they are making predictions and don't resume until after the tutor has left.

A similar interesting picture emerges when we examine patterns of monoglossic contributions over time. Despite the fact that the Indirect Agent condition was the highest on this, it is clear from the visualization over time that it did not start out that way. Monoglossia increases as more and more condition-specific prompts enter the conversation. The opposite is the case in the Direct Agent condition. Monoglossia starts out higher but is reduced as more of the condition specific prompts come in. In the unsupported condition, Monoglossia starts out at about the same level as in the Direct agent condition, but it does not reduce during the same period of time. The occurrence of Monoglossic contributions is relatively consistent within conditions, although distinct across conditions, between the time when the video is introduced and when the students were requested to return from the video.

Alex(Tutor)	In condition C, we made a modification to condition A. We replace the glucose solution with a starch suspension. In order to detect the presence of starch, we replace the distilled water with Lugol's Iodine solution.	m
Alex(Tutor)	You should now move on to discussing what will happen in Condition C and your explanation for this change.	he
s041	the world is going to end in 2012	m
Alex(Tutor)	S027, now would be a good time to ask S034 to build on what S041 is saying.	he
s034	im so confused!	m
s027	034, would you like to build on to what 041 is saying? and me too!	he
Alex(Tutor)	When you are in agreement, write down your predictions and explanations for Conditions A, B and C on your worksheet.	m
s027	who is 34?	na
s034	{s034}	m

Fig. 26.5 A frustrating interaction from the student perspective in the Indirect Agent condition

A substantial amount of heteroglossic discussion takes place in all but the Indirect Agent condition throughout that segment. After the students were instructed to return from the video, there is a period of intensive heteroglossic discussion both in the Unsupported condition and the Direct Agent condition, but not in the Indirect agent condition, and it occurred earlier in the Direct Agent condition.

What we see is that in the Direct Agent condition, students respond to the agent by getting on-task and contributing in a heteroglossic style. Both of these show engagement and positive attitude. The opposite is the case in the Indirect Agent condition. The visualizations tell us that the answer should be found relatively early in the discussions after the agent has asked the students to start making predictions. Since we know that the students in the Indirect Agent condition did show some indication of attempting to follow the agent's request for them to engage in Academically Productive talk, the question is what happened when they did.

Figure 26.5 illustrates two things that likely contributed to the effect of the Indirect Agent prompts. Note first that responding to an Indirect Agent prompt requires more from students than responding to a Direct Agent prompt. For the latter, a student only has to make one comment, possibly building on or expanding on what another student has said, and then it's over. But in the Indirect Agent condition, first one student needs to respond with an Academically Productive talk move, and then another student has to respond to that. Thus, it takes both more time and more coordination. In both conditions, the tutor prompts were sometimes generated inappropriately as we see in Fig. 26.5, however, in the Direct Agent condition, it was easier to disregard. Furthermore, we see an additional problem. Notice that s027 attempts to engage in Academically Productive Talk in response to the agent's inappropriate request. But then, to make matters worse, before the other student has time to respond, the tutor agent comes in with a task-related prompt, one that was scripted for a specific time in the interaction. Since these were given without regard for whether students had time to respond to the condition-specific prompts, the timed prompts were more interruptive in the Indirect Agent condition. Notice that after the task prompt, the students switched from their focus on the task, regardless of being confused, to off-task related talk.

The peaks in monoglossia in the Indirect condition were not always easy to explain, particularly for group C04 which did not experience any condition-related tutor prompts. For the other groups, these peaks were caused by four types of contributions: uncertainty about identities (e.g., "who is s011?" "I'm [s011], who are you"), expressions of confusion (e.g., "I don't understand this"), instructions (e.g., "write your predictions"), and giving answers ("condition [a] gained [weight]"). All four of these contribution types appear to have at their root the initial confusion about the task, exacerbated by the obliqueness of the indirect prompts which compound their lack of situatedness (asking students to perform Academically Productive Talk moves in inappropriate situations), the unfamiliarity of the situation (biology task, complex roles) and the problems over student identification. Distributing instructions and giving answers occur mainly when one of the students observes that his peers are struggling or confused and takes charge in resolving the situation. Lastly, the small heteroglossia peaks in the Indirect condition come from students following the tutor prompts and performing the requested Academically Productive Talk move, whether appropriate or not, increasing yet again student dissatisfaction and confusion.

Conclusions and Ideas for Redesign

One obvious problem with the condition-specific prompts was that they were sometimes triggered inappropriately, for example when students had uttered something in the form of a prediction that was actually off-task talk. However, this was the case both in the Direct Agent condition and the Indirect Agent condition. We do not see evidence that students responded the same way on average to the two kinds of prompts in the patterns of heteroglossia codes. Thus, it appears that a more serious issue was the lack of coordination between the condition-specific prompts and the timed task prompts, especially in the Indirect Agent condition where students were sometimes taxed by being interrupted while trying to follow the tutor's instructions. The realization of tutor interruptions conflicting with student actions is perhaps the most valuable insight that came out of the Souflé analysis, and provides the best direction for tutor design improvement.

From the summative statistical analysis, one might simply conclude that the Indirect Agent prompts were ineffective and annoying to students. However, a closer look with the Souflé analysis and time-based visualizations indicates that the real issue was a lack of coordination between types of prompts and a lack of consideration for how long it would take students to appropriately respond to the condition-specific prompts in the Indirect Agent condition. Considering the negative effect of the lack of coordination, it is no longer surprising that students showed a trend towards more APT in the Indirect Agent condition, but not a significant increase, and not the related increase we would have expected to see in explicit displays of reasoning that those moves are designed to elicit. Rather than drop the idea of the Indirect Agent condition, this analysis suggests that the timing and coordination problem are more important to address. This finding is consistent with recent work on triggering social prompts in agent supported CSCL environments where we have seen that trigger models that are trained using machine learning techniques on annotated data lead to significantly higher learning gains than other simpler trigger models (Kumar & Rosé, in press; Kumar et al., 2011).

The coordination issues revealed in this Souflé analysis were addressed in subsequent work on the Bazaar architecture for coordinating dynamic support for collaborative learning (Adamson & Rosé, 2012) and enabled subsequent positive results with APT agents (Adamson et al., 2013; Adamson et al., in press; Dyke et al., in press).

Acknowledgments This work was supported by the Pittsburgh Science of Learning Center and Graduate Training Grant from the Department of Education (#R305B040063).

References

Adamson, D., Ashe, C., Jang, H., Yaron, D., & Rosé, C. (2013). Intensification of group knowledge exchange with academically productive talk agents. *Proceedings of the 10th International Conference on Computer Supported Collaborative Learning*, Madison, WI, July 2013.

Adamson, D., Dyke, G., Jang, H., & Rosé, C. P. (In press). Towards adapting dynamic collaboration support to student ability level. *International Journal of AI in Education.*

Adamson, D., & Rosé, C. P. (2012). Coordinating multi-dimensional support in conversational agents. *ITS 2012 Proceedings of the 11th International conference on Intelligent Tutoring Systems*, Lecture Notes in Computer Science (Vol 7315, pp. 346–351) Springer Berlin Heidelberg.

Berkowitz, M., & Gibbs, J. (1983). Measuring the developmental features of moral discussion. *Merrill-Palmer Quarterly, 29*, 399–410.

Clarke, S., Chen, G., Stainton, K., Katz, S., Greeno, J., Resnick, L., Dyke, G., Howley, H., Adamson, D., & Rosé, C. P. (2013). The impact of CSCL beyond the online environment. *Proceedings of Computer Supported Collaborative Learning.*

Dillenbourg, P. (2002). Over-scripting CSCL: The risks of blending collaborative learning with instructional design. In *Three worlds of CSCL - Can we support CSCL?* Heerlen: Open Universiteit Nederland, pp. 61–91. http://dspace.ou.nl/bitstream/1820/1618/1/Three%20worlds%20of%20CSCL%20Can%20we%20support%20CSCL.pdf

Dyke, G., Adamson, D., Howley, I., & Rosé, C.P. (In press). Enhancing scientific reasoning and explanation skills with conversational agents. *IEEE Transactions on Learning Technologies.*

Dyke, G., Howley, I. K., Kumar, R., & Rosé, C. P. (this volume). Towards academically productive talk supported by conversational agents. In D. Suthers, K. Lund, C. P. Rosé, C. Teplovs, & N. Law (Eds.) *Productive multivocality in the analysis of group interactions*, Chapter 25. New York: Springer.

Dyke, G., Kumar, R., Ai, H., Rosé, C. P. (2012). Challenging assumptions: Using sliding window visualizations to reveal time-based irregularities in CSCL processes. *Proceedings of the International Conference of the Learning Sciences.* Sydney, Australia

Gweon, G., Jain, M., Mc Donough, J., Raj, B., & Rosé, C. P. (2013). Measuring prevalence of other-oriented transactive contributions using an automated measure of speech style accommodation. *International Journal of Computer Supported Collaborative Learning, 8*(2), 245–265.

Gweon, G., Rosé, C. P., Zaiss, Z., & Carey, R. (2006). Providing support for adaptive scripting in an on-line collaborative learning environment. *CHI 06: ACM Conference on Human Factors in Computer Systems.*

Howley, I., Mayfield, E., Rosé, C. P. (2011). Missing something? Authority in collaborative Learning. *Proceedings of the 9th International Computer Supported Collaborative Learning Conference*, 336–373.

Howley, I., Mayfield, E., & Rosé, C. P. (2013). Linguistic analysis methods for studying small groups. In C. Hmelo-Silver, A. O'Donnell, C. Chan, & C. Chin (Eds.), *International handbook of collaborative learning*. London: Taylor and Francis.

Howley, I., Mayfield, E., Rosé, C. P., Strijbos, J. W. (this volume). A multivocal process analysis of social positioning in study groups. In D. Suthers, K. Lund, C. P. Rosé, C. Teplovs, & N. Law (Eds.) *Productive multivocality in the analysis of group interactions*, Chapter 11. New York: Springer.

Howley, I., & Rosé, C. P. (2011). Modeling the rhetoric of human–computer interaction. In *HCII'11 Proceedings of the 14th International Conference on Human–Computer Interaction: Interaction Techniques and Environments* (pp. 341–350). Berlin: Springer-Verlag.

Kobbe, L., Weinberger, A., Dillenbourg, P., Harrer, A., Hämäläinen, R., Häkkinen, P., et al. (2007). Specifying computer-supported collaboration scripts. *International Journal of Computer-Supported Collaborative Learning, 2*(2), 211–224.

Kollar, I., Fischer, F., & Hesse, F. W. (2006). Collaborative scripts - a conceptual analysis. *Educational Psychology Review, 18*(2), 159–185.

Kumar, R., Beuth, J., & Rosé, C. P. (2011). Conversational strategies that support idea generation productivity in groups. *Proceedings of Computer Supported Collaborative Learning.*

Kumar, R. & Rosé, C. P. (2011). Architecture for building conversational agents that support collaborative learning. *Learning Technologies, IEEE Transactions on 4.1*, 21–34.

Kumar, R. & Rosé, C. P. (in press). Triggering effective social support for online groups. *ACM Transactions on Interactive Intelligent Systems.*

Martin, J. R., & Rose, D. (2003). *Working with discourse: Meaning beyond the clause*. Continuum International Publishing Group. London, UK.

Martin, J. R., & White, P. R. (2005). *The language of evaluation: Appraisal in English*. London: Palgrave.

Rosé, C. P., Jordan, P., Ringenberg, M., Siler, S., VanLehn, K., & Weinstein, A. (2001). Interactive conceptual tutoring in atlas - Andes. *Proceedings of AI in Education.*

Rosé, C. P., & VanLehn, K. (2005). An evaluation of a hybrid language understanding approach for robust selection of tutoring goals. *International Journal of Artificial Intelligence in Education, 15*(4).

Rummel, N., Spada, H., & Hauser, S. (2006). Learning to collaborate in a computer-mediated setting: Observing a model beats learning from being scripted. In S. A. Barab, K. E. Hay, & D. T. Hickey (Eds.), *Proceedings of the international conference of the learning sciences 2006* (pp. 634–640). Mahwah, NJ: Lawrence Erlbaum.

Sionti, M., Ai, H., Rosé, C. P., & Resnick, L. (2011). A framework for analyzing develpoment of argumentation through classroom discussions. In N. Pinkwart & B. McClaren (Eds.), *Educational technologies for teaching argumentation skills*. Oak Park, IL: Bentham Science.

Suthers, D. (2006). Technology affordances for inter-subjective meaning making: A research agenda for CSCL. *International Journal of Computer Supported Collaborative Learning, 1*, 315–337.

Teasley, S. D. (1997). Talking about reasoning: How important is the peer in peer collaborations? In L. B. Resnick, C. Pontecorvo, & R. Saljo (Eds.), *Discourse, tools, and reasoning: Situated cognition and technologically supported environments*. Heidelberg: Springer.

Veel, R. (1999). Language, knowledge and authority in school mathematics. In F. Christie (Ed.), *Pedagogy and the shaping of consciousness* (pp. 185–216), London, UK.

Wecker, C., & Fischer, F. (2007). Fading scripts in computer-supported collaborative learning: The role of distributed monitoring. *The role of distributed monitoring. In Proceedings of the 8th iternational conference on Computer supported collaborative learning* (pp. 764–772). International Society of the Learning Sciences.

Weinberger, A., & Fischer, F. (2006). A framework to analyze argumentative knowledge construction in computer-supported collaborative learning. *Computers & Education, 46*(1), 71–95.

Chapter 27
Successful Knowledge Building Needs Group Awareness: Interaction Analysis of a 9th Grade CSCL Biology Lesson

Ulrike Cress and Joachim Kimmerle

Introduction

Knowledge building is an emergent collaborative process in which people build on each other's ideas and create new knowledge (Bereiter & Scardamalia, 2003; Cress & Kimmerle, 2008). This process may be supported by using collaborative computer technology (Kimmerle, Moskaliuk, & Cress, 2011; Moskaliuk, Kimmerle, & Cress, 2009; Scardamalia & Bereiter, 2006). Technology that supports knowledge building should be explicitly designed in such a way that it facilitates collaboration between those who are involved in knowledge building activities. The more complex the tasks are that a group has to deal with (its primary purpose may be cooperative work, collaborative learning, development of new knowledge, etc.), the more important it is that the group and its members are supported in this process of collaboration. People who work collaboratively need adequate support in order to make sure that collaboration is able to demonstrate its potential advantages. Members of a collaborating group need to establish and maintain various types of knowledge about the other group members they are interacting with. People working collaboratively need to be enabled to rely on each other, on each other's behaviors, plans, activities, findings, understandings, and knowledge. For this purpose, they need to know some relevant aspects about their collaborators. Kreijns, Kirschner, and Jochems (2003), for example, have argued that social interaction in computer-supported collaborative learning (CSCL) environments will not automatically be a success simply because technology is available that allows some kind of social interaction. Instead, computer-supported social interaction needs to be facilitated

U. Cress (✉)
Knowledge Media Research Center, Schleichstr. 6, 72076 Tuebingen, Germany
e-mail: u.cress@iwm-kmrc.de

J. Kimmerle
Department of Applied Cognitive Psychology and Media Psychology,
University of Tuebingen, Tuebingen, Germany

D.D. Suthers et al. (eds.), *Productive Multivocality in the Analysis of Group Interactions*, Computer-Supported Collaborative Learning Series 16,
DOI 10.1007/978-1-4614-8960-3_27,

systematically, in order to avoid problems of interaction such as coordination or communication difficulties (Janssen, Erkens, Kanselaar, & Jaspers, 2007).

This chapter presents an analysis of chat protocols from four 9th grade biology classrooms with 50 students at a Public School in Pittsburgh, PA. Particular aspects of knowledge building processes in small computer-supported groups are described and explained. We provide examples from the chat protocols that hint to successful knowledge building and from which we can learn something about how the development of knowledge takes place. Moreover, we provide examples that illustrate why four types of group awareness (social, action, activity, and knowledge awareness) are crucial for collaboration, why a lack of group awareness may be detrimental to CSCL, and which strategies students will apply in order to establish group awareness and common ground. Concluding, we point to implications for future design processes of CSCL scenarios.

In the remainder of this introduction we provide a brief overview of the theoretical assumptions this chapter is based on, point out the purpose of analysis, and refer to the unit of interaction, the representations, and the manipulations the analysis builds on.

Theoretical Assumptions

This chapter is based on the assumption that groups can only be successful in exchanging knowledge and developing new understandings when their members have at least some minimal awareness about the other members of the group, that is, about their presence, their actions, their task-related activities, and about the other group members' knowledge.

Purpose of Analysis

The analysis points to social aspects in settings of collaborative knowledge building. With a focus on group awareness the analysis stresses the importance of users' subjective representations of a social situation. The analysis shows that the instructional setting as well as the technological environment should enable people to build valid representations about the social situation and to become aware of the knowledge and the activities of the other group members.

Unit of Interaction

The analysis considers sequences of utterances within the small groups as units of interaction. Such sequences provide examples of successful knowledge building as well as examples of misunderstandings, of missing common ground, or ignoring each other.

Representations

The analysis builds on the textual chat protocols. As the groups could only communicate via text chat, these protocols deliver all kinds of communicative acts that took place. The texts do not only provide the content of people's communication, but also their emotions and motivation. The chat protocols also show the distribution of activity among the learners, the processes of mutual (implicit or explicit) referencing and the temporal development within the communication of a group.

Manipulations

The analysis follows a deductive, theory-driven approach. It starts with theoretical assumptions about different types of group awareness and their influence on collaboration. Then the analysis identifies relevant examples that demonstrate the importance of these processes in the given data. In sum, the theoretically assumed types of group awareness turned out to be crucial indeed for successful communication and collaboration.

Group Awareness

Group awareness is a frequently used label for peoples' knowledge about their fellow group members (Bodemer & Dehler, 2011; Janssen, Erkens, Kirschner, 2011; Kimmerle & Cress, 2008). It is an important prerequisite for efficient computer-mediated communication and successful CSCL. Carroll, Neale, Isenhour, Rosson, and McCrickard (2003) have distinguished three different types of awareness in virtual settings: Social awareness, action awareness, and activity awareness. The authors point out that each type of awareness may be supported by specific awareness tools. Social awareness is a user's perception of the presence of others. It may be fostered by tools that visualize the presence of other group members in any form, for example, by providing photographs of the fellow team mates, or the like. Tools which support the development of action awareness present information about actions that are being carried out by group members at a certain point in time, for example, by displaying which (shared or unshared) resources—such as documents, videos, etc.—other group members are interacting with. So action awareness plays an important role in synchronous collaboration. The term activity awareness refers to the task that a group is supposed to accomplish. For the development of activity awareness, the activities of the group members are described with reference to the common task. So activity awareness tools should be able to provide, for example, feedback on the extent to which group goals have been achieved so far.

Another concept that has emerged in the CSCL literature recently and which plays an important role for the support of successful collaboration is an approach which has been referred to as knowledge awareness (Dehler-Zufferey, Bodemer, Buder, & Hesse, 2011; Engelmann, Dehler, Bodemer, & Buder, 2009). This concept addresses the issue that in a computer-supported group of collaborative learners it is not only overt behavioral information that needs to be provided technically, but also information that would be difficult to access in a face-to-face group as well, that is, the group mates' cognitive states, such as their beliefs, intentions, or their knowledge concerning a certain topic. So, in simple terms, knowledge awareness means that members of a group possess some knowledge about other people's knowledge. Tools that support knowledge awareness may take various forms. Some tools require that learners assess their own knowledge, while other tools look at content that is extractable, in order to allow some inference about what these people really know (cf. Engelmann et al., 2009). The concept of knowledge awareness is closely related to the so-called common ground in a communication process (Clark, 1996; Clark & Brennan, 1991): A certain degree of knowledge about other people's knowledge is necessary for an efficient course of communication. Common ground is an indispensable prerequisite for coordinating joint activities. It means that the partners involved in a joint activity share some information about their mutual activity (and about each other), and that they know that they share this information (cf. Engelmann et al., 2009). Common ground is necessary in order to allow a person to address other people purposefully and specifically, that is, according to those people's respective levels of knowledge. This ability and approach is referred to as audience design (Clark & Murphy, 1982).

Our view is that all these types of group awareness, and the opportunity to establish some common ground, are needed in order to allow successful interaction in computer-supported groups. In this chapter, we will provide a variety of examples which, we believe, can illustrate what makes group members successful (or not) in interacting with each other. We want to demonstrate that in certain situations certain types of awareness are necessary to facilitate learning and knowledge building, or, respectively, in which situations a lack of group awareness will lead to certain problems and difficulties. To demonstrate our ideas, we will refer to examples from chat protocols that were produced in the context of a biology course in a ninth grade in a Pittsburgh public school. This context of data collection is explained briefly in the following section. Subsequently, we will present the data analysis by demonstrating what successful knowledge building may look like, and by addressing both those aspects that indicate a lack of group awareness and those that show how people have, nonetheless, tried successfully to establish group awareness and common ground. The final part of the paper will conclude our findings and theoretical considerations, and suggest further implications for CSCL scenarios.

Data Collection and Methodology

The data which are analyzed in the following study were collected in a unit on homeostasis within the so-called "cell model microstudy," which took place in four 9th grade biology classrooms with 50 consenting students at a Public School in

Pittsburgh, PA, USA [for details see chapter by Dyke, Howley, Kumar, & Rosé (Chap. 25, this volume)]. In this lesson, students examined the function of semipermeable membranes in homeostasis. For this purpose, three students each were teamed together and had the opportunity to discuss their ideas through a computer text chat. In this process, several students of a class were sitting in the same room, each working with his or her own computer. The students did not know who of their classmates was in which group. Each group had a total 43 min for their computer-mediated knowledge exchange. The material that the students worked with consisted of a worksheet in which they were asked to write down their predictions, observations, and explanations in tables (each student worked individually with the same worksheet), a video that showed a variety of osmosis processes (each student started and watched the video on his or her own computer), reading materials about membranes, and a text chat tool. Apart from the three students in a group, there was also a program-operated tutoring agent (not a real person, but a software program) that participated in the chat and aimed at structuring the discussion process. This tutor had assigned to each of the three students a certain role (revoicer, challenger, explainer). The rationale for this role allocation was that one goal of the study was the application of "accountable talk" discussions (Resnick, O'Connors, & Michaels, 2007) in this CSCL scenario. But since most of the students did not take on those roles in a noticeable way, we will not go into this aspect any further in our analysis [detailed examinations of this aspect are provided in the chapters by Dyke et al. (Chap. 25, this volume); and by Howley, Kumar, Mayfield, Dyke, and Rosé (Chap. 26, this volume)].

The data analysis which is provided in the following section is based on the examination of the protocols of the groups' text chat discussions. In this way it, was possible to observe, describe, and explain particular aspects of knowledge building processes in small computer-supported groups. Small groups are the preferred unit of analysis in CSCL, and may indeed be considered [as has frequently been emphasized, e.g., Stahl (2006, 2009)] the crucial unit of analysis. The interaction analysis provided here allowed us to explore some processes of the students' unique, situated, sequential social interactions (cf. Cakir & Stahl, 2009; Stahl, 2006). A presentation that emphasizes the importance of the temporal sequentiality in these particular protocols can be found in the chapter by Stahl (Chap. 28, this volume).

Data Analysis

In this section, we will first provide examples from the chat protocols that hint to successful knowledge building and from which we can learn something about how the development of knowledge takes place. We will then provide some examples that illustrate why the four types of group awareness are crucial for collaboration, why a lack of group awareness may be detrimental to CSCL, and which strategies students will apply in order to establish group awareness and common ground despite unfavorable conditions.

Table 27.1 Explanations of insights

12.03.59	*s035*	*ook group what do we pedict*
12.05.03	*s042*	*I think that the water will seperate and travel to glucose. Because water has a lot of molecules and glucose is just sugar. So it would then balance out. What do you think?*
12.05.39	*s035*	*ook this is what i think...if we put the cell model in the water it will grow like the egg did....but if its in teh sugar it will shrink*
12.06.51	*s042*	*Oh I get it. But wouldn't it still expand and try to balance out?*

Table 27.2 Exchange of hypotheses

12.15.54	*s042*	*Okay, I think that when the distilled water is put into the glucose the glucose will get lighter and then well I don't know what then.....*
12.15.55	*s028*	*wouldnt for B, the water will once again shrink because its mixing water with glucose again?*
12.16.09	*s028*	*oh ok, or your prediction haha*
12.16.20	*s042*	*Oh I think it would, but it might expand at first then shrink*
12.16.47	*s028*	*Yeah*
12.16.51	*s028*	*that makes sense*

Examples of Successful Knowledge Building

The chat protocols provide some intriguing examples of social interaction and shared meaning making, in which students responded to each other's questions, took up each other's statements, and explained important insights to each other. An example (protocol *C02*) can be seen in Table 27.1.

Participant *s035* initiated this unit by asking the other group members for the hypotheses they were supposed to state concerning homeostasis. Participant *s042* answered by providing a prediction and an explanation for that prediction, as demanded by the worksheet, and closed with a question that invited others to give their opinion on this issue. The exchange continued by scrutinizing and challenging the previous thoughts and considerations.

Another example from the same protocol illustrates how two students exchanged hypotheses and reached agreement on their explanations (see Table 27.2).

Both examples demonstrate that the students' starting point for thinking and meaning making were the predictions that their worksheets had asked them to make. This is not surprising, as it is well known that the mode of presentation of a task in a collaborative scenario has a great impact on how learners will behave, or in other words, learners tend to adapt their behavior and thinking to the way in which information is presented to them. The meaning making that took place here was also strongly supported by the use of external representations, namely, the worksheet that all group members had to fill in on their own. On this worksheet, participants had to write down their predictions, observations, and explanations in respective tables (Fig. 27.1 shows an example of a completed worksheet). These tables seemed to provide affordances, that is to say, they inspired participants to swap ideas in a

Predicted results

Condition	Contents of internal environment	Contents of external environment	Predicted change	Explanation for prediction
A	Glucose Solution	Distilled Water	water will get absubed.	Water molecules
B	Distilled Water	Glucose Solution	The glucose will go in to the	
C	Starch Suspension	Distilled Water	The Glucose will mix with the starch	

Fig. 27.1 Example of a partly completed worksheet

Table 27.3 Finding explanations collaboratively

12.20.32	*s045*	*okkaaaaayyyy. do you guys know why this stuff happpned*
12.21.24	*s038*	*for a....it gained weight because th glucose was absorbed*
12.21.43	*s031*	*okay and then would it be the oppistie for b*
12.21.45	*s031*	*?*
12.21.56	*s038*	*umm...im not so sure*
12.22.08	*s045*	*iit lost weighht*
12.22.29	*s038*	*nevermind....yeah it would*
12.22.36	*s031*	*Okay*
12.22.39	*s038*	*what about for c?*
12.22.41	*s045*	*soooo it lost water from inside?*
12.22.57	*s038*	*yeah..weight was lost because the glucose didnt absord all the way*

meaningful way (cf. the influential explanations on *representational guidance* by Suthers, 2001).

The external representation provided certain constraints and made certain aspects salient (cf. Suthers, 2001). The comparison between the three conditions A, B, and C, as prompted by the worksheet, structured the groups' interactions implicitly. Knowledge was particularly shared and developed when the group members started to fill in the tables on their worksheets (cf. also Hron, Cress, Hammer, & Friedrich, 2007). The unit presented in Table 27.3 (protocol *C05*) provides a striking example of collaborative meaning making, in which all three people involved collaborated on finding explanations for the various conditions.

The tables on the worksheet used here were not really shared, but as all group members had the same tables, they could easily make references and comparisons. This led to a shared and coordinated workflow that made them state hypotheses, externalize their knowledge, ask concrete questions, and provide explanations.

But it might have been a further improvement if the setting had not only provided representations that had to be worked on individually, and in which people could not see what the others were doing and to what extent they had finished their task so far. If computer-based representational tools would have been available as shared external representations, knowledge building would possibly have been taking place in an even more successful way. It would then be easier for participants to see what the others are doing, to take up their activities, to refer to each other, etc. (see below).

Table 27.4 Lack of identifiability

08.09.03	*s003*	*who r u guys*
…	…	…
08.17.20	*s003*	*who is #11 and #7*
08.17.41	*s011*	*I'm {s011}, who are you*
08.19.23	*s003*	*{s003} lol*
08.19.38	*s011*	*haha whos number 07*

When the members of a small group are spatially separated from each other and work on their own computers, but are intended to collaborate with each other for knowledge building purposes, we are faced with a CSCL scenario in which communication and coordination are hindered. So this is a learning situation in which the four types of group awareness, as described above, are particularly essential. In the biology chat, however, the students had to deal with a CSCL setting that did not provide sufficient group awareness to the people involved. The interaction analysis yielded many examples that illustrate the point. But, at the same time, several cases of successful collaborative knowledge building occurred despite these adverse conditions. These observations allow some important insights into how groups enable themselves to deal with suboptimal situations and achieve shared meaning making.

Social Awareness

One can assume that a minimum of social awareness exists, if those involved understand that other people are "present" in the situation as well, and if they know who these people are. The first requirement was met in this biology lesson, but not the second one: Students were aware that two other students participated in the chat and in the learning task, because the chat tool indicated clearly when a participant entered the chat room. But who it was, which individual, who "joins the room," remained unclear (as the participants were only labeled with three-digit numbers), and it was obvious in virtually all protocols that the students were unhappy with this lack of identifiability. They spent considerable time and energy in finding out who the other participants were. An example (protocol *A03*) can be seen from Table 27.4.

As can be seen from Table 27.4, even after a few minutes (after the tutor has provided many instructions) the participants were not willing to start performing the task as long as they did not know whom they were supposed to work with.

The issue that the tool did not provide any information on the identity of other group members (which of the physically present classmates belonged to which group) was rather disturbing, as the tools that were used here treated the groups as if their members were distributed between different localities, although they were actually all in the same room. The result of our counting was that 23.14 % of all utterances (184 out of 795 contributions) referred to an effort to establish social awareness. In some groups, this applied to almost half the utterances ("empty" posts and tutor messages were not taken into account in this analysis).

Table 27.5 Confusion regarding the tutor

08.30.10	*s003*	*is alex a real person or a computer*
08.30.20	*s003*	*j jik*
08.30.47	*s011*	*Same but i didnt say why. And I dotn know. I think he was a computer thing.*
08.31.59	*s003*	*it was they just told me*
08.32.03	*s007*	*yeah he was & idk why*

Table 27.6 Physical identification

02.20.07	*s061*	*Who is s058*
02.20.45	*s064*	*the guy over there in the grey hoody*

Table 27.7 Struggle for action awareness

12.17.43	*s039*	*i didnt see the video for condition a though*
...	...	...
12.19.01	*s039*	*im gonna watch the video to see condition c*
12.19.21	*s032*	*Okay*
12.19.24	*s046*	*im gonna watch it too to understand whats going on*

Participants were also confused regarding the "social status" of the tutor "Alex." They were wondering if this was a human being or not (protocol *A03*) (Table 27.5).

What did people do in order to find out who the other participants were, or in other words, to establish at least a minimum of social awareness in their groups? Apart from asking each other for their names and indicating their own names (see above), they sometimes tried to utilize the fact that they were all located in one room (even though not assignable) by identifying themselves (or each other) physically. An example is provided in Table 27.6 (protocol *D03*).

Action Awareness

In order to accomplish action awareness, participants in a CSCL setting need to know what the other group members are doing at the present moment. In the biology chat, however, this remained unclear to the participants during large parts of the collaboration. For example, it was not recognizable whether the others were watching the video or had already finished watching it. So these participants had to spend additional time and effort to inform each other explicitly about what they were now doing and what they were intended to do next (Table 27.7; protocol *C06*).

The students used various strategies to find out what the other participants were doing just now. They wrote to each other what they had just done or not done, what they were planning to do next, and sometimes they requested others to do or omit

Table 27.8 Clarifying actions

12.06.29	*s037*	*m j :)*
12.06.33	*s044*	*Fag*
12.06.59	*s037*	*watch the videio..and that wasnt me..ha*
12.07.10	*s044*	*8=====>*
12.07.21	*s044*	*V*
12.07.22	*s037*	*MJ*
12.07.29	*s044*	*{s036} gay*
12.07.49	*s044*	*{s045} wrote all that*

Table 27.9 Mutual assistance

12.09.40	*s035*	*we need to watch the movie maybe?*
…	…	…
12.10.11	*s042*	*Yeah we should*
12.10.59	*s035*	*ook were do we go lol*
12.11.19	*s028*	*down at the bottom of the screen*
12.11.29	*s035*	*thnx found it*

Table 27.10 Targeted coordination

12.04.18	*s029*	*WHAT ARE WE SUPPOSED TO WRITE*
12.04.25	*s036*	*ok wat r we doin?*

some particular action as a next step. Often, they informed each other about things that were just going on, which had nothing to do with the actual task. For example, in the unit presented in Table 27.8, two students explained to each other that it was, allegedly, not they themselves who wrote certain contributions in their name—which is also related to the lack of social awareness as described above (protocol *C04*).

Activity Awareness

Activity awareness—information that refers to the task that the group has to accomplish—was difficult to establish in the biology chat. In general, participants struggled hard to gain task-relevant information in the first place: Often they did not know where the video was, what could be seen in the video, what to do with the worksheet, and so on. In such situations the participants helped each other. An example is provided in Table 27.9 (protocol *C02*).

Moreover, there are some relevant examples of processes of targeted coordination between the group members (e.g., protocol *C03*; Table 27.10).

What was a problem here was that participants could not see what the other group members were working on. Instruction was given verbally, but it seems that it was not sufficiently embedded into the learning setting. It would probably have helped if the tutor program had provided this kind of awareness, but its tutorial instructions did not respond adequately to the group's activities. So, instead of

Table 27.11 Interruptions by the tutor

12.06.03	*s039*	*for solution a i think water will be added to the cell*
12.07.10	*s039*	*my explanation is that the glucose will absorb the water causing the cell to expand*
12.07.23	*s032*	*i think the exact opposite*
12.07.33	*s039*	*why?*
12.08.10	*s039*	*lets go to the video*
12.08.28	*s032*	*to my knowledge glucose attractes water but when inside an ienvoronment itll push out water but i could be wrong*
12.08.43	*s032*	*so i like what you said*
12.08.45	*s032*	*Said*
12.08.50	*Alex (Tutor)*	*Are you guys back?*
12.09.58	*Alex (Tutor)*	*Now discuss what you observed in the video about conditions A and B. Compare what you observed to your predictions.*
12.10.53	*s039*	*alrite what do you think about condition B?*
12.12.59	*Alex (Tutor)*	*You should now move on to discussing what you observed in the video about condition C. Compare what you observed to your predictions. How is this different from what happened in condition A?*

providing awareness, the digital tutor often led to confusion. For example, as can be seen from Table 27.11, one group was doing rather well in discussing how a semipermeable membrane might work, but the tutor managed to interrupt these learners (protocol *C06*).

The tutor's first interruption occurred only 5 s after the last contribution of participant *s032*, so in the middle of a thriving discussion the team mates were prevented from continuing and potentially developing new insights. And shortly after *s039* had suggested discussing condition B, the tutor demanded to discuss condition C instead. The tutoring software not only failed in providing information on the students' progress with the task, but also failed to use the available progress information. Consequently, the groups disregarded the tutor to a large extent. There were only very few episodes in which people followed its instructions. But even if they did, this barely supported their knowledge building. In these cases, the students were simply following instructions from the tutor, without a noteworthy progress of insight.

Knowledge Awareness

When members of a small group pursue the aim of learning together and building knowledge jointly, they will need some information on what the other group members already know in general terms and about the relevant topic in particular. As group members in this biology course could not see each other's completed tables, it was not at all easy for them to become aware of the other students' knowledge. The following example (Table 27.12; protocol *C03*) shows a group in which

Table 27.12 Most knowledgeable student answers questions

12.04.18	*s029*	*WHAT ARE WE SUPPOSED TO WRITE*
12.04.25	*s036*	*ok wat r we doin?*
12.04.33	*s043*	*i think the water will travel into the tube with the gluecose because of osmosis*
12.04.57	*s029*	*for A?*
12.05.15	*s036*	*a? i thought it was #1?*
12.05.19	*s043*	*water will mix with the gluecose to try and balance out the water molecules*
12.05.26	*s043*	*yes for a*
12.06.03	*s036*	*for predicted change?*
12.06.42	*s029*	*wouldn't that et he same for B?*
12.07.06	*s036*	*i need help! ;(*
12.07.24	*s043*	*i think the water will be obsorbed*
12.07.35	*s043*	*for part b*
12.08.24	*s029*	*i think the glucose will dissolve*
12.08.33	*s043*	*there is more gluecose than water outside than inside so the water will come out*
12.08.43	*Alex (Tutor)*	*Is everyone back?*
12.09.01	*s036*	*ok can someone help me please!*
12.09.07	*s029*	*how do you know there's more glucose than water outside than inside?*
12.09.13	*Alex (Tutor)*	*Now discuss what you observed in the video about conditions A and B. Compare what you observed to your predictions.*
12.10.07	*s043*	*becasue the tube of water is siting inside the container of gluecose*

the participants' utterances provided many opportunities to find out what the others knew. Right from the beginning, *s029* and *s036* started the conversation by asking questions, whereas *s043* answered these questions and provided explanations. In addition, *s036* repeatedly asked for help during the chat.

A few minutes later, participant *s043*—who had been the most knowledgeable person until then—watched the video, and the two remaining students had to move forward on their own. The following part of the chat protocol (Table 27.13) shows that thereupon *s029* continued by asking questions, whereas *036* was now taking the role of an explainer. But s/he stills oriented on what *s043* had said before she left the interaction.

This sequence presented in Table 27.13 shows that both *s036* and *s029* recognized clearly that *s043* had provided much knowledge from which they could benefit. At first, this knowledge awareness led to a situation in which *s036* explicitly asked for help and acquired some knowledge of his or her own in this way. After *s043* had left the interaction, *s036* could make uses of this knowledge in order to explain it to *s029*. In this fashion, *s043*'s knowledge became *s036*'s knowledge as well: "*idk this is wat* she *was sayn*" in 12.19.13 emerged to "*i put for A the explantation prediction of what* i *said about A lol*" in 12.20.14.

Table 27.13 Another student takes the role of an explainer

12.18.12	*s036*	*b- water will obsorbed bc there is more gluecose/the water travel low water level to a higher source? ritght?*
12.18.33	*s029*	*wait, would osmosis be the explanation for A?*
12.18.59	*s036*	*yeah the A "the water will go into the tube bc of osmois"*
12.19.13	*s036*	*idk this is wat she was sayn*
12.19.25	*s029*	*so we're writing "Osmosis" in the box for Explanation?!*
12.19.43	*s036*	*but she is watching the video we are soo lost :(*
12.20.02	*s029*	*hahahh oh dear*
12.20.14	*s036*	*i put for A the explantation prediction of what i said about A lol!*

Discussion

In this chapter, we have described and explained in which respect and to what extent four types of group awareness (social, action, activity, and knowledge awareness) play a significant role in the context of successful learning and knowledge building in CSCL environments. The data presented here do not show completely what happened in the groups, but they illustrate some interesting cases of more or less successful collaboration sequences. In particular, the external representation on the worksheet, which had to be completed by the individual students, provided a useful scaffold for group interaction. It focused the activity of the group and was able, to a certain degree, to support the development of activity awareness. Moreover, in those cases in which groups managed to become aware of the knowledge of the others, examples of successful knowledge transfer could be found as well. The examples also demonstrated that a lack of social awareness and a deficit in action awareness hindered successful collaboration. Without all these types of group awareness, beneficial interaction was difficult to establish. The groups needed great effort and energy to achieve the required extent of group awareness before they could engage in a relevant and productive form of content-related interaction.

In future design processes of CSCL scenarios, the development of all these group awareness types should be supported: In many situations of computer-supported social interaction, the available tools should not only allow people to recognize that someone else is present, but they should also provide social awareness in terms of knowing who the person is that one is interacting with. Moreover, it is important that participants have the opportunity to recognize what the others in the group are doing and what resources they are concerned with in the present moment. Failing that, there may be confusion and problems in synchronizing group actions. Regarding activity awareness, tools are required that allow targeted coordination between the group members on the task that has to be accomplished. It should be possible to use shared tools that present information in such a way that they offer affordances that will structure interaction in a meaningful manner. Finally, in order to support the development of knowledge awareness in the collaboration process, it may be useful to apply tools that clarify who knows what in the group. This will give participants the opportunity to access additional knowledge resources as needed, and will facilitate knowledge transfer and emergence of new knowledge.

References

Bereiter, C., & Scardamalia, M. (2003). Learning to work creatively with knowledge. In E. De Corte, L. Verschaffel, N. Entwistle, & J. van Merriënboer (Eds.), *Powerful learning environments: Unraveling basic components and dimensions* (pp. 55–68). (Advances in Learning and Instruction Series). Oxford: Elsevier Science.

Bodemer, D., & Dehler, J. (2011). Group awareness in CSCL environments. *Computers in Human Behavior, 27*, 1043–1045.

Cakir, M. P., & Stahl, G. (2009). Interaction analysis of dual-interaction CSCL environments. In C. O'Malley, D. Suthers, P. Reimann, & A. Dimitracopoulou (Eds.), *Computer supported collaborative learning practices: CSCL2009 conference proceedings* (Vol. 1, pp. 3–12). International Society of the Learning Sciences (ISLS).

Carroll, J. M., Neale, D. C., Isenhour, P. L., Rosson, M. B., & McCrickard, D. S. (2003). Notification and awareness: Synchronizing task-oriented collaborative activity. *International Journal of Human-Computer Studies, 58*, 605–632.

Clark, H. (1996). *Using language*. New York: Cambridge University Press.

Clark, H., & Brennan, S. (1991). Grounding in communication. In R. M. Levine, L. B. Resnick, & S. D. Teasley (Eds.), *Perspectives on socially shared cognition* (pp. 127–149). Washington, DC: American Psychological Association.

Clark, H., & Murphy, G. (1982). Audience design in meaning and reference. In J. F. Le Ny & W. Kintsch (Eds.), *Language and comprehension* (pp. 287–299). Amsterdam: North-Holland.

Cress, U., & Kimmerle, J. (2008). A systemic and cognitive view on collaborative knowledge building with wikis. *International Journal of Computer-Supported Collaborative Learning, 3*, 105–122.

Dehler-Zufferey, J., Bodemer, D., Buder, J., & Hesse, F. W. (2011). Partner knowledge awareness in knowledge communication: Learning by adapting to the partner. *The Journal of Experimental Education, 79*, 102–125.

Dyke, G., Howley, I. K., Kumar, R., & Rosé, C. P. (this volume). Towards academically productive talk supported by conversational agents. In D. Suthers, K. Lund, C. P. Rosé, C. Teplovs, & N. Law (Eds.), *Productive multivocality in the analysis of group interactions*, Chapter 25. New York: Springer.

Engelmann, T., Dehler, J., Bodemer, D., & Buder, J. (2009). Knowledge awareness in CSCL: A psychological perspective. *Computers in Human Behavior, 25*, 949–960.

Howley, I. K., Kumar, R., Mayfield, E., Dyke, G., & Rosé, C. P. (this volume). Gaining insights from sociolinguistic style analysis for redesign of conversational agent based support for collaborative learning. D. Suthers, K. Lund, C. P. Rosé, C. Teplovs, & N. Law (Eds.), *Productive multivocality in the analysis of group interactions*, Chapter 26. New York: Springer.

Hron, A., Cress, U., Hammer, K., & Friedrich, H. F. (2007). Fostering collaborative knowledge construction in a video-based learning setting: Effects of a shared workspace and content-specific graphical representation. *British Journal of Educational Technology, 38*, 236–248.

Janssen, J., Erkens, G., Kanselaar, G., & Jaspers, J. (2007). Visualization of participation: Does it contribute to successful computer-supported collaborative learning? *Computers & Education, 49*, 1037–1065.

Janssen, J., Erkens, G., & Kirschner, P. A. (2011). Group awareness tools: It's what you do with it that matters. *Computers in Human Behavior, 27*, 1046–1058.

Kimmerle, J., & Cress, U. (2008). Group awareness and self-presentation in computer-supported information exchange. *International Journal of Computer-Supported Collaborative Learning, 3*, 85–97.

Kimmerle, J., Moskaliuk, J., & Cress, U. (2011). Using wikis for learning and knowledge building: Results of an experimental study. *Educational Technology & Society, 14*(4), 138–148.

Kreijns, K., Kirschner, P. A., & Jochems, W. (2003). Identifying the pitfalls for social interaction in computer-supported collaborative learning environments. A review of the research. *Computers in Human Behavior, 19*, 335–353.

Moskaliuk, J., Kimmerle, J., & Cress. (2009). Wiki-supported learning and knowledge building: Effects of incongruity between knowledge and information. *Journal of Computer Assisted Learning, 25*, 549–561.

Resnick, L., O'Connor, C., & Michaels, S. (2007). Classroom discourse, mathematical rigor, and student reasoning: An accountable talk literature review. http://www.learnlab.org/research/wiki/images/f/ff/Accountable_Talk_Lit_Review.pdf

Scardamalia, M., & Bereiter, C. (2006). Knowledge building: Theory, pedagogy, and technology. In K. Sawyer (Ed.), *Cambridge handbook of the learning sciences* (pp. 97–118). New York: Cambridge University Press.

Stahl, G. (2006). *Group cognition: Computer support for building collaborative knowledge*. Cambridge, MA: MIT Press.

Stahl, G. (2009). *Studying virtual math teams*. New York: Springer.

Stahl, G. (this volume). Interaction analysis of a biology chat. In D. Suthers, K. Lund, C. P. Rosé, C. Teplovs, & N. Law (Eds.) *Productive multivocality in the analysis of group interactions*, Chapter 28. New York: Springer.

Suthers, D. D. (2001). Towards a systematic study of representational guidance for collaborative learning discourse. *Journal of Universal Computer Science, 7*, 254–277.

Chapter 28
Interaction Analysis of a Biology Chat

Gerry Stahl

Analyzing Response Structure

This chapter takes a specific analytic approach, developed within the Virtual Math Teams (VMT) Project (Stahl, 2009). The VMT research team adapted video-based interaction analysis of face-to-face discourse (Jordan & Henderson, 1995) to analyze synchronous text chat by students in their mid-teens as they interact in the online VMT environment, discussing issues raised in school mathematics. We found that, from a structural viewpoint, the most important aspect of discourse is its temporal sequentiality; the field of Conversation Analysis has analyzed this extensively, beginning with (Sacks, 1962/1995) and summarized more recently by (Schegloff, 2007). We adapted such sequential analysis to student chat discourse in the VMT environment at the foundational level of "adjacency pairs" of mutually responsive postings (Stahl, 2006c)—which we take as the unit of interaction—and at the "longer sequence" level (Stahl, 2011)—which we feel is the key level of description for knowledge building in computer-supported collaborative learning (CSCL).

In this chapter, I apply the method of analyzing text-chat response structure that we developed in the VMT Project to chat among students discussing a biology experiment conducted in an early version of the environment formerly known as ConcertChat (now VMT). The text chat was integrated with class discussion, a worksheet and videos. In addition, the software was extended with a software agent, which interacted with the students as a chat participant. I ignore most of the larger context of the experiment (see chapters by Dyke et al., Chap. 25, and Howley et al., Chap. 26, this volume) and focus on what is visible in the chat log. I look at a representative case from each of the first two cycles of experimentation.

G. Stahl (✉)
Drexel University, Philadelphia, PA, USA
e-mail: Gerry@GerryStahl.net

D.D. Suthers et al. (eds.), *Productive Multivocality in the Analysis of Group Interactions*, Computer-Supported Collaborative Learning Series 16,
DOI 10.1007/978-1-4614-8960-3_28, © Springer Science+Business Media New York 2013

In undertaking this paper, I decided to do my own methodological experiment within the biology educational experiment. I wanted to see if sequential analysis could be used effectively as a quick-and-dirty method of evaluation within a design-based-research cycle. Design-based research is a widespread approach within educational research for designing technological and pedagogical interventions through iterative cycles of design, prototyping, user trial, analysis, and redesign. In the biology experiment, an intervention had been designed for biology classrooms; software agents had been prototyped within a version of the VMT collaboration environment; the intervention was tried in middle-school classrooms; and it was now time to analyze the results. While some experimenters may have been hoping that analysis would show the benefits of agent support or accountable-talk training, my aim was to discover what most needed redesign in the next cycle.

Although design-based research is a much used and discussed approach to educational research, there is no established method for conducting the analysis phase of the iterative cycles. Researchers both friendly to and opposed to Conversation Analysis (CA) have argued that CA sequential analysis is inappropriate in design-based research. Adherents of CA argue that CA cannot be applied to design efforts because it is interested in seeing what emerges of interest from an unguided analysis of the participants' discourse—which is unlikely to be relevant to a designer's goal-oriented concerns. On the other hand, researchers from other approaches, such as quantitative coding of discourse, insist that qualitative CA takes too long and is too costly to fit into the workflow and focused research questions of redesign cycles. My experiment was to see if I could conduct a quick sequential analysis that would cheaply and effectively point the way for redesign. That was the practical goal of my methodological experiment.

Theoretically, I was interested in understanding what "really" occurred in the interaction between students and agent. I wanted to "bracket out" the assumptions of the people who set up the biology experiment as well as assumptions about what went on in the heads of the students or the programs of the agent, based on reports from outside the discourse data. As a researcher of group cognition (Stahl, 2006a), I am interested in the effect of the intervention on the group processes, the interaction visible in the chat log. I wanted to see how much I could learn about the group process by viewing the structure resulting from sequential analysis. I wondered what I could fathom of the group knowledge building from micro-analysis of the discourse details, i.e., from how the participants articulated their responses to each other. The goal of accountable-talk training and support is presumably to change certain aspects of the talk by the students, and this is what I wanted to observe directly—not indirectly from statistical verification of hypotheses based on testing responses of individual students outside of the group-interaction context.

Obviously, the behavior of the students will be affected by countless factors, many of which could be studied in theory with various methods and data-collection efforts: the personalities and backgrounds of the students, the programming of the agents, the funding of the schools, the history of American education, prior testing results and future test schedules, etc. But I wanted to see how far I could get in making grounded redesign recommendations by just looking with some care at a small sample of interaction data.

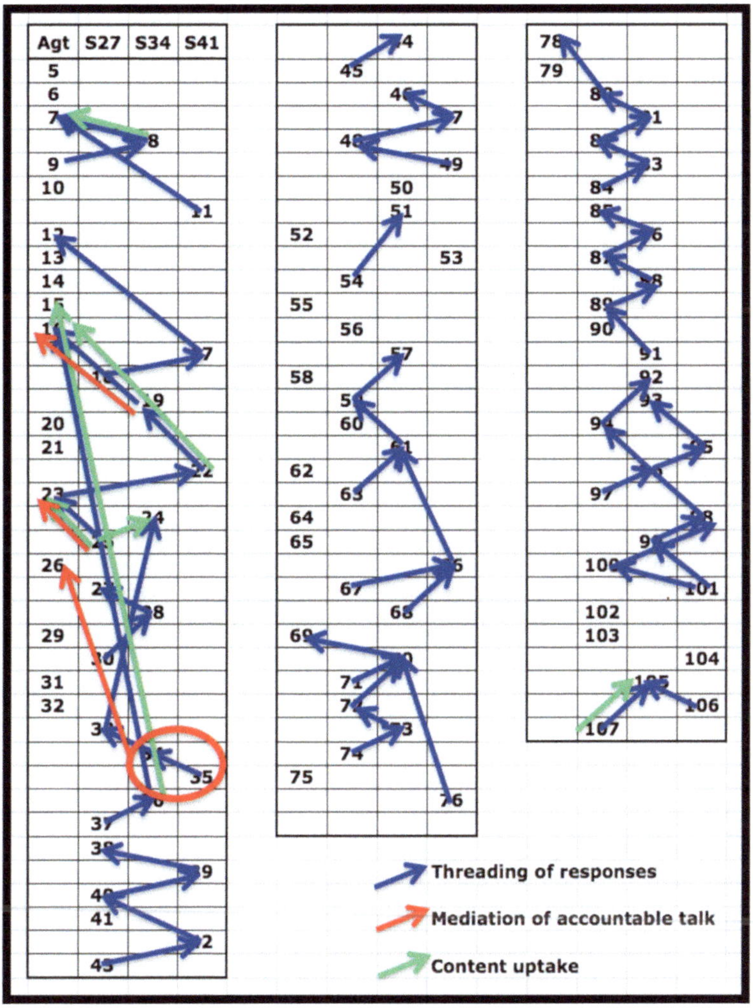

Fig. 28.1 Sequential response structure of chat C01. Note that only interactions between actors are represented, not instances of a posting by one actor building on his, her, or its own previous posting

Furthermore, I was only concerned about the group unit of analysis, that is the interactions among group members, not the status of any one individual member. Fortunately, because the group interaction for a period of time during the experiment was mediated by the VMT system, all group interaction among the students and the agent passed through the chat tool and was captured in the chat log exactly as it appeared to the participants. This gave me a complete and reliable log of the group interaction without all the complications and interpretive issues of videotaping and transcribing. As described below, I modified the chat log representation and then constructed a representation of the sequential interaction (Fig. 28.1). Simply looking at this representation allowed me to make some tentative conclusions about the nature of the interaction and to point these conclusions out to others. The conjectures based on this representation guided a careful look at the details of how the specific chat postings involved were designed by their posters, the groups of students.

The problematic aspects of interaction revealed in my quick response analysis of a student chat in the original intervention were taken into account in redesigning the intervention in a second cycle of design-based research a year later. I conducted a similar quick response analysis of a student chat in cycle 2 and was able to see a significant improvement in the behavior of the agent as well as in the discourse of the student group.

Method

1. Following the first classroom intervention, I was supplied with the logs of 16 chats, in spreadsheet format. The chats each lasted about a half an hour and contained the chat postings of three students and an agent. The 16 chats were divided among three conditions: in one condition the agent prompted students (indirectly) to ask each other to make specific accountable-talk moves; in a second condition the agent prompted students (directly) to make specific accountable-talk moves; in the final condition the agent did not make any accountable-talk prompts, but only guided the students through the steps of the assignment (as was also done in the first two conditions).
2. I read through each of the 16 chat logs that I was given and I wrote down a couple sentences of my initial reaction to the quality of the interaction. It struck me that similar patterns of interaction were arising in the 16 logs, and so I decided to analyze one chat in detail to get at key common patterns. I selected log C01 as representative and promising for illustrating the common patterns. This case was from the first condition, in which the agent gave indirect prompts. Clearly, other analyses with different research questions and approaches would want to contrast the different conditions (e.g., Howley et al., Chap. 26, this volume), but from my focus on response structure it seemed particularly useful to look closely at one typical example.
3. In order to make the interaction flow visible, I rearranged the spreadsheet to have the postings of each participant in its own column. The newer version of VMT produces logs in this format automatically for students, teachers and analysts. We often also have columns for time elapsed since the previous posting and time when a posting was starting to be typed. These figures sometimes help to determine which previous posting a new posting is responding to. In the current log, such detailed reasoning was not generally necessary.
4. I next sketched the response structure of the chat (see Fig. 28.1). I drew an arrow from each posting to the prior posting to which it was responding interactively, for instance to what question is an answer responding? This already gave a visual impression of some aspects of the patterns of responses. These patterns are central to the interactional dynamic of the group.

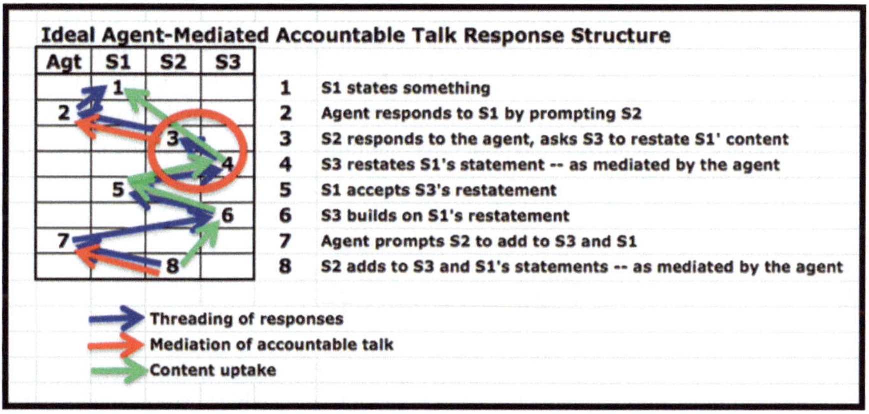

Fig. 28.2 Sequential response structure of accountable talk

5. An important phase of interaction analysis is the exploration of the data, line-by-line, in a *data session* with other researchers (Jordan & Henderson, 1995). This inherently dialogical or multivocal approach can bring in multidisciplinary perspectives and balance one-sided views. A data session can be most effective once some initial analysis has already been undertaken by one of the researchers. After the data session, suggestions have to be synthesized and followed up with further detailed data analysis. There can be multiple cycles of group and individual analysis. The data session for this chapter's analysis included experienced online educators from the Math Forum and two analysts from other chapters (Rosé and Goggins). The session suggested a more complex representation of the response structure, it refined interpretive details, and it situated the case study in a deeper understanding of the experimental context. In particular, the data-session discussion proposed the representation of response structure of accountable talk shown in Fig. 28.2, which was used in refining Fig. 28.1.
6. Once I had a preliminary view of the response structure of the discourse in the chat, I could start to formulate tentative observations about the case study. These observations led to looking at the textual content of the postings. This showed the nature of the group interaction in more detail. The evolving analysis (see next section) also revealed the understandings and reactions of the students to their situation. This highlighted the response of the student group to its given task and to the actions of the agent, to the accountable-talk training and to the software environment.
7. As I summarized my observations (see Sect. 4), I felt that they generally applied to the other chats as well. By grouping the problems in relation to different design decisions in the experiment, I was able to propose several general suggestions for future redesign (see Sect. 5). Other analysts, taking into account other data, additional knowledge of the constraints on the experiment, and alternative research questions will undoubtedly reach different—hopefully

complementary—conclusions. I was interested in seeing what insights an interaction analysis of a single case study could provide for the long-term design-based-research effort. I wanted to do this analysis strictly on the basis of the chat data from a single case study, without being concerned about the many constraints, practicalities, and concerns that influenced the experimental design in all its complexity.

Analysis of the Chat-Response Structure

Figure 28.2 shows a representation of the response structure of an ideal accountable-talk interaction, as hypothesized by the experimenters. The blue arrows indicate that the agent responds to the students (line 2 and 7) and that the students in turn respond to the agent (lines 3 and 8). There is also a sequence in which the students respond to each other (lines 3, 4, 5, 6). This produces a tight group interaction including the agent and the students. The green arrows indicate that subsequent postings often involve uptake of content from previous postings (e.g., lines 4, 5, 6, 8 by the students). The role of the agent does not involve content, but mediates the student uptake of content by means of accountable-talk prompts (lines 2 and 7, pointed to by the red arrows). Let us see the extent to which the data of actual interaction among students and the agent includes similar patterns of response.

Figure 28.1 indicates three instances of mediation of accountable talk (red arrows): (1) the response at line 19 to line 16, (2) the response at line 25 to line 23, and (3) the response at lines 34 and 35 to line 26. Let us consider each of these in turn.

1. The agent requests in line 16: "`Please discuss what you predict will happen in these two conditions.`" Student S034 complies after a lengthy two-and-a-half minutes of silence by asking the group, "`what do you think'ds going to happen?`" At this point, the agent interjects some information about a third condition and asks the students to move on to discussing that. The timing of this seems questionable if the goal is to encourage extended knowledge-building interaction among the students. Student S041 then ignores the agent's latest contribution and responds ironically to student S034's request for a prediction: "`the world is going to end in 2012.`"
2. The agent quickly picks up on S041's prediction by introducing the indirect prompting for accountable talk in line 23: "`S027, now would be a good time to ask S034 to build on what S041 is saying.`" This all confuses S034, who states, "`im so confused!`" But S027 dutifully instructs S034 to explain S041's remark by building on it and explaining it to S027: "`034, would you like to build on to what 041 is saying? and me too!`" The first part of this follows the script prompted by the agent, but S027 adds his sympathetic addendum, aligning with S034 by agreeing that he is also confused about what is being asked of them.
3. The final mediation is similar to the first. In line 26, the agent requests: "`When you are in agreement, write down your predictions and`

explanations for Conditions A, B and C on your worksheet." A minute later, after S027 complains again of not knowing what to do, S034 says, "someone predict something." Student S041 responds again to student 034: "THE WORLD IS GOING TO END IN 2012!."

As the green arrows indicate, almost all uptake of content is associated with these three mediated interactions. Line 8 merely introduces the student, repeating the word "name": S034 responds to the agent's "I didn't get your names yet" with "my name is [S034]." Line 107 responds to line 105's birthday greeting with "is it ur birthday?" These are not knowledge-building moves, but are social interactions, not directly relevant to accountable talk about curricular content.

There is some evidence that the agent is responding to student postings. The agent's line 7 succeeds in getting S034 to give his or her name and the agent then responds to that by assigning a role to S034. At line 23, the agent responds to a posting by S041 by asking S027 to ask S034 to build on what S041 said. This is an instance of the indirect mediation. While the timing is appropriate to ask S027 and S034 to discuss a posting by S041, the agent clearly fails to understand the significance of the posting. The agent assumes that S041 has made a prediction about the biology experiment, and not a sarcastic joke. This could have sent the group off on a distracting tangent, but in fact only confused the students about the agent's behavior and the meaning of the agent's requests.

If we look at the blue arrows in Fig. 28.1, we see that the only times that the agent responded to the students were in lines 9 and 23. In line 9, the agent started to assign roles that were ignored by the students. In line 23, the agent requested an accountable-talk script to build on a joke.

A look at the high-level visual structure for Fig. 28.1 indicates that the agent dominated the discussion in the early part, but then was ignored for most of the remainder of the chat. Toward the end, there was a significant pattern of interaction among the students, who seemed to be engaged as a group. A closer look at the content of the individual students' postings suggests that S034 is trying hard to accomplish the class task. S027 seems generally lost. S041 is not interested in the biology and is more oriented to clowning around. There is no apparent correlation of their individual behaviors to the roles assigned to them by the agent.

The period from posting 5–18 lasted about 4 min. This period is totally dominated by the agent, which posted over 260 words while the three students responded with a total of nine words, mostly just stating their names. The agent did not acknowledge their responses or appear to respond to them, except as noted above. Although delivering instructions to the students through the agent may have been motivated by an attempt to establish dialog between the agent and the students, it positioned the agent as an authoritative source of knowledge and commands, while positioning the group of students as a set of largely passive listeners, thus discouraging student discursive agency.

Of course, it made no sense for the agent to ask the students to "build on" to the sarcastic answer in line 22. This response by S041 shows that he/she already did not take the agent seriously. By not interacting with the students in a way that makes sense to them, the agent fails to establish itself as a serious participant in the group discourse. Caught in the middle between human interaction with the other students and obeying the authoritative orders of the agent, S027 follows the agent's command, but adds his protest against the agent's leadership in line 25.

S027 and the other students then stop orienting to the agent and the agent is ignored for the next 10 min until it again provides an unhelpful indirect prompt for accountable talk at line 69. Instead of responding to the agent prompt, S027 asks who is 34 and says "`ooh. hi`" when S034 responds. The students go on to work together to fill in the worksheet. One student provides the answers and the others try to figure out how to copy those answers into their own worksheets.

The agent continues to give commands, but they are generally ignored. When in line 69 the agent prompts once more for accountable talk, the students agree that the agent is being an insufferable nuisance. They evaluate the whole supported chat experience by agreeing that "`this would be so much easier just in a group`," meaning just sitting together without any computer or agent support and filling in their worksheets. Their only subsequent response to the agent is to celebrate when it leaves.

Discussion: Issues Observed

In the initial experiment, students were placed in small groups of three students and an agent in a chat room. This is a setting that calls for intense text-based interaction. The patterns in Fig. 28.1 are already visually suggestive. The agent does not significantly respond to (i.e., interact with) students. The student responses to the agent are problematic. After trying to be responsive, the students give up and start to engage in their own discussion. The later periods of student interaction show considerable back-and-forth responses as they elicit responses, provide responses, and then acknowledge the responses to each other in various ways. Student responses are tightly situated in the ongoing discourse, whereas the agent speaks like an academic textbook, with no sense of contextualization and little apparent attempt at interaction.

The educational experiment is an attempt to support collaborative learning with (a) the VMT software environment, (b) software helping agents (c) a social small-group setting, and (d) accountable-talk prompts. It is a CSCL intervention that aims to scaffold collaborative learning with these forms of computer support and communication structuring.

(a) The first problem is that the lesson design does not succeed in fostering collaboration. The students are each given their own worksheet to fill out and then they are each tested individually. There is no meaningful group task or group goal to be accomplished collaboratively. The questions to be addressed by the students are not open-ended issues to encourage group inquiry and discussion, but questions with instructor-defined correct answers that the students can solve individually. Consequently, there is little evidence of real knowledge building taking place collaboratively. The most that occurs is that a student who knows the correct answer will give it to students who do not know it. Rather than this taking place as accountable talk, it naturally takes place in the form of students copying each other's answers to fill in their individual forms, without caring much about understanding the science—i.e., a common school process understood by all as cheating rather than collaborating or learning. The VMT environment was designed for shared tasks, with a shared whiteboard provided as a shared external memory that can be even more important for communication and joint work than the text chat (Çakır, Zemel, & Stahl, 2009). Rather than this, the experiment uses the whiteboard to display once more a static cartoon of accountable talk, which appears to have been completely ignored by the students. The whiteboard could have contained the worksheet, to be filled out collaboratively by the team. That group artifact could then have been evaluated for the grading, rather than threatening the students with individual quizzes (causing expressions of test phobia). The shared whiteboard (or additional tabs with Web browsers or other whiteboards) could also have been used to present data of the biology experiment, rather than having the students have to start up other applications (causing further confusion).

(b) The second problem involves the design of the agent interventions. First of all, the agent was in effect non-interactive. The agent may have been carefully programmed to intervene in an interactive way, but it does not come off that way in a sequential analysis of the chat—which is more important than the intentions of the programmer. To the students, the agent's timing did not appear to be effectively coordinated with the student discourse or responses. Inevitably, the agent postings introduced confusion for the students rather than clear structure. They were incredibly verbose—within the chat medium, which is known for its conciseness of expression. It might have made more sense to explain the process in class before breaking into online chat groups. Helping agents should probably not be used to automate teacher-centric instructors, but should get out of the way of student interaction until the students express a need for help. When an agent does intervene, it has to know what is going on well enough to judge what kind of response might be helpful. The agent behavior programmed here was an extreme example of "over scripting" and the opposite of the recommended "SWISH approach" (Dillenbourg, 2002; Dillenbourg & Jermann, 2006).

(c) A third problem involves social identity. Teenage students are mainly learning social skills, despite teacher efforts to have them learn curricular content. So when they are put together to interact in small groups it is essential to them that they know as much as possible about each other. In the VMT Project, we tried to put together students with no prior knowledge of each other so that we researchers could know everything the students knew about each other, so that we could interpret their interaction logs on a par with their understanding of the group interaction. In this biology case study, the students knew each other very well and had well practiced relationships. By assigning the chat participants anonymous identifiers, the experiment interfered with their exercise of these important and motivating social relationships (see chapter by Cress & Kimmerle, Chap. 27, this volume). The students spent much time and attention in overcoming this circumstance (e.g., chat lines 17/18 and 27/28/30), positioning them in opposition to the conditions imposed upon their daily routines by this experimental intervention.

(d) Finally, accountable talk needs to take place at a sophisticated level of discourse. Like all effective discourse, it must be highly situated in the ongoing discussion. That is the skill of a teacher who has mastered accountable talk moves, to know just when and how to prompt. A complicated prompt cannot just appear out of the blue and hope to be helpful in building shared understanding. This poses a major technical challenge for software agents at many levels; it may require many cycles of design-based research to evolve an effective interaction behavior for helping agents that can effectively prompt for accountable talk by students.

Suggestions for Redesign

The biology experiment is cutting-edge research. The components that it brings together each require groundbreaking advances in the knowledge of their domain. It is not a matter of simply applying well-understood techniques.

(a) It took years of research by a large international, interdisciplinary team to develop the integration of pedagogy, problem, and technology for the Virtual Math Teams Project in the domain of collaborative online discourse of school mathematics—and there is still much investigation to be done there. Similar explorations will be needed for the domain of online discourse of school biology. A primary issue in guiding student inquiry in small online groups is how to avoid intruding in the important processes of small-group collaboration among the students; the case study just analyzed shows that there is a long way to go in achieving this with the approach tried. Our past research emphasizes how important yet difficult guidance or scaffolding of collaborative knowledge building is to achieve. In the VMT Project, we often had an adult facilitator in the chat room with the group of students. We trained the facilitators to avoid intervening too much in the interaction, mainly answering questions and helping with technology issues. A study of this showed the subtlety of supporting student group agency rather than interfering with it (Charles & Shumar, 2009).

(b) Involving software agents as participants in open-ended collaboration is quite different from the approaches that have been so successful in automated tutors of individual students being trained in well-defined algebra procedures within tightly constrained interfaces. In collaboration with Carolyn Rosè's research group, we started to explore the interaction of software agents with students in online discussions in the VMT environment with experiments in a mathematics classroom (Stahl et al., 2010). Here we discovered how invasive agents tend to be. Even with "wizard of Oz" experiments in which human researchers played the role of software agents, the presence of the "agents" radically transformed the online interaction. The students oriented their discussion to the agents instead of to each other and to the math problems. Much more experimentation seems necessary to design less invasive agent behaviors, even in theory. In addition, it may be necessary to study successful examples of accountable-talk prompts or interventions by skilled teachers, using the micro-analytic techniques of Conversation Analysis before trying to design software algorithms to replicate such expert behavior. In particular, we need to know how to effectively time interventions and how to adapt the linguistic structure of interventions to the ongoing discourse.

(c) Designing effective CSCL interventions and introducing new technologies to scaffold interaction is a complex undertaking. It requires many cycles of iteration. The data analyzed here functions as an initial, pilot iteration. It was probably premature to run multiple conditions and to expect to see effects in subsequent testing of individual students. If anything, the VMT environment, the software agents, and the accountable-talk prompts seem to have each done more to interfere with any possibility of collaborative discussion of biology than to promote it.

(d) The theory of accountable talk has intuitive appeal to scientifically well-trained, mature, rational adults, whose thinking is heavily influenced by explicit textual expression. However, theories relevant to CSCL stress the social, situated, and linguistic nature of cognition (Stahl, 2012). To introduce accountable-talk moves into the highly situated, socially interactive text-chat interaction of school children will involve much more than providing canned prompts of the form used in the case study. It will require understanding the situated, sequential, social, interactional character of student chat, developing agents that can follow these subtle processes through real-time analysis of cryptic, ironic, juvenile postings and can formulate agent postings that engage in the co-construction of shared understanding. It is even possible that actually accomplishing that would exceed the theoretical possibilities of artificial intelligence to engage in intersubjectivity with humans. But before we can reasonably speculate on that, it seems important to understand the nature of effective knowledge-building discourse and productive accountable-talk prompting; again, micro-analysis of prototypical examples of such interaction need to be carried out.

The point now is to take the lessons learned back to the drawing board for extensive redesign: (a) First, integrate more aspects of the biology experiment into the collaboration-support software environment by allowing the group to see the diffusion experiment results in a shared view and to embed its inquiry reasoning and its group conclusions in the VMT shared whiteboard. This can make better use of the collaboration tools of the software as a collaborative medium. (b) Second, develop the agents to follow the student discourse and to just intervene when needed. This involves real-time natural language processing of the student postings, which is a complex, subtle, and situated skill, which may exceed the current state of the art. (c) Third, encourage collaboration among friends by letting the students know each other's identities and having them work for a group product, rather than filling in individual worksheets and taking individual tests. This would transform the exercise from one focused on individual learning to collaborative knowledge building. (d) Fourth, figure out how accountable-talk prompts can be contextualized as part of natural verbal interaction. This will involve development of this approach beyond the current conceptualization of the technique.

Methodologically, this stage of research calls for observations of pilot studies in order to guide design in the various aspects of the project. A single case study, looking in detail at the interactions, can provide insight into what *group-cognitive processes* (Stahl, 2006a) take place empirically—in ways that quantitative comparisons of different conditions generally cannot. This can provide important correctives to what designers assumed would take place based on their best preconceptions. Statistical controlled comparisons and quantitative measures of changes in individual test results at this initial stage would likely produce results that would at best be confusing, but more likely be misleading when interpreted on the basis of researcher preconceptions of what transpires in student interaction. This response analysis from cycle one has tried to provide a detailed case study that analyzes the actual interaction (among humans and agents) to reveal processes that are fundamental to human interaction under such conditions and are therefore likely to take place in other cases. It has tried to show how interaction analysis focused on the response structure of interaction can provide insight into group-cognitive processes and can indicate how experimental interventions do or do not support the group interaction. It contributed to guiding the redesign of this design-based research effort at this early stage of educational design.

Cycle Two of Design-Based Research

Due to the practicalities of conducting an experiment in public schools and due to the level of redesign called for by the lessons of the analysis of the first cycle of user testing, it took a year before the next cycle's user testing could be conducted. In this section, I take a similar approach to seeing what a quick sequential analysis can yield with the data from the second cycle.

1. As described in Dyke et al. (Chap 25, this volume), the new intervention had students working in four conditions. I decided that the revoicing condition would be the most interesting. I wanted to see the effect of the agent prompting students to revoice their chat postings.
2. I read through each of the 5 chat logs in the revoicing condition and I wrote down a couple sentences of my initial reaction to the quality of the interaction. I selected log F01 as the one that seemed to have the richest student interactions.

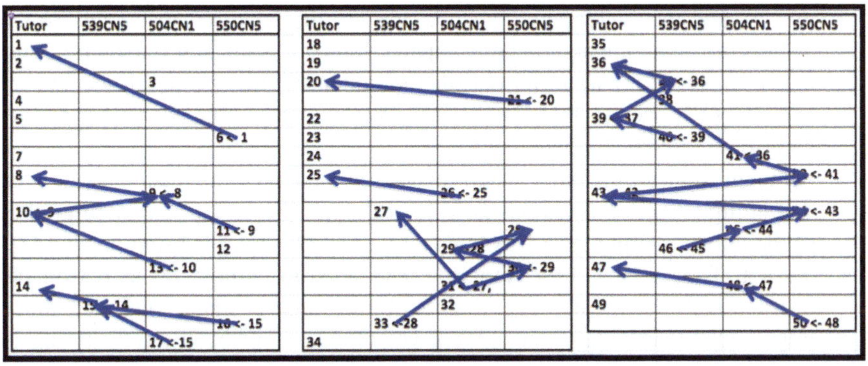

Fig. 28.3 Sequential response structure of chat F01

I wanted to see how the agent postings—particularly revoicing prompts—affected the accountable talk of the students.

3. I rearranged the spreadsheet to have the postings of each participant in its own column.
4. I next sketched the sequential response structure of the chat (see Fig. 28.3).
5. A visual scan of the response structure shows that the tutor (first column) is still very dominant in the discourse. Of 50 postings, now only 10 are by the tutor agent, but most of them are lengthy, whereas many of the student postings are only a word ("yes," "ok," or the student's name). Primarily, most of the student postings are in response—either directly or indirectly to the tutor. However, there are now several brief interactions among the students and even a couple of quite involved interactions (posts 27–33 and 41–50).
6. If we look at the content of the posts, we see that the whole discussion remains closely on-topic, following the agenda of the tutor. The tutor takes a strong instructionist teacher role. The students seem to accept this and respond to it much as they might to a classroom teacher. Although this was not the case in all of the chats, the one analyzed here seems quite successful in terms of student responses to the agent.
7. The student-to-student interaction (stimulated repeatedly by the tutor) progressed well. All the students participated (at least when prompted by the tutor), they discussed each other's proposals and they all agreed to a group answer after each of the extended interactions. This may have been encouraged by the formulation of the task, which was presented as a group task, to come up with an explanation that everyone agreed with.
8. The focus on accountable talk was reduced to the idea of revoicing—at least in terms of the tutor programming in this chat. The tutor only posted two explicit revoicing moves: postings 39 and 43. In both of these, the tutor proposed an alternative (and more scientifically formal) way of describing a biological

process and the student simply said, "yes" to the proposed revoicing. So the agent's move did not significantly expand the accountable discourse of the students. However, for whatever reason, the students in this group did seem to act in a generally accountable way by including and respecting each other and by describing biological phenomena.

9. Although some of the other groups expressed the kind of confusion about what was going on generally, about the role of the tutor and about the intelligibility of the tutor's postings that was rampant in the first year, the group in chat F01 did not. They accepted the tutor and responded to its postings as reasonable instructional statements. The timing of the tutor postings was also much improved. Student discussions were not often cut off by the tutor trying to follow a schedule. The tutor even seemed to react to student postings in ways the students could accept.
10. In conclusion, one cycle of redesign was adequate for eliminating the worst problems of agent intrusiveness, at least in the case of this one group, which I selected as most promising based on a skim of the logs. The ultimate goal of the theory of accountable talk is to have groups of students being accountable for their own discourse. It may be that at the level of ninth grade biology most students still need strong instructionist guidance and modeling before they can effectively adopt accountable talk practices in student-centered scientific discourse.

My quick analysis of a sample from the second cycle suggests that the major technical problems were adequately identified by my quick interaction analysis of the first cycle log and that they have been substantially addressed by the extensive redesign effort that it called for. The ground has now been laid for subsequent cycles exploring the complex issues of scaffolding group cognition among young students of science.

Issues for Further Multivocal Analysis

Design-Based Research for Designing Technology

Too often, research reports are written to give the impression that a well-defined hypothesis was tested and that everything went according to plan, resulting in the reported findings. The widespread popularity of design-based research in educational technology design is a testament to the fact that research in real classrooms rarely simply follows a preconceived experimental plan. Rather, understanding about how to design effective educational technology emerges gradually from iterative attempts to refine prototypes in response to unanticipated issues that only become apparent in messy trials. The initial attempt to promote accountable talk in a biology classroom through the use of conversational agents ran into myriad circumstances that modified the ideal experimental plan. Dyke et al. (Chap 25, this volume) listed some of these. Cress & Kimmerle (Chap 27, this volume) argued that the experimental situation, as actually implemented, did not support the social aspects of interaction that are so important to the students. The preceding sequential interaction analysis of one group's chat log from cycle one indicated that the agents were not very "conversational" in the resultant situation. Howley et al. (Chap 26, this volume) further investigated the social, linguistic and sequential structure of the chat interactions, both to see how the agents and students positioned each other as knowledge-building partners and to track the temporal unfolding of the chats. These analyses begin to inform the design of the software agents and of the educational intervention generally, suggesting approaches to be tried in cycle two and in subsequent iterations. Other types of analysis can no doubt offer additional suggestions for redesigning features of this multi-dimensional intervention.

Scripting of the Software Agents and Situated Interaction

Just as the experiment as a whole is situated amid the complex constraints on conducting experiments in typical public school classrooms, so the postings of the agent and students are situated in the unpredictable and subtle constraints of the social and linguistic interaction that unfolds in the chat room. In particular, each posting must make sense as following previous postings. Furthermore, when someone has difficulties making sense of the sequence of postings in this context then there is a need for "repair" processes. The sensitivity of a posting to preceding chat posts motivated my decision to look at the adjacency-pair structure, as a key indicator of the extent to which posts—particularly those of the agent—were meaningfully related to preceding and subsequent posts by students. My analysis revealed that agent posts in cycle one were not adequately situated in this sense. Furthermore, the agent showed no ability (or even inclination) to repair problems of meaning making when they arose.

In a chapter I wrote for a book on scripting (Stahl, 2006b), I cautioned that scripts should be conceptualized as situated resources rather than implementable plans for action. For instance, rather than scripting the agent to instruct the students to watch the video at precisely 8 min 15 s after the start of the chat, the agent should try to find an appropriate moment roughly 8 or 9 min into the chat for doing this, depending on what the students are doing at that point. I cited Suchman's (1987, p. 181) recommendation that computer support compensate for its limitations by: (1) extending its access to the actions and circumstances of the user; (2) clarifying for the user the limits of the computer's access to the users' rich interactional resources; and (3) providing a wider array of alternative resources, particularly to help the users respond to unforeseen breakdowns. Suchman was talking about the design of help systems for large copying machines. Compared to that, the conversational agents have the significant advantage of having access to all actions in the chat room—they have the same access that the students have to each other's actions. However, the agents have been programmed to project an anthropomorphic personality, pretending that they have meaning-making and language-understanding capabilities far in excess of what they can actually do. Suchman warned explicitly against doing this because it inevitably confuses the relationships and leads to misunderstandings and frustrations. As Cress & Kimmerly emphasized, a classroom is a highly social setting for the students, and introducing a new social partner with no social skills may not be an effective approach. Finally, the agent is designed to perform multiple roles, scripting the macro-level phases of work as well as the micro-level accountable-talk moves. When the students reject the agent, they are left to their own resources.

Sequential Interaction Analysis of Small Groups

While the design-based-research approach is often recommended for educational technology, this approach does not generally specify a method for analyzing the results of trials. In the past, I have suggested adapting Conversation Analysis to provide insight into how teachers and students are actually making use of a prototype, rather than quickly counting surface features of interactions or coding utterances based on the designer's or researcher's conceptualization of the intervention. Although we have found data sessions based on VMT sessions to provide quite useful design feedback in a matter of hours, many researchers claim that qualitative analysis is too time consuming to give timely feedback. That is why I tried in this paper to see how much insight into central problems of an intervention could be gleaned from a quick adjacency-pair analysis of one typical chat session.

For the data from cycle one, I skimmed through the chats and got a sense of the problematic nature of the sessions, much like the feelings that the authors of the related chapters expressed. I selected a chat session that seemed to have relatively clear examples of the problems. Specifically, I selected a session in the "indirect" condition, which was the condition of greatest interest for the experiment. I then sketched an initial version of Fig. 28.1. Based on the visual appearance of the figure and the content of the connected adjacency pairs of posts, I drafted an initial version of this chapter, arguing for the need for changes to the agents and to the intervention in subsequent iterations. During a data session with some of the other chapter authors, refining Fig. 28.1 and our understanding of what took place interactionally in the chat, we agreed on directions for further analysis and experimentation. In this way, the sequential interaction analysis with the graph of adjacency pairs provided a quick sense of where major issues lay, which needed to be addressed in redesign. Thus, it played a role similar to so-called "discount methods" in human–computer interaction, where designers need fast feedback at low cost.

Accountable Talk and Off-Task Student Practices

Throughout the history of CSCL, researchers have conducted educational interventions with expectations that the students would engage in knowledge building, inquiry, transactivity, collaborative learning, warranted argumentation and other lofty conceptions of scientific intellectual discourse. These expectations were operationalized so that research assistants could reliably interpret student utterances as falling into different coding categories. Inevitably, few utterances could be coded in the highest categories; a large percentage fell outside the scheme, and they were called "off topic."

To conclude this paper, I would like to raise the ethnomethodological question: what are the students doing when they are off topic? If they do not *do* being-a-student by engaging in recognizably accountable talk, how do they do it? Is it due to some personal characteristics of these students that they engage in "cheating" rather than in following the instructions of the agent? Perhaps if we break free of the conceptualizations imposed by the experiment's world-view, we can understand the off-topic behaviors in a positive light. As Cress & Kimmerle (Chap 27, this volume) suggest, the teenage students are engaged in social activity with one another. Their social relations support their discussions of curricular topics and their talk in the classroom feeds into their social relations. Any arrangements that interfere with their social relations—such as hiding everyone's identities—will interfere with the possibility of any kind of interaction and will generate attempts to repair the problem. In addition to the social practices involved in relating with their peers, the students are involved in established classroom practices, oriented largely around earning good grades. While the researchers were looking for accountable talk in the details of interaction, the students were oriented toward completing the individual worksheets and taking tests. Thus, many of the early interactions graphed in Fig. 28.1 involved finding out the identities of the chat room occupants and many of the later interactions involved filling out the worksheets. While the exchange of information needed for the worksheets did not fit the profile of accountable talk, it did contribute to a rather efficient sharing of worksheet answers. Although it did not meet the ideals of collaborative knowledge building or group cognition, the three students did effectively divide up the tasks and share the results, in the classic form of cooperation that most people tend to use. If the task had been to fill out a group worksheet, neither the students nor the researchers would have viewed this procedure as cheating. Classrooms are not blank slates waiting for a researcher's conceptualizations; they are richly structured with established practices that can only be discovered by analysis of interventions in the wild.

Appendix A: Chat Log of Group C01 from Cycle One

	time	Tutor	S027	S034	S041
1	11.54.45				
2	11.55.35				
3	11.55.41				
4	11.55.51				
5	11.56.01	Hi all! Today, each of you will be both the instructor and the student. Each of you will be responsible for looking for opportunities for a specific discussion strategy to be used in the conversation.			
6	11.56.06	Lets get started by introducing ourselves. I am Alex.			
7	11.57.04	S034,S041,S027... i didn't get your names yet.			
8	11.57.16			my name is {S034}	
9	11.57.16	Hi S034 you are the Revoicer. When an explanation or idea is given which would benefit from revoicing, please encourage one of your team mates to do so, for example, by asking them to explain what was said in their own words.			
10	11.57.24	S041, you are the Challenger. When you see a statement being made (whether it is right or wrong) which would benefit from being challenged, please encourage one of your team mates to do so, for example, by asking them whether they agree or disagree with what was said, and why.			
11	11.57.25				{S041}
12	11.57.31	Ok... So the Explainer is S027. When you see a place where it would be good to get more explanation, you can encourage one of your team mates to do that, for example, by inviting that person to add on to what the other student has already said.			
13	11.57.37	You should probably all mark an X opposite your role in the book, so you won't forget.			
14	11.57.43	Its nice to meet you all. :) Please make sure you have understood what you are responsible for.			
15	11.58.00	In condition A, we placed a glucose solution inside the cell model and immerse it in distilled water. In condition B, we did the opposite.			
16	11.58.05	Please discuss what you predict will happen in these two conditions.			
17	11.59.29				who is 27
18	12.00.06		{S027}		
19	12.00.46			what do you think'ds going to happen?	
20	12.01.02	In condition C, we made a modification to condition A. We replace the glucose solution with a starch suspension. In order to detect the presence of starch, we replace the distilled water with Lugol's Iodine solution.			
21	12.01.07	You should now move on to discussing what will happen in Condition C and your explanation for this change.			

(continued)

22	12.01.14				the world is going to end in 2012
23	12.01.21	S027, now would be a good time to ask S034 to build on what S041 is saying.			
24	12.02.00			im so confused!	
25	12.02.08		034, would you like to build on to what 041 is saying? and me too!		
26	12.02.32	When you are in agreement, write down your predictions and explanations for Conditions A, B and C on your worksheet.			
27	12.02.42		who is 34?		
28	12.02.53			{S034}	
29	12.03.00	You are now going to watch a video showing the cell in Conditions A, B and C.			
30	12.03.04		ooh. hi		
31	12.03.05	As you watch the video, write down your observations on your worksheet.			
32	12.03.11	Go to the Videos folder on the Desktop, and watch the video which is there.			
33	12.03.24		idk what to do.		
34	12.03.27			someone predict something	
35	12.03.56				THE WORLD IS GOING TO END IN 2012 !
36	12.04.05			oh i get it! what do you think is going to happen when you put the glucose in the distilled water?	
37	12.04.22		i think it will turn a different color.		
38	12.04.29		{S041}?		
39	12.04.45				ok dude what 27 said
40	12.05.04		in your own words		
41	12.05.10		?!?!?!		
42	12.05.26				you have to be more specific 27
43	12.05.43		ur one to talk!!!		
44	12.05.46			the answer to number one : the distilled water will move into the internal envorment of glucose solution	
45	12.06.36		kk		
46	12.06.45			and the explaination is glucose is denser	
47	12.06.57				huh?
48	12.07.05		just write what she[{S034}] said.		
49	12.07.20				for the explination

50	12.07.33			#2 the distilled water will move into the external enviroment	
51	12.08.14			watch the video	
52	12.08.38	Is everyone back?			
53	12.08.45				we have to do all the prediction , thats what NACHO said
54	12.08.48		ok, but where do i click to watch it?		
55	12.09.00	Now discuss what you observed in the video about conditions A and B. Compare what you observed to your predictions.			
56	12.09.40		nevermind i found it.		
57	12.10.01			#3 the water will move into the starch because starch absorbs it	
58	12.10.08	S027, now would be a good time to ask S041 to build on what S034 is saying.			
59	12.10.33		{S034} are these #'s all of the predictions or explanations?		
60	12.11.51		{S034}?		
61	12.12.01			yeaaaa	
62	12.12.02	You should now move on to discussing what you observed in the video about condition C. Compare what you observed to your predictions. How is this different from what happened in condition A?			
63	12.12.25		#3, is that the prediction?		
64	12.12.41	Lets spend the rest of the time we have discussing your observations from all the three conditions.			
65	12.12.47	Discuss explanations of what you observed in each condition and make sure you fill in your explanations on your worksheets.			
66	12.13.00				what is explanation for #2
67	12.13.07		idk.		
68	12.13.07			its both, the first part is the prediction the second part is the explanation	
69	12.13.13	@S027, can you ask S041 to build on what S034 is saying.			
70	12.13.34			i want to shoot myself in the foot -.-	
71	12.13.50		ikr.		
72	12.14.00		this is so stupid.		
73	12.14.11			this would be so much easier just in a group	
74	12.14.19		yep yep		
75	12.14.23	@S027, can you ask S041 to build on			

(continued)

		what S034 is saying.			
76	12.14.25				you dont have the balls to do it
77	12.14.36				
78	12.14.38	Ok, I gotta go.			
79	12.14.43	It was nice talking to you all. :-)			
80	12.14.55		our tutor left us!!!		
81	12.15.10			D:	
82	12.15.23		she is a baad tutor!!!		
83	12.15.44			im pretty sure its a he	
84	12.16.09		really? haah! oh well. idc. he/she/it		
85	12.16.31		i ddnt even watch the whole video		
86	12.16.49			i watched most of it	
87	12.16.58		do you know what happened?		
88	12.17.13			i just wrote the observations	
89	12.17.42		oh. what was it after 1 hour for condition A?		
90	12.19.17		{S041}, did you get everything so far?		
91	12.19.24			condition a: 620 glucose inside; 635 glucose inside; 630 glucose inside; explanation: the membrane will hold stuff for a certain amount of time	
92	12.20.05			condition b: 540 glucose outside; 525 glucose outside; 525 glucose outside; explanation: the glucose was not being absorbed by the water	
93	12.20.10			i think .	
94	12.21.45		k, i'll fast forward and try to find the last one, when im done writing the condition b.		
95	12.21.47				you mean i know !
96	12.22.02			yea, sure i do, whatever. hahahahahha	
97	12.22.15		lol		
98	12.22.19				lol what is C
99	12.22.27			i dont even think there was one	
100	12.22.33		i sd i'd try to find it if i could.		
101	12.22.51				there is & ok
102	12.23.04		aaah there is a quiz!!!!!!!!!!!!		

(continued)

103	12.23.19		im gonna fail!!!!!!!!!!!!!!		
104	12.23.57				stfu !
105	12.23.59			HAPPY BIRTHDAAAY {S041} :D :D :D :D :D :D :D :D :D	
106	12.24.10				lol thsnks
107	12.24.13		is it ur birthday?		
108	12.24.20			[blank line]	
109	12.24.23			[blank line]	
110	12.24.23			[blank line]	
111	12.24.23			[blank line]	
112	12.24.23			[blank line]	
113	12.24.23			[blank line]	
114	12.24.24		...		
115	12.24.25				
116	12.24.35		bye bye guys.		
117	12.25.01				
118	12.25.11				
119	12.28.34				

Appendix B: Chat Log of Group F01 from Cycle Two

No.	Time	Tutor	539CN5	504CN1	550CN5
1	13.57	Hi, my name is Sage. I'll be working with you in this activity. Take a minute to introduce yourselves			
2	14.42	I won't tell you if you're right or wrong, but I'll give feedback as you discuss things as a team			
3	14.49			Hi dajha!	
4	14.58	Let's look at a setup which is the opposite of the one we saw earlier—the iodine is now inside and the starch is outside			
5	15.24	On the whiteboard, you can see a picture of this new setup			
6	15.29				Vinny<3
7	15.38	Here's what happened in this setup over time			
8	15.59	Notice how the inside and outside environments changed. Why did these changes happen?			

(continued)

No.	Time	Tutor	539CN5	504CN1	550CN5
9	16.26			The iodine left the bag	
10	16.31	Okay, 504CN1—why would how we use indicators be relevant?			
11	17.39				The iodine was never in the bag the iodine went into the bag
12	17.48				
13	17.51			The weight is decreasing and the starch water changed colors	
14	17.54	Any input from 539CN5?			
15	18.59		If you look at the diagram it shows the big getting liter because the iodine left the bag		
16	19.21				Agreed…
17	19.26			Agreed…	
18	19.32	Okay, team—in a moment we'll move to another set of experiments			
19	19.56	For these experiments, we'll use the same membrane, but with pure water on the outside and a glucose solution on the inside			
20	20.27	Discuss the differences between glucose and starch molecules			

21	20.46				Glucose molecules weigh less
22	21.19	We'll use the tube weight as an indicator for water movement			
23	21.23	We'll also use glucose test strips to detect glucose in both environments			
24	21.37	On the whiteboard, you can see what happened in this setup over time			
25	22.01	The results with glucose are different than what you saw with starch—as a team, discuss and agree upon an explanation for these results			
26	22.50			The weight of the glucose solution increased over the time period	
27	22.59		The water has no sugar in it so the strip will stay yellow. the glucose solution has sugar so the strip will turn green		
28	23.03				I think the glucose molecules left the bag and went into the water in the process of that water molecules entered the bag

(continued)

No.	Time	Tutor	539CN5	504CN1	550CN5
29	23.41			Once again......agreed with dajha … haha backslash	
30	24.00				Always a smart thing to do
31	24.41			But also i agree with vicky too	
32	24.50				
33	25.32		And over time glucose cane out of the bag and got into the water and that is why after 1 and 24 h the strip that was in the water started to change		
34	25.38	Okay, everyone—let's move on to the last experiment			
35	26.00	In this setup we have glucose solution on the outside, pure water on the inside			
36	26.23	What changes do you think you'll see in this condition? Agree on a prediction for what you'll observe over time, and an explanation that everyone understands			
37	26.50		The water has not sugar so the strip is yellow		
38	27.36		and the glucose has sugar and the strip is green		

39	27.41	So is 539CN5 saying "a glucose test strip changes color when glucose is there?"			
40	27.51		Yes		
41	28.06			I think the glucose test strip will slowly change to a bright yellow color it is on the weighing scale. also i think the bag of distilled water will weigh more because over time of sitting in the glucose you will get some glucose molecules in the	
42	28.06				I don't think the sugar molecules will get in side the membrane i think they will be to big i think the water molecules will come out of the membrane and go into a concentrated equal environment
43	28.13	Would another way to say that be "molecules will move to the area of lower concentration if they can?"			

(continued)

No.	Time	Tutor	539CN5	504CN1	550CN5
44	28.38				Yes
45	29.41			Agreed........with dajha again!	
46	29.51		I agree to		
47	30.02	Make sure each of you is clear on your team's prediction and explanation, and write it in your worksheet			
48	30.09			We all agree!	
49	30.45	All right, we're wrapping up. Thanks, team!			
50	30.53				Just to restate… the sugar molecules will not go into the membrane the distilled water will come out and the test strip will turn yellow

References

Çakır, M. P., Zemel, A., & Stahl, G. (2009). The joint organization of interaction within a multimodal CSCL medium. *International Journal of Computer-Supported Collaborative Learning. 4*(2), 115–149. Retrieved from http://GerryStahl.net/pub/ijCSCL_4_2_1.pdf

Charles, E. S., & Shumar, W. (2009). Student and team agency in VMT. In G. Stahl (Ed.), *Studying virtual math teams* (pp. 207–224). New York, NY: Springer. Chapter 11.

Dillenbourg, P. (2002). Over-scripting CSCL: The risks of blending collaborative learning with instructional design. In P. Kirschner (Ed.), *Three worlds of CSCL: Can we support CSCL?* (pp. 61–91). Heerlen, The Netherlands: Open University of the Netherlands.

Dillenbourg, P., & Jermann, P. (2006). Designing integrative scripts. In F. Fischer, H. Mandl, J. Haake, & I. Kollar (Eds.), *Scripting computer-supported collaborative learning: Cognitive, computational and educational perspectives* (pp. 275–301). Dodrecht, The Netherlands: Kluwer–Springer.

Jordan, B., & Henderson, A. (1995). Interaction analysis: Foundations and practice. *Journal of the Learning Sciences. 4*(1), 39–103. Retrieved from http://lrs.ed.uiuc.edu/students/c-merkel/document4.HTM

Sacks, H. (1962/1995). *Lectures on conversation.* Oxford, UK: Blackwell

Schegloff, E. A. (2007). *Sequence organization in interaction: A primer in conversation analysis.* Cambridge, UK: Cambridge University Press.

Stahl, G. (2006a). *Group cognition: Computer support for building collaborative knowledge.* Cambridge, MA: MIT Press.

Stahl, G. (2006b). Scripting group cognition: The problem of guiding situated collaboration. In F. Fischer, H. Mandl, J. Haake & I. Kollar (Eds.), *Scripting computer-supported collaborative learning: Cognitive, computational and educational perspectives.* (pp. 327–335). Dodrecht, The Netherlands: Kluwer–Springer. Retrieved from http://GerryStahl.net/pub/scripting.pdf

Stahl, G. (2006c). Sustaining group cognition in a math chat environment. *Research and Practice in Technology Enhanced Learning (RPTEL). 1*(2), 85–113. Retrieved from http://GerryStahl.net/pub/rptel.pdf

Stahl, G. (2009). *Studying virtual math teams.* New York, NY: Springer.

Stahl, G. (2011). *The structure of collaborative problem solving in a virtual math team.* Paper presented at the Conference 2011. Seattle, WA. Retrieved from http://GerryStahl.net/pub/iconf2011.pdf

Stahl, G. (2012). Theories of cognition in collaborative learning. In C. Hmelo-Silver, A. O'Donnell, C. Chan & C. Chinn (Eds.), *International handbook of collaborative learning.* New York, NY: Taylor & Francis. Retrieved from http://GerryStahl.net/pub/clhandbook.pdf

Stahl, G., Rosé, C. P., O'Hara, K., & Powell, A. B. (2010). *Supporting group math cognition in virtual math teams with software conversational agents.* Paper presented at the First North American GeoGebra Conference. Ithaca, NY. Retrieved from http://GerryStahl.net/pub/geogebrana2010a.pdf

Suchman, L. (1987). *Plans and situated actions: The problem of human-machine communication.* Cambridge, UK: Cambridge University Press.

Chapter 29
Network Analytic Techniques for Online Chat

Sean P. Goggins and Gregory Dyke

Introduction

The purpose of this chapter is to evaluate the feasibility and validity of automated trace analysis and real time participant feedback in synchronous discourse through the application of multivocal methods. In the analysis of biology education synchronous chat data presented here, we experience multivocality as a framework for weaving new methods and perspectives into an existing mixed methods CSCL and CSCW research program. We contribute a *Group Informatics*[1] vocality to the book. Group informatics makes multivocal approaches to analysis of interactions more systematic. For example, weighted social networks are derived from interpretation of the conversations and analysis of contributions. The middle road between ethnomethodology and fully computational approaches is a good way of conceptualizing Group Informatics.

In this chapter, we describe two orthogonal analyses of biology students participating in a synchronous collaborative session in a chat environment scaffolded by conversational agents. Our first analysis contrasts the first author's network analysis techniques with Stahl's ethnomethodologically informed analysis (Stahl, Chap. 28, this volume) A key dimension of this first analysis is the reflexivity between the Group Informatics methods and the ethnomethodologically informed analysis of Stahl. Second, we explore multivocality as both a methodological and a data centered framework for gaining new perspectives. We present a descriptive study that illustrates methods and provides suggestions for future lines of inquiry.

---

[1] http://www.groupinformatics.org.

S.P. Goggins (✉)
The University of Missouri, Columbia, MO, USA

G. Dyke
University of Lyon, Lyon, France

D.D. Suthers et al. (eds.), *Productive Multivocality in the Analysis of Group Interactions*, Computer-Supported Collaborative Learning Series 16,
DOI 10.1007/978-1-4614-8960-3_29,

Theoretical Assumptions: Group Informatics

In past research at the intersection of small groups and ICT researchers faced difficulty defining the concept of "group," which inspired calls for abandonment of the concept of "group" as a construct for collaborative computing research (Schmidt & Bannon, 1992). Nonetheless, theory development related to ICT mediated groups continued as a relatively small thread within information science and CSCW research (Latour, 2007; Turner, Bowker, Gasser, & Zacklad, 2006). Reconsidering groups as a unit of analysis in information science is now justified by the diffusion of ICT beyond the work domain and into nearly every corner of life in the information society. Further motivation for this shift is supported by long standing analyses of social behavior that recognizes the central role small groups play in organizational and societal change (Fine & Harrington, 2004) and ICT adoption and use (Goggins, Laffey, & Gallagher, 2011; Mead, 1934, 1958; Stahl, 2006).

Group Informatics incorporates the practice of analyzing electronic trace data in concert with other methods. Network analyses focused on group emergence and leadership are grounded in data or salient theory to explicitly connect trace data with group phenomena occurring in a technologically mediated environment (Goggins, Mascaro, & Mascaro, 2012; Goggins, Mascaro, & Valetto, 2013; Goggins, Valetto, Mascaro, & Blincoe, 2013). The Group Informatics methodological approach overcomes deficiencies in existing applications of social network analysis methods to the analysis of electronic trace data. In this paper, we consider the role of electronic agents as members of synchronous, technologically mediated small groups and, in the second section, operationalize types of knowledge and action behavior as nodes in a network.

Network analysis of electronic trace data often overlooks validity issues arising from inconsistencies between data collection and research purpose, reliability issues associated with technical system failure and conceptualization issues related to how nodes (people) and edges (connections) are defined prior to analysis (Howison, Wiggins, & Crowston, 2012). For example, how and to what extent electronic agents—or humans masquerading as agents—in an electronic network should be evaluated as nodes in that network is not considered elsewhere.

Understanding how such analysis might be valid, reliable and appropriately conceptualized for chat interactions involving electronic agents and people in learning contexts is especially important. Such analysis could be immediately useful for understanding how similarly programmed agents interact with members in different sessions or help identify patterns of productive and nonproductive or hindering human–agent interaction. Productive and hindering interactions are especially important to distinguish in learning contexts where agents are used in order to prevent student frustration with agents (Stahl, Chap. 28).

Prior work detecting change in social networks composed entirely of agents or entirely of real people contributes to our understanding of the differences in these types of networks and limitations of network analytic techniques. McCulloh (2009; McCulloh & Carley, 2009) defined a set of statistical control charts capable of detecting statistically significant changes in social networks. However, these changes are difficult to tie back to concrete real-world events. In this study, we mix humans in the same environment with agents. Furthermore, the context of this study is synchronous, unlike much previous work. Social network analysis and qualitative research methods must work together in order to make sense of human–agent interactions in small socio-technical settings like learning-focused chat rooms. Application of trace data to increase awareness of participant structure and to provide insight to management or teaching leadership falls along a continuum between a preference for presenting interested parties (users and others) with raw, unprocessed representations of their trace data (Erickson, 2009) and viewing trace data as an incomplete view of ground truth that must be analyzed in a socio-technical context in order to be meaningful (Goggins et al., 2011). The role of time and the relative differences in time, connection type and explicit interaction type make a rich, weighted analysis of connections between members of a social network more closely reflect what interview, content analysis and ethnographic data suggest is "ground truth" (Goggins, Galyen, & Laffey, 2010).

Purpose of Analysis

The analyses and research questions presented in this chapter pertain to three overarching themes, the last of which is a focused application of the first two methodological issues, chosen because of its importance in Stahl's analysis.

- How can we derive Social Networks from chat, where explicit evidence of reading behavior does not exist?
- What utility can SNA provide for analysis of small groups?
- What role does the agent play in the group discussions of this dataset?

The analysis portion of this chapter is separated into two phases. Within these phases, each research question is related to the themes above.

1. The first phase is related to Stahl's analysis of a single group.
   (a) How can Stahls analysis inform a weighting heuristic for calculating explicit network ties?
   (b) What can the automatically derived network metrics tell us about the different groups and the roles of the agents within those groups?
   (c) How do the sociograms developed from Stahl's analysis compare with automatically derived sociograms?
2. In the second phase, we extend Stahl's analysis to a manual coding of all the groups.
   (d) How can we evaluate the validity of the automatically generated sociograms?
   (e) What can changes in sociograms over different phases of a discussion tell us about the group interaction?

(f) In labeled data showing the negotiation of knowledge and actions, how can the "label-o-gram" of the network of replies from one label type to another inform interpretation of the underlying interaction?

Representations

In this chapter, we use both tabular and sociogram representations to show social network analysis metrics between participants. In order to construct such representations, we must choose a unit of analysis, which lends itself to deriving the social network data. We then create an initial representation of the data showing the ties between events (determined by various methods, which we will compare). This representation is then transformed into the social network itself, for each group. Below, we describe our unit of analysis and the manipulations necessary to create our various representations in greater detail.

Unit of Analysis and Manipulations

Electronic trace data from conversationally focused asynchronous systems like discussion boards do not represent a complete record of user interactions because social connections derived from information or conversation displayed on a page are missing (Goggins et al., 2010, 2013; Goggins, Laffey, & Amelung, 2011). To overcome this obstacle, Goggins previously described methods for constructing such social connections from ordinary, found trace data and trace data that is designed to capture additional context, user reads and user posts (Goggins et al., 2010, 2011a, 2013). Figure 29.1 illustrates how log data and knowledge of the time sequence and user involvement in sequential page construction is used to construct the implicit social network data. Networks that incorporate implicit ties are particularly useful in multiuser environments where individual members are not able to explicitly address a specific prior message.

Knowing the basics of who posted each item is sufficient for a rudimentary social network, but it is still not complete. Weighting of ties and making connections for

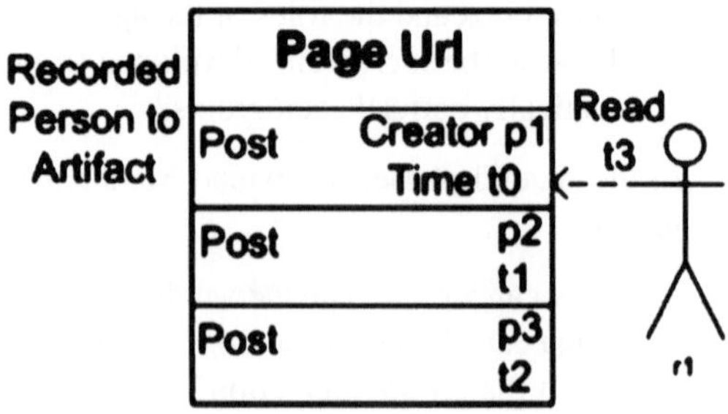

Fig. 29.1 Adding social relations for SNA

the appropriate number of prior items on a page is necessary when explicit ties are not expressed in the data. While often possible to produce explicit ties through a manual analysis of trace data, as Stahl does in the volume and Dyke does using Tatiana in our second phase, this is also more time consuming. The method described in phase one maintains an advantage of speed, and potential for synchronous feedback to participants.

Each component of implicit connection—how many posts back in a post sequence to imply a tie, and how to weight a tie—varies by sociotechnical context. In the first author's previous work on asynchronous data, the connections are determined by considering the number of previous posts visible to a user, and by using qualitative coding as a baseline for determining durations after which the likelihood that sequential posts are not closely related is low.

Rationale for Synchronous Weighting and Processing in This Case: Phase One

Synchronous trace data analysis is different in two ways from asynchronous analysis. First, the time between interactions is relatively stable between groups, assuming the same environment and an ongoing activity. Weighting according to time, then, takes a more linear form in synchronous data. Second, the interactions are displayed serially on the user interface, so what a user is taking in and how the user is processing the data is a different phenomenon than what is the case in asynchronous interaction. The question for the first phase analysis, then, is how to weight data and how to make implicit connections between the current user and prior user contributions in the same environment. To accomplish this goal we build on the ethnomethodologically informed analysis of Stahl.

Implicit connections rely on some assumptions, derived from more time consuming qualitative methods, about how many prior posts are being consumed by a user before they themselves post back. To determine how many posts back social network connections should be drawn from, we identify leverage Stahl's analysis of the same corpora. Based on Stahl's analysis (Chap. 28, this volume), we see that the average turn responds to a turn 2.4 times prior, with a standard deviation of 3.2, a max number of turns between of 20 and a minimum of 1. The mode is 1, and 85 % of interactions that Stahl analyzes occur within 3 turns. For this reason, we chose to calculate implicit connections going back three turns, and weighted those connections linearly according to seconds between responses.

Development of Explicit Networks Using Tatiana: Phase Two

In the second phase of analysis we use only explicit network ties identified by the second author's analysis of the interactions between students in each treatment using Tatiana. The tools a researcher has at their disposal influence what might be found.

By using Tatiana, we are able to rapidly build the connections for all 15 treatment groups. Though this analysis is not as deep as Stahl's ethnomethodological analysis (which informs the weighting in the heuristic presented above), it is sufficient for the development of explicit social networks.

An Introduction to Basic Social Network Analysis Measures

The analysis in this chapter leverages basic methods from social network analysis, including the use of sociograms to represent network structure. In prior analyses, we have shown the evolution of these structures over time. In this chapter, given the relative brevity and stability of participation (membership is quite stable in this data set compared with other trace data the first author previously analyzed), we present a single static view of each session. Key network analytic concepts include: *Degree Centrality*—a measure of how central individual actors are in discourse. *In-degree* centrality is greater when more turns are referring to an actor; *out-degree* centrality is greater when an actor makes more contributions.

The sociograms are colored according to in-degree and out-degree centrality. The vertices (nodes) are sized according to relative degree centrality within the session. We present comparisons of degree centrality across sessions in table form. The thickness of the edges in our sociograms reflects the strength of connection between two nodes (Fig. 29.2). The color red is used to represent out-degree centrality, which means that other users are referencing statements made by that user.

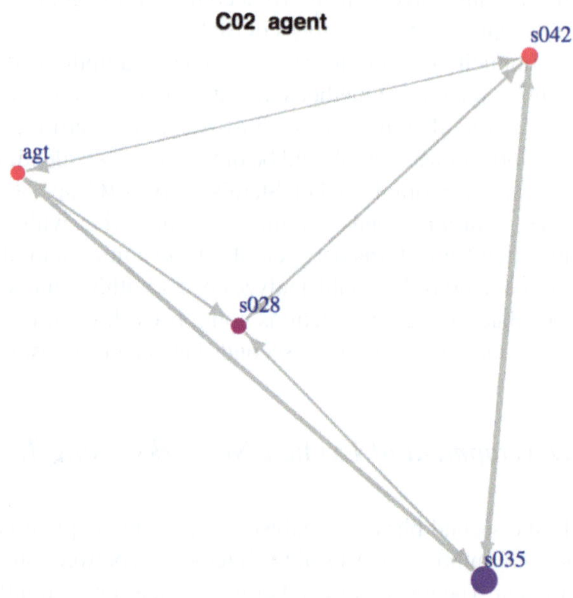

Fig. 29.2 Example of sociograms presented, using session C02 with agent as an example. Bluer vertices have greater relative in-degree centrality while redder vertices have greater out-degree centrality

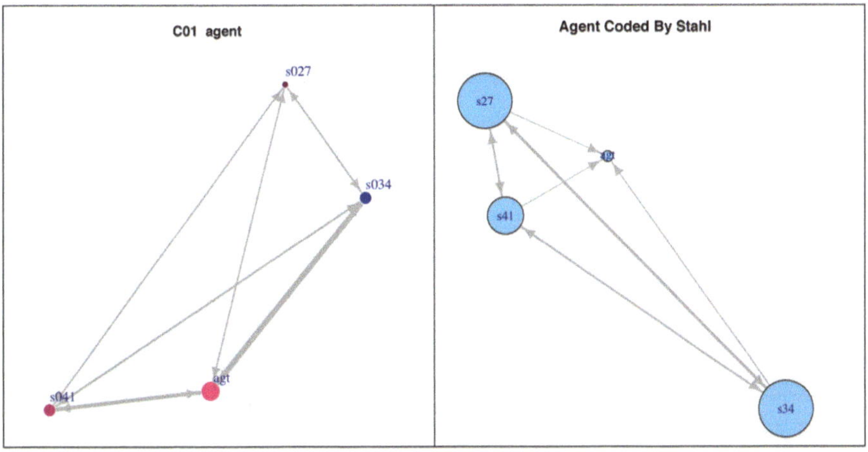

Fig. 29.3 C01 as analyzed with agents in the network. Using weighted trace analysis (*left*), based on Stahl's Conversation Analysis (*right*)

The color blue is used to represent in-degree centrality. Figure 29.3, for example, shows that user S035 has the highest out-degree centrality, while users agt and S042 have greater, relative in-degree centrality. All coloring is relative within the sociograms. The line thicknesses are calculated consistently across sociograms. Figure 29.3 shows the strongest connection between S035 and S042, for example.

Phase 1: Multivocality with Group Informatics and Ethnomethodology

In this chapter we explore multivocality at a meta level that corresponds with the origins of our methods. The first author has experience with analysis of asynchronous trace data, and our first attempt at multivocal analysis of this data involved serial analysis by multiple different team members; what we refer to as asynchronous multivocality. Next, following our workshop, our team met on a second occasion and the two authors continued their collaborative analysis in a synchronous—or at least synchronized—manner. We present both analyses, next, as they bring light to different dimensions of the data while illustrating multivocality within and between research teams. The first, asynchronous multivocal method, weights synchronous messages by time distance and proximity. Here, implicit connections are made based on a comparison with Stahl's initial analysis. In the second, synchronous multivocal method, the second author performs analysis to make the same kinds of connections Stahl made, but for every group. Here, we perform similar network analysis on the networks derived from these more precise measures.

Table 29.1 Degree centrality measures and degree centrality standard deviations between members in the biology data

With Agents			Group	Without Agents		
Degree Mean	In-degree Standard Deviation	Out-degree Standard Deviation		Degree Mean	In-degree Standard Deviation	Out-degree Standard Deviation
9.61	0.78	3.02	C03	5.94	1.34	2.02
9.77	3.61	3.05	A04	5.86	1.70	2.46
5.22	3.06	3.77	A02	5.60	1.86	3.88
7.97	2.74	3.38	C07	5.59	1.46	2.87
7.85	3.01	4.97	**D02**	5.28	3.72	3.57
7.15	3.55	1.98	C02	5.22	2.32	2.46
7.18	1.01	2.07	C06	5.03	0.70	1.56
5.48	1.81	2.02	A05	4.75	2.03	2.16
8.74	4.12	1.49	D01	4.57	2.66	2.53
8.82	3.29	3.59	**C01**	4.46	1.24	0.70
6.07	2.48	2.42	C04	3.67	1.23	1.29
7.91	3.14	0.78	**C05**	3.63	0.26	1.08
7.26	3.75	2.99	**A01**	3.50	0.36	0.36
4.14	1.20	1.83	A03	2.40	0.46	0.66

Interpretation of Heuristically Derived Social Networks in Small Group Situations

We compare network traits in all of the different treatments and groups using the heuristic for deriving weighted ties between messages, based on time distance and proximity. This analysis illustrates how SNA metrics can be interpreted in small group situations and focuses on the role of the agent, as a preparation for comparison with Stahl's analysis (Chap. 28).

There is substantial variance in the mean degree centrality and standard deviation of degree centrality averages across groups. This indicates that there are a number of different patterns of participation that are immediately made visible through automated analysis of session traces. Average weighted degree will be the same in both directions. It is a balanced equation. The standard deviation indicates the variability in the member participation. This is generally greater when the agent actions are figured into the network, as the degree centrality of the agent varies significantly compared with the people. When the agent actions are factored in, you can see that the ordering of the groups according to degree centrality will be different. For example, sessions C01 and D01 are much higher in rank order if we incorporate the agent in mean degree centrality. Session A02, in contrast would be much lower. Table 29.1 illustrates the two sessions with the highest and lowest standard deviations between member degree centrality using blue to indicate high standard deviation, and orange to indicate low standard deviation.

To understand member participation and relate that to other analysis of the biology data, a full set of member degree centrality by session is presented in Table 29.3 (Appendix). The highest in-degree centrality members are coded in blue, matching the color scheme in the sociograms. The highest out-degree centrality members are coded in red, also matching the color scheme in the sociograms. In cases where the top in-degree or out-degree members in a session are the same, with our without the agent in the analysis, those users are bolded. In 6 of 14 sessions, the agent has the highest in-degree centrality, and in 2 other sessions the agent has the highest out-degree centrality. So, in over half of the session, the agent appears to be either the most responded to or most active responder within three turns.

When the agent is removed from the analysis, members of each group in a session are the highest and lowest. There are 13 users who are highest in-degree or highest out-degree centrality with or without the agent's inclusion in the analysis. Deeper investigation of these users is likely to show that they do, in fact, play a central role in discourse, regardless of the presence of the agent. In sessions A02, C02, and D02, two members of the course sustain a highly central in-degree or out-degree role without regards to the presence of an agent. Sessions A02 and D02 also have some of the highest standard deviations between member degree centrality, indicating that some members, perhaps the two referenced or the agents, have significantly different participation patterns than the third human member.

This analysis suggests one of two potential situations. First, there are two members who are highly active with each other, and a third periphery member; and this has little to do with the agent. Or it is possible that the agent generates active responses from these two members in such a way that their centrality is sustained and overshadows the role of the agent. It is clear that, in some sessions, the agent plays a fairly central role. If Stahl's analysis is representative, it may be that agent interactions are heavy in the front of discussion consistently, which would suggest that future analysis explicitly address how the agent's behavior is intended, and whether or not it is focused on a particular part of the discussion.

Comparison of Stahl's Sociograms with Automatically Generated Sociograms

The raw SNA degree statistics and comparison of member variance across sessions provides a picture of several sessions, A02, C02 and D02 with participation traits that are distinct from the other sessions. It also suggests there are 13 members who are in central roles in these human triads + Agent configurations both with and without incorporating the agent in analysis. A comparison of the sociograms derived from Stahl's analysis with the sociograms built from our analysis of processed data is mechanism for understanding of the extent to which automated sociogram construction might begin to inform and guide which sessions are coded in an ethnomethodologically informed way, or might curry insight if analyzed using still other techniques. Figures 29.4 and 29.5 represent the automated sociogram generation

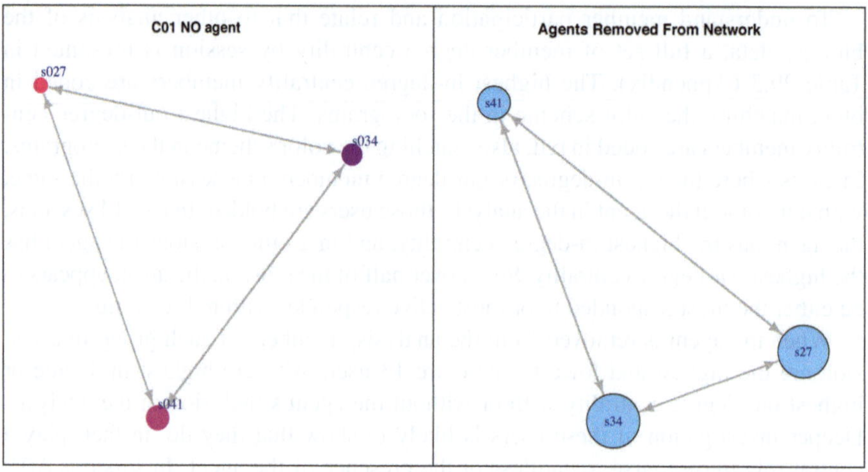

Fig. 29.4 Network representation of interaction network with agent actor removed. Using weighted trace analysis (*left*), using Stahl's conversation analysis (*left*)

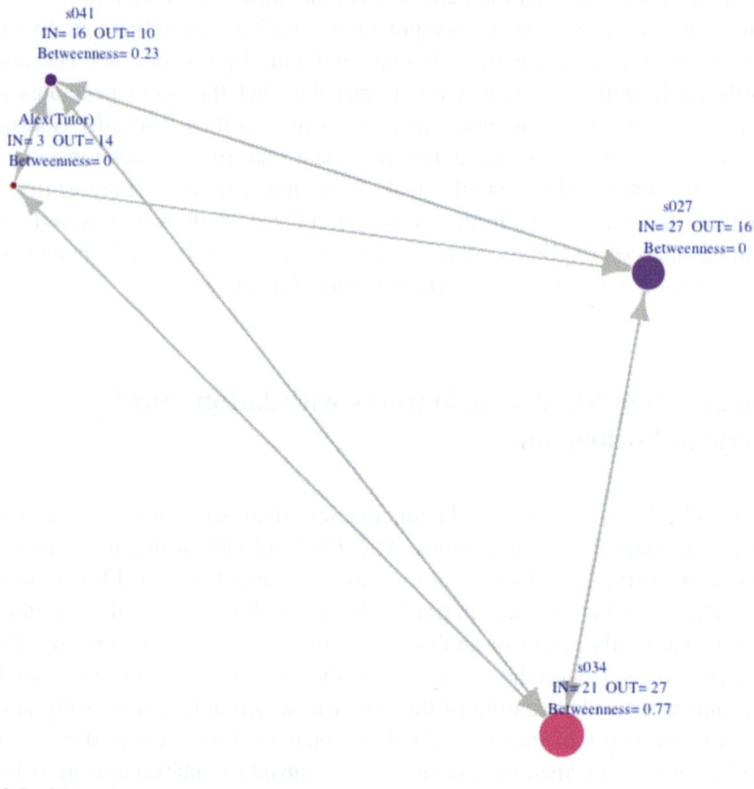

Fig. 29.5 C01 coded in Tatiana

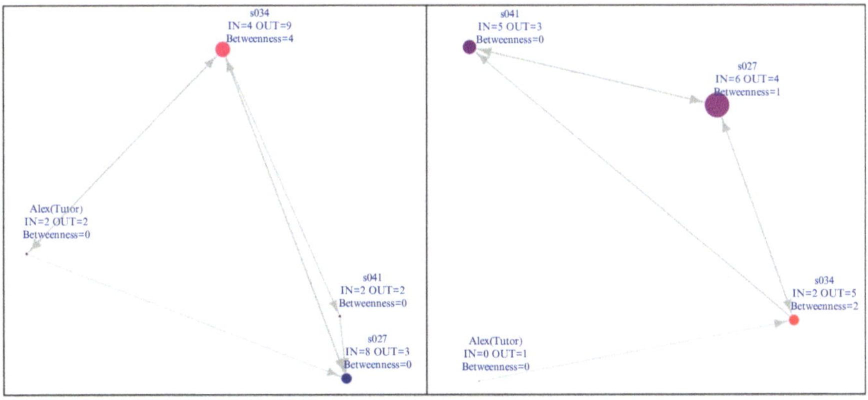

Fig. 29.6 CO1, explain phase (*left*), video phase (*right*)

and sociogram generation from Stahl's coding, with a network that includes the agent, respectively. Figures 29.6 and 29.7 represent automated sociogram coding and Stahl's coding without the agent in the network.

We see that the automated analysis puts the agent in a more central role than Stahl's analysis. The red color on the agent vertex in Fig. 29.4 indicates that this centrality is principally out-degree, which is consistent with a theory that the agent acts as a "broadcaster." There is little direct response to the agent identified in our analysis. This is not uniform across sessions. Eight of fourteen sessions show the agent's out-degree centrality is greater than in-degree. Coding of interaction types between the agent and the environment may reveal additional semantics and differences in agent interactions in future analysis.

Another contrast between the automated analysis and Stahl's analysis is between the centrality of S027 compared with S041. S027 appears less central in the automated analysis, compared with Stahl's; and S041 is shown as more central. This may be, in part, due to the nature of our weighted analysis. S034 is consistently in the middle of conversations between S027 and S041 in Stahl's analysis. We do not weight connections there, while the automated analysis gives greater weight to direct adjacency and short response times. More than one session would need to be analyzed in both ways in order to make any generalizable suggestions about the relationship between the automatically coded network and the ethnomethodologically informed network coding. S034 is a deeper blue than the other students, which is consistent with that student's role communicating directly with S027 and S041. The greater post count is reflected in the blue color.

With the agents removed from the network analysis, the degree weighting of their role in the network is removed from consideration, and we see the sizes of the vertices become more normalized. The agent was quite active in session C01, but not coded as a participant in a significant way by Stahl. In addition, the agent is not viewed as "communicating directly" with anyone in the ethnomethodological analysis, while the agent is viewed as "communicating to those three turns adjacent to it"

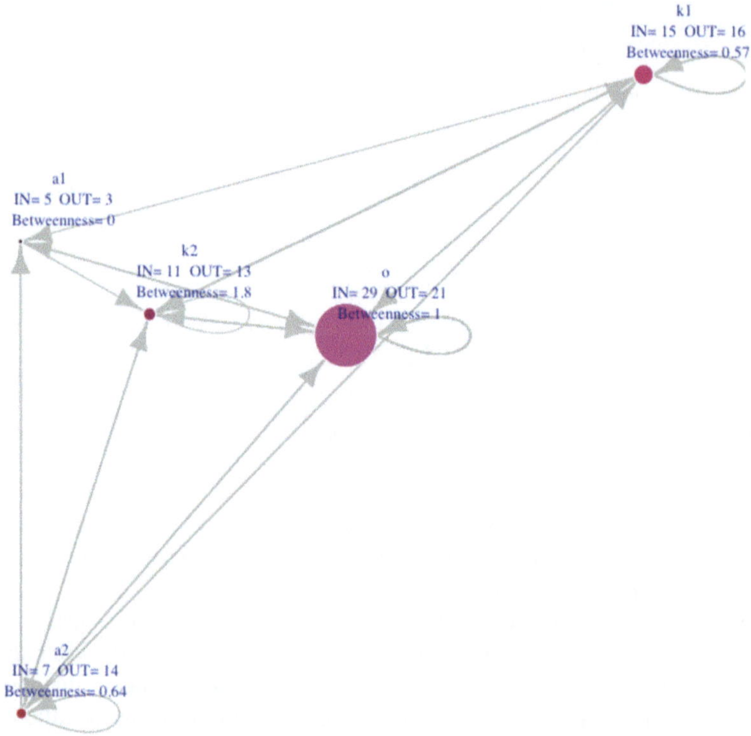

Fig. 29.7 C01, negotiation codes

in our analysis. When the agent is not included, this consideration is removed from the data. All three nodes take on a more reddish tone in the agentless diagram, compared to the diagram with agents. This is illustrated further in Table 29.3, where we can see that the relative degrees of all three students in the agentless analysis are more even than they are when heavy agent activity skews the network.

Phase 2: Comparing Implicit and Explicit Networks for All Groups and Operationalizing Knowledge Construction as a Network Node

Stahl's ethnomethodologically informed analysis identifies explicit interactions between actors, but is a time consuming analytical process that will not be widely adopted in practice for that reason. The estimation of interactions, informed by Stahl's results uses implicit interactions and weighting of those interactions to approximate understanding more rapidly. Tools like VMT could be instrumented to provide this kind of feedback in real time. Figure 29.7, showing Stahl's analysis

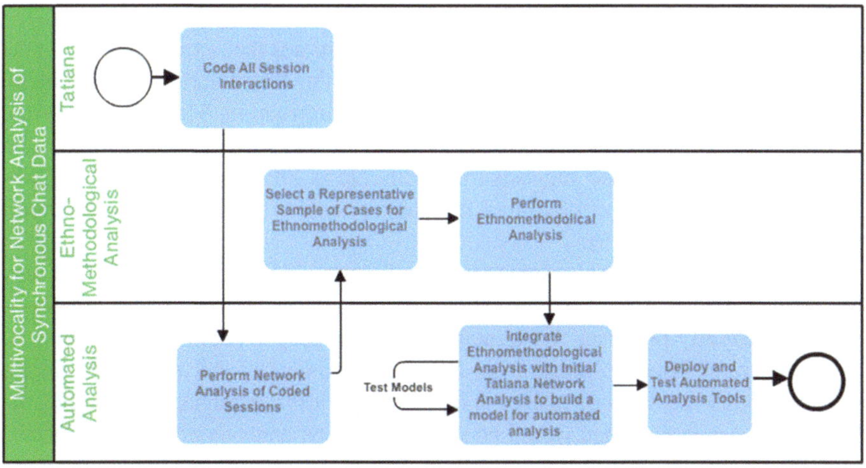

Fig. 29.8 An optimized multivocal candidate process

with the agent actor removed is most similar to phase one automated analysis where the agent is present. Since the session chosen by Stahl is one where the agents role was part of the indirect treatment, and there were few responses to the agent, it is not surprising that an automated analysis that reflects the relative position of the agent would most closely align with Stahl's analysis. In the second phase we seek a middle stance between ethnomethodologically informed analysis of electronic trace data and fully automated analysis, consistent with the stance of Group Informatics research about the importance of triangulating electronic trace data with other research methods. First, we use Dyke's (Dyke, Lund, & Girardot, 2009) Tatiana tool to rapidly code explicit interactions, and produce analyses from the results. Second, we take an orthogonal view of the data, and construct a network where the types of knowledge construction interactions—assertions (K1), queries (K2), narration (A1), and providing instructions (A2) are used to produce network diagrams. In this analysis, we show which types of moves become significant in technologically mediated discourse.

Figure 29.8, which also includes the agent role, illustrates a similar position for S027 and S034 in the discourse. It is more precise in its provision of key network statistics. Where Stahl's analysis produces relations and our automated analysis provides coloring to illustrate in-degree and out-degree centrality, in this analysis we include measures explicitly and add the network statistic of *betweenness*. This statistic is classically used to illustrate power brokerage relationships in analysis of physical networks. In this case, the statistic takes on a meaning that identifies whose actions, overall in the VMT tool, elicited response from more than one participant. Figure 29.8 shows that the level of participation for S027 and S034 is the highest of the three human participants. The *betweenness measure shows* that S041, while less prolific, did provide contributions that both other human participants responded to.

Dyke's Tatiana tool also facilitates the efficient coding of phases, and the construction of networks by phase. Figure 29.9 illustrates the "explain phase" in the study. The explain phase is when members are explaining their solution to the problem. In this phase, S034 and S027 made the greatest contribution. This is a part of the activity where members are negotiating a shared meaning. The limited participation of S041 during this phase suggests that their role in knowledge construction was more limited. Both degree centrality measures and betweenness measures support this conclusion (Fig. 29.5).

Figure 29.10 provides the same analysis for the video phase. Figure 29.6 now begin to tell a more complete story about the interaction practices of the members. S034 consistently produces more interactions than responses, while S027 consistently generates direct response from the other members of the triad. S041, in the key knowledge construction phases of interaction is less central, as illustrated by the betweenness measure, than would be suggested by an analysis of the entire period of interaction. Analysis by phase, therefore, brings the roles of members into a more clear focus. The high out-degree centrality of S034 in the Tatiana analysis is also consistent with the measures derived from the purely automated analysis, including implicit interactions, produced in phase one.

Table 29.4 (Appendix) illustrates that types of analysis produced by Stahl can be produced efficiently for all treatments through the combination of annotating conversation in Tatiana and applying network analytic techniques to the resulting data. Because the Tatiana coded interactions are all explicit, there are no implicit interactions represented here, and the network measures are therefore not weighted. We focus on session C in this analysis in order to highlight the contrasts in one session, having illustrated contrasts across sessions in Phase One.

The contrast between network analysis of the explicit connections from Tatiana with the automated analysis from phase one using implicit connections shows that the results are different. For example, Table 29.1 shows that, in session C01, S027 has the highest in-degree centrality based on explicit connections. Table 29.3 (Appendix) identifies S034 as having the highest in-degree centrality. Each approach to analysis of interactions in the treatments produces a nominally different result and identifies different members as having more or less central roles in the group. With the small data set we have to work with is not possible to know which method of analysis maps more closely to useful identification of members who play key roles in knowledge construction during these kinds of technologically mediated learning events. To direct focus in future studies, we include a measure of betweenness by member in Table 29.4 (Appendix). Individuals with high betweenness make statements that both other members of the team respond to and may therefore serve to direct future research questions and automated analysis.

Finally, in phase two we stepped back from our focus on understanding member structural position and role to examine the particular types of knowledge construction moves that were significant in discourse in VMT. These codes were produced using Tatiana as well. Figure 29.11 demonstrates that queries (K2) were the most responded to types of knowledge construction interaction in session C01. In addition, session C01 showed providing instruction (A2) and assertions (K1) were center

Table 29.2 Negotiation codes for all of classroom C with betweenness calculated

		K1	K2	A1	A2
Indirect	C01	0.57	1.8	0	0.64
None	C02	3.8	0.25	0	0
Direct	C03	5.7	0.33	0	0
Indirect	C04	0.75	0.25	0	0
None	C05	0	0	0	0
Direct	C06	11	0	0	0
Indirect	C07	1.4	0	0	0

points of interaction. We highlight C01 because that is the session that Stahl focused on, so it serves as a thread through our multivocal analysis in this chapter.

Ethnomethodological approaches are time consuming, and require the researcher to select specific sessions for close analysis. Automated analyses like those described in our two phases of multivocal analysis may aid the ethnomethodologically informed researcher in the selection of representative cases in future analyses. Table 29.2 highlights the extent to which C01 may not have been a representative choice for activities in session C. The betweenness calculations for all other sessions show that assertions (K1) are the most likely type of interaction to result in discourse among all three parties and that neither A1 nor A2 moves produce interactions among all the members in any of the seven cases in session C, except for C01.

Discussion

Network analysis provides a preview of interaction patterns that ethnomethodologists and other analysts may use as a guide for which groups to dive into more deeply. We see through the two phases of automated analysis that reflexively deriving implicit interaction methods from ethnomethodological analysis from a single case may not lead to a measure of network position and role that is representative of the explicit interactions across a range of cases. The use of Tatiana to build explicit connections for the full corpora surfaces the nature of C01 as an outlier in the data set. Together, phase one and two of our analysis suggests that a reflexive, multivocal strategy may produce a more efficient use of analytic resources. A first pass could perform a superficial analysis, such as that conducted with Tatiana. This could feed into an automated selection process, highlighting candidates for deeper investigation. And, in a final step in future experiments, we suggest reflexively incorporating findings from both Tatiana and the purposely directly ethnomethodological analysis to identify appropriate weighting strategies and factors for fully automated analysis. This kind of "next step" will enable progress toward a vision of automated, multivocal research and contribute to the emerging area of inquiry described by some as "learning analytics." One potential version of this process is illustrated in Fig. 29.8.

The role of the agent in different groups is made easily visible through this analysis, enabling quick adjustments to the agent's behavior during design and future research. The lack of direct interaction from human subjects back toward the agent is most clearly illustrated in the second phase of our work. Future studies should consider how the agent's connections are measured. Both Tatiana and the ethnomethodologically informed analyses of the data show, quite properly, that people do not respond directly to agents. However, with this sort of analysis we are making assumptions about agent impact that may not be appropriate. We opened the chapter with a question of whether or not the anthropomorphization of agents in analysis is more or less useful. A productive question for future analysis is whether or not there is some type of new kind of impact agents have on discourse that is not reflected by conversation analysis, explicit conversation connections or even implicit analysis. While Stahl and others have reported that agents in this study were regarded as somewhat of a nuisance by participants, and this analysis shows they were accordingly not addressed, we are concerned that the role and impact of agents on learning is not addressed by the voices of multivocal analysis reflected here. The voice of the agent is not human, nor is it purely nonhuman. Future ethnomethodological analysis of agent discourse is an appropriate voice for unearthing the new kinds of interactions prompted in people, by the existence and contributions of agents in discourse.

Appendix

Table 29.3 Complete listing of degree data

	With agents		Without agents		
Session	In-degree	Out-degree	In-degree	Out-degree	User
A01	11.92	6.62			agt
A01	8.64	11.62	3.16	3.87	**s001**
A01	4.61	5.90	3.46	3.16	s005
A01	3.87	4.90	3.87	3.46	s009
A02	1.48	2.59			agt
A02	5.00	2.98	4.48	2.34	s002
A02	8.97	4.57	7.75	4.57	**s006**
A02	5.43	10.73	4.57	9.89	**s010**
A03	2.63	6.75			agt
A03	5.42	3.19	1.96	2.15	s003
A03	3.83	2.63	2.88	1.89	s007
A03	4.68	4.00	2.34	3.14	s011
A04	14.42	7.11			agt
A04	5.88	14.12	4.81	8.64	**s004**
A04	10.39	8.40	7.81	4.94	s008
A04	8.40	9.46	4.94	3.99	s012
A05	3.55	4.10			agt

(continued)

(continued)

Session	With agents		Without agents		User
	In-degree	Out-degree	In-degree	Out-degree	
A05	7.89	3.76	6.61	3.17	s020
A05	4.92	8.17	2.58	7.21	**s021**
A05	5.56	5.87	5.06	3.87	s022
C01	8.90	13.01			agt
C01	5.42	7.25	3.09	4.92	s027
C01	13.26	4.74	5.51	3.66	**s034**
C01	7.71	10.28	4.78	4.81	s041
C02	4.44	8.50			agt
C02	7.37	6.03	5.03	3.88	s028
C02	12.10	4.96	7.63	3.72	**s035**
C02	4.70	9.11	3.01	8.06	**s042**
C03	10.20	10.39			agt
C03	9.64	5.74	7.30	4.33	s029
C03	10.09	13.04	4.62	5.29	s036
C03	8.50	9.26	5.91	8.22	s043
C04	5.10	8.14			agt
C04	5.01	3.62	4.49	3.62	s030
C04	4.40	8.15	2.25	4.99	**s037**
C04	9.75	4.35	4.28	2.41	s044
C05	12.51	8.26			agt
C05	7.34	8.58	3.47	4.71	**s031**
C05	5.72	8.03	3.49	2.55	s038
C05	6.08	6.79	3.93	3.63	s045
C06	7.29	6.34			agt
C06	8.54	5.09	5.08	3.90	s032
C06	6.64	9.96	4.31	6.80	**s039**
C06	6.24	7.31	5.70	4.38	s046
C07	10.15	4.96			agt
C07	3.95	8.76	3.91	7.55	s033
C07	8.89	12.40	6.31	6.92	s040
C07	8.88	5.76	6.54	2.30	s047
D01	14.12	7.14			agt
D01	5.19	9.21	2.27	3.73	s056
D01	5.85	10.58	3.97	7.41	**s059**
D01	9.81	8.04	7.48	2.57	s062
D02	7.70	7.85			agt
D02	5.86	14.87	2.93	9.40	**s057**
D02	5.68	4.62	3.34	3.34	S060
D02	12.15	4.05	9.57	3.10	**s063**

Highest in-degree centrality is noted in blue highlighting. Highest out-degree per session is noted in red. Users who are highest in one or both measures with or without the agent included in the analysis are bolded

Table 29.4 Key network statistics for Tatiana coded data

Node	In-degree	Out-degree	Betweenness
C01 (Indirect)			
1 Alex (Tutor)	3	14	0
s041	16	10	0.23
s034	21	27	0.77
s027	27	16	0
C02 (None)			
2 Alex (Tutor)	0	7	0
s035	13	12	0
s028	13	17	0
s042	20	10	0
C03 (Direct)			
3 Alex (Tutor)	2	5	0
s036	20	12	0
s029	17	18	0
s043	12	16	2
C04 (Indirect)			
4 Alex (Tutor)	0	9	0
s037	14	10	0.71
s030	7	5	0
s044	6	3	0.29
C05 (None)			
5 Alex (Tutor)	0	11	0
s038	21	31	0
s031	26	19	0
s045	24	10	0
C06 (Direct)			
6 Alex (Tutor)	2	14	0.27
s032	20	21	2
s046	4	4	0
s039	29	16	0.73
C07 (Indirect)			
7 Alex (Tutor)	3	11	0
s040	32	23	0.86
s047	14	16	0.14
s033	23	22	0

References

Dyke, G., Lund, K., & Girardot, J.-J. (2009). Tatiana: An environment to support the CSCL analysis process. *Proceedings of the 9th International Conference on Computer Supported Collaborative Learning* (Vol. 1, pp. 58–67)

Erickson, T. (2009). "Social" systems: Designing digital systems that support social intelligence. *AI & Society, 23*(2), 147–166.

Fine, G. A., & Harrington, B. (2004). Tiny publics: Small groups and civil society. *Sociological Theory*, *22*(3), 341–356. Retrieved from http://onlinelibrary.wiley.com/doi/10.1111/j.0735-2751.2004.00223.x/abstract

Goggins, S., Galyen, K., & Laffey, J. (2010). Network analysis of trace data for the support of group work: Activity patterns in a completely online course. *Proceedings from ACM Group 2010* (pp. 107–116) Sanibel Island, FL

Goggins, S., Laffey, J., & Amelung, C. (2011). Context aware CSCL: Moving toward contextualized analysis. In H. Spada, G. Stahl, N. Miyake & N. Law (Eds.), Connecting Computer-Supported Collaborative Learning to Policy and Practice: Proceedings of the 9th International Conference on Computer-Supported Collaborative Learning (CSCL 2011) (Vol. II, pp. 591–596). Hong Kong: International Society of the Learning Sciences.

Goggins, S. P., Laffey, J., & Gallagher, M. (2011). Completely online group formation and development: Small groups as socio-technical systems. *Information Technology & People, 24*(2), 104–133. doi:10.1108/09593841111137322.

Goggins, S., Mascaro, C., & Mascaro, S. (2012). Relief after the 2010 Haiti Earthquake: Participation and leadership in an online resource coordination network. *Proceedings from ACM Conference on Computer Supported Cooperative Work* (pp. 57–66). Seattle, WA

Goggins, S. P., Mascaro, C., & Valetto, G. (2013). Group informatics: A methodological approach and ontology for sociotechnical group research. *J Am Soc Inf Sci Tec, 64*(3), 1532–2890. http://dx.doi.org/10.1002/asi.22802. doi: 10.1002/asi.22802. 516–539.

Goggins, S., Valetto, P., Mascaro, C., & Blincoe, K. (2013). T Creating a model of the dynamics of socio-technical groups. *User Modeling and User-Adapted Interaction, 23*(4), 345–379. http://dx.doi.org/10.1007/s11257-012-9122-3. Springer Netherlands.

Howison, J., Wiggins, A., & Crowston, K. (2012). Validity issues in the use of social network analysis with digital trace data. *Journal of the Association of Information Systems, 12*(2)

Latour, B. (2007). *Reassembling the social: An introduction to actor-network theory.* Oxford

McCulloh, I. (2009). *Detecting changes in a dynamic social network.* Pittsburgh, PA, USA: Carnegie Mellon University.

McCulloh, I., & Carley, K. (2009). Longitudinal dynamic network analysis using the over time viewer feature in ORA. Retrieved from http://citeseerx.ist.psu.edu/viewdoc/summary?doi=10.1.1.161.3492

Mead, M. (1934). *Kinship in the Admiralty Islands.* Transaction Pub. Retrieved from http://books.google.com/books?hl=en&lr=&id=8Fyj0h0G_OUC&oi=fnd&pg=PR9&dq=mead+"small+group"+"M+mead"&ots=-7aKCAORBO&sig=IzsWex2tPvK2vmGGWBpfm-6OFXs

Mead, M. (1958). Cultural determinants of behavior. *Behavior and evolution.* (pp. 480–503)

Schmidt, K., & Bannon, L. (1992). Taking CSCW seriously. *Computer Supported Cooperative Work (CSCW), 1*(1), 7–40.

Stahl, G. (2006). *Group cognition: Computer support for building collaborative knowledge.* Boston, MA: MIT Press.

Stahl, G. (this volume). Interaction analysis of a biology chat. In D. D. Suthers, K. Lund, C. P. Rosé, C. Teplovs & N. Law (Eds.), *Productive multivocality in the analysis of group interactions*, Chapter 28. New York: Springer

Turner, W., Bowker, G. C., Gasser, L., & Zacklad, M. (2006). Information infrastructures for distributed collective practices. *Computer Supported Cooperative Work, 15*, 93–110.

Chapter 30
Multivocality as a Tool for Design-Based Research

Cindy E. Hmelo-Silver

This section of the book focuses on an initial enactment of accountable talk agents to support collaborative learning in an urban high school science laboratory (Dyke, Howley, Kumar, &, Rosé, Chap. 25, this volume). Much of the analysis focused on the initial enactment and at least two of the chapters also considered a second enactment that took into account what the researchers learned through the use of productive multivocality. This early stage in a design-based research (DBR) program is timely for understanding how a complex sociotechnical intervention affected collaboration.

In this chapter, I will present a brief overview of design-based research and consider why multiple perspectives offer a distinctive advantage (if not in fact being essential) in DBR. In the remainder of my discussion, I will address how multivocality has influenced the overall understanding of what was going on in the initial enactment and raise some conjectures as to how it might have influenced revision of the design. I will also address some ways in which the multivocal process was not as productive as it might have been, along with some caveats.

By way of disclosure, I have been both an early participant (ICLS 2008) and a late participant (ARV 2011) in the multivocality process. It has been challenging to try to get up to speed on a process that I missed a great deal of important progress in, with a new dataset that was also challenging. But taking on these challenges is critical if CSCL is to grow as a field that respects and learns from a range of research methods and theoretical frameworks.

C.E. Hmelo-Silver (✉)
Indiana University, Bloomington IN, USA
e-mail: hmelosi@indiana.edu

D.D. Suthers et al. (eds.), *Productive Multivocality in the Analysis of Group Interactions*, Computer-Supported Collaborative Learning Series 16,
DOI 10.1007/978-1-4614-8960-3_30,

Why Design-Based Research?

In her seminal article on DBR, Brown (1992) declared that she was engaged in the practice of engineering learning environments. But the goal for this engineering was to test theories of learning in the crucible of practice, a theme later echoed with some cautions by Shavelson, Phillips, Towne, and Feuer (2003). DBR is characterized by being a theory-driven iterative approach to research that seeks to understand an overall "learning ecology" (Cobb, Confrey, DiSessa, Lehrer, & Schauble, 2003).

The kinds of theories that DBR seeks to test are often what Cobb et al. (2003) refer to as humble theories that try to account for domain-specific learning processes (Cobb & Gravemeijer, 2008). DBR programs are thus accountable to both theory and pragmatic concerns as we try to understand how the interacting elements in a complex learning environment work, and how they might be adapted to other circumstances. As Cobb et al. (2003) note:

> "What works" is underpinned by a concern for "how, when, and why" it works, and by a detailed specification of what, exactly, "it" is. This intimate relationship between the development of theory and the improvement of instructional design for bringing about new forms of learning is a hallmark of the design experiment methodology (p. 13).

One of the lessons of DBR is often that it is not just the tools, but it is also the curriculum, the pedagogy, the participant structures, what the learners bring to the learning experience, how the teachers enact an innovative learning environment—in other words, a whole range of features (Hmelo-Silver, 2012). Thus, we need to think about learning at the micro level as well as at the larger social infrastructure (Bielaczyc, 2006; Hmelo-Silver, 2012).

By its nature, DBR is iterative. Between each iteration, multiple methods are needed to be able to understand the complex enactment context and how that relates to what learners take away from the design. One of the challenges is in having a research team with the requisite kinds of expertise to conduct these different kinds of analyses and to communicate with each other productively.

DBR is also concerned with building theory. To a great degree, the designs we construct are ways of enacting our theories about how people learn under a particular set of circumstances. When we revise our designs based on the feedback from different enactments, we also need to think about what the implications are for theory. Together this suggests that if multivocality is to be a tool for DBR, it also needs to be an ongoing process.

What Do Multiple Perspectives Bring to Design-Based Research?

One characteristic of complex systems is that they are composed of multiple interacting levels (Hmelo-Silver, Liu, Gray, Finkelstein, & Schwartz, 2007; Jacobson & Wilensky, 2006). Such systems can be viewed at both micro and macro levels, different units of analysis, and through a range of different lenses. If we are trying to understand the complexity of a learning ecology, there are many advantages to bringing in multiple perspectives that complement each other in understanding a complex CSCL environment (Hmelo-Silver 2003; Hmelo-Silver, Nagarajan, & Chernobilsky, 2009). This is the conjecture of productive multivocality. The multivocality process (Suthers, Chap. 1, this volume; Suthers et al., 2011) goes beyond having multiple perspectives that sit alongside each other—rather, the notion is to have an ongoing conversation that is mutually informative. As part of this process, analysts have participated in workshops, read each other's chapters and revised their analyses. My task in the remainder of this chapter is to consider how this can be a tool for DBR.

What Have Multivocal Analyses Revealed?

Not all of the multivocality process is explicit in the chapters themselves—some of it has happened between the lines and in face-to-face meetings. Several themes have emerged in these analyses. First and foremost, the intervention did not work as intended. That is neither a surprise for a complex intervention, nor is it a bad thing as there was much to be learned as Rosé and her team have taken on a difficult problem in trying to engineer a software agent to facilitate academically productive talk (Resnick, Michaels, & O'Connor, 2010). It is a problem well worth taking on if student-centered discourse-based approaches to instruction are to be widely implemented. The different analyses converged on some common findings as well as identifying unique issues to be addressed in the redesign, and indeed, some of these were successfully addressed.

These analysts represent a range of theoretical assumptions and frameworks, which necessarily lead to different approaches to analysis. Howley, Kumar, and Mayfield (Chap. 26, this volume) developed their design and analysis tool from the perspective of linguistics. As part of the multivocality process, the designers worked with Stahl (Chap. 28, this volume), who used group cognition as his theoretical framework and that informed his ethnomethodological analysis. Cress and Kimmerle (Chap. 27, this volume) took a very socially oriented framework that focuses on the importance of group members' awareness of social context, knowledge in the group, and actions and activities of other group members. Goggins and Dyke (Chap. 29, this volume) took a group informatics perspective that compared a completely bottom up social network analysis to analyses informed by both Stahl's ethnomethodological analysis and Dyke's Tatiana assisted coding (Dyke, Lund, & Girardot, 2009). This last analysis alone is a clear indication of multivocality in that it integrates across group cognition and informatics perspectives.

Moreover, different analysts had different purposes for their analyses. Howley et al. were interested in uncovering the type of support that would elicit academically productive talk moves. One can imagine that such a goal is entirely compatible with Cress and Kimmerle's stated purpose of identifying knowledge building episodes. Stahl took this dataset as an opportunity to see if sequential analysis could be used as a "quick and dirty" method of evaluation that could inform future design cycles. Goggins and Dyke's purpose was to provide descriptive information as well as considering the role of software agents in synchronous chat.

So far, we have different underlying theoretical frameworks and analytical goals, and consistent with these assumptions and goals, the different analysts had different units of analysis. These units ranged from individual utterances (e.g., Howley et al.) to adjacency pairs (Stahl) to knowledge building episodes (Cress & Kimmerle) to interaction networks (Goggins & Dyke) as well as considering pre and post tests (Howley et al.). The analysts used a range of manipulations to construct the representations that they operated on. Cress and Kimmerle operated on examples of chats and scanned worksheets as they coded and counted attempts at establishing different kinds of awareness. Stahl manipulated spreadsheets to show the sequential structure of the chat as he sketched who was responding to whom. This made salient the role of the software agents within the group. Howley et al. performed a variety of data manipulations as they scored tests of learning outcomes and coded utterances for transactivity and heteroglossia, calculated frequencies and inferential statistics across conditions. They then were able to create visualizations in Tatiana that allowed viewing the sequential nature of different coded utterances by student and agent. Goggins and Dyke computed social network statistics and constructed sociograms as well as engaging in sequential data analysis of codes in phase 2. In many of the chapters in the book, analysis focused around pivotal moments. That was not the case in this dataset. Some of the analysts focused on small pieces of the datasets or examples, whereas others considered the dataset as a whole—and that is consistent with the range of perspectives that that all brought to the table.

Despite all the differences, these analysts talked with each other. Goggins, Rosé, and Dyke met for the interaction analysis sessions that informed Stahl's chapter. Both Stahl and Dyke served as points of comparison for Goggins' analysis. Lessons learned from the Cress and Kimmerle analysis informed redesign in terms of making allowances for social awareness.

As in many initial design implementations, things did not go quite as the designers had intended. Howley et al. did not find effects on learning outcomes, though they did find some transfer to whole class discussions. Such findings need to be unpacked to understand the enactment in detail. It is rare that we get an inside look at where instructional innovations begin as there is much to be learned from those beginnings. What is lovely in this multivocality process is that a coherent picture emerged from these analyses that would prove to be extraordinarily fruitful in redesign as these analyses help with the needed unpacking.

From the social awareness perspective, Cress and Kimmerle identified some foundational problems in the group members not being aware of whom they were chatting with and what each other were doing. This led to difficulties in coordination and overall frustration in the anonymous conversation with people with whom they were otherwise familiar. Stahl identified the frustration with the anonymity as well. He argues that there were several concerns. First, with both the macroscripts and microscripts, the environment may well have been over scripted (Dillenbourg, 2002). Moreover, the agent was not really interactive—it did not fit into the conversation in ways that responded to the student discourse in the chat. Stahl also noted the lack of awareness, similar to Cress and Kimmerle. Another important concern was that the task was not designed in a way to take advantage of the collaborative environment. One of the particular affordances of the VMT environment is the shared whiteboard and the way it permits referencing. The whiteboard could have provided a focus for negotiations of ideas if it were used, as well as providing opportunities for students to be aware of what each other were doing. The network analyses of Goggins and Dyke made the agents' role in the interaction salient. It was informed by the work of both Dyke and Stahl. They noted that although human contributions were rarely ignored, the students were, not surprisingly, more likely to ignore the software agent than their other group members.

Despite these limitations, Howley et al. found that there were more APT moves in both supported conditions than in the control condition and that the use of APT in the chat was correlated with contribution to the larger whole class discussions. However, their Souflé analysis showed that something was going on in the indirect agent condition that was suppressing discussion. The Tatiana visualizations demonstrated the time course of monoglossic and heteroglossic utterances differed across conditions. As in the analyses by Stahl and Cress and Kimmerle, Howley et al. identified frustration with the software agent's contribution, and that this was greater in the indirect condition. They too attribute this to the lack of coordination between macroscripting timed prompts and microscripting prompts as they noted that tutor prompts often tended to disrupt student conversations, even when students were trying to follow an earlier direction from the agent.

In summary, the different analysts interacted with each other both directly at workshop and data sessions, as well as indirectly as the chapters were shared. They create a rich story of this initial enactment of a software agent to facilitate APT. Goggins and Dyke suggest that the SNA analysis could be used to pinpoint spots for in-depth analysis. Moreover, Stahl found that these quick sequential analyses could be used as part of a DBR process in a timely manner. I concur with the importance of being able to pinpoint where the action is for analysis—and often, one of the challenges in DBR is being able to engage in principled analysis in ongoing projects. The different perspectives brought out a range of different issues that figured into the redesign. However, one aspect that the multivocality process did not necessarily consider is how these different analyses were geared towards the questions that the original researchers were asking. So for the designers, the experimental manipulation represented an important research question but for the other analysts, it was not very relevant. An important lesson for multivocality in DBR is that it must address the designers' goals even if very different lenses are used for that. Because there are these dual research and practice goals, in DBR, it is important that the multivocal analysts understand how their analyses can be productive with respect to DBR.

Using Multivocal Analysis for Redesign: DBR Round 2

The instructional interventions in DBR are complex, and it is often difficult to tease out what the key factors are. Here, it was clear that there were issues in terms of when the agent intervened as well as with the students knowing who was participating and what others in their group were doing. The second iteration of the APT agents was able to take this into account.

First, the overall intervention was simplified to deal with attention focus issue as well as pragmatic issues related to teacher planning. Second, the role of the software agent was simplified to focus on the revoicing aspect of APT, and only used as a direct prompt. In addition, an agent provided positive feedback for using APT. The designers created better coordination between the revoicing prompts and third, students knew who they were talking to—consistent with the recommendations of the earlier multivocal analysis. It is clear that these changes were motivated by the multivocal analysis. The multivocality process continued to a small degree as two of the participants engaged in further analysis.

The design team evaluated learning outcomes in the second analysis in Howley et al. and found that the only APT revoicing manipulation affected learning outcomes and the quality of the co-constructed explanation. There was no effect of the manipulation on class discussion. Although the designers were disappointed that there was no prolonged effect of the manipulation, I would be encouraged by the local effect that they found. Changing discourse norms does not happen overnight—and in this short intervention, finding any effect is encouraging and worthy of further analysis. Indeed in a later study, the research team found that some of the some of the APT moves supported by the agents continued into the classroom discussion that followed (Clarke, Chen, Stainton, Katz, Greeno, Resnick, et al., 2013). Digging deeper into these data with the Soufflé style analysis may be fruitful in further unpacking what went well and where there is further room for improvement.

Some of this unpacking was accomplished with Stahl's sequential analysis. He found that there was more even involvement by the group members, but that the agent was still dominant. Moreover, the agent moves did not expand the accountable talk space but nonetheless, in the group studied, students were respectful and included each other. One question is how representative this group is and whether there were other discourse patterns. Stahl's analysis seems to confirm that the timing of the agent moves was more appropriate. Stahl raises the issue of what one can expect of students of this age. In my own consideration (and with my experience in the discursive context of problem-based learning), I would expect that it would take more time for the students to adjust to the norms of APT, even if they needed to be more directive in initial stages. This raises interesting questions about scaffolding contingencies and whether they can be faded—all of which seem to be promising directions for future DBR iterations. Nonetheless, in this later iteration, it provided evidence that the software agent was having positive effects.

Reflections on the Productive Multivocality Process and DBR

What is missing from the chapters are the discussions among the authors—the informal conversations that happened at dinner, on walks, over coffee and tea, and in hallways. Although some of this is reflected here, I have had many discussions with Rosé as have Stahl, Goggins, and Dyke. Cress offered important insights that were important in the redesign that were offered at the Alpine Rendez-vous workshop. These discussions were productive and respectful, despite coming from different theoretical commitments and analytic approaches.

It is clear from these chapters that multivocal analysis is not just for understanding learning but can be part of an iterative design process. Each analysis points out what went wrong but also suggestions for how this might be improved. This advanced the design process by providing detailed analysis giving different perspectives on both social and cognitive processes. What is still not clear is how the multivocality problem advanced theory here—about scripting, about the use of accountable talk, and software agents. It would be helpful to have seen the initial embodied conjecture (2004) and how that has been revised in terms of a small "t" theory of how APT can be supported with software agents.

Another important contribution of multivocality is in understanding learning as a complex system with interaction among different levels of the system. In these data and analyses, we have analyses at different units of analysis that connect what is going on in individual utterances, larger episodes, overall interactions, and student outcomes. What would be helpful would be to make some of these connections more explicit.

Multivocality or Cacophony?

Although we see different perspectives as multivocality was applied here, in talking with the provider of the dataset (Rosé), one concern was that the different analysts did not necessarily consider their analyses with respect to the questions that the researchers posed. There is the danger then of a cacophony if the different voices have different goals. But the owner of the data might also consider the extent to which the different questions raised have important considerations in design-based research that might complement the perspective of the original researchers.

One of the things that made this discussant task particularly challenging is that the research group did not reach shared agreement on what the pivotal moments might be. That made it challenging for me to see more broadly how these different perspectives might be complementary. Table 30.1 shows my rough parsing of how different voices dealt with some of the areas of common ground we were charged with identifying. This parsing shows that in addition to these differing assumptions, which we might assume would be productive; there were many challenges as well. Different analysts were focusing on different aspects of the data and one wonders

Table 30.1 Comparison of analyses

Perspective	Theoretical assumptions	Purpose	Unit of analysis	Data representations	Manipulations	Pivotal moments
Designers	Learning enhanced through APT scripting conversational agents	Teach productive class-room discourse specific strategies				
Cress & Kimmerle	Group awareness key to supporting successful CSCL	Identify instances of Knowledge building	Small group	Examples of chat, scanned worksheet	Counting contributions, attempts to establish social awareness	Many moments analyzed
Stahl, iteration 1, 2	Group Cognition	Sequential analysis as "quick and dirty" method of evaluation to inform next design cycle; effects of agents on group processes	Text pairs—adjacency pairs	Spreadsheets with response structure of chat	Spreadsheet arranged with participants in own column. Make response structure clear	C01 selected as representative F01 for iteration 2
Goggins & Dyke	Group informatics perspective: recognizes role of groups in organizational and societal change	Descriptive information complements ethno-methododologically informed analysis; role of agents as chat group members	Interaction networks, Dyke coded knowledge construction moves	Sociograms and Tabular representations	Inclusion and exclusion of agent, weighting of ties	C01 for initial analyses but then all groups
Howley et al., iteration 1	Linguistic perspective on how software agents support CSCL	How do linguistic interactions support CSCL; relation between facilitation moves and transactivity	Utterances, individual pre/post test gains	Tables comparing conditions, Tatiana visualizations	Coding of APT moves, reasoning moves, transactive moves. Inferential statistics	Several examples not identified
Howley et al., iteration 2		Hypothesis: dynamic APT microscripting to CSCL groups → increased learning	Pre/post test, group artifacts			

the extent to which this may have impeded productive multivocality for this DBR project. Another question relates to what the role of the facilitator is in the theoretical assumption. Like in other student-centered approaches to learning (e.g., Hmelo-Silver, 2004; Hmelo-Silver & Barrows, 2008), the teacher role is central in APT. It is however a different role from that of direct instruction. In APT, students are accountable to their community, to standards of reasoning and to disciplinary knowledge (Resnick et al., 2010). To be accountable to the community for example, student talk must build on the contribution of others. Being accountable to standards of reasoning means that talk involves logical connections, constructing explanations, and self-corrections. Accountability to knowledge refers to getting one's facts right and making evidence behind claims explicit. It includes being willing and able to challenge ungrounded assertions. But these kinds of discursive interactions do not happen in a vacuum—the role of the teacher in facilitating this accountability is central to APT. The teacher helps in positioning students as cognitively engaged. However, supporting this kind of APT discourse is challenging for teachers and part of the goal of research on software agents was to help support the teachers in creating these norms (Dyke et al., Chap. 25, this volume; Clarke et al., 2013).

These differences in theoretical assumptions were both broad and particular about the role of the teacher as facilitator. This was frustrating for the designers initially but they were able to take the multivocal feedback into account in their later designs.

So an important question for multivocality and DBR is whether we can harmonize these different perspectives, purposes, units of analyses and pivotal moments or are we stuck with a cacophony that is just noise. Being an optimist, I think the former is the case, but creating harmony is neither quick nor easy, but rather requires ongoing discussions. It might be useful to begin with boundary objects beyond the raw data itself. One possibility for such an object might be something that makes the designers' assumptions and intent in DBR really specific and where the analysts could clearly demarcate their analytic foci and differing assumptions.

One wonders what difference it would have made to the multivocal process if the design team had made their assumptions about the role of the teacher being embodied in the software agent clear and why they were doing their experimental manipulation explicit. It might have also helped to make what they were not doing clear—that is, that they were not testing the VMT chat software so much as using that as a medium to test the agents. The common procedures across the other datasets in this volume was to provide them to analysts without instructions or clarifications so as not to bias the analyses. But these other datasets were from more stable interventions—not as early in the design process as the data from the section. I return to this issue in the conclusion.

Another tack that one might take in understanding multivocality in this group builds on Sandoval's (Sandoval, 2004, 2013) notion of a conjecture map. Working through discussions with Rosé and with inferences from my reading of the chapters, I suspect a conjecture map for this project would look something like Fig. 30.1. Here the software agent was taking on the role of the teacher.

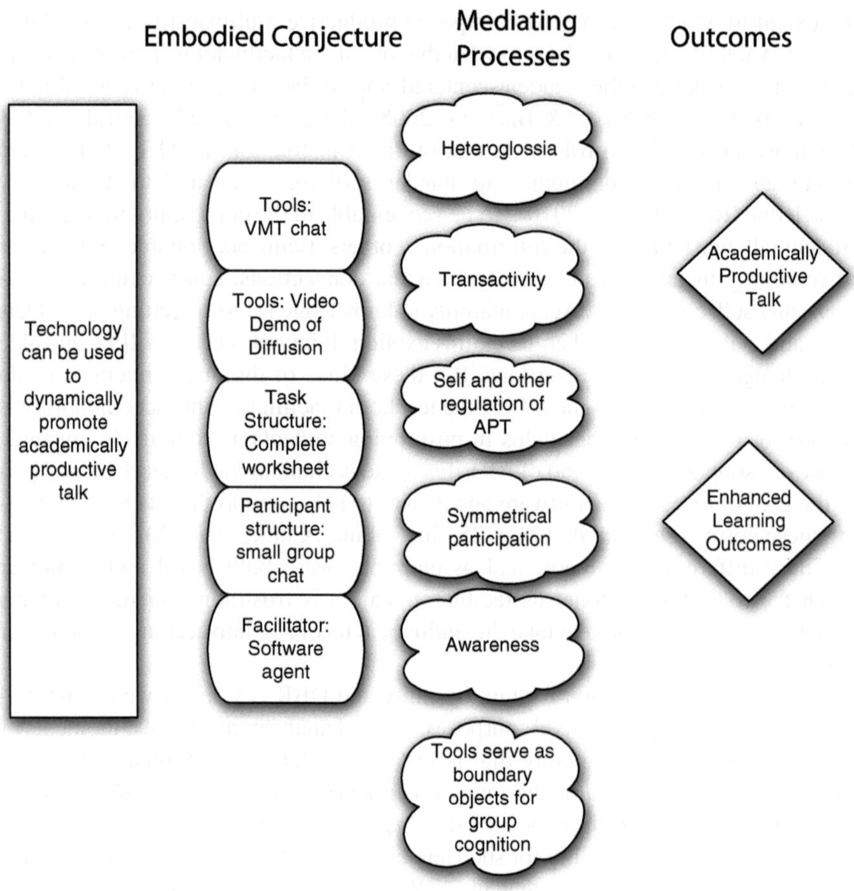

Fig. 30.1 Possible conjecture map for APT intervention

What this conjecture map shows is what the high level goal was and how I think it was embodied. What is not clear is that all the analysts shared an understanding of the goals of the research project or what the mediating processes might need to be to achieve the project outcomes. For example, Stahl focused on the affordances of the tools to support group cognition and was concerned about the agent interfering with that. Cress and Kimmerle added a mediating process to what the original designers may not have considered in terms of needing to be aware of what one's collaborators were doing with respect to the task. Goggins and Dyke focused on the change in the social network with and without the agent—relating to issues such as symmetry of participation and leadership. Howley et al. focused on explicit displays of reasoning, outcomes, and the collaborative dynamics involved in the leadership in the small groups. The Souflé analysis allowed them to examine dynamics in terms of authoritativeness, heteroglossia, and transactivity but also to pinpoint how these dimensions might be enhanced for both the agent and the group members. This helped the designers realize the issues in coordination with the agent (also clearly noted by the other analysts).

Finally, beyond being productive for this specific project, the respectfulness is important in promoting complementarity and synergy in the learning sciences. One of the key strengths of this field is its interdisciplinary nature, but often, we do not always take advantages of such affordances. In DBR, this multivocal process allows the whole to be more than the sum of the parts. One challenge to the multivocality process was that different analysts were not all looking at the whole elephant—that is, different analysts may have put their analytic focus on a particular tool (column 2 of Fig. 30.1) or mediating process (one of the clouds in column 3) without considering the overall embodied conjecture and its theoretical underpinnings. Thus, some discussion up front that established some shared goals among the analysts, and that had some resemblance to the designer's intent might have made this process smoother and more productive. However, it was also helpful for the designers to have fresh eyes to understand why things did not work as intended.

Concluding Thoughts

To be productive at different stages of research, the multivocality process must adapt. In a dataset from a mature project, as in many of the other sections of this volume, having analysts work with a dataset and talk later has been productive. In early phases of design, I would like to suggest that an initial discussion is important. The analysts need to have an understanding of the project goals. The designer needs to make their goals and intentions clear. The analysts need to have some initial conversation about how the different approaches can go beyond providing alternative lenses. It is important that these lenses contribute to moving the design forward. In this case, the role of the teacher/software agent was critical to helping to create norms for and facilitate APT. As the later design iterations demonstrate, the multivocality process was somewhat productive, if a little more frustrating that would be ideal. I do not mean to suggest that this process was not useful—but I think as in any design, the design of the multivocality process needs to considered after any enactment. The example here suggests ways that productive multivocality may be enhanced through considering where we are in the research and design process.

My last comment relates to how this experience might fulfill the DBR goal of contributing to theory. I think there is potential here as well. One key goal of DBR is answering the question "under what circumstances." The design of APT suggests that particular kinds of discourse moves promote learning and engagement. The results of the studies presented here suggest yes, that is true, but we also need to consider the timing—specifically the relations between the macroscript, microscript, and what the students are actually doing and how these discourse norms might be sustained. This is important for the design of software agents, but might also be helpful in professional development for teachers. This seems a promising direction for moving forward—and if that is the case, then the multivocality process has certainly been productive.

References

Bielaczyc, K. (2006). Designing social infrastructure: Critical issues in creating learning environments with technology. *Journal of the Learning Sciences, 15*, 301–329.

Brown, A. L. (1992). Design experiments: Theoretical and methodological challenges in creating complex interventions in classroom settings. *Journal of the Learning Sciences, 2*, 141–178.

Clarke, S. N., Chen, G., Stainton, C., Katz, S., Greeno, J. G., Resnick, L. B., Dyke, G., Howley, I., Adamson, D., & Rosé, C. P. (2013). The impact of CSCL beyond the online environment. In N. Rummel, M. Kapur, M. J. Nathan & S. Puntambekar (Eds.), To See the World and a Grain of Sand: Learning across Levels of Space, Time, and Scale: CSCL 2013 Conference Proceedings Volume 1 — Full Papers & Symposia (pp. 105-112). June 15-19, 2013, Madison, Wisconsin: International Society of the Learning Sciences.

Cobb, P., Confrey, J., DiSessa, A., Lehrer, R., & Schauble, L. (2003). Design experiments in educational research. *Educational Researcher, 32*, 9–13.

Cobb, P., & Gravemeijer, K. (2008). Experimenting to support and understand learning processes. In A. E. Kelly, R. A. Lesh, & J. Y. Baek (Eds.), *Handbook of design research methods in education* (pp. 68–95). New York: Routledge.

Cress, U., & Kimmerle, J. (this volume). Successful knowledge building needs group awareness: Interaction analysis of a 9th grade CSCL biology lesson. In D. D. Suthers, K. Lund, C. P. Rosé, C. Teplovs & N. Law (Eds.), *Productive multivocality in the analysis of group interactions*, Chapter 27. New York: Springer

Dillenbourg, P. (2002). Over-scripting CSCL: The risks of blending collaborative learning with instructional design. In P. A. Kirschner (Ed.), *Three worlds of CSCL* (pp. 61–91). Heerlen, Netherlands: Open Universitat Nederland.

Dyke, G., Howley, I. K., Kumar, R., & Rosé, C. P. (this volume). Towards academically productive talk supported by conversational agents. In D. D. Suthers, K. Lund, C. P. Rosé, C. Teplovs & N. Law (Eds.), *Productive multivocality in the analysis of group interactions*, Chapter 25. New York: Springer

Dyke, G., Lund, K., & Girardot, J. (2009). Tatiana: An environment to support the CSCL analysis process. In C. O'Malley, D. Suthers, P. Reimann, & A. Dimitracopoulou (Eds.), *Computer supported collaborative learning practices: CSCL 2009* (pp. 58–67). Rhodes, Greece: International Society for the Learning Sciences.

Goggins, S. P., & Dyke, G. (this volume). Network analytic techniques for online chat. In D. D. Suthers, K. Lund, C. P. Rosé, C. Teplovs & N. Law (Eds.), *Productive multivocality in the analysis of group interactions*, Chapter 29. New York: Springer

Hmelo-Silver, C. E. (2003). Analyzing collaborative knowledge construction: Multiple methods for integrated understanding. *Computers and Education, 41*, 397–420.

Hmelo-Silver, C. E. (2004). Problem-based learning: What and how do students learn? *Educational Psychology Review, 16*, 235–266.

Hmelo-Silver, C. E. (2012, March). Une Approche de Recherche Basée sur la Conception afin de Comprendre l'apprentissage Complexe: A Design-Based Research Approach to Understanding Complex Learning. Invited presentation. Seminaire sciences de la'apprendre. Institute Français de L'Education. Lyon, France.

Hmelo-Silver, C. E., & Barrows, H. S. (2008). Facilitating collaborative knowledge building. *Cognition and Instruction, 26*, 48–94.

Hmelo-Silver, C. E., Liu, L., Gray, S., Finkelstein, H., & Schwartz, R. (2007). *Enacting things differently: Using NetLogo models to learn about complex systems*. Paper presented at the Biennial meeting of European Association for Research on Learning and Instruction, Budapest, Hungary

Hmelo-Silver, C. E., Nagarajan, A., & Chernobilsky, E. (2009). Two sides of the coin: Multiple perspectives on collaborative knowledge construction in online problem-based learning. In K. Kumpulainen, C. E. Hmelo-Silver, & M. César (Eds.), *Investigating classroom interaction: Methodologies in action* (pp. 73–98). Rotterdam: Sense Publishers.

Howley, I. K., Kumar, R., Mayfield, E., Dyke, G., & Rosé, C. P. (this volume). Gaining insights from sociolinguistic style analysis for redesign of conversational agent based support for collaborative learning. In D. D. Suthers, K. Lund, C. P. Rosé, C. Teplovs & N. Law (Eds.), *Productive multivocality in the analysis of group interactions*, Chapter 26. New York: Springer

Jacobson, M. J., & Wilensky, U. (2006). Complex systems in education: Scientific and educational importance and implications for the learning sciences. *Journal of the Learning Sciences, 15*, 11–34.

Resnick, L. B., Michaels, S., & O'Connor, M. (2010). How (well-structured) talk builds the mind. In D. Preiss & R. J. Sternberg (Eds.), *Innovations in educational psychology: Perspectives on learning, teaching, and human development* (pp. 163–194). New York: Springer.

Sandoval, W. A. (2004). Developing learning theory by refining conjectures embodied in educational designs. *Educational Psychologist, 39*, 213–233.

Sandoval, W. A. (2013). Conjecture Mapping: An Approach to Systematic Educational Design Research. Journal of the Learning Sciences. DOI: 10.1080.10508406.2013.778204.

Shavelson, R. J., Phillips, D. C., Towne, L., & Feuer, M. J. (2003). On the science of education design studies. *Educational Researcher, 32*, 25–28.

Stahl, G. (this volume). Interaction analysis of a biology chat. In D. D. Suthers, K. Lund, C. P. Rosé, C. Teplovs & N. Law (Eds.), *Productive multivocality in the analysis of group interactions*, Chapter 28. New York: Springer

Suthers, D. (this volume). Multivocal analysis: Objectives and origins. In D. Suthers, K. Lund, C. P. Rosé, C. Teplovs & N. Law (Eds.), *Productive multivocality in the analysis of group interactions*, Chapter 1. New York: Springer

Suthers, D. D., Lund, K., Rosé, C. P., Dyke, G., Law, N., Teplovs, C., Chen, W., Chiu, M. M., Jeong, H., Looi, C.-K., Medina, R., Oshima, J., Sawyer, R. K., Shirozou, H., Strijbos, J.-W., Trausan-Matu, S., & van Aalst, J. (2011). *Towards productive multivocality in the analysis of collaborative learning*. In H. Spada, G. Stahl, N. Miyake & N. Law (Eds.), Connecting Computer-Supported Collaborative Learning to Policy and Practice: Proceedings of the 9th International Conference on Computer-Supported Collaborative Learning (CSCL 2011) (Vol. III, pp. 1015-1022). Hong Kong: International Society of the Learning Sciences.

Part VII
Reflections

Section Editor: Dan Suthers,
University of Hawai'i

This final section of the present volume steps back from the specific data corpora and their attendant issues to reflect on what our experience tells us about the possibilities and conditions for productive multivocality. In Chap. 31, "Achieving Productive Multivocality in the Analysis of Group Interactions," the case study editors (Suthers, Lund, Rosé, and Teplovs) offer a self-contained summary of the project and its major insights and lessons suitable (for example) for use in graduate seminars. After briefly reviewing the motivations and history of the project, they summarize the five data corpora, the analyses done on them, the challenges for productive multivocality that were encountered, and what was learned from these case studies. The chapter concludes with strategies for productive multivocality.

In Chap. 32, "Methodological Pathways for Avoiding Pitfalls in Multivocality," Rosé and Lund present a conceptual model for thinking about multivocality and how it relates to methodological traditions, and use this model to reflect on what we have learned about multivocality (particularly with respect to difficulties) in the five case studies of our project. With an eye towards dissemination to the next generation of researchers, they also compare our experience of multivocality in the work reported in this volume with the experience of a class of graduate students in their attempt to use multivocality in learning to do discourse analysis. From this they derive advice for avoiding the pitfalls.

Chapter 33, by authors Dyke, Lund, Suthers, and Teplovs, focuses on "Analytic Representations and Affordances for Productive Multivocality." Most analyses use various data representations and transform these into successive analytic representations. Authors examine how representations are used and given meaning in analysis, with examples derived from the case studies of this volume, and discuss the implications of representational affordances for multivocality. They conclude with strategies for effective use of representations in support of productive multivocality.

Multivocality can and should lead researchers to examine disciplinary epistemological foundations that might otherwise be left implicit. In Chap. 34, "Epistemological Encounters in Multivocal Settings," authors Lund, Rosé, Suthers, and Baker examine what happened when different epistemologies encountered each other in the case studies and discuss what could or should have happened (e.g., when the epistemologies did not engage with each other, or the engagement was not productive). The chapter shows how epistemological encounters can help to bridge between isolated traditions that work on similar objects of study.

In Chap. 35, "Implications for Practice," editor Nancy Law and colleague Therese Laferriere discuss what aspects of this project may have meaningful implications for educational practitioners such as teachers. They find that multivocal interaction analysis as exemplified in the five case studies is sometimes only of interest to researchers, but in other cases can contribute to two types of relevance to practice: informing immediate pedagogical decision-making, and providing more general insight and understanding to the processes and outcomes of learning and knowledge building in collaborative contexts.

Finally, two prominent researchers who were not involved in our data corpora analyses efforts, Tim Koschmann and Claire O'Malley, conduct a dialogue about our research collaboration in Chap. 36: "A Dialog on 'Productive Multivocality.'" Taking our case studies as their data, they critically examine whether our conclusions are warranted and discuss connections to other similar efforts.

Chapter 31
Achieving Productive Multivocality in the Analysis of Group Interactions

Daniel D. Suthers, Kristine Lund, Carolyn Penstein Rosé, and Chris Teplovs

This chapter summarizes the outcomes of a long-term research collaboration in the analysis of group interaction, reported in detail in the other chapters of the volume within which this chapter is contained (Suthers, Lund, Rosé, Teplovs, & Law, this volume). We call this collaboration the "productive multivocality project," as it involved an effort to bring the various "voices" of multiple theoretical and methodological traditions into productive dialogue with each other. This project had multilayered goals. In addition to individual participants' goals, our collective goals were to bring these various traditions to bear on the problem of understanding interaction in educational settings, while deliberately reflecting on and modifying our collaborative research practices to learn how multiple traditions might "speak to" each other in a manner that transcends yet leverages their differences. That is, our efforts at multivocal analysis of interaction not only produced research results concerning the interactions of students being studied, but also served as the setting for a research program concerning our own interactions as researchers, intended to inform other attempts at collaboration in multidisciplinary areas of study. Therefore, this chapter (and the volume within which it is contained) can be read for different purposes. It is primarily a report on what we learned from the productive multivocality project: how to bring different traditions into dialogue with each other in a manner that is beneficial to the participating researchers and to progress in the field. It also

D.D. Suthers (✉)
Department of Information and Computer Sciences, University of Hawai'i at Manoa, 1680 East West Road, Post 309, Honolulu, HI 96822, USA
e-mail: suthers@hawaii.edu

K. Lund
ICAR Research Lab, CNRS—University of Lyon, 15 parvis René Descartes, 69007 Lyon, France

C.P. Rosé
Carnegie Mellon University, Pittsburgh, PA 96822, USA

C. Teplovs
Problemshift Inc., Windsor, ON, Canada

D.D. Suthers et al. (eds.), *Productive Multivocality in the Analysis of Group Interactions*, Computer-Supported Collaborative Learning Series 16,
DOI 10.1007/978-1-4614-8960-3_31,

contains a condensed report of a number of studies of interaction in educational settings (32 distinct analysts conducting 17 analyses across five data corpora), so can be read to survey their research results. The full volume (which contains detailed descriptions of data and analytic methods) may also be of value to students and researchers who want to learn about the range of analytic approaches available for their own data, perhaps to expand beyond the disciplinary boundaries of their own training.

We begin our report in this chapter by establishing the context: who we are, what we were trying to accomplish, how we went about it, and to whom else this work might be of interest. The project comprises five collaborations, each consisting of several researchers analyzing a shared data corpus. The body of this chapter contains summaries from the editors who facilitated each of these data-focused collaborations. We then step back from these specific analytic efforts and consider the lessons learned for productive multivocality in multidisciplinary areas of study.

Motivations for Multivocal Analysis

The nearly 40 researchers involved in this project work in the areas of collaborative learning, technology enhanced learning, and cooperative work, and share an interest in understanding group interactions, including interactions mediated by various technologies ranging from paper and pencil to online environments. We approach this topic from a variety of disciplinary homes and theoretical and methodological traditions that converge in research communities such as the *learning sciences* (Kolodner, 1991; Sawyer, 2006b), the study of human learning and instructional innovations for furthering learning, and its subfield of *computer supported collaborative learning* (CSCL) (Koschmann, Hall, & Miyake, 2001; Stahl, Koschmann, & Suthers, 2006), the study of how interaction leads to learning with the support of designed artifacts. Representatives of diverse disciplines, such as education, psychology, computer and information sciences, applied linguistics, pragmatics, anthropology, sociology, and others, are found within these areas of study and their communities, and bring their associated research traditions. Methods include statistical analyses of experimental data, iterative design-based research, conversation analysis, grounded theory, and social network analysis, among many other approaches. Theoretical traditions such as cognitivism, ethnomethodology, socioculturalism, and others may be found side by side in the same journal or conference proceedings. This state of affairs is found in many other fields as well, particularly in the social and behavioral sciences in which no single tradition has established primacy. The challenge is to convert multidisciplinarity—disciplines contributing independently, in an additive manner, into interdisciplinarity—disciplines in discourse with each other, contributing in an integrative manner (Choi & Pak, 2006).

One can argue that interdisciplinarity is *essential* for fields that focus on how social settings foster learning. Consider the range of conceptions about learning in social settings (Suthers, 2006). Theories differ on who or what they identify as the agent that learns. The agent of learning may be individuals, small groups, or networks (including networked individuals, communities, cultures, and societies). Theories also differ in epistemologies or what they identify as the process of learning. Prominent epistemologies include learning as acquisition of knowledge or skills (e.g., Anderson, 1981; Wenger, 1987); learning as intersubjective meaning-making (Koschmann et al., 2005; Suthers, 2006) such as argumentation (Andriessen, Baker, & Suthers, 2003), co-construction (Weinberger & Fischer, 2006), transactivity (Sionti, Ai, Rosé, & Resnick, 2011), or group cognition (Stahl, 2006); or learning as the process through which communities expand their collective capital (Scardamalia & Bereiter, 1991) or sustain themselves through changes in social participation and identity (Lave & Wenger, 1991; Rogoff, 1995; Wenger, 1998). Although our research may focus on only one of these levels of agency and epistemologies at a time, it is highly plausible that individuals and collectives participate in the foregoing forms of learning simultaneously. This raises a larger question of how learning takes place through the interplay between individual and collective agency, a question that *requires* coordinated analyses and theorizing from multiple perspectives.

Yet, the need to make a multidisciplinary field interdisciplinary does not automatically make it so, or even possible. We are still faced with the question: is diversity of methodological and theoretical perspectives in multidisciplinary fields, such as the learning sciences and CSCL, a blessing or a curse? Are multidisciplinary areas of study doomed to be "balkanized," with independent strands of thought and investigation coexisting in journals and conferences based only on their common concern with a nominal phenomenon (such as "learning"), itself being variously conceived? Or is productive interaction among the traditions possible such that diversity becomes strength, and if so, through what strategies?

The "productive multivocality project" has taken on this challenge by deliberately attempting to bridge theoretical and methodological divides for the analysis of interaction in learning oriented settings. We take the term *multivocal* from Bakhtin (1981), who used it to describe the presence of multiple "voices" that can be discerned in texts (see also Koschmann, 1999). Here the "text" is the collective discourse of a field such as the learning sciences. Our working assumption is that scientific and practical advances in the social sciences (where there is often no one dominant paradigm as in the physical sciences) can be enhanced if researchers working in multiple traditions—including traditions that some assume to be incompatible—make a concerted effort to engage in dialogue with each other, comparing and contrasting their understandings of a given phenomenon and how these different understandings can either complement or mutually elaborate each other. We do not expect to eliminate our differences and achieve full unification, but rather hope to find productive tensions in this dialogue. "Productive" does not necessarily mean "agreement" (Matusov, 1996), and controversies can be "deployed" towards productive ends (Latour, 2005). Diversity enables us to explore alternate approaches to understanding learning in interaction. However, this diversity is advantageous only to the extent that there is sufficient commonality to support dialogue between the "voices" and reach some degree of coherence in our discourse. We need strategies and *boundary objects* (Star & Griesemer, 1989) that form the basis for dialogue between theoretical and methodological traditions applied to the analysis of learning in and through interaction. This project undertook to find what constitutes effective boundary objects and how they may be leveraged.

Evolution of the Project

The multivocality project developed over a period of 5 years through a series of workshops at the International Conference on the Learning Sciences (ICLS) in 2008 and 2010, the Computer Supported Collaborative Learning (CSCL) conference in 2009, and the STELLAR Alpine Rendez-Vous (ARV) in 2009 and 2011, as well as ongoing collaborations between the authors of this volume and other colleagues extending through 2013. Here we briefly consider the evolution of our thinking; a more detailed history of the project may be found in Chap. 1 (Suthers, this volume-b).

Initially we had different goals. In our first workshop (ICLS 2008), motivated by the observation that advances in shared representations, methods, and tools lead to progress in many scientific disciplines, we sought to establish a common conceptual model and abstract transcript that might also form the requirements for shared analytic software. Workshop participants presented analyses conducted on their own data, and we used a collection of dimensions (see below) to describe commonalities across our analytic approaches. Commonalities were difficult to identify, and we found that the dimensions were more helpful for describing ways in which our approaches *differed* from each other. Yet, participants were excited about the opportunity to compare analyses and tools.

In our second workshop (CSCL 2009), we decided to provide a stronger basis for comparison of our approaches by having analysts from different traditions analyze the same data (two corpora not represented in the present volume). We again tried to find "common objects" between our analyses along a refined set of dimensions, but again found that the dimensions highlighted how the analyses differed rather than their commonalities. Although we had hoped that multiple analyses of shared data corpora would provide a basis for dialogue, the analyses presented were disconnected in part because the analysts were approaching these corpora with entirely different questions (colleagues have reported that this is a common point of failure in other similar efforts). This observation led to the innovation of "pivotal moments" in the next workshop.

Our third workshop (ARV 2009) continued the prior strategy of having researchers from different theoretical and methodological traditions analyze shared data corpora. We used data from a Knowledge Forum discussion in education (the basis of the case study in Chaps. 20–24), and from a Japanese primary school mathematics class (Chaps. 4–8). As before, we deliberately paired up analysts from different methodological traditions, in some cases challenging them with forms of data to which they were not accustomed. Most importantly, we addressed the prior mismatch in analytic objectives by asking analysts to identify the *pivotal moments* in the interactions recorded in the data. The definition of pivotal moments was purposefully left unspecified, providing a projective stimulus (as it were) that drew out different researchers' assumptions and insights and in some cases led to exciting comparative and integrative discussion. As expected, analysts differed in their conception and identification of pivotal moments, but these differences (as well as some congruencies) generated productive discussion of how learning arises from interaction.

In this third workshop we first articulated our core strategy for productive multivocality: assign diverse analysts to shared corpora and charge them with analytic objectives that are deliberately open to interpretation (e.g., "pivotal moments"). Also, our emphasis shifted from seeking "common objects" to seeking *boundary objects* (such as the corpora and pivotal moments) that support dialogue between different traditions. Boundary objects "have different meanings in different worlds but their structure is common enough to more than one world to make them recognizable, a means of translation" (Star & Griesemer, 1989, p. 393). We also found it to be useful to align various analytic results (e.g., to find overlaps and differences in pivotal moments identified), so wanted to explore tools for juxtaposing analyses for comparison.

In our fourth workshop (ICLS 2010), we recruited new analysts and new data corpora from a Group Scribbles mathematics classroom in Singapore (subsequently replaced) and university level chemistry study groups in the U.S. (Chaps. 9–13). We replicated the core strategy of having deliberately diverse analysts identify pivotal moments in shared corpora, although another analytic objective, that of identifying "leadership," played a similar role across some of the chemistry corpus analyses. Also, we used a software tool (Tatiana; Dyke, Lund, & Girardot, 2009) to support data sharing and more detailed comparisons of analyses. The primary strategy again proved to be productive, surfacing issues and insights exemplified by the case studies; see for example (Dyke et al., 2011; Suthers et al., 2011).

The final formal workshop of this project (ARV 2011) brought in additional analysts and two more data corpora. At our request, our Singapore colleagues replaced the mathematics corpus with another corpus on learning about electric circuits with multimodal use of Group Scribbles and physical manipulatives (Chaps. 14–19). A final corpus was introduced involving iterative design of a software agent supporting accountable talk in discovery learning of ninth grade Biology (Chaps. 25–30). The end of the 2-day workshop was structured to identify themes common across the case studies and thus surface practical, methodological and theoretical issues and strategies for productive multivocality that are highlighted in the present volume (especially in Chaps. 32–35).

Subsequent collaborations continued beyond ARV 2011, resulting in a number of papers (e.g., Chiu & Fujita, 2014; Dyke, Adamson, Howley, & Rosé, 2013; Dyke et al., 2011; Dyke, Howley, Adamson, & Rosé, 2012; Dyke, Kumar, Ai, & Rosé, 2012; Howley, Mayfield, & Rosé, 2013; Jeong, Chen, & Looi, 2011; Medina & Suthers, 2013; Oshima, Matsuzawa, Oshima, Chan, & van Aalst, 2012; Oshima, Oshima, & Matsuzawa, 2012; Oshima, Oshima, Matsuzawa, van Aalst, & Chan, 2011; Reynolds & Chiu, 2012; Schwarz et al., 2010; Suthers et al., 2011; Wise & Chiu, 2011a, 2011b).

Dimensions

The dimensions we used for describing analytic approaches are as follows: details may be found in Chap. 2 (Lund & Suthers, this volume):

Theoretical assumptions underlying the analysis. What ontological and epistemological assumptions are made about phenomena worth studying, and how we can come to know about them?

Purpose of analysis. What is the analyst trying to find out about interaction?

Units of action, interaction and analysis. In terms of what fundamental *relationships* between actions do we conceive of interaction? What is the relationship of these units to the unit of analysis? A unit of *interaction* relates two actions (at some level of description) in a manner that constructs a model of interaction informative for the desired unit of analysis.

Representations of data and analytic interpretations. What representations of data and representations of analytic constructs and interpretations are used to capture these units in a manner consistent with the purposes and theoretical assumptions?

Analytic manipulations taken on those representations. What are the analytic moves that transform a data representation into successive representations of interaction and interpretations of this interaction? How do these transformations lead to insights concerning the purpose of analysis?

The last two dimensions essentially treat analysis as a form of distributed cognition (Hutchins, 1995) by describing how analyses are achieved through transformations of representations in a system of analysts and analytic representations. These dimensions will be referred to occasionally in the summary that follows.

Analytic Traditions and Data Corpora

Diversity of theoretical and methodological traditions is a necessity for a project on productive multivocality. The persons we were able to recruit use methods as diverse as various forms of content analysis, conversation analysis, polyphonic analysis, semiotic and multimodal analysis, social network analysis, statistical discourse analysis, computational linguistics, and uptake analysis. Theoretical traditions include cognitivism, constructivism, dialogism, ethnomethodology, group cognition or intersubjective meaning-making, knowledge building, progressive inquiry, semiotics, and systemic functional linguistics.

In selecting the data corpora (case studies) and analysts for this project, we were entirely dependent on what participants were willing and able to make available, but sought individual data corpora that had potential to show learning through interaction, and were compelling as evidenced by the desire and willingness of multiple analysts to spend time analyzing that data. We were also cognizant of collective criteria for the corpus as a whole, seeking diversity of age levels, settings (formal

Table 31.1 Summary of data corpora

Chapters	Topic	Age and institutional setting	Interactional setting and media
4–8	Mathematics	6th Grade Japanese classroom	Face-to-face with origami paper and blackboard
9–13	Chemistry	Undergraduate Peer-led team learning	Face-to-face with paper and whiteboard
14–19	Electricity	Primary school in Singapore	Primarily face-to-face with circuit components and Group Scribbles software
20–24	Education	Graduate Level in Toronto	Asynchronous discussions in Knowledge Forum
25–30	Biology	Secondary school in Pittsburgh	Mixed face-to-face and online with Concert Chat and conversational agents in support of collaborative learning

and informal learning in schools, workplaces, and elsewhere), interactional media (face to face, synchronous, and asynchronous computer mediated communication), and domains or topics of study. The result is summarized in Table 31.1. We were successful in obtaining a diversity of topics, age groups, and interactional media within formal educational settings. The emphasis is on science and mathematics. We are missing (and unsuccessfully solicited) case studies in informal settings or workplaces, and representatives of sociocultural traditions. Also, most of our data involves small group interactions rather than large-scale networks or communities of learners. Yet, we believe that we have sufficient diversity to have encountered and grappled with major issues in achieving productive multivocality in the analysis of interaction.

In the following five sections, we summarize each data corpus, the analyses undertaken for each corpus, and the most salient lessons learned from reflecting on and sometimes revising our attempts to engage in multivocal discourse about our analyses.

Case Study 1: Pivotal Moments in Origami

Section Editor: Kristine Lund

The earliest data corpus we used that survived to be represented in this project was the Japanese sixth grade fractions data corpus gathered by Shirouzu (this volume-b, Chap. 4). Shirouzu wanted to track conceptual change by exploring the diversity of the paths learners take and the goals they pursue while participating in a collaborative learning task of understanding fractions by folding origami paper. Specifically, Shirouzu (this volume-a, Chap. 5) analyzed (1) where personal foci of learners originated, (2) what happened next in the interaction once a learner focused on something (were learners “doing” or “monitoring” tasks?), and (3) what learning outcomes these foci and interactions led to, including what learners remembered 6 months later. The editors of this volume chose two other researchers to analyze Shirouzu’s corpus, on the basis that such data would challenge each of them (for different reasons). We hypothesized that getting researchers to work outside of their comfort zone could lead to innovative results.

The second analyst, Trausan-Matu had previously worked only with on-line chat data and was asked to use his approach on video data of face-to-face interactions (this volume, Chap. 6). He used a framework centered on the concept of “voice” (Bakhtin, 1981) to look for patterns of interaction where learners converged or diverged (called inter-animation patterns). His goals were (1) to understand how group interactions could scaffold individual learning, (2) to evaluate different collaborative situations in terms of what the inter-animation patterns revealed about the quality of collaboration, thus giving us an indicator for choosing them, and (3) to leverage these inter-animation patterns as a way for teachers to manage students’ activity.

The third analyst, Chiu, had developed his own quantitative method called Statistical Discourse Analysis (SDA). We challenged him to apply this method to a corpus that more typically lent itself to qualitative analyses (this volume-a, Chap. 7). Although Chiu’s specific goal for this corpus was to statistically model how cognitive and social metacognitive processes influence the likelihoods of new ideas, correct ideas and justifications, his statistical method is a general one that (1) identifies breakpoints that divide the data into distinct time periods according to changes in variables, (2) tests whether these variables are linked to greater or reduced likelihoods of dependent variables of interest, and (3) tests whether these links differ across time periods.

Two different lenses were used as ways to compare analysts’ results, and both catalyzed lessons for productive multivocality (Lund, this volume, Chap. 8): the five methodological dimensions, and the comparison of “pivotal moments.” The next two sections discuss each lens in turn.

Comparing Analyses on Methodological Dimensions

Theoretical Assumptions

Both Shirouzu and Trausan-Matu made inferences about learners’ intra and intermental activities in the contexts of their respective frameworks. This prompted discussion on the extent to which inferences and interpretations are substantiated only by direct observables, or may also be substantiated by a narrative that is compatible with direct observables. Another point of discussion dealt with causality: which types of explanatory schema are more substantiated in the data, those that attribute causes to individual characteristics or those that attribute causes to the unfolding situation? A third issue dealt with being influenced to redefine key analytical concepts in existing frameworks (e.g., “voices” in the Bakhtinian framework, “utterance” and "adjacency pair" (Schegloff & Sacks, 1973) in conversation analysis). For example, Trausan-Matu redefined utterances to be not only verbal, but also inferred thought as well as different types of actions; and instead of being essentially individual or co-elaborated, they could be group generated, such as when all students moved their chairs in chorus in order to get closer to their origami papers. Pairs of utterances were considered to be adjacent even if shared ordering could just be inferred, for example between an external utterance (talk or action) and an internal one that was presumed by the researcher to be "thought" by the learner. However, the ramifications of such redefinitions were not discussed with recognized representatives of those frameworks. Indeed, being influenced to modify important analytical concepts without the benefit of within-tradition scrutiny of these modifications may be a danger of multivocality.

Purpose of Analysis

Each analyst sought in part to understand the role of individual participants or their contributions in the group interaction. But the way each researcher chose to qualify those roles or contributions was different. Shirouzu assessed the role of the individual in terms of task doer or task monitor during group work. Trausan-Matu assessed the individual's contributions indirectly through "adjacency pairs" of utterances in interaction with interlocutors, which were classified as either converging or diverging. Chiu assessed the individual's contributions through how a particular utterance type can lead to other or the same utterance types coming from others in the group (or from the same individual). Taking these different ways of qualifying the nature of the individual contribution together gives a more complete picture and incites an integrative approach that remains open to still other definitions of the individual's roles or contributions within the group.

Unit of Analysis/Unit of Interaction

All three researchers shared a focus on sequences of related turns in their analysis of the learners' interaction, but they each had a unique approach to this focus. Redefinitions of utterance and adjacency pair were first considered, and discussions about collaborative utterances also prompted Chiu to reexamine the pivotal moments he had originally defined as single conversation turns that divided the interaction into distinct periods. He now looked at them as longer more contextualized moments that when analyzed qualitatively, could be more fully understood.

Data Representations and Manipulations

The group working on the fractions dataset benefitted from being one of our earlier data corpus collaborations, providing them with sufficient time to compare and refine analyses. They also benefited from the fact that the data gathered was adequate for the requirements of all the researchers. It is not surprising that our assumptions about the phenomena we study influence how we collect data to represent those phenomena. It may be somewhat more surprising that researchers who have different methodological and theoretical assumptions are satisfied with the way data is collected and readied for analysis, as was the case for the researchers of this section. However, being able to refer jointly to the interaction being studied helped us to tease out the differences in both theory and method. One result of this joint reference came about after the three researchers realized they had all defined the same moment (of varying duration) as pivotal. Chiu in turn realized that this pivotal moment had the greatest impact on producing justifications in his framework and this information enabled him to discover his model had additional analytical power.

Comparing Pivotal Moments

Shirouzu's definition of pivotal moments evolved as he was continuously taking into account the other researchers' points of view and integrating their results into his own (Chap. 5). For example he originally defined pivotal moments as occurring when the student monitor reflects upon externalized traces, but upon seeing other analysts' results, he modified that to when the student monitor or the student doer reflects upon externalized traces. Trausan-Matu identified pivotal moments as changes in the degree of inter-animation of voices as illustrated by converging and diverging utterances (Chap. 6). Chiu operationalized a pivotal moment as a conversation turn that separates a portion of the conversation into two distinct time periods (before and after) with substantially different likelihoods of the focal variable (e.g., correct ideas) appearing in each portion (Chap. 7).

We compared pivotal moments in relation to the theoretical concepts that were mobilized by each researcher. That meant we performed three comparisons: (1) conceptual change (Shirouzu) as compared to inter-animation patterns (Trausan-Matu) (2) conceptual change (Shirouzu) as compared to frequency of new ideas (Chiu), and (3) inter-animation patterns (Trausan-Matu) as compared to frequency of new ideas (Chiu). These comparisons first served to illustrate where researchers coincided with their definition of pivotal moments, even if for different reasons, and this in turn enabled discussion of how to build bridges between their underlying theoretical concepts while also solidifying their different views. Although pairs of researchers overlapped in defining pivotal moments, only one moment (of varying duration) was deemed pivotal by all three researchers. In this pivotal moment, the group of learners noticed conceptually that the solutions had the same areas, but that their shapes and production methods were different. The differential inter-animation pattern focused on "same" versus "differ," and this was also where there was a drop in new ideas in Chiu's analysis.

Multivocality Changes the Researcher

Shirouzu showed particular interest in understanding what other analysts had to say about his corpus, and throughout our collaboration strove to integrate their viewpoints into his own analysis, which kept evolving. This was possible because either he could reinterpret the results of others in his own framework or because he could appropriate their epistemological views on a different time scale. An example of the latter is when Shirouzu at first did not agree with Trausan-Matu that it was justifiable to use ability or character traits to explain behavior, preferring to rather explain behavior by the unfolding of the interaction between participants. However, he came to understand that perhaps ability or character traits could have more explanatory power if learners were followed over the long run, across multiple pedagogical tasks. Our collaboration also convinced Chiu that quantitative and qualitative methods can be used in concert to obtain a more complete understanding of group interactions. Finally, although Trausan-Matu interacted intensively with the other analysts, his research seemed to be most influenced by the new type of corpus with which he was confronted, provoking modifications to the definitions of his analytical concepts (i.e., "utterance" and "adjacency pair") and thus his epistemological stance. In general, researchers in the fractions section adopted an integrative stance, actively searching for ways in which the analyses of others could become either complementary to their own or illustrate the limits of their own approach.

Lessons Learned for Productive Multivocality

The collaboration around the fractions data corpus showed how multivocal analyses help researchers to gain new insights by modifying their epistemological presuppositions about human interaction (e.g., when could it make sense to explain human interaction by a learner's individual characteristics rather than by aspects of the situation?), their assumptions about learning (e.g., how does conceptual change come about?) and their analytical methods (what are new ways of measuring individual participation in the collective?). All analysts measured the quality of the collaboration in some way (Lund, 2011) but with different indicators and units of analysis, using both qualitative and quantitative methods that were adapted to the small size of the dataset. By comparing methodological dimensions and definitions of how particular moments of interaction were pivotal for learning over essentially 3 years of collaboration, the analysts in this section reconsidered both their theoretical and methodological positions, thereby surpassing the initial limits of their approaches.

Case Study 2: Peer Led Team Learning for Chemistry

Section Editor: Carolyn P. Rosé

We have learned through our experience on this project that multivocal approaches to analysis of collaborative learning interactions challenge our individual operationalizations of complex constructs such as social positioning, idea development, or leadership by revealing the ways in which they are each limited. Some of the most exciting insights that came from the multivocal analysis of the Peer Led Team Learning (PLTL) for Chemistry dataset were related to the realization that what might sound like very similar conceptual models and operationalizations from a high level may lead to very different codings and therefore different interpretations of the same data. These moments of insight provided the opportunity to challenge one another to think more deeply about the assumptions we were each making.

In the PLTL Chemistry section, we compared the discussion styles of two different PLTL groups (named the "Gillian group" and the "Matt group," after their leaders) as they solved a chemistry problem related to de Broglie's equation. PLTL is a collaborative learning approach that has been used on many college campuses, especially in large lecture classes in departments of chemistry. Prior work in the learning sciences community has shown that engaging in collaborative discourse contributes to deeper conceptual understanding and increased retention and transfer (Engle & Conant, 2002; Greeno, 2006; Sawyer, 2006a; Scardamalia & Bereiter, 2006). However, the challenge in promoting collaboration among undergraduates in large science lecture based courses is that such courses tend to be focused on transmission of knowledge to individuals (Seymour & Hewitt, 1994). Consequently, there has been a paucity of research on students' collaborative discourse practices in college science settings. PLTL has been designed to facilitate chemistry literacy and success for all students, including but not limited to chemistry majors, by supplementing the lecture with study group sessions that offer opportunities for active and collaborative learning (Gosser et al., 2001; Gosser & Roth, 1998; Sarquis et al., 2001; Siebert & McIntosh, 2001). Peer leaders are selected from undergraduate students who have successfully received an A in the class in an earlier semester. A peer leader is selected for each group of six to eight students who meet for 2 h once per week to solve chemistry problems designed by course instructors.

Several studies have offered impressive evidence that PLTL improves learning (Gafney & Varma-Nelson, 2008; Hockings, DeAngelis, & Frey, 2008; Tien, Roth, & Kampmeier, 2002). However, prior to this study, researchers had not done a detailed investigation of the discussion practices employed by peer leaders and students, and between the students themselves, that mediate the effect. To better understand the mechanisms that make PLTL work, Sawyer, Frey, and Brown (this volume-b, Chap. 9) videotaped three PLTL sessions for each of 15 veteran peer leaders over the course of one semester. The dataset we analyzed included transcripts of two PLTL groups as they solved the same problem. Analysis proceeded in two waves, with the multifaceted concept of leadership guiding the integration of findings from the first wave, and more of a stylistic approach to analysis of collaborative problem solving guiding the second wave.

The three analysis chapters in this section fall on three distinct places on the continuum between highly quantitative and highly qualitative. At the qualitative end, Sawyer and colleagues approached the conversations in a situated, turn-by-turn fashion. Their analysis in Chap. 10 (Sawyer, Frey, & Brown, this volume-a) includes many excerpts from the corpus in raw form, along with commentary and reflection on the substance of the interactions, including an assessment of the knowledge that was communicated and the manner in which it was communicated. At the opposite end of the spectrum, Oshima and colleagues (this volume, Chap. 12) applied a social network analysis approach to the same data. Connections are made between the evolution of the topology of the network representation over time and a theoretical framework from the Knowledge Building community. In between these two end points stands the coding and counting approach of Howley and colleagues (this volume, Chap. 11), which contains analyses from two different multidimensional coding schemes, including the Souflé analysis lead by Rosé and the CSM analysis lead by Strijbos. Both of these coding schemes consist of the same three dimensions, namely Cognitive, Relational, and Motivational, and were thus expected to reveal similar insights about the data, yet the frameworks actually brought out different insights from the data. Like the Sawyer chapter, the Howley chapter contains a number of example excerpts from the corpus. However, like the Oshima chapter, the inferences are made primarily based on quantitative measures as viewed through the lens of the theoretical frameworks that underlie the coding schemes. Only the Sawyer analysis includes an extended thick description of the two problem solving experiences. In their own way, all of the analysts investigated the ways in which participants made their reasoning public by articulating their reasoning, and commenting on and building on one another's reasoning. Each analytic approach made its own characteristic fine grained distinctions in the manner in which reasoning was articulated, shared, and integrated.

The rich contextualized turn-by-turn analysis of Sawyer and colleagues serves as a counter-point to the other analyses, which are quantitative in nature and attempt to draw conclusions from patterns found within a structure imposed on the data as an analytic lens. The qualitative and the quantitative analyses in this section of the book challenge one another's interpretations. Whereas the qualitative analysis has the benefit of contextual knowledge and human insight, it is limited with respect to the ability to distinguish between the typical and the idiosyncratic. On the other hand, the quantitative approaches provide the machinery to not only make but also quantify this important distinction, but they are prone to misinterpretation caused by over-generalization between instances of behaviors that are treated as the same type. The role of the qualitative analysis was to challenge the treatment of the significance of individual events within the three quantitative analyses. The corresponding role for the quantitative approaches was to challenge summative conclusions drawn within the Sawyer et al. analysis.

Most of the analysts agreed that the Gillian group was more conceptually oriented and interactive whereas the Matt group was more narrowly focused on problem solving. The Oshima, Strijbos, and Sawyer analyses all make this argument using their own style of analysis. Thus, these three analyses can be seen as providing a great deal of convergent evidence that this conclusion has support in the data. The Rosé analysis presents a slightly different view, however. In both the Gillian and Matt groups the same number of reasoning statements were uttered; however, this number is a smaller percentage of moves in the Gillian group than in the Matt group. Thus, while the articulation of reasoning might be framed differently in the two groups, both groups are equally open with one another about their reasoning, and the Matt group is more singularly focused on reasoning. In the Gillian group there is more packaging around the reasoning. This packaging might be what makes the conversation hang together better and appear more highly inter-connected. And yet, the fact that the same number of reasoning statements were uttered raises questions about what the conceptual versus procedural contrast signifies. When we probe more deeply, looking at transactive contributions (i.e., those that build on or comment on a previous reasoning statement) rather than simply reasoning statements, we see that a slightly higher proportion of reasoning statements in the Matt group are transactive. These mainly took the form of comparisons between problem solving approaches. While these comparisons were written off as simply procedural by many of the analysts, the Rosé analysis calls these out as places where the students are considering each other's approaches and comparing them with their own in order to determine how best to solve the problem.

The differences between what these analyses bring out about the group discussions raise questions about what is desirable in PLTL groups, and prompt further reflection on concepts as established as group knowledge integration. All of the analysts valued students sharing their reasoning and working together to refine that reasoning. Not all of the analysts agreed with what that should look like on the surface. Through this process the analysts became aware as a community that we have work to do before we will be able to assess important qualities of collaborative problem solving that we may have previously thought we already had a handle on as a community.

Case Study 3: Multimodality in Learning About Electricity with Diagrammatic and Manipulative Resources

Section Editor: Dan Suthers

The Electricity data corpus provided by Chen and Looi (this volume, Chap. 14) derives from a primary grade classroom in Singapore using the Group Scribbles collaborative whiteboard (Roschelle et al., 2007) and electrical components to learn about electric circuits. This data corpus is unique among those in this volume in that it mixes face-to-face interaction, collaborative physical manipulation of objects, and computer-mediated interaction. The corpus was analyzed by Looi, Song, Wen, and Chen (this volume, Chap. 15) using uptake and content analysis guided by a theory of progressive inquiry, Medina (this volume, Chap. 16) using uptake analysis with an ethnomethodological orientation towards unpacking group accomplishments, Lund and Bécu-Robinault (this volume, Chap. 17) focusing on coherence and conceptual change in translations between media and modes motivated by a theory of semiotic bundles, and Jeong (this volume, Chap. 18) using content analysis under her conception of "group understanding." The analyses focused on two major themes: what evidences understanding, and practices of multimodal interaction across various media. Understanding was analyzed via uptake structures, the coordination of multimodal acts in multiple media, and/or the contents of resulting artifacts in relation to canonical physics. Multimodality was understood in three ways: in terms of the unique affordances of each medium, conceptual coherence sustained through translations across media, and group accomplishments through simultaneous coordinated use of media.

Transcripts and Other Analytic Representations as Boundary Objects

Each analytic team created transcripts and other analytic representations as needed for their own purposes, and many analysts examined the six video records directly (synchronous viewing was made possible by Tatiana: Dyke et al., 2009). Subsequently, we tried to align the various analytic representations (including transcripts) so that we could compare results, but found it difficult to align the different units of analysis, which included participants' actions in different media; episodes defined by ideational, inscriptional, or other activities; the completion of artifacts; and translations between media/modes, among others. Discrepancies between the analytic teams' requirements for transcripts highlighted for us the purpose driven and hence theoretical nature of transcripts (Duranti, 2006; Ochs, 1979), and raised fundamental questions concerning the role shared transcripts play in a multivocal analytic collaboration. A shared transcript is a means to an end. If a shared transcript can be agreed on, it will be easier to compare analyses, but different disciplinary requirements placed on transcripts may preclude this agreement. In either case, the process of attempting to create aligned representations can lead analysts to become aware of dimensions of the data they might not otherwise have considered, and expose essential differences in viewpoints. Some of our own insights came when the group facilitator brought analyses into alignment for comparison and confronted the group with congruences and discrepancies, some of which are discussed below.

Comparing Pivotal Moments

As might be expected, definitions of "pivotal" differed across teams. For Looi, Song, Wen and Chen (Chap. 15), a contribution is pivotal if it shifts the direction of subsequent events, as evidenced by changes in content. Looi et al. were the only analysts who distinguished *manifest* pivotal moments: those that were actually taken up; and *latent* pivotal moments: those that had the potential to shift the direction but were not taken up. Perhaps more attention needs to be paid to the latter in educational research, as they may offer opportunities for interventions by practitioners. Lund and Bécu-Robinault (this volume, Chap. 17) work with Multimodal/Multimedia Reformulations (MMRs), or translations from one medium to another, as the units of analysis. An MMR is pivotal when it evidences conceptual change towards canonical physics, or progressions towards more complexity while sustaining the same level of coherence. Medina (this volume, Chap. 16) sought "pivotal sequences of interaction" rather than moments. A pivotal sequence is convergence of uptake in enacting an innovation. Jeong (this volume, Chap. 18) defined pivotal moments as "moments when changes in group understandings occurred both in terms of the development of the domain understanding and intersubjectivity," but found that in practice this is an incremental process. However, our most productive discussion came out of comparing interpretations of specific events rather than our general conceptions of pivotalness.

The Role of a Teacher Intervention

One issue surfaced by comparing analyses concerned a case where the teacher interrupted a group experiment with a two-battery, two-bulb configuration. The teacher wanted students to draw the circuits first, and then experiment to see whether they worked, and pointed out that there was "no draft" for the experiment in the Group Scribbles board. Medina, who was constructing an emic account of group accomplishments, initially saw the teacher's intervention as a disruptive move that "splintered the group's intersubjectivity and re-prioritized how they proceeded to manage their interaction." Looi, Song, Wen, and Chen also marked this intervention as pivotal, but in a positive light: it "changed the direction of the group inquiry path from trying to do a new experiment to reflecting and conceptualizing their working theories of how to light a bulb and the mechanism of the circuit." In our discussion, analysts came to agree that both views were correct. The teacher disrupted the group's indigenous activity for exogenous objectives, and the group did not get to explore "two batteries, two bulbs" until later, but by imposing the activity structure of the original lesson plans the teacher led the participants to represent their understanding in a different medium that exposed one individual's conceptual weak point, prompting other students to help him. Medina reports that this exchange led him to a more dimensional understanding of the pragmatics of individual and group dynamics.

Misconception or Innovation?

Another point of contention emerged concerning whether a student we call "Bruno" had a "canonical conception" of how to connect a wire to the negative end of a battery. Based on his diagramming (see inset for one example), Lund and Bécu-Robinault concluded that Bruno did not have a complete understanding. They wrote, "he does not clearly show that the wire going to the minus pole of the battery actually touches the pole and does not just end at the bottom left of the battery." Suthers (the facilitator) found the diagram to be inconclusive, because an effective strategy for attaching a wire to a battery at its flat negative end is to press the wire flat against the battery. He returned to the video record, and found that Bruno drew the wire into the middle of the first rectangular end of the battery rather than the corner. Examining Bruno's manipulation of physical batteries, three times Bruno clearly placed the wire flat against the flat end of the battery, establishing a solid connection. In one case, Bruno pressed the batteries down on the table with the wire flat underneath, thereby accomplishing with the table what would otherwise need a third hand. This innovation made the entire battery configuration more stable, supplying the physical contingencies for a subsequent innovation (two batteries, two bulbs) enacted by the group, as detailed by Medina's analysis. Thus, the video record of the actual production of artifacts suggests that Bruno had a good understanding of how to connect the wire, enabling an innovation upon which a subsequent group accomplishment was contingent. However, Lund and Bécu-Robinault are not making a claim about whether the drawing correctly showed the actual situation in the physical environment. Rather, they are concerned with whether Bruno has shown representational competence in diagramming circuits as a physics student trained in circuit modeling would. Our dialogue surfaced the essential issue: whether Bruno's understanding should be assessed based on domain standards for abstract representations that are external to the interaction, or based on an emic display of understanding by successfully lighting the bulb in a physical configuration that led to a group innovation.

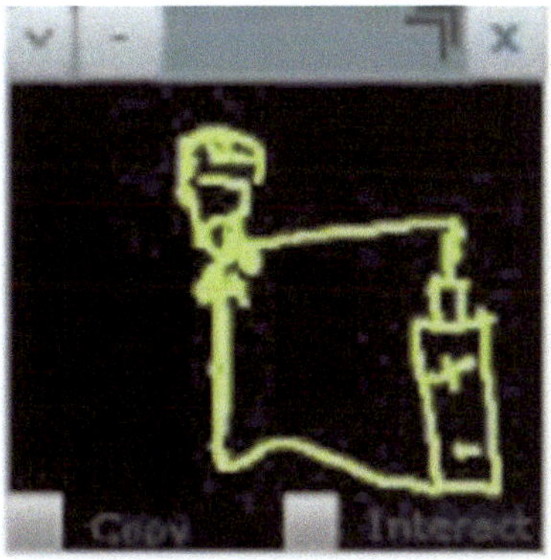

Agency and the Distribution of Activity Across Modalities

Stepping away from the interpretation of specific events, we also uncovered theoretical issues indicated by how the analyses construe the objects of study. Learning has been theorized as taking place at different granularities of *agency*, including individuals, small groups, or larger networks such as communities as the agent and locus of learning (Greeno, Collins, & Resnick, 1996; Suthers, 2006). Is the unit of agency individuals, the group acting or analyzed as a unit, or something in between, such as intertwined individual agencies from which group agencies emerge? A related question is how *activity* distributes across the available mode/media. Acts in different media can be treated as separate realms between which participants translate, or as simultaneously coordinated in a unified activity. For Jeong, the agent of interest is the group, but she does not assume that there is a singular group cognition or shared understanding. Jeong's approach of characterizing group understandings through the progression of individually produced conceptual artifacts is similar to how archeology characterizes a culture through examples of artifacts produced by individuals. Looi et al. analyze acts by individuals in different media separately, but identify shifts in these acts over time that evidence the group's progressive inquiry. Lund and Bécu-Robinault draw on a theory of semiotic bundles (Arzarello, 2004) that emphasizes coordination of multiple modes/media, but examine how translations across modes/media evidence canonical understanding ("coherence") in a manner following distributed cognition (Hutchins, 1995). The analysis is diachronic because they are concerned with conceptual change over time, and the ability to translate between representations is a key indicator of competence in the domain. Medina, influenced by Goodwin (2000), examines how simultaneously converging acts in multiple modalities/media are brought together in group accomplishments. For Medina, agency exists across the simultaneous coordinated actions of individuals, and activity is distributed across coordinated use of the modalities.

Lessons

The following lessons for achieving productive multivocality can be drawn from this experience. We found that interaction between multiple analysts with different viewpoints can drive advances in both analysts' individual and collective understanding. The potential value of collaborative data analysis has long been established (e.g., Jordan & Henderson, 1995), but there are additional opportunities as well as challenges for multivocality in multimethodological settings. Different analytic approaches make different demands on transcripts, and there are potential opportunities in the negotiation of shared transcripts as boundary objects. It is useful to attempt to map between analytic representations, or learn from the intrinsic incommensurabilities that prevent such a mapping. While abstractions such as transcripts, snapshots, and analytic structures play important roles in each analytic tradition, it may be necessary to go back to the original data record to resolve disputes. To do so, it is essential that the abstractions used by analysts index back to the data record in some shared coordinate system such as time. Not all of the benefits are found in the attempt to align and compare analyses. Some of the productivity of multivocality is found by comparing how analyses constitute the object of study, thereby making alternative theoretical conceptions explicit, such as in our discussion of the distribution of agency and activity across persons and media. Finally, a third party tasked with facilitating multivocal dialogue plays an important role in achieving the above.

Case Study 4: Knowledge Building Through Asynchronous Online Discourse

Section Editor: Chris Teplovs

This dataset provided by Fujita comes from an online graduate education course that was conducted exclusively online using Knowledge Forum at the University of Toronto. The instructor organized the course into 13 separate folders or "views" that corresponded to weekly topics. The discourse data consisted of 1,330 asynchronous online discussion messages or "notes" contributed by 17 graduate student participants, the researchers and the instructor. The dataset was part of a larger designed-based research study, the goals of which were twofold: to improve the quality of online graduate education in this particular instance, and to contribute to the theoretical understanding of how students collaborate to learn deeply and build knowledge through progressive discourse.

The dataset was first shared with other analysts at the 2009 Alpine Rendez-Vous, with the goal of gaining new perspectives on collaborative learning through the application of alternate analytic approaches. Not all analytic approaches undertaken in our workshops yielded promising results, and the three chapters included in this volume represent only the more successful analyses. It took several iterations of analyses and tool development to yield the current chapters. The multiple points of contact afforded by the series of workshops at multiple Alpine Rendez-Vous and several other conferences fostered the exchange of ideas and clarification of assumptions that resulted in a deeper understanding of collaborative learning.

The three analyses differed in their purposes: Teplovs and Fujita's analysis (this volume, Chap. 21) sought to examine the relationship between social interaction and the semantics of the written contributions of students. Law and Wong's analysis (this volume, Chap. 22) was motivated by the desire to create a dashboard designed for teachers that would represent students' progress. Chiu's analysis (this volume-b, Chap. 23) sought to use Statistical Discourse Analysis (SDA) to identify pivotal moments in the discourse, as he did with the Japanese Fractions corpus, but here taking on the additional challenge of data from asynchronous interaction.

Analyses by Teplovs and Fujita and Law and Wong share the data provider's theoretical underpinning of knowledge building, while Chiu's analysis applied a method whose theoretical assumptions are broadly compatible with other approaches. Furthermore, all three analyses shared the goal of identifying and exploring "pivotal moments," however broadly defined, and all three analyses shared a temporal aspect of analysis even if the specific units of analysis differed. It is perhaps these shared elements that allowed Fujita to reflect on the implications of productive multivocality for design-based research in her discussion (Fujita, this volume, Chap. 24).

The hallmarks of multivocality are attractive: multiple voices, commonality, and coherence. Whereas those may be the ultimate goals, the process by which multivocality is pursued can be difficult. In this collaboration, difficulty stemmed from a variety of sources. For example, asynchronous interaction does not result in a single record of interaction that is readily accessible (such as a video recording in the case of synchronous small group interaction). The need to produce high-fidelity representations of asynchronous data that could be understood and analyzed by researchers using different tools took considerable effort, and understanding what each other had accomplished as a result of their analyses was sometimes challenging. High-fidelity representations of asynchronous data, particularly those that convey information about the relationships between entities such as people and documents, often make use of network graphs. Network graphs are often difficult to interpret and care needs to be taken to explain what they show. More traditional representations, such as line graphs and bar charts of participation metrics tend to be more understandable by many researchers. Productive multivocality can be hindered by a lack of common understanding around representations.

In retrospect, it may be the case that each voice in our quartet paid little heed to other discrepant voices, but preferred to focus on aspects that resonated with existing beliefs and perspectives. This is not to say that important gains were not made: Chiu made improvements to SDA to facilitate its use with asynchronous discourse data, Teplovs and Fujita created new representations of interactions, and Law and Wong made progress toward their goal of an analytic tool for teachers. However, theoretical stances were for the most part reified rather than revised. The openness to change, the willingness to question one's assumptions, and the commitment to engage in productive dialogue and debate waned over time.

The authors of this section made repeated overtures indicating that they believe there is merit in pursuing the integration of approaches and techniques. For example, it was suggested that the analysis using KISSME (which visualizes patterns of interaction and community structure by combining social network analysis and latent semantic analysis; Chap. 21) could be repeated using week-by-week student models and the results compared to the findings made using other analyses. This sustained collaboration has failed to materialize at least in part due to lack of resources as well as the pioneering nature of multivocal analysis that goes against existing researcher practices. The problem of limited resources is pervasive and largely insoluble. However, researchers who are new to multivocal analyses can benefit from lessons learned in this volume.

Perhaps most important is a commitment to engage in an iterative process of research amongst the multivocal analysts. This commitment requires: (1) multiple attempts at analyses and (2) carefully considering the role of the data provider. Rather than following a process by which a dataset is presented to a number of analysts who may or may not collaborate with the data provider in its analysis, a more productive approach would include both the reporting back to the data provider and to other analysis the results of the first iteration of analyses as well as a commitment to at least one more round of analyses that would take into consideration feedback and reactions from the data provider and other analysts. By doing so, analysts would be more likely to pick up improvements from each other, and the workload of the data provider may also be decreased.

Case Study 5: A Data-Driven Design Cycle for Ninth Grade Biology

Section Editor: Carolyn P. Rosé

The unique focus of the fifth case study is using multivocality to enhance a data-driven design process by offering a multifaceted understanding of how interventions under development interact with group functioning. This raises unique challenges in sharing the task of data interpretation. While secondary data analysis is becoming more commonplace in the learning sciences, sharing pilot data is far less typical, especially pilot data from experiments gone awry. But the use of process analysis to inform iterative development of interventions for supporting collaborative learning is increasing and has great potential for impact within the field of CSCL. So within that scope, it is important to explore the potential value of multivocal analysis above and beyond a univocal process analysis approach. Do the benefits outweigh the costs in this type of setting? In this case study, four analysts offered their interpretation of what went right and what went wrong in a pilot evaluation of a new form of software-agent based support for scientific discovery learning. From this investigation we learned what we may or may not be missing in analysis of process data from prototypes by conducting the analysis from one specific theoretical and methodological lens. The discussant offers an interpretation of the multivocal process and its implications for a design-based research process (Hmelo-Silver, this volume, Chap. 30).

The study that provides the shared data for this case study is referred to as the Cell Model study, because it involved ninth grade biology student groups who were exploring how cell models work (Dyke, Howley, Kumar, and Rosé, this volume, Chap. 25). The broader project this study was part of builds on a large body of work that has shown that certain forms of classroom interaction, termed Academically Productive Talk (APT), are beneficial for learning with understanding (Resnick, Asterhan, & Clarke, in press). This work has also shown the crucial role of the teacher in facilitating these discussions. The academically productive talk form of classroom interaction is one in which a facilitator (or a software agent) poses a question that calls for a relatively elaborated response (e.g., both a solution and a reason for the solution), and then presses the group to build on or challenge these ideas, with the purpose of keeping student reasoning at center stage and increasing student ownership of ideas. The goal of the project is to increase the extent to which these APT based practices are used within typical urban classrooms, using professional development with teachers and Computer Supported Collaborative Learning (CSCL) experiences for small groups of students as tools for reshaping the classroom culture (Rosé & Tovares, in press).

The development goal of the project is a conversational software agent that provides support for collaborative learning by mimicking practices from the APT theory of classroom discussion facilitation (Adamson, Dyke, Jang, & Rosé, in press; Michaels, O'Connor, & Resnick, 2008). Researchers conducted two complete cycles of design development, deployment, and analysis over the 2-year project. The second year design drew on lessons learnt from the multivocal analyses presented in the chapters within this section, which were conducted after the first year study (Dyke et al., 2013). This development effort builds on an earlier history of successful deployment of intelligent conversational agents for support of small group learning. Technology for dynamic support for collaborative learning has matured both in terms of its ability to monitor collaboration through automatic collaborative learning process analysis as well as to offer context appropriate support for effective participation in groups (Kumar & Rosé, 2011). The novelty of the Cell Model study was an exploration of how one might design conversational agents that employ APT practices as scaffolding for on-line collaborative learning discussions, which were eventually successful at leading to increases in learning in the second year of the data collection effort (Adamson, Ashe, Jang, Yaron, & Rosé, 2013; Dyke et al., 2013).

Despite the critiques of analysts (below), the project did experience some success, even in the first year. Within the context of this district-wide design study, the focus was initially mainly on teacher training. Early on, a relatively slow rate of adoption by teachers led the researchers to consider alternative means to accustom students to APT in order that they might be more responsive to the teacher's classroom scaffolding. One approach they took was to introduce APT practices to students in small group activities facilitated by conversational agents. These activities played an important enabling role in the professional development effort: students came to whole class discussions better prepared and able to engage in intensive discussion after the CSCL activities, which then elevated the teacher's adoption of APT by 1.7 standard deviations (Clarke et al., 2013).

The analysts came to the task with a variety of disparate theoretical assumptions and methodological tools. Howley, Kumar, Mayfield, Dyke, and Rosé (this volume, Chap. 26) use a visualization tool to examine patterns of linguistic codes in a three dimensional analysis framework to show how linguistic evidence of social positioning within groups pinpoints negative student experiences. Cress and Kimmerle (this volume, Chap. 27) follow with an ethnographic study that examines the collaborative setting in terms of the desired (but missing) affordances for group awareness. Stahl (this volume, Chap. 28) and Goggins and Dyke (this volume, Chap. 29) focused on roles within the interaction, and specifically how the role taken by the agent may have limited the opportunities for leadership role taking of students within the group discussions. Stahl's analysis draws on ethnomethodology, and Goggins and Dyke integrate ethnographic analysis methods and social network analysis methods.

One major discrepancy between some of the analysts and the developers was in their assumptions concerning the ideal role for a facilitator in a collaborative learning encounter. All of the analysts agreed that the intervention did not work as intended. The biggest criticism from the analysts other than the developers was that the agent dominated the conversation too much—intervening too frequently, and with turns that were too long. This view of the ideal role of the facilitator that some of the analysts brought with them into the multivocal discussion was in contrast to the theory of APT based instruction, where the instructor plays a very integral role in the classroom discussion, and where much evidence exists to support the effectiveness of this form of classroom facilitation. Debates about the ideal role of a group discussion facilitator pervaded the entire multivocal process.

At the same time, each analyst focused on different questions, sometimes much different from the focus of the intervention designers. In particular, only the analysis conducted by the designers themselves focused on the specific ways in which the behavior of the conversational agents affected the interaction between students in intended versus unintended ways. Some analysts (Stahl and Cress) ignored the experimental manipulation altogether, not seeing any of the differences between conditions as relevant to their concerns, although significant effects of the manipulation were reported in the analysis conducted by the designers. Some of the points raised by analysts were issues that the designers were aware of but chose not to address in the first iteration. This points to special care that must be taken when an analyst participates in secondary data analysis, especially as part of a design-based research process where the designers are already aware of many obvious limitations of the intervention and would benefit more from insights related to limitations they are not already aware of.

On the other side, the other analysts challenged the designers to see beyond their own research questions to the ways in which infrastructure that was the foundation for the experiment itself was flawed, especially in the way it sometimes failed to avoid clashes between multiple aspects of support that were simultaneously active during a portion of the discussion. As Hmelo-Silver pointed out in Chap. 30, in these data and analyses there are many different units of analysis discussed and explored that connect what is going on in individual utterances, larger episodes, overall interactions, and student outcomes. A valuable contribution of multivocality in this context is that the combination of analyses provided insight into learning as a complex system, with interaction among different levels of the system. The feedback from the analysts to the designers was taken to heart in a redesign in terms of significantly extending the capabilities of the architecture for managing dynamic support as well as a major adjustment of the role of the agent in the conversation, which was ultimately successful. This success suggests that multivocal analysis can be valuable in challenging designers to break out of their own box and view their data more broadly. It can even lead to questioning the assumptions of their entire enterprise. In this sense, multivocality is "risky," as that is not what designers hope to gain from recruiting other analysts to their project. Nevertheless, it can be very valuable.

Reflections on Productive Multivocality

The five data-focused collaborations that we have just summarized were each individually and to varying degrees a locus for their own advances concerning the specific matters of research and practice at hand and the development of greater understanding among members of different analytic traditions. Yet they also collectively constitute the "data" for our larger enterprise, that of developing strategies for making data-focused dialogue between analytic traditions useful, and understanding implications of these attempts at "productive multivocality" for theory and practice. In the remainder of this chapter we summarize our conclusions concerning the larger enterprise. After positioning this project relative to the mixed methods literature, we offer a synopsis of strategies that we identified for productive multivocality that may be of use to others.

Mixed Methods

An obvious reference point for our project is "mixed methods" research (Frechtling & Sharp, 1997; Johnson & Onwuegbuzie, 2004; Tashakkori & Teddle, 2003). Mixed methods have been defined as "the class of research where the researcher mixes or combines quantitative and qualitative research techniques, methods, approaches, concepts or language into a single study." (Johnson & Onwuegbuzie, 2004, p. 17). As reflected in this quote, mixed methods has generally been conceived of as a mixture of "quantitative" and "qualitative" research. Johnson and Onwuegbuzie (2004) discuss the complementary advantages of these traditions, stating for example that qualitative methods add meaning to quantified results, while quantification adds precision to qualitative descriptions. Mixed methods can also increase generalizability of results by assessing ecological and phenomenological validity through qualitative studies while warranting causal claims through controlled experimental manipulation of variables. Achieving this complementary synergy depends on effective strategies for mixing the methods. Johnson and Onwuegbuzie (2004) elaborate on strategies for combination of quantitative and qualitative research, and offer a mixed research process model. Creswell (2003) discusses three strategies classified as (1) *sequential*, (2) *concurrent triangulation*, and (3) *concurrent nested*. For example, in a sequential strategy, qualitative methods can be used to derive and quantitative methods to test a grounded theory. Concurrent application of methods can provide stronger evidence for a conclusion through corroboration or convergence of findings. Similar strategies may be found in this volume. For example, in the Fractions case study, multiple analytic methods were applied to identify "pivotal moments" in the classroom session. Only one pivotal moment was corroborated by convergence, but the lack of convergence was also informative (Lund, this volume, Chap. 8). In the Biology case study, a deliberate effort was made to incorporate the insights of Stahl's conversation analysis into a social network analysis by Goggins and Dyke (this volume, Chap. 29), exemplifying a concurrent nested strategy in which the qualitative method is incorporated into the quantitative one.

Mixed methods have their share of critics. Issues listed by Johnson and Onwuegbuzie (2004) generally fall into difficulties with doing mixed methods and research community readiness. Due to the diverse knowledge and skill required to be sufficiently expert in multiple traditions and the sheer the amount of work to simultaneously conduct multiple analyses, it is difficult for one person to pull off. This points to the advantage of collaboration between multiple researchers, to which we return below. Other difficulties pertain to lack of acceptance of mixed methods in traditionally mono-methodological disciplines, and insufficient collective knowledge for combining methods. Although the qualitative/quantitative divide has long been questioned (Howe, 1988), some critics view mixed methods as incoherent (Yanchar & Williams, 2006). Another danger is that when an investigator working within a major paradigm mixes methods, the minor paradigm becomes a "handmaiden," not appreciated for its own value (Dourish & Button, 1998). Opportunities to challenge the assumptions of the major paradigm or exploit dialectics and synergies between two equal paradigms may be missed.

All of the claimed advantages and some of the potential disadvantages of mixed methods apply to productive multivocality, but there are some ways in which productive multivocality is not identical to mixed methods. Mixing can occur on dimensions other than the common quantitative versus qualitative research distinction. The productive multivocality project has found multiple voices even within a single tradition, for example, as seen in the different conceptions of "leadership" that developed in two methodologically very similar analyses in Chap. 11 (Howley, Mayfield, et al., this volume). More importantly, the objectives are different: mixed methods research is successful if the methods harmoniously work together towards a conclusion, but productive multivocality is also intended to surface conflicts, and can be considered successful when this happens as long as commonalities and essential differences are separated and well understood. From the point of view of our objectives, a limitation of the single-investigator mixed-methods approach is that there is only one agent representing the methods, and hence no true dialogue between the voices of different traditions, as the dialogue is entirely intrasubjective. This problem can be addressed by involving a committed representative of each tradition in a cooperative endeavor that yet has potential for genuine intersubjective argumentation. Our project took this approach, and elucidating argumentation resulted, for example, concerning the grounding of interpretations and causal claims in the Origami Fractions case study, the quality of Bruno's understanding in the Electricity case study, and the role of the agent in the Biology case study.

But multivocal analysis does not succeed simply by applying mixed methods distributed across multiple analysts: additional strategies are required to manage distributed agency. Other projects in the past have also attempted analysis of shared data by multiple traditions, with limited success. We encountered some of the same problems as these projects, but had the advantage of iteration over several years in which we were able to explore strategies for achieving productive multivocality. Some of these strategies are discussed below.

Strategies for Productive Multivocality

We have found the following strategies for coordinating multiple analysts to be useful, and offer them as a guide for future efforts at productive multivocality.

Analyze the Same Data

If each investigator takes on a different phenomenon and distinct sources of data, there may be no substantial basis for dialogue, and investigators' ad-hoc explanations might remain unchallenged. *Sharing data and comparing analyses* provides at least the possibility that alternative accounts are juxtaposed. Many efforts at research collaborations have involved this strategy (e.g., Koschmann, 2011; Stahl, 2009) and technologically oriented investigators have proposed or developed standards and metadata to enable data exchange (e.g., Harrer, Monés, & Dimitracopoulou, 2009; Reffay, Betbeder, & Chanier, 2012).

Analyze from Different Perspectives

Achieving epistemological multivocality requires *assigning analysts from different traditions to the same data* (although theoretical multivocality also can exist within a single tradition, as we saw in the Chemistry case study). When mixing traditions, analysts will encounter challenges in achieving agreement on what data is worth considering, and addressing differences in data needs. For example, some traditions would not find data from an experimental setting to be worth considering, or might consider it in an entirely different light than the experimenter (as happened with the Biology case study). Statistical methods based on sampling theory generally require more data (e.g., in terms of length or number of conversations) than microanalytic traditions, an issue we encountered in the Fractions case study. Issues may also arise concerning what constitute an adequate record of the phenomenon—such as what constitutes a "transcript"—as we saw in the Electricity case study. However, an important point is that these issues are not necessarily barriers to productive multivocality: they are opportunities to surface implicit assumptions of traditions and to bring them into dialogue with each other.

Push the Boundaries of Traditions Without Betraying Them

A related strategy is to *push analysts outside their comfort zone, while maintaining the integrity of their traditions*. We found that advances were made when analysts were asked to deal with data of a type they had not handled previously, as was the case with Trausan-Matu in the Fractions case study. However, concerns were also expressed that analysts so pushed were taking their traditions beyond what other members of the tradition would find comfortable. (Interestingly, these concerns were sometimes expressed by persons *not* in the traditions in question.) For example, we had discussions about whether Trausan-Matu (this volume, Chap. 6) should be generalizing adjacency pairs to include mental events, whether Chiu (this volume-a, Chap. 7) was being given enough data to meet the assumptions of statistical methods, and whether Stahl (this volume, Chap. 28) is violating the ethnomethodological tradition by attempting to generalize. Future efforts should seek a balance between pushing analysts to extend the value of their methods while staying grounded in their traditions, perhaps by discussing their innovations with other members of the traditions in question.

Begin with a Shared Pre-theoretical Analytic Objective

While needed, shared data is not enough. As we found in an early iteration of the project, analysts might ask entirely different questions about the data, resulting in analyses that are difficult to juxtapose because they construe the data differently. An additional strategy that provides further points of articulation is to *identify a shared but pre-theoretical concept as the analytic objective*. For example, beginning with the third workshop in our project, we posed analysts with the objective of identifying the *pivotal moments* in the collaboration. We left what constituted a "pivotal moment" unspecified, other than that such a moment (or event, episode, etc.) should be relevant to learning or collaboration. The deliberate vagueness solves a problem: over-specification of the analytic objective might privilege one tradition over another, as traditions differ in what they either consider worth investigating or what they are capable of identifying with their analytic toolkit. Left vague, "pivotal moment" served as a projective stimulus for the researcher/tradition. The different analyses that resulted could then be compared on the interesting questions of whether they identified the "same" moments, where and why they differed, and whether the moments identified by one tradition might lead to others to refine their approach. The concept (along with the data to which it was applied) served as a boundary object (Star & Griesemer, 1989)—an entity that could be understood and interpreted by different traditions, each in their own way, but with a shared referent (the data and the identified moments) that mutually ground dialogue. Another objective that played a similar role was that of identifying "leadership" in the Chemistry case study. Yet, shared data along with a vaguely specified shared objective may still not be sufficient. We found that further strategies for comparing the analyses that result are helpful.

Bring Analytic Representations into Alignment with Each Other and the Original Data

A straightforward yet powerful strategy is to *bring analytic representations into alignment*. It is difficult to compare results if analyses use entirely different representations and segment and describe the data in different ways. While each tradition will need to retain those representations that are essential to what it means to work in the tradition, efforts to identify where analytic representations address the same temporal, spatial and semantic spans of the original phenomenon will be rewarded with greater understanding of the relationships between approaches. Alignment need not necessarily be successful or even possible: the reward is as much in the effort to align and the discussion that results from this effort as in the aligned representations that result.

We found that this strategy can be facilitated by appropriate use of tools. For example, analyses of the Group Scribbles corpus on fractions (which was later replaced with the electricity corpus) were brought into alignment and compared using the Tatiana analytic software by Dyke and Lund at our fourth workshop (Dyke et al., 2011). It helps if the representations reference some common coordinate system and make the analytic interpretations salient: representational affordances for supporting dialogue amongst analysts are discussed in Chap. 33 (Dyke, Lund, Suthers, & Teplovs, this volume). It may be necessary to *return to the original data record* to resolve disputes (e.g., as in Chap. 19): mutual reference to the timeline of a shared data recording helps here.

Assign a Facilitator/Provocateur

The natural tendency for researchers is to focus on their own analyses, produce their results, and advocate for their viewpoint in communication with others. Researchers will put less effort into carefully examining others' analyses and performing comparisons. Of course, a commitment to do so is necessary for any meaningful collaboration of persons claiming to do multivocal analysis, but we must also acknowledge and plan for peoples' natural tendency as a barrier. A facilitator can serve various roles, including assisting in doing some of the work of aligning analytic results, and finding places where analysts disagree but may not have addressed their disagreement, as occurred in the Electricity case study (Suthers, this volume-a, Chap. 19).

Eliminate Gratuitous Differences

Efforts to align analytic representations will quickly make the need to *eliminate gratuitous differences* clear. Such differences include, for example, having chosen to analyze different temporal segments of a data stream, giving different names to the same entities (e.g., contributions), including or excluding private communications or nonverbal actions, etc. There are several examples in our project, not always successfully addressed. The Fractions and Chemistry corpora were fortunate in that analysts agreed on the temporal scope of data at the outset, and generally worked from the same transcripts, so did not need to iterate for this reason (Lund, this volume, Chap. 8). Analysts of the first Group Scribbles corpus on fractions differed on whether they looked at events in private workspaces as well as the public workspace, and on whether they considered nonverbal as well as verbal events (Suthers, this volume-a, Chap. 19). These differences were eliminated for a second pass, and analysts of the Electricity corpus did not differ in these ways, but the temporal scope of data considered varied widely for the latter corpus, resulting in only a small overlap where all analysts examined the same time period. Some differences are essential to the respective traditions participating, and must be respected. For example, Stahl (this volume, Chap. 28) ignored the experimental structure of the Biology data, choosing instead to perform an uptake analysis of a single chat session. Again, productive does not necessarily mean agreement, and the process of separating non-essential from essential differences will not only help make the later more salient but may also be rewarding in itself.

Iterate

Iteration is required to successfully realize the value of many of the other strategies. Gratuitous differences may emerge only after the first attempt to bring analytic representations into alignment, so some of the analyses may need to be reworked. Iteration is also useful to take advantage of what has been learned from the entire effort. For example, we have seen an analysis from one tradition spur an analyst working in a different tradition to consider a different conception of "pivotal moment" and redo his analysis (Lund, Chap. 8; Shirouzu, this volume-a, Chap. 5).

Attend to the Needs of the Data Providers

Data providers are providing a valuable service. It takes work to provide data to others: collaborations that share this work or otherwise provide resources for it are more likely to succeed. Analysts should be aware that data providers may have different objectives in the activity that produces the data, and it will not be perceived as helpful to critique that activity based on criteria that are not important to them or point out problems they are already aware of. This does not preclude making such critiques, but rather is a call to respect the other perspective in doing so. These points are illustrated by our experience with the Knowledge Forum (Fujita, this volume, Chap. 24) and Biology (Hmelo-Silver, this volume, Chap. 30) case studies: see also the summaries earlier in the present chapter. Data providers are taking a risk in exposing their activity to outside analysts, not only in exposing the details of their execution but also opening up the possibility that members of other traditions may question the value of the whole endeavor. Iteration in which analysts communicate their results candidly with data providers and then revise will more likely result in new understandings that are valued by both partners.

Reflect on Your Practice

The final and most important strategy we will note here derives from our argument that, while methods have biases, researchers have agency in applying them as tools and are not deterministically bound to the traditions those methods come from.

Methods are based on data and analytic representations and ways of manipulating those representations to derive new representations. Methods also include practices in using these tools, such as how to select questions worth asking and situations worth studying, how to map situations to data representations, and how to interpret the analytic representations. The argument has been made that methods intrinsically bring with them theoretical and epistemological commitments. While we agree that there are commitments, we believe that it is important to examine how these commitments are transmitted and which are non-negotiable, rather than accepting a methodological determinism dogmatically. But making this determination requires the final strategy.

The strategy is to remove one's methodological eyeglasses and view and *dialogue about methods as object-constituting, evidence-producing and argument-sustaining tools*. This dialogue requires careful consideration of what methods (understood as inscriptions and means of operating on inscriptions, with associated practices) intrinsically bring with them, and what teleological, epistemological, and theoretical commitments are made in the practices of applying these tools to a domain. It is our expectation that this reflection will be deeper when undertaken collectively rather than individually, and our hope that such collective reflection will help the community of researchers in multidisciplinary fields such as the learning sciences identify the conceptual centers of gravity that gives their work coherence, and identify and leverage the value of distinct disciplinary orbits around these centers for improving our understanding of the phenomenon as a whole.

Acknowledgements We are thankful to the efforts and perseverance of the many data providers and analysts who made this project possible. Your commitment to understanding each other bodes well for the future of the study of learning in interaction. The first author was partially supported by NSF Award 0943147 during the course of this project.

References

Adamson, D., Ashe, C., Jang, H., Yaron, D., & Rosé, C. P. (2013). Intensification of group knowledge exchange with academically productive talk agents. In *Proceedings of the 10th International Conference on Computer Supported Collaborative Learning*, Madison, Wisconsin.

Adamson, D., Dyke, G., Jang, H., & Rosé, C. P. (in press). Towards adapting dynamic collaboration support to support to student ability level. *International Journal of Artificial Intelligence in Education*.

Anderson, J. R. (Ed.). (1981). *Cognitive skills and their acquisition*. Hillsdale, New Jersey: Lawrence Erlbaum Associates.

Andriessen, J., Baker, M., & Suthers, D. D. (Eds.). (2003). *Arguing to learn: Confronting cognitions in computer-supported collaborative learning environments*. Dordrecht, The Netherlands: Kluwer.

Arzarello, F. (2004). Semiosis as a multimodal process. *Relime, Numero Especial*, 267–299.

Bakhtin, M. (1981). Discourse in the novel. In M. Holquist (Ed.), *The dialogic imagination* (pp. 259–422). Austin, TX: University of Texas.

Chen, W., & Looi, C.-K. (this volume). Group scribbles-supported collaborative learning in a primary grade 5 science class. In D. D. Suthers, K. Lund, C. P. Rose, C. Teplovs, & N. Law (Eds.), *Productive multivocality in the analysis of group interactions*, Chapter 14. New York, NY: Springer.

Chiu, M. M. (this volume-a). Social metacognition, micro-creativity and justifications: Statistical discourse analysis of a mathematics classroom conversation. In D. D. Suthers, K. Lund, C. P. Rosé, C. Teplovs, & N. Law (Eds.), *Productive multivocality in the analysis of group interactions*, Chapter 7. New York, NY: Springer.

Chiu, M. M. (this volume-b). Statistical discourse analysis of an online discussion: Cognition and social metacognition. In D. D. Suthers, K. Lund, C. P. Rosé, C. Teplovs, & N. Law (Eds.), *Productive multivocality in the analysis of group interactions*, Chapter 23. New York, NY: Springer.

Chiu, M. M., & Fujita, N. (2014). Statistical discourse analysis: Knowledge creation during online discussions. In S.-C. Tan, H.-J. So, & J. Yeo (Eds.), *Knowledge creation in education.* New York, NY: Springer.

Choi, B. C. K., & Pak, A. W. P. (2006). Multidisciplinarity, interdisciplinarity and transdisciplinarity in health research, services, education and policy: 1. Definitions, objectives, and evidence of effectiveness. *Clinical and Investigative Medicine, 29*(6), 351–364.

Clarke, S., Chen, G., Stainton, K., Katz, S., Greeno, J. G., Resnick, L. B., et al. (2013). *The impact of CSCL beyond the online environment.* Paper presented at the Proceedings of Computer Supported Collaborative Learning, Madison, Wisconsin.

Cress, U., & Kimmerle, J. (this volume). Successful knowledge building needs group awareness: Interaction analysis of a 9th grade CSCL biology lesson. In D. D. Suthers, K. Lund, C. P. Rosé, C. Teplovs, & N. Law (Eds.), *Productive multivocality in the analysis of group interactions*, Chapter 27. New York, NY: Springer.

Creswell, J. W. (2003). *Research design: Qualitative, quantitative, and mixed methods approaches.* Thousand Oaks, CA: Sage Publications.

Dourish, P., & Button, G. (1998). On "Technomethodology": Foundational relationships between ethnomethodology and system design. *Human-Computer Interaction, 13*, 395–432.

Duranti, A. (2006). Transcripts, like shadows on a wall. *Mind, Culture, and Activity, 13*(4), 301–310.

Dyke, G., Adamson, D., Howley, I. K., & Rosé, C. P. (2013). Enhancing scientific reasoning and explanation skills with conversational agents. *IEEE Transactions on Learning Technologies, 6*(3), 240–247.

Dyke, G., Howley, I. K., Adamson, D., & Rosé, C. P. (2012). Towards academically productive talk supported by conversational agents. In S. A. Cerri, W. J. Clancey, G. Papadourakis, & K. Panourgia (Eds.), *Proceedings of the 11th International Conference on Intelligent Tutoring Systems (ITS 2012), Chania, Crete, Greece* (pp. 531–540). Berlin, Germany: Springer.

Dyke, G., Howley, I. K., Kumar, R., & Rosé, C. P. (this volume). Towards academically productive talk supported by conversational agents. In D. D. Suthers, K. Lund, C. P. Rosé, C. Teplovs, & N. Law (Eds.), *Productive multivocality in the analysis of group interactions*, Chapter 25. New York, NY: Springer.

Dyke, G., Kumar, R., Ai, H., & Rosé, C. P. (2012). Challenging assumptions: Using sliding window visualizations to reveal time-based irregularities in CSCL processes. In J. van Aalst, K. Thompson, M. J. Jacobson, & P. Reimann (Eds.), *Proceedings of the 10th International Conference of the Learning Sciences* (Vol. 1, pp. 363–370). Sydney, NSW: International Society of the Learning Sciences.

Dyke, G., Lund, K., & Girardot, J.-J. (2009). Tatiana: An environment to support the CSCL analysis process. In C. O'Malley, P. Reimann, D. Suthers, & A. Dimitracopoulou (Eds.), *Computer supported collaborative learning practices: CSCL 2009 Conference Proceedings* (pp. 58–67). Rhodes, Greece: International Society of the Learning Sciences.

Dyke, G., Lund, K., Jeong, H., Medina, R., Suthers, D. D., van Aalst, J., et al. (2011). Technological affordances for productive multivocality in analysis. In H. Spada, G. Stahl, N. Miyake, & N. Law (Eds.), *Connecting computer-supported collaborative learning to policy and practice:*

Proceedings of the 9th International Conference on Computer-Supported Collaborative Learning (CSCL 2011) (Vol. 1, pp. 454–461). Hong Kong: International Society of the Learning Sciences.

Dyke, G., Lund, K., Suthers, D. D., & Teplovs, C. (this volume). Analytic representations and affordances for productive multivocality. In D. D. Suthers, K. Lund, C. P. Rosé, C. Teplovs, & N. Law (Eds.), *Productive multivocality in the analysis of group interactions*, Chapter 33. New York, NY: Springer.

Engle, R. A., & Conant, F. R. (2002). Guiding principles for fostering productive disciplinary engagement: Explaining an emergent argument in a community of learners classroom. *Cognition and Instruction, 20*(4), 399.

Frechtling, J., & Sharp, L. (Eds.). (1997). *User-friendly handbook for mixed method evaluations*. Arlington, VA: Directorate for Education and Human Resources, Division of Research Evaluation and Communication, National Science Foundation.

Fujita, N. (this volume). Critical reflections on multivocal analysis and implications for design-based research. In D. D. Suthers, K. Lund, C. P. Rosé, C. Teplovs, & N. Law (Eds.), *Productive multivocality in the analysis of group interactions*, Chapter 24. New York, NY: Springer.

Gafney, L., & Varma-Nelson, P. (2008). *Peer-led team learning: Evaluation, dissemination, and institutionalization of a college level initiative*. Dordrecht, The Netherlands: Springer.

Goggins, S. P., & Dyke, G. (this volume). Network analytic techniques for online chat. In D. D. Suthers, K. Lund, C. P. Rosé, C. Teplovs, & N. Law (Eds.), *Productive multivocality in the analysis of group interactions*, Chapter 29. New York, NY: Springer.

Goodwin, C. (2000). Action and embodiment within situated human interaction. *Journal of Pragmatics, 32*, 1489–1522.

Gosser, D. K., Cracolice, M., Kampmeier, J., Roth, V., Strozak, V., & Varma-Nelson, P. (2001). *Peer-led team learning: A guidebook*. Upper Saddle River, NJ: Prentice Hall.

Gosser, D. K., & Roth, V. (1998). The workshop chemistry project: Peer-led team learning. *Journal of Chemical Education, 75*(2), 185–187.

Greeno, J. G. (2006). Learning in activity. In R. K. Sawyer (Ed.), *Cambridge handbook of the learning sciences* (pp. 79–96). New York, NY: Cambridge University Press.

Greeno, J. G., Collins, A. M., & Resnick, L. B. (1996). Cognition and learning. In D. C. Berliner & R. C. Calfee (Eds.), *Handbook of educational psychology* (pp. 15–46). New York, NY: Simon & Schuster Macmillan.

Harrer, A., Monés, M., & Dimitracopoulou, A. (2009). Users' data: Collaborative and social analysis. In N. Balacheff, S. Ludvigsen, T. de Jong, A. Lazonder, & S. Barnes (Eds.), *Technology enhanced learning. Principles and products* (pp. 175–193). New York, NY: Springer.

Hmelo-Silver, C. E. (this volume). Multivocality as a tool for design-based research. In D. D. Suthers, K. Lund, C. P. Rosé, C. Teplovs, & N. Law (Eds.), *Productive multivocality in the analysis of group interactions*. New York, NY: Springer.

Hockings, S. C., DeAngelis, K. J., & Frey, R. F. (2008). Peer-led team learning in general chemistry: Implementation and evaluation. *Journal of Chemical Education, 85*(7), 990–996.

Howe, K. R. (1988). Against the quantitative-qualitative incompatibility thesis, or dogmas die hard. *Educational Researcher, 17*, 10–16.

Howley, I. K., Kumar, R., Mayfield, E., Dyke, G., & Rosé, C. P. (this volume). Gaining insights from sociolinguistic style analysis for redesign of conversational agent based support for collaborative learning. In D. D. Suthers, K. Lund, C. P. Rosé, C. Teplovs, & N. Law (Eds.), *Productive multivocality in the analysis of group interactions*, Chapter 26. New York, NY: Springer.

Howley, I. K., Mayfield, E., & Rosé, C. P. (2013). Linguistic analysis methods for studying small groups. In C. E. Hmelo-Silver & A. M. O'Donnell (Eds.), *International handbook of collaborative learning*. New York, NY: Routledge, Taylor and Francis Group.

Howley, I. K., Mayfield, E., Rosé, C. P., & Strijbos, J.-W. (this volume). A multivocal process analysis of social positioning in study groups. In D. D. Suthers, K. Lund, C. P. Rosé, C. Teplovs, & N. Law (Eds.), *Productive multivocality in the analysis of group interactions*, Chapter 11. New York, NY: Springer.

Hutchins, E. (1995). *Cognition in the wild.* Cambridge, MA: The MIT Press.
Jeong, H. (this volume). Development of group understanding via the construction of physical and technological artifacts. In D. D. Suthers, K. Lund, C. P. Rose, C. Teplovs, & N. Law (Eds.), *Productive multivocality in the analysis of group interactions*, Chapter 18. New York, NY: Springer.
Jeong, H., Chen, W., & Looi, C.-K. (2011). Analysis of group understanding in artifact-mediated discourse. In H. Spada, G. Stahl, N. Miyake, & N. Law (Eds.), *Connecting computer-supported collaborative learning to policy and practice: Proceedings of the 9th International Conference on Computer-Supported Collaborative Learning (CSCL 2011)* (Vol. II, pp. 786–790). Hong Kong: International Society of the Learning Sciences.
Johnson, R. B., & Onwuegbuzie, A. J. (2004). Mixed methods research: A research paradigm whose time has come. *Educational Researcher, 33*(7), 14–26.
Jordan, B., & Henderson, A. (1995). Interaction analysis: Foundations and practice. *The Journal of the Learning Sciences, 4*(1), 39–103.
Kolodner, J. L. (1991). The journal of the learning sciences: Effecting changes in education. *The Journal of the Learning Sciences, 1*(1), 1–6.
Koschmann, T. (1999). Toward a dialogic theory of learning: Bakhtin's contribution to understanding learning in settings of collaboration. In C. Hoadley & J. Roschelle (Eds.), *Proceedings of the 1999 Conference on Computer Support for Collaborative Learning (CSCL 1999).* Palo Alto, CA: International Society of the Learning Sciences.
Koschmann, T. (2011). *Theories of learning and studies of instructional practice.* New York, NY: Springer.
Koschmann, T., Hall, R., & Miyake, N. (Eds.). (2001). *CSCL II. Carrying forward the conversation.* Mahwah, NJ: Lawrence Erlbaum Associates.
Koschmann, T., Zemel, A., Conlee-Stevens, M., Young, N., Robbs, J., & Barnhart, A. (2005). How do people learn: Member's methods and communicative mediation. In R. Bromme, F. W. Hesse, & H. Spada (Eds.), *Barriers and biases in computer-mediated knowledge communication (and how they may be overcome)* (pp. 265–294). Amsterdam, The Netherlands: Kluwer Academic Press.
Kumar, R., & Rosé, C. P. (2011). Architecture for building conversational agents that support collaborative learning. *IEEE Transactions on Learning Technologies, 4*(1), 21–34.
Latour, B. (2005). *Reassembing the social: An introduction to actor-network-theory.* New York, NY: Oxford University Press.
Lave, J., & Wenger, E. (1991). *Situated learning: Legitimate peripheral participation.* Cambridge, UK: Cambridge University Press.
Law, N., & Wong, O.-W. (this volume). Exploring pivotal moments in students' knowledge building progress using participation and discourse marker indicators as heuristic guides. In D. D. Suthers, K. Lund, C. P. Rosé, C. Teplovs, & N. Law (Eds.), *Productive multivocality in the analysis of group interactions*, Chapter 22. New York, NY: Springer.
Looi, C.-K., Song, Y., Wen, Y., & Chen, W. (this volume). Identifying pivotal contributions for group progressive inquiry in a multi-modal interaction environment. In D. D. Suthers, K. Lund, C. P. Rose, C. Teplovs, & N. Law (Eds.), *Productive multivocality in the analysis of group interactions*, Chapter 15. New York, NY: Springer.
Lund, K. (2011). Analytical frameworks for group interactions in CSCL systems. In S. Puntambekar, G. Erkens, & C. E. Hmelo-Silver (Eds.), *Analyzing collaborative interactions in CSCL: Methods, approaches and issues* (pp. 391–411). New York, NY: Springer.
Lund, K. (this volume). A multivocal analysis of pivotal moments for learning fractions in a 6th grade classroom in Japan. In D. D. Suthers, K. Lund, C. P. Rosé, C. Teplovs, & N. Law (Eds.), *Productive multivocality in the analysis of group interactions*, Chapter 8. New York, NY: Springer.
Lund, K., & Bécu-Robinault, K. (this volume). Conceptual change and sustainable coherency of concepts across modes of interaction. In D. D. Suthers, K. Lund, C. P. Rose, C. Teplovs, & N. Law (Eds.), *Productive multivocality in the analysis of group interactions*, Chapter 17. New York, NY: Springer.

Lund, K., & Suthers, D. D. (this volume). Methodological dimensions. In D. D. Suthers, K. Lund, C. P. Rosé, C. Teplovs, & N. Law (Eds.), *Productive multivocality in the analysis of group interactions*, Chapter 2. New York, NY: Springer.

Matusov, E. (1996). Intersubjectivity without agreement. *Mind, Culture, and Activity, 3*(1), 25–45.

Medina, R. (this volume). Cascading inscriptions and practices: Diagramming and experimentation in the Group Scribbles classroom. In D. D. Suthers, K. Lund, C. P. Rose, C. Teplovs, & N. Law (Eds.), *Productive multivocality in the analysis of group interactions*, Chapter 16. New York, NY: Springer.

Medina, R., & Suthers, D. D. (2013). Juxtaposing practice: Uptake as modal transposition. In *Proceedings of the 10th International Conference on Computer Supported Collaborative Learning (CSCL '13)*, Madison, WI.

Michaels, S., O'Connor, C., & Resnick, L. B. (2008). Deliberative discourse idealized and realized: Accountable talk in the classroom and in civic life. *Studies in Philosophy and Education, 27*(4), 283–297.

Ochs, E. (1979). Transcription as theory. In E. Ochs & B. B. Schieffelin (Eds.), *Developmental pragmatics* (pp. 43–72). New York, NY: Academic.

Oshima, J., Matsuzawa, Y., Oshima, R., Chan, C. K. K., & van Aalst, J. (2012). Social network analysis for knowledge building: Establishment of indicators for collective knowledge advancement. In J. van Aalst, K. Thompson, M. J. Jacobson, & P. Reimann (Eds.), *The future of learning: Proceedings of the 10th International Conference of the Learning Sciences (ICLS2012)* (Short papers, symposia, and abstracts, Vol. 2, pp. 465–466). Sydney, NSW: International Society of the Learning Sciences.

Oshima, J., Matsuzawa, Y., Oshima, R., & Niihara, Y. (this volume). Application of network analysis to collaborative problem solving discourse: An attempt to capture dynamics of collective knowledge advancement. In D. D. Suthers, K. Lund, C. P. Rosé, C. Teplovs, & N. Law (Eds.), *Productive multivocality in the analysis of group interactions*, Chapter 12. New York, NY: Springer.

Oshima, J., Oshima, R., & Matsuzawa, Y. (2012). Knowledge building discourse explorer: A social network analysis application for knowledge building discourse. *Educational Technology Research and Development, 60*, 903–921.

Oshima, J., Oshima, R., Matsuzawa, Y., van Aalst, J., & Chan, C. K. K. (2011). *Network structure analysis for knowledge building: A macroscopic view of collaborative learning discourse.* Annual Meeting of Educational Research Association, New Orleans, LA

Reffay, C., Betbeder, M.-L., & Chanier, T. (2012). Multimodal learning and teaching corpora exchange: Lessons learned in five years by the mulce project. *International Journal of Technology Enhanced Learning, 4*(1/2), 11–30.

Resnick, L. B., Asterhan, C., & Clarke, S. (in press). *Socializing intelligence through academic talk and dialogue*. Washington, DC: American Educational Research Association.

Reynolds, R., & Chiu, M. M. (2012). Contribution of motivational orientations to student outcomes in a discovery-based program of game design learning. In J. van Aalst, K. Thompson, M. J. Jacobson, & P. Reimann (Eds.), *Proceedings of the 10th International Conference of the Learning Sciences* (Vol. 1, pp. 356–360). Sydney, NSW: International Society of the Learning Sciences.

Rogoff, B. (1995). Observing sociocultural activity on three planes: Participatory appropriation, guided participation, and apprenticeship. In J. V. Wertsch, P. D. Rio, & A. Alvarez (Eds.), *Sociocultural studies of mind* (pp. 139–164). New York, NY: Cambridge University Press.

Roschelle, J., Tatar, D., Chaudhury, S. R., Dimitriadis, Y., Patton, C., & DiGiano, C. (2007). Ink, improvisation, and interactive engagement: Learning with tablets. *IEEE Computer, 40*(9), 38–44.

Rosé, C. P., & Tovares, A. (in press). What sociolinguistics and machine learning have to say to one another about interaction analysis. In L. B. Resnick, C. Asterhan, & S. Clarke (Eds.), *Socializing intelligence through academic talk and dialogue*. Washington, DC: American Educational Research Association.

Sarquis, J. L., Dixon, L. J., Gosser, D. K., Kampmeier, J. A., Roth, V., Strozak, V. S., et al. (2001). The workshop project: Peer-led team learning in chemistry. In J. E. Miller, J. E. Groccia, & M. S. Miller (Eds.), *Student-assited teaching* (pp. 150–156). Boston, MA: Anker.

Sawyer, R. K. (2006a). Analyzing collaborative discourse. In K. R. Sawyer (Ed.), *The Cambridge handbook of the learning science* (pp. 187–204). New York, NY: Cambridge University Press.

Sawyer, R. K. (Ed.). (2006b). *The Cambridge handbook of the learning sciences*. New York, NY: Cambridge University Press.

Sawyer, R. K., Frey, R., & Brown, P. (this volume-a). Knowledge building discourse in peer-led team learning (PLTL) groups in first-year general chemistry. In D. D. Suthers, K. Lund, C. P. Rosé, C. Teplovs, & N. Law (Eds.), *Productive multivocality in the analysis of group interactions,* Chapter 10. New York, NY: Springer.

Sawyer, R. K., Frey, R., & Brown, P. (this volume-b). Peer-Led team learning in general chemistry. In D. D. Suthers, K. Lund, C. P. Rosé, C. Teplovs, & N. Law (Eds.), *Productive multivocality in the analysis of group interactions*, Chapter 9. New York, NY: Springer.

Scardamalia, M., & Bereiter, C. (1991). Higher levels of agency for children in knowledge building: A challenge for the design of new knowledge media. *The Journal of the Learning Sciences, 1*(1), 37–68.

Scardamalia, M., & Bereiter, C. (2006). Knowledge building. In R. K. Sawyer (Ed.), *Cambridge handbook of the learning sciences* (pp. 97–115). New York, NY: Cambridge University Press.

Schegloff, E. A., & Sacks, H. (1973). Opening up closings. *Semiotica, 8*, 289–327.

Schwarz, B., Wang, C., Chiu, M. M., Ching, C. C., Walker, E., Koedinger, K. R., et al. (2010). Adaptive human guidance of computer-mediated group work. In K. Gomez, L. Lyons, & J. Radinsky (Eds.), *Proceedings of the 9th International Conference of the Learning Sciences* (Vol. 2, pp. 149–156). Chicago, IL: International Society of the Learning Sciences.

Seymour, E., & Hewitt, N. (1994). *Talking about leaving: Factors contributing to high attrition rates among science, mathematics, and engineering undergraduate majors*. Boulder, CO: Bureau of Sociological Research, University of Colorado.

Shirouzu, H. (this volume-a). Focus-based constructive interaction. In D. D. Suthers, K. Lund, C. P. Rosé, C. Teplovs, & N. Law (Eds.), *Productive multivocality in the analysis of group interactions*, Chapter 5. New York, NY: Springer.

Shirouzu, H. (this volume-b). Learning fractions through folding in an elementary face-to-face classroom. In D. D. Suthers, K. Lund, C. P. Rosé, C. Teplovs, & N. Law (Eds.), *Productive multivocality in the analysis of group interactions*, Chapter 4. New York, NY: Springer.

Siebert, E. D., & McIntosh, W. J. (2001). *College pathways to the science education standards*. Arlington, VA: NSTA Press.

Sionti, M., Ai, H., Rosé, C. P., & Resnick, L. B. (2011). A framework for analyzing development of argumentation through classroom discussions. In N. Pinkwart & B. McClaren (Eds.), *Educational technologies for teaching argumentation skills*. Sharjah, UAE: Bentham Science.

Stahl, G. (2006). *Group cognition: Computer support for collaborative knowledge building*. Cambridge, MA: MIT Press.

Stahl, G. (Ed.). (2009). *Studying virtual math teams*. New York, NY: Springer.

Stahl, G. (this volume). Interaction analysis of a biology chat. In D. D. Suthers, K. Lund, C. P. Rosé, C. Teplovs, & N. Law (Eds.), *Productive multivocality in the analysis of group interactions*, Chapter 28. New York, NY: Springer.

Stahl, G., Koschmann, T., & Suthers, D. D. (2006). Computer-supported collaborative learning: An historical perspective. In R. K. Sawyer (Ed.), *Cambridge handbook of the learning sciences* (pp. 409–426). Cambridge, UK: Cambridge University Press.

Star, S. L., & Griesemer, J. R. (1989). Institutional ecology, 'translations' and boundary objects: Amateurs and professionals in Berkeley's Museum of Vertebrate Zoology. *Social Studies of Science, 19*(3), 387–420.

Suthers, D. D. (2006). Technology affordances for intersubjective meaning-making: A research agenda for CSCL. *International Journal of Computer-Supported Collaborative Learning, 1*(3), 315–337.

Suthers, D. D. (this volume-a). Agency and modalities in multimediated interaction. In D. D. Suthers, K. Lund, C. P. Rosé, C. Teplovs, & N. Law (Eds.), *Productive multivocality in the analysis of group interactions*, Chapter 19. New York, NY: Springer.

Suthers, D. D. (this volume-b). The productive multivocality project: Objectives and origins. In D. D. Suthers, K. Lund, C. P. Rosé, C. Teplovs, & N. Law (Eds.), *Productive multivocality in the analysis of group interactions*, Chapter 1. New York, NY: Springer.

Suthers, D. D., Lund, K., Rosé, C. P., Dyke, G., Law, N., Teplovs, C., et al. (2011). Towards productive multivocality in the analysis of collaborative learning. In H. Spada, G. Stahl, N. Miyake, & N. Law (Eds.), *Connecting computer-supported collaborative learning to policy and practice: Proceedings of the 9th International Conference on Computer-Supported Collaborative Learning (CSCL 2011)* (Vol. III, pp. 1015–1022). Hong Kong: International Society of the Learning Sciences.

Suthers, D. D., Lund, K., Rosé, C. P., Teplovs, C., & Law, N. (this volume). *Productive multivocality in the analysis of group interactions*. New York, NY: Springer.

Tashakkori, A., & Teddle, C. (Eds.). (2003). *Handbook of mixed methods in social & behavioral research*. Thousand Oaks, CA: Sage Publications.

Teplovs, C., & Fujita, N. (this volume). Socio-dynamic latent semantic learner models. In D. D. Suthers, K. Lund, C. P. Rosé, C. Teplovs, & N. Law (Eds.), *Productive multivocality in the analysis of group interactions*, Chapter 21. New York, NY: Springer.

Tien, L. T., Roth, V., & Kampmeier, J. A. (2002). Implementation of a peer-led team learning instructional approach in an undergraduate organic chemistry course. *Journal of Research in Science Teaching, 39*(7), 606–632.

Trausan-Matu, S. (this volume). Collaborative and differential utterances, pivotal moments, and polyphony. In D. D. Suthers, K. Lund, C. P. Rosé, C. Teplovs, & N. Law (Eds.), *Productive multivocality in the analysis of group interactions*, Chapter 6. New York, NY: Springer.

Weinberger, A., & Fischer, F. (2006). Framework to analyze argumentative knowledge construction in computer-supported collaborative learning. *Computers and Education, 46*, 71–95.

Wenger, E. (1987). *Artificial intelligence and tutoring systems: Computational and cognitive approaches to the communication of knowledge*. Los Altos, CA: Morgan Kaufmann.

Wenger, E. (1998). *Communities of practice: Learning, meaning and identity*. Cambridge, UK: Cambridge University Press.

Wise, A. F., & Chiu, M. M. (2011a). Analyzing temporal patterns of knowledge construction in a role-based online discussion. *International Journal of Computer-Supported Collaborative Learning, 6*, 445–470.

Wise, A. F., & Chiu, M. M. (2011b). Knowledge construction patterns in online conversation. In H. Spada, G. Stahl, N. Miyake, & N. Law (Eds.), *Connecting computer-supported collaborative learning to policy and practice: Proceedings of the 9th International Conference on Computer-Supported Collaborative Learning (CSCL 2011)*. Hong Kong: University of Hong Kong.

Yanchar, S. C., & Williams, D. D. (2006). Reconsidering the compatabilty thesis and eclecticism: Five proposed guidelines for method use. *Educational Researcher, 35*(9), 3–12.

Chapter 32
Methodological Pathways for Avoiding Pitfalls in Multivocality

Carolyn Penstein Rosé and Kristine Lund

Introduction

In his original work on multivocality in literary criticism (Bakhtin, 1981), Bakhtin argued that the novel as a form of literature offers more of an opportunity for multivocality than other narrowly and rigidly defined forms. In this spirit, we offer a perspective on research methodology that conceptualizes the analytic cycle in such a way that provides the opportunity for multiple perspectives to speak to one another and challenge one another as we examine data that are of common interest. The earlier chapters of this book have illustrated our own journey towards multivocality and have served the purpose of illustrating potential outcomes of productive multivocality. In Chap. 34 (Lund, Rosé, Suthers, & Baker, this volume), we explore the epistemological encounters that researchers had when they compared various aspects of their analyses. The perspectives of different researchers may either coexist in their natural, productive tension without being integrated, and thus remain limited with respect to their value to one another; or they may actively interact on theoretical and/or methodological levels, thus bringing new insight to the phenomenon being studied.

In this chapter we invite the reader to join us on our journey towards multivocality while we focus on the methods of analysis of collaborative interactions. We assume a diverse readership that may include expert analysts, steeped in their own

C.P. Rosé (✉)
Carnegie Mellon University, Language Technologies Inst. & HCI Inst.,
Gates Hillman Center, 5415, 5000 Forbes Avenue, Pittsburgh, PA 15213, USA
e-mail: cprose@cs.cmu.edu

K. Lund
ICAR Research Lab, CNRS—University of Lyon, 15 Parvis René Descartes,
69007 Lyon, France
e-mail: kristine.lund@ens-lyon.fr

D.D. Suthers et al. (eds.), *Productive Multivocality in the Analysis of Group Interactions*, Computer-Supported Collaborative Learning Series 16,
DOI 10.1007/978-1-4614-8960-3_32, © Springer Science+Business Media New York 2013

tradition, who want to forge new partnerships to embark on their own multivocal experience. Other readers may be students just learning about research methods in order to get a better grasp on the research landscape. Still other readers may be instructors of research methods who may be looking for ideas for how to use multivocality as a teaching paradigm in those courses. In all these cases, it is important to get insight into what makes multivocality challenging. Our goal is to offer these insights. In particular, we will examine what might be considered the "dark side" of the multivocal analysis process. We do not want to present an unrealistically rosy view of our own experiences for those who will follow in our footsteps. So here we discuss potential pitfalls of the multivocal analysis process and what might be some pathways towards working around them or avoiding them altogether.

First we place this discussion within a conceptual frame. Simply put, the important running theme is that multivocal research is an intensely team-oriented sport. Pitfalls come from a tendency for researchers to fall into isolation in one way or another in their work. Pathways in this chapter will thus consistently be framed as pressures to overcome these tendencies. Experienced researchers know that research is by its very nature social, just as literary criticism is social, in other words embedded within communities of practitioners. We see this when we consider that science is the accumulation of knowledge and that theories are storehouses for the collection and integration of knowledge gained through empirical investigations. We know that a single focused research contribution by itself, no matter how insightful or high quality, is too narrow to be of significant value. It becomes valuable as it is integrated with the results of other empirical investigations that have either already occurred or occur later and relate back to it as seminal work. Despite understanding this important truth, however, we still fall prey to a tendency towards isolation.

In Bakhtin's investigations we see that it is not something inherent in epics or novels that makes one multivocal and the other not; rather it is the way they are treated by communities of practitioners that make them that way. Students learning not only research methods but also the politics and sociology of research may be even more prone to isolation that runs counter to the goals of multivocality. As a running theme, then, we will compare two distinct types of potentially multivocal analysts. At one end of the continuum, we will examine students who are just beginning to engage in research practices and who may thus treat those practices as rigidly defined forms like epics in Bakhtin's explorations. We will compare this orientation with those of more senior researchers within the ecosystem who have gained facility with the practices and have earned a position that allows them to manipulate research practices in creative and productive ways.

Communication Flow of the Multivocal Analysis Process

Now we offer a schema for thinking about our conceptual frame from the perspective of communicative processes. We review the many diverse stakeholders who take part in the multivocal analysis process either directly or indirectly, and analyze

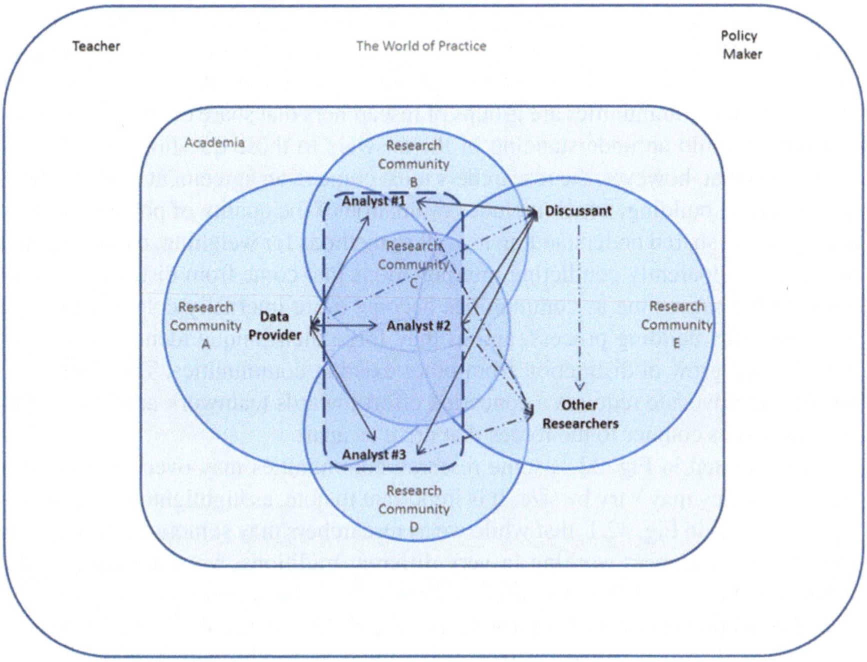

Fig. 32.1 Information flow in an iterative multivocal analysis process. Note that only relatively infrequent formal communication processes are indicated with explicit links. The more frequent and less formal communication and coordination between the analysts is signified by the *dashed, oblong shape* that joins them together

the information flow between them, since many of the pitfalls we address in this chapter occur as communication breakdowns or difficulties between stakeholders.

Most of the process of multivocal analysis occurs within the purview of Academia. Thus, in Fig. 32.1 all of the direct communication links, which are represented as unidirectional (dotted lines) or bidirectional links (solid lines), are within that area. However, as discussed in Chap. 35 (Law & Laferriere, this volume), ultimately the hope is that our research will impact the world of Practice, so it is important to consider practitioners as one audience for our work, including both teachers and policy makers. That communication is aided by researchers who span both worlds, such as the authors of Chap. 35.

Communication across stakeholders is the key to fueling the iterative multivocal process. In Fig. 32.1, informal processes of comparison between analysts are not displayed as lines. Instead their more frequent communication with one another is indicated by the oblong shaped dashed box that illustrates their joint status as an analytic team. Public presentations, which are one-way communications, are displayed with dashed lines, with an arrow indicating the direction of the communication. Direct communication involving the transfer of artifacts, including data, formal analyses, instructions, and feedback are displayed with solid lines. These communications in our work were typically two-way communications that involved discussion between both parties, and are thus indicated with bidirectional solid links.

In part because of the tendency towards distinction and isolation, the world of Academia consists of many overlapping research communities, each holding to their own epistemological and theoretical biases as well as methodological practices. Research communities are groups of researchers that share common questions and seek to build an understanding of the answers to those questions together. In order to do that, however, the researchers must come to an agreement on the criteria for consensus building, which includes evaluation of the quality of potential contributions to the shared understanding as well as methods for weighing, balancing, and reconciling apparently conflicting interpretations that come from distinct contributions. At the same time as communities become more internally coherent through this consensus building process, and as they forge their unique identity as a community, they grow in distinction from other existing communities. The multivocal process we advocate requires a concerted effort towards teamwork across communities that runs counter to the forces that drive us apart.

As illustrated in Fig. 32.1, some research communities may overlap more than others, and they may vary by size. It is important to note, as highlighted by the overlapping circles in Fig. 32.1, that while we as researchers may sometimes feel worlds apart from researchers working in very different traditions, we share more of the substance of our work than we might acknowledge. The similarity comes from the shared focus on the target of inquiry, in the case of this book, collaborative learning. It is the distinctions between these overlapping research communities, and the membership of the participants of that process within different ones of these communities, that makes multivocality what it is. That being said, although Fig. 32.1 places each stakeholder participating in the formal process within a single research community, the truth is that we frequently participate in different research communities at different times. Furthermore, Fig. 32.1 places each stakeholder in a different community from the other stakeholders, however, this need not be the case either as long as some stakeholders participate in the process as representatives of different communities than others in the process are representing at that time. Finally, Fig. 32.1 suggests that each participant plays only one role in the process, however, sometimes the Data Provider is also one of the Analysts or the Discussant.

The multivocal analysis process begins with the Data Provider, situated within his or her research community and typically aiming to serve some specific research agenda, who collects a set of data. That Data Provider may have collected the data with the intention of sharing it with other analysts, or may have collected it specifically for his or her own purpose, and then decided later to share it. In either case, the data will typically have had a history before it is shared with the other analysts. The data along with that history is communicated by the Data Provider to each Analyst, and often the Discussant as well. Each Analyst does his or her own analysis. The Analysts may share their analyses with one another informally or formally, or may wait to share them until a formal, public presentation, which was the function of many of the workshops that played a prominent role in our process. This sharing and comparing may lead to iteration in the analyses themselves. Most of this volume focuses on that iterative process, however, we see in Fig. 32.1 that process is only one piece of the bigger picture. The Discussant is among the audience of the public presentation as are Other Researchers who are consumers of the research. The Discussant plays a role in synthesizing and comparing the analyses, and sometimes challenging the Analysts. The Data Provider is also a consumer of the analyses produced by the Analysts and the meta-analysis produced by the Discussant.

This schema applies equally to the very different processes that took place within each of the five data sections in this volume. Thinking about the lifecycle of an iterative multivocal analysis process, process related questions become relevant once the analysis process is under way, and the chosen answers along the way influence the path the team will take. For example, how much iteration is desirable, and when should analysts decide to stop? What are the inputs to and outputs of each iteration that allow these multiple cycles to progress? When qualitative and quantitative researchers work together, will it logically and practically make more sense for them to work in parallel or in alternating iterations? The analysts on the Fractions and Knowledge Forum dataset spent the most time discussing and comparing analyses because those datasets were chosen early on in the collaboration. In the Group Scribbles dataset, researchers struggled with a common way of referring to the data, and this hindered time spent on discussion. In both the Chemistry and Biology datasets, there were relatively few rounds of formal analysis, but many rounds of reflection and discussion in between. The extensive discussion provided the analysts significantly different lenses through which to view the data. Thus, even with a small number of iterations, in both cases, the understanding that the analysts came away with were significantly altered by the process.

Multivocal Seedlings

As a comparison to our own processes of multivocality, we will explore the experiences of researchers in training as they first experience a similar process. The hope is that this comparison will spark inspiration among both researchers in training as well as instructors of research methods. The Computational Models of Discourse Analysis (CMDA) course was designed to offer primarily first year language technologies graduate students the opportunity to learn to do multivocal corpus research in teams. For many of the students who took the course, this was their first course in any kind of research methods. The situation of this course contrasts with the teams of seasoned researchers that worked together in the five data sections of this book in many respects, which makes this course interesting as a comparison case. For example, whereas the teams that worked together for this book were assigned by the editors to specific datasets for strategic reasons relating to their distribution of analytic expertise and preferences, the CMDA teams emerged through whole class discussions about possible research directions. Because they were new to research methods, they did not come in with strong preferences in terms of their analytic approach, but as part of their orienting process, they were encouraged by the instructor to pursue distinct, complementary paths that would provide useful fodder for challenging one another as they worked together over the 9 week project. In each team, all students played the role of Analyst. In most teams one, or sometimes two, students played the dual role of Analyst and Data Provider. Frequently the data came from their research outside the course. The instructor played the role of Discussant. It is important for those who plan to undertake a similar endeavor that the role of the instructor is both to teach the methods and to help the students learn the important teamwork skills that are the heart of the process.

Potential Pitfalls of the Multivocal Process

In the five Data sections of the book, we sought to emphasize the value of multivocality. Nevertheless, our path has not always been smooth and easy, especially as these team analysis efforts have been our sometimes fledgling attempts at accomplishing something new. In hindsight the difficulties are not surprising as we consider the extent to which multivocality can be viewed as a counter-culture, as we have hinted above. Throughout this project we were feeling our way as we went. As a result, sometimes we fell into pitfalls, which we see more clearly now in hindsight. In our efforts, we sometimes found pathways for moving forward, which we offer now as helpful hints. Here we outline the main types of pitfalls, and we expand on each along with examples from our own efforts later in the chapter.

The first two pitfalls can be thought of as operating at a more macro level, where the team is constituted, and where the team reaches out beyond itself. The other pitfalls operate at a within-team micro level, as the teams work together as teams. So we begin enumerating the pitfalls at the macro level, and then move on to the micro-level pitfalls.

The first type of pitfall, termed *Team Setup pitfalls*, occurs at the time that the multivocal analysis team is formed to analyze a specific dataset. For example, a team may have been selected to represent a specific distribution of analytic approaches. However, just as we have acknowledged that Analysts may have some expertise with multiple analytic approaches, they may choose to approach the data in a different way than was intended by the one who invited them to participate. From a different angle, a lack of understanding of one analytic approach's needs or assumptions might lead to a failure to meet the preconditions of a fully satisfactory application of a method to the data.

The next type of pitfall, termed *Public Presentation pitfalls*, occurs in the public presentations that occur throughout the process wherever there is formal communication indicated in Fig. 32.1 as links. These pitfalls can be characterized as a failure to manage the many different audiences for the public communication that exist within the full set of stakeholders. There may be a failure to respect the trust and vulnerability of the Data Provider, which is experienced as a loss of control on the part of the Data Provider. Lack of sensitivity in these public presentations may engender defensiveness or resentment that works against the intention of the multivocal process. Another Public Presentation pitfall is a failure to communicate research results clearly to those with very different expertise who may struggle to fully grasp some methods that are not familiar to them. A final Public Presentation pitfall occurs when an Analyst targets the presentation to some specific other stakeholder, rather than framing it in a way that is of general interest.

Another type of pitfall, this time a micro-level pitfall, are termed *Data Transfer pitfalls*. These occur from the perspective of the Data Provider in the transfer of data to Analysts. This includes potential failure to set expectations for the work with the Analysts that is mutually acceptable. For example, as will be illustrated below, a Data Provider may make assumptions about what is appropriate data, and these assumptions may not match those of the other analysts. Unspoken expectations and sometimes unconscious assumptions are almost sure to end in disappointment. A failure to adequately communicate important contextual information about the data, how it was collected, and how it was sampled or cleaned up prior to transfer may lead to misunderstandings that can negatively impact the ability of the Analysts to do their work. Or they may invest time and effort into analyses that they later regard as unmeaningful because they were conducted under faulty assumptions about the data. Here "cleaning up" refers to processes of reformatting the data and possibly removing some aspects that are deemed not pertinent for the analysis. Therefore, "cleaning up" only makes sense in certain analytic approaches. In others, it might invalidate the analysis altogether. Once we make the assumptions explicit, then we will know (more explicitly at least) whether the analyses that we want to do are possible or not. Another way these pitfalls may lead to a waste of Analyst time and effort is if Analysts spend time producing an analysis that has already in some ways been done before the data were shared because this prior knowledge and understanding was not communicated to them in the transfer.

Some culpability for miscommunication of context may also reside on the side of the Analysts for reading too much intentionality into choices made by the Data Provider, such as representativeness of the data selected for sharing. We refer to these breakdowns as *Analysis Transfer pitfalls*. Another Analysis Transfer pitfall may occur when an Analyst is selective in which contextual information provided by the Data Provider to take into consideration, such as ignoring heterogeneity within the dataset caused by an experimental manipulation or the inherent hierarchical nature of the data (e.g., students nested within groups, nested within classrooms, nested within teachers, nested within schools).

Within the Analytic team itself, there may also be a failure to engage productively with one another. These breakdowns are considered at length in Chap. 34, Epistemological Encounters in Multivocal Settings, and are thus not the focus of this chapter.

Pathways Around Team Setup Pitfalls

Maintaining a Diversity of Analytic Approaches

An important question to address at the inception of a multivocal research process is the composition of the analytic team. The five teams of expert analysts featured in this book were assembled based on interest in a particular dataset, diversity in research approach, and in a few cases deliberate positioning of researchers outside of their analytical comfort zones, as we understood them. In every team, we attempted to include a mix of qualitative and quantitative approaches. Within the set of participating quantitative researchers, we included ones that make use of a diversity of different tools in their work, some of which apply statistical methods or machine learning, some of which make use of visualizations and other representations, and still others who use network analytic techniques. It was important to us that the researchers within a team would challenge one another to reflect more deeply on their approach and conclusions. Thus, it was important to pick people that were not only familiar with different techniques, but each have some commitment to something distinct from what others in the team were committed to.

While we selected researchers who were known for their work using particular methods, we did not take any sort of heavy-handed approach to managing their analyses. Thus, the analyses they provided sometimes surprised us in terms of the approaches that were taken. For example, in the Biology team, although we selected the Cress and Kimmerle team (Chap. 27, this volume) because of their sophisticated expertise in the area of quantitative methods, because of the early stage of the research that produced the dataset, they did not feel that this approach was appropriate for analysis of this data, and thus took a qualitative approach instead. While this meant that the team as a whole was differently balanced than we as editors had intended, the qualitative analysis they provided was still distinct from the analyses provided by the other team members, and thus still played into the multivocal process illustrated in that data section (Chaps. 25–30).

The management of the CMDA teams was looser at the team formation stage but needed to be tighter in the process stage in order to keep the processes moving forward because the students were new to corpus research. CMDA students who shared similar topic interests or research questions gravitated towards one another through whole class discussions that took place in the initial 7 weeks of the 16 week course. Usually one student within each group ended up spearheading the project idea and had some idea of where to get appropriate data. Nevertheless, while the teams themselves chose a topic focus and data, they were all required to orient their analyses to three themes, which were used to structure the 9 week projects into phases that were naturally punctuated by check points in which the teams made public presentations to the whole class and received feedback. These checkpoints served a similar purpose to the workshops that provided the impetus for the long term effort from which this book emerged. The three themes that the students were required to orient their analyses to included: the self in relation to individual others, the self in relation to the community, and communities in relation to each other. The three themes were meant to serve two purposes. First, within phases of the project, they were meant to provide a common focus as a counterpoint to their separate analytic approaches, similar to the purpose of pivotal moments in our work. Across phases, they were meant to push the teams to see how rich interaction data is, and how it is possible to view the same data from very distinct vantage points. Within each group, the students were each required to contribute their own view of how to analyze their group's dataset from the standpoint of the theme using their own chosen methods. They were then supposed to work together to integrate their differing perspectives before presenting their analysis to the class in a public presentation. This is analogous to but different from the notion of pivotal moments, which provided a common thematic focus for many of the multivocal research teams featured in this book.

Altogether there were four student teams. The student teams were meant to be heterogeneous in terms of analytic technique and expertise. Team 1 was the most heterogeneous in terms of expertise and ability level. This team focused on the 2012 Republican debates as their corpus. An emergent research question, which tied together the three themes and gave their project a united focus, was the question of in what ways and to what extent each candidate succumbed to pandering to the public. Their experiences included significant exposure to qualitative research, varying levels of quantitative research in applied linguistics and discourse analysis, and an undergraduate with very little research experience. The undergraduate played mainly a supporting role in the work. The qualitative student took a mainly grounded theory approach. The other two students explored their research question through alternative computational techniques. They spent significant time comparing and contrasting these techniques as operationalizations of pandering, and then evaluating them in light of the qualitative approach.

Team 2 was less heterogeneous in terms of methodological approach, but their work was more grounded in theory, and each of the five team members adopted a distinct theoretical framework that guided their analysis and provided contrasting explanations for learning in their corpus, which they then worked to reconcile. Their research question was what properties of interaction accounted for learning gains in a corpus of peer tutoring interactions. The framing of their question was itself quantitative, and they each gravitated towards pursuing their question using an approach in which they developed a coding scheme, worked to achieve inter-rater reliability with another team member, coded some data, possibly by hand or partly by hand and partly using machine learning techniques, and then performed a quantitative analysis, either in terms of distributions of codes or in terms of sequences of codes, to identify patterns that predicted pre to post test gains. In that sense, their team did not achieve the kind of methodological multivocality that was initially of interest. However, we see that even among researchers that share a set of analytic tools, a productive multivocality along the theoretical dimension can still be achieved. Although the theories they chose as lenses were quite distinct in terms of the independent variables, because they shared a common dependent variable and unit of analysis, they were able to explore whether these perspectives were providing alternative views on the same learning, or accounting for different learning within the same interactions.

Team 3 was unique in that they shared a common theoretical perspective, namely, Good Death theory (Steinhauser et al., 2000), but explored two distinct datasets, one of which was an online cancer support forum and the other of which was a corpus of suicide notes. The questions they pursued were how concepts from Good Death theory are reflected through language behavior in the two corpora, and what is in common and distinct in the experience of death between cancer victims and suicide victims. Three of the members of the four student team were highly quantitative in their approach, and made heavy use of machine learning in their work. The fourth student adopted an approach that was very similar to that taken by the students in Team 2 but was preceded by significant time reading whole posting histories of long time participants in the cancer support forum and gaining a qualitative sense for what was happening in the data before adopting a quantitative approach. Team 3 produced a very interesting analysis of the experience of death in an online discussion forum, which was submitted for publication soon after the end of the semester. Eventually it came to light, however, that in order to achieve this positive result, the one student in the group with the most insight into the data really drove the whole process, which allowed the other analysts to take less of a leadership role. This highlights the importance of continually monitoring group processes for breakdowns in teamwork in a multivocal process.

The fourth team was extremely homogenous in terms of approach, all taking a very quantitative approach. This team of three students came in with the least expertise of any of the teams in terms of understanding and facility with theory driven research. In their project, which focused on a Supreme Court Hearing corpus, they took a strongly atheoretical approach to their analysis. Their engagement remained at a very superficial level when it came to integration of results. In order to address this issue, in every meeting, the instructor continued to pose challenges to them regarding the interpretation of their results in order to spark more intensive engagement between their perspectives.

Like the experiences of the five expert teams from this book, there were varying levels of success at achieving the kind of exploration and interaction that was intended within teams. At each stage the instructor played an integral role at scaffolding the teamwork. At the team setup stage, the instructor pressed each student to take responsibility for playing a distinctive role on the team, and yet the instructor continually challenged the members of each team to think about how the view they were seeing through their analytic lens could speak to the other students in the group. Part of this scaffolding was the reward structure for the class, where each student received a grade that was based in a group grade for the project, but which was adjusted based on individual contribution. However, as we see with Team 3, this was not always successful. We see that there are many productive paths towards multivocality that do not require diversity in all of the dimensions. However, in the case where readers plan to similarly use multivocality as a paradigm for research methods instruction, it is important to note the ways in which they may need to intervene and scaffold the process at every stage in order for it to remain productive.

Satisfying the Preconditions of a Diversity of Analytic Methods

In our experience working on the analyses that led to this book, two tricky questions come up related to how to prepare data for a multivocal research process. First, what is an effective process for preparing data for sharing, overcoming challenges in data sharing, and specifically challenges in communication about data for secondary analysis? Second, how can we deal effectively with the fact that different analysis methods have different data needs (i.e., quantitative approaches require larger datasets, qualitative approaches require more intimate knowledge of the context from an insider's perspective whereas quantitative approaches seek objectivity)? While setting up the data may be viewed as a mundane aspect of the analytic process, it is this early stage where seeds of pitfalls are sewn. The problem starts with the tendency towards isolation referred to above. Data analysis is not frequently undertaken as a team sport. Instead, researchers are more likely to retreat into their cave with their favorite analysis tools, doing the initial exploration of their data in tandem with data preparation. It is during the exploration phase that more plans for more extensive analysis form (apart from planned contrasts directly related to the experimental manipulation if any). However, this time of individual exploration and data manipulation might lead to work being done to set up the data that may need to be undone or redone differently later in order for the analysts to work together as a team later in the process. Important questions must be answered about how to meet the preconditions of the diversity of analytic approaches that will be used across Analysts. The later these questions are addressed within the team, the more likely it is that work done up front will have ended up including time and effort wasted on work that will need to be redone later or that was redundantly performed by more than one researcher.

Choosing and Preparing Data

The team that worked on the Chen and Looi Group Scribbles dataset provides the impetus for the first of these questions, which will give us the opportunity here to reflect on where assumptions about data gathering and preparing can come from. These reflections will give us the opportunity to observe how conceptual debates within the field can trigger forces towards isolation within teams that need to be overcome in order to achieve multivocaltiy. We will frame this discussion in terms of contrasts between "naturally occurring" or "authentic" data vs. "contrived" or "researcher provoked" data. In the case of Group Scribbles, the data provider made assumptions about how the data should be prepared that did not match with how the analysts wanted the data. We will see that the data provider was situated in the "naturally occurring" data gathering paradigm, and he had additional assumptions about preparing his data for analysis, that were not shared by the other analysts.

Let us consider the assumptions behind the debate concerning "naturally occurring" or "authentic" data vs. "contrived" or "researcher provoked" data. The former stem from human interactions that would have occurred even in the absence of the data-collection activity or in other words, they pass the "dead social scientist test." "If the researcher got run over on the way to the university that morning (Potter, 2004, p. 612)," would the interaction have nevertheless occurred and played out in the same way? The latter type of data are relatively contrived social science data sources such as surveys, interviews, and focus groups (Speer, 2002) and although there are many differences to be discussed between this and experimental data, for the purposes of this chapter we add to this latter type, data from experiments where participants perform tasks under controlled settings. The comparison of these two data types is meant to illustrate that a researcher gathers data in order to perform specific analyses (this means that other analyses could be difficult or impossible to do) and that the way in which the data is gathered and made ready is compatible with the researcher's assumptions (this means that the data gathered could be incompatible with other researchers' assumptions).

So, how can this distinction between "naturally occurring" and "artificially provoked" data help us illustrate our argument that data is collected and prepared according to underlying assumptions? Speer (op. cit.) gives an insightful overview of the views of conversation analysts that stem from its ethnomethodological origins on why experimental data is problematic. She goes on to argue that just because ethnomethodologists attribute their own assumptions and higher order goals to experimental data in order to explain why it is problematic does not mean that such data is problematic for researchers in an experimental paradigm who have their own assumptions. First of all, the ultimate objective for ethnomethodologists—as developed by Harvey Sacks and colleagues—was to produce an inventory of "recognizable social actions in this culture… the aim is to find it and provide an account of it empirically and precisely, not imaginatively or typically or hypothetically or conjecturally or experimentally, and to use actual, situated occurrences of it in naturally occurring social settings to control its description" (Schegloff, 1996a, p. 167). Given this, the assumption is that if researchers use "written texts, monologues, talk or writing produced under experimental or quasi-experimental conditions" (Schegloff, 1996b, p. 468) then since "these materials are not drawn from the naturally occurring interactional environments which seem to be the natural, primordial home for language use," the inventory of social actions will be compromised because the "primary and proximate interactional practices which undergird" the specific social action we are studying "may be largely or totally absent, often suppressed by specially designed circumstances of production" (Schegloff, 1996b, p. 468). So this is an argument that ethnomethodologists make against experimental data as analyzed in typical quantitative research methodologies, but according to their own assumptions.

If our goal as researchers is providing an inventory or a catalog of recognizable social actions as they occur naturally, it does not make sense to use imaginary, made-up examples of language use that are purported to be typical, based on the intuition that there is reason to doubt such conjecture. And it is certain that experimental conditions do *not* embody ordinary contingencies of interaction; instead, they "confront participants with quite distinctive, and potentially complicating, interactional exigencies" (Schegloff, 1999, p. 419). But researchers who are not ethnomethodologists may not be seeking to produce an inventory of recognizable social actions as they occur naturally, during ordinary conversation. Their goal may be to flesh out how experimental conditions do indeed affect language use within group interactions and to make probabilistic assertions about that, a goal foreign to conversation analysts (Golato, 2003). They may hypothesize that experimental conditions could *provoke new* language use, not usually present in ordinary interaction, but that may be beneficial for learning, for example. In that case, their data matches their assumptions and goals. Experimental data therefore escapes the criticism of not being naturally occurring, as it was never argued as being so and since experimental methods are used for different goals than conversational analysis, they can coexist, as long as experimental researchers do not treat their interventions as "neutral resources for accessing some truth or reality beyond or beneath the data" itself (Speer, op. cit.).

This discussion about how researchers in different paradigms have different assumptions about gathering and preparing data should have made clear that once it has been assured that gathered data is both appropriate for specific analyses to be carried out and compatible with a particular set of researchers' higher objectives and assumptions, difficulties will most likely be encountered when data is shared with researchers not party to the gathering. Although other researchers may share interest for a particular type of setting (e.g., group interaction), they can hold different assumptions about it, have different goals, focus on different aspects of it and as a result want to analyze data of a specific nature different from that provided by the data gatherers.

Though the issues the team examining the Looi & Chen (Chap. 14, this volume) data experienced grew out of seemingly unsolvable debates in the field, eventually the team worked out a productive solution. In particular, the Data Providers initially furnished a synthesis of their vision of the pedagogical interaction (concerning electricity) and not a transcription of the actual interaction. If this was suitable for their own goals and assumptions, it was not suitable for example, for Lund & Robinault (Chap. 17), who requested transcriptions. It was important for their research questions to have (1) complete transcriptions of talk, (2) talk as uttered and not modified (e.g., summarized) by the person doing the transcribing, (3) correct differentiation of turn taking—both in terms of content expressed and in terms of interactional chronology in relation to other turns, and finally (4) correct differentiation of speaker. In addition, since they analyzed the interaction from the view of physics didactics, it was also important for them to have knowledge of the sequence of learning activities of which the one classroom action was part: what kind of knowledge about electricity did the students have coming into the classroom? What were the specific learning goals for the classroom activity that was recorded? What kind of course was to follow? This information was not initially provided, perhaps because it was not all relevant to the data providers' own analysis. However, the needed information was eventually offered to the analysts who needed it when the problem came to light. This required some unanticipated effort on the part of the Data Providers, however.

The Interrelations of Methods and Data

The team that worked on the Fractions dataset provides the impetus for the question regarding the interrelations and constraints of applying any given method to a particular dataset, but also the method's relation to theory. The Fractions team was nicely poised for multivocality in terms of distribution of analytic expertise. In particular, Ming Ming Chiu (Chap. 7) was selected as an analyst because of the sophisticated statistical discourse analysis (SDA) technique that he contributed, which is striking in its contrast both to qualitative approaches as well as other, simpler quantitative approaches featured within the book that are not able to capture the sequential nature of collaborative discourse. However, a sophisticated statistical technique requires a large amount of data in order to be used appropriately and the fractions dataset was quite small; it was a transcript of one group discussion. While the typical tests for over-fitting using this approach did not show unequivocal signs of over-fitting, one must still use extreme caution when drawing conclusions from such a complex model applied to such a small amount of data.

Each type of method has its own constraints for application to data. Issues related to over-fitting are specific to quantitative approaches. A small amount of data can still provide the basis for insightful thick description using a qualitative technique. In fact, qualitative researchers may find the opposite challenge on multivocal teams. While there are well-established methods for sampling from a larger corpus to identify segments that can be approached from a qualitative standpoint, it still remains to be worked out how to integrate analyses across approaches when the quantitative analysis is applied to the whole corpus, whereas the qualitative analysis is applied only to a small portion. A qualitative analysis may be valuable even if it is not meant to illustrate a pattern that is claimed to be generalizable to the whole corpus. In fact, its value may be precisely because the scenario that is being highlighted is unusual. Therefore, the goals of the analysis from a qualitative standpoint and quantitative standpoint may be distinct, and thus the findings may require some creativity in order to integrate in a valid way. This exchange raises an important caveat that applies at the Analysis Transfer stage discussed below. Specifically, it is extremely important within multivocal teams at the time when analyses are shared across the Analysts that they take care to consider the limitations of what can be concluded from one another's analyses depending on the extent to which the preconditions for a felicitous application of those methodologies were met in the data. We observed difficulties in this regard, especially in connection with Ming Ming Chui's analysis in both Chaps. 7 and 23, because the other Analysts in the community found his approach to be beyond their level of technical expertise and somewhat mystifying, and thus they found themselves less capable of evaluating or challenging this work. A potential pathway towards addressing issues like this would be for Analysts within teams to offer short tutorials to one another to build common ground prior to exchanging analyses.

How Student Teams Dealt with Preparing Data and Choosing Methods

As mentioned earlier, there were four student teams in the CMDA course. The first question on choosing and preparing data was universally an issue for the student teams. While challenges with respect to data sharing and comparison of analyses came up for the student teams, none were insurmountable. In examining their experiences, we can learn to anticipate such issues and prepare for them in such a way that they can be dealt with efficiently. All of the datasets that students made use of required a substantial amount of time and effort to set up for the analysis. Thus, one important lesson we can learn is to make sure when constituting analytic teams that each includes someone with expertise in data cleansing and manipulation, and that time for that preparation is taken into account when the project is planned. Furthermore, that person should take responsibility to prepare the dataset for sharing before the other analysts begin their work. Such processes should not be entered into glibly, however, since as we discussed above, "cleaning up" the data is only a standard practice in some methodologies, and is inconsistent with others. So some serious discussion must be conducted with the team of analysts in order to agree on what makes sense in order to prepare for their joint endeavor that respects each represented methodology, or at least represents a compromise all of the Analysts agree to up front. Making sure this occurs effectively is the job of the instructor.

The second question on interrelating methods and data was most relevant for Team 1, where the strongest commitment to a grounded theory approach was found in one team member. The grounded theory method yielded two complementary sets of themes, one of which was related to the topics addressed by each candidate, and the other of which was related to the argumentation strategy that candidates adopted. Using these two sets of themes, when applied to debate transcripts, allowed the team to explore how candidates differed in terms of their associated distribution of strategies, but also how some kinds of strategies were more associated with some topics than others. As a comparison between the qualitative and quantitative approaches, they were able to determine whether the kinds of group level differences they saw with the automatically derived topics were similar to the ones they saw using the topics identified using a grounded theory approach. The challenge in that collaboration was that it did not become clear what the most valuable contrast between the qualitative approach and the quantitative approach would have been for constructing one integrated understanding of the dataset in the end of the project until relatively late in the process. In the case of this team, alternating between qualitative and quantitative analyses might have been a more strategic approach since the quantitative analysis cycles were quicker and were therefore able to encompass more data within shorter amounts of time.

Pathways Around Public Presentation Pitfalls

The Biology Data

The analysis of the Biology data is a good illustrative example of Public Presentation pitfalls. The analysis of this data was necessarily iterative since not all of the data included in the final analysis was available when the collaboration across groups began. The initial analysis also sparked a fair bit of controversy, including a question about whether multivocal analysis is even appropriate for data collected within an environment at an early stage of development.

A number of issues came up in the initial public presentation of the Biology dataset at the Alpine Rendez-Vous workshop at Garmisch-Partenkirchen in 2009. Some of these were symptomatic not only of Public Presentation pitfalls, but also of Data Transfer pitfalls, which will be the focus of the next section. The current Data chapter in the Biology section (Dyke, Howley, Kumar, & Rosé, Chap. 25, this volume) includes an explicit write up of the constraints the Data Providers were working within when collecting the data, which were not communicated to the Analysts adequately before they began their work. This information was included in the write up of that chapter as a response to this public discussion. Much of the discussion at this initial public presentation focused on what the researchers should have done differently, some of which were things the researchers did not have any control over given the context of their work, and some of which were issues they were aware of but were not the focus of their investigation. The time of this public presentation was not the appropriate time for these constraints to come to light for the Analysts. Time spent on discussion of issues the Data providers were already aware of and issues that were beyond their control took time away from what could have been a more productive intellectual exchange.

Another issue came up in Stahl's presentation and the subsequent write up of the Stahl analysis (Stahl, Chap. 28, this volume). Here the issue was that the chapter was written in the frame of communication from Stahl to the Data providers specifically, and focused largely on advice that would be useful to the Data providers in their process but might not provide value to other researchers not specifically involved in the design process because of its level of specific focus on the prototype intervention used to collect the data. Since Stahl and the Data providers were close colleagues for whom a frank exchange of views was the norm, and in fact was quite welcome in private communication, the main issue to be considered was whether the presentation was appropriate for public consumption. That analysis focused largely on lists of things Stahl would have done differently if he were setting up the experiment. The question here is whether the value in multivocality is in license to publically present unmitigated criticism, or whether there might be some more productive exchange that can take place in public settings when the object of analysis is data from a pilot experiment. This analysis contrasts with the Cress & Kimmerle analysis (Chap. 27), where a similar focus on what could be improved in the design was offered, but it was contextualized in a theoretical framework in a way that spoke to a broader research community.

As we see, the initial analyses were critical of the collaborative environment and the design of the study that produced the data in a way that the Data Providers found tangential to the questions the study was meant to address, and the way some of this was communicated in public presentations resulted in some angst. The analysis chapters in the Biology section preserve the issues with respect to Public Presentation pitfalls discussed here in order to provide visibility to researchers interested in embarking upon a similar journey in their own work. Nevertheless, it should be acknowledged that the initial divergence of perspectives did eventually lead to a series of productive interactions between analysts that challenged the Data Providers in their conception of the research as well as challenging the analysts in their conception of multivocality. In the context of the group discussion at the Garmisch-Partenkirchen workshop, the Data Providers came to understand the underlying conceptual differences in assumptions about the ideal role of the facilitator in discussion groups that the different Analysts highlighted. This distinction eventually became the lens through which the diverse group of Analysts was able to debate and build consensus. This teaches us that while these pitfalls are prone to occur, and though they may cause some temporary friction within an analytic team, they need not cause irreconcilable difficulties in collaboration. Perhaps a take away is that Data Providers should come in to the process with an expectation that some thickness of skin and perseverance will be necessary.

After rounds of discussion and reflection, an additional analyst (Goggins & Dyke, Chap. 29, this volume) was added to contribute an extensive network analysis, to bridge the coding and counting approach in the original Howley et al. analysis (Chap. 26, this volume) and the qualitative Stahl analysis. After this interaction, a second data collection effort in an updated version of the environment provided a complementary set of data, and then eventually a third. After much reflection and discussion, the team converged in their understanding of multivocality and its role in iterative, design based research, and the multivocal process resulted in a number of observations that led to a successful redesign of the intervention. The successful intervention then produced new knowledge for the field about how conversational agents can be used to support group discussion. And the resulting agent design represented insights drawn from diverse perspectives on appropriate support for group discussion in the field from those who may not have had the opportunity to work together apart from a desire to engage in a multivocal process.

The Student Teams

The student teams struggled with different issues in their public presentations than the expert teams. In particular, because of their relatively early stage of familiarity with the methods they were using in their analyses, they found it challenging to clearly articulate their analyses and findings and to compare and contrast with one another in preparation for these public presentations. Thus, although the teams presented together, and although they were instructed to present not only their own analyses but also lessons learned from interaction within their teams, many presentations came across as a patchwork. This left more work for the instructor who was acting as a Discussant to engage the teams, sometimes privately in advisory sessions with the groups prior to public presentations on some iterations, but also in the context of whole class discussion. The purpose of this scaffolding was to clarify what each analysis demonstrated and how the alternative analyses may challenge one another. These facilitated discussions served to scaffold the communication and coordination between analysts that often occurred outside of public view in the more experienced teams featured in this book. Thus it is important to note that when incorporating multivocality in an instructional setting, these public presentations may serve as valuable opportunities to learn the multivocal process itself rather than simply opportunities to communicate with a broader audience.

Pathways Around Data Transfer Pitfalls

The Chemistry Data

The analysis of the Chemistry data provides a convenient example of Data Transfer pitfalls. The analysis of this data proceeded in three phases. In a first phase, an initial version of all of the analyses was completed by individual researchers. All but the Sawyer analyses were presented at a workshop in 2010 at the International Conference of the Learning Sciences (ICLS). During the discussion at the workshop, the leadership theme emerged and then became a consistent thread in all subsequent analytic work by the team. Eventually, both the Rosé and Strijbos teams revised their characterization of their respective multidimensional coding frameworks. The workshop sparked a collaboration between the Rosé team and the Strijbos team, which proceeded in terms of informal discussions over more than a year, and then finally a formal reanalysis in time to write a chapter about their integration (Howley, Mayfield, Rosé, & Strijbos, Chap. 11, this volume). Discussions with the other analysts proceeded in parallel. Elaboration of both of those other analyses ensued, until finally the discussant used the emergent leadership theme to contrast the findings across the four analyses.

The Data Transfer pitfall that came up in this process occurred immediately subsequent to the 2010 workshop. The culpability here in the transfer may have been on the side of the Analysts some of whom may have read more in to the data sampling process than was warranted. The Data provider had chosen two discussion groups whose style provided an interesting contrast from the standpoint of the theoretical framework that motivated the data collection in the first place. The concept of leadership was not central to the contrast that the Data Providers were necessarily interested in. And the Data Providers never asserted that the particular problem solving episode that was selected for examination in order to compare the two groups was necessarily representative of every aspect of the collaboration within those groups. As the discussion comparing analyses across Analysts turned towards leadership, it was tempting to draw inferences about relationships between the students within the groups from the small amount of data that was provided. Interest in pursuing the issue further led to a request for the Data Provider to offer more transcripts from the same groups solving other related problems. Only one Analyst actually examined this larger corpus in detail. That extended analysis revealed that the encounters examined jointly by the multivocal team were not in fact representative of consistent leadership taking within the groups. This additional analysis was eventually dropped since it was conducted only by one analyst. In hindsight it was not surprising that while the selected examples served the original intention of the Data providers in their own analysis, it was not necessarily ideal from the standpoint of analyses that focused on different issues. This simply raises a note of caution to take into consideration during the process of transferring data for multivocal analyses. If the analysts had asked more questions up front, they might have focused their questioning of the data in a direction that was more consistent with the considerations used in selecting the sample, or they might have negotiated for different sampling criteria in the data sharing process.

The Student Teams

With the student teams, issues with respect to data transfer came up primarily for teams 1 and 2. In the case of team 1, the Republican Debates, not all of the students within the team shared the same amount of expertise regarding the American political process, and thus some insights about how the data might be productively questioned were not shared. For example, some students who did not grow up in the United States were far less aware of the important role of the region in which a debate took place would play in terms of what could be assumed about the audience the debaters were presenting themselves to. Eventually more knowledgeable students within the groups shared their expertise with the less knowledgeable students as part of the process of working towards integration of findings. However, in hindsight, scheduling in time to have these discussions at the very beginning of the process might have allowed the team to use their time more efficiently. A more difficult issue came up for Team 2, who was using a dataset that none of them had participated in collecting. In the case of the expert teams, there was a plan for the data to be shared with other analysts, and written documentation to facilitate the sharing was provided in addition to the public presentation of the data that occurred at the workshops at a key stage. In the case of the student teams, they received the data from a researcher who was remote from the process and did not anticipate what the student team would need to know in order to do its work. Thus, the team engaged in a lot of guess work about the data up front. The time lost in the process due to the guess work highlights the importance of taking the data transfer stage seriously as a critical part of the multivocal process and thus actively engaging the Data Provider.

Pathways Around Analysis Transfer Pitfalls

The Fractions Data

Engagement between researchers is the underlying concern of Analysis Transfer pitfalls. The Fractions team was exemplary in terms of the level of engagement of the analysts in the multivocal process. The sharing and analysis of the Fractions data began previous to the 2009 Alpine Rendez Vous in Garmisch-Partenkirchen, where three initial analyses were presented by individual researchers: Shirouzu, Chiu, and Trausan-Matu. These analyses continued to evolve, already influenced by each other and rediscussed informally at a workshop in Chicago, 2010 at ICLS and more formally in preparation for a symposium in Hong, Kong, 2011 at the Conference for Computer Supported Learning (CSCL) conference. At the Alpine Rendez Vous 2011 in La Clusaz, France, analyses evolved still further, with Shirouzu, the data provider taking a kind of integrative stance, recognizing that different units of analysis and frameworks could be complementary, and both Chiu and Trausan-Matu making changes to their views as well. Discussions led to so many changes in analyses that Lund, the discussant for the Fractions section had to rewrite the chapter (Lund & Bécu-Robinault, Chap. 17, this volume) many times in order to account for them. Nevertheless, what we see here is evidence that continued engagement between analysts can lead to progressive refinement of constructs, ideas, and conclusions over an extended period of time.

The Chiu analysis (Chap. 7, this volume) in particular highlights other relevant challenges in sharing results across very different analytic approaches. Although Chiu was indeed responsible for assuring that his method was applicable to the fractions data, certain contributions of his analysis within the context of the fractions analysis team lay on a different level than his results, per se. This can lead to challenges in communication across Analysts. For example, the pivotal moments that Chiu found prompted Shirouzu (Chap. 5) to give meaning to them in his own framework, illustrating that the results of one method can be reinterpreted within an alternative theoretical framework. Chiu claims his SDA method to be in a sense *atheoretical,* able to be used with a variety of theoretical frameworks (although this was questioned by Fujita, Chap. 24). In the fractions section, he looked for micro-creativity, but he could search for patterns of any type. These issues are not insurmountable however, they must just be carefully considered and explicitly discussed among Analysts.

The Group Scribbles Data

The Group Scribbles experience stands in contrast especially with that of the Fractions dataset where there was a notable intensive exchange between analysts over multiple iterations. With the Group Scribbles team, the analysts required some prompting, sometimes by discussants, to engage deeply with the distinctions between their analyses. Data providers Chen & Looi (Chap. 14) shared the Group Scribbles data at the end of 2010 and initial analyses on the data concerning electric circuits were presented in March of 2011 at the Alpine Rendez Vous. Contrary to the other datasets, the data providers did not present their data in person, but rather sent the group documents describing it. Suthers presented the dataset at the ARV2011, and he became the discussant for this section (Chap. 19, this volume). Jeong (Chap. 18), Lund & Becu-Robinault (Chap. 17), and Medina (Chap. 16) all contributed analyses, in addition to an analysis offered by colleagues (Wee, Song, & Looi) of the data providers. The team had first drafts of all analysis chapters at the end of June 2011, after some analysts obtained partial transcriptions that they put together from the videos. Some of the group members met Lund at CSCL2011 in Hong Kong in early July to discuss how our conclusions compared, but this was difficult as the analysts were analyzing different empirical material and so did not have a simple way of referring to places in the dataset that would be easy to map from one analysis to another. Until January 2012, discussion continued over e-mail through part of August 2012 when Suthers posted the discussion chapter and Jeong and Lund commented on it. But, it was only after this that Lund & Becu-Robinault succeeded in aligning the transcript they had greatly modified (in order to respect their epistemological constraints) and their pivotal moments with the data providers' original synthesis of the interaction. Unsurprisingly, the lack of a common reference to the data greatly hindered the exchange between analysts.

The Knowledge Forum Data

Like the Fractions dataset, work on the Knowledge Forum data began at the 2009 Alpine Rendez Vous in Garmisch-Partenkirchen, where two separate analyses of the data were presented, namely, Teplovs and Fujita as one, and Tscholl and Dowell as the other. The first round of analyses revealed some challenges with the multivocal process. In particular, the Tscholl and Dowell analysis was eventually discarded because the sample selected for up close analysis was felt to be nonrepresentative by the data provider, and thus the conclusions drawn from the analysis did not have face validity from the data provider's perspective. The original Teplovs and Fujita analysis focused more heavily on the methodology and less heavily on the data than the current Teplovs and Fujita chapter (Chap. 21). This prompted a discussion by Rosé that similarly focused on methods rather than substantive conclusions that might have fed back into the design based research process that produced the data. We see here that a multivocal process can get off to a slow start.

Ultimately, the outcome of this first round was disappointment to the data provider that the analyses and discussion did not necessarily further the research goals of the project nor fully appreciate the complexities of the data. In response, the Teplovs and Fujita chapter that is included in the book addresses the research goals of the project more explicitly and clearly than the initial analysis presented in 2009, and an additional analysis was invited, this time coming from within the Knowledge Building community itself by researchers who were able to fully appreciate the larger goals of the project, namely, Law and Wong (Chap. 22). A further analysis was invited by Chiu (Chap. 23), which provides a purely quantitative sequential analysis of the data, in the same spirit as the Chiu analysis provided for the Fractions dataset. This second round provided a richer multivocal experience. Nevertheless, convergence is difficult. In the end, as seen in the Discussant chapter written by the data provider herself (Chap. 24), the value in the multivocal process was attributed to the impact it had on her view of the data, but not in its contribution to the design goals that prompted the data collection. In this case, one might conclude that further iteration would be required in order to provide that needed convergence in order to inform design.

The Student Teams

As mentioned, the CMDA students were required to orient their analysis to three related themes, each associated with a major presentation the groups were required to give to the whole class, which provided answers to these questions for them to follow. At each check point, the instructor acted as the discussant, giving the teams suggestions for how they might push their individual analyses further as well as explore comparisons and contrasts between analyses. Each check point presentation also involved time for group discussion.

Iteration was really essential for the student teams in order to spur them to do thorough analyses, since they were just learning what it meant to do a rigorous analysis. This was, of course, not an issue with the experienced researchers we were privileged to partner with in creating the five data sections featured in this book. When the student teams reached their first check point, none of the groups had done the integration. They spent until the wee hours of the morning of the presentation doing their own analyses, so each team presented a patchwork that was not integrated. The instructor sent them back with the feedback that at the next check point they should strive for better integration. When the teams presented their second theme analyses, the results were much stronger in every group. All but one group had several interesting stories to tell. But they still did not leave time before the presentation to think about the integration. Again, they were up until the wee hours the morning of the presentation finishing up their own analyses. The instructor then decided that since most of the groups had substantial raw material for a multivocal analysis by then, they were allowed to abandon the three themes in order to place their emphasis on the integration question for the third iteration. By the final presentation, most of the teams had come to a point of seeing value in integration and had spent substantial time working to reconcile the alternative perspectives offered by the distinct approaches their teams had pursued.

In hindsight, one might argue that trying to teach multivocality to students simultaneously as they are learning their own analytic methods is just too high of cognitive load and that this should be regarded as an advanced method, not to be entered into until one has some facility with at least one research method. An alternative option may have been to take a jigsaw approach, however, this would have been difficult in the CMDM course due to resource constraints from the side of instructional time and teacher resources on the one hand and breadth of learning objectives on the other. Nevertheless, the CMDM course indeed struggled along the way with concepts related to conducting theory driven research at all, and found it difficult to manage their time in order to balance doing their own analysis with integrating with those of their team mates, many students commented after the fact about having benefitted from working with their team and being challenged by their team. Thus, it would certainly be possible to argue that despite the cognitive load demands, there was benefit from the additional struggle because of the broader perspective it provided. Furthermore, in working in a group on the analyses, students were able to be more ambitious in their goals for the analysis.

Lessons Learned: Reflections on Methods for Multivocality

In this chapter we have worked to abstract away from the specific processes that the five teams in the book engaged in that were illustrated within the five data sections, and have explored some practical questions in light of lessons learned from these processes. As a comparison case, we have contrasted the expert teams that worked with us on this book with a set of four student teams just learning how to do theory driven research and engage in a multivocal process in tandem.

In addition to the rules of thumb and practical suggestions that have been offered in this chapter, we can draw some conclusions in reflection. What we see is that in both the expert teams as well as the student teams, there were ways in which the multivocality proceeded successfully as planned, ways in which it did not work out so well, and ways in which it worked out differently than planned, but successfully nonetheless. It is the last of these three points that is potentially the most important, because we see that it is possible to benefit from a multivocal process even when it is not perfectly planned out to begin with, and even if it doesn't play out exactly as anticipated. Thus, one should not shy away of multivocality for fear of making mistakes. Even the student teams who were just fledgling researchers, for the most part, benefitted from being challenged to look at their data from multiple perspectives. Many of them were surprised in the end that they found out how brittle an analysis conducted from only one perspective might be and how subtle differences in operationalization even of constructs that seem to be identical when conceived from a conceptual standpoint dramatically change the claims one feels comfortable making as a result of the analysis.

Perhaps the most valuable lesson learned in all of this is the contrast between the real benefit of a multivocal analysis and more standard mixed methods approaches. Whereas there is increasing consensus about the benefits of a mixed methods approach for research findings, strength of the conclusions, and depth of insight into the target phenomena, there is still something missing from mixed methods research that is gained through multivocal methods. Whereas mixed methods approaches benefit the research, multivocality benefits the research community, forging new connections in terms of relationships and publications between researchers and their respective communities that did not exist before. This sentiment is echoed in the words of the discussant of the Knowledge Forum section, who is a researcher well experienced in mixed methods approaches prior to participation in the multivocal research process that produced this volume (Fujita, Chap. 24, this volume). In mixed methods there is one agent, so the methods are not likely to challenge each other deeply. In multivocal analysis, there is a different agent representing each method, so the dialogue can be more genuine, "multivocal," and there can be more substantial challenges.

Acknowledgments This work was supported by the Pittsburgh Science of Learning Center and Graduate Training Grant from the Department of Education (#R305B040063) and by the French CNRS.

References

Bakhtin, M. (1981). Discourse in the novel. In M. Holquist (Ed.), *The dialogic imagination: Four essays by M. M. Bakhtin*. Austin: University of Texas Press.

Chen, W. & Looi, C.-K. (this volume). Group Scribbles-supported collaborative learning in a primary grade 5 science class. In D. D. Suthers, K. Lund, C. P. Rose, C. Teplovs & N. Law (Eds.), *Productive multivocality in the analysis of group interactions*, Chapter 14. New York: Springer

Chiu, M. M. (this volume). Social metacognition, micro-creativity and justifications: Statistical discourse analysis of a mathematics classroom conversation. In D. D. Suthers, K. Lund, C. P. Rosé, C. Teplovs & N. Law (Eds.), *Productive multivocality in the analysis of group interactions*, Chapter 7. New York: Springer

Chiu, M. M. (this volume). Statistical discourse analysis of an online discussion: Cognition and social metacognition. In D. D. Suthers, K. Lund, C. P. Rosé, C. Teplovs & N. Law (Eds.), *Productive multivocality in the analysis of group interactions*, Chapter 23. New York: Springer

Cress, U., & Kimmerle, J. (this volume). Successful knowledge building needs group awareness: Interaction analysis of a 9th grade CSCL biology lesson. In D. D. Suthers, K. Lund, C. P. Rosé, C. Teplovs & N. Law (Eds.), *Productive multivocality in the analysis of group interactions,* Chapter 27. New York: Springer

Dyke, G., Howley, I. K., Kumar, R., & Rosé, C. P. (this volume). Towards academically productive talk supported by conversational agents. In D. D. Suthers, K. Lund, C. P. Rosé, C. Teplovs & N. Law (Eds.), *Productive multivocality in the analysis of group interactions,* Chapter 25. New York: Springer

Fujita, N. (this volume). Critical reflections on multivocal analysis and implications for design-based research. In D. D. Suthers, K. Lund, C. P. Rosé, C. Teplovs & N. Law (Eds.), *Productive multivocality in the analysis of group interactions,* Chapter 24. New York: Springer

Goggins, S. P., & Dyke, G. (this volume). Network analytic techniques for online chat. In D. D. Suthers, K. Lund, C. P. Rosé, C. Teplovs & N. Law (Eds.), *Productive multivocality in the analysis of group interactions,* Chapter 29. New York: Springer

Golato, A. (2003). Studying compliment responses: A comparison of DCTs and recordings of naturally occurring talk. *Applied Linguistics 24/1*, 90–121, Oxford University Press

Howley, I. K., Kumar, R., Mayfield, E., Dyke, G., & Rosé, C. P. (this volume). Gaining insights from sociolinguistic style analysis for redesign of conversational agent based support for collaborative learning. In D. D. Suthers, K. Lund, C. P. Rosé, C. Teplovs & N. Law (Eds.), *Productive multivocality in the analysis of group interactions,* Chapter 26. New York: Springer

Howley, I. K., Mayfield, E., Rosé, C. P., & Strijbos, J.-W. (this volume). A multivocal process analysis of social positioning in study groups. In D. D. Suthers, K. Lund, C. P. Rosé, C. Teplovs & N. Law (Eds.), *Productive multivocality in the analysis of group interactions,* Chapter 11. New York: Springer

Jeong, H. (this volume). Development of group understanding via the construction of physical and technological artifacts. In D. D. Suthers, K. Lund, C. P. Rose, C. Teplovs & N. Law (Eds.), *Productive multivocality in the analysis of group interactions*, Chapter 18. New York: Springer

Law, N., & Laferriere, T. (this volume). Multivocality in interaction analysis: Implications for practice. In D. D. Suthers, K. Lund, C. P. Rosé, C. Teplovs & N. Law (Eds.), *Productive multivocality in the analysis of group interactions*, Chapter 35. New York: Springer

Law, N., & Wong, O.-W. (this volume). Exploring pivotal moments in students' knowledge building progress using participation and discourse marker indicators as heuristic guides. In D. D. Suthers, K. Lund, C. P. Rosé, C. Teplovs & N. Law (Eds.), *Productive multivocality in the analysis of group interactions*, Chapter 22. New York: Springer

Lund, K., & Bécu-Robinault, K. (this volume). Conceptual change and sustainable coherency of concepts across modes of interaction. In D. D. Suthers, K. Lund, C. P. Rose, C. Teplovs & N. Law (Eds.), *Productive multivocality in the analysis of group interactions*, Chapter 17. New York: Springer

Lund, K., Rosé, C. P., Suthers, D. D., & Baker, M. (this volume). Epistemological encounters in multivocal settings. In D. D. Suthers, K. Lund, C. P. Rosé, C. Teplovs & N. Law (Eds.), *Productive multivocality in the analysis of group interactions*, Chapter 34. New York: Springer

Medina, R. (this volume). Cascading inscriptions and practices: Diagramming and experimentation in the group scribbles classroom. In D. D. Suthers, K. Lund, C. P. Rose, C. Teplovs & N. Law (Eds.), *Productive multivocality in the analysis of group interactions*, Chapter 16. New York: Springer

Potter, J. (2004). Discourse analysis. In M. Hardy & A. Bryman (Eds.), *Handbook of data analysis* (pp. 607–624). London: Sage

Schegloff, (1996a). Confirming allusions: Toward an empirical account of action. *American Journal of Sociology*, *102*(1), 161–216

Schegloff, (1996b). Some practices for referring to persons in talk-in-interaction: A partial sketch of a systematics. In B. A. Fox (Ed.), *Studies in anaphora*. (pp. 437–485). Amsterdam: Benjamins Publishing Company

Schegloff, E. (1999). Discourse, pragmatics, conversation analysis. *Discourse Studies, 1*(4), 405–435.

Shirouzu, H. (this volume). Focus-based constructive interaction. In D. D. Suthers, K. Lund, C. P. Rosé, C. Teplovs & N. Law (Eds.), *Productive multivocality in the analysis of group interactions*, Chapter 5. New York: Springer

Speer, S. (2002). "Natural" and "contrived" data: A sustainable distinction. *Discourse Studies, 4*, 511–525.

Stahl, G. (this volume). Interaction analysis of a biology chat. In D. D. Suthers, K. Lund, C. P. Rosé, C. Teplovs & N. Law (Eds.), *Productive multivocality in the analysis of group interactions*, Chapter 28. New York: Springer

Steinhauser, K., Clipp, E., McNeilly, M., Christakis, N., McIntyre, L., & Tulsky, J. (2000). In search of a good death: Observations of patients, families, and health care providers. *Annals of Internal Medicine, 132*, 825–832.

Suthers, D. D. (this volume). Agency and modalities in multimediated interaction. In D. D. Suthers, K. Lund, C. P. Rosé, C. Teplovs & N. Law (Eds.), *Productive multivocality in the analysis of group interactions*, Chapter 19. New York: Springer

Teplovs, C., & Fujita, N. (this volume). Socio-dynamic latent semantic learner models. In D. D. Suthers, K. Lund, C. P. Rosé, C. Teplovs & N. Law (Eds.), *Productive multivocality in the analysis of group interactions*, Chapter 21. New York: Springer

Chapter 33
Analytic Representations and Affordances for Productive Multivocality

Gregory Dyke, Kristine Lund, Daniel D. Suthers, and Chris Teplovs

Introduction

In each of the sections of this book, researchers have had to overcome a variety of obstacles in their effort to engage productively in multivocality. The abstract nature of the theoretical constructs discussed in the early days of the productive multivocality effort convinced us of the necessity to ground these discussions not only in shared data but also in shared objectives. In order to create a shared basis for dialogue, the *pivotal moment* was proposed as a boundary object. As discussed in other chapters, these strategies went some way towards providing focus to methodological and theoretical discussions.

A large portion of the analytic knowledge derived when producing statements about and interpretations of the data is inscribed in representations. Therefore, another strategy for helping the various researchers engaged in a multivocal effort to speak to each other is to leverage the representations they create to juxtapose their statements and interpretations. Dyke et al. (2011) showed that it may be beneficial to identify representations that are specifically relevant to the multiple vocalities of multiple analysts, and thus that assist in this juxtaposition, rather than representations that are relevant only to a single analysis. Therefore we will delve into the

G. Dyke (✉) • K. Lund
ICAR Research Lab, CNRS—University of Lyon, 15 parvis René Descartes, 69007 Lyon, France
e-mail: gregdyke@gmail.com

D.D. Suthers
Department of Information and Computer Sciences, University of Hawai'i at Manoa, 1680 East West Road, POST 309, Honolulu, HI 96822, USA

C. Teplovs
Problemshift Inc., Windsor, ON, Canada

D.D. Suthers et al. (eds.), *Productive Multivocality in the Analysis of Group Interactions*, Computer-Supported Collaborative Learning Series 16,
DOI 10.1007/978-1-4614-8960-3_33, © Springer Science+Business Media New York 2013

representations used in this project as recorded in the present volume to better understand their value for productive multivocality.

In this chapter, we describe the variety of representations present in the different parts of this book, the purposes representations can serve within a single analysis, and strategies for further manipulating these representations for productive multivocality across analyses. We believe this survey to be useful, first to guide the design of tools to support productive multivocality, and second, to guide researchers on choosing tools and representations with multivocality in mind. The remainder of this chapter is structured in four sections. The first two sections discuss representations for analysis in general (and may be of limited interest to researchers not interested in designing tools and visualizations for analysis); the last two sections discuss the role of representations in a multivocal context. First, we lay some conceptual foundations for discussing representations and their role in "analytic discourse": the dialogue between analysts and their data (Thomas & Cook, 2005). Second, we describe the representations used in this book: what information is made salient in each representation and how do we move from one representation to another? Third, we propose nine strategies to achieve productive multivocality from a representational standpoint, based on examples taken from each of the data parts. Last, we conclude by discussing the requirements for an analytic framework to coordinate between analytic representations and assist in applying the proposed strategies.

Representations in Analytic Discourse

Thomas and Cook (2005) describe analysis as an iterative process: information is gathered, represented in a way that assists interpretation, new insight is developed, and a new representation is created to *crystallize* this insight. They define *analytic discourse* as the "technology-mediated dialogue between an analyst and his or her information to produce a judgment about an issue" (p. 38). As an analyst uses representations in order to interact with information via technology, we are well advised to examine the role of representations in the analytic process in preparation for understanding the role of representations in the multivocal process. Although there is substantial work on the role of representations in science (e.g., Koschmann & Zemel, 2009; Latour, 1990; Roth, 2003), and also on the roles of representations ("external" as well as mental) in cognition and learning (e.g., Ainsworth, 2006; Collins & Ferguson, 1993; Lajoie & Derry, 1993; Suthers & Hundhausen, 2003; Suthers, Vatrapu, Medina, Joseph, & Dwyer, 2008), our understanding of how representations are applied in collaborative learning research is limited. Quite a lot has been published on specific methods (e.g., Hmelo-Silver, 2003; Puntambekar, 2006; and other articles in the special issues in which these were published), but these works generally advocate for the utility of their methods rather than taking the process by which they are applied as their object of study. Harrer et al. (2007) derived a model of analysis from multiple analyses of the same dataset performed within a project. For the purpose of tool design, and understanding the role of representations in multivocal analysis, however, this description does not sufficiently disambiguate between questions of methodology—theory-driven questions of what representations need to be created and in what order, and representational questions—methodology-driven questions of what data should be encoded in a given representation, how new knowledge can be encoded and, more generally, how to move from one representation to another. Dyke, Lund, and Girardot (2009) propose a framework called Tatiana for creating and coordinating analytic representations, but make very few commitments as to how and why representations are created.

Medina and Suthers (2013) explain that *inscriptions* (markings in some medium) acquire meaning and become *representations* in the context of some practice. (The term *notation* can be used instead of inscriptions when it is important that the inscriptions follow some rules for their construction within a notational system; see also Suthers, 1999.) We will first examine some properties of inscriptions that make them useful for analytic practice, and then summarize a well-known account of the practices.

Analytic Inscriptions

A long tradition within artificial intelligence, cognitive psychology and education has investigated how the choice of problem representation influences individual effectiveness in problem solving and learning (e.g., Amarel, 1968; Koedinger, 1991; Larkin & Simon, 1987; Utgoff, 1986). In interaction with our perceptual-motor systems, inscriptions make certain things visible and certain actions possible (Zhang, 1997). Suthers (1999, 2000) observes that inscriptions differ on two major dimensions: constraints and salience.

Constraints pertain to the *expressiveness* of a notational system: what can be "written" with it. For example, a notational system may provide a set of categories, constraining one's representations to the corresponding ontology (e.g., "nodes" and "ties" in social network analysis). Expressive flexibility may range (for example) from being restricted to undifferentiated binary relations to the full expressiveness of natural language text.

Salience pertains to *heuristic adequacy*, or what information is easily available once it has been written into a notational system. Some visual notations make certain information more available, for example, a node and link diagram makes connections between nodes immediately visible, and one can quickly work out connections in a tabular representation by indexing by rows and columns. But salience also includes the salience of information that is missing. For example, in a matrix representation it is obvious when a cell is empty; in a node and link visual diagram one can quickly pick out nodes that are isolated (not connected to any others); but to find relationships that have not been considered in a textual narrative requires a careful reading. The salience of missing elements leads to the related concept of *prompting*: users of a representation are in a sense prompted to add the elements that are saliently missing.

Motivated by work in machine learning showing the importance of such "representational biases" (Utgoff, 1986), Suthers (1999, 2001) called these propensities of notational systems to enable their users to notice certain relationships and act in certain ways *representational guidance*. Representational guidance is related to the concept of visible affordances (Norman, 1999) in that the notational system offers certain potentials for action that are likely to be noticed and may (or may not) be taken up by the actors involved. We can thus understand the utility of representations as tools for expressing certain kinds of information and making certain aspects of that information salient. Representational guidance is not solely an individual phenomenon: later we discuss how representations support and influence collaborative meaning-making. Representations may differ in the extent to which they provide the guidance necessary to compare and contrast analyses, particularly in a multivocal setting. Yet, this account is only half the story: inscriptions become representations through practices (Medina & Suthers, 2013), and in a multivocal setting some analysts may not understand the practices behind other analysts' representations.

Analytic Practices

Although we are lacking empirical studies of analytic practice in the analysis of interaction data, Sanderson and Fisher (1994) describe a variety of techniques for the analysis of sequential data (i.e., where the chronology of the data is meaningful). Sanderson and Fischer describe two nested loops that make up what they call Exploratory Sequential Data Analysis (ESDA). In an external loop, the scientific process consists of asking questions, collecting and analyzing data and producing statements, all within an epistemological framework that constrains the nature of questions that can be asked, the data to be collected, and the acceptability of the generated statements. An inner loop, for data analysis, concerns the iterative creation of artifacts that have been transformed from the recorded data (or from artifacts created in a previous iteration). In this inner loop, analysts attempt to "*[smooth] the data—manipulating it so that its essential structure becomes apparent*" (Fisher & Sanderson, 1996, p. 28). When bringing together multiple analyses in a multivocal process, the question at hand (as addressed in other chapters) is the extent to which different epistemological frameworks allow the steps of the outer loop in each analysis to speak to each other. Further iterations within the inner loop will enable the answering of this question.

Sanderson and Fischer describe eight "smoothing" operations that they call the "eight Cs": chunks, comments, codes, connections, comparisons, constraints, conversions, and computations. *Chunks* concern the segmentation and re-segmentation of data, potentially in a hierarchical structure. *Comments* are unstructured, informal annotations, which are attached to data, to chunks or to other intermediary products. *Codes* are a way of abstracting the data, capturing its meaning while reducing the variability and lack of precision of its vocabulary. *Connections* describe relationships between the data such as temporality, the implicit or explicit structure of the data (e.g., the implicit reply structures in a chat and the explicit reply structures in a threaded forum), or the relationship between different types of media (e.g., video and digital traces). *Comparisons* examine the data through differentiating lenses such as the same coding by different analysts, data from different experimental conditions or between a predictive behavior model and the described behavior. The *constraints* applied to the data can be used to filter it so as to only show, for example, a selected interval, a particular medium or data coded with a certain keyword. *Conversions* transform data in order to present it in a different way—a different coding scheme, a different graphical representation, a different grain of analysis. Finally, *computations* reduce the data to summary representations: statistics, average values, etc. Fisher and Sanderson (1996) note that, "By putting these fundamental operations together in different ways, according to need, rather than by slavishly following a particular technique, the analyst has a good chance of crafting a new methodological approach that will meet his or her needs." (p. 30).

Sanderson and Fisher make explicit the operations that are applied to the data, but they do not describe the objects (or representations) that are the inputs and outputs of these operations. As the operations are intended to be put together in different ways, it is important to describe these inputs and outputs so as to know whether any step can be applied at any point or whether there are additional constraints. Dyke et al. (2009) base their Tatiana framework on a single abstract representation type, called a *replayable*, that is composed of events, and to which the different operations can be applied within a coherent framework. These operations cover the same analytic space as the eight Cs but distributed over different operations. *Transformation* creates new replayables. These new replayables can contain new events (chunks), or include only a subset of events (constraints). *Visualization* creates concrete visual representations (conversions) of the abstract replayables; *Synchronization* allows multiple visualizations to be replayed and synchronized in time, affording comparison (to a limited extent) between different representations and allowing the relationship between events in time to be explored (a subset of connections). Last, *enrichment* allows various kinds of analytic knowledge to be layered on top of existing data, allowing the creation of codes and comments on top of events and connections between events. Of these four operations, transformation, visualization, and enrichment can be applied to abstract replayables, and synchronization can be applied to their concrete visual representations. The outputs of these operations are new replayables for transformation and enrichment, and visual representations for visualization and synchronization. This guarantees that the output of an operation within this framework can always be the input of a new operation. However, this framework does not in general describe the result of computations. In general a computation removes the temporal aspect producing a result that is not a replayable (and to which the framework's operations no longer apply). Dyke et al. further note the difficulty in describing the result of comparisons beyond opening replayables side by side. While connections can be created within this framework, exploiting different kinds of connections requires new kinds of synchronization: sometimes events that happen at the same time are not actually related, while it is difficult to observe the link between events that are related across time. The advantages of this Tatiana framework lies in describing manipulations in terms of their inputs, providing better guidance for the software implementation of an extensible analytic environment, and making explicit the representations that are created at each step and the means by which these different representations can be coordinated.

Describing the link between theory and representations, Suthers, Dwyer, Medina, and Vatrapu (2010) make explicit the difference between three foundations of an analytic representational framework: the empirical foundation comprises the real-world observable objects the data is composed of (typically events and relationships between events); the representational foundation comprises how the empirical foundation is modeled and subsequently visualized (e.g., as a table or as a graph); the conceptual foundation comprises the mapping of epistemological concepts (e.g., ideas and uptake) onto the other two foundations. For example, in the Traces analytic framework (Suthers et al., 2010), empirical foundations are reflected in a "contingency graph" of observed relationships between acts and events, and conceptual foundations are mapped to an "uptake graph" of postulated interactional intentions behind these acts: translation from contingency to uptake graphs is an act of analysis. Suthers and colleagues suggest that the abstract representational foundation they propose, while being ideally suited to the related conceptual foundation of their framework, might also be generic enough to be used as a boundary object for a number of other related conceptual foundations. Dyke and colleagues (2011) propose a near identical abstract representational foundation and provide a notational space that affords both a variety of notations and the means to transition knowledge from one notational system to another. For example, in a tabular notation it is possible to enrich the data by adding codes. These codes are then available to affect a graphical timeline visualization, created automatically from the same data. Similarly, in a box and arrows representation, links between events can be created, which are then available for computations based on these links (e.g., social network analysis).

Other practices not discussed here (as they are too varied and voluminous) are the tradition-specific practices by which representations are used. For example, consider the varying practices of a conversation analyst, a grounded theorist, and an experimentalist with respect to a corpus of transcriptions of talk. What the representation means and offers to each is deeply bound up with the subsequent transformations they will undertake to create other analytic representations and interpretations. By better understanding these transformations, we provide analysts from different traditions with the means to document their analyses in a way that offers hints to other analysts as to the analytic leverage they obtain by creating a given representation.

Notational Systems and Manipulations

As we saw in the previous section, analytic discourse—the dialogue between analyst (and theory) and data—takes place through representations and the ways in which these representations are interpreted. In preparation to examining representational affordances for group analytic discourse in a multivocal setting, in this section we describe the notational systems that have been used in the present volume in order to understand their role within individual analyses. We break this down into two parts. First, we catalogue the aspects of the data that are made salient through the various notational systems (drawing from the 16 analysis chapters and two data chapters that contain notational systems not found elsewhere: Shirouzu, this volume-b, Chap. 4; Fujita, this volume, Chap. 20). Second we examine the manipulations that lead from one notation to another.

Constraints and Saliences of Notational Systems

Analysts tend to use a given notational system to show a collection of entities of the same nature. These entities can be events (such as actions and time periods), participants, groups, codes (such a experimental conditions, and various categories and tags attributed to other entities), artifacts created as a result of the events, or links between entities. In general a notation will show all of the dataset (i.e., all of the data that was collected about a given collaborative learning process, or whatever subset of that data is available in a given medium, such as a transcript), but in some cases it will be constrained to a specific group (e.g., Goggins & Dyke, this volume, Chap. 29), episode (all of the transcript excerpts), or condition (e.g., Howley, Kumar, Mayfield, Dyke, & Rosé, this volume, Chap. 26). This affords comparison between different subsets of the data, and the ability to only see data relevant to the current analysis.

Many of the notational systems that feature events (both actions and time periods) attempt to make the relationship between these events salient, be it relationships of sequence (Y happened after X), relationships of positioning and duration in time (Y started 2 min after X, lasted 10 min and overlapped with X for the first 5 min) or explicit links between events (such as links showing uptake and other forms of interaction). When sequence is most relevant, *event based tabular representations* with one event per row (one such a representation is a *transcription*) tend to be used, providing the additional ability to encode a lot of textual information describing the attributes of each event (participant, various codes, etc.). This verbosity leads to long tabular representations, only episodes of which tend to be reported in chapters. Such tabular representations also tend to not explicitly show links between events, relying instead on the implicit links provided by sequence of adjacent rows.

When affordances for making other kinds of relationships salient are needed, *timelines* can be used, making it easier to place links (as arrows) and using proximity as an indication of relative positioning and size of the event to indicate duration (explaining why). This can further provide both relevant and misleading guidance. For example different participants can be shown on different rows, helping identify contributions by the same participant, but also misleadingly showing the top- and bottom-most participants' events as being further away (and therefore possibly less related) than the middle participants' events.

Using graphical timelines is a tradeoff between their succinctness—showing an overview of many events and the relationships between them—and the completeness of information about their attributes. With the additional cost of needing a legend and still being difficult to read, some information can be encoded as graphical properties such as color, shape and size (e.g., Looi, Song, Wen, & Chen, this volume, Chap. 15; Lund & Bécu-Robinault, this volume, Chap. 17). One can include extra text as in Looi et al. (this volume, Chap. 15) and Teplovs and Fujita (this volume, Chap. 21), but the bigger the graphical objects representing objects are due to their verbosity, the more difficult it is to use size and position to make relationships salient. Affordances of digital notations to zoom in and out can overcome these problems; another possibility is to provide details either through a popup, or give a referential index to the event (such as a name, number, or timestamp) and provide details in another representation. Examples of this can be found in all chapters that refer to events in narratives, and in Stahl (this volume, Chap. 28), where the links between different events are shown in a graphical timeline and additional information about events is provided in a tabular representation.

Timelines can also be exploited for their affordances to see the evolution over time of other entities than events, for example showing summary statistics for groups (Oshima, Matsuzawa, Oshima, & Niihara, this volume, Chap. 12), codes (Law & Wong, this volume, Chap. 22), or the distribution of codes (Howley et al., this volume, Chap. 26).

Other notational systems remove the salience of time and sequence. Non-temporal *Graphs* show the relationships between entities such as codes (Goggins & Dyke, this volume, Chap. 29; Oshima et al., this volume, Chap. 12) or participants (Goggins & Dyke, this volume, Chap. 29; Oshima et al., this volume, Chap. 12; Teplovs & Fujita, this volume, Chap. 21), again using shape, color and size to encode information. In these cases, change over time can be made salient by comparing graphs at different time points (Goggins & Dyke, this volume, Chap. 29; Oshima et al., this volume, Chap. 12).

Affordances for comparison of different entities (groups, codes, participants) are provided by descriptive statistics, which summarize frequencies, proportions or indicators for the entities or the events pertaining to them. *Statistical models,* which encode consistent sequence relationships between participants and codes, can be shown by path diagrams (Chiu, this volume-a, this volume-b, Chaps. 7 and 23). Other relationships such as correlation or co-occurrence can be made salient by statistical tests (e.g., Howley et al., this volume, Chap. 26; Oshima et al., this volume, Chap. 12).

Manipulations

The representations described in this chapter are those found in the book. A far larger number of intermediate representations probably exist, which the authors have not included in their chapters. Furthermore, the practices involved in the notations *becoming* representations are mostly not apparent and would have required recording of analysts at work to better understand. Where these practices can, to a small extent, be found, is in the choice of manipulations reported in the various "manipulations" sections of each chapter. From an overview of these, it seems that the eight Cs are an accurate report. They can, however, be further broken down in terms of the representations (i.e., the notational systems and their representational guidance) that are created, and the reasons for creating these representations (the constructs that they seek to make salient).

Chunks separate the data into different units of analysis (or adds in new units on top of the existing ones). Analysts typically create events, phases, episodes, etc. so as to have the correct unit of analysis for their further manipulation. This can result in discrepant units between analysts (cf. the effort in Part 4 to identify a common transcript on which to work: see Chap. 19). Chunking into phases can illustrate how the events of a phase belong to it and are different from its neighbors (e.g., Chiu's breakpoints, this volume-a, Chap. 7), how that phase articulates with its neighbors within a session (e.g., Lund & Bécu-Robinault, this volume, Chap. 17), how ideational and inscriptional episodes relate to each other (Medina, this volume, Chap. 16); or highlight important parts of the data, which have been selected for deeper analysis. Systematic chunking can provide uniform units for coding (Jeong's initial analysis, this volume, Chap. 18), or help illustrate trajectory from one time period to another (e.g., the time slices in Howley et al., this volume, Chap. 26 and Oshima et al., this volume, Chap. 12) or the consistency of sequences within a time period (e.g., the lag variables in Chiu, this volume-a, this volume-b, Chaps. 7 and 23).

Comments can be applied to specific events, but more generally can take the form of narratives attached to participants, sessions or artifacts. As these narratives increase in completeness, the scope of entities that they encompass increases in quantity and quality. When narratives are no longer keyed to a specific entity, the salience of individual entities is reduced and it is more difficult to follow referential indices in either direction (narrative to entity or entity to narrative) to align and compare analyses and interpretations. For example, Looi et al. (this volume, Chap. 15) refers in narrative to individual contributions shown in a graph, to relations between contributions shown in the graph, and also to the role of the individual contributions within the interaction as a whole. Understanding this narrative requires going back and forth between the narrative itself and the graph it refers to. When going from narrative to graph, the uniqueness of the contribution in the graph makes it relatively easy to find the entity referred to (although tools could provide better affordances for this, with hyperlinks to navigate from the narrative to the graph). Going from graph to narrative, however, is difficult as a contribution can be referred to multiple times, either explicitly, or implicitly when it is part of the context for describing another contribution. In a setting where researchers attempt to compare different narratives, this issue is even more problematic.

Codes can also be attached to any kind of entity, leading to similar issues in reference and alignment as discussed for comments. Codes have a variety of provenances. For example, Looi et al. (this volume, Chap. 15) and Lund and Bécu-Robinault (this volume, Chap. 17) code actions according to the medium in which they happen (an empirical concept, although for Lund, these codes are then interpreted in relation to theoretical concepts), while Howley (Chaps. 11 and 26) codes actions according to a set of theoretical concepts. Codes can also be inherited (or computed in an inheritable way): the summary statistic value of a participant for a given code (e.g., data section 2) is the number of events that have that code.

Constraints occur in two situations. First, they occur when deciding that only a subset of the data should be made salient in a given representation (e.g., only a given session, episode or participant). Incompatible constraints can make analyses become incomparable, sometimes unnecessarily: see for example discussion of the choices to analyze public vs. private spaces and verbal vs. visual contributions in the first Group Scribbles corpus, Chap. 19. Second, they occur when partitioning the data so that different subsets of it can be compared, either in an event-based representation (e.g., comparing conditions in Howley et al., this volume, Chap. 26) or in summary statistics.

Conversions involve choosing a notational system (such as those described in the previous subsection) that provides adequate representational guidance to obtain the information necessary for subsequent manipulation. Some notational systems are standard representations (e.g., summary statistics) whereas others, such as those of Shirouzu (this volume-a, this volume-b, Chaps. 4 and 5), attempting to illustrate novel and complex relationships between various entities, both theoretical and empirical, will be more ad hoc. The creation of ad-hoc notational systems provides a challenge, particularly in multivocal settings, as it is difficult to foresee how existing or different analytic practices will bias interpretation.

Connections, as highlighted in the description of the Tatiana framework (Dyke et al., 2009), encompass several different practices. Synchronization and replay in time allows the connections between simultaneously occurring events in different representations to be made salient (necessary for connecting transcript and video). Indexical synchronization is similar but highlights entities (events, participants, etc.) that are identical and occur in different representations (e.g., using labels to identify events, Stahl, this volume, Chap. 28). Creating links shows connection between entities (sequence, causality, uptake, or some other relationship; e.g., Chaps. 15, 16, and 28), while the ability to follow links enables the interpretation of these connections, or the folding of links (e.g., transforming an actor–code network into and actor–actor network, Teplovs & Fujita, this volume, Chap. 21). When subsuming (chunking) events into phases, an implicit connection of belongingness is created between the events and the phase. A similar connection is created when an entity-type attribute is assigned to another entity (e.g., when a participant is assigned to a group, or an event to a condition). Last, notational systems with variable properties (e.g., the size of a sliding window in Howley et al., this volume, Chap. 26) can be connected to facilitate comparison of multiple notations.

Computations allow the creation of new entities and the more specific computation of various attributes of these entities. They can exploit various connections (e.g., Chiu, this volume-a, this volume-b, Chaps. 7 and 23; Goggins & Dyke, this volume, Chap. 29; Oshima et al., this volume, Chap. 12; Teplovs & Fujita, this volume, Chap. 21) and partitions that exist to establish sets of like objects for which a similar computation can be established. Such computations include sums (of event counts or duration), calculating these sums as percentages, averages, differences, significance tests, etc.

Comparisons make salient the differences between partitions of the data. This can happen in a variety of ways. Significance tests are a computation that affords the comparison of different kinds of entity. Multiple representations using the same notational system can exploit various forms of connection (particularly synchronization and connection within multiple instances of the same notational system) or be opened side by side to facilitate comparison. Some notational systems can show multiple data subsets within the same representation (e.g., multiple group trajectories on the same plot, Oshima et al., this volume, Chap. 12). We were not able to identify an example of comparisons that was not achieved using the previously described manipulations.

The description of these manipulations, many of which are expressed in terms of other manipulations (or of their results) illustrates the need to better understand how manipulations can be integrated within a framework, both for purposes of planning and describing analyses, and for purposes of designing tools to help perform and share them. Some of these manipulations are optional within a given analysis but are particularly useful when trying to achieve productive multivocality. In the next section, we look at how representations were manipulated in order to achieve this goal.

Analytic Discourse for Multivocality

In this section, we examine how representations and manipulations can support the achievement of multivocality. We first examine how analytic discourse described earlier in this chapter translates to a group setting, before describing how multivocality panned out, in each chapter, from a representational standpoint.

Group Analytic Discourse

Within Sanderson and Fisher's outer loop, analysts choose questions, settings, data, manipulations and statements, all of which are constrained, to some extent, by their theoretical standpoint and teleological objectives. A first goal therefore lies in ascertaining the amount of commensurability, particularly in terms of the statements produced at the end, and how they will speak to each other, along with the acceptability of common data and of the analytic manipulations performed on that data. Within the inner loop, analysts create and manipulate representations. This representational foundation mediates between the empirical and conceptual foundations. The points of agreement and disagreement on each of those foundations must be established in order to achieve multivocality in analytic discourse. To assist in this, within each vocality the mapping between the different foundations must be made explicit. Once these points of commensurability have been established, researchers can choose representations that adequately further both their own analyses, and also the meeting of these analyses in a multivocal setting.

The effects of constraints and salience on groups collaborating with representations are not just the sum of the effects on individual analysts; they also operate at the level of group processes in ways that are not reducible to individuals. The theory of representational guidance identified the following effects among others (Suthers, 2006; Suthers & Hundhausen, 2003):

Negotiation potentials. When two or more persons are collaborating on constructing a representation, they may feel obligated to discuss potential modifications to that representation with each other. Constraints and salience affects what potential actions are suggested by the affordances of the notational tool (e.g., the possibility of selecting a category for a statement one is about to make), and by the salience of "missing" elements in a particular inscription constructed in that tool (e.g., the possibility of connecting two objects in a graph or filling in the cell of a matrix). Similarly, analytic methods and their associated potentials for action will affect what collaborating researchers consider. For example, in Goggins and Dyke (this volume, Chap. 29), Goggins and Dyke collaborated based on the gap afforded by Goggins' method for social network analysis, which had been applied to implicit ties automatically calculated from the data. Stahl (this volume, Chap. 28) had manually coded one session, providing a reply structure from which SNA could be derived in the same way. Dyke filled the gap by manually coding the explicit ties in the other sessions so that the same method could be applied and the results compared.

Referential Resources. As highlighted in the previous section, one theme common across many notational systems is the ability to make the identity of various entities salient, allowing representational artifacts to function as indexical resources for conversation. If certain aspects of the represented are more visible, or reified in singular inscriptional constituents that are easy to point to or refer to, then it will be easier to talk about ideas associated with those constituents (perhaps through prior discourse as well as representational conventions), as one can invoke the ideas with simple deixis. In Part 2 (Chaps. 4–8), the unique referential indices of events (and the use of the same event set in all vocalities) allowed researchers to negotiate meaning by referring to these events in their narratives. In Part 4 (Chaps. 14–19), lack of a shared referential framework was a barrier to comparing analytic results.

Other dimensions along which representations can influence collective activity include (im)mutable mobiles, recruiting others to one's analytic agenda and enable remote participation; reflecting subjectivity, providing implicit awareness of the activity and attention of others; and integration of group activity over time and space (Suthers, 2006).

Strategies for Productive Multivocality

Because of the real-world constraints of these multivocal efforts, in most cases spending a large amount of time to engage with each others' representations by applying further manipulations was impractical, so there may be as yet unexplored opportunities for multivocality that we cannot report upon. It may be a consequence of this (because of representations not having enough negotiation potential or providing insufficient referential resources), or it may be a characteristic of multivocal work, but in all cases, analysts' narratives and interpretations were key representations for multivocality. Researchers interact on Sanderson and Fisher's outer loop, producing statements and examining the extent of their commensurability, turning to the inner loop of data manipulation to perform only comparisons in order to ascertain the extent to which the interpretations are based on compatible empirical and representational foundations. In the following paragraphs, we present strategies based on our experiences. They are not offered as advice that *must* be followed, but rather as ways of moving forward in a situation for which the path to multivocality is not immediately apparent.

In Part 2 (Fractions), multivocality was grounded in the iterative definition and position of pivotal moments. None of the analysts performed any new chunking of the data, allowing the position of the pivotal moments to be identified by the referential resources common to all representations. Shirouzu's complex representations of artifact evolution were not taken up by the other two analysts, but this may be more related to conceptual than representational issues. This part illustrates two strategies for productive multivocality:

1. Referring to the same entities and providing common referential indexes to these entities will maximize the potential of using representations as referential resources and reestablishing the context in which a given event occurred.
2. Iteration will maximize the opportunities to refine a multivocal analysis, allowing multiple strategies to be applied.

In Part 3 (Chemistry), the theoretical focus on the roles of participants and a convergence towards representations presenting descriptive statistics naturally led to side-by-side comparison of the narratives and the profiles of individual students and groups according to different coding schemes. This illustrates how certain representations lend themselves to comparison more easily than others. Event-centered representations tend to be difficult to compare because of the variety of ways in which theoretical constructs can be projected onto events—e.g., should a "new idea" code be applied to the moment when an individual first starts creating it in their private space, when the idea is completed, or when it is published to the public space? As per the strategy described in the previous paragraph, reducing the number of entities to a fixed number of students and groups makes comparison much more tractable, with less negotiation needed to achieve common referential resources (the identities of the students and groups). This part also illustrates two further strategies.

3. Using similar notational systems will afford greater potential for comparison.
4. Using notational systems that already support comparison within a vocality will make comparison easier in a multivocal setting.

These two strategies do not suggest that analyses should be shoehorned into easily comparable representations for comparability's sake. They also do not suggest that having immediately comparable representations removes the need to delve deeper into the empirical and conceptual foundations. Rosé (this volume, Chap. 13) explores whether the interpretations are referring to a commensurable narrative of the underlying events, and to a compatible multivocal interpretation of the different theoretical lenses through which leadership can be construed.

In Part 4 (Electricity), the issues highlighted by Suthers (this volume, Chap. 19) illustrate the difficulty in using representations to achieve multivocality when there isn't a basis for juxtaposing analytic representations. We can expect that each analyst will come up with different chunks, codes, and comments, segment the data into different episodes, etc.: this is normal. But it is much easier to compare how these different derived analytic representations characterize the phenomenon if they can be brought into alignment via a shared coordinate system such as a timeline. Yet, some of the analytic representations produced did not even reference a timeline, let alone a shared transcript. Nevertheless the different analysts appear to have independently arrived at similar views to achieve agreement on the empirical level. The difficulty in comparing representations (which seems like a missed opportunity, given the relative similarity in notational systems across all four analyses) led the discussant to engage deeply with the different representations in order to eliminate gratuitous differences (such as different events being used to refer to the same artifact) and bring into alignment the different analyses. This alignment involved understanding the relationships between empirical, conceptual and representational foundations for each analysis, identifying points of agreement, disagreement and apparent disagreement (i.e., gratuitous differences) between analyses, and producing representations on which this alignment could be discussed. This provided focus for the analysts to understand and discuss to what extent their narratives and analytic statements were coherent and commensurable. As well as providing a further example of strategy 1 (positioning events on the same timeline to provide referential indexicality in the absence of ability to refer to common events) this illustrates two further strategies:

5. Aligning analyses (identifying points of agreement and disagreement and eliminating gratuitous points of apparent disagreement) will lead to the creation of new representations (typically narratives) with negotiation potentials for multivocal discussion.
6. Sharing complete and explicit analytic inscriptions, including the ability to return to the original data, (as opposed to the products that make it into paper publications, which make salient only aspects relative to the story told in the paper) makes it easier to achieve alignment.

In Part 5 (Education), the widest divergence in notational systems can be found. This led to a discussion about the affordances of the representations to identify different theoretical constructs, in turn helping to discuss these constructs. As the statements produced aim to help iteration in design-based research, they need only suggest noncontradictory improvements. In this part, the strategy of aligning analyses (at least from an empirical and representational standpoint) was difficult to apply, among other reasons because the strategy of using similar notational systems was also not applicable. Furthermore, this part highlighted the difficulty in understanding notational systems when the practices necessary to turn them into representations are not widespread, or immediately suggested by representational guidance. Chiu's statistical discourse analysis (this volume-a, this volume-b, Chaps. 6 and 23) provides quantified sequential or contextual relationships (strengths and significance levels) between different codes (or contexts, such as the gender of the participants). Because of the widespread practice of interpreting correlations, this notational system provides immediate affordances for interpretation. However, the complexity of the manipulation to produce these correlations makes it difficult to immediately identify the caveats that must be applied in interpretation—although these are systematically discussed within the chapters in which they appear. Teplovs' method for describing similarity between participants (Chap. 21) is difficult to interpret, not because of a lack of representational guidance, but because of the lack of established practices in using similarity between participants to analyze their collaboration. This suggests a strategy that was followed to some extent in this part, but might have been carried further:

7. Identifying, sharing and developing the practices needed to understand different notational systems and their relations to different conceptual foundations will enable researchers from different analytic traditions to better understand each other's interpretations.

In Part 6 (Biology), the most integrative representation-based work can be found. Goggins and Dyke (this volume, Chap. 29) take link information from Stahl's coding of a sample group (this volume, Chap. 28) to produce an automated heuristic for producing links. They further take coding information from Howley et al. (this volume, Chap. 26) and integrate it with their own links to identify links between different codes. This work used the notational flexibility of Tatiana (Dyke et al., 2009), illustrating how much easier it is to engage with others' representations if the data encoded in those representations is easily transferable to another notational system. This exemplifies the last two strategies presented in this chapter:

8. Enabling the capitalization of the manipulations used to produce a notation (i.e., documenting the necessary inputs, operations, and outputs), will make it possible to rerun the same manipulation on different underlying data (such as the output of another analysis). This requires that knowledge inscribed in a representation should be encoded in a reusable data format.
9. When creating notations, choosing tools that can support the application of the various strategies will minimize the overheads of the already costly process of productive multivocality.

As we have seen in this brief overview, there exist a variety of strategies for using representations to help achieve multivocality. In fact, whereas we had previously thought that the result of applying these strategies was a precursor to multivocality, it is actually the engagement in these strategies that *is* the productive multivocality from a representational standpoint: using the various affordances of the representations to discuss the points of contention and commensurability of different analyses. Several of these strategies involve practices that are not needed within a single vocality (even when using mixed methods). In particular certain aspects of analysis that are often left implicit must be made explicit in order to work effectively with interlocutor analysts. What is the link between the empirical, representational and conceptual foundations of the analysis? What practices are necessary to give meaning to the notational systems used in the analysis? What manipulations are used to produce a notation in a given notational system? What is the complete representation on which the product shared in a publication was based? What are the entities present in this representation and how can we use them as referential resources? What are the relationships between entities that are considered relevant and have they been made explicit?

Conclusion: Affordances to Support Productive Multivocality

In this chapter, we attempted to summarize the lessons learned in using representations for achieving multivocality. We first turned to the literature to examine the role that is played by representations in analysis, identifying the cycles through which analyses happen, the manipulations that are applied in these cycles, the kinds of representations they produce, the reasons for which they get created, and the ways they get interpreted. We then examined the notational systems used in the various parts of this book, the entities they feature, the aspects of the data they make salient, and the manipulative practices used to create them. This description of manipulative practices refines descriptions found in the existing literature by explaining more clearly how different manipulations are related to each other. Last, we described the instances of representation-based productive multivocality, presenting nine strategies for researchers seeking to engage in productive multivocality. We now draw upon the three sections of this chapter to summarize the affordances needed for productive multivocality, in order to guide the design and choice of tools (strategy 9) to support analysis.

All the affordances needed for single vocality analysis will be present in multivocal analysis: the ability to create representations in a variety of notation systems (making salient aspects such as sequence, links and differences) and the ability to apply different manipulations to transition from one representation to another, or exploit the various existing representations and the guidance they afford. Through these affordances it must also be possible to apply the strategies we identified for multivocality. This calls for a specific focus on manipulations that provide these affordances, which are indispensible in a tool to support multivocality, particularly as we know that iteration (strategy 2) is of key importance.

The referential resources of representations are of importance to support discussing points of difference and agreement (alignment of analyses, strategy 5), and provide context. These include the ability to referentially index various entities (strategy 1) and to exploit various forms of connections: through synchronization and replay, the context across multiple modes can be provided; through indexical synchronization, the connections between multiple narratives and representations can be drawn and followed (i.e., it should be possible to easily follow the link from discussion of an entity in one narrative to discussion of that entity in all other narratives); by making explicit the links between entities, it is then possible to trace these links from source to target and back, to further interpret their role.

Inscription of new analytic knowledge, such as chunks, links, codes and comments should produce objects that can be reused in other analyses (particularly multivocal analyses), which can be capitalized upon and exploited. This sharing of complete and explicit inscriptions (strategy 6) will assist in alignment. Furthermore, in combination with describing manipulations such as conversions and computations in such a way that they can be rerun on this new analytic knowledge (strategy 8), this will make it easier to transform representations into notational systems that best enable comparison (strategy 3). This highlights the need to better understand the results of computations, how we can transition seamlessly from notational systems featuring events and entities to notational systems featuring groups, participants, and codes and how not event-centric notations can be further exploited.

Comparisons and notational systems that afford comparison (strategy 4) need to be better understood. In the Manipulations section, we showed how comparisons exploit all of the other manipulations: partitioning the data into comparable subsets of entities, producing notations in systems that have affordances for comparison, and coordinating the visual properties of these notations so that, for example, a set of timelines uses the same colors and shapes to encode the same information. Also, in a multivocal setting, tools for juxtaposing not easily comparable sets of entities are needed. For example, a shared coordinate system such as a timeline greatly assists in comparing how analyses have parsed the event stream into different episodes and mapped them to analytic concepts (as done successfully in Part 2 and problematically with effort in Part 4).

Last, the practices for making meaning of notations (strategy 7), and the narratives and statements that are the results of an analysis must also be supported within tools for sharing and coordinating representations, so that the complete set of representations is available at each iteration in the analysis cycle.

As the paragraphs above indicate, certain kinds of manipulations (connections, comparisons, and computations) need further investigation. We hope that tool designers will both take up this challenge, and use this chapter as guidance in terms of how different kinds of manipulations and representations can interact within a coherent framework. Beyond describing strategies for multivocality to help guide researchers seeking to engage in multivocality, this chapter argues that certain affordances are needed, supported by tools for the creation and manipulation of representations, in order to maximize the potential of multivocality.

Acknowledgements The authors are grateful to the ASLAN project (ANR-10-LABX-0081) of Université de Lyon, for its financial support within the program "Investissements d'Avenir" (ANR-11-IDEX-0007) of the French government operated by the National Research Agency (ANR).

References

Ainsworth, S. E. (2006). DeFT: A conceptual framework for considering learning with multiple representations. *Learning and Instruction, 16*(3), 183–198.

Amarel, S. (1968). On representations of problems of reasoning about actions. *Machine Intelligence, 3*, 131–171.

Chiu, M. M. (this volume). Social metacognition, micro-creativity and justifications: Statistical discourse analysis of a mathematics classroom conversation. In D. D. Suthers, K. Lund, C. P. Rosé, C. Teplovs, & N. Law (Eds.), *Productive multivocality in the analysis of group interactions*, Chapter 7. New York, NY: Springer.

Chiu, M. M. (this volume). Statistical discourse analysis of an online discussion: Cognition and social metacognition. In D. D. Suthers, K. Lund, C. P. Rosé, C. Teplovs, & N. Law (Eds.), *Productive multivocality in the analysis of group interactions*, Chapter 23. New York, NY: Springer.

Collins, A., & Ferguson, W. (1993). Epistemic forms and epistemic games: Structures and strategies to guide inquiry. *Educational Psychologist, 28*(1), 25–42.

Dyke, G., Lund, K., & Girardot, J.-J. (2009). Tatiana: An environment to support the CSCL analysis process. *Proc. CSCL 2009, Rhodes, Greece* (pp. 58–67).

Dyke, G., Lund, K., Jeong, H., Medina, R., Suthers, D. D., van Aalst, J., et al. (2011). Technological affordances for productive multivocality in analysis. In H. Spada, G. Stahl, N. Miyake & N. Law (Eds.), Connecting Computer-Supported Collaborative Learning to Policy and Practice: Proceedings of the 9th International Conference on Computer-Supported Collaborative Learning (CSCL 2011) (Vol. I, pp. 454–461). Hong Kong: International Society of the Learning Sciences.

Fisher, C., & Sanderson, P. (1996). Exploratory sequential data analysis: Exploring continuous observational data. *Interactions, 3*(2), 25–34.

Fujita, N. (this volume). Online graduate education course using knowledge forum. In D. D. Suthers, K. Lund, C. P. Rosé, C. Teplovs, & N. Law (Eds.), *Productive multivocality in the analysis of group interactions*, Chapter 20. New York, NY: Springer.

Goggins, S. P., & Dyke, G. (this volume). Network analytic techniques for online chat. In D. D. Suthers, K. Lund, C. P. Rosé, C. Teplovs, & N. Law (Eds.), *Productive multivocality in the analysis of group interactions*, Chapter 29. New York, NY: Springer.

Harrer, A., Zeini, S., Kahrimanis, G., Avouris, N., Marcos, J. A., Martinez-Mones, A., et al. (2007). Towards a flexible model for computer-based analysis and visualisation of collaborative learning activities. *Proc. CSCL 2007, New Jersey, USA*.

Hmelo-Silver, C. E. (2003). Analyzing collaborative knowledge construction: Multiple methods for integrated understanding. *Computers and Education, 41*, 397–420.

Howley, I. K., Kumar, R., Mayfield, E., Dyke, G., & Rosé, C. P. (this volume). Gaining insights from sociolinguistic style analysis for redesign of conversational agent based support for collaborative learning. In D. D. Suthers, K. Lund, C. P. Rosé, C. Teplovs, & N. Law (Eds.), *Productive multivocality in the analysis of group interactions*, Chapter 26. New York, NY: Springer.

Jeong, H. (this volume). Development of group understanding via the construction of physical and technological artifacts. In D. D. Suthers, K. Lund, C. P. Rose, C. Teplovs, & N. Law (Eds.), *Productive multivocality in the analysis of group interactions*, Chapter 18. New York, NY: Springer.

Koedinger, K. R. (1991). On the design of novel notations and actions to facilitate thinking and learning. In *Proceedings of the International Conference on the Learning Sciences* (pp. 266–273). Charlottesville, VA.

Koschmann, T., & Zemel, A. (2009). Optical pulsars and black arrows: Discoveries as occasioned productions. *The Journal of the Learning Sciences, 18*, 200–246.

Lajoie, S. P., & Derry, S. J. (Eds.). (1993). *Computers as cognitive tools*. Hillsdale, NJ: Lawrence Erlbaum Associates.

Larkin, J. H., & Simon, H. A. (1987). Why a diagram is (sometimes) worth ten thousand words. *Cognitive Science, 11*, 65–99.

Latour, B. (1990). Postmodern? No simply amodern. Steps towards an anthropology of science. An essay review. *Studies in the History and Philosophy of Science, 21*, 145–171.

Law, N., & Wong, O.-W. (this volume). Exploring pivotal moments in students' knowledge building progress using participation and discourse marker indicators as heuristic guides. In D. D. Suthers, K. Lund, C. P. Rosé, C. Teplovs, & N. Law (Eds.), *Productive multivocality in the analysis of group interactions*, Chapter 22. New York, NY: Springer

Looi, C.-K., Song, Y., Wen, Y., & Chen, W. (this volume). Identifying pivotal contributions for group progressive inquiry in a multi-modal interaction environment. In D. D. Suthers, K. Lund, C. P. Rose, C. Teplovs, & N. Law (Eds.), *Productive multivocality in the analysis of group interactions*, Chapter 15. New York, NY: Springer.

Lund, K., & Bécu-Robinault, K. (this volume). Conceptual change and sustainable coherency of concepts across modes of interaction. In D. D. Suthers, K. Lund, C. P. Rose, C. Teplovs, & N. Law (Eds.), *Productive multivocality in the analysis of group interactions*, Chapter 17. New York, NY: Springer.

Medina, R. (this volume). Cascading inscriptions and practices: Diagramming and experimentation in the group scribbles classroom. In D. D. Suthers, K. Lund, C. P. Rose, C. Teplovs, & N. Law (Eds.), *Productive multivocality in the analysis of group interactions*, Chapter 16. New York, NY: Springer.

Medina, R., & Suthers, D. D. (2013). Juxtaposing practice: Uptake as modal transposition. *Proceedings of the 10th International Conference on Computer Supported Collaborative Learning (CSCL '13), Madison, WI.*

Norman, D. A. (1999 May–June). Affordance, conventions, and design. *ACM Interactions, 6*, 38–42.

Oshima, J., Matsuzawa, Y., Oshima, R., & Niihara, Y. (this volume). Application of social network analysis to collaborative problem solving discourse: An attempt to capture dynamics of collective knowledge advancement. In D. D. Suthers, K. Lund, C. P. Rose, C. Teplovs, & N. Law (Eds.), *Productive multivocality in the analysis of group interactions*, Chapter 12. New York, NY: Springer.

Puntambekar, S. (2006). Analyzing collaborative interactions: Divergence, shared understanding and construction of knowledge. *Computers and Education, 47*, 332–351.

Rosé, C. P. (this volume). A multivocal analysis of the emergence of leadership in chemistry study groups. In D. D. Suthers, K. Lund, C. P. Rosé, C. Teplovs, & N. Law (Eds.), *Productive multivocality in the analysis of group interactions*, Chapter 13. New York, NY: Springer.

Roth, W.-M. (2003). *Toward an anthropology of graphing: Semiotic and activity-theoretic perspectives*. Dordrecht, The Netherlands: Springer.

Sanderson, P., & Fisher, C. (1994). Exploratory sequential data analysis: Foundations. *Human-Computer Interaction, 9*(4), 251–317.

Shirouzu, H. (this volume). Focus-based constructive interaction. In D. D. Suthers, K. Lund, C. P. Rosé, C. Teplovs, & N. Law (Eds.), *Productive multivocality in the analysis of group interactions*, Chapter 5. New York, NY: Springer.

Shirouzu, H. (this volume). Learning fractions through folding in an elementary face-to-face classroom. In D. D. Suthers, K. Lund, C. P. Rose, C. Teplovs, & N. Law (Eds.), *Productive multivocality in the analysis of group interactions*, Chapter 4. New York, NY: Springer.

Stahl, G. (this volume). Interaction analysis of a biology chat. In D. D. Suthers, K. Lund, C. P. Rosé, C. Teplovs, & N. Law (Eds.), *Productive multivocality in the analysis of group interactions*, Chapter 28. New York, NY: Springer.

Suthers, D. (2000). Initial Evidence for Representational Guidance of Learning Discourse. In *Proceedings of International Conference on Computers in Education*, Taipei, Taiwan, 21–24.

Suthers, D. D. (1999, January 5–8). Representational support for collaborative inquiry. In *Proceedings of the Hawai'i International Conference on System Sciences (HICSS-32), Maui, Hawai'i (CD-ROM)*. Maui, Hawai'i: Institute of Electrical and Electronics Engineers, Inc. (IEEE).

Suthers, D. D. (2001). Towards a systematic study of representational guidance for collaborative learning discourse. *Journal of Universal Computer Science*, *7*(3). http://www.jucs.org/jucs_7_3/towards_a_systematic_study.
Suthers, D. D. (2006). Technology affordances for intersubjective meaning making: A research agenda for CSCL. *International Journal of Computer-Supported Collaborative Learning, 1*(3), 315–337.
Suthers, D. D. (this volume). Agency and modalities in multimediated interaction. In D. D. Suthers, K. Lund, C. P. Rosé, C. Teplovs, & N. Law (Eds.), *Productive multivocality in the analysis of group interactions*, Chapter 19. New York, NY: Springer.
Suthers, D. D., Dwyer, N., Medina, R., & Vatrapu, R. (2010). A framework for conceptualizing, representing, and analyzing distributed interaction. *International Journal of Computer-Supported Collaborative Learning, 5*(1), 5–42.
Suthers, D. D., & Hundhausen, C. (2003). An experimental study of the effects of representational guidance on collaborative learning. *The Journal of the Learning Sciences, 12*(2), 183–219.
Suthers, D. D., Vatrapu, R., Medina, R., Joseph, S., & Dwyer, N. (2008). Beyond threaded discussion: Representational guidance in asynchronous collaborative learning environments. *Computers and Education, 50*(4), 1103–1127.
Teplovs, C., & Fujita, N. (this volume). Socio-dynamic latent semantic learner models. In D. D. Suthers, K. Lund, C. P. Rosé, C. Teplovs, & N. Law (Eds.), *Productive multivocality in the analysis of group interactions*, Chapter 21. New York, NY: Springer
Thomas, J. J., & Cook, K. A. (Eds.). (2005). *Illuminating the path: The research and development agenda for visual analytics*. Los Alamitos, CA: IEEE Press.
Utgoff, P. (1986). Shift of bias for inductive concept learning. In R. Michalski, J. Carbonell, & T. Mitchell (Eds.), *Machine learning: An artificial intelligence approach* (Vol. II, pp. 107–148). Los Altos, CA: Morgan Kaufmann.
Zhang, J. (1997). The nature of external representations in problem solving. *Cognitive Science, 21*(2), 179–217.

Chapter 34
Epistemological Encounters in Multivocal Settings

Kristine Lund, Carolyn Penstein Rosé, Daniel D. Suthers, and Michael Baker

Introduction

What makes it difficult for a heterogeneous group of differently trained researchers to collaborate? They may not share the same assumptions about the nature of scientific knowledge, the role of theory, their research object, or how to gather and represent data. They may not agree on the appropriate relation between researcher and data, about how a particular analytical construct should be defined, or about what constitutes relevant units of analysis. They may not share the same bases for value judgments they make about their data. They may also differ in their opinion on which methods to apply, how to apply them or about how to validate their results. These assumptions and others are anchored in a researcher's epistemology. An epistemology is a logos or reasoned discourse on *epistémé* or the nature of knowledge and how it can be acquired. An epistemology is learned in the early years of research training, and it is often learned implicitly; its consequences are not always explicitly laid out (cf. Lund & Suthers, this volume, Chap. 2). Sometimes a researcher may not be aware of his or her epistemology and only later discover its constraints

K. Lund (✉)
ICAR Research Lab, CNRS—University of Lyon, 15 parvis René Descartes, 69007 Lyon, France
e-mail: kristine.lund@ens-lyon.fr

C.P. Rosé
Carnegie Mellon University, 5000 Forbes Avenue, Pittsburgh, PA 15213, USA

D.D. Suthers
Department of Information and Computer Sciences, University of Hawai'i at Manoa, 1680 East West Road, POST 309, Honolulu, HI 96822, USA

M. Baker
LTCI Research Lab, CNRS—Telecom ParisTech, 46 rue Barrault, 75013 Paris, France

D.D. Suthers et al. (eds.), *Productive Multivocality in the Analysis of Group Interactions*, Computer-Supported Collaborative Learning Series 16,
DOI 10.1007/978-1-4614-8960-3_34,

through discussions with researchers with different epistemologies. This chapter reflects on epistemological encounters in a 5-year project of multidisciplinary collaborations in the analysis of interaction (Suthers, Lund, Rosé, & Teplovs, this volume, Chap. 31).

How Can Different Epistemologies Encounter Each Other?

Typically, if researchers have contrasting epistemologies, we think of them as being from different disciplines (e.g., linguistics and psychology), but they can also be from the same general discipline (e.g., generative linguistics and interactional linguistics). Situations in which researchers desire to collaborate, but where they have contrasting epistemologies, are multivocal and have the potential to be productive. van den Besselaar and Heimeriks (2001) take a narrow and homogenous view of a disciplinary research field where "a group of researchers [work] on a specific set of research questions, using the same set of methods and a shared approach (op. cit., p. 706)". They argue that there are different ways of combining elements from various disciplines (and frameworks within disciplines) in order to get them to productively interact. Such interactions can take many forms, ranging from communicating and comparing ideas, exchanging data, methods and procedures to mutually integrating concepts, theories, methodologies, and epistemological principles (op. cit., 2001). However, since researchers most often gravitate towards interactions within their disciplinary tradition's boundaries, it is not uncommon that different communities of researchers work separately on similar objects of study without benefitting as much as they could from each other's efforts. We maintain that this is not an efficient use of resources at the community level. Our efforts towards multivocality in this book project were meant to explore ways of engaging communication between different traditions.

In this chapter, we argue for maintaining diversity of epistemological approaches while either achieving complementarity within explanatory frameworks on different levels or maintaining productive tension. We then present the extent to which researchers in the five case studies comprising our project encountered each other's epistemologies when they compared their analyses of shared corpora. We also include two examples from a similar project. Some comparisons in specific contexts (e.g., some research results, tool use,, or the alignment of representations) did not lead to engagement between epistemologies whereas other types of comparisons in other contexts (e.g., other research results, units of analysis, analytical objective, whether or not categorizing interaction is seen as acceptable, analytic constructs, the notion of what counts as data, the relation between researcher and data, validation of research) did lead to engagement between epistemologies and in one case to radicalizing incommensurable stances. These epistemological encounters (whether they led to epistemological engagement or not) were productive in either a seamless way, in a way that began with difficulties or in ways that could be construed as missed opportunities.

Relating Epistemologies Within a Multivocal Approach

In research on group interactions, as elsewhere, the nature of a researcher's epistemology affects his or her dialogue with other researchers. So, what happens when people holding different theories confront their epistemologies around the analysis of a shared artifact? If all of the differences that define research traditions were starkly contrasted and if all researchers held to their epistemologies with the utmost tenacity, then discourses within disciplinary traditions would be incommensurable, and our efforts towards multivocality fruitless. What we have found, however, is that if researchers are open to engaging with others over epistemological issues, then they can find ways to gain insights from each other—or from the process of engagement—even if they agree to disagree.

We lead you now on this journey with us. Why, you might ask, would we want to take the trouble to try and relate one epistemology with another? Researchers who study group interactions come from a large variety of disciplines, use many epistemological frameworks and don't necessarily interact in the same research communities, so it makes sense to develop a forum within which to examine epistemological frameworks that strive to take into account similar empirical data or seem to manipulate similar concepts. The premise of our project is that it is productive to compare epistemologies through multiple analyses of shared corpora, without necessarily having a specific common goal. Doing comparative work between theoretical frameworks generates fundamental questions for the scientific communities involved, for example how do theoretical assumptions drive research? In addition, combining theoretical perspectives has the potential benefit of bridging between similar areas of literature that are traditionally isolated, and showing why it is interesting for two research traditions that don't usually communicate to work together.

Keeping the Diversity

Our goal is not to merge neighboring theoretical perspectives into a single "super theory" that would account for all aspects of group interactions in a coherent way. On the contrary, diversity in theoretical assumptions and methodological approaches is an unavoidable and desirable characteristic of a multidisciplinary research community. Indeed, diversity is one of the strengths that we wish to maintain since dialogues about analytical constructs between researchers that differ in their ontology and epistemology are particularly enlightening (Abend, 2008). Stahl (2011) points out that the study of group interactions reveals distinct phenomena at different levels of description, and that because these phenomena interact with each other in complex ways—notably in Computer Supported Collaborative Learning (CSCL) settings—it is highly likely that CSCL requires multiple theories so that different aspects of the interaction can be studied at different levels (micro, meso, macro) and at different time scales with methods of investigation that are appropriate for the questions posed.

Different Explanatory Frameworks Can Be Complementary

Sometimes working at these different levels of description means that we only accept a particular type of explanation as valid and that we are not sensitive to other types of explanations at other levels of description. Indeed our historically anchored disciplinary training teaches us to distinguish between better or worse explanations and to prefer certain types of explanatory schemata (Morange, 2005) in which these explanations exist, but sometimes, "competing" explanations that occur at different levels can both be true. For example, there are two possible explanations for migratory phenomena. A bird migrates because climatic or daylight changes trigger physiological modifications in the bird's organism. It also migrates because moving elsewhere will bring it more food, thus favoring its survival and reproduction capacities. The first is a proximal cause, understood by mechanical explanatory schemas from biochemistry, molecular biology, and physiology and the second is an ultimate cause, understood by natural selection and Darwinian explanatory schema. Both explanatory schemas contribute to understanding the phenomenon of migration, but give a more complete picture when combined. Similar combinations of explanatory schemas can be done for the study of group interactions.

The rest of this chapter is organized as follows. We analyze the extent to which researchers engage with each other's epistemologies when they compare aspects of their analyses of group interactions in a multivocal setting, both in this project and in another similar setting. We discuss what might be behind the fact that certain comparisons lead to a researcher engaging (or not) with another's epistemology. The majority of comparisons in various contexts led to engagement between epistemologies and some of these epistemological encounters were productive and glitch free, others had difficulties, but still led to productivity, while still others led to missed opportunities and in one case to radicalizing incommensurable stances. A minority of comparisons in other contexts did not lead to researchers encountering each other's epistemologies, but could either still be fruitful or not productive at all and these are the examples we take up first, in the next section.

Comparisons that Do Not Lead to Epistemological Encounters

Three examples are discussed in this section, one from the Origami Fractions section (Part 2), one from the Asynchronous Knowledge Building section (Part 5), and one from the Multimodal Electricity section (Part 4). The first two examples are productive interactions, despite the fact that they do not lead to epistemological encounters and the third was a very problematic interaction, precisely because epistemological concerns were not addressed.

Reinterpretation of Another's Results Is Compatible with One's Own Epistemology

In this first example, Shirouzu (this volume-a, this volume-b) (see Chaps. 4 and 5 for presentation of data and the data provider Shirouzu's analysis) is able to enrich his view of group interaction by reinterpreting Chiu's results (this volume-a, Chap. 7) in his own framework. First, Chiu performed new analyses with his Statistical Discourse Analysis (SDA) method that focused on the class discussion activity phase of the paper-folding pedagogical task on fractions (2/3×3/4=?) after understanding that Shirouzu had a special interest in it. Shirouzu then demonstrated that he was able to match new meanings to Chiu's interpretations of pivotal moments (occurring in Chiu's framework) that were relevant to Shirouzu in his own framework. For example, Chiu viewed a particular pivotal moment as indicating the end of a period of frequent ideas, occurring just after a teacher acknowledgment. First, Shirouzu noticed that this moment could also be considered as a collective display of new understanding, something that he had not seen previously. Indeed at the moment when collective understanding is reached, a drop in new ideas could occur because learners are consolidating their knowledge in terms of concepts already expressed. Secondly, reexamining this moment in terms of Chiu's definition of ideas as "new" or "old" led Shirouzu to suggest that in his own framework, new ideas could correspond to conceptual or procedural changes of how to view the solutions (e.g., as a physical area or as an algebraic equation), progressing potentially towards a collaborative pivotal moment. Third, Shirouzu noticed that Chiu's five breakpoints corresponding to frequency of new ideas also corresponded to when and how the pedagogical designer's intentions were actualized by students' behavior. This example shows that when researcher A studies moments considered pivotal by researcher B in B's framework, but that A does not originally consider as pivotal, researcher A can appropriate these new pivotal moments to be meaningful in his own framework. Such appropriation proceeded without difficulty because there were no fundamental differences in the researchers' epistemologies.

Implicit Epistemological Compatibility Allows Researchers to Focus on Tool Integration

In the Asynchronous Knowledge Building section (Part 5), there were no fundamental epistemological incompatibilities between the three groups of analysts. Teplovs and Fujita (this volume, Chap. 21) and Law & Wong (this volume, Chap. 22) are representatives of the knowledge building community. Chiu (this volume-b, Chap. 23) argues that his Statistical Discourse Analysis (SDA) can be used with multiple theoretical frameworks, but his methodological assumptions show that he subscribes to the theoretical assumptions of social metacognition, likened by Fujita (this volume, Chap. 24) to the knowledge building principle of collective cognitive responsibility. All three groups strove to innovate new techniques to inform future work. Researchers were also stretched to imagine how the other analytic tools used in the section could inform their understanding of their own approaches. For example, Law & Wong could imagine how the KISSME analytic tool (Chap. 21) could serve to better identify semantic markers of interest to them. Conversely, Teplovs imagined how the use of keyword sets derived from the Law & Wong approach could be used to improve the analytic capabilities of KISSME. Chiu's SDA could be enhanced by using the latent semantic analysis capabilities of KISSME to aid in ensuring notes were correctly tagged, and Chiu's SDA could be used to identify breakpoints for examination from the perspective of discourse markers (Law & Wong) or network analysis (Teplovs & Fujita). It seems reasonable to argue that this example illustrates how the analysts' epistemological compatibility ensured that their diverse higher-order goals could be combined together, all involving ways of improving knowledge building.

Difficulty in Aligning Analytic Representations Due to Different Transcript Needs

As seen in the Group Scribbles data section (Chaps. 14–19, this volume), there is a tension between the need to align analytic representations for comparison purposes and for each analysis to have representations optimized for their own purposes. For example, some analyses need uniform sampling such as 30s intervals, some work at the granularity of recognizable acts, and some also work at granularities of episodes defined by participant activity but also in relation to the focus of analysis (e.g., inscription-construction, artifact-manipulation, or multimodal reformulation episodes). These different representations may be aligned for comparison purposes if there is a common dimension of reference, such as time in a shared video data source. But even if so aligned, conclusions may be attached to units of activity that do not coincide exactly. However, as discussed in Chap. 19 (Suthers, this volume), one can learn as much from the mismatches between analysis-specific representations as from comparisons concerning the conclusions of representations that have been carefully aligned. That being said, had the researchers engaged early on in relation to what they considered as adequate data for the assumptions they held about learning interactions, perhaps all the troubles they experienced concerning the transcript and the subsequent difficulties in aligning the representations that were inferred from them could have been avoided. They would have then been freer to engage in discussions about what could possibly be considered as more fundamental issues, such as how to qualify the learning going on.

Summary of Comparisons that Do Not Lead to Epistemological Encounters

In conclusion, in this section, we illustrated both how multivocality could be productive even when researchers do not specifically engage with others' espistemologies, but also how not engaging can lead to difficulties that prevent researchers from collaborating effectively on deeper issues. In the first example, a researcher's reinterpretations of another's results were possible, largely because his epistemology allowed them, even if this coherence was never made explicit. In the second example, implicit or latent epistemological compatibility allowed for seamless reflection on integrating tools. In the third example, a lack of discussion concerning what researchers needed from the data had the consequence of making analyses and comparison difficult, thus taking energy away from discussion involving comparisons. However, the struggles to get into a position to compare—for the most part led by the discussant of the section—led that discussant (and not the analysts themselves) to pinpoint discrepancies in analyses (cf. the sections below entitled "Interrogating the underlying epistemological assumptions of claims about learner agency and learner activity" and "Interrogating the underlying epistemological assumptions of claims about the evaluation of learning").

Comparisons that Do Lead to Epistemological Encounters

Eight examples are discussed in this section. The first two come from the Origami Fractions section (Part 2), the third comes from Part 5: Designing Biology, the fourth from Part 3: Peer Teams Chemistry, the fifth and sixth examples are from Part 4: Multimodal Electricity, and the last two examples are from a project different from the one in this book, yet very similar called the "MOSAIC" project, described when we get to those examples.

Missed Opportunities for Debating Changes to Key Analytic Constructs

In this first example, we show how the comparison of Shirouzu's (this volume, Chap. 5) and Trausan-Matu's (this volume, Chap. 6) pivotal moments lead to Trausan-Matu integrating aspects of Shirouzu's viewpoints into his own *problématique*[1] in two different ways. The collaboration described in this book introduced Trausan-Matu to the analysis of transcribed oral conversations, encompassing both talk and gesture, a type of corpus he had not focused on before. Adding gesture to his analysis of human interaction amounted to extending the domain of application of his PolyCAFe tool but more importantly to extending the concept of Bakhtin's "voices" to include gestures (both communicative and technical, the latter referring to manipulation of the origami paper); this was the first way he modified his

[1] "Problématique" is a French word used to name a coherent set of problems and assumptions. It provides a coherent framework to express problems, why it is interesting to solve them and how the current research described is able to do so. This term is not a synonym for the English word "problematic".

problématique. We interpret this reconceptualization of "voices" to mean that when Trausan-Matu was confronted with a corpus that presented forms of interaction he was not used to analyzing (i.e., gestures), he was able to reconsider the types of data he took into account as important for understanding learning and to integrate them into his theoretical and methodological framework. This change in conceptualization illustrates how closely related our theoretical frameworks are to the nature of the data we analyze. We argue that in this example theoretical convergence occurs in that Trausan-Matu widened Bakhtin's framework in order to take into account new types of corpora and by doing so, came closer to Shirouzu's framework. However, there was no one else in the fractions group who was expert in Bakhtin's framework and so the act of modifying the concept of "voices" to include gestures was not debated from an epistemological standpoint.

As a result of engaging with the patterns Shirouzu defined in his analyses, Trausan-Matu modified the parameters of two other conceptual terms, but in the framework of conversation analysis; this was the second way he modified his *problématique*. Trausan-Matu extended the definitions of both "utterance" and "adjacency pair" to reflect the inner speech that Shirouzu included in his analyses. Including inferred inner speech that fit with the context of the interaction (e.g., talk, gestures, manipulation of origami paper, and writing on the blackboard) allowed both Shirouzu and Trausan-Matu to constitute coherent stories about how the interaction played out. For Trausan-Matu, utterances were now not only verbal, but could also be inferred thought as well as different types of actions and instead of being essentially individual or co-elaborated, they could be group generated, such as all students moving their chairs to move closer to their origami papers, in chorus. Pairs of utterances were considered to be "adjacent" even if shared ordering could just be inferred, for example between an external utterance (talk or action) and an internal one that was presumed by the researcher to be "thought" by the learner. In contrast to the modification of the concept of "voice" in the Bakhtinian framework, both Shirouzu and Trausan-Matu were challenged in the discussions going on in that section to explain how they backed up their claims about inferred speech. However, submitting such a radical change of definitions of utterance and adjacency pair to the larger conversation analysis community is another matter. Both this and the previous example illustrate a danger that multivocality may lead to the modification of existing analytical concepts without explicitly taking into account the epistemological assumptions that underlie these terms. It seems likely that changing the definition of analytical terms already in widespread use will hamper researchers' ability to effectively communicate.

Comparing Pivotal Moments Leads to Epistemological Modifications that Enrich Analyses

In this second example, after discussing how each of them defined pivotal moments with both Shirouzu and Trausan-Matu, the author Chiu (this volume-a, Chap. 7) decided to look at the context around which his pivotal moments occurred (thus enlarging his unit of interaction). He extended his notion of a pivotal moment to beyond the single turn, seeing that a single turn could be interpreted in relation to what comes before and after. This extension had two consequences. The first is that Chiu broadened his understanding of why his breakpoints were pivotal using qualitative analyses to supplement his quantitative Statistical Discourse Analysis approach and the second is that he thus became more convinced of how quantitative and qualitative methods can be used in concert to obtain a more complete understanding of group interactions. In a sense Chiu maintained two visions of a pivotal moment, one that encompassed a single turn, which he needed in order to perform his analysis that located breakpoints in an interaction, and a second that took into account the context around a breakpoint in order to better understand it qualitatively. Chiu saw that a detailed qualitative analysis of the interaction around the pivotal moment could uncover hypothesized mechanisms that change the interaction; these could be searched for across case studies in order to determine their hypothesized robustness and then specified through operationalized variables in order to be statistically tested. This example illustrates how a researcher can adapt mixed methods for deeper understanding.

Opening Up to Engagement with Different Epistemological Assumptions

In the Biology section (Part 6), a variety of methods were applied to the analysis of a data set produced in a first iteration of a Design-Based Research cycle. In order to test their hypotheses, researchers who provided the data applied quantitative summative evaluations and coding and counting process analyses. Other researchers applied ethnographic, ethnological, and SNA methods to the analysis. At the start of the multivocal process, while the researchers were all aware of the epistemological underpinnings of their approach, researchers were less aware of a much more subtle difference in theoretical assumption, namely the assumed idealized role of a facilitator in a collaborative learning interaction. This difference played a much greater role in the interaction between researchers.

Quite orthogonal to the diversity of methodological approaches represented by the analytic team, the collection of analysts brings together two different research communities, one from classroom discourse where we get the Academically Productive Talk (APT) framework (Resnick, O'Connor, & Michaels, 2007) developed largely from research on primary school learners and whole classroom interactions facilitated by teachers, and another from CSCL where we get theories of problem-based learning (Hmelo-Silver & Barrows, 2006) and group cognition (Stahl, 2006); both of which have largely been developed from analyses of older learners, and in the case of problem-based learning, largely advanced learners (i.e., medical school students). Between the two communities, much is shared in terms of desired characteristics of the student interactions. Correlational analyses, largely from within the collaborative learning research community, offer empirical support for the value of certain characteristics of interactions between students for triggering learning processes. In the CSCL community, there has also been a large amount of research on how to achieve these kinds of interactions in small groups of students (e.g., scripted collaboration as well as study of expert PBL facilitation techniques). The PBL facilitation research is the easiest to directly compare with APT since it involves characterization of rhetorical moves used by humans to engage groups of students in discussion. Much can be found in common, for example, between the Hmelo-Silver work on PBL facilitation and specification of APT moves. However, if one looks deeper into the assumptions about the ideal positioning of the facilitator, one finds that within the PBL facilitation viewpoint, as in the Group Cognition framework, the idealized role of the facilitator is much more minimal than in an APT classroom where the instructor plays an integral role throughout the discussions, sometimes offering nearly half of the contributions that make up the discussion, while carefully self-locating outside of the interaction between the students so that student reasoning is kept at the center, and students maintain an authoritative footing within the interaction. Whereas in Stahl's view, the teacher should get out of the way, in APT, the teacher is constantly an integral part of the interaction. This view of the idealized role of the teacher colors the way three of the chapters evaluate the work.

The work on the Biology conversational software agents (see Dyke, Howley, Kumar, & Rosé, this volume, Chap. 25) did not begin with the goal of addressing the question of the ideal role of a facilitator. In fact, the data providers were not fully aware of the extent of the distinction in views until the multivocal process had begun. Instead, their goal was to accomplish several things within the theoretical framework of APT, a goal that was not successfully communicated to the other analysts at the inception of the multivocal process.

Empirical validations of the theory of APT show that when expert teachers use APT facilitation with whole classes, the classes achieve high test scores across subject areas and maintain their advantage for years (Resnick, Asterhan, & Clarke, in press). Around these studies, a belief about the mechanism for the effect has emerged but never validated through careful experimentation. The treatment has always been complex, involving multiple facilitation moves used with whole classes, where a human teacher insightfully decides when and with whom to use each move. Thus, it is not clear whether there are differential effects across the individual facilitation moves, or whether there are preconditions either at the group or individual level for their effects. Furthermore, there is a belief about the connection between results on achievement and impact on identity, motivation, and affective dimensions, although this had never been formally tested. The series of controlled studies with small groups of students that involves APT agents was meant to fill this empirical gap, giving the data providers the opportunity to carefully manipulate how and when each move was used, and investigating the effect on cognitive, motivational, and social dimensions and how those effects interact with individual differences between students and with group composition characteristics.

In the Howley, Kumar, Mayfield, Dyke, & Rosé, this volume analysis (Chap. 26), the role of the agent as an APT instructor is taken for granted, and the evaluation is with respect to how successful that intervention was in achieving the desired end in the interaction itself, with the goal of answering the research questions as outlined above. The Stahl and Goggins & Dyke analyses (this volume, Chaps. 28 and 29, respectively) took a different view—rather than taking the role of the agent for granted, they assumed an idealized minimal role for the agent and then evaluated the interactions in terms of how minimal the role of the agent was. The Stahl analysis ignored the differences between conditions, whereas the Goggins & Dyke analysis did look for differences between conditions. The Cress & Kimmerle, this volume analysis (Chap. 27) ignored the agent altogether and instead focused on the environment and task setup. So that chapter did not deeply participate in the multivocal discussion about the role of the agent, although the analysis was nevertheless useful for informing the redesign.

In connection with the issues we began this section with, consider in particular Stahl's criticism of the software agents as dominating the group discussions and getting in the way of student interactions (Chap. 28). In Stahl's analysis, the data was first formatted in the columnar representation suggested by Ochs (1979), and then analyzed as an uptake network structure with the goal of illustrating how participation in the conversation was distributed. Interpretation of the agent as "getting in the way" came from a count of the number of words in the agent column relative to those in the student columns as well as a representation of the focus of attention layered on Stahl's hand constructed graph representation. The analysis was conducted in a generic way, without separating out phases of the discussion that were set apart in the task design for different purposes and where students and the tutor agent were intended to play different roles. Arguments in Stahl's analysis are made as interpretations of interactional patterns within uptake networks and the particular layout of the transcription, both of which are displayed in Stahl's chapter. Although only a single transcript was selected for the main analysis presented by Stahl, a causal interpretation of the interactional structure, i.e., that the agent's role taking as interpreted by Stahl was the cause of other observed patterns in student interactions, is inferred in Stahl's analysis, where this would not be warranted in other epistemologies. These interpretations begin with mostly unstated assumptions about the meaning behind interactional structure (i.e., in terms of value judgments about the facilitator based on positioning in the uptake network). Thus, identification that the desired pattern is not present is then interpreted by Stahl as a failure and a cause of other undesirable behavior identified within the same interaction, where it might be interpreted differently using other analytic lenses.

At first, the data providers were frustrated with the chapters from other analysts because they seemed to be operating within a different theoretical plane—the other analysts were not trying to answer the data providers' questions and did not offer them insights related to those specific questions. But what they did do was question the theoretical assumptions the data providers were making about what the role of the facilitator should be. In the case of Stahl and Goggins & Dyke, there was a direct challenge to the teacher to "get out of the way". In the case of Cresse, it was more a process of questioning whether the environment was conducive enough to learning and collaboration that we could get a meaningful read on whether the manipulation worked in the first place. One could see all of these chapters as questioning whether the data providers were really asking the right questions to begin with.

In the end, the data providers reluctantly took the hard feedback to heart. Their own analysis did point to places where their intervention was getting in the way through poor timing. Problems with the agent's timing were indeed pointed out in multiple places throughout the analysis chapters. Addressing these issues with a new architecture (Adamson & Rosé, 2012) was one major technical advance between the first study, which provided the shared dataset, and the subsequent studies, which were more successful in terms of producing learning gains and other positive effects. In the data providers' redesign, they also scaled back the extent to which the agents were intervening in the discussion in the subsequent, much more successful trials. Even with the scale back, however, they have hung on to the basic moves from the APT theoretical framework (Dyke, Adamson, Howley, & Rosé, 2013; Adamson, Jang, Ashe, Yaron, & Rosé, in press; Clarke et al., in press) and have not changed their commitment to an APT style ideal facilitator. As the data providers have continued to investigate the impact of the facilitation moves on learning and motivational variables like self-efficacy (Howley et al., 2012), they have found results that validate the line of questioning they began with. In particular, the results have called into question the simplicity of the APT theory's model of facilitation as increasing self-efficacy through positively positioning students as authoritative in interaction. Instead, they find differential effects of the moves depending on student ability and self-efficacy. In connection specifically with multivocality, this example shows that while it was initially frustrating to get advice and feedback that seemed to be ignoring the theoretical framework the data providers were working within, researchers coming from other theoretical frameworks were able to look at what they were doing with less of a tunnel vision and were therefore in a better position to push them to reconsider things that might have hindered them from answering their questions if they had not stopped to make adjustments to their experimental setting.

Alternative Operationalization Brings Out Different Aspects of a Complex Analytical Construct

In the Chemistry section (Part 6), we observe a similar situation to the Biology section with respect to fine-grained distinctions in operationalizations, but one that played out more smoothly. Here we compare the two processes and work towards understanding why they played out so differently.

The team of analysts who worked on the Chemistry dataset was diverse methodologically in the same way as the Biology team. In particular, one analysis team (led by Keith Sawyer, this volume, Chap. 10) took an ethnographic approach. Another team (led by Jun Oshima, this volume, Chap. 12) took a social network analysis approach. And two teams (led by Carolyn Rosé and Jan-Willem Strijbos, respectively, and reported in Howley, Mayfield, Rosé, & Strijbos, this volume, Chap. 11) took a coding and counting approach. Despite these similarities between the two multivocal processes in terms of team composition, however, the circumstances of the data collection for the Chemistry effort were quite distinct from the design-based research process that provided the context for the Biology data collection. This distinction in turn led to a very different dynamic in terms of the communication between analysts and data provider.

In particular, the goal of the Chemistry data collection was to explore a process that was tried and true (i.e., Peer Led Team Learning) and to understand better how it was working rather than to evaluate the first instantiation of an intervention that was at an early stage and determine how to make it better. Thus it was natural for the process to be framed as a casual discussion about what different lenses applied to the data might teach us about the reasons why the intervention may have been working. The data providers for the Chemistry data had already completed their analysis and were satisfied with the answers they had found to their own research questions. From the data they collected and had examined, they selected two focal groups for the purpose of making an interesting contrast rather than evaluating the efficacy of the intervention. Because of the small size of the dataset and the absence of any experimental manipulation, there was no implicit invitation for analysts to provide a value judgment on the intervention. This stands in contrast to the Biology data effort where the purpose of the study was clearly to evaluate an intervention that was manipulated experimentally in the data, and the whole corpus was shared, with all flaws exposed. Considering this contrast, it is not surprising that most of the analyses of the Biology corpus were framed as evaluations of the quality of the intervention, and the focus of many of the write ups was on what went wrong and what the authors thought the data providers should have done differently.

Among the analysts of the Chemistry data, a common lens used to facilitate discussion among groups was the idea of leadership and role taking within groups. Similar to the Biology collaboration, the analysts brought into their work unvoiced assumptions about leadership roles and what they look like. These subtle distinctions in unvoiced underlying assumptions went unnoticed at first when discussions were at the level of broad strokes discussion about the style of interaction within the two student teams whose discussions were the focus of the analysis. However, when conclusions about individual students and their role taking within the interaction were compared across analyses, differences in conclusions led to line by line comparisons, which in turn eventually led to insights about distinctions in definitions. Upon reflection, the distinctions that were revealed led to interesting discussions about how expansive and complex the idea of leadership is.

The interaction around the questions of what leadership really is, including comparisons across operationalizations, were productive. The differences did not lead to conflict or friction among the analysts. The contrast to other case studies might have stemmed from the fact that evaluation of who was taking a leadership position in groups was not as value laden in this analysis. The distinctions in operationalization did not carry implications for design, per se, and did not reflect positively or negatively on anyone's work. Operationalizations themselves are not right or wrong. Instead they are either successful or not at capturing particular phenomena faithfully. There are always choices to make in operationalization, and differences across alternative operationalizations offer the opportunity to triangulate. A researcher may evaluate the validity of another researcher's operationalization. However, while discussions surrounding the patterns found in the Chemistry dataset centered on the idea of leadership in a number of interactions between analysts, none of the analysts were particularly invested in convincing others to see the role taking in the interactions in any specific light or saw the value of their constructs in terms of the extent to which they could convince others to view leadership taking the same way. The fact that one researcher's alternative operationalization brings out different aspects of a complex construct like leadership does not necessarily detract from the value of an alternative operationalization that brings out other facets that are also interesting. Thus, the discussion may proceed comfortably for all despite the disagreements that come up.

The contrast between what happened when researchers compared analyses from the previous Biology example and in this Chemistry example highlights the importance of taking care in how data is shared in a multivocal process, with clear communication about goals and expectations conducted up front so that the exchange can be comfortable and productive for all.

Interrogating the Underlying Epistemological Assumptions of Claims About Learner Agency and Learner Activity

Epistemological comparisons need not only be grounded in comparisons between the conclusions of analyses. They can also be grounded in comparisons of representational devices and empirical claims that evidence how analyses constitute the object of study in the first place. For example, the discussion of the Group Scribbles analyses (Suthers, this volume, Chap. 19) examined the units of agency and activity about which analysts made claims. Such groundings begin close to the data, but uncover epistemological assumptions that may not be discussed explicitly by researchers. Some analyses (e.g., Lund & Becu-Robinault, this volume, Chap. 17) focus on individuals' acts (reformulations across modes and media) to characterize their conceptual understanding: they clearly focus on the individual as the agent of learning. Others (e.g., Jeong, this volume, Chap. 18) do not track individuals at all, taking artifacts produced by individuals as evidence for collective or group understanding: the group is clearly the agent of learning. Some interesting nuances can be found by comparing how analyses characterize sequences of events across media. For example, Looi, Song, Wen, & Chen, (this volume, Chap. 15) and Lund & Becu-Robinault both examine how content changes as it is expressed in one medium and then another, implying that activity takes place in a given modality and diachronic features (change over time) are most important for understanding learning; while Medina (this volume, Chap. 16) discusses how events taking place synchronically or nearly simultaneously in multiple modes converge to constitute a single activity, implying that there is agency at the group level and activity is simultaneously distributed across modalities. Some of these analysts did not state these epistemological positions explicitly in their chapters: rather, we uncovered these positions by comparing the units and relations of their analysis. We should add that these comparisons draw conclusions about analyses not about researchers. For example, other discussion and citations by Lund & Becu-Robinault show that they hold a synchronic view of activity as well as the diachronic view exemplified by their analysis. This example shows how the units of agency and activity that researchers used in their analyses actually *embody* their underlying epistemological assumptions. The first assumption concerns where the researcher is looking for learning—at the individual or group level, or using a combination of both. The second assumption concerns how the researcher conceptualizes learning. This is discussed more specifically in the next section.

Interrogating the Underlying Epistemological Assumptions of Claims About the Evaluation of Learning

Comparisons of analyses of two different data corpora from the Group Scribbles setting (Chen & Looi, this volume, Chap. 14) both exposed epistemological differences on the part of analysts concerning criteria for evaluating the quality of participant activity. Analysts first analyzed a Group Scribbles corpus on fractions that was ultimately not used for this book. As discussed in Chapter 19 (this volume), an analyst (van Aalst) approaching the data from the theoretical perspective of Knowledge Building found the data uninteresting, as students' verbal discussion did not display evidence of students taking control of the learning opportunity. Simultaneously, another analyst (Medina) influenced by ethnomethodological and conversation analytic traditions analyzed students' actions in the workspace to show how a graphically expressed proposal was contingent upon the setting in a manner evidencing the development of shared representational practices. These different assessments are not merely due to one analyst having analyzed only talk, while the other analyzed workspace actions. There is a fundamental difference in whether student interaction should be approached with theoretically driven standards, or whether instead this interaction should be taken on its own terms as displaying organized group participation. A similar epistemological encounter occurred in the electricity data corpus analyzed for this volume. Lund and Becu-Robinault (Chap. 17) evaluated student actions in terms of whether individuals displayed understanding of conventions, such as for diagramming electrical circuits, that are taught in school and used by professionals, while Medina (Chap. 16) evaluated student actions in terms of how group activity led to innovations. Consequently and as discussed in Chap. 19, their assessments of one particular individual, "Bruno", differed. Lund and Becu-Robinault found that Bruno failed to exhibit a canonical representation of how a wire connects to a battery in a diagram, while Medina found that Bruno's innovation of stacking batteries on top of a wire was one of the various contingencies enabling a group innovation, "two batteries, two bulbs". Again, the difference is not only on whether we focus on diagramming or building circuits but also in analysts' fundamental epistemological criteria: should student acts be evaluated in terms of pedagogically driven criteria or by the internal logic of their collective accomplishments? This example delves more deeply into how a researcher's way of evaluating learning reveals his or her epistemological assumptions. One possibility shown in this dataset (cf. the Lund and Becu-Robinault's analysis) is to define learning as change over time where the focus is on the learner's evolving capacity (but as a member of a group that may or may not exhibit this capacity collectively) to be able to translate domain concepts from one representation to another while respecting taught domain knowledge. This desire of the researcher to track such an evolving capacity does not deny that learners also use different modalities simultaneously when they (co)construct knowledge in groups. A second way of conceptualizing learning in this dataset (cf. Medina's analysis) focuses more specifically on the group and describes how converging acts in multiple modalities and media are brought together to accomplish the group's activity, without evaluating the knowledge being constructed from any domain standard.

Integrating Methods on the Basis of Compatible Epistemologies

In the remaining two sections, we take examples from the MOSAIC project, financed by the French National Research Agency. This project was also multivocal, involving psychologists (of different specializations) and linguists (working on conversation analysis). Although its primary goal was to understand collaborative design processes in the domain of architecture (see analyses in Détienne & Traverso, 2009), its secondary goal was to specifically confront methodologies around an analysis of shared data. The researchers involved intended to construct a bridge between theoretical and descriptive research on interaction carried out in the language sciences on the one hand and studies of cognition and dialogue in complex collective design activities carried out in ergonomics on the other. During one collaboration within the MOSAIC project, it proved possible to integrate a socio-cognitive interactionist approach with a cognitive ergonomics approach, and this was to a great extent because of similar epistemological positions. In that collaborative effort, a discursive dimension accounting for argumentative and enunciative[2] activities was analyzed together with an epistemic dimension that accounted for the intermediate design products as well as the knowledge mobilized during these activities of elaboration and reconstruction (Baker, Détienne, Lund, & Séjourné, 2009). Both parts of this integrated analysis were built up from coding the interaction,

[2]Ducrot defines an enunciator as the instance of *the source of a viewpoint* expressed in the propositional content of an utterance (Brandt, 2013).

based on a priori categories gleaned from the literature and that were confronted with the data. In this example, although the psychologists represented different specialties, they all had a similar epistemological approach to studying group interactions and so integrating their approaches was seamless. For instance, they all agreed that researchers could define analytical categories, in relation to theory and research questions and then refine them in relation to part of the corpus. They also agreed that the researcher's task is to observe and analyze data, in order to elaborate theories and models of the data set. They also agreed that the validation of research concerns so-called objective markers, indicators of categories, and intersubjective agreement between independent coders.

Recognizing Incommensurability Radicalizes Researcher Positions but Also Makes Researchers More Aware of Their Constraints

In another collaborative effort within the MOSAIC project, it proved impossible to actually integrate the interactional linguistics approach with the cognitive ergonomics approach, largely because of their differing epistemological positions. For instance, whereas for the cognitive ergonomists, data selection was in large part determined by theory and model, the interactional linguists attempted to take into account the minute details of interaction in a way that was not conditioned by prior theorization. The two approaches do not agree on what constitutes "the corpus" and it is arguable that agreeing on what constitutes the corpus in the first place can more quickly allow researchers to compare their respective analyses (and access deeper conceptual issues) because they will be able to collectively refer to the same parts of the empirical material. On the other hand, the very act of deciding what the corpus should be obligated the researchers to be specific about their epistemological positions regarding what they needed to answer the questions that interested them and that were pertinent to ask in their respective theoretical frameworks. In general, the act of comparing their respective methods led to the cognitive ergonomists and interactional linguists detailing the very specific differences that illustrated the consequences of foci stemming from epistemological positions and these led to defining "zones of maximal analytical vigilance" (Traverso & Visser, 2009: p. 169), where researchers had to be particularly careful in respecting their methodological constraints. For example, the interactional linguist worked to make her description of the interaction coincide with how (*she understood that*) the participants themselves interpreted and demonstrated the interaction, and although the cognitive ergonomist's descriptions were also formed from the activities of the participants, she recognized that her analysis was a personal construction (differing perhaps from other colleagues' descriptions), colored by theories and models she would render explicit. Her descriptions included inferences that were based on activities that were implicitly present within the interaction, but that could be argued to be present, based on observables (much like Shirouzu and Trausan-Matu argued in the Origami

Fractions section, Part 2). The crucial question here is the extent to which researchers are able to substantiate their analytic claims. Both cognitive ergonomists and interactional linguists claim to base their interpretations on observables, but the difference seems to occur on two intersecting planes. The first is the extent to which an object, event, or phenomenon can be considered observable. Is being "observable" some kind of proxy for "objective" or is it impossible to separate observing human interaction from our own human experience so that we necessarily both view and make inferences about it? And the second is the extent to which analyses of human action are effectively grounded in what is observable (i.e., are interpretations about human interaction (the interactional linguistics position) more grounded in observables than inferences about human interaction (the ergonomist position))?

These last two examples from the MOSAIC project show the crucial roles that differing epistemologies played in bringing about productive multivocality. Here, underlying epistemological assumptions determined whether or not methods could be integrated and when they could not, the comparison of such assumptions served to determine how researchers grounded their analyses and interpretations.

Summary of Comparisons that Do Lead to Epistemological Encounters

In the first example (Origami Fractions section, Part 2), modifications of major analytical concepts of two theoretical frameworks were carried out and functioned locally, but an opportunity was missed to discuss the ramifications of these modifications with representatives of the relevant research communities, thus potentially leading to a confusing use of terms. If researchers are unilaterally redefining analytical constructs that are in widespread use, then how can other researchers evaluate work done with those constructs, if they don't mean the same thing anymore? On the other hand, if the innovative changes in definition respect the epistemological constraints of the framework in which they are supposed to function, then perhaps being free of having to conform to community norms is what will allow for scientific progress.

In the second example (also Origami Fractions section, Part 2), an epistemological modification led to mixed methods being incorporated in a coherent way. There is a well-known danger to mixing different methods, but Chiu did not fall victim to it. The *incompatibility thesis* critiques naive methodological eclecticism (in other words, mixing methods without examining their underlying assumptions) with the claim that methods from diverse traditions are based on incompatible philosophical assumptions, and so cannot be combined without incoherence (Yanchar & Williams, 2006). However, identifying the specific ways in which methods intrinsically carry assumptions while also defending the agency of researchers in choosing and combining methods allows researchers to escape methodological determinism and apply methods coherently in different theoretical frameworks than in which they originated.

The third example (Designing Biology, Part 5) shows that it can be fruitful to open up to engagement with the differences in underlying assumptions between our own theories and methods and those of others. Researchers may come to a consensus about how they might strategically relax or reformulate some of their methodological or theoretical assumptions in order to find some common ground on which to exchange insights. Alternatively, they may not come to such a compromise, but at least their perspectives will have been broadened by honestly asking the relevant questions. It should also be noted that even though data providers did not sufficiently communicate their context to analysts, thus leading to misunderstandings about the data, this ultimately led to insights about how to improve the pedagogical situation through Design-Based Research.

In the fourth example (Peer Teams Chemistry, Part 3), unvoiced assumptions about the analytic construct of leadership were discovered through differences in conclusions that led to line by line comparisons of coding categories by the analysts and eventually to distinctions in definitions. Alternative operationalizations are beneficial, but also make for a harmonious researcher interaction when they can bring out complementary aspects of an analytical construct without taking away from another operationalization.

The fifth and sixth examples (Part 4: Multimodal Electricity) show how the units of agency and activity that researchers used in their analyses actually *embody* their underlying epistemological assumptions about where learning takes place and how to track it. In addition, this interrogation led to an analysis of how researchers use different criteria to evaluate learning. There is a tension between describing in detail the ways in which groups work together and how individuals contribute to the collective, and doing that in addition to evaluating the individual and group activity on the basis of whether or not it reflects an understanding of taught (or canonical) domain knowledge.

In the last two examples from the MOSAIC project, researchers maintained their original assumptions, coherent with their own frameworks, but they specifically sought out the comparison of epistemological assumptions with researchers of different traditions to see how the assumptions affected methods in order to gain insight about collaborative processes. We showed that when epistemological frameworks were compatible, it was easy to combine methods but also that when frameworks were not compatible, attempting to combine methods only served to radicalize the researchers involved. At the same time, explicating their epistemological positions also had the effect of those radicalized researchers more clearly understanding how they were expected to respect the constraints of their respective frameworks and so made them more careful researchers.

Conclusions

In this chapter we have explored what occurs when researchers encounter each other's epistemologies in multivocal settings. We have argued that one of the major reasons that researchers from different traditions may find it difficult to collaborate is that they do not share the same epistemological assumptions about the nature of scientific knowledge and how it can be acquired. We further argued that researchers in the same community, but from different traditions may work separately on similar objects of study without benefitting from each other's efforts and that this is not effective at the community level. We reflected on how multivocal contexts could remedy that by studying the extent to which researchers who compared aspects of their analyses of shared corpora had encounters with other researcher's epistemologies. Our hypothesis was that engaging with researchers with different epistemologies could help bridge between traditions and make for more effective collaboration at the community level. Sometimes the comparisons researchers made about aspects of their analyses led to epistemological engagement and sometimes it did not.

In the cases where comparisons of aspects of analyses did not lead to epistemological engagement, multivocality could still be productive. For example, when the underlying epistemologies of researchers were already coherent (so there was no pressing impetus to engage), they were able to reinterpret each other's analyses in their own frameworks and focus on integrating each other's tools. On the other hand, when the underlying epistemologies of researchers were not coherent and they did *not* engage, this proved problematic for obtaining the kind of data researchers needed to do their analyses.

In the cases where comparisons of aspects of analyses did lead to epistemological engagement, multivocality was either (1) productive and glitch free, (2) productive but difficult, or (3) led to missed opportunities. For example, during productive glitch-free epistemological engagement, researchers successfully mixed qualitative and quantitative methods, used multiple operationalizations of an analytical construct in a harmonious way in order to bring out the complementary aspects and combined methods of analysis from different traditions once it was ascertained that epistemologies were compatible.

During productive but difficult epistemological engagement, analysts experienced misunderstandings about data due to insufficient communication about its context by the data provider, and although in this latter's view, analysts were not treating the issue at hand (analysts even questioned the legitimacy of the pedagogical situation), the data provider ended up gaining insights from their analyses. In two other examples of productive but difficult epistemological engagement, the discussant was doing the engaging on behalf of the researchers who had expended their energy on attempting to get the data they needed and then attempting to align their analyses without an obvious common empirical basis. Despite the difficulties, the discussant was able to draw out differences in learner agency, learner activity, and evaluating learning, but unfortunately not all researchers engaged with the discussant, so this example also could be classed in part in the category of missed opportunities.

During epistemological engagement that led to missed opportunities, a researcher—influenced by a colleague's analysis—modified the meanings of analytical constructs that were in widespread use; on the one hand he could be construed as innovating, but on the other, he missed an opportunity to debate those changes within the larger community. Our final example also has both positive and negative aspects. Researchers from different traditions were not able to integrate their methods due to incommensurable differences in epistemologies. They became more entrenched in their respective stances and perhaps lost faith in the multivocal process. However, the act of engaging made them more reflective.

We are now in a position to offer up some concluding remarks. First, comparing aspects of analyses can be productive without leading to engaging with other epistemologies. That being said, many more examples were productive once researchers did engage. At the same time, some situations of productive epistemological engagement also led to difficulties or to missed opportunities. In the end, it all comes down to the researcher's will to participate in multivocality. When there were difficulties, they were due to communication breakdown (see Rosé, & Lund, this volume, Chap. 32), to a non-willingness to put in the necessary effort, but also to epistemological entrenchment or in other words a non-willingness to strategically relax or reformulate some of their methodological or theoretical assumptions in order to find some common ground on which to exchange insights. Researchers who are willing to engage in multivocality are innovators. They are on the cusp of interdisciplinarity, beginning to mutually integrate concepts, theories, methods, and epistemological principles (van den Besselaar & Heimeriks, 2001). Our hypothesis was that engaging with researchers with different epistemologies could help bridge between traditions and make for more effective collaboration at the community level. This chapter illustrates that we have taken some steps towards that goal for disciplinary traditions interested in collaborative learning.

Acknowledgements The first author is grateful to Benoit Habert for his insightful comments on a draft version of this chapter. Our thanks also go to Nancy Law for provoking a key pivotal moment useful in elaborating the argument developed here. This work was supported by the French CNRS and by the Pittsburgh Science of Learning Center and Graduate Training Grant from the Department of Education (#R305B040063).

References

Abend, G. (2008). The meaning of 'Theory'. *Sociological Theory, 26*(2), 173–199.

Adamson, D., Jang, H., Ashe, C., Yaron, D., & Rosé, C. P. (in press). Intensification of Group Knowledge Exchange with Academically Productive Talk Agents. *Proceedings of Computer Supported Collaborative Learning.*

Adamson, D., & Rosé, C. P. (2012). Coordinating Multi-Dimensional Support in Conversational Agents. *ITS 2012 Proceedings of the 11th International conference on Intelligent Tutoring Systems,* Springer-Verlag, pp 346–351.

Baker, M. J., Détienne, F., Lund, K., & Séjourné, A. (2009). Etude des profils interactifs dans une situation de conception collective en architecture. In F. Détienne & V. Traverso (Eds.), *Méthodologies d'analyse de situations cooperatives de conception: Corpus MOSAIC* (pp. 183–220). Nancy: Presses Universitaires de Nancy.

Brandt, L. (2013). *The Communicative Mind. A Linguistic Exploration of Conceptual Integration and Meaning Construction.* Cambridge: Cambridge Scholars Publishing.

Chen, W., & Looi, C.-K. (this volume). Group Scribbles-supported collaborative learning in a primary grade 5 science class. In D. D. Suthers, K. Lund, C. P. Rose, C. Teplovs & N. Law (Eds.), *Productive multivocality in the analysis of group interactions, Chapter 14.* New York, NY: Springer.

Chiu, M. M. (this volume-a). Social metacognition, micro-creativity and justifications: Statistical discourse analysis of a mathematics classroom conversation. In D. D. Suthers, K. Lund, C. P. Rosé, C. Teplovs & N. Law (Eds.), *Productive multivocality in the analysis of group interactions, Chapter 7*. New York, NY: Springer.

Chiu, M. M. (this volume-b). Statistical discourse analysis of an online discussion: Cognition and social metacognition. In D. D. Suthers, K. Lund, C. P. Rosé, C. Teplovs & N. Law (Eds.), *Productive multivocality in the analysis of group interactions, Chapter 23*. New York, NY: Springer.

Clarke, S., Chen, G., Stainton, K., Katz, S., Greeno, J., Resnick, L., Howley, H., Adamson, D., Rosé, C. P. (in press). The Impact of CSCL Beyond the Online Environment, *Proceedings of Computer Supported Collaborative Learning.*

Cress, U., & Kimmerle, J. (this volume). Successful knowledge building needs group awareness: Interaction analysis of a 9th grade CSCL biology lesson. In D. D. Suthers, K. Lund, C. P. Rosé, C. Teplovs & N. Law (Eds.), *Productive multivocality in the analysis of group interactions, Chapter 27*. New York, NY: Spring

Détienne, F., & Traverso, V. (Eds.). (2009). *Méthodologies d'analyse de situations cooperatives de conception: Corpus MOSAIC*. Nancy: Presses Universitaires de Nancy.

Dyke, G., Adamson, D., Howley, I., & Rosé, C. P. (2013). Enhancing scientific reasoning and explanation skills with conversational agents. *IEEE Transactions on Learning Technologies, 6*(3), 240–247.

Dyke, G., Howley, I. K., Kumar, R., & Rosé, C. P. (this volume). Towards academically productive talk supported by conversational agents. In D. D. Suthers, K. Lund, C. P. Rosé, C. Teplovs & N. Law (Eds.), *Productive multivocality in the analysis of group interactions, Chapter 25*. New York, NY: Springer.

Fujita, N. (this volume). Critical reflections on multivocal analysis and implications for design-based research. In D. D. Suthers, K. Lund, C. P. Rosé, C. Teplovs & N. Law (Eds.), *Productive multivocality in the analysis of group interactions, Chapter 24*. New York, NY: Springer.

Goggins, S. P., & Dyke, G. (this volume). Network analytic techniques for online chat. In D. D. Suthers, K. Lund, C. P. Rosé, C. Teplovs & N. Law (Eds.), *Productive multivocality in the analysis of group interactions, Chapter 29*. New York, NY: Springer.

Hmelo-Silver, C. E., & Barrows, H. S. (2006). Goals and strategies of a problem-based learning facilitator. *Interdisciplinary Journal of Problem-based Learning, 1*, 21–39.

Howley, I. K., Kumar, R., Mayfield, E., Dyke, G., & Rosé, C. P. (this volume). Gaining insights from sociolinguistic style analysis for redesign of conversational agent based support for collaborative learning. In D. D. Suthers, K. Lund, C. P. Rosé, C. Teplovs & N. Law (Eds.), *Productive multivocality in the analysis of group interactions, Chapter 26*. New York, NY: Springer.

Howley, I. K., Mayfield, E., Rosé, C. P., & Strijbos, J.-W. (this volume). A multivocal process analysis of social positioning in study groups. In D. D. Suthers, K. Lund, C. P. Rosé, C. Teplovs & N. Law (Eds.), *Productive multivocality in the analysis of group interactions, Chapter 11*. New York, NY: Springer.

Howley, I., Adamson, D., Dyke, G., Mayfiled, E., Beuth, J., & Rosé, C. P. (2012). Group Composition and Intelligent Dialogue Tutors for Impacting Students' Self-Efficacy, ITS 2012 Proceedings of the 11th International conference on Intelligent Tutoring Systems, Lecture Notes in Computer Science Volume 7315, Springer-Verlag, pp 551–556.

Jeong, H. (this volume). Development of group understanding via the construction of physical and technological artifacts. In D. D. Suthers, K. Lund, C. P. Rose, C. Teplovs & N. Law (Eds.), *Productive multivocality in the analysis of group interactions, Chapter 18*. New York, NY: Springer.

Jun Oshima, J., Matsuzawa, Y., Oshima, R., & Niihara, Y. (this volume). Application of network analysis to collaborative problem solving discourse: An attempt to capture dynamics of collective knowledge advancement. In D. D. Suthers, K. Lund, C. P. Rosé, C. Teplovs & N. Law (Eds.), *Productive multivocality in the analysis of group interactions, Chapter 12*. New York, NY: Springer.

Law, N., & Wong, O.-W. (this volume). Exploring pivotal moments in students' knowledge building progress using participation and discourse marker indicators as heuristic guides. In D. D. Suthers, K. Lund, C. P. Rosé, C. Teplovs & N. Law (Eds.), *Productive multivocality in the analysis of group interactions, Chapter 22*. New York, NY: Springer.

Looi, C.-K., Song, Y., Wen, Y., & Chen, W. (this volume). Identifying pivotal contributions for group progressive inquiry in a multi-modal interaction environment. In D. D. Suthers, K. Lund, C. P. Rose, C. Teplovs & N. Law (Eds.), *Productive multivocality in the analysis of group interactions, Chapter 15*. New York, NY: Springer.

Lund, K., & Bécu-Robinault, K. (this volume). Conceptual change and sustainable coherency of concepts across modes of interaction. In D. D. Suthers, K. Lund, C. P. Rose, C. Teplovs & N. Law (Eds.), *Productive multivocality in the analysis of group interactions, Chapter 17*. New York, NY: Springer.

Lund, K., & Suthers, D. D. (this volume). Methodological dimensions. In D. D. Suthers, K. Lund, C. P. Rosé, C. Teplovs & N. Law (Eds.), *Productive multivocality in the analysis of group interactions, Chapter 2*. New York, NY: Springer.

Medina, R. (this volume). Cascading inscriptions and practices: Diagramming and experimentation in the Group Scribbles classroom. In D. D. Suthers, K. Lund, C. P. Rose, C. Teplovs & N. Law (Eds.), *Productive multivocality in the analysis of group interactions, Chapter 16*. New York, NY: Springer.

Morange, M. (2005). *Les secrets du vivant: contre la pensée unique en biologie [The secrets of the living: against a singular paradigm in biology]*. Paris: Editions La Découverte.

Ochs, E. (1979). Transcription as theory. In E. Ochs & B. B. Schieffelin (Eds.), *Developmental Pragmatics* (pp. 43–72). New York, NY: Academic.

Resnick, L., Asterhan, C., Clarke, S. (in press). *Socializing intelligence through academic talk and dialogue*. Washington, DC: American Educational Research Association.

Resnick, L., O'Connor, C., and Michaels, S. (2007). *Classroom Discourse, Mathematical Rigor, and Student Reasoning: An Academically Productive Talk Literature Review*, unpublished manuscript.

Rosé, C. P., & Lund, K. (this volume). Methodological pathways for avoiding pitfalls in multivocality. In D. D. Suthers, K. Lund, C. P. Rosé, C. Teplovs & N. Law (Eds.), *Productive Multivocality in the Analysis of Group Interactions, Chapter 32*. New York: Springer.

Sawyer, K., Frey, R., & Brown, P. (this volume). Knowledge building discourse in peer-led team learning (PLTL) groups in first-year general chemistry. In D. D. Suthers, K. Lund, C. P. Rosé, C. Teplovs & N. Law (Eds.), *Productive multivocality in the analysis of group interactions, Chapter 10*. New York, NY: Springer.

Shirouzu, H. (this volume-a). Focus-based constructive interaction. In D. D. Suthers, K. Lund, C. P. Rosé, C. Teplovs & N. Law (Eds.), *Productive multivocality in the analysis of group interactions, Chapter 5*. New York, NY: Springer.

Shirouzu, H. (this volume-b). Learning fractions through folding in an elementary face-to-face classroom. In D. D. Suthers, K. Lund, C. P. Rosé, C. Teplovs & N. Law (Eds.), *Productive multivocality in the analysis of group interactions, Chapter 4*. New York, NY: Springer.

Stahl, G. (2006). *Group cognition: Computer support for building collaborative knowledge*. Cambridge, MA: MIT Press.

Stahl, G. (2011). Theories of cognition in collaborative learning. In C. Hmelo-Silver, A. O'Donnell, C. Chan & C. Chinn (Eds.), *International handbook of collaborative learning*. New York, NY: Taylor & Francis. Web: http://GerryStahl.net/pub/clhandbook.pdf.

Stahl, G. (this volume). Interaction analysis of a biology chat. In D. D. Suthers, K. Lund, C. P. Rosé, C. Teplovs & N. Law (Eds.), *Productive multivocality in the analysis of group interactions, Chapter 28*. New York, NY: Springer.

Suthers, D. D. (this volume). Agency and modalities in multimediated interaction. In D. D. Suthers, K. Lund, C. P. Rosé, C. Teplovs & N. Law (Eds.), *Productive multivocality in the analysis of group interactions, Chapter 19*. New York, NY: Springer

Suthers, D. D., Lund, K., Rosé, C. P., & Teplovs, C. (this volume). Achieving productive multivocality in the analysis of group interactions. In D. D. Suthers, K. Lund, C. P. Rosé, C. Teplovs &

N. Law (Eds.), *Productive Multivocality in the Analysis of Group Interactions, Chapter 31*. New York, NY: Springer.

Teplovs, C., & Fujita, N. (this volume). Socio-dynamic latent semantic learner models. In D. D. Suthers, K. Lund, C. P. Rosé, C. Teplovs & N. Law (Eds.), *Productive Multivocality in the Analysis of Group Interactions, Chapter 21*. New York, NY: Springer.

Trausan-Matu, S. (this volume). Collaborative and differential utterances, pivotal moments, and polyphony. In D. D. Suthers, K. Lund, C. P. Rosé, C. Teplovs & N. Law (Eds.), *Productive multivocality in the analysis of group interactions, Chapter 6*. New York, NY: Springer.

Traverso, V., & Visser, W. (2009). Co-élaboration de solutions et role du graphic-gestuel: bilan de la comparaison méthodologique. In F. Détienne & V. Traverso (Eds.), *Méthodologies d'analyse de situations cooperatives de conception: Corpus MOSAIC* (pp. 169–182). Nancy: Presses Universitaires de Nancy.

van den Besselaar, P., & Heimeriks, G. (2001). Disciplinary, Multidisciplinary, Interdisciplinary: Concepts and Indicators. In M. Davis and C.S. Wilson (Eds.), *ISSI 2001, 8th international conference of the Society for Scientometrics and Informetrics*, Sydney: UNSW 2001. pp. 705–716.

Yanchar, S. C., & Williams, D. D. (2006). Reconsidering the compatibility thesis and eclecticism: Five proposed guidelines for method use. *Educational Researcher, 35*(9), 3–12.

Chapter 35
Multivocality in Interaction Analysis: Implications for Practice

Nancy Law and Thérèse Laferrière

Introduction

There has been increasing interest in the adoption of social constructivist approaches (Bransford, Brown, & Cocking, 1999; Sawyer, 2006) to organizing teaching and learning in formal educational settings. While the study of interactions has been an important area of research (Kumpulainen & Wray, 2002), interactions among students in collaborative learning contexts are by nature very different from those between teachers and students. In contrast to traditional instructional approaches, collaborative learning generally values interactions among students as an important input to the learning process, and gives more scope for students to take responsibility and make decisions in the learning process. Whereas IRE (Initiate-Response-Evaluate, Cazden, 1988) has been identified as the most prevalent form of interaction in traditional, teacher-centered classrooms, the nature and impact of peer to peer interactions are much more varied and less well understood. Research on peer interactions in collaborative learning contexts is becoming an important field in education research, but few metrics or tools have been offered to help teachers make sense of students' collaborative process and to make decisions on whether and what interventions should be made to advance the learning of groups and individuals.

With the increasing accessibility of the Internet and popularity of web 2.0 applications, many teachers have also introduced collaborative learning mediated through synchronous and/or asynchronous information communication technologies (i.e., computer-supported collaborative learning, CSCL for short) into their day-to-day pedagogical repertoire. CSCL brings rich, diverse possibilities to collaborative learning that would not have been possible otherwise, as illustrated through the

N. Law (✉)
University of Hong Kong, Pokfulam Road, Hong Kong
e-mail: nlaw@hku.hk

T. Laferrière
Laval University, Quebec City, QC, Canada

D.D. Suthers et al. (eds.), *Productive Multivocality in the Analysis of Group Interactions*, Computer-Supported Collaborative Learning Series 16,
DOI 10.1007/978-1-4614-8960-3_35, © Springer Science+Business Media New York 2013

three datasets generated in CSCL settings in the present volume. The Biology data set introduces conversational agent technology into the CSCL interactions to enact facilitation for an approach called Academically Productive Talk. The Group Scribbles environment associated with the Electricity dataset allows students to construct their own digital artifacts in the form of drawings and text, and to share those with others within and outside of their own small groups. Just for a small group of four students, the collaboration data collected in the Electricity dataset consisted of five videos of the tablet screens, one for each of the student's individual screen and one for the shared screen, and this is in addition to the video of the actual physical activities of the group and the audio recording of their verbal discourse. The Knowledge Forum® Education dataset contains discourse data for a master level 13-week course conducted totally online through an asynchronous forum and the data accumulated amounted to more than 200K words! Such advances in learning technology provide unprecedented opportunities for rich peer collaborative learning opportunities, but also poses serious challenges to the teacher as facilitator on how to even be aware of what the students are discussing, let alone whether they are making progress, whether they are facing difficulties with the subject matter or the socio-dynamics.

While the focus of this book is generally to explore whether multivocal analysis of the same dataset can lead to productive interactions among researchers, and the possible theoretical and/or methodological developments that this may bring about (see Chaps. 1 and 31–34), this chapter explores whether such multivocality would have meaningful implications for practice. Would the different analyses reported in this volume offer insight as to whether and how such analytical methods and tools may be helpful to teachers in helping them to understand how students respond to different collaborative learning designs, pedagogical settings and strategies, even though these methods are underpinned by disparate theoretical assumptions and analytical approaches?

The five datasets forming the basis for the multivocal analyses in this book are all collected in formal educational settings, spanning primary, undergraduate and postgraduate education. The modes of interaction are also various, ranging from entirely online synchronous, entirely online asynchronous, face-to-face augmented with online platform for individual and shared artifact construction, to entirely face-to-face. Further, the pedagogical principles underpinning the learning designs in the five settings also differ greatly. This provides a welcome diversity for an initial exploratory study reported here.

In this chapter, we will be addressing the following questions:

- Do pivotal moments identified by researchers have implications for pedagogical practice?
- Would teachers be interested in the multivocality or the pivotal moments identified by researchers?
- Would teacher's understanding of their own practice and students' learning be enriched by the multivocality of the researchers?
- Is it likely that such analytical processes would be considered for productive adoption on a routine basis?

We invite the reader to inquire with us in (1) the value of such analysis as a diagnostic tool for the teacher as opposed to analyses that help to answer specific research questions, and (2) the feasibility of the analysis being easily conducted by the teacher on the basis of ease of administration.

Background

Classroom interaction and talk have been inquired into for several decades (Fisher, 1993; Flanders, 1970; Kumpulainen, 1996; Lemke, 1990; Mercer, 1996; Resnick, Levine, & Teasley, 1991). A major thrust of such studies has been to make teaching more information-based (or data-driven) through classroom observation schedules. However, researchers' analytical methods, results, and tools do not easily find their way into teachers' regular practice. One exception has been Moreno's (1934) sociogram. It was and remains a popular tool for teachers. However, much of the complexity of sociometry has been reduced. The same with Gardner's theory of multiple intelligences, which has a broad impact on teachers. Related educational resources and professional development workshops simplified Gardner's contribution. Large-scale dissemination is often at the expense of complexity, and rich analytical methods and tools applied in a rigid, technical manner.

To counter technical rationality in the education of professionals, Schön (1983) put forward the reflective practitioner approach, which has become a dominant approach in professional education. Within this paradigm, it is a common activity to collect data on one's own teaching through means such as videos, interviews, students' work, etc. to act as foci for analysis and dialogue. In the early days of sociometry, teacher educators who engaged pre- and in-service teachers in its derived techniques did not have such a purpose. Neither did they have the technology available today—to provide practitioners with rich descriptions through visualizations that help develop deeper understanding of specific learning environments.

UNESCO's (2011) three ICT competencies for teachers, especially deep understanding and knowledge creation, invite teachers to such an exercise. It is our working hypothesis that the data–information–knowledge trio can be achieved in teacher education and professional development. For this to happen, researchers are providing tools (conceptual, methodological, and technological) to transform data into information for evidence-based knowledge in support of pedagogical decision-making. Writing this chapter, which focuses on peer interaction and talk, we have in mind teachers, including preservice teachers and teacher educators, whose circumstances could allow them to analyze their own learning environments for improvement purposes.

Given that the analyses in this book are on human interactions for learning purposes in which peer interactions are important, the assumption is that the methods or tools can only be of relevance to pedagogical practices that are at least broadly social constructivist in nature. We are also aware that there are great diversities across different social learning theories and so different foci in terms of analysis for researchers interested in the different theories. Teachers also have been exposed to a variety of theoretical perspectives and pedagogical methods. They hold different sets of values and beliefs but are open to new tools that have resonance for their practice.

Methodology

We developed an analytical grid for exploring the chapters in terms of their pedagogical relevance for practice. Four themes were identified as follows:

- Analytical focus or Pivotal moments.
- Relevance for pedagogical practice.
- Mechanisms/tools for detection.
- Potential for automatic detection to interactively inform pedagogical decision-making.

First, we looked at the analytical focus of the chapter and including the pivotal moments, if applicable. The special attention given to pivotal moments was to align with the fact that they are considered boundary objects for those researchers participating in the writing of this book (as described in Chaps. 1 and 31). Some chapters pinpointed pivotal moments in their analysis of the same set of data (e.g., Oshima, Chap. 12; Lund and Bécu-Robinault, Chap. 17; Shirouzu, Chap. 5; and Trausan-Matu, Chap. 6) and others not. Some researchers found more pivotal moments than others (e.g., Trausan-Matu, Chap. 6 and Chiu, Chap. 7). This is no surprise as the definition of a pivotal moment had been left open on purpose. Thus, some researchers identified pivotal moments, while others presented a different interpretation regarding the same dataset (e.g., the Electricity data set). Looi, Song, Wen and Chen (Chap. 15) identified seven pivotal contributions, one of them being the following one: "The teacher's intervention (T12) to ask the group to draw their electrical circuit of lighting one bulb using two batteries in GS was considered pivotal to shape the students' inquiry to a higher level for conceptualization". Medina (Chap. 16) interpreted pivotal moments differently, as he posited that "a pivotal sequence of interaction occurring in the later half of the activity in which one member of the group proposes an innovation for illuminating two light bulbs in a single circuit". Other researchers distanced themselves from pivotal moments as boundary objects. Their analytical foci did not point directly to pivotal moments (e.g., Stahl, Chap. 28; Teplovs and Fujita, Chap. 21).

Second, we read the chapters with an eye to their relevance for pedagogical practice. Were these results (content or process) relevant to teachers? Two kinds of relevance were distinguished: results that presented general relevance (generalities about collaborative learning) and results that had more specific relevance as they pertained to specific learning contexts.

Third, we gave attention to the unit of interaction and mechanisms/tools for identification of pivotal moments or other points of analytical focus. We reckon that the complexity (or simplicity) in the way the unit of interaction is defined and operationalized would play a role in teachers' ease of understanding the focus of the analysis and hence their readiness to make use of the analytical results. Furthermore, some analyses are very refined and the mechanisms used for conducting them may not be easily understood conceptually by practitioners, which may influence a teacher's readiness to incorporate such analyses into their pedagogical decision-making, even if the analyses are done for them.

Fourth, in our analysis, we also wished to identify analyses that have strong possibilities for using technological support to inform teachers' pedagogical decision-making in interactive ways. This is important because teachers have less time than researchers to devote to interaction analysis. Technology support to analyze online talk would be most welcome, particularly those results that point to interactions indicative of particular states or transitions in the collaboration process.

The current availability of timesaving analyses of mechanisms/detection tools limits how far practitioners can take accounts of the complexity of classroom talks (contexts, affects, and the like). Complexity remains a most challenging issue. Therefore, automatic detection would be considered a plus. The challenge of uncovering pivotal moments with the support of automatic detection kept being at the center of our exploration. We know it is complex but have confidence that advances could be made.

Results

Based on the methodology described above, we reviewed and analyzed all the analysis chapters in the five data sections in this volume according to the four themes identified. Table 35.1 presents a brief overview of the findings for 13 of the data analysis chapters, within which we see strong promise of the analysis described to have pedagogical relevance. In the remainder of this section, we will present our key findings and explicating the contents of the table in the process.

Focus and Purpose of Analyses

The focus and purpose of the analyses have been found to be of paramount importance to whether the analysis has pedagogical relevance. The contributors of the five datasets were all contributing one analysis chapter. These data contributors also played a major role in the pedagogical design of the collaborative learning settings. It is hence not surprising that the purposes and foci of their analyses chapters have a clear pedagogical connection with the respective data collection context, and have potential pedagogical relevance to teachers from that perspective. For example, Shirouzu's analysis of his own Origami-fractions data focused on identifying students' collaborative advancement as well as their individual progression, which

Table 35.1 A brief overview of the findings for a sample of the data analysis chapters

Analytical focus or pivotal moments	Relevance for pedagogical practice	Units of analysis and/or mechanism of identification	Potential for automated analysis to interactively inform pedagogical decision-making
FRACTIONS—video of a group of primary children exploring fractions multiplication by paper folding			
Shirouzu (Chap. 5)—collaborative advancement and individual progression	For deepening teachers' understanding of children's thinking about the topic	Focus-based constructive interaction analysis—at the group level and individuals' changes of focus	Analysis too complex for yielding direct intervention
Trausan-Matu (Chap. 6)—convergence and divergence of voices, inter-animation patterns and changes in rhythm	For general pedagogical awareness of pairs of verbal and nonverbal utterances	Identification of a differentiated pattern following a series of repetitions	Difficult, though NLP might offer a way to automatic detection
CHEMISTRY—video data of two groups of undergraduate students in peer-led team learning tutorial			
Sawyer (Chap. 10)—knowledge building	For training peer leaders and for improved design of questions in team learning	Adjacency pairs/episodes	Narrative approach taken in analysis, not applicable for machine analysis
Howley (Chap. 11)—social positioning in study groups	Potentially relevant but need further research to build pedagogical model	Role analysis of peer leader: transactivity, heteroglossia/monoglossia	Possibility of automated analysis is claimed
Oshima (Chap. 12)—collective knowledge advancement	Peer leader's prompting role	Semantic SNA—degree centrality coefficient of nodes in a network of words	Good technological potential, require teacher to have deep understanding
ELECTRICITY—video and audio record of small group physical (expt.) and online (GroupScribbles) work			
Looi (Chap. 15)—pivotal contributions: shift direction of subsequent events	Manifested and latent pivotal contributions as lens to study pedagogical design	Sequential traces of uptake through conversational turns	Too complex to be automated in the near future
Lund (Chap. 17)—conceptual change identified as action/concept reformulation	Identify two types of change: naïve to canonical concept, retain canonical coherency and argumentative complexity	Semiotic bundles of mode/medium couplet(s) and MMR (multimedia multimodal reformulations) in between	Not immediately obvious

Jeong (Chap. 18)—increase in domain understanding and intersubjectivity	Collective activity may not imply intersubjectivity, group understanding as cognitive state	Categorization of fixed time segments and combination to form meaning-based episodes	NA—difficult to construct episodes and distill meanings automatically
ASYNCHRONOUS KNOWLEDGE BUILDING—discussion data of a totally online Master course on Knowledge Forum®			
Teplovs and Fujita (Chap. 21)—identification of learners with similar content interest	Identification can help to bring learners with similar content interests to work together	Adjacency matrix based on intensity of shared reading events using KISSME, referred to as latent semantic learner models	The analysis and visualizations are already automatically generated
Law and Wong (Chap. 22)—progression in sociometacognitive dynamics and subject matter understanding	Quick overview of likely pivotal moments in students' knowledge building progress	Detection through participation and discourse marker indicators: provide overview of nature and depth of students' engagement	Most of the analysis is already performed automatically using computational tools
BIOLOGY—chat data of online group tutorials under different accountable talk tutor agent conditions			
Howley et al. (Chap. 26)—reasoning events observed in transactive moves	For identifying support to encourage articulation of reasoning, selection of intelligent elearning tools	Conversation analysis using *Soufflé* framework	Potential for automatic analysis to improve performance of online tutor
Stahl (Chap. 28)—response structure of conversation	Highlighting potential damage of tutor-centric intelligent agents	Adjacency-pair analysis of response threading, content uptake, and mediation	Automation for conversational turns display, not yet response structure identification
Cress/Kimmerle (Chap. 27)—group awareness: social, activity, action, knowledge	Highlight online CL may be obstructed by lack of group awareness in environment	Identify chat sequences showing successful knowledge building, misunderstanding, etc.	N/A for pedagogical decision-making

was the purpose of the study for which the data was collected. Likewise, Sawyer's analysis of his own Chemistry discussion data was to identify which kind of peer leader role would be more conducive to knowledge building emerging through group discussions in peer-led team learning contexts. Looi's analysis of the Electricity data looked for pivotal contributions that shifted the group of students' foci and subsequent action/understanding in their attempts to connect bulbs and batteries. The purpose of Teplovs–Fujita's analysis of the Knowledge Forum® discourse data was to explore the usefulness of the KISSME tool in generating and testing predictive models of learner interactions to optimize learning. Howley et al.'s analysis of the Biology chat data was to identify what type of online tutor agent interaction would encourage students to articulate their reasoning and to listen and respond to the reasoning of others. It is also worth noting that of these five analyses, Sawyer's and Teplovs and Fujita's are primarily concerned with developing a generalizable model about particular aspects of collaborative learning, while the other three analyses also have specific relevance pertaining to the particular learning contexts.

While the five datasets were all collected by researchers with pedagogically related analytical goals, the analyses by researchers other than the data provider may be motivated by very different research goals. We find that irrespective of the analysts' theoretical or methodological constructs, whether the work has pedagogical relevance depends largely on the purpose and focus of the analysis. This can be illustrated using the analyses reported in Part 4 on the Electricity data. Lund et al.'s analysis was grounded on the science education literature and the goal was to look for instances of conceptual change, which is very different from a theoretical standpoint from Looi's identification of uptake that was grounded on the theory of intersubjectivity. However, these differences between the analyst and the data provider will not stop science teachers from finding meaning in Lund et al.'s analysis to track the students' transformations of conceptual content from the physics domain perspective as they communicate by talk, gestures, and drawings in the GS interface and through manipulations of the experimental apparatus. On the other hand, the purpose of Medina's analysis, which was also grounded on the identification of uptake based on the theory of intersubjectivity, was an academic one: to explore how sequential structures of multimodal interactions, including the availability of persistent artifacts generated on inscription devices, influence joint meaning making processes. While the findings from such research may have implications for understanding collaborative learning involving multimodal interactions, these are rather more distant from the immediate concerns of the practitioner faced with achieving the set curriculum goals through designing and facilitating collaborative learning.

Pivotal Moments

While all the data analysts were asked to identify pivotal moments (as defined by the analyst concerned), which are to be used as boundary objects for scaffolding productive multivocality, not all analysis chapters identified pivotal moments.

Where pivotal moments have been explicitly defined in studies that have pedagogically relevant analytical purposes, the pivotal moments may serve as important conceptual artifacts and scaffold understanding that are very helpful to practice. For example, in Part 5, while the chapter of Teplovs and Fujita and the chapter of Law and Wong were similarly grounded on the theory of knowledge building and had a similar purpose of providing pedagogically relevant analytical visualizations of learner interactions, they differ in that the former did not identify pivotal moments while the latter did. The KSV tool used by Teplovs and Fujita is a very innovative one integrating latent semantic analysis and social network analysis, and providing very flexible graphic visualizations to the user. They used the tool to identify students who shared similar latent semantic learner models, found that many of these students did not have high-level interactions, and went on to hypothesize that the tool may be of value to teachers as a basis for purposefully promoting interactions among these students. The validity of this hypothesis is yet to be substantiated. The Law and Wong paper focused on two types of pivotal moments. The first was to look for "pivotal weeks" during which the statistical interaction parameters indicate having reached a social dynamic condition illustrative of some of the knowledge building principles. The second type of pivotal moments was breakthroughs in students' understanding of key concepts based on semantic analysis of the note contents. These pivotal moments link directly with the concerns of the teacher.

Pivotal moments that are directly linked to the subject matter domain being studied are likely to be easily appreciated by teachers as relevant to their practice. For example, the analysis of the electricity data by Lund and Bécu-Robinault to identify instances of action/concept reformulation as a specific group of students engages in collaborative learning would be enlightening to and very much appreciated by science teachers.

However, not all pivotal moments have direct relevance to pedagogical practice. For example, in Chiu's analysis of the Origami-fractions data, a pivotal moment is a "conversation turn that separates a portion of the conversation into two distinct time periods (before and after) with substantially different likelihoods of the focal variable (e.g., correct ideas)". This formulation of a pivotal moment does not link directly to the practice concerns of the teacher on a day-to-day basis.

Unit of Analysis and Mechanism of Pivotal Moment/Event Detection

As discussed in the Methods section above, analyses in which the units of analysis or mechanisms of identification that are complex to understand and/or to operationalize would face more challenges in convincing teachers of their pedagogical relevance. For example, Shirouzu's analysis of the Origami-fractions data (Chap. 5) identified two units of analysis, the group and the individual. For the former, the focus of analysis was to look for those collaborative utterances that are indicative of constructive interaction, while the latter looked for changes in personal focus. These parallel teachers' interest in knowing about students' individual gains in understanding, as well as in how the collaborative process might have contributed to their advances in understanding. Chiu's analysis of the same dataset (Chap. 7) defined his unit of analysis as a conversation turn, which is simple to understand, but the unit of interaction was defined as a sequence of one type of action following another, with the actions being "microcreativity" that is to be identified through coding of the argumentative attributes of each conversation turn, and the units of interactions to be identified through statistical discourse analysis. While some of the pivotal moments identified by these two analyses are the same, Shirouzu's analysis would be more accessible to teachers, and hence this type of analysis is more likely to have impact on practitioners.

We suspect that the complexity of the units of analysis/interaction as well as the mechanisms for identification of the analytical point of interest does not only impact on the uptake of the related analysis by practitioners but other researchers as well. For example, to construct the uptake graphs in Looi et al.'s analysis of the Electricity dataset (Fig. 15.1) requires such detailed analysis and meticulous construction of the visualization that it is doubtful whether members of this research team concerned will repeatedly conduct the same analysis after the research advance targeted has been achieved. On the other hand, the visualization of Jeong's analysis of the emergence of group understanding of circuits as demonstrated through the physical and digital artifacts students constructed (Fig. 18.4) has a simplicity in its ease of construction and clarity in communication that other interaction analysts may wish to learn from.

Potential for Automated Analysis to Interactively Inform Pedagogical Decision-Making

Automatic capturing, processing, and analysis of interaction data is not one of the common themes for the present volume on productive multivocality in interaction analysis. However, we would like to argue that the exploration and sharing of automated tools to facilitate digital processing, analysis, and visualization of interaction data is one valuable potential outcome to be achieved as productive multivocality for a project of this nature. Automated, or even semiautomated analysis of interactions, would be particularly relevant in pedagogical situations where the analysis will provide information on the behavior or performance patterns of the specific group or individuals, as such information would be able to scaffold further pedagogical decisions.

Of all the analyses reported in this volume, only Teplovs and Fujita (Chap. 21) conducted their entire analysis through automated tools, the KSV and KISSME. Law and Wong (Chap. 22) identified the pivotal weeks using the ATK tool built into

Knowledge Forum®, while their identification of pivotal breakthroughs in students' understanding of key concepts was achieved through an automatic selection of sentences based on keyword search, followed by critical reading and qualitative analysis of the selected sentences to identify the critical advances. Howley et al. (Chap. 11) mentioned explicitly that the analysis of social positioning transactive interactions has been successfully automated in their team's earlier work, though the specific analysis reported was done manually. For the other chapters, no explicit mention has been made on the issue of analysis automation.

From the perspective of providing just-in-time analysis results to support teachers' pedagogical decision-making, the format of the interaction data also matters. Cases where the entire set of data can be captured digitally and ready for processing and analysis as in the case of the asynchronous discussion data on Knowledge Forum® consisting entirely of text data, or the synchronous chat log in the Part 6 Biology dataset offer a relatively low threshold for automation. Participation statistics and easily computed interaction patterns such as Social Network Analysis still offer valuable insight to teachers, despite their limitations. With advances in text-to-speech technology, it may be possible in the near future for the kind of network analysis of words reported by Oshima et al. (Chap. 12) to be carried out relatively easily in a timely fashion to inform teachers of the progress in students' collective knowledge advancement. With advances in natural language processing, it is anticipated that some of the analyses that are primarily grounded on looking for linguistic features/patterns in discourse would also be candidates for possible automation, such as the identification of convergence and divergence through detecting changes in repetition, etc. in Trausan-Matu's analysis of the Origami-fractions data (Chap. 6) and Howley et al.'s *Souflé*-based conversation analysis to identify students' reasoning behavior in their analysis of the Biology data (Chap. 26).

Further Observations

We are very heartened by the findings from this preliminary study as it provides substantial evidence that the multivocality in interaction analysis can be productive in providing valuable insight and pedagogical support to teachers interested in implementing collaborative learning in their everyday practice. The five different sets of data analyzed in this volume are very diverse in the contexts from which they were collected, not only in terms of their level and subject domain of study but also in terms of the use or otherwise of technology in the collaborative learning process. Our analysis demonstrates at length the presence of collaborative learning and the feasibility of at least some of the analyses being productive in identifying important issues for practice, despite the fact that some of the cases involved multimodal multimedia data (e.g., the electricity dataset), while others involved only monotonic textual data. Our findings also illustrate the potential of interaction analysis as a productive method for teachers' evaluation of the suitability of specific CSCL environments when they wish to select one to realize particular collaborative processes, as illustrated by the analysis of the Biology data.

Our analysis also demonstrates that a meaningful analysis from the practice perspective can be made by researchers who do not themselves generate the data, and using analytical methods that are grounded on theoretical frameworks different from the ones underpinning the pedagogical practice contexts from which the data were collected. In fact, the relevance to practice appears to depend largely on the goal and focus of the analyst.

It is interesting to note that there is some consistency in the triangulation of the analyses results, for example, although the conceptualization of pivotal moments in the three analyses of the Origami-fractions data were different, some of the pivotal moments identified by them referred to the same moments. Whether such pivotal moments are particularly significant ones in the students' collaborative learning process has to be confirmed on a case-by-case basis. However, when the results from one analysis are reinforcing the results of another, there is productive multivocality by providing a complementary perspective to the validation, and offer clear targets for teachers to consider as a first priority.

Examining analysis results on the same dataset that do not provide triangulated validation could also be productive from a practitioner perspective as the difference could promote reflection from multiple perspectives. For example, Howley et al.'s analysis of the Biology data (Chap. 26) was grounded on the assumption that heteroglossic conditions would be more conducive to students' adoption of accountable talk. When their experimental hypothesis that the Indirect Agent condition would be more conducive to accountable talk behavior was not supported by their analysis of the aggregate data collected from all the groups, they examined some of the interaction segments between the tutor agents and the students, and came to the conclusion that a crucial problem was the lack of coordination between the condition-specific prompts and the timed task prompts from the agents, and that the Indirect Agent should be improved with respect to timing and coordination. Stahl chose to analyze in detail the data of only one of the groups from the same dataset (Chap. 28) and constructed a visualization of the response structure of the group to show the threading of responses, mediation of accountable talk, and content uptake. This enabled him to home in on three instances of the Indirect Agent successfully mediating accountable talk. The in-depth analysis of these "pivotal moments" led him to rather different conclusions. On the basis of the holistic trace of an entire chat log, Stahl identified issues of lesson design, overscripting of the agents, and masking of the social identity of students. He proposes that accountable talk is a sophisticated level of discourse, which needs the skills of a teacher with mastery of this pedagogical approach and not simply the canned interventions from automated agents. Such fundamental conclusions arising from multivocality in analysis will also help teachers in their evaluation and selection of CSCL environments.

Discussion

The Potential of Interaction Analyses and Analytical Tools to Contribute to Pedagogical Practice

Classroom interaction analysis is something all teachers practice informally whenever direct interaction with students is involved. The focus and level of sophistication of the analysis, however, greatly varies according to the time devoted to such analysis, and its objective, means, and context. For those willing to go a step beyond the free-flow of thoughts regarding what is going on, during or after a class, observation/analysis tools that harmonize with the acquisition metaphor (AM, Sfard, 1998) are available. However, they are rarely used by teachers and remain in the toolkit of supervisors. The social psychology theory of power relationship and the critical theory of education argue for giving space for encouraging participation. But teachers moved by the participation metaphor (PM) have less direct access to what students are talking about or doing than when the AM metaphor applies. Questions as simple as "Are some students working and others not? Are students simply adding up individual contributions without much discussion?" are haunting ones for teachers. Therefore, reflective practitioners who engage students in collaborative learning may be more appreciative of ready-to-use analytical tools likely to inform them about what is going on. The analytic tools provided in this book are a most significant contribution to the PM repertoire of analytical tools.

Written electronic conversations have the great advantage of providing readily accessible traces of student participation. An online tool usually offers a teacher the possibility to glance through what is going on, be attentive to author names and turns, time and length of messages, and the like. Teachers can monitor, scaffold, evaluate. Depending on the functionalities of the online tool and related analytical tools (e.g., chat, GroupScribbles, Knowledge Forum), the teacher as a reflective practitioner can proceed to further analyses. See, for instance, the analysis of keywords in Law and Wong's chapter. For analysis of the epistemological/conceptual foci of the discourse (KB), KSV and KISSME provide visualizations of group or classroom talk. To interpret results, however, a teacher has to be acquainted with Scardamalia and Bereiter's (2003) perspective on knowledge building. In other cases, as with the use of TATIANA, knowing a theoretical perspective such as Michaels, O'Connor, and Resnick's (2008) Accountable Talk brings meaning to a tool that would otherwise remain opaque to a teacher.

To fully interpret results, however, would require an understanding of the metrics beneath. Few teachers are likely to be interested. Some teachers may be inclined to trust the metrics and engage students in reflecting upon automated analytical results. (This means that automated data analysis could have implications for student learning—for teachers to use the tools to support student self-reflection, peer collaboration on reflection and whole-class reflection, on own performance, and on the collaborative process, thereby building their own capacity as self-directed learners or autonomous inquiry groups.) The community of practice theory, which argues for giving more space/permission for participation, and the power relationship theory may be instrumental as a theoretical basis for creating this opportunity in the classroom. Here, visualizations that highlight pertinent aspects of the interactions may prove especially attractive to groups and classrooms. Students' identification of pivotal moments in onsite/online conversations would be an important metacognitive act for them to perform.

Therefore, the very introduction of the pivotal moment concept and its spread in analytical activity may be one of the most important outcomes of this book. This is an advance in terms of analyzing human interactions in the classroom. Most teachers are likely to have the conceptual understanding (conceptual tools) to be able to see the pedagogical relevance of identifying pivotal moments during onsite/online group/classroom talk. The writers of this book make a major contribution by showing that collaborative learning does take place in the classroom, and that it leads to deeper understanding and knowledge building.

At a time when UNESCO (2011) is fostering deep understanding and knowledge creation as key teacher competencies in the digital age, the pedagogical implications of the previous chapters' results extend to teacher educators. They are the ones most likely to grasp the value of pivotal moments during classroom talk, and consider them as an important feature of classroom/group talk that is student focused. While planning learning experiences for pre- and in-service teachers that will help them develop this kind of understanding, teacher educators may want to begin with pivotal moments that point to changes in knowledge and understanding, and patterns of interaction, as these are of general concern to teachers. At an experiential level we suggest to combine synchronous (onsite/online) and asynchronous (online) talk on questions that matter to teachers. The exercise of identifying pivotal moments in their own conversations would be valuable for teachers. Moreover, it could become a nice addition to the practice of group reflective analysis, an innovation in teacher professional development going back to the eighties and nineties (Schön, 1983) and one that stresses the importance of multiple perspectives (Valli, 1992).

Although in this chapter we are seeing opportunities for pedagogical practice and arguing that multivocal analyses of CL interactions are relevant for practice, whether this is really the case remains to be empirically explored, and should best be done as an interdisciplinary collaboration between learning analytics researchers and education/pedagogy researchers and teachers. It is unlikely that even those understanding and valuing multivocality will apply it on a regular basis, at least not in the near future.

Suggestions for Learning Analytics Researchers with an Interest in Supporting/Influencing Practice

Learning analytics researchers interested in classroom interaction analysis may want to engage teachers in discussing the analytical results from their practice, thus uncovering whether teachers may find these to be helpful to them in understanding students' learning, in orchestrating/facilitating learning, and if so, in what formats or what kinds of visualizations would be more helpful. They may have already done so with local teachers as they developed or tested their metrics.

Researchers may also want to consult teachers and/or learners about whether providing such results to learners would be meaningful and helpful to the learning process. Teachers could share with researchers (1) ways in which they would introduce learning analytics results to students, and (2) ways in which they have attempted to share such results with students. It would help researchers to know the circumstances within which teachers may find multivocality results informative for immediate pedagogical action.

Researchers may have a graduate seminar that would allow them and their students to give special attention to pivotal moments during classroom talk, thus bridging their research and teaching activities. Pivotal moments could serve as scaffolds for teachers and learners for understanding and self-direction in steering learning. In the context of supporting assessment for learning, and in particular assessment for collaborative learning, it is important that analytical results not only be composed of discrete learning outcomes, but provide more nuanced understanding of how these come about as an integral part of the interactional process, which would give teachers and learners greater sociometacognitive agency in CL. For instance, as an integral part of a graduate seminar, circumstances of use could be grasped, results could be interpreted with a sense of the "whole". The closest illustration of such a process is Sawyer's attempt (Chap. 10) to better understand the Peer-led team learning discourse practices used by peer leaders and students, and among the students themselves, that give rise to an enhanced understanding of the chemistry content.

The Potential of Multivocality in Interaction Analysis to Contribute to Pedagogical Practice

The results presented in Table 35.1 indicate that the goal of using technology to support teachers in almost real-time pedagogical decision-making would not be feasible in most cases. Such a "pessimistic" conclusion probably reflects not only limitations in the current state of development of analytical and visualization technology but also the complexity of many of the analyses and our inadequate expertise in this area. On the other hand, the multivocal analytical methods and tools presented in this volume may probably become a pathway or scaffold for realizing the use of technology tools as supports to teachers in assessment for learning, which could have significant implications for practice.

We also feel that there is good potential for some of these analyses and tools to be used for the purposes of professional development and teacher learning (teacher education), especially in supporting teacher reflection on the impact of different pedagogical designs and facilitation on the processes and outcomes of collaborative learning. This offers a pathway for teachers (pre- and in-service) to engage in and hence to learn about social constructivist models of learning. The kind of data and analysis made available in this volume is a very attractive and appropriate resource to support more open, exploratory modes of learning in teacher education as the multivocality contained therein ensures that these would not be used as "ideal" or "authoritative" analysis, but as stimulus for further discussion.

There is also the potential of interaction analysis to illuminate which kind of learning technology is likely to be supportive of more open, collaborative, and knowledge-building-oriented approaches to learning and teaching, and which ones are likely to restrict it (e.g., the analyses of the biology data). The following kinds of analysis could help practitioners to understand and assess the appropriateness of a learning technology for CSCL:

- Analysis that reveal whether the interaction between technology and learners resemble teacher-/instructor-centered interactions which tend to obstruct students' knowledge building or encourage more open interaction and exploration among learners.
- Analysis that reveal and encourage students' agency in learning, such as the generation of good inquiry questions and sustained efforts in improving understanding, which are important if knowledge building is to be achieved through CSCL. Such analysis would also help teachers to differentiate software platforms that provide external agents to direct student learning from ones that foster student agency to take responsibility for and to monitor their own learning progress.

As mentioned in the introduction, multivocal interaction analysis can contribute to two types of relevance to practice: those that can inform more immediate pedagogical decision-making and those that provide more general insight and understanding to the processes and outcomes of learning and knowledge building in collaborative contexts. The literature on interaction analysis has provided a scientific basis for more nuanced understanding of collaborative learning, which clearly has pedagogical implications. However, our analysis reveals that research that offers greater alignment between the analytical goals and learning outcomes or processes of importance to the daily milieu of a teacher's practice is more likely to contribute to advances in CSCL practice in educational settings.

References

Bransford, J. D., Brown, A. L., & Cocking, R. R. (Eds.). (1999). *How people learn: Brain, mind, experience, and school*. Washington, DC: National Academy Press.

Cazden, C. B. (1988). *Classroom discourse: The language of teaching and learning*. Portsmouth, NH: Heinemann.

Fisher, E. (1993). Distinctive features of pupil-pupil classroom talk and their relationship to learning: How discursive exploration might be encouraged. *Language and Education, 7*, 239–257.

Flanders, N. (1970). *Analyzing teacher behavior*. Reading, MA: Addison-Wesley.

Kumpulainen, K. (1996). The nature of peer interaction in the social context created by the use of word processors. *Learning and Instruction, 6*, 243–261.

Kumpulainen, K., & Wray, D. (2002). *Classroom interaction and social learning*. London: RoutledgeFalmer.

Lemke, J. L. (1990). *Talking science: Language, learning and values*. Norwood, NJ: Ablex.

Mercer, N. (1996). The quality of talk in children's collaborative activity in the classroom. *Learning and Instruction, 6*, 359–377.

Michaels, S., O'Connor, C., & Resnick, L. B. (2008). Deliberative discourse idealized and realized: Accountable talk in the classroom and in civic life. *Studies in Philosophy and Education, 27*(4), 283–297.

Moreno, J. L. (1934). *Who shall survive? A new approach to the problems of human relations*. Washington, DC: Nervous & Mental Disease Publishing Co.

Resnick, L. B., Levine, J. M., & Teasley, S. D. (Eds.). (1991). *Perspectives on socially shared cognition*. Washington, DC: American Psychological Association.

Sawyer, R. K. (Ed.). (2006). *Cambridge handbook of the learning sciences*. New York, NY: Cambridge University Press.

Scardamalia, M., & Bereiter, C. (2003). Knowledge building. In *Encyclopedia of Education* (2nd ed., pp. 1370–1373). New York, NY: Macmillan Reference.

Schön, D. A. (1983). *The reflective practitioner: How professionals think in action*. New York, NY: Basic Books.

Sfard, A. (1998). On two metaphors for learning and the dangers of choosing just one. *Educational researcher, 27*(2), 4–13.

UNESCO. (2011). *ICT competency framework for teachers*. Paris: Author. Retrieved from http://unesdoc.unesco.org/images/0021/002134/213475e.pdf

Valli, L. (Ed.). (1992). *Reflective teacher education: Cases and critiques*. Albany, NY: SUNY Press.

Chapter 36
A Dialog on Productive Multivocality

Timothy Koschmann and Claire O'Malley

An Inquiry into Inquiries

We are grateful to the editors for inviting us to comment on this large and fascinating project. Thomas Kuhn (1970), in his well-known treatise on scientific revolutions, argued that science advances when a community of researchers comes to agree upon a research question and a set of methods for addressing it. Taken together, the question and the associated methods represent what Kuhn termed a "paradigm." Despite some early, overly optimistic proposals (e.g., Koschmann, 1996), the research arena that we have come to know as CSCL has yet to develop its own identifying paradigm. It is a scholarly community with a shared interest in learning in settings of collaboration, but with neither an agreement on a focusing question nor a common methodology. The current volume engages this as its central problematic.

Current research in CSCL is incredibly diverse. Some work involves engineering new technologies, some the evaluation of instructional innovations, and some focuses on better understanding practices of sense-making and collaboration. The Productive Multimodality (PM) Project focuses on the latter, but even within this restricted subdomain, a rich variety of research traditions and disciplines are represented. The project organisers sought to engender conversations across these traditions. Their strategy was to invite workers with different theoretical backgrounds to look at a common set of data together. In this way, the PM Project is reminiscent of

---

Prepared for inclusion in *Productive Multivocality in the Analysis of Group Interactions* (Edited by Dan Suthers, Kristine Lund, Carolyn Rosé, Chris Teplovs, and Nancy Law).

T. Koschmann (✉)
Department of Medical Education, Southern Illinois University, Springfield, IL, USA
e-mail: tkoschmann@siumed.edu

C. O'Malley
School of Psychology, University of Nottingham, Nottingham, UK

D.D. Suthers et al. (eds.), *Productive Multivocality in the Analysis of Group Interactions*, Computer-Supported Collaborative Learning Series 16,
DOI 10.1007/978-1-4614-8960-3_36, © Springer Science+Business Media New York 2013

various earlier efforts in the learning sciences to conduct "collaborative analyses" (Koschmann, 2011, p. 9). These would include special journal issues organised around a single set of analytic materials (e.g., Koschmann, 1999; Sfard & McClain, 2002) and book-length projects such as those organised by Cobb (1995), Koschmann (2011), Maher (2011) and Stahl (2009). The current volume reports on five such analytic exercises involving five different data sets and five different sets of analysts. The recruited analysts were charged with locating "pivotal moments" (PP. ch1–4) within the materials provided to them. The project, in this way, is designed to encourage the analysts to focus on the same phenomena and be explicit about their research practices and about the assumptions that inform their work. It is, in short, an inquiry into their methodic inquiries. Is that about the way you read it, Claire?

Yes. It is interesting to note that in every case, one of the analysts was the person responsible for generating the data and representing it—in several senses—to the other analysts. This seems to have been quite a deliberate strategy, since one goal is to compare (and/or contrast) what would have been the primary researcher's analysis with that of other analysts. So the difference between the primary analyst and the secondary analyst is not merely one of different methodological approaches or theoretical orientations, but also the difference in their relationship to the data (see also Rosé & Lund, Chap. 32). This clearly became both a practical challenge for the project at certain times—and indeed caused frustration, where other researchers disagreed with the way that data were collected or recorded—but also a source of potentially productive multivocality, often for the primary researchers themselves. Another aspect of this difference in relationship with the data surfaces in the observations by the editors about the (usually only tacitly acknowledged if at all) ontological commitments implicit in (a) the data themselves (i.e., *what is collected/measured entails some assumptions), (b) how they are recorded* (i.e., *transcripts already encode certain primary analyses/assumptions) and (c) represented (as in higher order coding schemes). One wonders whether the PM project would have been more or less productive if none of the analysts had a relationship to the primary data?*

The PM Project is very reminiscent of those earlier projects you outline, Tim, but it also connects to certain broader themes in educational research today. Firstly, there is the idea of having multiple researchers all working on a common set of data, often quite large. Over the years there have been many examples of this including, for example, TIMSS and PIRLS,[1] *PISA,*[2] *the US National Household Education Surveys,*[3] *the UK Household Longitudinal Study.*[4] *Secondly, there is a great deal of interest in so-called "big data"—not just in science, but also with respect to social*

[1] http://timssandpirls.bc.edu

[2] http://www.oecd.org/pisa

[3] http://nces.ed.gov/nhes

[4] http://www.esrc.ac.uk/funding-and-guidance/tools-and-resources/research-resources/surveys/understanding-society.aspx

science—data collected by various government agencies, companies, and via interactions through social media. The distinction between the first and second kinds of efforts are that with the second kind, the data are usually collected indirectly and are potentially capable of being mined for all sorts of purposes not originally planned for. Thirdly, there is the idea of aggregating diverse bodies of research in the interest of identifying convergent findings. This lies at the heart of the many meta-analyses and systematic reviews conducted in recent years.

There has been a long tradition in educational research of so-called mixed-methods research, usually taken to be a combination of qualitative and quantitative methods. Although some have argued for this approach to be superior to single method approaches (Jonson & Onwuegbuzie, 2004), others have been more critical (Symonds & Gorard, 2010).

The PM Project shares the aim of the first of these themes—several analysts sharing the same set of data. However, it is at a much smaller scale, both in terms of the size of the dataset and in terms of the number of analysts. Perhaps another difference is that it is probably true to say that, for most of the kinds of examples given under the first theme (i.e., datasets), although not all users of those data would come from the same disciplines or theoretical orientations, they are more likely to agree about method than disagree (e.g., for one thing, all these examples are quantitative datasets). The difference in scale also raises a question about how well the PM approach can "scale up," both in the sense of the size of the datasets and in terms of the numbers of analysts. It seems fairly fundamental to the approach that the different analysts needed a great deal of face-to-face interaction in order to achieve the level of shared understanding that might lead to productive multivocality—and one has to remark that 5 years is a long period of time for the development of such shared understandings. So it is worth asking whether the approach could generalise to larger-scale scientific programmes, or, indeed, whether the outcomes of a multivocal analysis endeavour can be generalised—i.e., perhaps you always have to "be there" and go through the process first-hand?

The PM Project shared some common features with the second theme, in that the whole endeavour was to see whether new understanding could be obtained from analyses for which the data were not originally intended, and we see throughout the book various examples of where secondary analysts saw different phenomena in the data, whether it was because they focused on the nonverbal interactions, or because they aggregated data at higher levels of scale, thus revealing emergent properties (see also Lund et al., Chap. 34). One other possible similarity with this second theme, and a slightly different project to the PM Project, would be where different analysts collect different data from the same context, then try to link those data—or even "swap" their data—in order to perform different analyses.

Finally, the PM Project shares some commonality with the aim of the third theme—i.e., to attempt to converge on common findings from diverse analyses. However, what is distinctive about meta-analyses and systematic reviews is that it is essential to compare like with like—similar data, collected from studies employing similar methods. The whole point about the PM Project is at least to deliberately use different methods of analysis and on the same data. This also raises what is perhaps a missed opportunity of the PM Project, and that is, to use similar data across each of the knowledge domains. As it is, not only do methods of analysis vary but so do knowledge domains, types of data, educational contexts and age groups. This is discussed in the book as a deliberate strategy in order to maximise coverage. The downside is that there is so much diversity it is difficult to know always whether any difficulties in agreement, or even productive disagreement, are due to the primary concern (different methods of analysis) rather than any secondary features (e.g., context of data collection).

Yes, the theme of pursuing mixed-method inquiry is also often described in the literature in terms of "triangulation" and the method receives passing reference here. For example, Suthers et al. in their discussion chapter mention "concurrent triangulation" (PP. ch31-19) and Law and Laferrière discuss "triangulated validation" (PP. ch35-9). Denzin (1970), who is often credited with introducing the term, spoke of four different kinds of triangulation: *data* triangulation, *investigator* triangulation, *theory* triangulation and *methodological* triangulation. In the way in which this Project was structured we can see the latter three all coming into play. Data triangulation, on the other hand, which involves applying fixed precepts to different datasets is an exact inversion of the strategy employed in this project. Central to the PM Project was the notion of holding a set of data in common.

Five Orienting Questions

As explained in Chap. 2 by Lund & Suthers, each recruited analyst was asked to address five questions within their respective chapters: (1) "What ontological and epistemological assumptions are made about phenomena worth studying, and how can come to know about them?," (2) "What is the analyst trying to find out about interaction?," (3) "In terms of what fundamental relationships do we conceive of interaction? That is, what is your 'unit of interaction'?," (4) "What representations of data and representations of analytic constructs and interpretations capture your research purposes and units in a manner consistent with the theoretical assumptions?" and (5) "What are the analytic moves that transform a data representation into successive representations of interaction and interpretations of this interaction? And, how do these transformations lead to insights concerning the purpose of analysis?" These were presented as "dimensions along which to describe analytic methods" (PP. ch2-1). In other words, depending on how individual analysts choose to answer these questions, their contributions can be positioned within a space of possible approaches to studying interaction.

These are interesting questions, but I could see how they might be difficult for chapter authors to answer forthrightly. The first, for example—where do you even start? As learning scientists, I suppose, we might begin with our theories of learning. Learning, after all, is our most cherished phenomenon. But what is *its* ontological status and how do we come to know about it? Our various and sundry theories of learning have lots of assumptions built into them and bringing these assumptions into the daylight is probably a good start toward building common foundation for the field.

Let's take the first question concerning ontological and epistemological assumptions. Data are "theory-laden." One issue that arises is what counts as the primary data. Indeed what data are captured (talk, video, data logs) is arguably already suffused with theoretical and methodological assumptions (cf., Hall, 2000*; Ochs,* 1979*). Indeed there was at least one instance in the PM Project where there was significant dispute between analysts about whether the data were the "right" data to collect to begin with! The way in which data are represented is also not neutral to the analysis. A transcript already carries framing assumptions related to both theory and method. What is left out of a transcript (e.g., pauses, timings, accompanying nonverbal behaviour) is as significant as what is represented (words). Several authors also had to grapple with this issue. It isn't necessarily that, for productive multivocality one should agree on how to represent the primary data, just that it ought to be a conscious and explicit choice, recognising that this will affect the analysis.*

The second question pertains to the purpose of the analysis. But within the current project, the purpose seems to be dictated, at least in part, by the requirement to locate "pivotal moments" within the supplied data sets. As Suthers recounted in his introduction, in their first attempts to produce collaborative analyses, the organisers were disappointed to find that the analysts were "'talking past' each other" (PP. ch1-4) because each analyst was pursuing a different research question. To avoid this in their subsequent efforts at collaborative analysis, the organisers imposed a requirement that the analysts identify "pivotal moments." What actually makes a moment pivotal, however, was left "explicitly vague" (Garfinkel, Lynch, & Livingston, 1981).

Yes, the vagueness of the notion of "pivotal moment" is regarded as a strength, since the authors claim that it is important that these instances bring out what is pivotal in the eyes of the analyst—indeed at times the authors seem to use this concept interchangeably with the concept of a "boundary object," which is not defined objectively, but surfaces at points of difference between analysts/methodologies. However, at other times the use of the term "pivotal moment" seems more to refer to what is seen as a significant "shift" in the interaction, whether it is new forms of learning or changes in talk or behaviour, which is not quite the same thing as a boundary object. So I wonder whether the concept of "pivotal moment" really did succeed in overcoming the "talking past each other" problem.

Several, if not all, the analysts in this book did use the concept of "pivotal moments," which were usually higher level segments of talk or action that for some reason denoted significant or foregrounded activity—e.g., points at which insights were achieved. Some analysts chose common pivotal moments, but others highlighted different ones.

The third question posed to the recruited analysts had to do with the "units of interaction" that they utilised in their respective analyses. As I will discuss a bit later, there are some fundamental differences between spoken language and textually mediated interaction such as chat or blogging, one of these having to do with how the interaction is segmented. Segmentation of computer-mediated communication (CMC) is done explicitly by the producer when hitting the "send" button. Face-to-face (F2F) communication, on the other hand, is a little more complicated and it is difficult to formulate any strict rule for what constitutes a turn at talk. If a speaker starts, pauses for a period, and then restarts, is that one turn or two? If one speaker seizes the floor, talking for an extended period without break, is that one turn or several? A number of chapters in the book deal with what have been termed in the CA literature "collaborative turn sequences" (Lerner, 2004). Here, one speaker initiates a statement that is then completed by another. Again, for the purposes of defining a "unit of interaction," does this represent one turn or two? This is not to say that we cannot differentiate turns at talk, because we clearly can. We do so routinely whenever we engage in conversation. But difficulties arise when we have to say exactly how we accomplish this.[5] None of the chapters of this book actually address this problem, but it would appear to be a foundational one for anyone wishing to understand the fundamental organisation of interaction.

At some points in the book the authors talk about "gratuitous differences in data considered." An example they discuss is where analysts differed in terms of whether they analysed "private" as well as public activity. In other words, I take it, some analysts worked with inferred data or units of analysis, whilst others worked only with what could be seen by other researchers in the data presented. This is a related point to yours Tim, in the sense that what constitutes a turn can sometimes be emergent in the interaction. But it is also a slightly different point for me. That is, the authors at some points talk about how it is important to reflect upon differences in what is regarded as the units of interaction, or "appropriate" data—these, after all, provide the "boundary objects" they seek in the PM Project. On the other hand, at times they want to eliminate such differences and it is never really made clear what counts as "gratuitous" versus perhaps what we might call "fortuitous" differences.

The fourth orienting question has to do with representations of interaction and this generally involves transcripts. Let me begin by making a simple observation: transcripts are never complete. This is true for a couple of reasons. First, there is always a certain amount of slippage that occurs in putting spoken language on paper—sometimes there are problems with the quality of the recording, people don't always enunciate clearly, when the subjects are using another language, there may be vagaries of translation, etc. So there will inevitably be varying degrees of certainty attached to every line entered into a transcript. Second, when we begin to include multimodal aspects of conduct such as gaze, gesture, posture, facial expression, etc., the number of things that could potentially be noted is essentially unbounded. This is also true when we just consider all of the possible aspects of delivery (e.g., intonation, stress, pronunciation, dialect, etc.) that could be rendered. Since having a transcript is a prerequisite to doing an analysis, we may not know,

[5] But see Sacks, Schegloff, and Jefferson (1974) for one attempt to provide a formal account.

when making the transcript, how much detail will be required. So, as a practical matter, transcript elaboration and analysis often proceed hand-in-hand. Some things may be initially included, but later dropped when determined to have no apparent value for the analysis. These same elements, however, may turn out to be crucially important to someone else doing some other kind of analysis. This is related to what you were describing previously as the "theory-ladeness" of transcripts. This creates a particular problem in a comparison study in which the secondary analysts are supplied with a transcript readymade. Ideally, one would want to structure the task in such a way that all of the analysts are required to construct their own from scratch. Under such circumstances, the transcripts too would be objects of comparison, just as the analytic findings are. The things these transcripts make visible and the things they elide will determine what eventually gets noticed in the various analyses (see also Suthers' comments on this in Chap. 19).

The point above about all transcripts being incomplete has one exception and that exception arises in computer-mediated, textual interaction. Here what you see is exactly the same as what the subjects saw, with no need for correction or additional annotations. In this case we effectively get a perfect transcript and we get it for free. This, in fact, was the case in two of the five data sets studied in this volume. This seemingly represents a big break for researchers who do this kind of research, but it should also raise a caution—this fundamental difference between transcripts used in studying F2F interaction and CMC interaction ought to give one pause when attempting to make generalisations about and across these very different kinds of interaction. Others (e.g., Garcia & Jacobs, 1999) have made a similar point, but formulated it in different terms.

At times it feels rather difficult to follow the distinctions drawn between data, units of analysis, units of interaction and representations. For some data, for some representations, it is fairly clear (e.g., the data are turns of talk, the representations are codes attached to those turns), but for others, the distinction between primary data and secondary representation is not so clear. Is the transcript the data or a representation of the data, as discussed above. The difficulties in making this distinction are highlighted in the discussion about the use of software tools to "align" different representations. However, software tools also make ontological commitments as much as other (non-software) representations. It is not as if a toolkit is neutral representationally, so if the analysts use a common set of tools to align other data or representations, it behoves the researchers to be explicit about what assumptions are built into the tools.

In the chapter describing the methodological dimensions, Lund and Suthers describe interactional analysis as an iterative process of transforming different one representation of the data into another. The fifth and final question deals with manipulations and analytic moves that transform one data representation into another. This works out in quite different ways, however, in different kinds of research (see also Dyke et al., Chap. 33).

The transformation of one representation into another is most evident in research that entails coding of interactional conduct, research that I will term, for simplicity of reference, *discourse analytic* (DA).[6] Here units of text or talk are extracted to form a tabulation vector, the transcript giving way to the coded tabulation, which, in turn, may be further worked into additional representations such as correlation tables or comparison charts. These transformations are essentially irreversible because certain information (e.g., sequence, temporality, speaker attribution) is discarded at each turn. DA methods are used in many of the analyses presented here including the chapters written by Chiu (Chaps. 7 and 23); Sawyer, Frey and Brown (Chap. 10); Howley, Mayfield, Rosé and Strijbos (Chap. 11); Law and Wong (Chap. 22); and Howley, Kumar, Mayfield, Dyke and Rosé (Chap. 26). Rather than simple coding, another strategy is to represent the interaction as a graph or network. Here again we see a procedural operation being used to transform one representation (i.e., a transcript or log of posts) into another. Examples of analyses that utilise graph-based representations can be found in the chapters by Oshima (Chap. 12); Teplovs and Fujita (Chap. 21); and Goggins and Dyke (PP. Chap. 29). Again, the transformation is irreversible.

But the view of analysis as iterative transformations of various forms of representation receives its biggest challenge from studies that seek to give direct accounts of the organisation of collaborative interaction. Here the final account is derived through some sort of hermeneutic processing of the transcript and primary data materials. Examples would include Shirouzu (Chap. 5) and Trausan-Matu's (Chap. 6) analyses of the origami classroom, the various "Group Scribbles" studies (Looi et al., Chap. 15; Jeong, Chap. 16; Medina, Chap. 17; and Lund and Bécu-Robinault, Chap. 18), and the analyses of the "cell models experiment" prepared by Stahl (Chap. 27) and Cress and Kimmerle (Chap. 28). As I mentioned earlier, rather than simply serving as a starting point for an analysis, transcripts here tend to evolve in concert with it, serving as a storage medium for all the useful noticing made along the way. In this way, it is not simply used and discarded, but is rather continuously improved and brought into better alignment with the goals of the analysis. The way in which the representation is transformed, therefore, is fundamentally different from that seen in DA or graph-based forms of analysis. This constitutes a tension between these two very different ways of approaching the task of understanding interaction in collaboration. The nature of that tension lies at the heart of what needs to be examined in this "inquiry into inquiries."

A related point is that, for most, if not all analytic methods, the experience and the skills of the analyst are key to producing "good" research. This is especially so for the more qualitative methods that rely on interpretation and a good deal of sophistication in terms of sensitising constructs—it is probably a hallmark of the more critical theoretical approaches, including for example "discursive psychological" approaches. It is also true of the more quantitative and positivistic end of the

[6] My choice of term is admittedly one that invites confusion. 'Discourse analysis' is used in a wide variety of ways in the literature (cf., Brown & Yule, 1983; Cicourel, 1980; Gee, 1999; Fairclough, Mulderrig, & Wodak, 2011; Potter, 2004; Sinclair & Coulthard, 1975). Here I am using it specifically to denote those methods for studying interaction that apply categorization reductionistically.

continuum. There is a skill—arguably an art—certainly a craft, in designing a good experiment, in spotting patterns in data, in knowing which differences are meaningful and which are not. Hence the frustration often with less experienced researchers who are slave to "significant differences" and don't take account of effect size—and even where there are measurably large effects sizes, knowing what is "important" as opposed to "substantial." This sense of the analyst as professional tends to be overlooked in the PM Project. An assumption underlying the whole endeavour is that it is the analysis (method) that is important, and not the analyst.

You comment, Tim, that the transformations of one representation into another are "essentially irreversible," but perhaps one goal of the PM Project should be to develop methods for translating between representations so that it is always possible to backtrack, zoom in and out of levels of data analysis, so that a researcher encountering a higher level of representation might be able to recover the intermediate stages and assumptions that led to the abstraction. This would only be possible if, in the process of translating between representations, every step would be made explicit to a third party. The project has taken a significant step along the way to being able to do this, but some further work is needed to make more explicit the translation steps between the various representations employed by the analysts in the different case studies presented here.

Being Multivocal "Productively"

Claire, a couple of decades ago we were both active in helping to form a community of scholars focusing on issues related to fostering learning in settings of collaboration. The research questions in CSCL drew on different kinds of expertise—pedagogical, psychological, sociological, technological—and, so, the enterprise was from the git-go an interdisciplinary one. The challenge in any such undertaking is achieving some sort of synergy across the different disciplinary perspectives represented. Fundamentally, it is a challenge of creating a conversation in which all parties participate as full members, no disciplinary contribution beholden to another. Over the last 20 years CSCL has grown into a vital research community, one with its own scholarly journal and a biennial international conference. Nevertheless, the problem of managing conversations across disciplinary and theoretical boundaries remains. The PM Project is to be applauded for attempting to address this problem.

It could be observed that all research is to some degree multivocal in that within any publication the voices of many (co-authors, reviewers, past collaborators, former mentors) can often be heard. But in seeking "productive" multivocality, the organisers of the current volume seek to achieve a more radical form of conversation. Here participants are obliged to work with materials that they did not personally gather, to be articulate about the assumptions underlying their methods, and, in some cases, to adopt methods that are foreign to their usual research practices. But there seems to be some ambiguity with regard to how the PM Project is to be taken, as a template for how research should be conducted in the future, as a tutorial for methodology, or just as an instructive exercise.

I agree. There is also an issue about just what is productive about multivocality—this is never really fully articulated in the book. How does one recognise success? Does it matter as long as you, the researcher, feel it is productive? It could be that the multiple forms of analysis reveal new insights. This could be due to the second analysis providing additional findings to the first, or that two different methods combined reveal emergent findings at some second order level. Or it could be that a second method exposes gaps or problems in the first or reveals underlying assumptions that would otherwise be tacit.

There are clearly some practical issues related to adopting this as a model for future research in CSCL. The researchers involved in this volume are all concerned with documenting aspects of interaction in learning settings. While this is clearly a focus of some work in CSCL, it does not span the full scope of research being done under the CSCL banner, much less the wider range of educational research. Much of the work in education focuses not on process, but rather, following in the Thorndikean tradition (Koschmann, 2011), on learning outcomes. This remains the prevailing paradigm in education research and somehow we need to find a way of bringing it into the conversation as well. There are other issues, as well. Is it feasible to expect every project to take on a team of analysts? Would all principal investigators be willing to expand their projects in this way?

Yes, it isn't entirely clear whether the organisers are truly being prescriptive about this approach and arguing for a new agenda in educational research (as in the design-based research initiative, or those who have advocated mixed-methods research as the gold standard). Some seem to be a little more tentative (e.g., Chap. 30).

At times, however, it seems as though the primary goal of the project was to provide a useful source for research methods training. One way in which this might be taken forward is in further articulating the differences in the different methods and analytic traditions employed in this project. As you say, it is clear that what is represented here in terms of range of approaches is not exhaustive or even representative of the whole range of traditions in the fields of learning sciences more broadly, and CSCL in particular. The authors do not claim that they have done this and are clear that they worked with what they had, for whatever reasons. Nonetheless, it does represent a diverse set of perspectives. It would be interesting to develop this with a crisper overview and categorisation of the methods employed here. The authors enumerate a variety of methods/traditions in both Chap. 1 (PP. ch1-7) and Chap. 31 (PP. ch31-5) (content analysis, conversation analysis, polyphonic analysis, semiotic and multimodal analysis, social network analysis, statistical discourse analysis, computational linguistics, uptake analysis, knowledge building analysis, systemic functional linguistics and so on). Perhaps a useful next step would to build upon this project in order to develop dimensions along which these analyses vary, which methods are convergent, which divergent.

At first blush, several of these approaches might be amenable to grouping into a smaller set of categories. However, it does raise the question—why is there such a seeming diversity of labels for what might, to outsiders, seem to be just minor variants of the same approach? Here I am tempted to comment upon the broader issue of traditions of professional practice in the learning sciences community. So, for researchers whose primary focus is research methodology—i.e., that is their object of study—there is perhaps a disincentive to hone a particular method and a positive incentive to differentiate one's own method from that of other researchers. After all, originality and distinctiveness is a key criterion for professional advancement in the field. If you are a researcher into methods, there are fewer rewards for incremental advances and more rewards potentially for achieving a paradigmatic shift (as in more citations, associating your name with method x, etc.) So, the goal then is not so much contributing to the field by improving method y, but showing that method x is "better." Perhaps this in part explains the proliferation in the field of analytic methods that have different labels but actually, underneath, may well be just minor variants of the same class of methods. (The same might be said for software tools for analysis—a case of "not invented here.") It would have been interesting to see some reflection of this kind in the book and, in practice, there are hints at this issue at times where the discussants comment upon points of disagreements between analysts about what constitutes as "proper data" or appropriate representations. To use this text as a source book on method, however, would require careful scaffolding. The analyses presented here are somewhat complex and the conclusions are not always clear-cut. Added to this, there is the issue that the primary data are, for all intents and purposes, not available; so evaluating each approach is somewhat difficult.

Yes, that will be a problem for all readers, both novice and expert. Not having access to the primary materials upon which the analyses were made really made it difficult for me to sort out the differences between the various approaches. I am reminded of McDermott, Gospodinoff, and Aron (1978) dictum that all descriptive accounts must minimally provide sufficient access to the primary analytic materials to enable the reader/auditor to evaluate the claims being made. Given that they wanted to include five datasets and seventeen analyses, space was no doubt at a premium, but leaving the data sources out will result in a great loss to readability.

Perhaps, in the end, the greatest contribution of the PM Project will be as an instructive exercise. The editors argue that what is different in multivocality (compared with mixed-methods research) is that it is a reflexive practice (Chap. 32), where the researcher or analyst gains insight into methods themselves. So, one wonders whether it is important to actively engage in multivocal research first-hand and that such insights cannot be gained second-hand—for example through reading this book. On the other hand, what the book contains is a very rich resource for other researchers—whatever their level of experience—for reflecting seriously and systematically on how their approaches, from the framing assumptions underlying their approach to data collection to the assumptions underlying their analyses and interpretations, compare and differ with others in the field.

In conclusion, I feel that this has been a hugely interesting project to undertake and should serve to stimulate much discussion in the field. Although I might be accused of relying on positivist assumptions, if I were to take this project forward,

I would articulate more clearly the range of methods, and take care to ensure that they are representative of the diverse traditions of the learning sciences. I would adhere to the policy of holding the dataset constant, but I would make more explicit what counts as criteria for success (of the multivocality enterprise)—for example, the effectiveness of the different methods for revealing insights into learning outcomes and effective teaching strategies, efficiency of methods (perhaps in terms of resources and in terms of generalisability). Or, apply the same methods to different datasets in order to test the assumptions of the methods.

The authors and the many researchers involved in this project should be congratulated for their courage in embarking on such an ambitious project—one which has already generated several papers in high quality outlets—and in their patience in seeing it through over a long period of time, involving many research teams internationally, and in synthesising a very complex body of work. The field is very much in need of such an ambitious programmatic enterprise and, even if the project has revealed as many problematic issues as "successes," it is still to be counted as productive in my view, for opening up these debates as much as anything else.

References

Brown, G., & Yule, G. (1983). *Discourse analysis*. New York: Cambridge University Press.

Cicourel, A. V. (1980). Three models of discourse analysis: The role of social structure. *Discourse Processes, 3*, 101–130.

Cobb, P. (1995). Mathematical learning and small-group interaction: Four case studies. In P. Cobb & H. Bauersfeld (Eds.), *The emergence of mathematical meaning: Interaction in classroom cultures* (pp. 25–130). Hillsdale, NJ: Lawrence Erlbaum Assoc.

Denzin, N. K. (1970). Triangulation: A case for methodological and combination evaluation. In N. K. Denzin (Ed.), *Sociological methods: A sourcebook* (pp. 471–475). Chicago: Aldine.

Fairclough, N., Mulderrig, J., & Wodak, R. (2011). Critical discourse analysis. In T. A. van Dijk (Ed.), *Discourse studies: A multidisciplinary introduction* (2nd ed., pp. 357–378). London: Sage Publications.

Garcia, A., & Jacobs, J. (1999). The eyes of the beholder: Understanding the turn-taking system in quasi-synchronous computer-mediated communication. *Research on Language and Social Interaction, 32*, 337–368.

Garfinkel, H., Lynch, M., & Livingston, E. (1981). The work of discovering science construed with materials from the optically discovered pulsar. *Philosophy of Social Science, 11*, 131–158.

Gee, J. (1999). *An introduction to discourse analysis: Theory and method*. New York: Routledge.

Hall, R. (2000). Videorecording as theory. In E. Kelly & R. Lesh (Eds.), *Handbook of research design in mathematics and science education* (pp. 647–664). Mahwah, NJ: Lawrence Erlbaum.

Jonson, R. B., & Onwuegbuzie, A. J. (2004). Mixed methods research: A research paradigm whose time has come. *Educational Researcher, 33*(7), 14–26.

Koschmann, T. (1996). Paradigm shifts and instructional technology: An introduction. In T. Koschmann (Ed.), *CSCL: Theory and practice of an emerging paradigm* (pp. 1–24). Mahwah, NJ: Lawrence Erlbaum Assoc.

Koschmann, T. (1999). The edge of many circles: Making meaning of meaning making. *Discourse Processes, 27*, 103–117.

Koschmann, T. (2011). Theorizing practice. In T. Koschmann (Ed.), *Theories of learning and studies of instructional practice* (pp. 3–17). New York: Springer Science+Business Media.

Kuhn, T. S. (1970). *The structure of scientific revolutions* (2nd ed.). Chicago, IL: University of Chicago Press.
Lerner, G. (2004). Collaborative turn sequences. In G. Lerner (Ed.), *Conversation analysis: Studies from the first generation* (pp. 225–256). Amsterdam, The Netherlands: John Benjamins.
Maher, C. A. (2011). The longitudinal study. In C. A. Maher, A. B. Powell, & E. B. Uptegrove (Eds.), *Combinatorics and reasoning: Representing, justifying and building isomorphisms* (pp. 3–8). Dordrecht: Springer.
McDermott, R., Gospodinoff, K., & Aron, J. (1978). Criteria for an ethnographically adequate description of concerted activities and their contexts. *Semiotica, 24*, 245–275.
Ochs, E. (1979). Transcription as theory. In E. Ochs & B. B. Schieffelin (Eds.), *Developmental pragmatics* (pp. 43–72). New York: Academic.
Potter, J. (2004). Discourse analysis. In M. A. Hardy & A. Bryman (Eds.), *Handbook of data analysis* (pp. 607–624). London: Sage Publications.
Sacks, H., Schegloff, E., & Jefferson, G. (1974). The simplest systematics for the organization of turn-taking for conversation. *Language, 50*, 696–735.
Sfard, A., & McClain, K. (2002). Analyzing tools: Perspectives on the role of designed artifacts in mathematics learning. *Journal of Learning Sciences, 11*(2/3), 153–138.
Sinclair, J. M. H., & Coulthard, M. (1975). *Towards an analysis of discourse: The English used by teachers and pupils*. London: Oxford University Press.
Stahl, G. (Ed.). (2009). *Studying virtual math teams*. New York: Springer.
Symonds, J. E., & Gorard, S. (2010). Death of mixed methods? Or the rebirth of research as a craft. *Evaluation & Research in Education, 23*, 121–136.

Index

D.D. Suthers et al. (eds.), *Productive Multivocality in the Analysis of Group Interactions*, Computer-Supported Collaborative Learning Series 16,
DOI 10.1007/978-1-4614-8960-3,

D

E

F

Q

R

MIX
Papier aus verantwortungsvollen Quellen
Paper from responsible sources
FSC® C105338
www.fsc.org

If you have any concerns about our products,
you can contact us on
ProductSafety@springernature.com

In case Publisher is established outside the EU,
the EU authorized representative is:
Springer Nature Customer Service Center GmbH
Europaplatz 3, 69115 Heidelberg, Germany

Printed by Libri Plureos GmbH
in Hamburg, Germany